THEORETICAL CHEMISTRY

From the Standpoint of Avogadro's Rule & Thermodynamics

MAXWELL PRESS

THEORETICAL CHEMISTRY

From the Standpoint of Avogadro's Rule & Thermodynamics

By

WALTHER NERNST Ph.D

Winner of Nobel Prize for Chemistry

Georg August University of Göttingen

MAXWELL PRESS

Chennai Trichy NewDelhi

MAXWELL PRESS

This edition has been published in india by arrangement with Carson Books, UK

ISBN 978-93-90063-43-7 Maxwell Press

No. 44, Nallathambi Street,
Triplicane, Chennai 600 005

MJP 890

Publisher : C. Janarthanan

Publisher's Note

The legacy of a country is in its varied cultural heritage, historical literature, developments in the field of economy and science. The top nations in the world are competing in the field of science, economy and literature. This vast legacy has to be conserved and documented so that it can be bestowed to the future generation. The knowledge of this legacy is slowly getting perished in the present generation due to lack of documentation.

Keeping this in mind, the concern with retrospective acquiring of rare books has been accented recently by the burgeoning reprint industry. Maxwell Press is gratified to retrieve the rare collections with a view to bring back those books that were landmarks in their time.

In this effort, a series of rare books would be republished under the banner, "Maxwell Press". The books in the reprint series have been carefully selected for their contemporary usefulness as well as their historical importance within the intellectual. We reconstruct the book with slight enhancements made for better presentation, without affecting the contents of the original edition.

Most of the works selected for republishing covers a huge range of subjects, from history to anthropology. We believe this reprint edition will be a service to the numerous researchers and practitioners active in this fascinating field. We allow readers to experience the wonder of peering into a scholarly work of the highest order and seminal significance.

MAXWELL PRESS

THIS BOOK IS DEDICATED TO

Prof. ALBERT v. ETTINGHAUSEN

OF GRAZ

IN LOYAL MEMORY OF

HE AUTHOR'S YEAR OF STUDY AND TRAVEL

TRANSLATOR'S PREFACE

THE first edition of the German text was translated by Prof. C. S. Palmer of Colorado, soon after its publication, and long remained the only English version of Nernst's valuable treatise. Meanwhile, the original has passed through three further editions, has grown some two hundred pages in bulk, and has gained in usefulness by repeated careful revision. Under these circumstances the English publishers, in 1903, asked me to re-edit their translation. In doing this, I have translated all the new matter contained in the fourth German edition, and revised certain parts of the old, in the endeavour to free the book from any mistakes and ambiguities that a first edition is so liable to contain. The bulk of the old text, however, it was.

ROBERT A. LEHFELDT.

December, 1904.

PREFACE TO THE FOURTH EDITION

THE quantity of valuable new investigations now made in the domain of theoretical chemistry, especially in chemical statics and dynamics, is so great that no attempt at completeness of reference could be made, as in the first edition, and much valuable work has been left unmentioned, in order to keep the length of the book, as seemed desirable, approximately the same.

In the short interval since the last edition, the most remarkable phenomenon of radio-activity has come to be more closely investigated; in this, chemical processes of quite a different order to those previously known, seem to be dealt with,—a matter of the highest consequence to the theoretical chemist. Though the phenomena are still very obscure I have attempted to present in a new chapter, "The Atomistic Theory of Electricity," the development of electron theory, which is more and more seen to be a powerful extension of atomistic method.

I have to thank Dr. F. Krüger and Mr. A. Siemens for their help in correcting the proofs.

W. N.

GÖTTINGEN, *November* 1903.

PREFACE TO THE SECOND EDITION

In the five years since the publication of the first edition of this book, theoretical chemistry has hardly suffered any deep change, or rearrangement in principle; but the assiduous researches that have been made have accumulated a great quantity of new material, which the fruitful conceptions of the new theoretical chemistry have put in a clear light. Accordingly I have hardly changed the plan of the work in re-editing it, but have attempted to weave the latest results of experiment into the text, as illustrations of the leading theoretical conceptions.

W. N.

Gottingen, *September* 1898.

PREFACE TO THE FIRST GERMAN EDITION

THE following presentation of Theoretical Chemistry is a development of an "introduction" prepared by the writer two years ago for the *Handbuch der Anorganischen Chemie* (*Handbook of Inorganic Chemistry*), edited by Dr. O. Dammer (vol. i. pp. 1 to 358). In keeping with the broader requirements of an independent text-book — and not merely an introduction to a special work—it has been considerably rearranged and extended. It aims to consider the very latest investigation, and while this requires no change in the fundamental nature of the more recent theories, yet it implies a surprising development of them.

I believe that at present there has come a period of quiet but fruitful toil for the investigator of physical chemistry. The ideas are not only at hand, but they have attained a certain maturity. It is fortunate that new thoughts are always fruitful, in that they are followed by a period of enthusiastic activity, and thus at present we see the powers of investigation of many cultured nations occupied with rare unanimity in the hearty and successful development of the general system of theoretical chemistry.

At such times there is especial need of a statement of the guiding ideas, which shall not only give instruction to the student, but which shall also give advice to the investigator. As is well known, this need was completely satisfied in every respect by the two most excellent treatises of Ostwald, viz. the short *Outlines* and the larger *Manual of General Chemistry* (*Lehrbuch der Allgemeinen Chemie*), the latter of which has passed to a new edition.

Therefore I entertained serious doubts as to the advisability of a new work on the same subject from the same point of view, at least as regards all the essential questions. But in such a case this adage seemed to be apposite, viz. "When many come, much comes"; and my scruples were finally completely removed by the direct encouragement of Professor Ostwald in the preparation of this separate edition of the introduction to Dammer's Handbook; and I was supported in this undertaking not only by the editor, Dr. Dammer, but also by the kindness of my publisher, Mr. F. Enke [of Stuttgart].

Of course in a treatise on theoretical chemistry, the different chapters of physics and chemistry must find their place. And the essential contents of this will invariably imply so much of chemistry as shall be indispensable for the physical investigator, and so much of physics as shall be indispensable for the chemical investigator; and in all of this the physicist must conduct himself as a specialist of physics, and the chemist as a specialist of chemistry. Thus the development of physical chemistry as a special branch of natural science means—and I would lay particular emphasis on this—not so much the shaping of a new science, but rather the co-operation of two sciences which hitherto have been, on the whole, quite independent of each other.

In the selection of the material which has been furnished by the physical or chemical investigator, I have not been guided by the aim to make this as complete as possible,—a task to which I do not feel equal,—but I rather wish to describe thoroughly, only those experimental data which possess a universal significance, or which give promise of achieving it; only those hypotheses which have already proved themselves to be helpful; and finally, only those applications capable of being used *systematically*, whether their nature is that of calculation or of experiment. And this latter aim implies the description and illustration of several important pieces of apparatus for the laboratory. I attach great importance to the incidental description of some simple experiments for the lecture-desk which I have tested for several years in my capacity as Docent of Physical Chemistry.

Inasmuch as, in giving the science a form which is not rendered hazy by the accident of historical development, I have primarily sought to represent it as it appears at the present, rather than as it has been in the past—therefore of necessity the historical element has been repressed, and this resulted in more particular reference to the later literature. I trust that this method of treatment will not be interpreted as implying any lack of respect for those remarkable men who, by their work, laid the very foundation on which we build at present.

In working this out, it continually became clearer to me that, in the theoretical treatment of chemical processes,—which constitutes the chief part of my problem, — the most important foundations are: firstly, the *Rule of Avogadro*, which seems to be an almost inexhaustible "horn of plenty" for the molecular theory; and, secondly, the *Doctrine of Energy*, the authority of which is recognised by all the processes of nature. I considered that this view should be emphasised in the title of my book.

In noting corrections I have been greatly aided by the friendly assistance of Dr. C. Hohmann.

W. N.

GÖTTINGEN, *April* 1893.

CONTENTS

INTRODUCTION TO SOME FUNDAMENTAL PRINCIPLES OF MODERN INVESTIGATION

Empirical facts and their hypothetical generalisation, 1. Measurement, 4. The indestructibility of matter, 5. The transmutability of matter, 6. Simple and compound matter, 6. The indestructibility of energy, 6. Measurement of energy, 9. Equations of motion of a particle, 12. The convertibility of energy, 15. Isothermal process, 18. Transfer of heat, 19. Summary of the two laws, 22. Analytical formulation of the maximal work, 24. Free energy, 26. Conditions of thermodynamic equilibrium, 28. Physical mixtures and chemical compounds, 29. The law of constant and multiple proportions, 30. The molecular hypothesis, 31. The atom and the molecule, 32. The table of the elements, 33. Classification of natural processes, 34. The principle of arrangement, 35.

BOOK I

THE UNIVERSAL PROPERTIES OF MATTER

CHAPTER I

THE GASEOUS STATE OF AGGREGATION

The universal properties of gases, 37. The laws of gases, 38. The hypothesis of Avogadro, 39. The content of energy of a gas, 42. The specific heat of gases, 42. The specific heat of gases at very high temperatures, 43. The dependence of the specific heat of gases on the temperature, 45. The thermodynamics of gases—first law, 46. The ratio of the specific heats of a gas, 49. The thermodynamics of gases—second law, 52. The behaviour of gases at higher pressures, 53.

CHAPTER II

THE LIQUID STATE OF AGGREGATION

The general properties of liquids, 56. Surface tension, 57. Vapour pressure and the heat of vaporisation, 58. The form of the vapour pressure curve, 61. The boiling-point, 65. The critical phenomena, 66.

CHAPTER III

THE SOLID STATE OF AGGREGATION

General properties of solid bodies, 69. Melting-point and pressure, 69. The vapour pressure of solid substances, 72. The crystallised state, 74. The fundamentals of geometrical crystallography, 75. The classification of crystals according to symmetry, 79. Regular system, 83. Hexagonal system, 83. Tetragonal system, 84. Orthorhombic system, 85. Monoclinic system, 86. Triclinic system, 86. Twinning and growths of crystals, 87. The physical properties of crystals, 88. The optical properties of crystals, 90. The physical properties of lower symmetry, 92. The determination of the crystallographic symmetry, 94. Polymorphism, 95. Liquid crystals, 97. Amorphous state, 98.

CHAPTER IV

THE PHYSICAL MIXTURE

General observations, 100. Gas mixtures, 100. The physical properties of liquid mixtures, 105. The optical behaviour of mixtures, 107. One-sided properties, 109. The vaporisation of mixtures, 110. The theory of fractional distillation, 112. The thermodynamic treatment of liquid mixtures, 114. The critical point of mixtures, 119. Isomorphous mixtures, 120. The physical properties of mixed crystals, 122. Fusion and solidification of mixtures, 123. The thermodynamics of isomorphous mixtures, 128. Adsorption, 129.

CHAPTER V

DILUTE SOLUTIONS

General remarks, 130. Osmotic pressure, 130. The direct measurement of the osmotic pressure, 132. Indirect methods for the measurement of osmotic pressure, 136. Separation of the pure solvent from the solution, 136. Separation of the dissolved substance from the solution, 144. The law of osmotic pressure, 146. Osmotic pressure and concentration, 146. Osmotic pressure and temperature, 146. Osmotic pressure and the heat of dilution, 147. Osmotic pressure and the nature of the substance dissolved, 148. Osmotic pressure and gas pressure, 148. The law of absorption of Henry and Dalton, 153. The nature of the solvent, 154. Molecular state of bodies in solution, 155. Osmotic pressure and hydro-diffusion, 156. Osmotic pressure in mixture, 159. Vapour pressure of concentrated solutions, 164. Ideal concentrated solutions, 166. The extent of the domain of the laws of solutions, 168. Solid solutions, 168.

BOOK II

ATOM AND MOLECULE

CHAPTER I

ATOMIC THEORY

Combining weight and atomic weight, 172. The rule of Avogadro, 174. The law of Dulong and Petit, 175. Isomorphism, 177. The periodic system of the elements, 182. Physical properties of the elements, 189. Melting-point of elements on the absolute scale, 190. Significance of the periodic system in the table of atomic weights, 191. The spectra of the elements, 193. Regularities in the distribution of the spectral lines of the elements, 195.

CHAPTER H

THE KINETIC THEORY OF THE MOLECULE

General observations, 200. The kinetic theory of gases, 201. The rule of Avogadro, 205. The ratio of the specific heats, 206. The mean length of the free path, 208. The behaviour of gases at higher pressures, 210. The equation of van der Waals, 212. The kinetic theory of liquids, 215. The reduced equation of condition, 223. Application of the theory of corresponding states, 224. The coexistence of liquid and vapour, 227. The demonstration by Young, 230. The heat content of compressed gases and liquids, 236. Critique of the results, 239. The kinetic theory of the solid state of aggregation, 240. The kinetic theory of mixtures, 242. The kinetic theory of solutions, 243.

CHAPTER III

THE DETERMINATION OF THE MOLECULAR WEIGHT

The molecular weight of gaseous substances, 247. Regnault's method, 248. Dumas's method, 248. The Gay-Lussac-Hofmann method, 249. Victor Meyer's method by the displacement of air, 249. The determination of the vapour density at very high temperatures, 251. The determination of the vapour density at diminished pressure, 254. Calculation of atomic weight from gas density, 255. The determination of the molecular weight from the osmotic pressure of the dissolved substance, 256. The depression of the freezing-point, 257. Freezing-point of very dilute solutions, 259. The lowering of the vapour pressure, 261. The investigation of volatile substances, 265. The lowering of the solubility, 266. The distribution of a substance between two solvents, 267. The *rôle* of the solvent, 267. The molecular weight of strongly-compressed gases and of liquids, 269. Molecular weight of solids, 272.

CHAPTER IV

THE CONSTITUTION OF THE MOLECULE

Allotropy and isomerism, 273. The constitution of the molecule, 274. The chemical forces, 275. The doctine of valence, 277. The dualistic and the unitary view, 279. The changeableness of chemical valence, 280. Molecular compounds, 282. The compounds of carbon, 283. The peculiarities of the carbon compounds, 284. The methods for the determination of the constitution, 286. The benzene theory, 288. The stereochemistry of carbon, 291. Optical isomerism, 293. Geometrical isomerism, 294. The stereochemistry of nitrogen, 296.

CHAPTER V

PHYSICAL PROPERTIES AND MOLECULAR STRUCTURE

General observations, 299. Specific volume and molecular volume, 300. The density of solid bodies, 304. The refractive power, 304. The molecular refraction of organic compounds, 306. Dielectric constants, 313. Magnetic rotation of the plane of polarisation, 316. Magnetism, 316. The heat of combustion, 317. Regularities in the boiling-points of organic compounds, 321. The boiling-point curves, 325. The critical data, 326. Heat of evaporation, 326. The melting-point, 328. Internal friction, 330. The natural rotation of the plane of polarisation, 331. The decomposition of a mixture of optical isomers, 334. Quantitative relations of the rotatory power, 335. The absorption of light, 336. The theory of colouring substances, 338. Fluorescence, 340. Crystal form, 340. The systemisation of the physical properties, 342.

CHAPTER VI

THE DISSOCIATION OF GASES

Abnormal vapour densities, 344. Dissociation, 345. The extent of dissociation, 345. The physical conduct of dissociated gases—effusion, 346. Colour, 347. Specific heat, 347. Thermal conductivity, 348. The condition of dissociation, 349.

CHAPTER VII

ELECTROLYTIC DISSOCIATION

Dissociation in solutions, 350. Water solutions, 352. Electrolytic conduction, 353. The free ions, 355. The determination of the degree of electrolytic dissociation,

357. Hittorf's transportation values and Kohlrausch's law of the independent migration of ions, 360. The characteristics of electrolytic dissociation, 364. The diffusion of electrolytes, 365. The friction of the ions, 371. Other solvents, 372. Werner's theory of molecular compounds, 377. Incandescent gases, 378. Historical observations, 379.

CHAPTER VIII

THE PHYSICAL PROPERTIES OF SALT SOLUTIONS

The necessity of the additive relation, 381. The density of salt solutions, 381. The refractive power of salt solutions, 383. Light absorption and colour, 384. The natural power of rotation, 386. Ionic mobility, 387. Systematic résumé of the properties of ions, 387.

CHAPTER IX

THE ATOMISTIC THEORY OF ELECTRICITY

General, 390. Ion and electron, 391. Free electrons, 392. Action of electrons on gases, 394. Other methods of formation of gas ions, 395. Mobility of the gas ions, 396. Diffusion of gas ions, 397. Condensation by gas ions, 399. Positive and negative elements, 399.

CHAPTER X

THE METALLIC STATE

General, 401. Metallic solutions, 402. Metallic alloys, 404. Electric conductivity of alloys, 405. The theory of metallic conduction, 407.

CHAPTER XI

COLLOIDAL SOLUTIONS

Colloids and crystalloids, 408. Osmotic pressure, 409. The freezing-point and the vapour pressure, 410. The molecular weight of colloids, 411. Gelatination, 412. Dialysis, 414.

CHAPTER XII

THE ABSOLUTE SIZE OF MOLECULES

The superior limit, 416. The space occupied by molecules, 416. The density of molecules, 419. The size of molecules, 420. The number and weight of molecules, 420. The electric charge of an ion, 421. Determination of the absolute charge of an ion by J. J. Thomson, 422.

BOOK III

THE TRANSFORMATION OF MATTER

(THE DOCTRINE OF AFFINITY, I)

CHAPTER I

THE LAW OF CHEMICAL MASS-ACTION

The aim of the doctrine of affinity, 423. The condition of chemical equilibrium, 425. Reversible reactions, 426. The kinetic development of the law of mass-action, 428. The history of the law of mass-action, 432.

CHAPTER H

CHEMICAL STATICS—HOMOGENEOUS SYSTEMS

Equilibrium between gases, 434. The formation of hydrogen iodide, 435. The dissociation phenomena of gases, 438. The dissociation of nitrogen dioxide, 440. The dissociation of the compound of hydrochloric acid and methyl ether, 441. The influence of indifferent gases, 442. The influence of an excess of the dissociation products, 442. The frequency of dissociation phenomena, 443. The decomposition of carbon dioxide, 444. Decomposition of water vapour, 446. Equilibrium in homogeneous liquid systems—esterification, 447. The influence of the nature of the reacting components, 449. The dissociation of esters, 450. Equilibrium in solutions, 451. The distribution of hydrochloric acid between alkaloids, 452. Dissociation in solutions, 455. Esterification in benzene, 456. Evidence of chemical action from the osmotic pressure of solutions, 458. The influence of the solvent, 458. The participation of the solvent in the reaction, 459. Equilibrium in solid systems, 461.

CHAPTER III

CHEMICAL STATICS—HETEROGENEOUS SYSTEMS

The kind of heterogeneity, 462. General law regarding the influence of the relative proportions, 462. Complete and heterogeneous equilibrium, 464. Phases of variable composition, 466. Equilibrium between a gaseous phase and a solid substance—sublimation, 467. The dissociation of a solid substance which produces only one gas, 467. The production of one gas from several solid substances, 470. The dissociation of one solid substance which produces several gases, 470. Reaction between any arbitrary number of gases and solid substances, 473. Explanation of the preceding general case by means of sublimation and dissociation, 474. The equilibrium between a liquid phase and solid substances—the solubility of solid substances, 475. Solubility of hydrates, 477. Equilibrium between solids and solutions, 477. Several phases of variable composition—the vapour pressure of solutions, 479. The law of distribution, 481. The simultaneous evaporation of the solvent and of the dissolved substances, 481—the mutual solubility of liquids, 484. The distribution of a substance between two solvents, 485. Application of the law of partition to determination of chemical equilibrium, 487. The freezing of solutions and crystallisation from solutions, 488. The freezing-point of dilute solutions, 489. The crystallising out of dissolved substances, 490. The so-called cryohydrates, 490. The equilibrium between liquid and solid solutions, 490. The most general case, 493. Applications, 494.

CHAPTER IV

CHEMICAL EQUILIBRIUM IN SALT SOLUTIONS

The reaction capacity of ions, 496. Electrolytic dissociation and chemical nature, 499. The strength of the affinity of organic acids, 500. The mixture of two electrolytes containing the same ion, 503. The equilibrium between any selected electrolytes, 507. Electrolytic dissociation of pure water, 508. The most general case of homogeneous equilibrium, 510. The distribution of one base between two acids, 511. The strength of acids and bases, 516. Hydrolytic dissociation, 519. The theory of indicators, 521. The so-called "Hydrate Theory," 525. The mutual influence of the solubility of salts, 526. Anomalies due to formation of complex ions, 531. Application of the law of mass action to strongly dissociated electrolytes, 532. The reaction between any number of solid salts and their solution, 533. Normal and abnormal reactions, 536. Formation and solution of precipitates, 538. Distribution of an electrolyte between water and a second phase, 539.

CHAPTER V

CHEMICAL KINETICS

General observations, 541. The inversion of sugar, 543. The catalytic action of hydrogen ions, 544. The catalysis of esters, 547. Formation of sulphuretted hydrogen from its elements, 549. Unimolecular reactions, 549. Bi-molecular reactions—the saponification of esters, 550. The theory of saponification, 554. Further bi-molecular reactions, 558. Tri- and multi-molecular reactions, 559. The course and the mechanism of a reaction, 562. The reaction-velocity and the constitution, 565. Catalysis, 566. Auto-catalysis, 567. The progress of incomplete reactions, 568. Multi-rotation of milk sugar, 572. Tautomerism, 572. Complications in the course of the reaction, 574. The influence of the medium, 577. The kinetics of heterogeneous systems, 579. The linear velocity of crystallisation, 582. The kinetic nature of physical and chemical equilibrium, 583.

BOOK IV

THE TRANSFORMATION OF ENERGY

(THE DOCTRINE OF AFFINITY, II)

CHAPTER I

THERMO-CHEMISTRY I

THE APPLICATIONS OF THE FIRST LAW OF HEAT

General remarks, 586. Heat of reaction, 587. The thermo-chemical method of notation, 588. The law of constant heat summation, 591. The influence of temperature upon the heat of reaction, 594. Thermo-chemical methods, 594. Gases and solutions, 597. The changes of the state of aggregation, 597. The heat of solution, 598. The heat of formation, 600. The heat of combustion of organic compounds, 601. The thermo-chemistry of electrolytes, 603.

CHAPTER II

THERMO-CHEMISTRY II

TEMPERATURE AND COMPLETE CHEMICAL EQUILIBRIUM

The application of the second law of the mechanical theory of heat—historical, 608. Gibbs's "phase rule," 609. The point of transition, 612. The equilibrium between the different phases of water, 613. The equilibrium between water and sulphur dioxide, 616. The hydrates of ferric chloride, 622. Detection of chemical compounds by means of melting curves, 628. Systems of three components, 629. The thermodynamics of complete equilibrium, 630. Condensed systems, 631. Allotropic transformation, 633. The melting of salts containing water of crystallisation, 635. The formation of double salts, 636. The double decomposition [metathesis] of solid salts, 637. The vapour pressure and the solubility at the temperature of transformation, 638. The determination of the temperature of transformation, 640.

CHAPTER III

THERMO-CHEMISTRY III

TEMPERATURE AND INCOMPLETE EQUILIBRIUM

Thermodynamics of incomplete equilibrium, 642. The reaction isotherm, and the reaction isochore, 643. The derivation of the reaction isotherm, 645. The derivation of the reaction isochore, 649. Vaporisation, 650. The dissociation of solid substances, 651. The dissolving of solid substances, 652. The dissociation of solid substances by dissolving, 653. The dissolving of gases, 654. The dissociation of gaseous substances, 655. The dissociation of dissolved substances, 658. The most general case, 660. The process of the blast furnace, 662. The law of the mutuality of phases, 662. Influence of the extent of a phase, 663. The influence of the temperature and of the pressure upon the state of chemical equilibrium, 664. Effect of non-uniform pressure, 667. Thermodynamic potential, 668.

CHAPTER IV

THERMO-CHEMISTRY IV

TEMPERATURE AND THE REACTION VELOCITY

The acceleration of chemical reactions by means of an elevation of the temperature, 670. Non-reversible decomposition, 671. The application of thermodynamics, 673. Explosions and combustions, 675. The reaction capacity of gases, 677. Reactivity of oxygen, 679. The catalytic action of water vapour, 681. The velocity of the transmission of explosions, 682.

CHAPTER V

THERMO-CHEMISTRY V

HEAT AND CHEMICAL ENERGY

The "Principle of maximal work," an erroneous interpretation of heat of reaction, 685. The standpoint of thermodynamics, 687. The standpoint of kinetics, 688. The results of experiments, 688. The measure of affinity, 691. Methods for the determination of affinity, 693.

CHAPTER VI

ELECTRO-CHEMISTRY I

GENERAL FACTS

The conveyance of electrical energy, 696. Electrolytic conduction, 697. Electrolysis, 698. The development of electrical energy by chemical systems, 699. Special electro-chemical reactions, 701.

CHAPTER VII

ELECTRO-CHEMISTRY II

THERMODYNAMIC THEORY

Electrical work, 703. Application of the first law of thermodynamics, 703. Reversible elements, 706. Law of conversion of chemical energy into electrical energy,

709. Galvanic polarisation, 713. Thermodynamic calculation of electromotive forces, 713. Application of thermodynamics to the lead accumulator, 715. Electromotive force and chemical equilibrium, 717. The galvanic element considered as a chemical system, 718.

CHAPTER VIII

ELECTRO-CHEMISTRY III

OSMOTIC THEORY

The mechanism of current-production in solutions, 720. The dissolving of metals, 723. Theory of the production of the galvanic current, 725. Concentration cells, 726. Normal and abnormal potentials, 730. Gas batteries, 731. Oxidation and reduction cells, 731. Theory of electrolysis, 732. Chemical application of osmotic theory, 737. Theory of galvanic polarisation, 740. General theory of contact electricity, 740.

CHAPTER IX

PHOTO-CHEMISTRY

The action of light, 743. Actinometry, 745. Photo-chemical extinction, 748. Photo-chemical induction, 748. The latent light-action of silver salts, 750. The laws of photo-chemical action, 753. The theory of photo-chemical action, 735.

INDEX OF SUBJECTS 757

INDEX OF AUTHORS 765

INTRODUCTION

TO SOME FUNDAMENTAL PRINCIPLES OF MODERN INVESTIGATION

Empirical Facts and their Hypothetical Generalisation.—As the most common and immediate aim of physics and chemistry, we may propose to answer as thoroughly and as simply as possible, and for as many cases as possible, this question, viz. in a given system of limited dimensions, what events will transpire, and what will be the condition of this system after the lapse of a definite time? For the solution of this problem it appears necessary, first of all, to realise and to trace out the history of that system whose future we would follow. With what success this problem is to be attacked depends upon the skill and the resources of the observer, and results will grow with the art of experimentation.

But the unlimited variety of the systems occurring in nature to interest us on the one hand, and on the other the unlimited pains which human endeavour must apply to the explanatory investigation of any one of the complex and changeable systems, these difficulties would frighten back the discouraged student from a systematic investigation of natural phenomena, were there not furnished another assistance, in addition to the immediate impressions of the senses. This assistance is the theoretical realisation of the experiences derived from different systems, and consists primarily and solely in this, viz. that by means of a conclusion of analogy the perceptions obtained from one system may be transferred to another. If we have studied the case of the falling of a heavy body at one point of the earth's surface, then it is at once possible to transfer some of the observed phenomena to other systems; for example, to the falling of a heavy body at other points of the earth's surface. The successful scrutiny of the student of nature reveals itself in finding out that which phenomena, apparently very diverse, have in common, and the results are the more brilliant, as the parallelised phenomena appear at the outset more diverse.

The transference of observations from one case to another is naturally at first attended with uncertainty, but with the repeated corroboration of the experiences in question, there follows an ever-growing credibility, until this transference finally attains to the rank of an empirical law of nature. The discovery of such a law—as, for example, that which permits the calculation of the specific heat of solid compounds, from certain numerical coefficients of the elements, viz. the so-called atomic-heats—invariably indicates great and undoubted progress, as it embodies many empirical facts and allows the anticipation of many new ones previously unknown.

The history of the exact sciences teaches us that we may discover new laws of nature in two essentially different ways, one of which may be designated as the empirical, the other as the theoretical. Thus in one way by suitable observations, one collects abundant material, capable wherever possible of mathematical expression, concerning the phenomena between which he suspects a connection; and then by a repeated and purely empirical grouping of the data so obtained, he seeks to approach the desired goal: in this way, for example, were found certain relations between the properties of the elements and their atomic weights. The second way, on the other hand, leads from suggested conceptions regarding the nature of certain phenomena, through pure speculation to new information, the correctness of which must be determined by subsequent research; thus by kinetic considerations of the combination and dissociation of substances reacting on each other, the law of chemical mass action was discovered.

The first of these two methods, the empirical, can be proposed in all cases, and will invariably lead, after work which indeed is very painstaking, to definite results. As regards the value of a law of nature so won, in the first place its applicability will be authoritative, and the esteem which men render it will be so much the greater, as the natural phenomena concerning which the law affords information are more numerous and varied. Therefore, to offer what is unquestionably the most brilliant example of a law of nature empirically proven, viz. the doctrines of thermodynamics, these are applicable to every occurrence which we see transpiring in nature, and therefore demand attention in the scientific investigation of every particular natural phenomenon. On the other hand, to be sure, the comprehension of such a far-reaching law of nature will be the more difficult, and its treatment will require more practice, as it is more universal, and in the given cases, the difficulties of a correct and complete application to a particular phenomenon are commonly very great; and though the successful transference of universal principles to special cases possesses undoubted scientific significance, yet the result so obtained in utilising a universal principle may offer nothing new.

Although this purely inductive method has always had, and will always have great consequence for the advancement of the natural sciences, yet we undoubtedly penetrate more deeply into the nature of the phenomena considered, when by means of the second way, viz. on the basis of suggested conceptions and their subsequent realisation, we arrive at new laws of nature; and therefore this latter way appears the more fascinating. Obviously it is only by the fortunate choice of conceptions, fundamental in theoretical considerations, that we can arrive at safe conclusions. But now, according to the nature of the case, it is commonly true that we cannot subject these fundamental conceptions to the direct test of experiment, in order to determine their value or worthlessness; and the investigator who rashly works in this way, is continually in danger of being led astray by the *ignis fatuus* of principles unfortunately chosen.

Conceptions of this sort, incapable of direct proof by experiment, are called hypotheses: as, for instance, the assumption of a space-pervading luminiferous ether which, as though itself of imponderable mass, avoids our senses which are identified with matter; or the assumption that all matter consists of very small but not infinitesimal particles, incapable of further subdivision, and which, on account of their smallness, are directly intangible to the senses. The introduction of these hypotheses, as above observed, is necessitated in order to obtain a deeper knowledge of natural phenomena, which in turn leads to new and legitimate results. These are accessible to experiment, and good results prove, not the correctness of the hypothesis itself indeed, but its utility; while a miscarriage shows beyond a doubt not only the unsuitableness, but also the incorrectness of the conceptions from which we have started.

In suitable ways the hypothesis is a very important assistance to science; it has no self-interest at all, at least not for the student of the exact sciences, but must rather adduce proof for its own right of existence, in that like a bridge it unites known empirical facts with one another, or it brings us to new ones. The advantage of a good hypothesis consists essentially in deepening and broadening our knowledge of natural phenomena, *i.e.* in doing the same as an empirical law of nature. That the human mind has turned with predilection at all times, though in different degrees, to the elaboration of hypotheses, is to be ascribed to the circumstance that the mental satisfaction afforded in finding a new law of nature is greater when it is inferred deductively from universal generalisations, than when inferred inductively from knowledge tediously acquired.

We may say then that the speculative activity of the investigator must be directed, not only to observation and measurement of phenomena, but to the discovery of the most general laws and the most useful hypotheses. When these theorems are put into words or formulæ others than the discoverer can take part in testing them; and any really sound new theorem brings the power of foretelling a whole sheaf of detail. "He who learns the law of phenomena wins not only learning, but the power of entering into the course of nature and of working on it further according to his will and need. He wins insight into the future course of these phenomena. He wins indeed faculties that in superstitious times were looked for in prophets and magicians" (Helmholtz, Goethe Lecture, 1892).

At present we possess some empirical laws and hypotheses which have the widest application in every branch of natural science; these, accordingly, deserve a preliminary treatment in a detailed didactic manner, and demand a special consideration in showing the present condition of theoretical chemistry. The law of the indestructibility of matter was first clearly demonstrated by chemical investigation. The law of the indestructibility of energy has called into being a special branch of chemistry, viz. thermochemistry; and the fruitfulness of the law of the convertibility of heat into external work has perhaps never had a more brilliant illustration than in its application to chemical processes. Finally, the atomic and molecular hypotheses appear to be indispensable for a comprehension of the nature of chemical compounds.

Measurement.—It must be the constant endeavour of the investigator to give his observations a quantitative form. A description of phenomena is usually unintelligible and misleading, repetition of observations is made greatly more difficult, when data as to the magnitudes concerned are not given. For didactic treatment of natural phenomena, reference to quantitative relations is equally indispensable.

The choice of units is arbitrary; various practical and historical accidents have led to the fundamental units of length, time, mass, and temperature. Similarly other quantities that arise in our growing knowledge of nature give occasion for new units, many of which are in use; for example, the "atmosphere," "candle-power," horse-power," calorie, etc.

It was consequently a great advance when Gauss (1832) and Weber (1852) showed in the case of magnetic and electric units that this arbitrariness could be, if not quite done away with, at least much restricted. Their method was to *use the laws of nature to define new units.*

Thus instead of comparing electric currents in any arbitrary way, and so restricting themselves to *relative* measurements, they made use of the electrodynamic action between currents to refer current strength to the fundamental units, and defined the unit current *absolutely* as such that two portions, each 1 cm. long, and at a great distance L apart on the same axis, exercise a force $1/L^2$. The unit of resistance thus follows immediately as that in which unit current produces in unit time an amount of heat equivalent to the unit of work; and electromotive force is defined by means of Ohm's law, so that the potential difference between the ends of a conductor of unit resistance is unity, when unit current flows through it.[1]

This method of referring new units to old, and comparing new quantities by reference to those previously known, is not free from the arbitrary; in the foregoing case, the unit of current might equally well be defined by means of elements of a different shape or position; or, as Gauss and Weber pointed out, electrostatic forces instead of electrodynamic might be made the basis of the system of measurement. Still the arbitrariness is much reduced by the principle of Gauss and Weber. But it is even more important that the absolute system does away with the numerical factor in the expression of many laws of nature, so that they assume a simpler form; thus this system, like a good theorem, gives the physicist a quantity of detailed knowledge of the art of measurement, and of the most varied character.

The four fundamental units so far generally used, and mentioned above, are not necessary. Mass may be referred to length and time by means of Newton's law of gravitation (Maxwell): The gas equation $pv = RT$ may by putting $R = 1$ be used to define temperature in mechanical measure, as a quantity of energy. For various reasons such definitions are not convenient, at least at the present time, and it will be well to wait for the discovery of new laws of nature before reducing the number of fundamental units. The latter should of course be so chosen that they can be as accurately measured as possible; this condition is satisfied by length, mass, time, and temperature to a high degree, but not by energy, which for that reason is not acceptable as a fundamental unit.

The Indestructibility of Matter.—Numerous investigations have shown that neither by physical change of a substance, as, for example, by pressure, heating, magnetisation, etc., nor by chemical decomposition, does there occur a variation of its mass, as measured by the attraction of the earth (Lavoisier). Innumerable chemical analyses and syntheses speak for the correctness of this law; in spite of the mighty chemical processes occurring in the sun, its attraction of the planets remains unchanged—an extraordinarily sharp proof that in these processes the total mass of the sun itself remains unchanged.

The question whether the weight of a product of reaction is equal to that of the reacting bodies has lately been tested by H. Landolt, with great accuracy (*Ber. d. d. chem. Ges.* [1893], **26.** 1820; *Zeitsch. f. phys. Chem.* [1893], **12.** 1). It appeared that in the cases investigated the change of weight due to chemical reaction was at most a

[1] Further details in F. Kohlrausch, *Leitfaden der prakt. Physik, Anhang.*

millionth part, probably much less, and in no case was certainly proved at all.

The Transmutability of Matter.—The properties of a substance vary with the external circumstances under which we study it; but, nevertheless, for a slight change of external conditions (especially of pressure and temperature) there corresponds only a slight change in the *physical* properties of a substance. On the other hand, if we bring together different substances, as, for example, sugar and water, or sulphur and iron, even when the same external conditions are maintained, there commonly occurs a deep-seated change in their properties, producing substances which on comparison with the original are very different in many respects. Thus it is possible for the same substance under the same external conditions to assume entirely different external properties: the substance is convertible into another.

But according to our experiments hitherto, the convertibility of matter is limited to certain conditions. The law of the indestructibility of matter furnishes the first limitation, viz. that in any event, this change in physical property concerns only *identical masses* of the substances [*i.e.* that in the change there is neither gain nor loss of mass]. Further experience gained in this direction,—the result of a vast amount of painstaking work of the chemical laboratory, from the attempts of the alchemists to change base metals into gold, to the wonderful syntheses performed by our organic chemists of the present,—this experience has brought the further knowledge that, in general, even identical masses of substances *essentially different* are not convertible into each other.

Simple and Compound Matter.—Innumerable investigations, which have had for their object, on the one hand, by chemical analysis, to reduce compound matter into simpler, and, on the other hand, by chemical synthesis, to bring together different substances to make a new one, have led to the conviction that, in decomposing the substances occurring in nature, one always comes to a number of substances incapable of further decomposition, the so-called "ground-stuffs" or elements; of these we can thus far isolate about seventy. Every attempt at the further decomposition of these "ground-stuffs" has thus far been fruitless; but from these ground-stuffs" can be made synthetically all the substances collectively known to us. Only those substances are convertible into each other which contain the same elements, and indeed each element in the same proportion.

The Indestructibility of Energy. (The first law of thermodynamics).—Many fruitless attempts to find a *perpetuum mobile*—*i.e.* a machine which of itself is able to perform external work continuously, and to an unlimited degree,—have finally led to the conviction that such a thing is impossible, and that the fundamental notion of making such a machine is in opposition to some law of nature. This law of nature is stated in the following way:—If one subjects any selected system to a reversible process, *i.e.* if he allows any changes whatsoever to occur so that finally it shall return to its original condition, then the external work A, performed by the system during the reversible process, is proportional to the amount of heat, W, absorbed at the same time, *i.e.*

$$A = JW \quad . \quad . \quad . \quad . \quad . \qquad (a)$$

The coefficient J, the mechanical equivalent of heat, is independent of the nature of the system selected, and its numerical value varies only with the scale of measurement employed to determine the amount of heat and external work.

If any system whatever is subjected to any desired changes, these are, in general, identified with the following changes in energy: firstly, a certain amount of heat is either absorbed or given out; secondly, a certain amount of external work is either performed by the system or is performed against it; thirdly, the internal energy of the system will either diminish or increase. In general in any event the diminution of the internal energy U must be equal to the external work A accomplished by the system, minus the amount of heat Q absorbed: *i.e.* the following relation exists—

$$U = A - Q \quad . \quad . \quad . \quad . \quad . \qquad (b)$$

In this equation all the quantities must be expressed in terms of the same unit of energy, *e.g.* heat must be put into work-units.

Of course each of these three quantities may be negative: thus when by the changes there occurs a development of heat Q is negative, when there occurs an increase of internal energy U is negative, and when there occurs an introduction of external work A is negative. When the system considered consists of substances capable of reaction, and when the change is a chemical decomposition, then respectively $-Q$ signifies the heat of reaction, U the change of the substances' internal energy occasioned by the decomposition, and A the external work performed by the reaction, and which consists in overcoming the external pressure, and, as above, is positive when the reaction is accompanied by an increase of volume of the system, but negative when accompanied by a diminution of volume. When, as often happens, the external work is negligibly small, U is equal to the heat evolved in the reaction.

If we bring a system that has suffered any change back to its original state, the work done by the system is, according to equation (*a*), equal to the heat supplied. Hence by (*b*) U must be zero, or the system possesses the same content of energy as before the change; the energy contained at any moment is therefore completely determined by the state of the system at that moment.

U must therefore be a single valued function of the variables characterising the system; and dU be capable of being put in the form of a complete differential. If, for example, the only external work done is against an external pressure, the system is in general completely defined by the temperature T and volume v, and we may put

$$dU = \frac{\partial U}{\partial v} dv + \frac{\partial U}{\partial T} dT.$$

As already stated, equation (*b*) is applicable to every occurrence; for it is the direct analytical expression of the law of the *conservation of energy*. A change of the internal energy, or, as also designated, "the total energy" of a system can occur in very many ways; as partly by a simple change of temperature, partly by isothermic changes of condition, and partly by the two together. In the first case, the change of energy, which is usually determined easily and exactly, is measured by the product of the heat capacity of the system and the change of temperature; in the second case, by a certain quantity of energy (as the latent heat, heat of reaction, and the like), + the external work; the third case, finally, can always be ascribed to the first two cases, as the following consideration shows:—

Let a system suffer any desired change, and at the same time let its temperature change from T to T + t. Now, let us think of this process as conducted in the two following ways: in the first way, the process is finished at constant temperature T, whereby the change of energy amounts to U_T, and then the system is warmed to T + t, whereby it needs the introduction of Kt calories, if K denotes the heat capacity of the system after it has suffered the change; in the second way, the system is warmed at once from T to T + t, whereby it needs an introduction of K_0t calories of heat, if K_0 denotes the initial heat capacity, and after this, the process is finished which is associated with a change of energy amounting to U_{T+t}. In both ways, we pass from the same initial to the same final condition; then, according to the law of the conservation of energy, the changes of energy in both cases must be equal: in the first case, the diminution of total energy amounts to $U_T - Kt$, in the second, to $U_{T+t} - K_0t$, and therefore we have the following equation—

$$U_T - Kt = U_{T+t} - K_0t,$$

or

$$K_0 - K = \frac{U_{T+t} - U_T}{t}.$$

The right-hand expression is the increase in the energy change per degree rise of temperature, and for small temperature differences may be written $\frac{dU}{dT}$; this, according to the above law, is equal to the difference between the thermal capacities of the system before and after the change. If, for example, we consider the process of fusion, the proposition given above declares that the heat of fusion of one gram of a solid substance increases as much for every degree of temperature elevation as the specific heat c, of the fused substance, is greater than that c_0, of the solid.

If a homogeneous substance be heated through dT, at constant pressure, the heat required is $c_p dT$, where c_p means the thermal capacity at constant pressure; the process may, however, be conducted in this way: raise the temperature through dT at constant volume, with absorption of heat $c_v dT$ (c_v = thermal capacity at constant volume), then allow it to expand isothermally by the amount dv, for which the quantity $-\frac{\partial U}{\partial v}dv + pdv$ will be required. Put dv equal to the expansion that would be caused by rise of temperature dT at constant pressure, and we arrive in each case at the same final, from the same initial state, so that the quantities of heat absorbed must be the same—

$$c_p dT = c_v dT + \left(p - \frac{\partial U}{\partial v}\right)dv,$$

or

$$c_p - c_v = \left(p - \frac{\partial U}{\partial v}\right)\frac{\partial v}{\partial T}.$$

The law of the conservation of energy, above all other laws, has introduced a new epoch of investigation of nature; it was clearly stated for the first time by Julius Robert Mayer (1842), but was first recognised in its full significance, and in its legitimate employment was first applied to the most various phenomena, by Hermann von Helmholtz, in his paper "On the Conservation of Force," 1847.[1] It received its first quantitative confirmation by the fundamental investigation of Joule (1850) on the conversion of work into heat, which in turn led to the determination of the mechanical equivalent of heat.

The share taken by these investigators in the common work is well characterised by Mach (*Prinzipien der Warmetheorie*, Leipzig, 1896, p. 268): "Mayer brought out clearly the dependence on law, and showed its applicability to all subjects. To Helmholtz is due complete critical working out in detail, and the relation to previous knowledge. Joule introduced the new method and conception, in the most varied manner, into the region of quantitative experiment."

Measurement of Energy.—Since we shall have much to do with energy, some special remarks on its measurement may be in

[1] Ostwald's *Klassiker*, No. 1. Leipzig, 1889. *Ueber die Erhaltung der Kraft.*

place here. The absolute system gives as unit the work that is done in moving the point of application of unit force throughout 1 cm. The unit force called the dyne is that which in one second gives to one gram the unit velocity (1 cm./sec. called "cel" from celeritas); it is, moreover, nearly the weight of a milligram (more exactly $\frac{1}{980 \cdot 6}$ gm. in middle latitudes). The corresponding unit of work is called the "erg" (ἔργον), and is of course equal to the kinetic energy $\left(\frac{m}{2}v^2\right)$ of two grams moving with a speed of 1 cel.

This unit of work is often inconveniently small, and for a long time past other units, suited to particular purposes, have been in use. The "kilogrammeter" is used in technology, *i.e.* the work done in lifting one kilogram through one meter, the unit of length being here taken as the meter, and that of force as the weight of a kilo. But as work can be done in increasing a volume against pressure, or causing electricity to flow against a difference of potential, units of work are suggested for such cases in the form of products of pressure by volume, and quantity of electricity by potential. If, as is customary in scientific calculations, the C.G.S. system is adhered to, the unit of work is of course always the same; but if, as we shall occasionally do for clearness, the conventional measures are used, the work unit will of course be different in different cases.

The unit of heat is determined in principle by the law of conservation of energy as being equivalent to the unit of work. But here also for practical reasons the reduction is often avoided, and a special heat unit, in closer connection with the methods of measurement, adopted; as such we shall always use the gram calorie (cal.), *i.e.* the heat required to raise one gram of water through 1° on the air thermometer scale.[1] But as the specific heat of water is appreciably variable, it is necessary to complete this definition by stating the temperature of the water. Now, by far the most calorimetric measurements, especially in thermochemistry, are made by observing the rise of temperature produced in water at atmospheric temperature by the heat to be measured; so that it is best for our purposes to choose as unit that quantity of heat which will raise one gram of water at 15° through 1° Cels.; and between 15° and 20° the specific heat of water may for most purposes be regarded as constant.

Besides the calorie given above, there are also the so-called "mean calorie," *i.e.* $\frac{1}{100}$ of the quantity of heat which is necessary to warm 1 g. of water from 0° to 100°; the so-called "zero-calorie," *i.e.* the quantity of heat which is necessary to raise 1 g. of water from 0° to

[1] For reduction of a Jena glass mercury thermometer to the air scale, see Wiebe, *Z. S. f. analyt. Chem.*, **30.** 1; *Chem. Centralbl.*, 1891, **1.** 249; *Z. S. f. Instrumentenk.*, **10.** 233, 435 (1890).

1° ; and also a number of other calories which refer to temperatures arbitrarily or casually chosen.[1]

It appears that the variation of the specific heat of water has been determined recently with satisfactory exactness, so that different calories can be compared with one another; and since we must frequently calculate the reduction of the valuation of calorimetric measurements of very different observers, there are arranged in the following table the recent results of some observers of the specific heat of water. In the first column are the figures of Rowland,[2] obtained by Joule's method of the change of friction into heat at different temperatures; in the second column the figures of Bartoli and Stracciati,[3] obtained by the method of mixing; in the third those of Lüdin [4] by the same method; in the fourth those of Callendar and Barnes,[5] by a continuous electric calorimeter.

	I.	II.	III.	IV.
0°	...	1·0080	1·0051	1·0044
5°	1·0054	1·0046	1·0027	1·0027
10°	1·0019	1·0018	1·0010	1·0012
15°	1·0000	1·0000	1·0000	1·0000
20°	0·9979	0·9994	0·9994	0·9990
25°	0·9972	0·9997	0·9993	0·9982
30°	0·9969	1·0000	0·9996	0·9977
35°	0·9981	...	1·0003	0·9973

The zero calorie is consequently 1·005 times the usual calorie; the value of the mean calorie according to the older measurements of Dieterici is 1·013, whereas Lüdin found the two units exactly equal.

For the mechanical equivalent of the usual calorie at latitude 45°, Pernet (*l.c.*) reckons 42,555 gram centimeters from the experiments of Joule, 42,547 from Rowland, and 42,637 from Miculesen; we shall adopt 42,600. The meaning of this number is that if in latitude 45° a gram be allowed to fall through 42,600 centimeters the kinetic energy acquired converted into heat would suffice to raise the temperature of a gram of water at 15° by 1° on the air thermometer scale.

[1] Considerable confusion unfortunately exists here, and it must therefore be noted as inadmissible for one to publish or to apply numerical values on the mechanical equivalent of heat, on the heat of fusion, etc., without at first emphasising *which calorie* is employed as the unit in the method of observation. Greater care on this point is the first step toward making the accuracy of calorimetric measurements approach the accuracy attained long ago in optical and electrical measurements.

[2] *Mechanical Equivalent of Heat, etc.* Cambridge, 1880.

[3] *Calore specifico dell' aqua.* Catania, 1892.

[4] *Dissertation*, Zurich, 1895; see also a critical discussion by J. Fernet. *Vierteljahrschrift der nat. Ges.*, Zürich, **41**, Jubelband II. 1896.

[5] Brit. Ass. Rep. Dover, 1899, p. 624; *Z. S. f. Instrumentenk.*, **20**. 270 (1900).

In the absolute system of measurement the value of the ordinary calorie is

$$42{,}600 \times 980{\cdot}6 = 41{,}777{,}000 \text{ ergs}$$ [1]

(980·6 = acceleration of gravity in latitude 45°).

Very often, and in calculations which are of especial importance for the chemist, the problem is given to express, in units of heat, the work performed in overcoming the pressure on a definite volume. For example, in a cylinder with a movable piston of a cross-section of 1 sq. dcm., on which the atmosphere presses with a pressure of one atmosphere, let the piston be raised 1 dcm. so that the atmospheric pressure will be forced back from the space of a liter. This unit of work, after the analogy of the "meter kilogram," we appropriately call a "liter-atmosphere." The pressure of an atmosphere on a sq. cm., as is known, is 1·0333 kg., and on a sq. dcm. 103·33 kg.; the work performed in raising the piston is as great as though 103·33 kg. were raised 1 dcm., or as though 1 g. were raised 1,033,300 cm. Therefore the work sought, in calorimetric units, is

$$1 \text{ Litre-atmosphere} = \frac{1{,}033{,}300}{42{,}600} = 24{\cdot}25 \text{ g.-cal.}$$

Equations of Motion of a Particle.—Some remarks on this point will be inserted here, partly to elucidate the law of conservation of energy, partly because we shall later have to do repeatedly with the movement of particles. If a particle of mass m move in a direction which may be defined as that of the axis of x in a system of co-ordinates, under the influence of a force X, and t is the time, then the fundamental laws of dynamics are expressed by

$$m\frac{d^2x}{dt^2} = X \quad . \quad . \quad . \quad . \quad (1)$$

(or mass × acceleration = effective force). Multiplying (1) by the identity

$$\frac{dx}{dt}dt = dx,$$

and remembering that

$$m\frac{dx}{dt}\frac{d^2x}{dt^2}dt = d\left[\frac{m}{2}\left(\frac{dx}{dt}\right)^2\right],$$

and finally replacing the velocity $\frac{dx}{dt}$ by the symbol v we have

$$d\left(\frac{m}{2}v^2\right) = Xdx \quad . \quad . \quad . \quad . \quad (2)$$

or in words: the increase in kinetic energy of the particle during any element of time is equal to the work spent by the force.

[1] According to Griffiths (*Phys. Zeitsch.*, **4**. 176, 1902-3), 41840000 ergs.

If the particle is travelling in the direction s which at the moment considered makes with the axes of a rectangular co-ordinate system the angles β, γ, then

$$\cos \alpha = \frac{dx}{ds}, \quad \cos \beta = \frac{dy}{ds}, \quad \cos \gamma = \frac{dz}{ds}.$$

If forces X, Y, Z act along the axes, the force T exerted on the point is

$$X \cos \alpha + Y \cos \beta + Z \cos \gamma,$$

and equation (1) becomes in this case

$$m\frac{d^2s}{dt^2} = X\frac{dx}{ds} + Y\frac{dy}{ds} + Z\frac{dz}{ds},$$

or, by a similar transformation to that previously used

$$d\left(\frac{m}{2}v^2\right) = Xdx + Ydy + Zdz,$$

in which the velocity

$$v = \frac{ds}{dt}.$$

If we have a system of particles, acted on by no external forces, the work done depends only on the forces between the particles; and if the right-hand expression is a complete differential, its integral is called the *potential*, and in this case the work is done exclusively at the cost of the potential energy. Then the increase of kinetic energy is equal to the decrease of potential energy, or the total energy of the system is constant (law of conservation of energy in mechanical systems).

If the particle suffers friction in its course, equation (1) requires a limitation. The friction is to be regarded as a force acting in the direction opposite to that of the motion of the particle at the moment, and in many cases is proportional to the velocity v. Hence there acts on the particle the force $X - kv$ where k is the opposing force for unit velocity. Then (1) becomes

$$m\frac{d^2x}{dt^2} = X - kv \quad . \quad . \quad . \quad . \quad (3)$$

and (2) becomes

$$d\left(\frac{m}{2}v^2\right) = (X - kv)dx \quad . \quad . \quad . \quad (4)$$

When X is constant, v must obviously increase if $X > kv$, and decrease in the contrary case, so that in both cases $X - kv$ approaches zero, and v approaches the limit

$$v_0 = \frac{X}{k} \qquad . \quad . \quad . \quad . \quad . \qquad (5)$$

After a certain (often immeasurably short) time the velocity is equal to the force divided by the frictional resistance; consequently in the case of movement under sufficiently great friction it is not, as in pure dynamics, the acceleration, but the velocity that is proportional to the applied force.

Writing (4) in the form

$$d\left(\frac{m}{2}v^2\right) + kvdx = Xdx,$$

we see that the work spent by the (internal or external) forces, Xdx, goes partly to produce kinetic energy, partly to do work against friction. The larger the latter portion is compared to the former, the more nearly true is the very simple equation (5).

Writing (5) in the form

$$kv_0 = X,$$

multiplying by dx and integrating from 0 to x and 0 to t we get

$$\int kv_0dx = \int kv_0^2dt = \int Xdx,$$

or

$$kv_0^2t = Xx \qquad . \quad . \quad . \quad . \quad . \qquad (6)$$

But Xx is the work done on the particle in time t; this frictional heat is equal to the product of resistance (k), velocity squared, and time. The resemblance of (5) to Ohm's law and (6) to that of Joule is obvious.

To calculate the time after which (5) may be regarded as true, we will write (3) in the form

$$\frac{dv}{X - kv} = \frac{dt}{m}$$

and integrate

$$-\frac{1}{k}\log(X - kv) = \frac{t}{m} + \text{const.}$$

But for $t = o$

$$-\frac{1}{k}\log X = \text{const.}$$

whence

$$v = \frac{X}{k}\left(1 - e^{-\frac{m}{k}t}\right). \qquad . \quad . \quad . \quad . \quad . \qquad (7)$$

The expression in brackets increases rapidly with time towards the limit 1, *i.e.* (7) passes over into (5); and the more rapidly the larger the friction k and the smaller the mass m.

The Convertibility of Energy. (The second law of the mechanical theory of heat).—While the law of the conservation of energy furnishes us with the quantitative relations which must necessarily be satisfied in changing into each other the various forms of energy, as external work, heat, internal energy, etc., the so-called "second law of the mechanical theory of heat" teaches us the limitations of the convertibility of the different forms of energy. Its qualitative side may be fairly expressed in the following statement: *External work, such as the kinetic energy of moving bodies, may be transformed in many ways and completely into another form, into heat* (most simply in this way, that external work may be applied to moving ponderable masses, and this may be spent by friction and converted into heat, as by applying brakes to a railway train); *but conversely the reconversion of heat into work is not at all, or only partially, possible* (*principle of Carnot and Clausius*).

The order of thought by which one is led to sift this question, viz. how far the different forms of energy are convertible into each other, and to the conjecture that here we have a restricting law of nature, is essentially the following:—As fruitless as have been the endeavours of innumerable inventors to construct a machine which should be able to do work continually without ever needing an expenditure of force to keep the machine in motion, so brilliant is the knowledge which explains this miscarriage by a law of nature. On the ground which was richly fertilised by the ruined fancies of unlucky inventors, there flourished the tree of the knowledge of the law of the indestructibility of energy, the golden fruit of which was plucked by Mayer and Helmholtz. Yet, in the judgment of some inventors who are completely permeated with the accuracy of this law, it is by no means regarded as impossible to construct a machine which should be able to furnish work as desired and free of cost. According to this law, external work and heat are equivalent to each other—both are manifested forms of energy. But energy in the form of heat is in abundance, so that it only needs an apparatus in which one shall apply it in driving our machine, to use up the energy of its environment. Such an apparatus, for example, might be sunk in a great water reservoir, whose enormous quantity of energy could be changed into useful work; it would, for example, make the steam-engines of our ocean steamers unnecessary, and would keep the screw of a ship in motion as long as desired, and at the cost of the immeasurable store of heat in the sea. Such an apparatus would be in certain respects a *perpetuum mobile*, and yet no impossible thing to contradict the first law of thermodynamics, since, acting according to this law, it would extract the heat of its environment and give it back again as external work, which, after its expenditure, according to rule—in this case, as a result of the friction of the screw —would change itself back again into heat, to enter on the cycle anew.

Unfortunately such an apparatus, which would make coal worthless as a source of energy, appears to be a chimera, exactly as was the *perpetuum mobile* of the inventor of the last century; at least many fruitless attempts have made this more than probable. Thus, as we sum up the numerous abortive endeavours, we come, in a way analogous to that which led to the knowledge of the first law of thermodynamics, to the proposition that an apparatus which could continually change the heat of its environment into external work, is in contradiction to a law of nature, and therefore an impossibility. Although, by recognising this law, the human spirit of invention may be poorer by one problem, yet natural investigation is compensated for it by a principle of almost unlimited application.

This result, which at first appears to have an entirely negative character, can be applied in finding the quantitative statement of the law, only by the aid of certain considerations, and by inference from some empirical facts; this law which limits the convertibility of energy is comprehended under the name of the "*second law of thermodynamics.*"

The fundamental considerations which led to this law were stated with great clearness by Carnot[1] as far back as 1824, before the law of conservation of energy was precisely understood; but to bring the second law into the form of a universally applicable and fruitful law of nature, and to give it exact mathematical expression, was the undying work of Clausius.[2] The following is a brief deduction of the fundamental formulæ:—

The earliest applications were technical, but to give the result of experience in the form most suitable as a physical principle we may say:—

I. *Every process which takes place in a given system by itself*, i.e. *without application of energy in any form, can, when properly used, yield a finite amount of work.*

By process is meant a change in the system by which it passes from an initial state to a different final state. If application of energy from without is not excluded, a system can naturally yield an indefinite amount of work, *e.g.* an electromotor supplied with sufficient current may be regarded, at least in principle, as an inexhaustible source of work.

We may ask now what is the best use, *i.e.* how the process may be made to give the maximum amount of work. For this it is obviously necessary that the apparatus used to gain the work must be technically

[1] *Réflexions sur la puissance motrice du feu.* Paris, 1824. German translation, Ostwald's *Klassiker*, No. 37. Leipzig, 1892.

[2] The separate papers from 1850 on were collected by Clausius in his book, *Mechanische Wärmetheorie.* Braunschweig, 1876.

efficient, so that loss of external work due to secondary defects is avoided (friction and similar causes, leakage of a piston in a cylinder during expansion or compression of gases, faults of insulation in electrical circuits, loss of heat in thermal machines, etc.); but, next, the change must be so conducted that force and resistance are almost equal at every stage of the process. If the resistance is made smaller, the process will go in one sense; if greater, in the opposite; and as all losses are avoided, as much work is gained in the one case as is absorbed in reversing it. In such cases the process is called reversible; we shall find cases in which this ideal limiting state can, at least in theory, be approached. We shall assume that this is general, and postulate that—

II. *A process yields the maximum amount of work when it is conducted reversibly.*

We may easily see that I. and II. are identical with the law that no arrangement is possible by which work shall be continuously performed at the expense of the surrounding heat. According to I., since a process can yield only a finite amount of work, such an arrangement must be a periodically acting machine, which, after a certain time, returns to its initial state; according to II., such a machine at the best (ideally perfect construction), after a period, has absorbed no external work; it can never yield any, because, working reversibly in order to avoid losses, the work given out during the outward process must be equal to that taken in in the return.

Examples of self-acting processes in the sense of I. are: the falling of a stone to earth, mixing of two liquids or gases, solution and diffusion of solids in a solvent, and all chemical reactions that take place of their own accord. The problem of calculating for each case the maximum work obtainable, *i.e.* when the process is conducted reversibly according to II., is of the highest importance, and its solution, in particular instances, has led to fruitful discoveries.

When bodies at different temperatures are put in contact, a transfer of heat from hotter to colder occurs. This process, it is well known, takes place of itself, for no work need be done to make it go; from I. it follows that it is possible to gain work by means of this interchange of heat, and, on the other hand, that work is required to reverse it, *i.e.* to transfer heat from the cold to the hot body.

Clausius stated the last result as a separate principle, "that heat cannot of itself, *i.e.* without compensation by means of external energy, pass from a colder to a hotter body." But this principle is really only a case of a more general law.

We shall now, in order to elucidate these general discussions, apply them to two specially important cases: first, isothermal processes; second, processes consisting essentially in transfer of heat (equalisation of temperature differences).

1. Isothermal Process.—Suppose the system undergoing a change to be throughout at the same temperature, and to be immersed in an indefinitely large bath at the same temperature; further, that all processes are conducted so slowly that any heat evolved is given up to the bath, and any required is absorbed from it, without appreciable temperature differences being set up. The system is then clearly not isolated, for it is in thermal connection with the bath; but we may treat the system and the bath together as a new isolated system, to which the preceding considerations are applicable.[1] Also let it be possible to conduct the processes in question reversibly.

The last condition is essential. Moreover, the problem of conducting a process isothermally and reversibly has not been solved in all cases in which it appears possible. The expansion of a gas or evaporation of a liquid can be performed in this manner easily by means of a cylinder and piston. Certain voltaic cells, such as the Daniell, can be used isothermally and reversibly by combining them with a well-constructed, *i.e.* efficient, electromotor; if the cell works it causes the motor to rotate and so does external work; if the motor be reversed by application of the same amount of work from without, it causes a current to flow through the cell in the opposite sense, which reverses the chemical process that produced the original current. But in other cases such arrangements are not known; thus, it is not yet possible to conduct isothermally and reversibly the combustion of many organic compounds, or the radiation of a phosphorescent body.

Let A be the external work that a process in a given system is capable of doing, by means of an arrangement working isothermally and reversibly. Imagine now another arrangement, which, under the same conditions, but by a different mechanism, performs an amount of work A′ while the same system passes through the same change; and to fix the ideas let $A > A'$. Then, by combining the two, we can construct an arrangement of this kind; by the first, we allow the process to take place with production of external work A; by the second, we reverse it, with expenditure of work A′. The system has then passed through an isothermal and reversible cyclic process, leaving it in its initial state; this may be repeated any number of times, and each time a net amount of work

$$A - A'$$

is performed. This would be a machine capable of yielding an indefinite amount of work at the expense of the heat bath—an impossibility. We conclude, therefore, that A′ cannot be different from A or

$$A = A' \qquad . \quad . \quad . \quad . \qquad (c)$$

[1] Strictly, the system + heat bath must fall in temperature in proportion to the external work done, according to the law of conservation; but if the capacity of the bath is great enough, the fall of temperature is negligibly small.

This equation, of which we shall make repeated use, expresses **that**—

III. *The external work that can be done by a given process when most efficiently used is independent of the arrangement by which the work is obtained;* or, more briefly, *the work done in an isothermal, reversible, cyclic process is zero.*

During such a cyclic process, certain quantities of heat are, in general, given to, and taken from, the heat bath; according to the law of conservation of energy the sum of these must be zero.

2. Transfer of Heat.—The calculation of the work that can be done by the passage of heat from higher to lower temperature is of great importance; we can easily solve the problem for a simple case—a cyclic process performed with a perfect gas—since here the quantities of energy involved are known from the laws of gases.

Let there be two reservoirs, at different temperatures, say in the form of large masses of water, which we can draw on for supplies of heat. For simplicity of calculation we will suppose the two temperatures to differ only infinitesimally, and that they are (on the absolute scale) T and T + dT. We have then to think out a mechanism by which heat may be withdrawn from the second reservoir (at T + dT) and given to the first, reversibly, and so with the greatest useful effect.

For this purpose we take a cylinder, closed by a piston and containing a certain quantity of a gas—say one gram-molecule or "mol" (32 g. of oxygen or 28 g. of nitrogen, etc.). The apparatus is put in contact with reservoir I., *i.e.* immersed in the large vessel of water at T, and the gas, originally occupying a volume v_1, compressed to v_2. Work is thus spent to the amount

$$A = RT \ln \frac{v_1}{v_2},$$

where R is the gas constant,

$$R = \frac{p_0 v_0}{273}$$

(p_0 and v_0 the pressure and volume at the temperature of melting ice). The heat Q required for this, measured in the same unit as A, is

$$Q = RT \ln \frac{v_1}{v_2},$$

given to the first reservoir. Now bring the cylinder in contact with reservoir II. so that it is warmed to T + dT; the heat thus absorbed is KdT, where K is the thermal capacity of the cylinder and its contents. During the heating the volume v_2 is to be kept constant, so that no external work is done. Now, on allowing the gas to expand from v_2 to v_1, we gain the external work

$$A + dA = R(T + dT) \ln \frac{v_1}{v_2},$$

and withdraw from reservoir II. the equivalent quantity of heat

$$Q + dQ = R(T + dT) \ln \frac{v_1}{v_2}.$$

Finally, put the apparatus in contact with the colder bath, and keeping its volume constant at v_1, allow it to give up the quantity KdT of heat, and so fall to T. It is then in the initial state again.

The sum of the work done by the gas in the cycle is

$$dA = RdT \ln \frac{v_1}{v_2},$$

and at the same time the quantity of heat

$$Q + KdT$$

is carried from reservoir II. to I., and has consequently fallen in temperature from T + dT to T, whilst the quantity of heat dQ = dA is converted into useful work. KdT is an infinitesimal quantity that may be neglected by comparison with Q.

If the thermal capacity of the cylinder and its contents were zero, instead of Q + KdT only Q of heat would have been transferred from T + dT to T, and this arrangement would increase the useful effect, it is true, but only by an infinitesimal percentage. Hence we may say, without appreciable error, that the quantity Q of heat has fallen from T + dT to T during the cyclic process; comparing this with the work done we find

$$dA = Q\frac{dT}{T} \quad . \quad . \quad . \quad . \quad (d)$$

i.e. of the quantity Q *of heat transferred from reservoir* II. *to* I., *the part* $Q\frac{dT}{T}$ *is turned into external work.*

We can now easily show that this result, though derived from a special case, is general. Assume that some other reversible cyclic process exists, by which the quantity Q′ of heat falling from T + dT to T accomplishes dA′ of work. Then we may combine this with the former process, choosing the quantity of gas so that in it also Q′ of heat suffers the temperature change. Now let the two systems each pass through a cycle, in opposite senses, so that the temperature of the quantity Q′ of heat falls in the one by dT but rises in the other by the same amount, and the thermal transport is thus neutralised; then there can be no external work done, otherwise we should have constructed a machine capable of converting heat into work indefinitely, *i.e.* we must have

$$dA' = Q'\frac{dT}{T} \quad . \quad . \quad . \quad . \quad . \qquad (d')$$

Hence—

IV. *If a process consists merely in transfer of heat, and a quantity* Q *be carried from one body at temperature* T + dT *to another at* T, *the work that can be accomplished is* $Q\frac{dT}{T}$ *in whatever way this work is obtained, provided it is a reversible way.*

There are few results so fruitful as those contained in (*d*) and (*d'*), which form the quantitative expression of the convertibility of heat into work, and have been applied, especially to chemical processes, with important results.

It will be worth while to remark further on the meaning of the quantity dA (or dA'). It must not be supposed that in the equation

$$dA = Q\frac{dT}{T} \quad \text{or} \quad \frac{dA}{dT} = \frac{Q}{T}$$

dA is the work done in raising the temperature of the system by dT. This is not the case, for we have carried out the cyclic process in such a way that no work is associated with the rise of temperature. Rather dA is the *excess* of work spent in reversing at T + dT the same process that has been performed at T, so that $\frac{dA}{dT}$ is the temperature coefficient of the availability of the process for work. We should obtain a wrong value for dA if a positive or negative amount of work were associated with the change of temperature of the system, and so a change in its availability for work took place, of a secondary character with regard to the present argument.

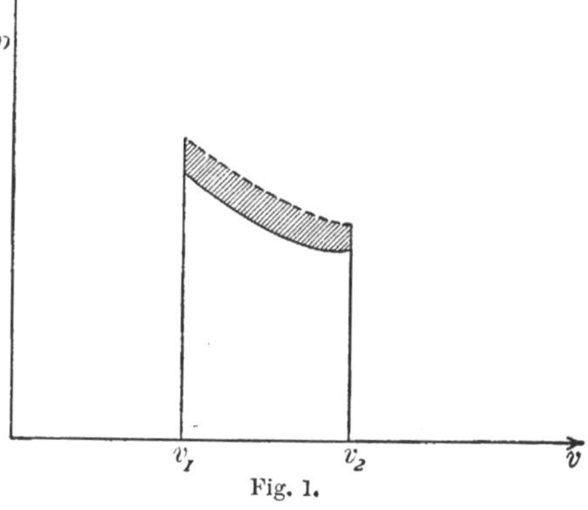

Fig. 1.

Clapeyron's graphical method is very instructive in this matter. The work done in an expansion $\int_{v_1}^{v_2} pdv$ is expressed in the figure by the area between the curve p = f(v), the two ordinates, and the axis of abscissæ. When the temperature is raised this curve is changed to

$$p = f(v) + \frac{\partial f(v)}{\partial T}dT,$$

and we thus obtain a neighbouring line (dotted in the figure). Then clearly the shaded portion expresses dA. If in any case other than expansion the work is given by $\int_{\mathfrak{w}_1}^{\mathfrak{w}_2} \mathfrak{K} d\mathfrak{w}$, then $\mathfrak{K}$ must be used as ordinate, $\mathfrak{w}$ as abscissa.

All results so far known in thermodynamics can be deduced by means of cyclic processes; the laws to be stated in the four following sections contain nothing essentially new, they are merely another way of expressing the relations already dealt with in this section. This must be expressly stated, as many authors, after establishing results by means of cyclic processes, repeat them by means of the entropy principle, thermodynamic potential, free energy, etc., and seem to regard this as a more convincing proof. In a pædagogic sense this may be true, but usually the only thing that is proved is that the author has studied the work of his predecessors with intelligence. There is no independent scientific value to be attributed to such discussions.

Which method is to be preferred—the explicit discussion of a cyclic process, or its abstraction in the theory of functions—is a matter of taste; it should be remarked, however, that while the prophets *ex eventa* of thermodynamics, with one accord, prefer the thermodynamic functions, investigators have mostly used the original method of Carnot and Clausius, as in establishing new results an experimentally realisable cyclic process offers more protection against error.

Still, any one who wishes to go deeper into the problems of the theory of heat may be recommended to a careful study of thermodynamic functions, especially in the admirably clear exposition of M. Planck (*Thermodynamik*, Leipzig, 1897).[1]

Summary of the Two Laws.—The results arrived at in the foregoing section may be summarised as follows:—

1. Every natural process can, by the most advantageous use, yield a definite amount of work, and by the expenditure of the same amount of work, it is theoretically possible to reverse the process; here by "process" we mean any change of a system that can occur without application of work from without.

2. If the process takes place isothermally, the maximal work A depends only on the initial and final states, and not on the path between.

3. If the process consists in an equalisation of temperature by which a quantity of heat Q falls from the absolute temperature T + dT to T, it is capable, whatever the nature of the system, of the work

$$dA = Q\frac{dT}{T}$$

as a maximum. (A and Q in the same units.)

[1] English translation by Ogg (London 1903).—Tr.

Now, the first law gives the relation (p. 7)

$$U = A - Q.$$

Eliminating Q from the two last equations, we have

$$dA = (A - U)\frac{dT}{T},$$

or

$$A - U = T\frac{dA}{dT} \qquad . \quad . \quad . \quad . \qquad (e)$$

Here A is the maximum work of an *isothermal* process at temperature T, U the simultaneous decrease in the energy of the system A – U (called Q in the foregoing); or the heat withdrawn from the surroundings during the process, to which the old term "*latent heat*" may conveniently be given.

Hence, according to (*e*), *the excess of the maximal work of an isothermal process over the decrease in total energy—the latent heat—is equal to the absolute temperature multiplied by the temperature coefficient of the maximal work.*

This is the most intelligible statement of the second law, because it contains only quantities with a direct physical meaning; for the change in total energy is always simply measurable by allowing the process to take place, without production of work, in a calorimeter. The maximal work is more difficult to measure, because for that the process must be conducted reversibly; but in many cases this quantity also can be directly recorded (*e.g.* by a Watt indicator, or a Siemens electrodynamometer), so that there can scarcely be any great difficulty in making its meaning clear in new instances. The above forms of statement is especially to be preferred in application to chemical and electrochemical reactions.

Remembering, further, that on account of the relation between A and U, both[1] these quantities are independent of the path, and completely defined by the initial and final states, and further, that A – U is the latent heat of the system, equation (*e*) may be regarded as including both fundamental laws.

When the work done by the process consists merely in overcoming a pressure during an expansion of the system, we can put A in the form

$$A = \int_{v_1}^{v_2} p\,dv,$$

where v_1, v_2 are the initial and final volumes respectively. Frequently

[1] U is so by the first law.

the pressure remains constant during this change (evaporation, fusion, dissociation of a solid, etc.) and then

$$A = (v_2 - v_1)p; \quad dA = (v_2 - v_1)dp,$$

and

$$A - U = Q = (v_2 - v_1)T\frac{dp}{dT},$$

where Q is the heat absorbed during the expansion from v_2 to v_1.

When p varies during the expansion (evaporation of a mixture, expansion of a gas or liquid, etc.), we may regard it as constant during an infinitesimal expansion $v_2 - v_1 = dv$, and obtain from the last equation

$$dQ = T\frac{dp}{dT}dv.$$

Here it is to be noted that dp is the increase of pressure at constant volume, but dQ means the heat absorbed when at constant temperature the volume of the system increases by dv (and external work is done). This is indicated by the symbols

$$\left(\frac{dQ}{dv}\right)_T = T\left(\frac{dp}{dT}\right)_v \quad \text{or} \quad \frac{\partial Q}{\partial v} = T\frac{\partial p}{\partial T}.$$

Analytical Formulation of the Maximal Work.—The greatest difficulty in applying the fundamental equation (*e*) lies in obtaining an analytical expression for the maximal work; for there are few cases in which A and $\frac{dA}{dT}$ can be measured by direct experiment, and otherwise the quantities must be determined by values characteristic of the initial and final states, and these values differentiated with respect to temperature, under the condition that during the temperature change no external work is done.

In most cases there is only one set of variables to express the initial and final states of a system when the temperature is given; but in case of choice the following is important. The quantity A of work can in general be decomposed into two factors, such as force × distance, pressure × volume, thermodynamic potential × mass, electric potential × quantity of electricity, and so on, of which the first is in general a temperature function, but the second can be made independent of temperature; the latter should therefore, with the temperature, be chosen as independent variable for expressing A.

This remark needs elucidation. Let A_0 be the expression characterising the initial state, F the final; then

$$A = F - A_0. \qquad (1)$$

Let A_0 and F differ only infinitesimally, so that A becomes the differential dA. If we can find a variable to the changes of which dA is proportional, we may put

$$dA = \mathfrak{K} d\mathfrak{w} \text{ (T const.)} \quad . \quad . \quad . \qquad (2)$$

in which the factor $\mathfrak{K}$ is equal to the partial differential coefficient of A with respect to $\mathfrak{w}$ and A itself, when $\mathfrak{K}$ is known as a function of $\mathfrak{w}$ can be arrived at by an integration from the initial to the final state

$$A = \int_{\mathfrak{w}_2}^{\mathfrak{w}_1} \mathfrak{K} d\mathfrak{w}.$$

The simplest case is when a force X acts at a point; if the point of application moves through the element dx

$$dA = Xdx,$$

i.e. for small displacements the changes of A and x are proportional. Now the analysis (2) appears to be always possible, so that by analogy we may call $\mathfrak{K}$ the *force* and $d\phi$ an *element of path.*

If, then, the system has several ways of doing work,

$$dA = \mathfrak{K}_1 d\mathfrak{w}_1 + \mathfrak{K}_2 d\mathfrak{w}_2 + \ . \ . \ . \text{ (T const.)} \qquad (3)$$

and $\mathfrak{K}_1$, $\mathfrak{K}_2$ are the partial differential coefficients of A with respect to $\mathfrak{w}_1$, $\mathfrak{w}_2$, etc. Thus the derivative of A with respect to volume is pressure, with respect to quantity of electricity is electromotive force, etc. So that, in general, $\frac{\partial A}{\partial \mathfrak{w}}$ is the force with which the system suffers a certain change; but as we shall be concerned almost exclusively with changes involving great friction, by analogy with (5) (p. 14), the velocity of the corresponding change is equal to the quotient of force by friction. As the latter depends essentially on the particular system, thermodynamics gives no information with respect to it.

The evaluation of $\frac{dA}{dT}$, the temperature coefficient of the maximal work, is now easy. After the value of

$$A = F - A_0$$

is arrived at by integration, we have to differentiate A partially with respect to T; since according to equation (3), when $\mathfrak{w}$ is constant no work is done, the condition (p. 21) that no work should accompany rise of temperature is necessarily satisfied.

Again, supposing F and A_0 to differ only infinitesimally, the maximal work becomes

$$\mathfrak{K}_1 d\mathfrak{w}_1 + \mathfrak{K}_2 d\mathfrak{w}_2 + \ . \ . \ .$$

where the quantities $d\phi$ represent small, but definite displacements. Applying the fundamental equation (*e*) to the process so defined, we have

$$\mathfrak{K}_1 d\mathfrak{w}_1 + \mathfrak{K}_2 d\mathfrak{w}_2 + \ldots - dU = T\left(\frac{\partial \mathfrak{K}_1}{dT} d\mathfrak{w}_1 + \frac{\partial \mathfrak{K}_2}{\partial T} d\mathfrak{w}_2 + \ldots\right)$$

or

$$\Sigma\left(\mathfrak{K} - T\frac{\partial \mathfrak{K}}{\partial T}\right) d\mathfrak{w} = dU,$$

in which, therefore, dU is the change of total energy due to the displacements $d\phi$. The equation on p. 24 is a particular case of this ($\mathfrak{K} = p$, $\mathfrak{w} = v$).

There is a relation between the values of $\mathfrak{K}$ that we will develop in the case of two kinds of external work; then

$$dA = \mathfrak{K}_1 d\mathfrak{w}_1 + \mathfrak{K}_2 d\mathfrak{w}_2 \qquad (T \text{ const.})$$

Since dA in this equation is a complete differential (independent of the "path") we must have

$$\frac{\partial^2 A}{\partial \mathfrak{w}_1 \partial \mathfrak{w}_2} = \frac{\partial^2 A}{\partial \mathfrak{w}_2 \partial \mathfrak{w}_1},$$

i.e.

$$\frac{\partial \mathfrak{K}_1}{\partial \mathfrak{w}_2} = \frac{\partial \mathfrak{K}_2}{\partial \mathfrak{w}_1},$$

a relation which in its applications coincides with equation (*c*) of p. 18. The generalisation for more than two variables is obvious, but has so far hardly found any application.

Finally, it may be remarked that the assumption of reversibility is implicitly made in all equations (such as (3)) which express the external work by means of variables, and do not contain the time.

Free Energy.—The quantity A expresses that part of the energy change involved in any process that can be turned into external work, kinetic energy, etc., without limitation—a *fortiori* into heat—and is therefore freely convertible. Helmholtz [1] gave to this the name "*change of free energy*," and to A – U, *i.e.* the difference between the changes of free and total energy, the name "*change of bound energy.*"

Physical meaning is, of course, only to be given to the *change* of free or total energy. The absolute value of these quantities is unknown, and without interest for us, since probably it does not influence the course of surrounding phenomena. In the same way, we speak only of relative movements of the bodies surrounding us, since their absolute velocities are unknown and, so far as we know, indifferent.

[1] *Sitzber. d. Berl. Akad.*, 1882; *Ges. Abh.*, **2.** 958. See also Massieu, *Journ. de Phys.*, **6.** 216 (1877), and W. Gibb's *Trans. Connecticut Acad.*, **3.** 1875-78. German translation by Ostwald, Leipzig, 1892.

Thus, writing $\mathfrak{U}$ for the absolute value of the total energy, and A for the free energy, we measure (calorimetrically)

$$U = \mathfrak{U}_2 - \mathfrak{U}_1,$$

and (*e.g.* by a Watt's indicator)

$$A = \mathfrak{A}_2 - \mathfrak{A}_1,$$

but the actual values of $\mathfrak{U}$ and $\mathfrak{A}$ escape us. We may, however, count U and A from a fixed initial condition, and speak of them as the "total" and "free" energy; then (*e*) would be expressed in words as: *the difference between the free and total energy* (also described as the bound energy) is equal to the absolute temperature multiplied by the temperature coefficient of the free energy.

According to the first law the total energy remains constant in a system isolated from communication of energy; according to the second law the same would be true of the free energy in a system at constant temperature in which all the processes were reversible. In reality it always decreases, for this ideal limiting case can never be strictly realised, but friction and similar processes constantly convert free into bound energy.

Clausius introduced a function which is useful for many purposes, viz. the entropy—

$$\mathfrak{S} = \frac{\mathfrak{U} - \mathfrak{A}}{T},$$

i.e. the difference between free and total energy divided by the absolute temperature. This is especially useful in dealing with *adiabatic* processes (*i.e.* when no heat is taken in or given out). Clausius showed that this function is increased by all irreversible processes that occur in an isolated system, so that the second law may be put in the form of the *principle of the increase of entropy.* This conception is not, however, more general than that of the decrease of free energy; the former principle is directly applicable only to isolated systems, the latter only to systems at constant temperature; but as from each we may deduce the impossibility of the engine described on p. 16, each of these special principles may be separately regarded as the most complete expression of the second law.

The combination of

$$\mathfrak{S} = \frac{\mathfrak{U} - \mathfrak{A}}{T} \quad \text{and} \quad \mathfrak{U} - \mathfrak{A} = -T\frac{d\mathfrak{A}}{dT}$$

gives

$$\mathfrak{S} = -\frac{d\mathfrak{A}}{dT},$$

i.e. the entropy is the negative temperature coefficient of the free energy. We shall henceforth always work with the intelligible notion of the maximal work (free energy), and not with its negative temperature coefficient (entropy); moreover, as Helmholtz pointed out, it is preferable to use the integral function (free energy) rather than its derivative (entropy), because $\mathfrak{U}$ can be determined from equation (*e*), and $\mathfrak{S}$ can easily be expressed in terms of $\mathfrak{A}$ by the preceding equation.

Conditions of Thermodynamic Equilibrium.—Finally, it may be remarked, the fundamental equations yield an easy test whether a system is in equilibrium or not. It was noticed above that for a reversible process force and resistance must be equal; if one slightly exceeds the other, a change takes place in one or the other sense. Hence follows that—

A system is in equilibrium at every phase of a reversible process; and for a system to be in equilibrium it is sufficient that the conditions should be so chosen that only reversible processes are possible.[1]

Thus, a mixture of chemically indifferent gases in a closed vessel is in equilibrium if the temperature and composition are the same throughout, for otherwise the irreversible processes of diffusion and thermal conduction will occur. Further, there is the condition that the containing vessel shall not burst under the pressure of the gas, for that would introduce a new irreversible process, and so on.

Since no system is in equilibrium in which there are temperature differences, we may confine ourselves to considering isothermal systems; for on bringing a system at uniform temperature into a bath at the same temperature, no existing equilibrium is destroyed nor any new equilibrium created. But as by p. 27, if all changes are accompanied by an increase of A, irreversible processes are excluded, we find that—

A system at constant temperature is in equilibrium when its free energy is a minimum.

The analytical condition is

$$\delta\mathfrak{A} = 0 \qquad (\text{T const.}) \qquad . \quad . \quad . \qquad (1)$$

Since the entropy is increased by all irreversible processes in an isolated system (p. 27), we find the corresponding condition that in a system of constant energy all changes involve an increase of the entropy $\mathfrak{S}$, or that $\mathfrak{S}$ is a maximum

$$\delta\mathfrak{S} = 0 \qquad (\text{U const.}) \qquad . \quad . \quad . \qquad (2)$$

The two equations of equilibrium (1) and (2) are however identical, for whether the equilibrium of a system is disturbed by

[1] Conversion of potential into kinetic energy is an irreversible process in all systems in nature, since friction is never entirely absent.

change of temperature, or by introduction of heat, comes to the same thing.

Physical Mixtures and Chemical Compounds.—Closely connected with the empirical laws given above, and which form at once the basis of both physical and chemical science, there is another law which concerns compound substances, and which leads us at once into the region of chemistry. Many different elements can unite in many ways so as to form new *homogeneous* substances, *i.e.* such as are everywhere alike, and which when examined with the most powerful microscope, present the same properties at all points. Experience teaches us that in all cases the properties of compound substances vary more or less with the composition, but that in no case does composition alone determine the properties. Thus "knall-gas,"[1] water vapour, liquid water, and ice, are all substances having the same composition, but having very different properties even when compared under the same external conditions of temperature and pressure.

The thorough study of the relations of compound substances regarding their dependence upon the composition, has led to their division into two classes. By the mixture of two gaseous elements, as of hydrogen and iodine vapour, for example, we can obtain, according to the conditions, two very different gaseous mixtures, each of which appears homogeneous in all physical and chemical respects, but which yet offer marked chemical differences. In one mixture we can easily recognise the properties of each particular element: the iodine vapour by its colour, and the hydrogen by its enormous diffusibility through porous partitions. In the other mixture many properties of both components are totally changed; the colour of the iodine vapour has vanished, and we seek in vain for the characteristic diffusibility of hydrogen. Further, the mixtures conduct themselves in a notably different way on condensation to the liquid or solid form: from the first, by cooling or by suitable compression, there is at once separated from the gas mixture solid iodine, a substance of entirely different composition from the gas remaining uncondensed; from the second mixture, by similar treatment, we obtain a homogeneous liquid of the same composition as the uncondensed gas.

We call the first of these gaseous admixtures, a *physical mixture;* the second, a *chemical compound*, of hydrogen and iodine, viz. hydriodic acid; and we are forced to conclude that the union of the two gases is much more intimate in the second case than in the first.

We make the same distinction in the union of different compound substances with each other, as with elements; here we often meet with homogeneous aggregates which can be liquefied, evaporated, solidified, recrystallised, etc., without change of composition, and whose pro-

[1] Knall-gas, the explosive mixture of two parts of hydrogen and one of oxygen; it might be well for us to adopt this concise and convenient term in English.—Tr.

perties are in many respects totally different from those of their respective components; these, beyond all doubt, can be affirmed to be chemical compounds. Other aggregates are of such a sort as to change their composition very easily by condensation, volatilisation, etc., their particular components can be reobtained in many ways without difficulty, and in the mass we can recognise many properties possessed by the components when separated. These complexes, with all certainty, are affirmed to be only physical mixtures.

Further, as a rule, the formation of chemical compounds is attended with considerable changes of volume and energy, quite different from the case of simple union in a physical mixture; moreover, the external work which we must apply in order to separate the components of the mixture, is much greater in the case of the chemical compound than in that of the physical mixture. All this means that the union of the components of the first, *i.e.* the physical mixture, is much less intimate than in the second, *i.e.* the chemical compound.

Nevertheless, the distinctions between physical mixtures and chemical compounds are only relative, and we find in nature all gradations between these extremes. Thus we are inclined to speak of the solution of a salt in water as a physical mixture, on account of the ease with which the components can be separated from each other; thus we can remove pure water from the solution both by evaporation and by freezing; and, on the other hand, we can extract the pure salt without difficulty by crystallisation. At the same time, many properties of salts may be changed in a very pronounced way by the process of solution: this happens in a very manifest way, for example, on dissolving anhydrous copper sulphate in water; it is a greyish-white substance which in solution assumes an intense blue colour. These and other phenomena suggest emphatically that there is a chemical process associated with solution. Nevertheless, on the other hand, in the union of various substances to form a new homogeneous complex, which for good reasons is to be regarded as a chemical compound, we find that certain properties of the components are unchanged. Thus, for example, in the union of iodine and mercury to form mercuric iodide, we find that the heat capacity of the newly-formed complex is almost exactly the same as before the union; thus the specific heat of each element remains unchanged in the compound.

The Law of Constant and Multiple Proportions.—The quantitative investigation of the amounts in which the different elements are present, has led to a much sharper distinction between physical mixtures and chemical compounds: the former usually have a variable composition according to the method of preparation, but the composition of the latter is constant. John Dalton (1808) found a wonderfully simple law to express the relative quantities according to which the particular elements are contained in compounds of constant composition; this is called the *law of constant and multiple* proportions. This states that one can find for every particular element, a certain number which we will designate as the *combining weight*, which is the standard unit for the quantity of the element entering into *all its various compounds. The quantities of the various elements in their respective compounds are either in the exact ratios of their combining weights, or else in simple multiples of these.*

This law is the foundation of chemical investigation; thus innumerable analyses, and also especially the determinations of the combining weights of the elements, conducted in the most various ways and with the greatest possible care, show that this law holds good with practically absolute exactness, for the union of the elements in all compounds which are shown to be genuine chemical compounds.

The Molecular Hypothesis.—Although it appears entirely feasible to construct a methodical chemical system on the basis of the empirical laws of the exchange of matter and energy as given above, in which the empirical data can be comprehensively arranged, yet in addition to these, and indeed earlier than they, there was introduced a hypothesis concerning the constitution of material aggregations; although this was advanced in ancient times, yet it was first practically employed for a deeper and clearer conception of chemical processes by Dalton and Wollaston at the beginning of the present century, and since then it has remained as the guiding principle of Chemistry and Physics. In the light of this hypothesis, a material aggregate does not fill, continuously and in all points, the gross space occupied by it; but the material aggregate is composed of particles which, though very small, are yet of *finite* dimensions; these are situated more or less distantly from each other, and are called the *molecules* of matter. The fact that matter appears to fill space continuously, that the gaps between the separate molecules escape us, and that these, as such, are inaccessible to our immediate knowledge, much more to our unbiassed senses, all this is easily explained by the smallness of the molecule, and by our inability to grasp such tiny dimensions.

Whether the molecular hypothesis can be squared with the actual facts, or whether it merely owes its origin to our existing inability at present to come at a deeper knowledge of natural phenomena from any other view—whether, perchance, the further building up of the doctrine of energy will lead to another and a clearer conception of matter, this is not the place nor the time to discuss. As a matter of fact, this is the most important and decisive verdict, viz., that the molecular hypothesis, more than any other theoretical speculation, has given powerful and varied assistance to every branch of natural science, and to chemistry in particular. Therefore, in the following presentation of theoretical chemistry, the molecular hypothesis will receive special consideration, and also in some cases it will be introduced where, perhaps, one can advance as far without it as with it, but where by the introduction of molecular conceptions the demonstration gains in interest and brevity of expression. Even to our day, the further extension of the molecular hypothesis has borne remarkably great and unexpectedly fine fruit for the positive enrichment of our science; therefore why should we not ever strive to make our conception of the molecule more tangible, and at the same time furnish our eyes with increasingly powerful microscopes for the consideration of this molecular world?

The Atom and the Molecule.—The first great result of the assumption of a discrete parting of matter in space, was the simple and obvious explanation of the law of constant and multiple proportions, by its discoverer Dalton. This hypothesis, by one effort of modern science, arose like a phœnix from the ashes of the old Greek philosophy.

The formation of a chemical compound from its elements, in the light of the molecular hypothesis, may be most easily conceived in this way, viz., that the smallest parts of the element enter into the molecules of the compound. Therefore these molecules must be divisible, and such a separation takes place when a compound decomposes into its elements. Thus we reach the assumption that a molecule does not continuously fill the space appropriated by it, but is a discrete aggregate of separable particles in its total space. These particles we call "*atoms.*" These atoms, by the union of which, molecules are made, are all alike if they belong to the molecule of one element, but different if they belong to the molecule of a chemical compound. Only in the first case [*i.e.* of an element] can it happen that a molecule may consist of a single atom. The force which binds the atoms in molecular union, we call the "chemical affinity" of the atom. To the action of this force we must primarily ascribe the fact that the properties of the atom vary so much according to the molecular compound to which they belong, and that their properties in the compounds are commonly so different from their properties in the free state.

Many experimental facts are in favour of the obvious assumption that the atoms of one and the same element are equally heavy, and that the molecules of a unitary compound have the same composition. Therefore the relative weights with which the elements enter into a compound must be those according to which the elements enter into the molecule; and as a definite number of atoms always unite to form a molecule of a compound, so will the composition of this molecule always be definite. Further, since molecules of all kinds of compounds always contain a whole number of atoms, and usually not many, therefore *the relative weights with which the elements enter into union to form the different compounds, must be either in the same ratios as the atomic weights, or simple multiples of these.* Experiment confirms these demands of the atomic theory to the greatest degree: the last statement contains the law of constant and multiple proportions, but it is an essential extension of this empirical statement, in that the combining weights, which are only obtained by experiment, are given a more obvious meaning. In the light of the atomic theory, the combining weights and the atomic weights must obviously stand in simple ratios to each other; these latter, [*i.e.* the atomic weights,] cannot be ascertained without further experimental data, not always free from arbitrariness, and only by the comparison of the atomic weight determinations, is the way assured which leads to the adoption of the relative value of the atomic weights, at present assumed as safe, within the limits of the error.

The Table of the Elements.—Since we cannot obtain the absolute atomic weights from the stoichiometric combining ratios, but only the relative figures, there is some opportunity for the choice of unity. By Dalton, hydrogen was taken as the basis of the atomic weights, as its figures are the smallest of all the elements. But as the exact determination of the relative combining weights of the other elements respectively with hydrogen, offers experimental difficulties, and furthermore, as most combining weights of the elements are made from oxygen compounds, Berzelius made the atomic weight of oxygen the basis, and placed it at 100, in order to have no atomic weight less than unity. Recently there has been a return to Dalton's unit for many reasons; but the misfortune exists as before, that the ratio in which hydrogen and oxygen unite to form water has not yet been satisfactorily determined—at least to one part in a thousand; and therefore, after every new determination of this ratio there must follow a recalculation of all the atomic weights. How unsatisfactory it is to work with atomic weights of a selected value is obvious. The fact that it is so very hard to obtain pure hydrogen, and to weigh it with satisfactory accuracy, explains the great difficulty of a sharp determination of its combining weight. Therefore from different sources[1] a proposal has been made by way of compromise, the universal acceptance of which will be only a question of a short time. The ratio of the atomic weights of hydrogen and oxygen is nearly 1 : 16; if then we assume the atomic weight of oxygen as the normal, and not 1, but

$$O = 16{\cdot}000,$$

then the atomic weight of hydrogen will be nearly but not exactly 1, and we unite the advantages of the units chosen by both Dalton and Berzelius; thus we are entirely relieved from the necessity of changing the atomic weight of all the other elements, after every new determination of the composition of water.

In the following table the elements are arranged in alphabetical

[1] Compare Ostwald, *Alleg. Chem.*, **1**. 20. Leipzig, 1891; and especially the Report of the Commission on Atomic Weights (*Ber. d. deutsch. Chem. Ges.* 1898, p. 2761).

order with their symbols and their atomic weights according to the Commission :[1]—

Substance.	Symbol.		Substance.	Symbol.	
Aluminium	Al	27·1	Neon	Ne	20·0
Antimony	Sb	120·2	Nickel	Ni	58·7
Argon	A	39·9	Niobium	Nb	94
Arsenic	As	75·0	Nitrogen	N	14·04
Barium	Ba	137·4	Osmium	Os	191
Beryllium	Be	9·1	Oxygen	O	16·00
Bismuth	Bi	208·5	Palladium	Pd	106·5
Boron	B	11	Phosphorus	P	31·0
Bromine	Br	79·96	Platinum	Pt	194·8
Cadmium	Cd	112·4	Potassium	K	39·15
Cæsium	Cs	133	Praseodymium	Pr	140·5
Calcium	Ca	40·1	Radium	Ra	225
Carbon	C	12·00	Rhodium	Rh	103·0
Cerium	Ce	140	Rubidium	Rb	85·4
Chlorine	Cl	35·45	Ruthenium	Ru	101·7
Chromium	Cr	52·1	Samarium	Sm	150
Cobalt	Co	59·0	Scandium	Sc	44·1
Copper	Cu	63·6	Selenium	Se	79·2
Erbium	Er	166	Silicon	Si	28·4
Fluorine	F	19	Silver	Ag	107·93
Gadolinium	Gd	156	Sodium	Na	23·05
Gallium	Ga	70	Sulphur	S	32·06
Germanium	Ge	72·5	Tantalum	Ta	183
Gold	Au	197·2	Tellurium	Te	127·6
Helium	He	4	Terbium	Tb	160
Hydrogen	H	1·008	Thallium	Tl	204·1
Indium	In	114	Thorium	Th	232·5
Iodine	I	126·85	Thulium	Ta	171
Iridium	Ir	193·0	Tin	Sn	119·0
Iron	Fe	55·9	Titanium	Ti	48·1
Krypton	Kr	87·8	Tungsten	W	184·0
Lanthanum	La	138·9	Uranium	U	238·5
Lead	Pb	206·9	Vanadium	V	51·2
Lithium	Li	7·03	Xenon	X	128
Magnesium	Mg	24·36	Ytterbium	Yb	173
Manganese	Mn	55·0	Yttrium	Y	89·0
Mercury	Hg	200·0	Zinc	Zn	65·4
Molybdenum	Mo	96·0	Zirconium	Zr	90·6
Neodymium	Nd	143·6			

Classification of Natural Processes.—Natural changes have long been grouped into physical and chemical, in the former the composition of matter usually plays an unimportant part, whereas in the latter it is the chief object of consideration. From the point of view of molecular theory a physical process is one in which the molecules remain intact, a chemical process, one in which their composition is altered. This classification has real value, as is shown by the

[1] *Ber. d. d. Chem. Ges.* 1903, p. 8.

customary separation of physics and chemistry, not only in teaching, but in methods of research: a fact that is all the more striking as both sciences deal with the same fundamental problem, in reducing to the simplest rules the complicated phenomena of the external world. But this separation is not altogether salutary, and is especially an obstacle in exploring the boundary region where physicists and chemists need to work in concert.

But since thermodynamic laws appear to be applicable to all the phenomena of the external world, a classification based on them suggests itself. The fundamental formula

$$A - U = T\frac{dA}{dT}$$

indicates the following special cases:—

1. $U = A$; the changes in free and total energy are equal. Then the temperature coefficient of A and therefore also of U is zero, *i.e.* temperature does not influence the phenomenon in question, at least not as regards its thermodynamic properties. Conversely, if the last condition is fulfilled $A = U$. This behaviour is shown by all systems in which only gravitational, electric, and magnetic forces act; these can be described by means of a function (the potential) which is independent of temperature. Also in most chemical processes A and U are approximately equal (see Book IV. chap. v.).

2. $U = 0$, so that $A = T\frac{dA}{dT}$, or A is proportional to the absolute temperature. The expansion of a perfect gas and mixture of dilute solutions are instances of this behaviour, in which the influence of temperature comes out the most clearly (gas thermometer).

3. $A = 0$, and therefore $U = -T\frac{dA}{dT}$.

This condition can only occur at single points of temperature; but A can be small in comparison with U over a considerable range of temperature. As then the percentage variation of A must be large, the influence of temperature must be very marked in such cases (evaporation, fusion, dissociation), *i.e.* all properly "physico-chemical" phenomena.

Case 3 is of course not so strikingly simple, and does not bring to life such important hypotheses as Case 1, which introduced attractive forces into science, and Case 2, which was decisive for the development of the molecular theory.

The case $U = 0$ and $A = 0$ would not be a process in the thermodynamic sense. Such cases, however, exist and are of importance (movement of a mass at right angles to the direction of gravity, passage of one optical isomer into the other, etc.), so it appears that though thermodynamics suggests important points in the classification of phenomena, it is too narrow [1] to cover the whole.

The Principle of Arrangement.—As we have thus obtained a general glimpse in this introductory sketch, a few words may here be premised, touching the division of the material to be considered in the following pages. It is to be treated in a series of four books. In the *first* we will consider the *universal properties of matter*, devoting ourselves almost exclusively to the basis of facts experimentally verified from all sides; the doctrine of energy will also here do useful service, inasmuch as it will not only throw light upon the empirical data, but also very often will broaden and deepen the results of observations. The *second* book is devoted essentially to developing the *molecular hypothesis*, and, accordingly, more than the first, which is on the whole physical, leads us into the region of questions characteristically chemical; here will be specially considered the relations between chemical composition and physical behaviour.

As we shall have studied the systems thus considered, in the light respectively of their properties of matter or of energy, we will then direct our attention to the changes which they undergo under the action of chemical forces, and the conditions under which these forces find themselves reduced to equilibrium. The last two books are, accordingly, devoted to the *doctrine of affinity*, and according to the twofold nature of chemical decompositions, relate respectively to the properties of matter and of energy: in the *third* book we will consider the *transformations of matter*; in the *fourth*, the *transformations of energy*.

[1] This of course is due to the fact that the two laws of thermodynamics are insufficient as an explanation of nature (*e.g.* they take no account of the course of phenomena in time); unlike the molecular theory, in which such a limitation has not so far been shown to exist.

BOOK I

THE UNIVERSAL PROPERTIES OF MATTER

CHAPTER I

THE GASEOUS STATE OF AGGREGATION

The Universal Properties of Gases.—Matter occurring in the gaseous state has the property of completely filling all available space, and, if not acted upon by any external force, as gravity for example, of filling it at all points in equal density. The gas particles are easily movable against each other, and the very small internal friction makes it easy to give a gas mass, at constantly maintained volume, any desired form without noticeable expenditure of work. For the vessel in which the gas is kept, by its shape and capacity, determines the form and density of the enclosed gas.

But a change of volume, and the associated change in density, calls for an important expenditure of work. Now, since gases have the tendency to occupy the greatest amount of space possible, or, in other words, as a result of this tendency, to exert a pressure on the walls of the enclosing vessel; therefore by diminishing their volume external work must be performed, and by an increase of their volume external work is gained.

In their reciprocal relations, gases are characterised by unlimited reciprocal permeability: one can mix with each other gases most dissimilar in composition, in all proportions, and for this there is needed no expenditure of external work; but it is rather a process which takes place "of its own accord," and therefore by the right utilisation, it can perform external work.

A further characteristic, at least under the conditions commonly employed for working with gases, is the very slight density exhibited by matter in the gaseous state; thus 1 c.cm. of atmospheric air, under atmospheric pressure at 0°, weighs only 0·001293 times as much as 1 c.cm. of water. But this is not an essential characteristic of the gaseous condition; since under certain conditions, by raising the external pressure, we can bring a gas to any desired density; but by reason of the great tenuity of matter, which we meet with in the case of gases not too strongly compressed, there results a clearly expressed simplicity of the laws which control both their physical and their chemical relations,—laws, moreover, which are of fundamental significance in the development of our conceptions of matter and of energy, and also for our conceptions of chemical processes. As we will make continual use of these gas laws, they will be enumerated below, and in a form suited for subsequent application; in a later section, reference will be made to the changes which these laws experience in the case of gases compressed to a great density.

The Laws of Gases.—1. *The pressure* p, *and the volume* v, *at constant temperature, are inversely proportional to each other, i.e.* $pv = a$ *constant* (law of Boyle and Mariotte). If we bring successively into the space of one liter, 1, 2, 3, . . . to n grams of oxygen, then we obtain 1, 2, 3, . . . to n-fold pressure respectively; every gram of oxygen presses as much on the enclosing vessel as though it were there alone. We can express this result thus:—the pressure of a gas is proportional to its concentration, provided its temperature be constant.

2. *As a result of an elevation of temperature, at constant volume, the pressure of a gas increases a certain amount which is independent of the nature of the gas and of the initial temperature:—or what is almost identical in the light of the first law of gases, at constant pressure, the volume increases;* the increase of pressure corresponding to a rise of 1° C. is $\frac{1}{273}$ [= 0·003663] of that pressure which the gas exerts at the same volume at 0°. Accordingly, at the temperature t°, if p_0 and v_0 denote respectively the pressure and volume at 0°, then the pressure which the gas exerts at the constant volume V_0, is

$$P = p_0\left(1 + \frac{t}{273}\right) = p_0(1 + 0{\cdot}003663t);$$

and the volume which the gas would assume, at constant pressure p_0, would be (law of Gay Lussac)

$$V = v_0(1 + 0{\cdot}003663t).$$

When pressure and volume both change during the heating, if at t° they amount respectively to p and v, then, according to the first law,

$$p = P\frac{v_0}{v} \quad \text{and} \quad v = V\frac{p_0}{p};$$

and if by one of these equations we eliminate P and V in the equations given above, we obtain

$$pv = p_0v_0(1 + 0{\cdot}003663t),$$

and if we reckon from $-273°$ instead of from $0°$, we obtain

$$pv = \frac{p_0v_0}{273}T,$$

in which $T = 273° + t°$, and its factor is a constant for any given quantity of gas.

3. *When different gases unite chemically, the volumes of the reacting gases, measured of course under similar conditions, bear simple ratios to each other, and the volume of the resulting compound, if gaseous, also stands in a simple ratio to the volumes of its components* (law of Gay Lussac).

4. *The total pressure exerted by a gas mixture on the walls of a vessel, is equal to the sum of the pressures which each gas singly would exert* (Dalton's law).

The laws given above are the immediate expression of experiment; as to their universal validity, it can at once be shown that they are not absolutely correct, but that they suffer variations (at most very slight), which, however, increase, the greater the pressure and the lower the temperature. Moreover, they are not free from great exceptions, inasmuch as there are gases, as for example nitrogen di-oxide, iodine vapour, etc., which at elevated temperatures certainly cannot be condensed proportionally to the external pressure, and which do not expand in proportion to their distance from the absolute zero. But in all these cases it has been shown, as we will see by the laws of dissociation, that the change of volume or temperature is accompanied by a chemical decomposition, and that if we take the influence of this into consideration, the laws of gases preserve their full validity, at least with variations of small amount.

The Hypothesis of Avogadro.—To explain the laws of gases, and especially the fact that the union of gases always occurs according to very simple proportions by volume, Avogadro (1811) advanced a hypothesis which, after much opposition, has come to be recognised as an important foundation of molecular physics, as well as of all chemical investigation. *According to this hypothesis, all gases under the same conditions of temperature and pressure, in unit volume, have the same number of molecules.* Thus the densities of gases, measured under the same external conditions, have the same ratios to each other as their molecular weights.

Naturally this hypothesis can no more be absolutely proved than the molecular hypothesis; but it appears at the outset very plausible, and it explains in the simplest way the rationale of the third law of gases. Further, the numerous results which have been obtained in consequence of its application, are telling witnesses in its favour; the molecular weights inferred from investigations of a purely chemical character, are commonly in surprising accord with those reckoned from the vapour density of gases. The kinetic theory of gases, as will be further shown in detail, leads, in an entirely independent way, to the same assumption. The utility of the hypothesis will be placed in such a clear light that, as will be shown in the theory of solutions, it can be applied to the case of substances occurring in very dilute solution, and furnishes proof thereto, the applicability of which was formerly entirely unsuspected. Finally, a fact which speaks convincingly in favour of the hypothesis, is that it has been very successful in a case which was formerly supposed to be a noted exception, viz., abnormal vapour densities, as will be explained in the description of the dissociation phenomena of gases.

By the help of Avogadro's rule, the laws of gases[1] can be summed up in the following form:—

If we take a gram-molecule of the various gases into consideration, *i.e.* the molecular weight expressed in grams, as for example, 2g. H_2, 32g. O_2, 18g. H_2O, etc., then for every gas mixture there exists the following simple relation between p, v, and the temperature T, counting from $-273°$, thus,

$$pv = \frac{p_0 v_0}{273} T = RT,$$

in which the factor R *is only conditioned by the unit of measure chosen, but is independent of the chemical composition of the gases in question.*

According to measurements of the density of various gases the pressure which a gram-molecule, or *mol* (as it is now called) of a gas at 0° would exercise is 22·42 atmospheres, *i.e.* 22·42 × 760 mm. of mercury at 0° measured at sea-level and latitude 45°, when the gas occupies one litre. Hence

$$pv = \frac{22{\cdot}42T}{273} = 0{\cdot}0821T.$$

The numerical factor occurring in this equation,—an equation which finds abundant application in the most varied calculations, is attended with a slight uncertainty which is too small to be regarded for most purposes. The cause of this lies not so much in the measurements from which the numerical factor is derived as in the deviation that the various gases show from the laws according to which the formula was deduced—a deviation that is perceptible within the limits of pressure of the measurements. Thus oxygen, hydrogen, etc. give different values of R, and again the values obtained at other temperatures differ from those for 0°. But, indeed, if we employ the measurements obtained from gases of normal behaviour, the variations of this numerical factor are only some fractions of a per cent.

[1] Horstmann, *B. Berichte*, **14.** 1243 (1881).

The following observations serve for calculating the gas constant. According to the measurements of Regnault, Jolly, Leduc, and Rayleigh the densities of the following gases at 0° and 760 mm. at sea-level and 45° latitude are :[1]—

Gas.	M.	Density.	p_0.
Hydrogen	2·016	$89·88 \times 10^{-6}$	22·43
Oxygen	32·00	1429·1	22·39
Nitrogen	28·08	1250·7	22·45
Nitric oxide	30·04	1342·65	22·37
Carbon monoxide	28·00	1250·7	22·44
Nitrous oxide	44·08	1970·6	22·37
Methane	16·04	714·64	22·44
Ammonia	17·07	762·1	22·39

Of course only such gases are available for calculating R as follow the laws of gases closely, and in particular have the normal coefficient of expansion.

Dividing 1000 times the density (*i.e.* mass of a litre) by the molecular weight M we obtain the numbers given under p_0, *i.e.* pressure in atmospheres of a mol contained in one litre. As at latitude 45° the acceleration of gravity is 980·6 and the density of mercury at 0° is 13·596 the atmosphere is

$$76 \times 980·6 \times 13·596 = 1,013,250 \text{ absolute units.}$$

The densities of the first three gases have been measured with remarkable agreement by a number of experimentalists[2] (Regnault, Jolly, Leduc, Rayleigh, Morley, etc.), the rest are from the observations of Regnault. The mean of the first three is $p_0 = 22·42$ of the others 22·40. We shall adopt the value **22·42.**

The density of atmospheric nitrogen (containing argon) is $1257·1 \times 10^{-6}$ from which $p_0 = 22·34$. The difference between this value and 22·40 or 22·42 was sufficient—in view of the exactness of the gaseous laws—to indicate an impurity in the nitrogen, and so lead to the discovery of argon.

If n mols of different gases are under the pressure P, and if they fill the volume V at temperature T, then, of course, in the sense of Dalton's law

$$PV = nRT.$$

The counting of the temperature from − 273° is called the "absolute temperature scale," and T the "absolute temperature." If we regard the gas equation as available even for very small values of T,

[1] For the gravity correction see Landolt and Börnstein, *Physikalische-Chemische Tabellen*, 2nd ed., p. 115, table 59.

[2] See Lord Rayleigh, *Chem. News*, **67**. 183, 198, 211 (1893).

and whether this is allowable or not is generally irrelevant in its practical application, we arrive at the result that a gas cooled to $-273°$ Celsius, would exert no pressure on the walls of the containing vessel: this point, viz. $-273°$, is called the "absolute zero."

The formula given above, by means of the mechanical theory of heat, is applicable to several of the most diverse processes; thus the gas law is employed to ascertain the convertibility of heat into external work (p. 20). The gas equation is implicitly contained in all thermodynamical formulæ, in which the "absolute temperature" almost always plays an important part. Scarcely any empirical law of nature has as yet been applied to such an extent as the law expressed by this gas equation.

The Content of Energy of a Gas.—If one connects two vessels containing gas, but at different pressures, so that an equalisation of pressure between the vessels takes place, but with no performance of external work, *then in this process heat will be neither developed nor absorbed* (Gay Lussac, Joule, Thomson); one will obtain the same result if one vessel is exhausted by an air-pump.

This result, expressed most simply in the language of thermodynamics, states that *the content of energy of a gas is independent of its volume.*

If one allows two vessels which contain two different gases to communicate with each other, these gases begin at once to diffuse into each other, and the final condition consists in a complete equalisation of the difference in composition. In this process, as well as in the mixing of two gases, neither development nor absorption of heat is observed, provided, of course, that no chemical action takes place. *The content of energy of a gas mixture is accordingly equal to the sum of the contents of the ingredients.*

These laws are of fundamental importance for the thermochemistry of gases; they are, like the laws mentioned above, available only in the case of ideal gases, and become inexact with lowering temperature and increasing pressure of the gas in question.

The Specific Heat of Gases.—Experience has shown that the quantity of heat which must be imparted to a definite quantity (by weight) of a gas, in order to raise its temperature 1°, varies accordingly as the heating takes place at constant pressure, or at constant volume; in the first case the volume increases from the increase in temperature according to the laws of gases (p. 38), and in the latter case the pressure of the gas increases $\frac{1}{273}$ of the pressure exerted at 0°.

Accordingly we must distinguish between the specific heat of a gas at constant pressure, and that at constant volume. Regarding the experimental determination, the simplest way to measure the specific heat at constant pressure, is by a process patterned after the method of mixture, and which consists in leading the warmed gas through a spiral calorimeter tube, which is surrounded with water, and so determining the quantity of heat which is given out by the cooling. Delaroche and Bérard first worked with this method (1811); then Regnault very extensively (1853); and recently (1876) E. Wiedemann, who considered especially the influence of temperature on the specific heat of gases, and Lussana,[1] who investigated the influence of pressure.

For the direct determination of the specific heat of a gas at constant volume, one must enclose it in a vessel, warm it to a measured temperature, and then cool it by dipping it in a calorimeter; but the attempt to determine it exactly, is thwarted by the fact that the heat capacity of the vessel must be subtracted from that of the system, in order to obtain that of the contained gas, and the heat capacity of the vessel is much greater than that of the contained gas.[2] But with the aid of thermodynamics, we can at once derive a simple formula which allows the calculation of the specific heat at constant volume, from that at constant pressure on the one hand, and on the other we will be led to certain experimental measures which show how the ratio of the two specific heats may be determined.

Instead of the specific heat itself, *i.e.* the heat capacity of 1 g., we will commonly work with the heat capacity of 1 g.-mol, the so-called "molecular heat." If c_p and c_v denote respectively the specific heats of a gas of mol wt. M, at constant pressure and constant volume, then the two molecular heats are respectively

$$C_p = Mc_p \quad \text{and} \quad C_v = Mc_v,$$

and between these two we find the relation

$$C_p - C_v = 1{\cdot}99 \text{ g.-cal.},$$

as will be further shown below.

The Specific Heat of Gases at Very High Temperatures.—A number of French investigators have recently succeeded in determining the specific heats of gases *at very high temperatures* with quite a considerable degree of accuracy. Mallard and Le Chatelier[3] exploded a gas mixture of known composition in a closed iron cylinder, and determined the *maximal pressure* developed thereby: this last measurement was performed at first with a Bourdon's manometer, which recorded its data by means of a needle on an evenly rotating cylinder. Later these same investigators employed "the crushing manometer" constructed by Sarreau and Vieille,[4] on which the pressure was

[1] Lussana, *Nuov. Cim.* (3), **36** 5, 70, 130 (1894).

[2] But Joly (*Proc. R. S. Lond.*, **55.** 390, 1894) showed that for sufficiently condensed gases this difficulty could be overcome.

[3] *Compt. rend.*, **93**, 1014, 1076 (1881).

[4] *Ibid.*, **95.** 26 (1882).

measured by the permanent deformation of a small, solid, copper cylinder placed between an anvil face and a piston on which the pressure to be determined acted; also here the time occupied in developing the pressure was determined by means of a needle and a rotating cylinder. Thus by measuring the maximal pressure of the explosion, one can calculate the maximal temperature, and since the heat developed by the explosion is known from the thermochemical data, there is given at once the *heat capacity* of the gas mixture. A correction must be made to account for the heat given off to the walls of the explosion bomb, which on account of the quick occurrence of the explosion is inconsiderable; this correction can be calculated from the velocity of cooling observed by the decrease of pressure after the explosion, or it can be determined quite accurately in an experimental way by the use of receiving vessels of varying size.

From the fact that the maximal pressure of a gas mixture, for example, "knall gas," when exploded with admixture of equal volumes of nitrogen, or oxygen, or hydrogen, or carbon monoxide, experiences the same diminution of temperature, it follows *that these gases possess the same heat capacity up to the maximal temperature of explosion*, i.e. *up to* 2700°. This is true, provided that these gases have the same coefficient of expansion up to these temperatures, a thing which can scarcely be doubted after the measurements of V. Meyer and Langer.[1]

These results have been confirmed as well as broadened at several points by Vieille,[2] and also by Berthelot and Vieille.[3] It was shown later that nitrogen and carbon monoxide have the same heat capacity, by the explosion of a mixture of cyanogen and oxygen, which led to the formation of carbon monoxide and nitrogen; thus

$$C_2N_2 + O_2 = N_2 + 2CO;$$

from the heat developed by this reaction and the maximal temperature observed, there could be calculated the specific heats of nitrogen and carbon monoxide. Finally, from further researches which were conducted with an excess of nitrogen, there could be stated the following formula for the molecular heat of N_2, H_2, O_2, and CO at constant volume, and which can be regarded as good from 1600° to 4500°; viz.

$$4{\cdot}76 + 0{\cdot}00244t.$$

For the latter temperature, 4500°, this formula gives 15·7; *i.e. the molecular heat of the permanent gases increases very considerably with the temperature.*

From observations on the maximal pressure in the explosion of hydrogen and carbon monoxide, in the presence of varying quantities

[1] *Pyrochem. Unterss.* Brunswick, 1885.
[2] *Compt. rend.*, **96.** 1358 (1883).
[3] *Ibid.*, **98.** 545, 601, 770, 852 (1884).

of nitrogen, Berthelot and Vieille calculated the molecular heat of water vapour to be

$$16 \cdot 2 + 0 \cdot 0038(t - 2000),$$

and for that of carbon dioxide

$$19 \cdot 1 + 0 \cdot 0030(t - 2000).$$

Both formulæ refer to the molecular heat *at constant volume.*

Later Mallard and Le Chatelier[1] used in a very similar way the maximal pressures of various explosives as observed by Sarreau and Vieille, to calculate the specific heats of gases, and thus obtained results showing undoubtedly that the molecular heats of the "permanent" gases increase strongly at higher temperatures.

For the mean molecular heats at constant volume between 0° and t°, they found—

Carbon dioxide . . .	6·50 + 0·00387 t
Water vapour	5·78 + 0·00287 t
Permanent gases . . .	4·76 + 0·00122 t

It has recently been thought, for various reasons, that the specific heat of gases does not increase so rapidly with temperature as this; further measurements of these highly important data are most desirable.

The Dependence of the Specific Heat of Gases on the Temperature.—It has been already repeatedly shown that the specific heat of gases *steadily decreases* with decreasing temperature.[2]

As Le Chatelier[3] noticed, this decrease *indicates a tendency for the specific heats of the most different gases to approach the closer, the more the temperature sinks, and at absolute zero the values of the molecular heats at constant pressure seem to converge at about* 6·5. The observations, therefore, can be represented by the formula

$$C_p = 6 \cdot 5 + aT,$$

in which T represents the absolute temperature, and "a" represents a coefficient which is greater according as the molecule is *more complex.*

The following values have been calculated for "a":—

H_2, N_2, O_2, CO . .	0 0010	C_2H_5Br . . .	0·0324
NH_3	0·0071	C_3H_6O . . .	0·0403
CO_2	0·0084	C_6H_6	0·0510
N_2O	0·0089	$CH_3COOC_2H_5$ . .	0·0674
C_2H_4	0·0137	$(C_2H_5)_2O$ . . .	0·0738
$CHCl_3$ . . .	0·0305		

If, for example, one calculates the *mean* specific heats measured by

[1] *Wied. Beibl.*, **14.** 364 (1890).

[2] The apparent exceptions exhibited by the vapours of acetic acid and of nitrogen dioxide at certain temperature intervals, have been referred to dissociation phenomena.

[3] *Zeitschr. phys. Chem.*, **1.** 456 (1887).

E. Wiedemann between 20° and 200°, with the help of the formula given above, he will find very good agreement. Also the values found by Le Chatelier at very high temperatures (see above) are in good agreement, since the true molecular heats at constant pressure, calculated for - 273°, for the following substances are

CO_2	6·39
H_2O	6·22
H_2, N_2, O_2, CO	6·10

All are nearly equal and not far removed from 6·5.

The Thermodynamics of Gases: First Law.—As we have thus in the preceding sections familiarised ourselves with the most important *empirical facts*, to a knowledge of which we have been led by the experimental study of the behaviour of gases, we will now consider, from the standpoint of thermodynamics, that which will lead to a broader and deeper knowledge. It must be emphasised here that the results so obtained are not to be regarded as inferior in accuracy to the empirical facts on which they are based.

The law of the conservation of energy insists that the change in total energy, which a system experiences in passing through any chosen process, is independent of the way by which it passes from the one to the other condition. Let us imagine 1 g.-mol of any selected gas enclosed in a vessel of volume v, which can be brought into communication with another [empty] vessel of volume v′; and let us conceive that this system passes in the two following ways from the same initial to the same final condition. In the first way, the g.-mol of gas, which originally occupies the volume v, is warmed 1° in temperature, and then a communication established between the two vessels. In the second way, communication is established at once between the two vessels, and then the gas, which now occupies the volume v + v′, is warmed 1°. In neither case is the flowing over of the gas from the full to the empty vessel accompanied by any change of energy—neither the performance of external work, nor the development of heat—since the content of energy of a gas is independent of its volume (see p. 42); and in both cases the change of energy consists solely in the amount of heat which must be introduced to raise the temperature of the gas 1°. But in both cases the quantity of heat is equal to the molecular heat of the gas at constant volume, and the only difference is that in the first way the gas is warmed at the volume v, in the second at the volume v + v′. The two quantities of heat must be equal to each other, *i.e. the molecular heat, and of course the specific heat also, of a gas at constant volume, is independent of the volume at which the warming occurs.*

If 1 g.-mol of an ideal gas is changed from the volume v to that of v + v′ in this way, viz. that a vessel of volume v which holds the g.-mol is connected with an empty vessel of volume v′, then there occurs no change in the total energy; but, as in the case of every automatic process, there is a decrease of the free energy, and this can therefore be used to do external work. The question is then to find such a mechanism as shall furnish the maximal work as is necessary for the application of the second law.

Such a mechanism would be very simple in this case: a cylinder closed at one end by a strong head, and shut at the other by a movable air-tight piston, will satisfy the demands. When we raise the piston the enclosed gas does work because it presses on the piston from within; when we lower it we must overcome the gas pressure. If the cylinder is only capable of sufficient expansion we can bring the enclosed gas to any desired volume; and that by means of this apparatus we really obtain the *maximal* work capable of being obtained in the expansion, we may know in this way, viz. that in a compression we must apply the same amount of work which we obtain in the dilation. Thus the mechanism described is "reversible."

When a gas expands without performing external work, one observes no thermal effect; but if, by the apparatus above described, *e.g.*, one uses the expansive force of a gas to do external work, then the gas must lose the equivalent amount of energy; and conversely, if one compresses the gas, it must absorb an amount of energy equal to the work of compression. Therefore, by the application of the law of the conservation of energy to the ideal gas, we find the following result, viz. a gas when expanding absorbs heat equivalent to the external work it performs, and when suffering compression generates heat equivalent to the external work spent on it.

Let us now imagine 1 g.-mol of a gas, enclosed in the cylinder above described, to be warmed 1° by the addition of heat; in general, the pressure exerted by the gas, as well as the volume occupied by the gas, will vary with the temperature; the amount of heat necessary for a definite elevation of temperature, will also vary according to the manner in which these values change, as is shown both by experiment, and by a consideration of some suggestions afforded by the law of the conservation of energy. Most easily considered are the two following limiting cases:—Firstly, we will add to 1 g.-mol of a gas the amount of heat C_p, required to raise it 1° in temperature, the warming being so conducted that the *pressure* remains constant:—or, Secondly, we will add to 1 g.-mol of a gas the amount of heat C_v, required to raise it 1°, the warming being so conducted that the *volume* remains constant. These quantities of heat, C_p and C_v, as already observed, are called respectively, the *molecular heat at constant pressure*, and the *molecular heat at constant volume*; by division by the mol. wt. of the gas in question, we obtain respectively, the *specific heat at constant pressure*, and the *specific heat at constant volume*.

In order to obtain the relation between these two specific heats, on the one hand we heat the g.-mol of gas, whose volume is v, at constant pressure, 1°; this requires C_p g.-cal. But since the gas expands during the heating, and, indeed, overcomes the constant pressure p which is weighting it down, so at the same time external work will be performed, which is found by the product of the pressure and the increase of volume; and since the former is p, and the latter is $\frac{v}{T}$, it amounts to $\frac{pv}{T}$. If, on the other hand, we heat the gas at constant volume, there occurs no performance of external work, and it requires only the addition of C_v g.-cal.

Now, according to the law of the conservation of energy,

$$C_p - \frac{pv}{T} = C_v,$$

or if we combine with it the gas equation (p. 40),

$$pv = RT,$$

we shall have

$$C_p - R = C_v.$$

If the molecular heat is expressed in g.-cal. the constant R must be also so reckoned.

Now, if we express p in atm. and v in litres, then

$$pv = 0{\cdot}0821T,$$

in which

$$R = 0{\cdot}0821;$$

the left side of the equation, which represents the product of pressure and volume, is a quantity of energy which we estimate according to the method given above in liter-atmospheres (p. 12); in order to transform it to g.-cal. we remember that

$$1 \text{ litre-atmosphere} = 24{\cdot}25 \text{ g.-cal.},$$

and we shall have

$$pv = 0{\cdot}0821 \times 24{\cdot}25T = 1{\cdot}991T,$$

and consequently

$$R = 1{\cdot}991.$$

The difference of the two molecular heats, therefore, amounts to

$$C_p - C_v = 1{\cdot}991\left[\frac{\text{g.-cal.}}{\text{degrees Cel.}}\right],$$

and the difference of the specific heats will we

$$c_p - c_v = \frac{1{\cdot}991}{M}.$$

Since, for the calculation of litre-atm. in g.-cal., one needs to know the mechanical equivalent of heat, so conversely one can determine experimentally this equivalent of heat from the difference between these two specific heats of a gas. As is well known, this is the way used by Mayer in 1842.

The relation between C_p and C_v can also be derived from the equation of p. 9—

$$C_p - C_v = \left(p - \frac{\partial U}{\partial v}\right)\frac{\partial v}{\partial T}.$$

If this be applied to one mol of gas, and it is remembered that

$$\frac{\partial U}{\partial v} = 0, \quad v = \frac{RT}{p}, \quad \frac{\partial v}{\partial T} = \frac{R}{p},$$

it follows that

$$C_p - C_v = R.$$

The Ratio of the Specific Heats of a Gas.—The application of the first law of thermodynamics to the ideal gas, has led further to two very important experimental methods, which make possible quite an exact determination of the quotient

$$k = \frac{C_p}{C_v} = \frac{c_p}{c_v}.$$

1. The Method of Clément and Desormes.—In a large glass globe is enclosed the gas to be investigated under the pressure P_1, which is made a little greater than the atmospheric pressure P. The globe is opened for an instant so that the pressure in the interior of the vessel sinks to that of the atmosphere, and then is closed again as quickly as possible. In consequence of the expansion of the gas it is somewhat cooled: then heat will flow in from without, and consequently the pressure in the interior of the vessel will increase a little over that of the atmospheric pressure, to the value P_2.

We develop the formula for the case when P_1 (and of course P_2 also) is only very slightly different from P, so that if we make

$$P_1 = P + p_1,$$
$$P_2 = P + p_2,$$

then p_1 and p_2 are very small in comparison with P, a proviso which must be realised in practical experiment.

If V is the volume of the globe, then on opening it at atmospheric pressure P, there escapes the volume $V \times \frac{p_1 - p_2}{P}$ which accordingly performs the work $V(p_1 - p_2)$ against the atmospheric pressure. This amount of work refers to the cooling of the gas in consequence of the outflow: on the supposition that, at the moment of the outflow, the quantity of heat introduced to the gas from without vanishes, or, as we say, that the dilation occurs "adiabatically," then the amount of undercooling, by which the gas sinks below the temperature of the experiment T, can be calculated from the equation

$$\frac{t}{T}=\frac{p_2}{P},$$

since, in consequence of the subsequent reheating from $T-t$ to T, which occurred from the heat coming in from without after closing the vessel, the pressure of the gas increased from P to $P+p_2$. This reheating took place at constant volume; now as there were contained in the vessel

$$n=\frac{PV}{RT}\text{ g.-mol (p. 41)};$$

so accordingly there was introduced

$$t\frac{PV}{RT}C_v=\frac{p_2V}{R}C_v\text{ g.-cal.},$$

and this must be equal to the work performed by the gas, and calculated above, [viz. $V(p_1-p_2)$]; hence, if we express this in calorimetric units we have

$$p_1-p_2=\frac{p_2}{R}C_v.$$

This equation will serve for the experimental determination of C_v; if we multiply it by the equation developed above on p. 49, viz.,

$$C_p-C_v=R,$$

we will have

$$k=\frac{C_p}{C_v}=\frac{p_1}{p_1-p_2}$$

In consequence of the unavoidable fact that, during the outflow, the undercooled gas extracts heat from the walls of the vessel, the value found for t, and accordingly that for p_2, and also that for k, will come out too small. One can reduce these sources of error to a minimum by operating with as large a globe as possible, and also by using as small differences of pressure as possible. In this way Röntgen (1870) found for dry air $k=1{\cdot}4053$, Lummer and Pringsheim (1894) $k=1{\cdot}4015$, Maneuvrier and Fournier (1896) $k=1{\cdot}395$.

2· The Method of the Velocity of Sound Waves (Dulong, Kundt).—According to a formula first developed by Laplace, the velocity u, of transmission of a wave is given by the following equation :

$$u = \sqrt{\frac{p}{d}k},$$

in which d is the density of the gas in question. The ratio of the wave velocities of two gases u_1 and u_2, with the densities d_1 and d_2, when measured under the same conditions of pressure and temperature, is

$$\frac{u_1}{u_2} = \sqrt{\frac{k_1 d_2}{k_2 d_1}},$$

or if we replace the densities of these gases by their respective mol weights (M_1 and M_2),

$$\frac{u_1}{u_2} = \sqrt{\frac{k_1 M_2}{k_2 M_1}}.$$ [1]

Now the ratio of two wave velocities is capable of an exact measurement by a method given by Kundt.[2] If one rubs a clamped glass tube (see Fig. 2) with a slightly moistened cloth till it sounds its longitudinal tones, the air contained within it will be thrown into fixed waves which can be rendered visible to the eye by the introduction of

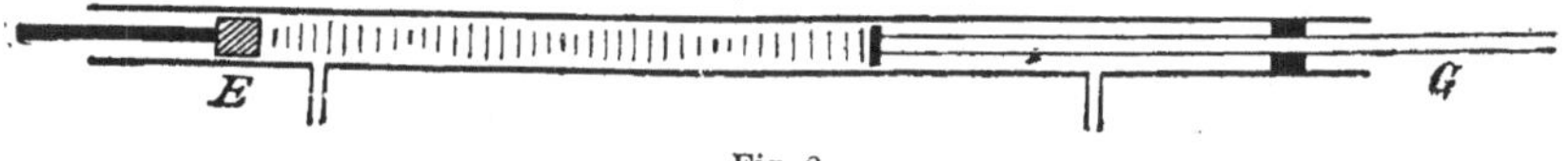

Fig. 2.

fine dust, as cork-powder, precipitated silica, etc., and thus can be directly measured. By adjusting the stops E, closing the end of the tube G, one can easily find the point at which the stationary waves are most sharply defined. The distance E between two nodes is proportional to the velocity of sound in the gas.

The ratio for the specific heat of any desired gas, of mol. wt. M, is given by comparison with the similarly determined value for atmospheric air (viz. 1·400), from the equation

$$k = 1{\cdot}400 \frac{M \, . \, l_1^2}{28{\cdot}88 l_2^2},$$

when l_1 and l_2 are the node intervals for the gas in question and for air respectively, measured under the same conditions of temperature and pressure; instead of $\frac{M}{28{\cdot}88}$, we may write the vapour density.

We will return later to the results obtained experimentally by these methods.

[1] Since pressure and density are proportional, the velocity is independent of pressure; in the above formula, therefore, only equality of temperature is assumed.

[2] *Pogg. Ann.*, **127.** 497 (1866); **135.** 337, and 527 (1868).

The Thermodynamics of Gases: Second Law.—The application of the second law of thermodynamics will in this case teach us nothing new, since we derived the quantitative side of this law from the relations of an ideal gas to pressure and temperature, on learning that one could at once generalise the experiments conducted in any one system; it is here advisable to turn this about, and thus to convince one's self how the following universal equation is handled, viz.

$$A - U = T\frac{dA}{dT}.$$

The process which we will consider will be the flowing over of a gas from a full to an empty space, and let the volume of 1 g.-mol increase from v_1 to v_2: the change of total energy is equal to zero (p. 42)

$$U = 0.$$

The maximal external work to be obtained in this process, a knowledge of which is essential for many purposes, is easily found; for it is equal to the work (calculated from the considerations on p. 47), required for the compression of 1 g.-mol at constant temperature, from v_2 to v_1.

If we diminish the volume v of a gas, at pressure p, by an amount dv, then the work pdv must be done; if we diminish the volume v_2 of 1 g.-mol. of a gas to v_1, then accordingly the work $\int_{v_1}^{v_2} pdv$ must be done; if we observe that pv = RT, then we will have

$$A = \int_{v_1}^{v_2} pdv = RT\int_{v_1}^{v_2}\frac{dv}{v} = RT \ln\frac{v_2}{v_1},$$

in which ln denotes the natural logarithm. If the pressure of 1 g.-mol at the volumes v_1 and v_2 respectively, amounts to p_1 and p_2, then according to the law of Boyle and Mariotte

$$\frac{v_2}{v_1} = \frac{p_1}{p_2},$$

and therefore

$$A = RT \ln\frac{p_1}{p_2}.$$

The maximal external work to be obtained by the increase in volume of 1 g.-mol of an ideal gas from v_1 to v_2, or as we have already defined it, the diminution of free energy associated with this process is, accordingly—

1. Only dependent on the relative difference between the initial and final volumes and pressures respectively, but is independent of their absolute magnitudes.
2. Proportional to the absolute temperature.
3. Equally great for the various gases.

Of course n fold work is necessary for the compression of n molecules. If we choose, as unit of work, that which will be spent, when the pressure of one atm. works during the increase in volume of one litre, then according to the above (p. 40),

$$A = 0{\cdot}0821T \ln \frac{v_2}{v_1} = 0{\cdot}0821T \ln \frac{p_1}{p_2} \text{ litre-atm.}$$

On the other hand, if we choose, as the unit of work, as is frequently more practical, the work performed when 1 g. is raised 1 cm., against the earth's attraction, then it must be noticed that the pressure of one atmosphere, *i.e.* of a column of mercury 76 cm. high, amounts per sq. dcm. to

$$100 \times 76 \times 13{\cdot}596 = 103{,}330 \text{ g. in weight};$$

and that therefore, as this pressure weighs down the volume of a litre, and as the force mentioned above works in an opposite direction a distance of 10 cm., then A assumes the value

$$A = 84{,}800T \ln \frac{v_2}{v_1} \text{ cm.g.}$$

Finally, in order to obtain A in our practical scale of energy, viz., the g.-cal., we must divide this last expression by the mechanical equivalent of heat, 42600, and we obtain

$$A = \frac{84{,}800}{42{,}600} T \ln \frac{v_2}{v_1} = 1{\cdot}991T \ln \frac{v_2}{v_1} \text{ g.-cal.}$$

Now, observing that

$$U = 0; \quad A = RT \ln \frac{v_2}{v_1}; \quad \text{and} \quad \frac{dA}{dT} = R \ln \frac{v_2}{v_1},$$

then the equation of the second law of thermodynamics becomes what is in this case an identity, viz.,

$$A - U = T\frac{dA}{dT}.$$

The Behaviour of Gases at Higher Pressures.—When one works with strongly-compressed gases, then the laws which we have thus far studied can be applied only with precaution; and at high pressures, and also at great densities, they lose their value entirely.

Slight concentration, the dissemination of gaseous material throughout a relatively great volume, give the conditions for the simple and regular behaviour of the ideal gas.

The compressibility of highly-condensed gases has been investigated by Natterer, Regnault, Cailletet, Andrews, and with special thoroughness recently by Amagat. As the behaviour of strongly-compressed gases has given occasion to some remarkable investigations on the molecular theory, this subject will be treated more thoroughly in the second book. Here will be given a series of observations by Amagat[1] on nitrogen at 22°, as follows:—

p in Atm.	pv	p in Atm.	pv
1·00	1·0000	126·90	1·0015
27·29	0·9894	168·81	1·0255
46·50	0·9876	208·64	1·0520
62·03	0·9858	251·13	1·0815
73·00	0·9868	290·93	1·1218
80·58	0·9875	332·04	1·1625
90·98	0·9893	373·30	1·2070
109·17	0·9940	430·77	1·2696

The compressibility is at first much greater than required by the Boyle-Mariotte law, reaches a minimum, then increases considerably, and at about 124 atm. pv regains and passes its normal initial value. This behaviour is general [*i.e.* with other gases].

The *thermodynamic treatment* of highly-compressed gases offers corresponding complications. If we consider again the expansion of 1 g.-mol. of a gas from v_1 to v_2, it must be at once observed in the application of the equation

$$A - U = T\frac{dA}{dT} \qquad . \quad . \quad . \quad . \qquad (1)$$

that U, the heat developed by expansion without the expenditure of external work, is no longer equal to zero, but has a value by no means inconsiderable, and, after the analogy of the heat of evaporation, it is usually negative. A is again given by the integral

$$A = \int_{v_1}^{v_2} p dv \qquad . \quad . \quad . \quad . \qquad (2)$$

to calculate which one must determine p as a function of v in any given case by experiment. Finally, the first law of thermodynamics gives for change of U with the temperature (p. 9),

[1] *Ann. chim. phys.* [5], **19.** 345 (1880).

$$\frac{dU}{dT} = C_{v_1} - C_{v_2} \quad . \quad . \quad . \quad . \quad . \quad (3)$$

in which C_v denotes the molecular heat at volume v. The differentiation of the equation (1), and also the direct application of the equation developed (on p. 24) give

$$\frac{\partial Q}{\partial v} = T\frac{\partial p}{\partial T} \quad . \quad . \quad . \quad . \quad . \quad . \quad (4)$$

in which $Q = A - U$, which denotes the heat absorbed in isothermal expansion with expenditure of work. Equation (3) can also be written in the form

$$\frac{\partial^2 U}{\partial T \partial v} = -\frac{\partial C_v}{\partial v} \quad . \quad . \quad . \quad . \quad . \quad . \quad (3a)$$

as we assume $v_2 - v_1$ to be very small. Now

$$\frac{\partial Q}{\partial v} = \frac{\partial(A - U)}{\partial v} = p - \frac{\partial U}{\partial v},$$

or combined with (4)

$$\frac{\partial U}{\partial v} = p - T\frac{\partial p}{\partial T} \quad . \quad . \quad . \quad . \quad . \quad (5)$$

and this differentiated for T gives

$$\frac{\partial^2 U}{\partial v \partial T} = -T\frac{\partial^2 p}{\partial T^2}$$

or finally with the aid of (3a)

$$\frac{\partial C_v}{\partial v} = T\frac{\partial^2 p}{\partial T^2} \quad . \quad . \quad . \quad . \quad . \quad (6)$$

Finally, the equation of p 9

$$C_p - C_v = \left(p - \frac{\partial U}{\partial v}\right)\frac{\partial v}{\partial T}$$

in combination with (5) yields

$$C_p - C_v = T\frac{\partial p}{\partial T} \cdot \frac{\partial v}{\partial T} \quad . \quad . \quad . \quad . \quad . \quad (7)$$

Although hitherto, applications of these equations have been advanced[1] only in a disjointed way, nevertheless it cannot be doubted that they are of great value in proving and extending investigations conducted on highly-compressed gases. Moreover, it should be emphasised that *these equations also hold good for the compression of homogeneous bodies, whether liquid or solid.*

[1] Compare, for example, Margules, *Wiener Sitzungsber.*, **97.** 1385 (1888), and especially Amagat, *J. de Phys.* [3], **5.** 114 (1896).

CHAPTER II

THE LIQUID STATE OF AGGREGATION

The General Properties of Liquids.—If one can diminish at pleasure the available volume of a definite amount of a simple gas, by increasing the external pressure, the pressure exerted by the gas on the surrounding walls grows continually with the diminution of volume: if one works at a temperature sufficiently reduced, there suddenly comes a point at which, by diminishing the volume, the pressure experiences no increase, but remains constant. At once it is observed that the material contained in the vessel, although of the same composition at all points, occupies its space in twofold form, distinguished by density, refractive index, etc. The substance has passed from the gaseous state of aggregation to a *much greater state of condensation.* Thus matter has the property of existing at the same time in two forms together, associated with certain conditions of temperature and external pressure; and for every temperature there is a certain pressure, the so-called "vapour pressure" (also called the vapour tension, maximal pressure, etc.), at which the gas and the more condensed form can exist side by side, or, as is said, *they exist in equilibrium with each other.* If we diminish this external pressure, the whole mass goes over into the gaseous state; if we increase the pressure, it goes over to the condensed condition. After all the gas is condensed at constant pressure, then by a further increase of pressure one can observe, in a qualitative way, the same phenomena which appeared before the beginning of condensation. A gradually diminishing volume corresponds to an increase of the opposing pressure; only now the diminution of volume, corresponding to the same degree of increase of pressure, is much smaller than before the condensation.

Now it is possible, according to the nature of the gas, by means of the experimental method described above, to condense the substance into two essentially different forms; these appear, accordingly, as they are under the influence of external forces, gravity especially, either in the spherical form or else crystallised in geometrical forms; in the first case the substance is called a *liquid*, in the latter a *solid*. Matter in the liquid state, like that in the gaseous state, has a relatively easy mobility of its particles, though the work which must be performed in this reciprocal mobility, measured by the "internal friction," and which appears in the form of heat, is always much greater in the liquid state than in the gaseous. It is not yet shown that that which is common to both the gaseous and liquid states, as contrasted with the solid, is to be ascribed to any lack of a tendency to assume a definite shape; but rather a liquid when left to itself, as when suspended in another liquid of the same density, and so removed from the action of gravity, by virtue of those inherent forces, which we will presently consider under "surface tension," takes the sharply-defined form of a sphere.

The special properties of the solid state will be dealt with in the following chapter.

Surface Tension.—The characteristic tendency of substances in the liquid state, when left to themselves to take a spherical form, to reduce the free surface to a minimum under all circumstances, can be clearly explained by the method of Young (1804), in considering that certain characteristic forces are active in the free surface of liquids. The superficial layer behaves like a tightly-drawn elastic skin which strives to contract. The force exerting itself normal to a section 1 cm. in length of the surface, is called the surface tension. A strip of the superficial layer of water of 1 cm. in breadth, for example, exerts a contracting force equal to 0·082 g. weight. Contrary to the behaviour of an elastic skin, this force, of course, is unchanged with unchanging temperature, despite an increase or decrease of the surface.

The absolute value of the superficial tension can be directly measured in several ways; thus, for example, if one dips a moistened cylindrical tube of radius ν vertically into the surface of a liquid, then the superficial tension works in the section cut off, and in a direction at right angles to the tube, with the following force, viz. the length of the section times the superficial tension $= 2\pi\nu\gamma$, if we denote the latter by γ; this force will raise the liquid in the tube to a sufficient height to compensate for the action of gravity; it will be

$$2\pi\nu\gamma = \nu^2\pi hs;$$

and if h is the height of the liquid, and s its specific gravity, then the right side of the equation represents the weight of the liquid raised. Therefore the surface tension, expressed in terms clearly and easily definable, is

$$\gamma = \frac{1}{2}h\nu s.$$

Vapour Pressure and the Heat of Vaporisation.—If we bring a simple liquid into a vacuum, evaporation takes place at once, till the pressure of the gas formed has reached a definite maximal value, viz. the corresponding vapour pressure. This pressure increases with the temperature, and usually very rapidly. In the presence of another, but an indifferent gas, evaporation takes place till now the *partial pressure* of the resulting vapour is equal to the vapour pressure (law of Dalton).

The passage from the liquid to the gaseous state, under the pressure of the saturated vapour, is attended with very important changes of energy: firstly, external work is performed; secondly, heat is absorbed from without.

Thus, firstly, let there be enclosed in a cylinder with a movable air-tight piston, a simple liquid, the vapour of which has the same composition as the liquid itself; the vapour pressure which it exerts at the temperature T in question, in the absolute scale, will amount to p. Now we will raise the piston so far that 1 g.-mol of the substance shall pass over to the state of a gas; if this amount as a liquid occupied the space v', but as a gas under the pressure p, the volume v, then the external work performed in its evaporation is $p(v - v')$. When p is not too great, we can, firstly, consider v' as a very small fraction of v, and which can be neglected; and, secondly, we can apply the gas equation, $pv = RT$, in the form where as usual the volume can be reckoned in litres and the pressure in atmospheres. The external work which can be gained by the evaporation of 1 g.-mol of any selected liquid under these conditions, amounts to 0·0821T litre-atmospheres, or 1·991T g.-cal. (p. 48); *it* [i.e. *the external work*] *is independent of the nature of the liquid, and is proportional to the absolute temperature.*

Secondly, the evaporation of 1 g.-mol is associated with the absorption of heat. That quantity of heat, which is identified with the evaporation of 1 g. of a substance, is called the *heat of vaporisation;* and that identified with the evaporation of 1 g.-mol of a substance is called the *molecular heat of vaporisation.* They vary with the substance; their dependence on temperature allows of their calculation on thermochemical principles (p. 8), from the difference of the specific heats of the liquid and of the saturated vapour, in the following way.

According to what has been said above, the diminution U of total energy in the evaporation of 1 g.-mol is equal to the external work performed RT, minus the heat absorbed, *i.e.* the *molecular heat of vaporisation* λ; accordingly, if we assume that we can calculate the external work from the gaseous laws

$$U = RT - \lambda,$$

and

$$\frac{dU}{dT} = R - \frac{d\lambda}{dT} = Mc - C_v,$$

if we denote the molecular heat of the vapour at constant volume by C_v, and the heat capacity of 1 g.-mol of the liquid by Mc. If, according to p. 49, we make $C_p - C_v = R$, and if we substitute the value of R, we will have

$$\frac{d\lambda}{dT} = C_v + R - Mc = C_p - Mc,$$

or if l denotes the ordinary heat of vaporisation,

$$\frac{dl}{dT} = c_p - c \quad . \quad . \quad . \quad . \quad (1)$$

Now, in all cases hitherto investigated, the specific heat of a substance is greater in the liquid than in the gaseous state; therefore *the heat of vaporisation always decreases as the temperature increases.*

The specific heat of liquid benzene was found by Regnault to be 0·436 between 21° and 71°; that of benzene vapour at the mean temperature (46°) can be calculated according to p. 45, as

$$\frac{6{\cdot}5 + (273 + 46) \times 0{\cdot}051}{78} = 0{\cdot}292,$$

so that

$$c_p - c = -0{\cdot}144.$$

Actually Griffiths and Marshall[1] found for the heat of evaporation of benzene between 20° and 50°

$$l = 107{\cdot}05 - 0{\cdot}158t,$$

the difference between 0·144 and 0·158 lies within the limits of experimental error. The specific heat of liquid water at 100° is 1·03, that of water vapour (at constant pressure) 0·48, whence the latent heat ought to fall off by 1·03 − 0·48 = 0·55 per degree, whilst Regnault found the decrease to be 0·7 per degree. It should be remarked that water vapour deviates from the laws of gases appreciably on account of polymerisation (formation of double molecules).

The second law of thermodynamics gives an important connection between the heat of vaporisation and the change of vapour pressure with the temperature: in the same way in which we considered the flowing of a gas into an empty space (p. 47), we will now apply to the evaporation of 1 g.-mol of a liquid in a vacuum, the equation

$$A - U = T\frac{dA}{dT};$$

the volume, which is at the disposal of the evaporating liquid, we will

[1] *Phil. Mag.* (5), **41**. 1 (1896).

choose as equal to the difference between the specific volumes of the saturated vapour V and of the liquid V′; then we will have

$$A = p(V - V'); \quad \frac{dA}{dT} = \frac{dp}{dT}(V - V'); \quad U = p(V - V') - l,$$

and accordingly

$$l = T\frac{dp}{dT}(V - V'), \quad . \quad . \quad . \quad (2)$$

an equation first developed by Clausius, and also given by the immediate application of the equation on p. 24.

Regnault (1847) found for water, for example, the heat of vaporisation at 100° to be 536·5 g.-cal.

The vapour pressure of water at 99·5° is 746·52, at 100·5° is 773·69 mm. (Wiebe, *Tables of Water Vapour Pressure*, Brunswick, 1894); hence[1] at 100°

$$\frac{dp}{dt} = 773{\cdot}69 - 746{\cdot}52 = 27{\cdot}17 \text{ mm.} = 0{\cdot}03570 \text{ atm.}$$

The specific volume V of saturated water vapour[2] at 100°, amounts to 1658 c.cm., somewhat greater than that calculated from the theoretical vapour density of water (viz., 0·623 compared with air), the volume V′ of liquid water being 1·0 cm.; therefore we find

$$V - V' = 1{\cdot}657 \text{ litre,}$$

and

$$l = 373 \;.\; 0{\cdot}03570 \;.\; 1{\cdot}657 = 22{\cdot}065 \text{ litre-atm.,}$$

or

$$l = 22{\cdot}065 \times 24{\cdot}25 = 535{\cdot}1 \text{ g.-cal.,}$$

a result in satisfactory agreement with observation.

As in the example given above, one can very often neglect V′ in comparison with V, and can calculate V with satisfactory accuracy from the laws of gases. The specific volume v, of 1 g.-mol of saturated vapour amounts to

$$v = MV = \frac{RT}{p},$$

and therefore

$$lM = \lambda = \frac{RT^2}{p}\frac{dp}{dT},$$

or transformed and expressed, like R, in g.-cal. (p. 48),

[1] On determining differential coefficients in such cases see Nernst and Schönflies, *Einleitung in die math. Behandl. der Naturw.*, 2nd edit., München, 1898, p. 266 ff.

[2] Wüllner and Grotrian, *Wied. Ann.*, **11.** 545 (1880).

$$\lambda = 1{\cdot}991T^2 \frac{d\ln p}{dT} \text{g.-cal.}, \quad . \quad . \quad . \quad (3)$$

a formula well suited for practical applications.

Within small temperature limits $T_2 - T_1$ we can regard λ as constant; then equation (3) suitably transformed

$$d \log p = \frac{\lambda}{1{\cdot}991} \cdot \frac{dT}{T^2}$$

can be integrated over this interval, giving

$$\lambda = \frac{1{\cdot}991 T_1 T_2}{T_2 - T_1} \log \frac{p_2}{p_1},$$

in which p_1, p_2 are the vapour pressures at T_1, T_2, introducing common logarithms

$$\lambda = 1{\cdot}991 \times 2{\cdot}303 \frac{T_1 T_2}{T_2 - T_1} \log \frac{p_2}{p_1}.$$

For benzene, Kahlbaum and v. Wirkner (*Dampfspannkraftmessungen*, **2.** 17, Basel, 1897) found, in good agreement with the earlier measurements of Regnault and Young,

$$T_1 = 273 + 20 \qquad p_1 = \ \ 75{\cdot}0 \text{ mm.}$$
$$T_2 = 273 + 30 \qquad p_2 = 118{\cdot}0 \text{ mm.}$$

whence the molecular latent heat λ is calculated at 8010 cal., whilst at the mean temperature (273 + 25) the measurements referred to on p. 59 give ($C_6H_6 = 78$)

$$(107{\cdot}05 - 0{\cdot}158 \times 25)78 = 8040.$$

For the temperatures

$$T_1 = 273 + 70 \qquad p_1 - 546{\cdot}5$$
$$T_2 = 273 + 80 \qquad p_2 = 750{\cdot}0,$$

whence $\lambda = 7640$ and the observed value is 7426. The appreciable difference in this case is due to benzene vapour being more dense at these pressures than accords with the gaseous laws.

The Form of the Vapour Pressure Curve.—The question, how the vapour pressure of a liquid changes with the temperature, or, in other words, what is the form of the *vapour pressure curve*, has been the subject of much experimental and theoretical investigation. At first the purely empirical result was found that the vapour pressure increases with the temperature, and indeed very rapidly, and accordingly the vapour pressure curve is convex downwards and highly bent; its upper extremity is found at the critical point (see below); *its lower* *extremity in all probability is not found till the absolute zero*, where gases cease to be capable of existence, and the corresponding vapour pressure becomes zero. If, on the other hand, one chooses the vapour pressures as the abscissæ, and the corresponding temperatures as the ordinates, he obtains the *curve of boiling-points* of the liquid.

There have been a very great number of researches to find out a universal law which should express the dependence of the temperature of boiling, on the external pressure, with satisfactory accuracy. Theoretically, indeed, this problem was solved by Clausius in the formula derived above, viz. :—

$$l = T\frac{dp}{dT}(V - V'),$$

in so far that, by means of this formula, the change of the boiling-point with the external pressure can be calculated with all strictness, and doubtless also with the greatest accuracy, if the heat of vaporisation, and the specific volumes of the liquid and of its saturated vapour in their dependence on the temperature, are known.

Thus, if the vapour pressure p_0 at some one temperature is known, we can calculate the value of V belonging to it, and thence $\frac{dp}{dt}$ by the above equation. For a temperature higher by the small amount Δt, the vapour pressure is $p_0 + \Delta t\left(\frac{dp}{dt}\right)_{p=p_0}$, whence we may find V again, and so on. It is therefore possible to calculate the vapour pressure curve by successive steps.

If the gaseous laws are applicable to the vapour (p. 60)

$$\lambda = RT^2\frac{d \log p}{dT} \quad . \quad . \quad . \quad . \quad (1)$$

and, by p. 59,

$$\frac{d\lambda}{dT} = C_p - C \quad . \quad . \quad . \quad . \quad . \quad (2)$$

where C_p is the molecular heat of the vapour at constant pressure, and C the molecular heat of the liquid ; we can then integrate this over an interval of temperature such that C_p and C can be regarded as constant, *i.e.* λ is a linear function of the temperature

$$\lambda = \lambda_0 - (C - C_p)T \quad . \quad . \quad . \quad (3)$$

(1) and (3) give

$$\frac{d \log p}{dT} = \frac{\lambda_0}{RT^2} - \frac{C - C_p}{RT},$$

from which, by integration,

$$\log p = -\frac{\lambda_0}{RT} - \frac{C - C_p}{R} \log T + \text{const.},$$

or

$$p = Ae^{-\frac{B}{T}} T^{-\frac{C - C_p}{R}},$$

where A and B are two constants that can be calculated by means of two pairs of values of p and T. This equation is suitable over intervals of temperature that are not too great and are well removed from the critical point.[1]

In general, it is not possible to apply the thermodynamic calculation of the vapour pressure curve for want of data. Another method, also difficult in application, is that of Van der Waal's theory (Book II. chap. ii.).

Comparison of the vapour pressure curves of different substances has led to results of importance that are very convenient in application. Dalton[2] pointed out such a relation, and by reckoning the boiling-point from the absolute, instead of from a conventional, zero, more striking regularities have been discovered.

As was shown in numerous examples by Ramsay and Young,[3] whose extended observations on the evaporation of liquids gave important information on many points, the ratio of the boiling-points of two substances *chemically related*, is nearly constant when measured at equal pressures and absolute temperatures, *i.e.* reckoned from $-273°$; in the comparison of substances chemically different, this ratio changes with the temperature.

In the following table, the figures of which are taken from the measurements of Schumann,[4] are given the absolute boiling-points with the corresponding pressures, for a number of esters, *i.e.* [ethereal salts] of closely related composition:

[1] For an application see Hertz, *Wied. Ann.*, **17.** 193 (1882); *Ges. Abh.*, **1.** 215.

[2] See also Duhring, *Wied. Ann.*, **11.** 163 (1880); **52.** 556 (1894), and a criticism on Duhring's rule by Kahlbaum and v. Wirkner, *Ber. d. d. chem. Ges.*, **27.** 3366 (1894).

[3] *Phil. Mag.* [5], **20.** 515; **21.** 33, 135; **22.** 32; and *Zeitschr. physik. Chem.*, **1.** 249 (1887), and S. Young, *Phil. Mag.* (5), **34.** 510 (1892); Groshans, *Wied. Ann.*, **6.** 127 (1879).

[4] *Wied. Ann.*, **12.** 58 (1881).

[TABLE

Substance.	T_1 760 mm.	T_2 200 mm.	$\frac{T_1}{T_2}$.
Methyl formate	305·3	273·7	1·115
Methyl acetate	330·5	296·5	1·115
Methyl propionate	352·9	316·7	1·114
Methyl butyrate	375·3	336·9	1·114
Methyl valerate	389·7	350·2	1·113
Ethyl formate	327·4	293·1	1·117
Ethyl acetate	350·1	314·4	1·114
Ethyl propionate	371·3	333·7	1·113
Ethyl butyrate	392·9	352·2	1·116
Ethyl valerate	407·3	365·3	1·115
Propyl formate	354·0	318·0	1·113
Propyl acetate	373·8	336·1	1·112
Propyl propionate	395·2	355·0	1·113
Propyl butyrate	415·7	374·2	1·111
Propyl valerate	428·9	385·6	1·112

According to the rule of Ramsay and Young, the ratio of the boiling-points in the vertical columns for any two substances must be equal; therefore, if one writes the quotient $\frac{T_1}{T_2}$ it must give an almost constant value for all cases, as is in fact strikingly shown in the last column.

One finds other relations on comparing the boiling-point curves of two substances of unlike chemical composition, as, for example, mercury and water. In the following table are given some boiling-points, on the absolute scale, corresponding to the pressure p:—

p	T Hg	T H_2O	Ratio.
34·4	495·15	304·5	1·6262
157·15	553·2	334·2	1·6553
760·83	631·68	373·03	1·6934
2904·5	721·0	415·36	1·7359

The ratio of any two boiling-points at equal pressures is here by no means constant, but increases with the temperature; but if one obtains the quotient of the increase of this ratio, divided by the corresponding rise of the boiling-point of mercury, he finds for the three intervals of the preceding table, 0·00098; 0·00098; 0·00100, *i.e.* the ratio of the absolute boiling-points corresponding to the same pressures, increases *linearly with the temperature.* If one calculates the relative increase corresponding to the rise of the boiling-point of water, he finds 0·00098; and one can calculate the absolute boiling-point T of mercury for any desired pressure, from the absolute boiling-point T of water at the same pressure, from the equation,

$$T = 1{\cdot}6934T_0 + 0{\cdot}00098(T_0 - 373)];$$

in this $1{\cdot}6934 = \frac{T}{T_0}$, when $T_0 = 373$, the vapour pressure is equal to one atmosphere.

The Boiling-Point.—A vapour is constantly ascending from the surface of a liquid except when the partial pressure of the vapour is greater than, or equal to, the tension of the liquid: in the former case, a reverse condensation takes place; while only in the latter case can there be an equilibrium between vapour and liquid. Its surface is always weighted down with a pressure, which, according to Dalton's law, is equal to the sum of the partial pressures of the vapour and of the other gases present. If one heats the liquid so high that its vapour pressure begins to overcome the external pressure weighting it down, then vapour bubbles begin to form in its interior, and one observes the phenomenon of *boiling*: the *lowest* temperature at which a liquid can remain *steadily* boiling is called the *boiling-point* corresponding to that particular pressure. That corresponding to the pressure of 760 mm. is called the *normal boiling-point*, or simply the *boiling-point*; its experimental determination is so simple and so common that we may dispense with a thorough treatment of it, but attention should be called to some important corrections.

Thus, to wit, the fact should be noticed that the boiling-point conducted at the prevailing atmospheric pressure, must be reduced to the normal pressure. The change dT, which the boiling-point experiences by a variation of the pressure amounting to dp, can be calculated, at least theoretically, from the formula of Clausius (p. 60), viz.:—

$$dT = T\frac{V - V'}{l}dp;$$

but since the heat of evaporation l is as a rule unknown, it is proposed to introduce a method recently given by Crafts,[1] which is based on the laws of boiling-points found by Ramsay and Young (p. 63), namely, that the absolute boiling-points, corresponding to the same pressures, of two substances chemically related, stand *in a constant relation*. In the following table are given the changes of boiling-point, divided by the absolute boiling-point of the respective substances at normal pressure; the changes refer to a change in pressure of 1 mm. of mercury, calculated from the change of boiling-point which the substances experience in varying the pressure from 720 to 770 mm.

[1] *Berl. Ber.*, **20.** 709 (1887).

Water	0·000100	Carbon disulphide	0·000129
Ethyl alcohol	0·000096	Ethylene bromide	0·000118
Propyl alcohol	0·000096	Benzene	0·000122
Amyl alcohol	0·000101	Chlor-benzene	0 000122
Methyl oxalate	0·000111	m-Xylene	0·000124
Methyl salicylate	0·000125	Brom-benzene	0·000123
Phthalic acid anhydride	0·000119	Oil of turpentine	0·000131
Phenol	0·000109	Naphthalene	0·000121
Aniline	0·000113	Diphenyl-methane	0·000125
Acetone	0·000117	Brom-naphthalene	0·000119
Benzophenone	0·000111	Anthracene	0·000110
Sulfobenzide	0·000104	Triphenyl-methane	0·000110
Anthraquinone	0·000115	Mercury	0·000122

If one has determined the boiling-point of a substance at a pressure removed from the normal, he can correct for it approximately, by the addition of the absolute temperature 273, and then selecting that substance *most closely resembling it* from the preceding list: the auxiliary factor is multiplied by the absolute boiling temperature, by which is obtained the correction which must be applied to the observed variation per mm. from the normal pressure.

The Critical Phenomena.—If one heats a liquid in contact with its saturated vapour, the density of the saturated vapour increases very quickly, since the vapour tension rises rapidly with the temperature. But the density of the liquid, which is expanding in consequence of the rise in temperature, is conversely, continually diminishing. Now the question arises, Does there exist a point at which the densities of the liquid and of the saturated vapour may be equal to each other? The study of this question led to the discovery of the *critical phenomena*, which have assumed great significance in our conceptions of the nature of the liquid state.

As was discovered by Cagniard de la Tour,[1] and later thoroughly investigated by Andrews,[2] in the compression of a gas, or in heating a liquid enclosed in a vessel, the following is observed. If one compresses a gas, carbon dioxide for example, at a suitably high pressure and low temperature, the contents, originally homogeneous, separate into two parts with a dividing surface sharply defined; or in other words, it is partially liquefied. The pressure at which this happens is, of course, the corresponding maximal pressure of the separated liquid, and it increases very considerably with the temperature. Now the question arises whether liquefaction can occur at all temperatures with a suitably high pressure,—a question which has been decided in the *negative* by the work of the investigators just mentioned. Thus, for example, carbon dioxide is convertible into a liquid by the application of a pressure of 70 atmospheres below 30·9°; but above that temperature, one can increase the pressure at will, without observing that the gas loses its homogeneity, and without liquefaction taking place.

[1] *Ann. chim. phys.* [2], **21.** 121, 178; **22.** 411 (1821).
[2] *Trans. Roy. Soc.*, **159.** 583 (1869).

When, on the other hand, one heats a glass tube filled with liquid and gaseous carbon dioxide, evaporation takes place gradually because the vapour pressure of the liquid increases faster, with a rise of temperature, than the pressure of the gaseous part. But at 30·90°, when the vapour pressure has reached 70 atmospheres, suddenly there occurs the evaporation of all the liquid part, the meniscus which separated the liquid from the gas, and which had already begun to flatten out, vanishes at this temperature, and the contents of the tube are completely homogeneous. When it is again cooled, at the same temperature a fog appears, which quickly gathers as a liquid in the lower part.

These very remarkable phenomena are called "critical." That temperature, above which the liquid ceases to be capable of existence, is called the "*critical temperature*" (or the "absolute boiling-point"); the vapour pressure of the liquid at this point is called the "*critical pressure*," and its specific volume the "*critical volume*." These three magnitudes are the characteristic *critical* data for every simple liquid, and, as will be shown in the second book, they are typical of the entire relations between gases and liquids.

The critical phenomena make it possible to change a liquid into a gas *in one continuous process*, *i.e.* without its losing its homogeneity, by partial evaporation during the process, and the reverse. One can thus heat a liquid till above its critical temperature, with the external pressure increasingly greater than its vapour pressure, and finally greater than its critical pressure; then if one allows the volume to increase, the original liquid mass remains homogeneous to any desired tenuity; it has been converted continuously into the gaseous state. In order to convert a gas continuously into a liquid, one must raise its temperature above the critical point, while the pressure is kept below that required for condensation; then it must be compressed above its critical pressure, and cooled below its critical temperature, while the external pressure is kept greater than the corresponding maximal pressure of the liquid; if now one allows an increase of volume, then the original gaseous mass loses its homogeneity, gives off vapour, and is to be regarded as a liquid.

The designation "absolute boiling-point" took its origin from Mendelejeff;[1] but since we ordinarily understand that to be the boiling-point of a liquid reckoned from $-273°$, there is here good reason for a change, and therefore the designation "critical temperature" is to be given the preference, and particularly so as it is now universally accepted.

On account of the results of the measurement of critical data, for which we are indebted to Cagniard, Andrews, Pawlewski, Dewar, and others, and recently to Ramsay and Young especially, reference should here be made to the very complete compilation which Heilborn[2] has

[1] *Lieb. Ann.*, **119.** 1 (1861). [2] *Zeitschr. physik. Chem.*, **7.** 601 (1891).

brought out. I will only mention here a simple lecture experiment, which I consider very instructive for the demonstration of critical phenomena. A small glass tube with thick walls, partially filled with liquid sulphur dioxide, is placed in a larger test-tube which contains paraffin. If one carefully heats this, which is in an inclined position, from the side, with a Bunsen burner, so that the upper part of the inner little tube shall be heated above the critical temperature of sulphur dioxide (155·4°), while the lower part, surrounded by unmelted paraffin, remains considerably cooler, then the meniscus vanishes. Under these conditions we have without doubt *gaseous* sulphur dioxide in the upper part of the little inner tube, and *liquid* in the lower part, *yet one sees nowhere a parting surface.* Thus this spectacle shows how liquid and gas can pass into each other continuously, in the circuit over the critical point. One will do well, while conducting this experiment, to protect himself against danger from explosion, by a thick plate of glass.

If the liquid is not quite pure the vapour will in general have a somewhat different composition, and as owing to slowness of diffusion the state of equilibrium is only very gradually reached, the meniscus will disappear on heating at a different temperature to that at which it appears on cooling, unless the operations are very carefully conducted. Some experimentalists, in recent times, have been so deceived by this that they doubted the existence of a well-defined critical point. The careful experiments of Ramsay,[1] Young,[2] and Villard[3] have, however, put this beyond question.

[1] *Zeitschr. phys. Chem.*, **14.** 486 (1894). [2] *Trans. Chem. Soc.* 1897, p. 446.
[3] *Ann chim. phys.* (7), **10.** 387 (1897).

CHAPTER III

THE SOLID STATE OF AGGREGATION [1]

General Properties of Solid Bodies.—If we condense a substance occurring in the gaseous state, at a sufficiently low temperature,—*i.e* below its fusing-point,—or if we cool a liquid substance to its hardening-point,—then the matter appears in the solid state. This, in common with the liquid, and in contrast to the gaseous state, has this property, viz. that *a change of volume* by a pressure from all sides, is resisted by an extraordinarily great force. But a property belonging to the solid state alone, must be mentioned here, viz., that *a change of form* without compression encounters the so-called *elastic forces.* Work which appears again as heat, must be expended indeed in a change of form of gases and liquids, in consequence of their internal friction. But in the case of solids, it happens that by a relative distortion (not too great) of the particles, the system is placed in a state of tension which is associated with the storage of considerable potential energy. When the action of the deforming force ceases, then the body resumes again its original form.

Melting-Point and Pressure.—In the same way that at a given temperature a substance can coexist in both the liquid and gaseous states, only at a certain definite external pressure, so a solid substance can exist in equilibrium with its molten product, only at very definitely related values of pressure and temperature. We find this quantitative distinction, viz. that while the boiling-point varies very much with the external pressure, the melting-point varies very little; so that for practical purposes one can usually regard this as unchangeable, and for a long time it was not observed.

This was established by William Thomson [2] in 1850 for the case

[1] In the preparation of the crystallographic part of this chapter I have enjoyed the valuable co-operation of Dr. Pockels.

[2] At present honoured and recognised by the title of Lord Kelvin; this correction may serve once for all in all references to this great investigator in this book. This is from no wish to disregard the present title, but simply because most of his work is to be found in literature under his former title.—TR.

of water, though James Thomson had previously predicted it theoretically. The process of melting can be treated thermodynamically, similarly to that of vaporisation.

The maximal work to be obtained by the fusion of 1 g. of a solid substance, is of course equal to the product of the increase of volume $V - V'$ (when V is the specific volume of the liquid, and V' that of the solid substance), and the pressure p, at which both states are in equilibrium; then will

$$A = p(V - V'), \quad \text{and} \quad dA = dp(V - V').$$

The diminution of total energy U is equal to the work performed A, minus the heat absorbed r, the so-called heat of fusion, *i.e.*

$$U = p(V - V') - r,$$

the change of which with the temperature (p. 8), neglecting the insignificant external work, is calculated to be

$$-\frac{dU}{dT} = \frac{dr}{dT} = c' - c''$$

when c' and c'' denote respectively the specific heats of the substance in the liquid and solid states. The specific heat of ice is 0·5; that of water 1·00; therefore the heat of fusion of ice increases per degree of rise in temperature, by an amount equal to $1{\cdot}00 - 0{\cdot}5 = 0{\cdot}5$ g.-cal.

The equation,

$$A - U = T\frac{dA}{dT}$$

in this case assumes the form,

$$p(V - V') + r - p(V - V') = T(V - V')\frac{dp}{dT},$$

or

$$r = T(V - V')\frac{dp}{dT},$$

or, as is also given by the direct application of the equation on p. 24,

$$\frac{dT}{dp} = \frac{T(V - V')}{r}.$$

An increase of pressure corresponds to a positive value of dT, *i.e.* a rise of the melting-point, if $V - V'$ is positive, *i.e.* when the change to the liquid state is accompanied by an increase of volume; on the other hand, it corresponds to a fall of the melting-point when the body contracts in melting, as is the case with ice.

Carrying out the calculation for water, as $T = 273°$; $v = 0{\cdot}001$;

and v′ = 0·001091 litre; the heat of fusion r = 80·3 cal., then the work is equal to

$$r = \frac{80 \cdot 3}{24 \cdot 25} \text{ litre-atmospheres,}$$

and therefore

$$\frac{dT}{dp} = -0 \cdot 0077^\circ,$$

i.e. an increase of an atmosphere in the external pressure corresponds to a lowering of the melting-point of water, of about 0·0077°. William Thomson (1851) showed that, as a matter of fact, by raising the external pressure about 8·1 and 16·8 atmospheres, the temperature of melting ice sank from 0° to − 0·059° and − 0·129° respectively, while these calculated from the formula given above, should be − 0·062° and − 0·127° respectively.

On the other hand, that the melting-point is raised by applying external pressure, in the case of substances which melt with increase of volume, was first shown by Bunsen (1857) for spermaceti and paraffin. Recently Batelli[1] and Demerliac[2] has found Thomson's formula to be well established for a number of organic compounds. Further, L. E. O. de Visser[3] has succeeded in measuring with great accuracy the change in melting-point with the pressure, *for acetic acid;* since he did not, as was done before, determine the melting-point corresponding to a definite pressure, but the pressure corresponding to a definite temperature. The simple apparatus used by de Visser, and which he named a *Manocryometer,* consisted of a large thick-walled thermometer; this was inverted, its capillary tube bent upwards, and then horizontally (Fig. 3). According to the temperature of the surrounding bath, it would assume that pressure at which the substance in A, partly liquid, partly solid, would find itself in equilibrium; the capillary tube filled with mercury served as a closed manometer with which to measure this pressure. The direct measurements of de Visser showed that

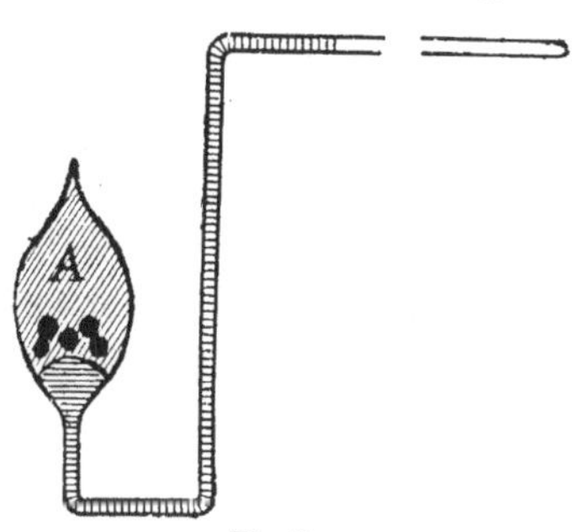

Fig. 3.

$$\frac{dT}{dp} = 0 \cdot 02435^\circ.$$

If the heat of fusion is 46·42 g.-cal., the fusing-point T, 273° + 16·6

[1] *Atti del R. Ist. Ven.*, (3) **3.** (1886). [2] *Compt. rend.* **124.** 75 (1897).
[3] Dissertation, Utrecht, 1892; ref. *Zeitschr. physik. Chem.*, **9.** 767 (1892).

$=289{\cdot}6°$, and the increase of volume $V - V' = 0{\cdot}0001595$ litre, then calculation gives

$$\frac{dT}{dp} = 0{\cdot}0242° ;$$

the coincidence is remarkable.

Recently the effect of pressure has been studied over a very wide range by G. Tammann.[1] The curve of fusion for benzene was traced to pressures of more than 3000 atmos.; the heat of fusion was determined directly and shown to be constant within 1% between 1 and 1200 atmos. The volume change $V - V'$ was also measured as a function of pressure along the curve of fusion. The following table gives the results; the pressure p is in kg/sq. cm.; $V - V'$ in c.c. per gram of substance :—

t	p	V − V′	r
5·43	1	0·1307	29·2
10·12	161	0·1272	30·0
20·13	533	0·1118	29·6
29·59	925	0·1053	30·9
42·06	1455	0·0919	30·6
55·02	2040	0·0770	29·0
66·00	2620	0·0738	30·6
77·96	3250	0·0693	31·8

To calculate r from the foregoing formula the empirical equation

$$p = 34{\cdot}4(T - 5{\cdot}43) + 0{\cdot}150(T - 5{\cdot}43)^2$$

was obtained from the observations, $\frac{dp}{dT}$ deduced and then r for various temperatures. The heat of fusion appeared constant, in harmony with the result of direct measurement; the mean calculated value was 30·0, the mean of the calorimetric observations 30·4.

The fact that $V - V'$ falls off rapidly while r remains constant seems to be general, and indicates that the two quantities do not vanish at the same point.

The Vapour Pressure of Solid Substances.—As in the case of liquids, so also for every solid substance, at a given temperature, there is a correspondingly definite vapour pressure, although, indeed, in most cases it is so extraordinarily small as to escape a direct measurement. The vaporisation of a solid substance is called *sublimation*. Sublimation, like evaporation, takes place gradually from a solid substance in contact with the free atmosphere under all conditions, but it is especially rapid if the *sublimation pressure* exceeds that of the atmosphere. If

[1] *Ann. d. Phys.*, **3.** 161 (1900).

this point, comparable with the boiling-point of the liquid substance, lies below the melting-point, then when heated in the open air it will sublime without melting. It is only by heating the substance in a closed vessel that it is possible to heat the substance to the melting-point, and thus to accomplish its liquefaction.[1] But ordinarily the sublimation pressure of solid substances at the melting-point is much smaller than the atmospheric pressure.

The "*heat of sublimation*," *i.e.* the quantity of heat absorbed in the vaporisation of 1 g. of the solid substance, can be derived from the change of vapour pressure with the temperature $\frac{dp}{dT}$, and from the specific volumes V and V′ of the vapour and substance respectively, in the same way that was used for the calculation of the heat of vaporisation (p. 58), since the considerations there advanced hold good whether a solid or a liquid substance is evaporated, and accordingly

$$s = T\frac{dp}{dT}(V - V').$$

On account of the insignificance of the vapour tension of the solid substance, V′ can, almost always, be neglected in comparison with V.

At the melting-point, the heat of sublimation is equal to the heat of fusion + the heat of vaporisation of the melted substance, *i.e.*

$$s = r + l = T\frac{dp}{dT}V;$$

and further for the vaporisation, the following equation holds good :—

$$l = T\frac{dP}{dT}V,$$

if by P we denote the vapour pressure of the liquid substance in the neighbourhood of the melting-point. Subtracting these two equations gives

$$r = TV\left(\frac{dp}{dT} - \frac{dP}{dT}\right).$$

The part in the parenthesis indicates the angle which the pressure curves of the substance in the solid and the liquid states, make with each other at the melting-point. But now the vapour pressures of the two states must be *equal* at the melting-point, because it is the point at which the solid and liquid states of the substance are in equilibrium with each other. Otherwise there would occur an isothermal distilla-

[1] A simple but instructive experiment to show this, is to heat a little iodine (which usually barely melts in the air) in a test-tube with say 2 cm. of H_2SO_4, when the melted iodine is clearly seen below the transparent acid, which increases the external pressure to the required point.—TR.

tion process which would cease only with the disappearance of that state which would have the greater vapour tension, *i.e.* the two states of aggregation would not be in equilibrium. Therefore the pressure curves *intersect* at the melting-point, as is shown in Fig. 4. Also the dotted line is the pressure curve of the *under-cooled* liquid substance, and forms the continuation of the pressure curve of the liquid. Probably the two curves would intersect asymptotically at abs. zero,[1] if one could under-cool a liquid so far. The preceding formula, which was developed by W. Thomson (1851), and also again independently by Kirchhoff (1858), was later proved by experiment by Ramsay and Young (1884) for benzene, and by W. Fischer (1886) for water; also recently by Ferche[2] (1891) for benzene. From the latter's measurements are taken the following numbers. In accordance with theory at the melting-point (5·6°), the pressure of solid and liquid benzene had the same value, viz. 35·5 mm. Hg. The measurement gave

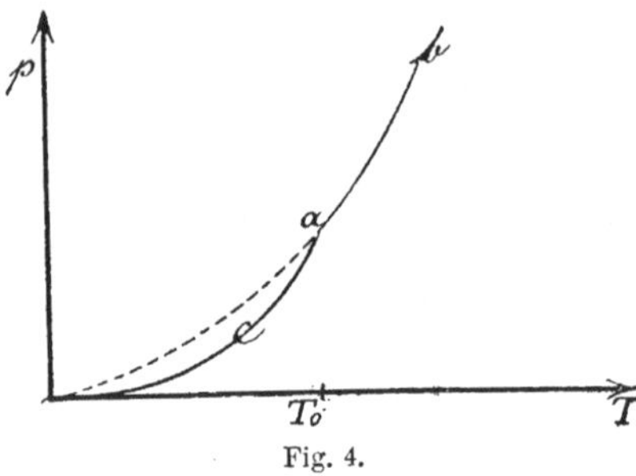

Fig. 4.

$$\frac{dp}{dT} - \frac{dP}{dT} = 2{\cdot}428 - 1{\cdot}905 = 0{\cdot}523.$$

On the other hand, this value can be calculated with the aid of the formula given above, from the heat of fusion, which, reduced to litre-atmospheres, was 30·18 g.-cal.; and thus according to the laws of gases we write

$$TV = \frac{0{\cdot}0821(273 + 5{\cdot}6)^2 \cdot 760}{35{\cdot}5 \cdot 78},$$

and, finally, this pressure in atm. when reduced to mm. of Hg, is

$$\frac{dp}{dT} - \frac{dP}{dT} = \frac{30{\cdot}18 \cdot 35{\cdot}5 \cdot 78}{24{\cdot}25(273 + 5{\cdot}6)^2 \cdot 0{\cdot}0821} 0{\cdot}541,$$

a result in satisfactory accord with experiment.

The Crystallised State.—Most solid substances separate by condensation from the gaseous state, or by congealing from a state of fusion, or by precipitation from solutions, in regular polyhedral forms, in case unfavourable circumstances do not interfere with their normal formation, *i.e.* they *crystallise.* All the physical properties have the

[1] There is no evidence in favour of the view, often expressed, that there is a temperature below which solids and liquids do not evaporate at all.

[2] *Wied. Ann.*, **44.** 265 (1891).

closest connection with the external form; both the external form and the physical properties are conditioned by the *structure* of the particular body. Thus a *crystal* can be defined as a *homogeneous body, in which the different physical properties conduct themselves differently, in the different directions radiating from one of its points.*

The proviso of *homogeneity*, which is always tacitly assumed in the following sketch, declares that the physical properties depend *only on the direction, not on any particular portion*, of the crystal, and accordingly they are the same for all *parallel directions.* It is well to notice that this definition of a crystal does not consider the geometrical form of the limiting surface; this latter is the most obvious external sign of the crystallised condition, but it occurs only when the formation is undisturbed.

Crystallised bodies are contrasted with the *amorphous*, in which all directions are alike respecting the physical properties. Yet certain facts, and especially their elastic behaviour,[1] have led to the suspicion that many amorphous solid substances are composed of crystal fragments too small for detection. Should this be established in future for all such substances, then a crystal structure would be regarded as characteristic of the solid state of aggregation.

If two or more directions radiating from one point of a crystal are *equivalent*, we say that the crystal possesses *symmetry.* The symmetry of a crystal, as regards the different physical properties, varies according to the nature of the physical process considered. Experience has shown that the lowest grade of symmetry is always shown in the processes of *growth* and *disintegration* of crystals, especially in the *external polyhedral form* attending undisturbed growth. Therefore the latter, which lends itself easily to observation, is especially suited for the characterisation and classification of crystals. We will consider next, on the aforementioned basis, the laws which regulate the *forms* of crystals.

The Fundamentals of Geometrical Crystallography.—As is known, normally-formed crystals are bounded by flat planes, which make *convex polyhedra, i.e.* such polyhedra as are cut by any straight line in only two points at most; [or we may say that convex polyhedra have no re-entrant angles]. It follows from the definition of a crystal as a *homogeneous* body, that planes having the same direction are equivalent. It must be noticed that in a plane the two sides are to be distinguished, and therefore two planes of the bounding surface can be similarly directed only when the direction of their *external* normals are coincident. Therefore, in the following sketch, it must always be borne in mind that, in studying the form of a crystal,

[1] See W. Voigt, *Wied. Ann.*, **38.** 573 (1889); *Gött. Nachr.*, 1889, 519, and 1890, 154.

the bounding planes can always be imagined to be moved at will parallel with themselves.

The first fundamental *law* of geometrical crystallography is the "*law of the constancy of interfacial angles*" (discovered by Steno in 1669); this declares that *the inclination of two definite crystal planes to each other, for the same substance, and measured at the same temperature, is constant, and independent of the size and development of the planes.*

The abundant data obtained in crystal measurement by means of the reflection goniometer, have taught us that this law has only an *approximate* value, for not infrequently on good crystals of the same chemically pure substance, which have apparently formed under the same conditions, and even on the same crystal, the corresponding angle is liable to variations of over 0·5°.

By the term "*zone*" is understood a set of planes which intersect with *parallel edges;* the common direction of the latter is called the "zonal axis." A zone is fixed by any two of its planes. Conversely, if it is known concerning a plane that it lies in two known zones, it is completely determined, for it is parallel to the two zonal axes, viz. two straight lines, which intersect. Experience has led to the following law, formulated by F. E. Neumann (1826), which is called the "*law of zones*," and which is the *characteristic fundamental law of crystallography; viz. all planes which can occur on a crystal are related to each other in zones;* or, in other words, *from any four planes, no three of which lie in one zone, all possible crystal planes can be derived by means of zones.*

Presumably there might be an unlimited number of bounding planes on a crystal; but by no means can every arbitrary geometrical plane be a face of a crystal. We must mention two other forms of this aforementioned law which will make this point clearer. Regarding this law of zones, it must be considered that it insists on the zone relation only for all *possible* crystal faces. The planes then actually occurring on a single crystal need not necessarily stand in complete zone relation to each other.

In order to state properly, *i.e.* by numbers, the position of a face of a crystal, it can be compared with any system of co-ordinates fixed relatively to the crystal. There are now selected as the co-ordinate planes (on grounds to be explained below), any three crystal faces not lying in the same zone (such faces, for example, as are remarkable for their prevailing occurrence, or as cleavage planes), and to correspond to the co-ordinate axes, three crystal edges, viz. the lines of intersection of the three planes, OX, OY, and OZ in Fig. 5. In general this system of co-ordinates will be *oblique angled.* Then the position of any given fourth crystal plane (E_1, E_2, E_3), is determined by the length of the intersections OE_1, OE_2, and OE_3, which it makes on the co-ordinate axes; if a, b, and c respectively are these lengths, the equation of the plane is

$\frac{x}{a} + \frac{y}{b} + \frac{z}{c} = 1$. As only the *direction* of the face is in question, we are only concerned with the *ratios* of a : b : c, which remain constant if the plane is moved parallel to itself. Usually one selects as *units*, the axial intersections a, b, and c, of any face (the unit plane), which does not lie in the same zone with two of the co-ordinate planes (fundamental planes); and by means of these units, the intersections of all other faces are expressed. These "axis units" a, b, and c, are usually unequal, and stand in irrational proportions to each other. Now if ma, nb, and pc, are the axial intersections (parameters) of any fifth face of a crystal (H_1, H_2, H_3, Fig. 5), one calls three numbers h, k, l, which are proportional to $^1/_m$, $^1/_n$, $^1/_p$, respectively, the *indices* of this latter plane, and so, after establishing the fundamental planes and the unit planes, the direction of any other will be determined by the ratios of these three numbers h, k, l. Now it can be easily shown that the law of zones leads to the conclusion that *the ratios of the indices of all possible bounding planes occurring on a crystal are rational numbers.* This *law of the rationality of indices*, the principle of which was first advanced by Hauy in 1781, of course holds good for any arbitrary choice of fundamental planes and unit planes, provided that these are crystal faces. The arbitrary common factor to h, k, l is so chosen that h, k, l shall be the smallest possible *whole* numbers; in the case of a plane which is parallel to one or two co-ordinate axes, one or two indices respectively will be equal to zero. It is to be observed that the indices are, in almost all cases, the *lowest whole* numbers (usually 0, 1, 2, 3, 4, etc.), and from this the law of the rationality of indices derives its significance; for if they were indeed irrational, one could always express the indices in sufficiently large whole numbers with an accuracy corresponding to that of the angle measurements.

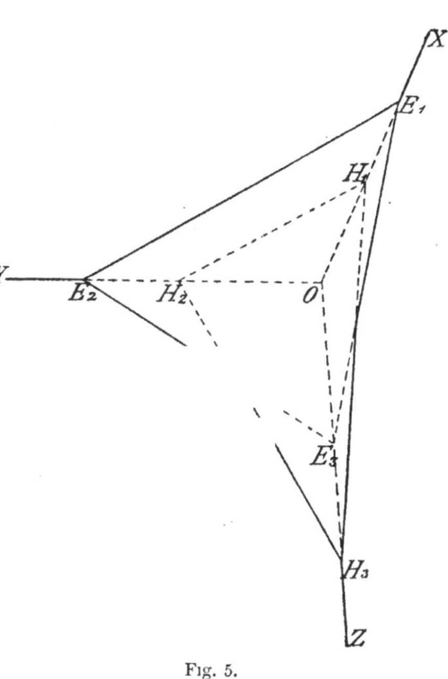

Fig. 5.

The calculations of the indices from the angle measurements usually give at first irrational proportions. But according to the preceding empirical law, one is seldom left in doubt as to which whole numbers are to be regarded as the true indices; for one selects those whole numbers whose ratios are nearest to the irrational values. In general, for the determination of the indices of a face, it is necessary to measure two angles (see the paragraph on "The Determination of Crystal Symmetry," p. 94); but if one zone is known in which the plane lies, then the measurement of only *one* angle is necessary; and if one knows two zones in which the plane lies, then the indices are fully given without the measurement of any angle.

Since this last case occurs very frequently in the determination of crystals, the course to be pursued will be somewhat elucidated. For this purpose we must first consider the method of determining the *direction of the edges.* From any four crystal edges (*three fundamental edges* and *one unit edge*), one can develop all other possible ones, in accordance with the law of zones, by finding out successively the lines of intersection of the planes which connect two zones already known. Now if α, β, γ, are the co-ordinates of a point of the unit edge, referred to the fundamental edges, and if $\alpha\xi$, $\beta\eta$, and $\gamma\zeta$, are the co-ordinates of a point of any fifth crystal edge, then it follows from the preceding law that the ratios $\xi:\eta:\zeta$ are rational. We call ξ, η, ζ, the *indices* of the edge or zone, and designate this latter by the symbol $[\xi, \eta, \zeta]$. Now if one selects for the fundamental edges the intersection lines of the fundamental planes, and for the unit edge the resultant from the axial units a, b, c, so that $\alpha : \beta : \gamma = a : b : c$, then to express the relation that a plane h, k, l passes through an edge $[\xi, \eta, \zeta]$, or lies in a zone $[\xi, \eta, \zeta]$, we have the equation

$$h\xi + k\eta + l\zeta = 0.$$

It is evident from this how the indices of a plane lying in two given zones, or how those of the intersection edge of two given planes, can be calculated from the solution of two linear equations, and also at the same time how the law of the rationality of indices follows from the law of zones.

A third method of expressing the fundamental law of crystallography, and coming from Gauss (1831), is the *law of the rationality of compound ratios;* this declares that between four crystal planes lying in one zone, or between four crystal edges lying in one plane, there exists such a connection that their compound ratio is rational. By the compound ratio of four such straight lines or planes α, β, γ, δ, is meant the following quotients of the sines of their included angles, viz.—

$$\frac{\sin(\alpha \cdot \gamma)}{\sin(\beta \cdot \gamma)} : \frac{\sin(\alpha \cdot \delta)}{\sin(\beta \cdot \delta)}.$$

That this quotient must be rational for the planes and edges of a *crystal*, is easily shown by the fundamental principles of projective geometry according to the law of zones, or according to that of the rationality of indices. In consequence of this relation, if we know three planes in a zone, or three edges in a plane, then all the planes or edges possible for the crystal, lying in this zone or plane, are completely determined. Since the compound ratio of four planes lying in a zone can be expressed by their indices, one obtains formulæ which are of fundamental importance for the calculation of crystals, since they give the solution of the two following problems, viz. :—1. If we know the indices of four planes of a zone, and also two of the six angles between these planes, to calculate the other four angles; 2. If we know the angles between four planes of a zone, and the indices of three of them, to find the indices of the fourth plane.[1]

In order to describe the geometrical form of a crystal it is clear from the preceding statements that one must specify as follows:—1. The three angles between the fundamental edges or between the fundamental planes; 2. The relative lengths of the units of the axes; 3. The indices of the planes (or edges) occurring on the crystal. The quantities given under 1 and 2, which are the so-called axial units or the geometrical constants of the crystal, are, in general, characteristic for the crystallised substance in question (compare the last sections in this chapter, p. 86). The choice of these, as already observed, is to a certain extent arbitrary. At the same time, one must consider not only the fundamental law of crystallography, but also the relations of the crystal regarding its *symmetry* (see also p. 80).

As Hauy first observed, in its limitation the *symmetry* of crystals holds good to this extent, viz. that the differently directed planes on a well-developed crystal, which are regularly related to each other, are always of equal value and physical behaviour (as shown by the glance, striation, etc.). The relations of symmetry can be established much more certainly by angle measurements; for the angles between two pairs of equivalent planes must of course be identical. If the symmetry were clearly expressed in the external form of the crystal polyhedron, we must suppose all equivalent planes to be moved parallel, each with itself, till they are all at the same distance from a fixed point in the interior of the crystal. This ideal geometrical shaping, which occurs in nature only in very favourable cases, must always be presupposed in the study of the symmetry of a crystal polyhedron. Let it be noticed here for subsequent use, that the related set of equivalent faces is called a *crystal form* in distinction from a *combination*, which latter is a crystal polyhedron made up of dissimilar planes. It is not necessary for a simple crystal form to be a closed polyhedron; but it can consist merely, for example, of a pair of parallel planes, or even of one single plane.

The Classification of Crystals according to Symmetry.—We

[1] Compare Th. Liebisch, *Geometrische Krystallographie*, chap. iv., Leipzig, 1881.

now turn to the more detailed study of the properties of the symmetry that a crystal polyhedron can have.

Symmetry is conditioned by the occurrence of the three following sorts of *symmetral elements*, viz.: 1. A centre of symmetry; 2. One or more axes of symmetry; 3. One or more planes of symmetry. These symmetral elements have the following meaning in crystal structure and geometrical form.

A Centre of Symmetry occurs when every two opposite directions are equivalent. In that case the crystal polyhedron is bounded by pairs of parallel planes which, in ideal construction, are at an equal distance from a fixed [interior] point.

An Axis of Symmetry is thus defined, viz., that by a rotation through a certain fraction of 360° about this axis, every direction is transferred to an equivalent one, and the crystal polyhedron again coincides with itself. If $\frac{1}{n} \cdot 360°$ is the smallest angle of rotation needed, then the axis is described as n-fold. According as its two opposite directions are congruent or not, this axis is said to be "*two-sided*" [bi-lateral], or "*one-sided*" [hemimorphic]. Three sorts of "one-sided" [hemimorphic] axes are to be distinguished according as the groups of faces lying about their ends are enantiomorphic directly, or only after a rotation, or as they are wholly different; in the latter case, the axis of symmetry is styled "*polar*." From the fundamental laws of crystallography it follows also, among other deductions, that an axis of symmetry is always a possible crystal edge, and the plane normal to it a possible crystal face; and further, that only 2-, 3-, 4-, and 6-fold axes are possible.

A Plane of Symmetry is one in reference to which the two equivalent directions lie, like an object and its reflected image, and by this plane the crystal polyhedron is divided into two equal halves resembling each other as object and image. One sees easily that a plane of symmetry is always parallel to some possible crystal face.

When several of the symmetral elements occur in certain combinations with each other, then the occurrence of certain other symmetral elements is conditioned thereby. Those symmetral elements which condition the others are styled "*generative.*" An important example of this is the law, that a centre of symmetry, a plane of symmetry, and an even-fold axis of symmetry normal to the latter, constitute a set of elements such, that any two condition the third absolutely.

Simple forms which cannot be brought into coincidence with their reflected image, and which therefore possess neither a centre of symmetry nor a plane of symmetry, are said, after Marbach, to be "turned back on themselves." Such a crystal form and its reflected image are styled *enantiomorphic* (Naumann), and are distinguished as right and left forms.

As such forms are of considerable theoretical interest in stereochemistry drawings are here given of two enantiomorphically similar, but not congruent quartz crystals.[1]

From the different possible combinations of the generative symmetral elements, there comes about *the division of crystallised substances into thirty-two groups*,[2] which will be considered below. Of these groups, each of which is characterised by special symmetral elements, those are included in one *crystal system* which can be referred to *the same system of crystallographic axes*. By a *system of crystallographic axes* is meant such a system of co-ordinate axes as shall, on the one hand, satisfy the law of the rationality of indices, *i.e.* such axes as are parallel to possible crystal edges; and, on the other hand, such as express the symmetral behaviour of the crystal, in that this crystal is always brought to coincide with itself by all those operations which transfer a direction over to an equivalent one. In reference to such

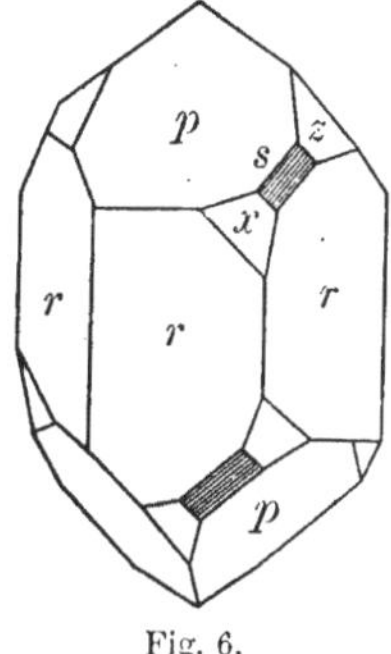

Fig. 6.

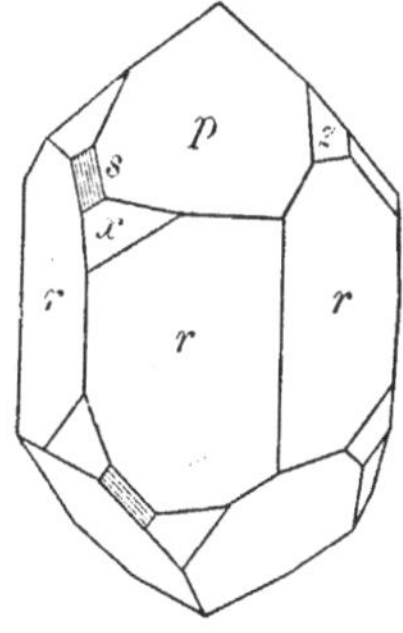

Fig. 7.

a system of axes, all equivalent faces receive indices which differ from each other only in their order of succession and their sign. Therefore one can concisely designate *a simple crystal form*, by a statement of the indices of any one of its planes (such indices are enclosed in parentheses to distinguish them from the symbols of particular faces). (For the numerous other symbols of crystal forms, which have been introduced, see the text-books of Crystallography.)

Before we proceed to the enumeration of the thirty-two groups, and of the crystal forms belonging to each, we must call to mind certain geometrical relations which exist between the simple forms of the groups belonging to the same system. The symmetral elements

[1] From Bodländer, *Lehrbuch d. Chemie*, p. 427, Stuttgart, 1896.

[2] This was developed by J. F. C. Hessel, 1830; A. Bravais, 1850; A. Gadolin, 1867; P. Curie, 1884; B. Minnegerode, 1886; a thorough presentation is given in Th. Liebisch's *Physikalischer Krystallographie*, pp. 3-50, Leipzig, 1891; also in the work of A. Schoenfliess, *Krystallsysteme und Krystallstruktur*, Leipzig, 1891. [For a brief but clear presentation of this somewhat difficult subject, see chapter ix., "Deduction of all Theoretically Possible Classes of Crystal Forms" in *Elements of Crystallography*, 3rd edit., by G. H. Williams, New York (1892).—Tr.]

of the groups of lower symmetry of every system, form a part of the symmetral elements of the groups of highest symmetry of the same system; hence from the simple forms of the latter groups (*holohedral*), we can geometrically derive all the other groups by causing one-half, or three-fourths, or seven-eighths of the planes of the holohedral forms to disappear, according to well-defined laws. The forms so obtained are called respectively *hemihedral*, *tetartohedral*, *ogdohedral*, and those which have *one* polar axis of symmetry *hemimorphic*. It must be noticed that not all the simple forms of the hemihedral and tetartohedral groups are geometrically different from the corresponding holohedral forms; thus, for example, the hexahedron [cube] and the dodecahedron appear in all the groups of the regular system. But the lower symmetry of such crystals can always be recognised by their physical behaviour, and especially by the position and shape of the *etched figures* on their faces. The notation of the hemihedral forms is accomplished by prefixing certain Greek letters, as (κ, π, γ, τ, ρ), before the symbols of the holohedral forms from which they are derivable; in the case of tetartohedral forms, two such letters are prefixed, since these forms can be developed by the combined application of two sorts of hemihedrism. The correlated complementary forms of the hemihedral and tetartohedral groups respectively, corresponding to the holohedral forms, which are always physically different, but only geometrically different in case they are enantiomorphic, these are designated by prefixes, or by arrangement of the indices. In the accompanying abstract of the thirty-two groups, the following abbreviated symbols will be used for the symmetral elements, viz. :—

C, for a centre of symmetry;

for m equivalent, n-fold, two-sided axes of symmetry, mL_n (in which the opposite directions are both counted, so that m at least is equal to 2);

for m equivalent, one-sided axes, and their opposites, mL_n and ml_n, mL_n* and ml_n*, and $m\bar{L}_n$ and $m\bar{l}_n$ respectively, accordingly as they are one-sided of the first variety, or of the second variety, or polar; and, finally,

for a plane of symmetry which is normal to no axis of symmetry, P; and for one which is normal to an n-fold axis, P_n.

Unequal axes of symmetry or planes of symmetry of the same sort should be distinguished by accents. Of the simple crystal forms of the groups of lower symmetry, only those will be mentioned which differ geometrically from the corresponding holohedral forms. Further, the forms of the hemimorphic groups will not be especially mentioned, since it is clear in these cases that they can be derived from the holohedral forms by allowing the planes grouped about one end of the polar axis to disappear. Those groups of which no examples are known, will be given in brackets without stating the forms; in the other cases there will be instanced a number of the most important minerals and artificially-prepared substances belonging to them.

I. Regular System

Crystallographic system of axes; three equal axes at right angles to each other (parallel to the edges of the hexahedron [or cube]).

1. Holohedrism. C; $6L_4$, $8L_3$, $12L_2$; $3P_4$, $6P_2$.

Hexakis-octahedron (hkl) h>k>l; triakis-octahedron (hhl); icositetrahedron (hll); tetrakis-hexahedron (hk0); octahedron (111); rhombic-dodecahedron (110); hexahedron (100).

Examples: P, Si, Fe, Pb, Cu, Ag, Hg, Au, Pt; PbS, Ag_2S; As_2O_3, Sb_2O_3; NaCl, AgCl, AgBr, CaF_2, K_2PtCl_6; Fe_3O_4; $MgAl_2O_4$, and the other spinels: garnet, analcite, perofskite, sodalite, etc.

2. Tetrahedral Hemihedrism. $6L_2$, $4\bar{L}_3$, $4\bar{l}_3$; 6P.

Hexakis-tetrahedron, κ(hkl); deltoid-dodecahedron, κ(hhl); triakis-tetrahedron, κ(hll); tetrahedron, κ(111).

Examples: Diamond; ZnS as zinc-blende; tetrahedrite; boracite; helvite.

3. Plagiohedral Hemihedrism. $6L_4$, $8L_3$, $12L_2$.

Pentagonal ikosa-tetrahedron, γ(hkl).

Examples: Cu_2O; KCl, $(NH_4)Cl$.

4. Pentagonal Hemihedrism. C; $6L_2$, $4L_3^*$, $4l_3^*$; $3P_2$.

Dyakis-dodecahedron, π(hkl); pentagonal dodecahedron π(hk0).

Examples: FeS_2 (as iron pyrites), $CoAs_2$, etc.; SnI_4; the Alums.

5. Tetartohedrism. $6L_2$, $4\bar{L}_3$, $4\bar{l}_3$.

Tetrahedral pentagonal dodecahedron, $\kappa\pi$(hkl); the other forms as in 2, excepting (hk0), which appears as the pentagonal dodecahedra.

Examples: $NaClO_3$, $NaBrO_3$, $Ba(NO_3)_2$, $Sr(NO_3)_2$, $Pb(NO_3)_2$; $Na_3SbS_4 + 9H_2O$; $NaUO_2(C_2H_3O_2)_3$.

II. Hexagonal System

System of crystallographic axes; one principal [vertical] axis, and three others [lateral] normal to it: these are equal to each other, and intersect each other at an angle of 120°. Each plane has four indices, viz. i, h, k, l, three of which, referring to the lateral axes, are related thus—$i + h + k = 0$.

6. Holohedrism. C; $2L_6$, $6L_2$, $6L_2'$; P_6, $3P_2$, $3P_2'$.

Dihexagonal pyramids (ihkl), i>h, $-k = \bar{k} = i + h$; hexagonal pyramids, 1st order ($i0\bar{i}l$), and 2nd order ($i.i.\overline{2i}.l$); dihexagonal prisms (ihk0); hexagonal prism, 1st order ($10\bar{1}0$), and 2nd order ($11\bar{2}0$); basal plane (0001).

Examples: Beryl, Milarite.

7. Hemimorphic Hemihedrism. $\bar{L}_6$, $\bar{l}_6$; 3P, 3P′.

Examples: ZnO; ZnS (as Wurtzite), CdS (Greenockite); $KLiSO_4$.

[8. Trapezohedral Hemihedrism. $2L_6$, $6L_2$, $6L_2'$.][1]

9. Pyramidal Hemihedrism. C; L_6, l_6; P_6.

Hexagonal pyramids, 3rd order, π(ihkl); hexagonal prisms, 3rd order, π(ihk0).

Examples: $Ca_5Cl(PO_4)_3$, $Pb_5Cl(PO_4)_3$, $Pb_5Cl(AsO_4)_3$, $Pb_5Cl(VdO_4)_3$.

10. First Hemimorphic Tetartohedrism. $\bar{L}_6$, $\bar{l}_6$.

Examples: $Sr(SbO)_2(C_4H_4O_6)_2$ and $Pb(SbO)_2(C_4H_4O_6)_2$.

[11. Sphenoidal Hemihedrism. $2L_3$, $3\bar{L}_2$, $3\bar{l}_2$; P_3, 3P.]

[12. Sphenoidal Tetartohedrism. L_3, l_3; P_3.]

13. Rhombohedral Hemihedrism. C; $2L_3$, $3L_2$, $3l_2$; $3P_2$.

Skalenohedron, ρ(ihkl), rhombohedron, ρ(i0il).

Examples: P, Te, As, Sb, Bi, Mg, Pd, Os; H_2O, Al_2O_3, Fe_2O_3, Cr_2O_3, $Mg(OH)_2$; $NaNO_3$, $CaCO_3$ (as calcite), $MgCO_3$, $FeCO_3$, $ZnCO_3$, $MnCO_3$; eudialite, chabazite, etc.

14. Second Hemimorphic Tetartohedrism (Hemimorphism of Group 13). $\bar{L}_3$, $\bar{l}_3$; 3P.

Examples: $Ag_6Sb_2S_6$ and $Ag_6As_2S_6$ (ruby silver), tourmaline; $NaLiSO_4$; tolyl-phenyl-ketone, $C_{14}H_{13}O$.

15. Trapezohedral Tetartohedrism. $2L_3$, $3\bar{L}_2$, $3l_2$.

Trigonal trapezohedron, $\rho\tau$(ihkl); rhombohedron (1st variety), $\rho\tau$(i0$\bar{i}$l); trigonal pyramids, $\rho\tau$(i.i.2$\bar{i}$.l); ditrigonal prisms, $\rho\tau$(ihk0), trigonal prisms, $\rho\tau$(11$\bar{2}$0).

Examples: SiO_2 (as quartz); HgS; $K_2S_2O_6$, $PbS_2O_6 + 4H_2O$, $SrS_2O_6 + 4H_2O$, $CaS_2O_6 + 4H_2O$; benzil $C_{14}H_{10}O_2$, matico-stearoptene, $C_{10}H_{16}O$.

16. Rhombohedral Tetartohedrism. C; L_3*, l_3*.

Rhombohedron, 3rd order, $\rho\pi$(ihkl), 1st order, $\rho\pi$(i0$\bar{i}$l), and 2nd order, $\rho\pi$(i.i.2$\bar{i}$.l); hexagonal prisms, 3rd order, $\rho\pi$(ihk0).

Examples: $CaMg(CO_3)_2$ (dolomite), titaniferous iron; H_2CuSiO_4, Zn_2SiO_4, Be_2SiO_4.

17. Ogdohedrism. $\bar{L}_3$, $\bar{l}_3$. Hemimorphism of 15 or 16.

Example: $NaIO_4 + 3H_2O$.

III. Tetragonal System

System of crystallographic axes; one chief [vertical] axis to which the third index refers, and two equal lateral side axes at right angles to it and to each other.

18. Holohedrism. C; $2L_4$, $4L_2$, $4L_2'$; P_4, $2P_2$, $2P_2'$.

Di-tetragonal pyramids (hkl), tetragonal pyramids, 1st order (hhl),

[1] According to recent work, *tri-methyl tri-mesitate* is an example of trapezohedral hemihedrism.—Tr.

and 2nd order (h0l); di-tetragonal prisms (hk0), tetragonal prisms, 1st order (110), and 2nd order (100); basal plane (001).

Examples: B, Sn; SnO_2, TiO_2 (rutile and anatase), $ZrSiO_4$; Hg_2Cl_2, HgI_2, $Hg(CN)_2$; $MgPt(CN)_6 + 7H_2O$; $NiSO_4 + 6H_2O$; KH_2PO_4, $(NH_4)H_2PO_4$, $Pb_2Cl_2CO_3$; vesuvianite, melilite, gehlenite, apophyllite.

19. Hemimorphic Hemihedrism. $\overline{L}_4$, $\overline{l}_4$; 2P, 2P′.

Example: Iodo-succinimide. $C_4H_4O_2NI$.

20. Trapezohedral Hemihedrism. $2L_4$, $4L_2$, $4L_2'$.

Tetragonal Trapezohedron, τ(hkl).

Examples: Guanidine carbonate, Strychnine sulphate.

21. Pyramidal Hemihedrism. C; L_4, l_4; P_4.

Tetragonal pyramids, 3rd order, π(hkl), and prisms, 3rd order, π(hk0).

Examples: $CaWO_4$, $PbMoO_4$; scapolite; erythrite, $C_4H_{10}O_4$.

22. Hemimorphic Tetartohedrism. $\overline{L}_4$, $\overline{l}_4$.

Example: $Ba(SbO)_2(C_4H_4O_6)_2 + H_2O$.

23. Sphenoidal Hemihedrism. $2L_2$, $4L_2$; 2P.

Tetragonal di-sphenoid, κ(hkl), and sphenoid, κ(hhl).

Examples: $CuFeS_2$ (copper pyrites); CH_4N_2O (urea).

[24. Sphenoidal Tetartohedrism. L_2*, l_2*.]

IV. Orthorhombic System

System of crystallographic axes; three unequal axes at right angles to each other.

25. Holohedrism. C; $2L_2$, $2L_2'$, $2L_2''$; P_2, P_2', P_2''.

Orthorhombic pyramids (hkl); prisms (hk0), (h0k), and (0hk); pairs of planes (pinacoids) (100), (010), and (001) the basal plane.

The great majority of minerals, and especially also of artificially-prepared substances, crystallise partly in this group and partly in group 28.

Some examples are: S, I, $HgCl_2$, HgI_2; FeS_2 (marcasite), Cu_2S; Sb_2S_3; TiO_2 (brookite); $BaCO_3$, $CaCO_3$ (aragonite), $SrCO_3$, $PbCO_3$; KNO_3, $AgNO_3$; $CaSO_4$, $BaSO_4$, $SrSO_4$, $PbSO_4$; K_2SO_4; Mg_2SiO_4, Fe_2SiO_4; topaz, andalusite; $MgSiO_3$.

26. Hemimorphism. $\overline{L}_2$, $\overline{l}_2$; P, P′.

Examples: $Zn_2(HO)_2SiO_3$, $Mg(NH_4)PO_4 + 6H_2O$; resorcin, $C_6H_6O_2$, tri-phenyl-methane $(C_6H_5)_3CH$.

27. Hemihedrism. $2L_2$, $2L_2'$, $2L_2''$.

Orthorhombic sphenoid, κ(hkl).

Examples: $MgSO_4 + 7H_2O$, $ZnSO_4 + 7H_2O$; $K(SbO)C_4H_4O_6$, $KHC_4H_4O_6$, $KNaC_4H_4O_6 + 4H_2O$, $NH_4NaC_4H_4O_6 + 4H_2O$, $C_3H_8O_3$, asparagine, mycose, etc.

V. Monoclinic System

System of crystallographic axes; two inclined axes, and another (ortho-axis) normal to the plane of the first two; the middle index refers to this ortho-axis: all three axes are unequal.

28. Holohedrism. C, L_2; l_2; P_2.

Prisms (hkl), (hk0), (0kl): pairs of planes out of the zone of the ortho-axis (h0l); special orthopinacoid (100), and basal plane (001); clinopinacoid (010).

Examples very numerous; among others: S (two modifications), Se; AsS; As_2O_3; Sb_2O_3; $NaCl + 2H_2O$; $KClO_3$; $Na_2CO_3 + 10H_2O$; $BaCa(CO_3)_2$; $(CuCO_3)_2 + H_2CuO_2$; $Na_2SO_4 + 10H_2O$; $CaSO_4 + 2H_2O$ (selenite), $FeSO_4 + 7H_2O$; $MgK_2(SO_4)_2 + 6H_2O$; and its isomorph, $CaNa_2(SO_4)_2$; $PbCrO_4$; $(MnFe)WO_4$ (wolframite); $Fe_4(PO_4)_2 + 8H_2O$; $Na_2B_4O_7 + 10H_2O$; $CaSiO_3$ (wollastonite); augite, hornblende, euclase, epidote, orthite, datholite, orthoclase, mica, titanite, heulandite, harmotome, etc.; most organic substances, as $C_2H_2O_4 + 2H_2O$; $KHC_2O_4 + 2H_2O$; $C_7H_6O_3$; $C_6H_4O_2$, $C_{10}H_8$, $C_{14}H_{10}$.

29. Hemimorphism. L_2, l_2.

Examples: Tartaric acid, quercite, cane sugar, milk sugar, etc.

[30. Hemihedrism. P.][1] Example: skolezite.

VI. Triclinic System

System of crystallographic axes; three obliquely inclined, unequal axes.

31. Holohedrism. C.

All the simple forms are only pairs of [parallel] planes.

Examples: $B(OH)_3$; $CuSO_4 + 5H_2O$, $K_2Cr_2O_7$, $CaS_2O_3 + 6H_2O$; $MnSiO_3$, Al_2SiO_5 (disthene), axinite, microcline, albite and anorthite, racemic acid, $C_4H_6O_6 + 2H_2O$.

[32. Hemihedrism. No element of symmetry.]

Example: d. monostrontium tartrate, $Sr(HC_4H_4O_6)_2 + 2H_2O$.

One can also deduce these thirty-two groups, without assuming the law of the rationality of indices, by starting with the conception of homogeneity. This method of treatment is considered in the "theories of crystal structure,"[2] which rest on molecular conceptions, and according to which there occurs an arrangement of the molecules of a crystal about its centre of gravity *in a regular system of points.* All the possible regular *point-systems* can be developed by the interpene-

[1] *Pyroxene* is an example of *monoclinic hemihedrism.* See G. H. Williams, *Crystallography*, pp. 168 and 195.—Tr.

[2] Frankenheim, 1835; Bravais, 1850; Camille Jordan, 1868; Sohncke, 1876; Fedorow, 1890; and with special thoroughness in Schoenfliess' *Krystallsysteme und Krystallstruktur.*

tration of congruent, parallelopipedal *space-gratings*, of which latter there are fourteen varieties, and according to their symmetry, the point-systems arrange themselves in the well-known thirty-two groups. This theory leads also, in an obvious way, to the law of the rationality of indices, by regarding the faces of a crystal as *net-planes* of the point-system.

Twinning and Growths of Crystals.—We must now consider those regular growths called *twins*, [threelings, fourlings, etc.], which consist respectively of two or more individuals, and which allow the production of forms of apparently higher symmetry than corresponds to their crystal structure. These growths can, as a rule, be recognised by the presence of re-entrant angles, which never occur on simple crystal individuals, or else by a varying behaviour of different parts of the same face.

We distinguish between twins with parallel axes, and those with axes which are not parallel. The former are growths of two hemihedral (or tetartohedral) forms respectively which bear the same relative position to each other as in the [theoretical] derivation from their holohedral (or hemihedral) forms. They are called complementary twins. Examples are the interpenetration of two pentagonal dodecahedra of iron pyrites, or the intergrowths of right- and left-handed quartz crystals. For the second class of twins [viz. with axes not parallel], which are by far the more common, the regularity is that the two crystal individuals lie symmetrically with reference either to one of their *possible faces or one of their edges.* In the case of those twins which are symmetrical to a plane, this plane is called the *twinning plane*; in the case of those twins which are symmetrical with reference to a straight line, the plane normal to this line is the *twinning plane*; the line normal to this plane is the *twinning axis*; in crystals which are centrally symmetric it is a symmetric axis of twinning. The two twinning individuals may either penetrate each other (interpenetration twins), or they may touch each other along a plane (contact twins); and, as is often the case, it is not necessary that this plane should be the twinning plane.

Repeated twinning growths frequently occur in parallel planes; in this case the alternating individuals (often as thin sheets), occur in parallel arrangement; in this way arise the *twinning lamellæ*, for instance, of calcite, aragonite, and especially of the triclinic feldspars (polysynthetic twinning). Twins or fourlings may show forms which apparently belong to crystal systems of a higher symmetry than that possessed by the single individuals; important illustrations are the pseudohexagonal prisms (or pyramids) of orthorhombic aragonite and witherite; also the crystals of chrysoberyl, harmotome, and phillipsite; such forms are called *pseudo-symmetric* or *mimetic*.

It is worthy of notice that twinning growth may occur not only in the original make-up of a crystal, but also in a secondary way from external pressure; calcite is a remarkable example of this.

The *growth of crystals, i.e.* the relative development of the different faces, is very variable, and often irregular, so that equivalent faces often have a very different size; these are distortions. The conditions controlling these distortions are not well understood. P. Curie (1885) and others have suggested as a controlling factor a certain "surface energy," analogous to that on the surface of liquids, which would be different for the unequal planes of the same crystal; but except for the observation that larger crystals grow at the expense of smaller ones, when placed in the same saturated solution, this hypothesis has led to hardly any results in harmony with experiment.

The solubility of a crystal is not the same in all directions. This can be shown by treating the surfaces for a short time with a solvent. Regularly bounded marks are thus formed (*Aetzfiguren*) whose arrangement shows the same symmetry as the ideally developed crystal, and often offers the only means of determining the crystallographic group.[1] Of two dissimilar faces of the same crystal, one may show the greater solubility in one solvent, the other in another. This may explain the varying *habitus* of crystals of the same substances, *e.g.* sodium chloride, which usually crystallises in cubes, comes down with octahedral faces from a solution containing urea.

The Physical Properties of Crystals.—In the case of all physical properties which are directed quantities, there appears in crystals a dependence on direction which, in the elementary laws of the processes in question, is expressed by a number of constants characteristic for each crystal. As already stated, the symmetry, which in the behaviour of a crystal appears correlated with some physical action, is always the same as that of its geometrical form, or higher; and the division into groups which one obtains by a classification based on any physical property, is always in harmony with the classification previously developed from the crystalline form. In the following section we will consider the physical properties particularly in their relations to symmetry; and we will consider the physical laws themselves only in so far as they are of especial importance for the determination of crystallographic symmetry. For anything more, reference must be made to the text-books of physics; and here special mention should be made of Th. Liebisch's *Physikalische Krystallographie.*

The physical properties of crystals may be divided into two groups, according as they possess a higher or lower symmetry. But the symmetry of the physical properties is never lower than that shown by the processes of growth and solution. Higher symmetry is shown by all those physical properties whose elementary laws in crystals can

[1] Liebisch, *Grundriss den physikalischen Krystallographie*, p. 43, Leipzig, 1896.

be expressed by an ellipsoid (for which, therefore, regular crystals behave as isotropic); lower symmetry by those for which this is not possible.

To the first class belong *thermal expansion* and *compression by equal pressure from all sides;* the *conduction* of *heat* and *electricity, dielectric* and *magnetic polarisation,* and finally the *thermo-electric phenomena.*

The significance of the ellipsoid in these cases is essentially as follows:—

In thermal expansion the inner part of the crystal, conceived to be originally a sphere, is distorted to an ellipsoid, the principal axes of which are proportional to $1 + \lambda_1 t$, $1 + \lambda_2 t$, $1 + \lambda_3 t$; in these formulæ λ_1, λ_2, and λ_3 respectively are the principal coefficients of expansion, and t is the amount of the change of temperature. If this *ellipsoid of dilatation* is given for a definite temperature interval t, then one can calculate the changes of the dimensions and of the angles of the crystal.

In the case of the phenomena of conduction, suppose that at a point of a crystal (of unlimited extension), there is a source of heat or of electricity; then the surfaces of equal temperature (or potential respectively) will be similar and similarly arranged ellipsoids, the principal axes of which are proportional to the square roots of the principal conductivities.

Finally, the relations between the intensity and the direction of di-electric polarisation (or of the magnetic polarisation) of a crystal, in a homogeneous field, on the one hand, and the direction of the lines of force on the other hand, can be shown by the aid of an ellipsoid; its principal semi-axes are the square roots of the reciprocals of the principal di-electric coefficients (or magnetic constants respectively). Similarly, an ellipsoid serves to show the connection between thermo-electric force and the direction of the greatest fall in temperature.

Inasmuch as six magnitudes must be determined in order to fix the directions and lengths of the principal axes of an ellipsoid of un-equal axes, so processes of this sort, in the most general cases, depend on *six* physical constants, the values of which will vary with the temperature (and the pressure). Thus, from the consideration of each of the physical properties enumerated above, there results the classification of crystals into the six following groups:—

I. Regular System: the ellipsoid is a sphere; no distinction from the behaviour of amorphous bodies. 1 constant.

II. Hexagonal and Tetragonal Systems: the ellipsoid is one of rotation, where the chief crystallographic axis is the axis of rotation; the length of this axis and of the equatorial semi-diameter are alone to be determined; therefore 2 constants. For every physical property one can make two subdivisions of this class, according as the ellipsoid of rotation is oblate or prolate.

III. Orthorhombic System: the ellipsoid has three unequal axes having a fixed direction, but of variable lengths ; these are the crystallographic axes. 3 constants.

IV. Monoclinic System : one of the three unequal axes of the ellipsoid coincides with the ortho-axis ; the directions of the two others vary. 4 constants.

V. Triclinic System : all three axes of the ellipsoid vary in length and direction. 6 constants.

In relation to the properties considered here, the crystals of group I. are *isotropic*, those of group II. have *one axis of isotropy*, those of groups III., IV., and V. have *no axis of isotropy*.

The Optical Properties of Crystals.—A somewhat greater multiplicity of the relations of symmetry appears in the *optical properties* of crystals. Here also, as will be seen below, the behaviour of the great majority of *transparent* crystals can be explained by the aid of an ellipsoid, but there are two groups of transparent crystals, viz. the "optically active," for which the ellipsoid is not suitable ; also in the case of the absorption of light by crystals, the relations are very complicated. The latter subject we must pass by with a reference to P. Drude's[1] presentation of the subject, only it should be observed that an absorbing crystal has always twice as many optical constants as a [perfectly] transparent one of the same grade of symmetry.

The division of crystals, then, according to their optical behaviour, is as follows :—

A. Optically Isotropic Crystals

I. *Singly refracting :* regular system.—The wave-surface is a sphere ; 1 constant. (Very often regular crystals also show weak, irregular double refraction ; this belongs to the so-called "optical anomalies," the cause of which, though certainly of a secondary nature, has not been satisfactorily explained.)

II. *Having the power of rotation* [rotary polarisation].—A part of the crystals of group 5 belong here. The wave-surface consists of two concentric spheres. 2 constants.

B. Optically Anisotropic Crystals with One Axis of Isotropy

III. *Optically Uni-axial without Circular Polarisation* (hexagonal and tetragonal system).—The wave-surface (Huyghens') consists of an ellipsoid of rotation (the rotation axis of which coincides with the chief crystallographic axis), and of a sphere which touches the former at the poles. The optically uni-axial crystals are subdivided into

[1] *Wied. Ann.*, **40.** 665 (1890).

positive and negative varieties, according as the ellipsoid of rotation is prolate or oblate. 2 constants.

IV. *Optically Uni-axial with Circular Polarisation* (groups 15, 17, and 20).—The wave-surface consists of two rotation-surfaces, one differing but little from an ellipsoid, and the other but little from a sphere. 3 constants.

C. Optically Anisotropic Crystals without Axis of Isotropy

The wave-surface (Fresnel's) is universally a surface of the fourth degree, with three twofold axes of symmetry at right angles to each other. It is cut by its planes of symmetry in a circle and in an ellipse, and has two pairs of singular tangential planes touching it in circles. The normals to these tangential planes are *the optic axes.* Distinctions arise here if we take into consideration the crystallographic orientation of the optic axes of symmetry, and their connection with the temperature and with the wave-length of light. There are the three following classes:—

V. *Optically Bi-axial Crystals of the Orthorhombic System.*—Fixed directions of the optic axes of symmetry, coinciding with the crystallographic axes. 3 constants.

VI. *Optically Bi-axial Crystals of the Monoclinic System.*—Only one optical axis of symmetry is fixed, viz. that having the same direction as the ortho-axis. 4 constants.

VII. *Optically Bi-axial Crystals of the Triclinic System.*—All three of the optic axes of symmetry have variable positions. 6 constants.

The law of double-refraction in inactive anisotropic crystals, or those of categories III., V., VI., and VII., can be easily developed in the following way with the help of the so-called "*index-ellipsoid.*" Its semi-axes coincide in direction with the optic axes of symmetry, and their values can be easily derived from the three or two chief indices of refraction as follows.

One obtains the velocity of transmission, and the polarising directions of those plane waves which advance in a given direction, by passing, through the middle point of the index ellipsoid, a plane normal to this direction, and then determining the principal axes of the resulting elliptical section. These axes are the polarising directions sought, and their reciprocals are the velocities of transmission, compared with these in air as unity; here it is to be observed that the velocity of transmission of every wave is determined by that ellipse axis which is at right angles to the plane of polarisation. The two directions for which the ellipse section is a circle, when the polarisation direction is indefinite and the two wave velocities are equal to each other, are the *optic axes.* On this method is based the derivation of many phenomena of double refraction, which cannot be entered into here, but which will be found thoroughly presented, for example, in Liebisch's *Physikalische Krystallographie.*

Optical activity is superimposed on ordinary double refraction, in the case of crystals of category IV.; this leads to the result that in the direction of the axis of isotropy there occurs *circular polarisation,* but *elliptic polarisation* in all other directions inclined to this. An analogous superposition of optical activity on ordinary double refraction, might occur in the case of those optically *bi-axial* substances which have crystal forms "turned back upon themselves"; but such instances have not been recognised hitherto, perhaps because the circular polarisation may be obscured by the double refraction, in all directions highly inclined to the optic axis. Optical activity is theoretically possible in all groups having neither centre nor plane of symmetry, *i.e.* apart from those mentioned, in the groups 3, 8, 10, 22, 24, 27, 29, and 32. Experience shows that in the case of crystals "turned back upon themselves," it does *not* follow conversely that there must be circular polarisation; instances of this are found in the regular tetartohedral nitrates of Pb, Ba, and Sr. It is worthy of remark that those (organic) substances which are optically active in solution always crystallise in the inverted forms (Pasteur, 1848); but most crystals with circular polarisation (for example $NaClO_3$, $NaBrO_3$) give *inactive* solutions. Optical activity in both crystallised and dissolved states is shown only by a few organic compounds such as strychnine sulphate, rubidium tartrate, etc.[1]

The Physical Properties of Lower Symmetry.—The most important property by which all regular crystals are distinguished from amorphous substances is *elasticity.* According to the theory of elasticity, the potential energy of unit volume of a deformed homogeneous crystal is a quadratic function of the moduli of deformation (as named by Kirchhoff); the coefficients occurring here are the so-called *elasticity constants* of the crystals, provided that the crystallographic axes are chosen as the co-ordinate axes. As Minnegerode[2] first showed, according to the symmetry of their behaviour respecting elasticity, the following nine classes of crystals are to be distinguished:—

(*a*) Regular system; groups 1–5, with 3 elasticity constants.
(*b*) Hexagonal system; groups 6–12, with 5 elasticity constants.
(*c*) Hexagonal system; groups 13–15, with 6 elasticity constants.
(*d*) Hexagonal system; groups 16 and 17, with 7 elasticity constants.
(*e*) Tetragonal system; groups 18, 19, 20, and 23, with 6 elasticity constants.
(*f*) Tetragonal system; groups 21, 22, and 24, with 7 elasticity constants.

[1] See H. Traube, *Jahrb. f. Mineralogie,* 1896, p. 788.
[2] Minnegerode. *Gött. Nachr.,* 1884.

(*g*) Orthorhombic system; groups 25–27, with 9 elasticity constants.

(*h*) Monoclinic system; groups 28–30, with 13 elasticity constants.

(*i*) Triclinic system; groups 31 and 32, with 21 elasticity constants.

That no more groups occur is due to the fact that the elastic pressure, as well as the amount of deformation, are magnitudes *centrally* symmetric. The elastic symmetry is expressed with especial clearness in the dependence of the *coefficient of elongation* (*i.e.* the elongation effected by the pull 1, in the direction of this pull) upon the direction. This relation can be shown by means of a closed surface, by plotting through a fixed point each pull-direction corresponding to the coefficients of elongation. Similarly the *torsion coefficient* of a circular cylinder can be represented as a function of the crystallographic orientation of the axis of the cylinder. (Reference should be made to the numerous researches of W. Voigt, a summary of which is given in Liebisch's *Phys. Kryst.*, chap. ix., and especially the account in § 7 of Voigt's *Kristallphysik*, Leipzig, 1898). It is especially worthy of mention that, in the case of the groups of the hexagonal system, under (*b*) in the preceding classification, the elasticity is the same in all directions having the same inclination to the principal axis. This does not hold good for the other groups of the hexagonal system, nor for those of the tetragonal system. Moreover, there are very remarkable variations of the coefficients of elongation and of torsion respectively, in different directions, as appears from the observations on calcite, dolomite, tourmaline, and barite, for example.

The properties associated with cohesion, as *cleavage* and *hardness*, in respect to which the regular crystals also seem to be anisotropic, have not as yet yielded to mathematical treatment.

A physical process, which *is not centrally symmetrical*, and which therefore leads to an entirely different classification of crystals, is that of *electrical excitement by elastic or thermal deformation* (piezo-electricity and pyro-electricity). As long as one considers only homogeneous deformation, these phenomena can only occur on crystals *without a centre of symmetry.* Pyro-electric excitement by uniform heating or cooling, is particularly characteristic of the groups having a characteristic polar axis of symmetry, viz. 7, 10, 14, 17, 19, 22, 27, and 29 (groups 30 and 32 have not been observed as yet); this property, therefore, is a good sign whereby to recognise these groups, apart from their crystalline form and etched figures. Remarkable examples are tourmaline, calamine, struvite, cane-sugar, and tartaric acid; also quartz, in which the three side-axes, and boracite, in which the four threefold axes are polar, these can take electrical charges only by *uneven* heating or cooling, with + and − alternately on the ends of the polar axes.

The Determination of the Crystallographic Symmetry.— If one is studying transparent crystals he will properly begin with an optical test. It should first be determined, by a polarisation apparatus for parallel light, or by a microscope fitted with polarising and analysing nicols, whether the crystal is optically isotropic (singly refracting or circularly polarising), or doubly refracting; and in the latter case, how the directions of extinction are orientated with reference to the bounding crystal faces: (these extinction directions are the directions which must be parallel to the polarising planes of the crossed nicols, when the crystal appears dark, and they are also the polarisation planes of the waves transmitted in the direction of observation).

Next follows an examination in converging polarised light (with the Nörremberg's apparatus, or, in the case of small objects, with a microscope fitted for converging light), when the light is passed through crystal sections with parallel planes, which are either naturally suited for this, or artificially prepared, by grinding or splitting. Usually it will be advantageous and save polishing if one immerses the crystal in a liquid having nearly the same refractive index. The following is a description of the characteristic phenomena which one first observes in homogeneous light; it will be noticed whether the crystal is optically uni-axial or bi-axial, and in the latter case how the optic axes lie. In the case of uni-axial crystals, in plates cut perpendicularly to the optic axis, between crossed nicols one sees a series of concentric circles, alternately light and dark, which are intersected by two dark bands which are at right angles to each other; in plates cut parallel to the axis, the rings become equilateral hyperbolas; a section, cut from a crystal in any other direction, shows excentric systems of ellipses or hyperbolas, according to its inclination to the optic axis. In the case of crystals optically bi-axial, in sections normal to the "*first middle line*" (*i.e.* the line bisecting the acute angle of the optic axes), in case the angle is not too great, one sees a system of *lemniscates* which are crossed by two dark bands at right angles to each other, or by two equal dark hyperbolic brushes, accordingly as the plane of the axes coincides with the principal section of one nicol (normal position), or as it makes an angle of 45° with it (diagonal position). With sections normal to "the second middle line" (*i.e.* the line bisecting the obtuse angle of the optic axes), or parallel to the plane of the optic axes, there are seen systems of concentric hyperbolas which are exactly symmetrical in the latter case, but not so in the former. Finally, it is to be noticed that sections normal to one axis of a bi-axial crystal show an interference figure consisting of almost circular curves which are crossed by *one* dark band. Interference curves are visible in white light, but only in sufficiently thin sections, or such as are nearly normal to one optic axis; the symmetry of the distribution of colour observed in the normal or diagonal position of the plate, and which depends on the position of the middle lines of each particular colour, allows one to determine whether the crystal system is orthorhombic, monoclinic, or triclinic. We distinguish in the case of monoclinic crystals three varieties of the "dispersion of the optic axes of symmetry," viz. 1, inclined dispersion when the optic axes lie in the plane of symmetry; 2, horizontal, and 3, crossed dispersion when the plane of the optic axes is at right angles to the plane of symmetry, and the first and second middle lines respectively lie parallel to this latter plane.

After one has thus made himself certain by the study of these optical phenomena, or at least has gained some idea regarding the symmetry of the crystal to be determined,—as will be of great value,—especially in complex combinations with the irregular development of planes,—he will next proceed to the measurement of angles on a reflection goniometer, having previously made a perspective sketch showing, as far as possible, the arrangement of the particular planes, each of which is marked with a letter.

The angular measurements will now afford more disclosures respecting the symmetry of the geometrical form (because equal angles lie between similar planes); and, further, they will serve to determine the elements of the axes (*i.e.* the geometrical constants) of the crystal. In the case of monoclinic and triclinic crystals, the choice of fundamental planes is purely arbitrary (for the former, 100 and 001, and also for the latter 010), but one can give to any fourth plane not lying in a zone of any two of the first planes, arbitrary indices, for example 111, and can calculate from the inclination to the fundamental planes, directly or indirectly obtained, the axial units a : b : c. The angles between the axes are already known by establishing the fundamental planes. To determine the indices of the other crystal planes, one needs at most two angles, measured from different zones; but it is the custom to have more, if possible, in order to have a control of the calculated elements of the axes and indices.

If the occurrence of the faces gives rise to doubts as to which group of the system a crystal belongs, we must proceed to study the etched figures, the pyro-electric properties, etc. Finally, to ascertain the symmetry and the geometrical constants, one must test still further the characteristic physical properties, especially the character (*i.e.* whether positive or negative), and the strength of the double-refraction must be determined; in the case of crystals optically bi-axial, *the angle between the optic axes;* and in monoclinic and triclinic crystals, the *orientation of the optic axes of symmetry* must be determined for the different colours; also, when it is possible, one must measure the principal indices of refraction by the method of total reflection, or by the prism method, and eventually one must study the absorption phenomena (pleochroism).

Polymorphism.—Although, in general, each substance occurs in a definite crystallographic symmetry and form, which is characteristic for itself, yet many instances occur where *the same substance exhibits different crystal forms.* The appearance of one and the same substance, having not only the identical composition, but also the identical constitution, in two or more crystal forms, *i.e.* forms with different symmetries, and also with different elements of the axes, this is called *dimorphism* or *polymorphism* respectively. This phenomenon was first recognised by E. Mitscherlich, in the salt $NaH_2PO_4 + H_2O$ (1821), and in sulphur (1823). The different kinds of crystals of a polymorphous substance are to be regarded as *different modifications* analogous to the different states of aggregation; they are therefore designated as *physical isomers*, in contrast to *chemical isomers*.

The following are among the important examples of polymorphic substances. For the particular group to which each modification belongs, see the preceding summary of the thirty-two groups.

C, S, Se, Sn; Cu_2S, ZnS, HgS, FeS_2; As_2O_3, Sb_2O_3, SiO_2 (quartz and tridymite), TiO_2; AgI, HGI_2, ICl; $CaCO_3$, KNO_3, $NaClO_3$, $KClO_3$, NH_4NO_3; K_2SO_4, $NiSO_4 + 6H_2O$, $MgSO_4 + 7H_2O$, $FeSO_4 + 7H_2O$, $NaH_2PO_4 + H_2O$, boracite, Al_2SiO_5 (andalusite, disthene, and sillimanite), the humite group, zoisite and epidote, leucite, potassium feldspar; also many organic compounds, as chlor-m-di-nitro-benzene, chlor-o-di-nitro-benzene, benzo-phenone, β-di-brom-propionic acid, mono-chlor-acetic acid, mono-nitro-tetra-brom-benzene, benzoïn, carbon trichloride, hydroquinone, malon-amide, m-nitro-para-acet-toluide.

Polymorphism (unlike chemical isomerism) is restricted to the solid state; on sublimation, polymorphic modifications yield the same vapour; on solution, the same liquid, just as the vapour from ice or water is identical, and the solution of ice or water in alcohol gives an identical aqueous alcohol.

But the vapour pressure (and solubility) of two polymorphic modifications is, in general, different; hence, in general, two polymorphic forms are not in equilibrium together, for vapour would pass from the form with the higher to that with the lower pressure, and the one would grow at the cost of the other. Equilibrium can occur only at the temperature at which the vapour pressure of the two forms is the same, *i.e.* at the point of intersection of the two vapour pressure curves, just as (p. 74) coexistence of the solid and liquid states depends on equality of vapour pressure.

The point at which the two modifications coexist is called the *transition temperature;* above that, only one, below it only the other, form is stable, just as below 0° water freezes, and above 0° ice melts. But an important difference is that solids cannot be heated above their melting-point without melting, whereas most polymorphs can be kept for some time above their transition temperature before being entirely converted, and some even show no tendency to be converted. Thus, calcite and aragonite can exist together through a wide range of temperature, and others, such as diamond and graphite, have not so far been converted by mere change of temperature. The transition is usually facilitated by contact with substance that has been already converted, just as the solidification of an undercooled liquid is brought about by contact with the frozen substance.[1]

The energy of two modifications is usually considerably different, as with the solid and liquid states; the evolution of heat in passing from one into the other is called the *heat of transition.*

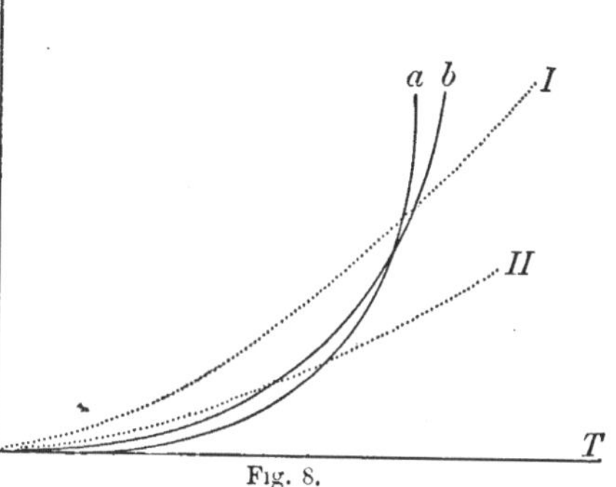

Fig. 8.

These relations can best be studied with the aid of Fig. 8 which, like Fig. 4 (p. 74), shows the vapour pressure curves of the different modifications. The point of intersection of curves a and b which represent the polymorphic forms A and B is the transition point, for there the vapour pressure of both is the same. Below that point A has the smaller vapour pressure, and is stable; above, the conditions are reversed, and B is stable.

The melting-point of each modification is at the intersection with the vapour pressure curve of the liquid. Two cases are possible.[2] Either the latter curve cuts a and b above the transition point (I, Fig. 8) or below (II). Only in the former case is the transition point actually attainable; in the latter, both modifications would melt first. Sulphur is an example of the first case; rhombic sulphur on heating to 95·6 passes over into the monoclinic form, and the latter on cooling reverts to rhombic. Benzophenone is an instance of the second; two modifications with different melting-points are known, and, as appears from Fig. 8, that with the higher melting-point must be the more stable. In this case, then, it is not possible to pass from either modification to the other by mere heating or cooling; the unstable modification has to be prepared by crystallisation from suitably undercooled liquid.[3]

Liquid Crystals.—To explain crystalline structure, we must obviously assume that forces act between the molecules which compel their regular arrangement. The stronger these forces, the more rigid the molecular structure will be, and the more the crystal will resist deformation. If, on the other hand, these forces become very weak, it is conceivable that deformation should occur even under the

[1] See "Synopsis of Literature and Researches," by W. Schwarz, *Umkehrbare Umwandlungen polymorpher Körper, Preisschrift der phil. Fakultät*, Gottingen, 1892.
[2] Ostwald, *Allg. Chem.*, 1st ed., **1.** 695 (1885); 2nd ed., **1.** 948 (1891).
[3] K. Schaum, "Arten der Isomerie," *Habilitationsschrift*, Marburg, 1897.

influence of gravity or capillarity, *i.e.* we should have a liquid crystal.

Actually Reinitzer,[1] some years ago, observed a peculiar liquid, but cloudy, modification of cholesterylbenzoate, which appeared bright between crossed nicols, and is therefore doubly refracting. The optical behaviour of this substance, and of some others, discovered by Gattermann,[2] which show similar phenomena, was thoroughly studied by O. Lehmann.[3] It appeared that these compounds are liquid, and that free drops show the structure of sphero-crystals, and like sphero-crystals give a black cross between crossed nicols.[4] We may perhaps assume a radial arrangement of molecules in such liquid drops.

On heating, the cloudiness disappears at a definite temperature, and with it the double refraction. This may be described as the melting-point of the crystal, *i.e.* the arrangement of molecules breaks down there.

Recently, however, G. Tammann[5] has put forward objections that appear well founded against these observations and conceptions. More careful observation of cholesterylbenzoate showed that the milky liquid slowly deposited small solid sphero-crystals and became clear, so that the "liquid crystal" was nothing more than a mixture of isotropic liquid and innumerable small crystals. In other cases, according to Tammann, formation of layers does occur on melting, *i.e.* the liquid crystal is an emulsion, possessing, however, the property of appearing light between crossed nicols. But these phenomena are the less noticeable the more carefully the preparation is purified. On the hypothesis of an emulsion, it has to be assumed that clear crystals separate into parts on melting—a hitherto unknown phenomenon. We must then await further investigation, especially whether it is possible to reproduce the phenomena of "liquid crystals" artificially by appropriate mixtures.

Amorphous State.—On cooling a homogeneous liquid sufficiently, it acquires, in general, the property of crystallising; at points of its interior, nuclei arise which grow more or less fast, and eventually cause the whole liquid to become crystalline. A liquid can therefore be *undercooled* (*i.e.* cooled below the melting-point) the more easily the fewer the nuclei in it, and the more slowly they grow.

If a liquid were suddenly cooled to the absolute zero, the facility for crystallising would not occur; for then molecular movement would

[1] *Monatsschrift f. Chem.*, **9.** 435 (1888).

[2] *Ber. d. d. Chem. Ges.*, **23.** 1738 (1890).

[3] *Z. S. phys. Chem.*, **4.** 462; **5.** 427 ff.; *Ber. d. d. Chem. Ges.*, **23.**, 1745; *Wied. Ann.*, **40.** 401 (1890).

[4] See also R. Schenck, "Die Kristallinischen Flussigkeiten," *Habilitationsschrift*, Marburg, 1897.

[5] *Ann. d. Phys.*, **4.** 524 (1901); **8.** 103 (1902).

cease, and the favourable collisions of molecules (Book II., chap. ii.) needed for the formation of a crystal nucleus would be wanting; nor could a nucleus grow, since the velocity of crystallisation must vanish too. It appears, then, that at the absolute zero the state of under-cooled liquid must be practically stable; it is further obvious that, with the cessation of molecular movement, the easy mobility of the parts of a liquid must cease too, whereas its isotropy must remain (in contrast with the crystalline state).

Numerous observations show that such a state is possible for many substances at atmospheric and higher temperatures, *i.e.* great under-cooling can be relatively stable; this condition has long been called *amorphous*[1] (in opposition to crystalline).

The characteristics of the amorphous state follow from what we have said; externally it has the properties of a solid, owing to great viscosity and considerable rigidity, produced by strong mutual action of the molecules. An amorphous body differs from a crystal, however, in its complete isotropy and absence of a melting-point; on heating, it passes continuously from the amorphous to the usual liquid state, as its properties show steady change with rise of temperature, and no breaks anywhere.

Since amorphous bodies (such as glasses) show rigidity, it would follow from this hypothesis that ordinary liquids, and even gases, must also, though to a very slight extent; so far, this has not been proved, possibly owing to the extreme smallness of the property.

Quartz, silicates (glasses), and many metallic oxides, are good examples of the amorphous state. Amorphous precipitates, difficult to wash and apt to bring down impurities with them, are frequent in chemical work and disadvantageous.

Sometimes crystallisation of amorphous bodies cannot be observed within ordinary limits of time; sometimes it occurs slowly, as in the devitrification of glass; occasionally with almost explosive violence. Tammann (*loc. cit.*) quotes a good example of this in Grove's discovery (1855) of explosive antimony. This metal is deposited in amorphous form (always containing some chloride) by electrolysis from solutions of antimonious chloride; on passing over into the crystalline state an evolution of 21 calories per gram occurs. If a part of the antimony be heated to 100 – 160°, rapid crystallisation takes place, and this, on account of the heat developed, is propagated like an explosive wave.

[1] This view of the amorphous state is developed and discussed by G. Tammann, *Schmelzen und Kristallisieren*, Leipzig, 1903 (Ambrosius-Barth).

CHAPTER IV

THE PHYSICAL MIXTURE

General Observations.—In the preceding chapters we have considered chiefly the behaviour of substances having a simple chemical composition: in the following chapter will be considered the most important properties of the physical mixture. By the term *physical mixture* we mean a complex of different substances, which is at all points homogeneous, both in a physical and a chemical sense. According to the state of aggregation, we must distinguish between gaseous, liquid, and solid mixtures; in special cases we call them gas-mixtures, solutions, isomorphous mixtures, alloys, and the like.

The physical mixture must not be confounded with the coarse mechanical admixtures of solid or liquid substances, as powders, emulsions, and the like, the various ingredients of which can be separated from each other without much difficulty, as by detection with the microscope, separation by gravity, by washing out, etc. They originate by the mutual molecular intermingling of the various substances of which they are composed, and the separation of their components is for the most part conditioned by the expenditure of a very considerable degree of work, a knowledge of which is of the very greatest importance. In terms of the molecular hypothesis, physical mixtures differ from chemically simple substances, in that the latter consist of the same kind of molecules, while the former consist of different kinds of molecules.

The gases alone possess collectively the property of unlimited miscibility. The solvent power of different liquids for each other, and also of liquids for solids, is much more limited; and still more rarely are crystallised substances able to form mixed crystals in all proportions.

Gas Mixtures.—As would be expected, the most simple relations are found in the case of mixtures arising from the mutual interpenetration (diffusion) of different gases. In those cases where there is no chemical action associated with the intermingling, the properties of each particular gas remain unchanged in the mixture, *i.e.* each gas conducts itself as though the others were not present; the pressure exerted upon the containing walls, the capacity of absorbing and refracting light, the solubility in any selected solvent, the specific heat, etc., all of these properties experience no change in the intermixture. Thus one can predict the [physical] properties of a gas mixture of known composition, if he has previously determined the properties of each particular ingredient. These relations should be regarded as characteristic only of ideal gases. As all the laws of gases are only approximate rules, one will find, presumably, by exact measurement, small deviations from the rules given above. At least it is already proven that the pressure exerted by a gas mixture does not strictly obey the law of the sum of their pressures.

If one mixes two gases, at the same time allowing them to diffuse into each other, at a volume which is kept constant, there occurs no development of heat, and also no expenditure of external work; consequently *the total energy* [U], remains unchanged in this process. But, nevertheless, the process, if rightly directed, can do external work, as it may be already inferred that the mutual intermingling of two gases is a process occurring automatically; and as is universally true, so here, the *maximum* of external work may be obtained by making the intermingling *reversible*.

Now, as a matter of fact, a very simple mechanism may be contrived to work in the desired way, and which depends upon a property possessed by certain partition walls, *of being permeable for certain substances, but not for others;* partitions of this sort are called "*semi-permeable*"; they have recently proved of immense service in the advance of both theory and experiment. Let there be

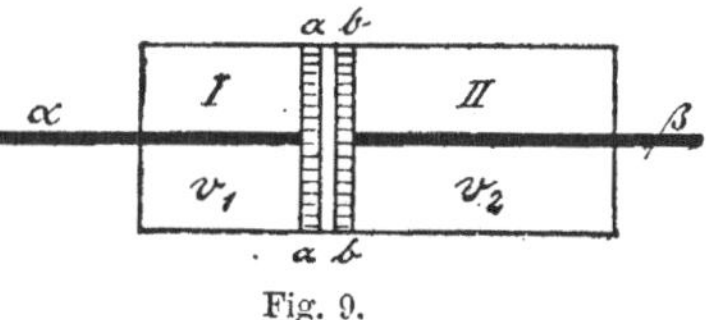

Fig. 9.

n_1 g.-mol of a gas in the division I of a cylinder (Fig. 9), the volume of which is v_1, and in division II, the volume of which is v_2, n_2 g.-mol of another gas. During the process the system is maintained at the constant temperature T. In the cylinder are two closely-fitting, movable pistons a and b; let piston a be permeable for the *first* gas, but not for the second; and, conversely, piston b should give free passage for the *second* gas, but not for the first. Two piston rods, α and β, which pierce the two air-tight cylinder heads, serve to carry the motion of the contents to the outside. At first the pistons should be pushed up close to each other; then the first gas will exert a pressure on piston b, but not on a, since it can freely pass through the latter; and conversely for a similar reason, the second gas, by virtue of its expansive force, will tend to move only the piston a. If we offer no resistance to the piston rods, as a result of this pressure action, there will occur a movement of the pistons to the ends of the cylinder, as is shown in Fig. 10; at the same time there has occurred the *intermingling* of the two gases. If, conversely, we press in the two piston rods, the first gas will offer resistance only to the movement of piston b, and the second gas only to the movement of piston a; if we push back the pistons to their original positions, there occurs a *separa-*

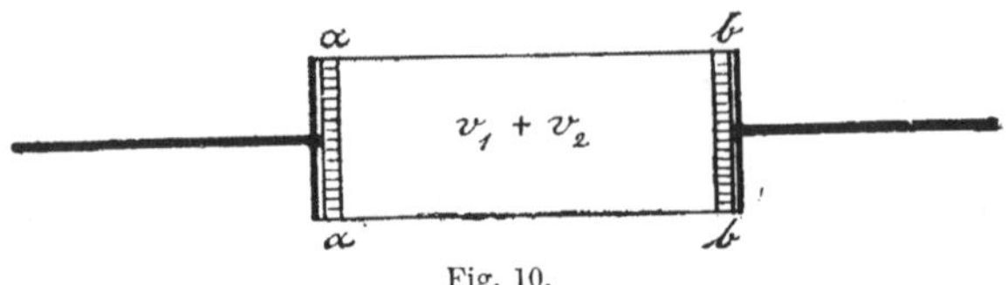

Fig. 10.

tion of the gases, and, at the same time, we bring the two gases back to their original volumes. The process is thus completely reversible.

Now the work which can be obtained by the intermingling is easily calculated: during the expansion each gas presses with its actual partial pressure on the piston opposing its passage, and accordingly performs exactly the same work, *as though the other gas were not present.* Since the n_1 molecules of the first gas expand from v_1 to $v_1 + v_2$, the work performed by them according to p. 52 is

$$A_1 = n_1 RT \ln \frac{v_1 + v_2}{v_1},$$

and for the second gas the work is

$$A_2 = n_2 RT \ln \frac{v_1 + v_2}{v_2},$$

and for the sum of the two

$$A = A_1 + A_2 = RT\left(n_1 \ln \frac{v_1 + v_2}{v_1} + n_2 \ln \frac{v_1 + v_2}{v_2}\right) \quad . \quad . \quad \text{(I.)}$$

This formula was first developed by Lord Rayleigh,[1] but afterwards independently and more thoroughly by Boltzmann.[2] Of course, if the system were not maintained at constant temperature, it would be cooled during the performance of the work, since there results no heat from the simple intermingling of the gases; under these circumstances, the equivalent quantities of heat would be absorbed from the environment, exactly as in the expansion or compression of a single gas.

From the equation (p. 23), viz.—

$$A - U = T\frac{dA}{dT},$$

[1] *Phil. Mag.*, **49.** 311 (1875). [2] *Wiener Sitzungsber.*, 1878, II. p. 733.

since in our case U is equal to zero (p. 52), it follows that

$$A = T\frac{dA}{dT},$$

or integrated

$$\ln A = \ln T + C,$$

in which C denotes an unknown constant; from this it follows that

$$A = cT,$$

in which c denotes a new constant.

Thermodynamics also establishes the correctness of the result given above, inasmuch as it requires a ratio between A and the absolute temperature, which, in fact, coincides with the formula obtained by the method described.

Though the derivation given above is obviously correct, yet it must not be overlooked that *it cannot be used in demonstration* until it has been proven that the semi-permeable walls used in the process *can be realised* in the actual cases: thermodynamic considerations must be employed, not in fictitious cyclic processes, but only in those which are possible in nature, if they are to attain to the rank of scientific proof, and not merely to that of arbitrary speculation. I would emphasise this the more, because hitherto the question of the practicability of the process in question has been disposed of rather too easily. Now, inasmuch as the "semi-permeable partitions" have played a very important rôle in many lines of research in recent times, and as their introduction into calculation has simplified and strengthened the proof, a few words will be offered concerning their legitimate use in thermodynamic conclusions.

For certain cases there are undoubtedly partitions having the quality desired; thus, palladium foil at high temperatures possesses the property of allowing hydrogen to pass through easily, but of hindering other gases. A gas very soluble in water can easily diffuse through a membrane moistened with water, while a gas which is soluble with difficulty in water cannot so diffuse. Thus if one ties a bit of moist pig's bladder over the mouth of a funnel, fixed with the mouth upwards, and with the tip in connection with a mercury manometer, and then leads ammonia gas under a jar held inverted over the funnel, he will at once see, as I have demonstrated many times, a rise of the manometer amounting to a tenth of an atmosphere, occasioned by the partial pressure of the diffused ammonia.[1] Thus the experiment shown in Figures 9 and 10 is realised, if the question concerns the mixture of ammonia and hydrogen. Thereby at the same time it becomes probable in the highest degree that the formula (I), obtained on p. 102, holds good universally; and since it is strongly proven for one

[1] This simple experiment is recommended for the lecture desk.

case, it can hardly be assumed that a nature law of that kind is dependent, so to speak, upon the casual question as to whether we possess the desired partition or not for all special cases.

Further, we next observe the more striking argument, in that another cyclic process has been devised by which the intermingling can be conducted isothermally and reversibly, and which affords the same final result, as was thoroughly proven by Boltzmann in 1878. A cyclic process of this sort can be represented in a much simplified form, as seems to me. Let us suppose that we have a solvent which easily dissolves one of two gases which are to be mixed, but the other with great difficulty. Thus water, for example, is a solvent having the desired properties for the gas pair, nitrogen and hydrogen sulphide, and still better for nitrogen and hydrogen chloride.

Suppose we are to consider the separation of hydrogen chloride and nitrogen: we first bring the mixture of gases to a large volume V, and then connect it with a large volume of a water solution of hydrogen chloride, choosing such a concentration for the latter that the partial pressure of its hydrochloric acid is equal to the partial pressure of this gas in the gas mixture of volume V. Then when we bring the gas mixture and the hydrochloric acid solution into contact with each other, hydrogen chloride will neither enter nor leave the solution. [Or rather, the quantities of hydrochloric acid entering and leaving the solution will be equal, and thus will balance each other.—Tr.] Now we will compress the gas mixture, which is in constant contact with the solution, to a volume which is very small in comparison with the original volume V; during this the hydrogen chloride will practically go completely into solution. Then we will separate the nitrogen, which remains behind in a state of almost complete purity, from the solution, and then will liberate from the solution the amount of hydrogen chloride dissolved; thus the desired separation has been effected. If we neglect the vapour tension of the water, which can give rise to no doubt if we work at low temperatures, and if we consider that the water vapour mixed with the two gases may be almost completely removed by compression, and if we also neglect the nitrogen dissolved in the solution, then it can be easily proven that the work expended in the separation of the gases is given by formula (I).

If we select a solvent in which the solubility of nitrogen cannot be neglected, but one in which the solubility of the two gases is very different, we can effect the separation of the two gases isothermally and reversibly by the repeated application of a cyclic process similar to the one just described, *i.e.* by "fractionation of the solution"; and it is not necessary to follow out the calculation, to conclude from the second law of thermodynamics, that the work to be applied can be calculated from formula (I.), since its efficiency has been already proven for the mixture of nitrogen and hydrogen chloride.

Now we can take any arbitrary pair of gases, and can find some suitable solvent which will dissolve the two gases in different degrees, for otherwise they would be chemically identical; and thus a cyclic process can always be realised which can be calculated by formula (I.). *This formula therefore holds good universally, and thereby it is proven, at the same time, that it is permissible to work with semipermeable partitions in all cases concerning the mingling or the separation of two gases which are chemically different.*

The Physical Properties of Liquid Mixtures.—The relations of liquid and solid mixtures are much more complicated than in the case of gases. Here, in the case of liquids, there is usually associated with the mingling, a change of the properties belonging to the particular ingredients of the mixture. The volume of the mixture of two liquids is in general different from the sum of the volumes of the individual ingredients, since with the mingling there is associated a contraction or a dilation which, to be sure, is not very considerable; and, moreover, the colour, refractive index, specific heat, etc., of the mixture are slightly different from these properties of the ingredients, calculated as though they persisted in the mixture. But one commonly obtains results at least approximately coincident, by calculating the properties of a physical mixture from those of the ingredients, as though these remained unchanged in the mixing.

In accordance with a proposal of Ostwald, we will call properties of this sort *additive;* the essential idea will be best made clear by examples.

Many liquids can unite with each other, as also many solid bodies with liquids, to form homogeneous liquid aggregates, the volumes of which are very nearly equal to the sums of the volumes of the respective ingredients. If we denote the volume of the mixture by V, and the volumes of the particular ingredients by V_1, V_2, etc., then it is approximately true that

$$V = V_1 + V_2 + \dots .$$

Let the weights of the particular ingredients be m_1, m_2, etc.; let us denote the volume of the unit of mass of the mixture, the so-called "specific volume," by v, and the respective volumes of the particular ingredients by v_1, v_2, etc.; then the preceding equation becomes

$$v(m_1 + m_2 + . .) = v_1 m_1 + v_2 m_2 + . . ,$$

or

$$v = v_1 \frac{m_1}{m_1 + m_2 +} + v_2 \frac{m_2}{m_1 + m_2 +} + \dots$$

The specific volume of the mixture can also be calculated from the specific volumes of the particular ingredients, by the so-called "partnership" calculation, viz. as though each particular ingredient entered the mixture with the specific volume which it had in the free state. The specific volume under these circumstances is an additive property. Of course, on the basis of the purely algebraic nature of the specific gravity, the reciprocals of the aforesaid values cannot be an additive property;[1] and this example teaches that it is only necessary to effect an algebraic transformation of values determined experimentally, to meet with much simpler relations.

The *heat capacity* of a liquid mixture is likewise as a rule equal to the sum of the heat capacities of the ingredients. If we denote by c the specific heat of a mixture containing m_1 g. of one ingredient of a specific heat c_1, and m_2 g. of another ingredient of a specific heat c_2, then according to the preceding law

$$c(m_1 + m_2) = c_1m_1 + c_2m_2;$$

but since $\frac{100\, m_1}{m_1 + m_2}$ denotes the value of the weight per cent of the first component, we obtain, as the formula to calculate c—

$$c = c_1\frac{p}{100} + c_2\frac{100 - p}{100}.$$

As a matter of fact, in the case of particular mixtures, as for example chloroform and carbon disulphide, this calculation proves very good, the differences between the observed and the calculated values being less than one per cent; but in other cases, as in the mixture of acetic acid and water, the results are only tolerable. Indeed, we find in the case of a mixture of alcohol and water, that the real specific heat is greater than it should be according to the preceding formula; thus a mixture of equal parts does not have a mean of the specific heats, thus,

$$\frac{1{\cdot}000 + 0{\cdot}612}{2} = 0{\cdot}806, \text{ but the specific heat is } 0{\cdot}910.$$

Mixtures of salt and water have as a rule a considerably smaller heat capacity than the sum of the ingredients, and sometimes smaller than what should correspond to the amount of water contained in the mixture (Thomsen, 1871, Marignac, 1873). In the case of solutions not too concentrated, one can estimate the heat capacity as nearly equal to that of the contained water, a very important rule for practical thermochemistry.

The *optical relations* of mixtures have been investigated with especial thoroughness; these will be dealt with in the next chapter. It need only be noticed here that excepting one case, when one works

[1] That is not, if as above the calculation is performed according to per cent by weight; but it will be additive if calculated according to per cent by volume.

with great exactness, he always finds differences exceeding the limits of error, between the results of observation and those calculated on the assumption of a simple additive relation. From all that we know up to date, in the case of liquid and solid mixtures, there is only one strictly additive property, viz., that of *mass*, which of course both in a simple mixture as well as in the case of a chemical compound, must remain unchanged, according to the law of the indestructibility of matter.

There can be no doubt that the deviations observed in the simple additive relations of liquids, will some day be of great significance in determining the nature of the force concerned in the mingling of two liquids, and of the kind of mutual action occurring. That commonly, and especially in the use of water in particular as a solvent, there occurs a *chemical action*, *i.e.* the formation of new molecules or the splitting of those previously existing,—this is well established beyond all question; it will be especially treated of, in the section on incomplete reaction (see the Doctrine of Affinity); but hitherto it has been impossible to prove this with certainty. Perhaps this much may be maintained, viz. that in cases where the properties of the mixture differ very considerably from the mean value [of those of the ingredients], a chemical action is probable; and in the case of such mixtures, a certain parallelism can be asserted in the variations attending their different properties; at least in those mixtures where we find a strong contraction, there being a deviation from the simple additive relations of the specific volumes, the specific heat seems to be considerably removed from the mean value. Certainly the fact on the other side, viz. that certain compounds which are undoubtedly chemical, show properties having an additive character,—this warns us not to be premature in drawing conclusions like the above. (See Second Book, Chap. V.)

The Optical Behaviour of Mixtures.—The *specific refractive power* of a mixture, by which is meant the quotient of the refractive index minus one, divided by the density [*i.e.* $\frac{n-1}{d}$], can be determined from the corresponding values of the particular components in exactly the same way as the specific volume; and as Landolt[1] has shown, the calculated value of the specific refractive power of the mixture coincides so well with that determined experimentally, that one conversely, can usually determine with great certainty the composition of a mixture from its optical behaviour. This example shows how sometimes, by a skilful combination of two physical properties (in the preceding case these being the refractive index and the specific gravity), one can arrive at simple and universal relations.

[1] *Lieb. Ann.*, Suppl., **4.** 1 (1865). Also compare Gladstone and Dale, *Phil. Trans.*, 1858, p. 887.

If we denote by n, the refractive power of a mixture, measured for any selected line of the solar spectrum, and if its specific gravity is d, and if it contains p per cent by weight of one ingredient, and 100 − p of the other; and moreover, if the corresponding values for these ingredients when pure are n_1 and d_1, and n_2 and d_2 respectively; then the specific refractive power of the mixture can be calculated by the formula

$$\frac{n-1}{d} = \frac{n_1-1}{d_1}\frac{p}{100} + \frac{n_2-1}{d_2}\frac{100-p}{100}.$$

As an example of the accuracy with which this formula corresponds to observations, there is given the following series of measurements of mixtures of ethylene bromide and propyl alcohol, for the temperature 18·07° and the sodium line.[1]

The Refractive Powers of Mixtures of Ethylene Bromide and Propyl Alcohol.

Ethylene Bromide per cent (p).	Sp. Gr.	Refraction Coefficient.	Calc.—Obs.	
			Diff. I.	Diff. II.
0	0·80659	1·386161	0·00000	0·00000
10·0084	0·86081	1·391892	+ 32	− 1
20·9516	0·92908	1·399136	+ 66	− 3
29·8351	0·99300	1·405958	+ 95	− 5
40·7320	1·08453	1·415815	+ 128	− 12
49·9484	1·17623	1·425748	+ 155	− 18
60·0940	1·29695	1·439013	+ 171	− 34
70·0123	1·44175	1·455063	+ 180	− 46
80·0893	1·62640	1·475796	+ 169	− 57
90·1912	1·86652	1·503227	+ 116	− 59
100	2·18300	1·540399	0·00000	0·00000

In the fourth column are given the differences between calculation and observation, which one obtains if he develops the values of n according to the preceding formula, from the experimentally determined values of p, n_1, n_2, d_1, d_2, and d. The differences all lie on the same side, and almost reach two units of the third decimal place, thus being a hundredfold greater than the errors of observation, which can be assigned as ± 0·00002. We conclude from this example that the law of mixtures just described, by no means represents a strict law of nature, but yet that as an approximate rule it frequently finds practical application.

[1] Schutt, *Zeitschr. phys. Chem.*, **9.** 349-377 (1892). For an extended form of the law of mixtures, from the introduction of a new constant, compare Pulfrich, *ib.* **4.** 161 (1889), and Buchkremer, *ib.* **6.** 161 (1890).

Now it has been investigated, as to whether another function of the refractive power may be better suited to represent the phenomena, and in fact H. A. Lorentz and L. Lorentz (1880), guided by certain theoretical considerations, found that the expression $\frac{n^2-1}{n^2+2} \cdot \frac{1}{d}$ is more accurate than the expression $\frac{n-1}{d}$, although it is not so simple. In the last column of the preceding table are given the differences between calculation and observation, which one obtains by solving the value of n from p, n_1, n_2, d_1, d_2 and d, by the formula

$$\frac{n^2-1}{n^2+2} \cdot \frac{1}{d} = \frac{n_1{}^2-1}{n_1{}^2+2} \cdot \frac{p}{100d_1} + \frac{n_2{}^2-1}{n_2{}^2+2} \cdot \frac{100-p}{100d_2};$$

as a matter of fact the result so obtained coincides much better with experiment.

These results have a practical interest to this extent; viz. that by means of the result of mixtures,

$$R = R_1\frac{p}{100} + R_2\frac{100-p}{100},$$

in which the specific refraction R denotes either $\frac{n-1}{d}$, or $\frac{n^2-1}{n^2+2} \cdot \frac{1}{d}$, *one is enabled from the specific refraction of a mixture to calculate the specific refraction of one ingredient, provided he knows the composition of the mixture and the refraction of the other ingredients.* If this method of calculation should give only approximate results, even then the relation between the specific refraction and the chemical composition (see this section) will be of great value in ascertaining these values for substances which one cannot study pure in the liquid condition, but only in the presence of a solvent.

Finally, it has been attempted to obtain results harmonising better with experiment, by introducing into the formula for the rule of mixtures certain constants suitably chosen for each particular case; thus, for example, instead of $\frac{n^2-1}{n^2+2} \cdot \frac{1}{d}$, let $R = \frac{n^2-1}{n^2+C} \cdot \frac{1}{d}$. But however useful such an equation might be as an exact interpolation formula in special cases, it would not possess universal interest, until from the deviations from simple additive relations, given by the numerical value of these constants, it could furnish a deeper insight into the nature of the mixture to be investigated.

One-sided Properties.—The study of the relations of mixtures is of especial importance for those physical properties which belong to only one compound of a mixture; if, for example, one mixes a coloured with an uncoloured substance, or one optically active with one optically inactive, he can obviously answer for such cases with all safety the question whether a change of property occurs by the mixture or not; and therefore in such cases one would have the best chance to reach a universal point of view. If, for example, we find that the refractive power of a mixture is the same as that calculated on the supposition of a simple additive relation, although at first sight it appears so, yet in reality there is no sharp proof that the specific refractive power of each component in the free state remains unchanged in the mixture, since it is possible that the property of one component might have increased to the same degree to which that of another has diminished. But if, for example, the specific rotatory power of an active substance in the mixture is exactly the same as in the free state, then the conclusion is unavoidable that every molecule rotates as strongly in the mixture as in the free state. In fact, "one-sided properties" of this sort, as they are called, are of the greatest importance in the investigation of solutions; besides the rotation and the absorption of light, there should be mentioned particularly the electrical conductivity and the osmotic pressure of dissolved substances.

As shown in the case of the two optical properties just referred to, so also is the absorption of light greatly altered by intermixture with colourless liquids; as a rule, but not always, the absorption bands of a dissolved substance are displaced the more toward the red end, accordingly as the refractive power of the solvent is greater (Kundt's rule).[1] Also, as a rule, the optical rotation of active molecules is greatly influenced by the presence of inactive molecules; and this influence, which may consist in a strengthening or a weakening of the rotatory power, varies with the quantity of the inactive substance added;[2] a universal conformity has not as yet been found. Further, the surmise is advanced that every important influence of this sort in mixtures may be ascribed to the formation of new molecular groupings in the way of incomplete reactions.

The Vaporisation of Mixtures.—One property of liquid mixtures, which changes with the proportions of the ingredients in a much more complicated way than those hitherto described, is their *vapour pressure.* The vapour emitted by a mixture will in general have the same components as the liquid remaining behind: therefore we must notice the *partial pressures* of the ingredients, the sum of which is the vapour pressure of the mixture. Sometimes, of course, the volatility of one ingredient is so very small that its pressure can be neglected; thus the vapour pressure of water solutions of salts is simply equal to the pressure of the water vapour in equilibrium with the solution.

1 Compare P. Stenger, *Wied. Ann.*, **33.** 577 (1888).

2 For further particulars see Landolt, *Optisches Drehungsvermögen organischer Substanzen*, Brunswick, 1879, p. 50 and following.

The following law holds good universally, and is of fundamental significance: *the partial pressure of each component of a mixture is always smaller than its vapour pressure in the free state (liquid or solid) at the same temperature.* If this were not the case the vapour of the component whose partial pressure was greater than in the pure state would be supersaturated; it would deposit from the vapour, and we should thus have spontaneous separation of the mixture. But this would mean that a self-acting process—the formation of a mixture—was reversed without any compensation, and this is not possible (see also below on the thermodynamic treatment of liquid mixtures). In the following chapters we shall see that the vapour pressures of dilute solutions follow very simple laws.

Regarding the vapour pressure of mixtures of *two liquids which dissolve each other in all proportions,* on the basis of the law just given, the following may be stated:—If we add to a liquid A a small quantity of another liquid B, then on the one hand the vapour pressure of A will be diminished, and on the other hand the total vapour pressure of the resulting solution will be increased by the circumstance that the quantity of B dissolved also gives out vapour; and, moreover, the partial pressure of B in the vapour existing in equilibrium with the resulting solution, will be so much greater as *the coefficient of solubility* of the vapour of B in the liquid A, is smaller. Accordingly as the first or second action of the added liquid preponderates, the vapour pressure of the solution will be smaller, or greater than that of A. Of course the same holds good for the solution arising from the addition of a slight quantity of A to B, which likewise, according to circumstances, can have a smaller or a greater vapour pressure than the liquid B in the pure state. But now since the properties of mixtures must always vary with the composition, the three following characteristic cases must be distinguished, regarding the dependence of the vapour pressure upon the varying proportions of the two liquids:—

I. The vapour of A is *easily soluble* in B, and that of B in A. Then in the case of the addition of a slight amount of A to B, as also of B to A, the vapour pressure of the resulting solution will fall. If we start with the pure solvent A (the vapour pressure of which at the temperature employed amounts to p_1), and add in succession small but increasing quantities of B, then the vapour pressure will at first diminish, will reach a *minimum,* then will begin to increase, and, finally, on the addition of large quantities of B will approach the vapour pressure p_2, that of the pure solvent B.

II. The vapour of A is only *slightly soluble* in B, as well as that of B in A. Then by the addition of a slight amount of A to B, as well as that of B to A, the vapour pressure of the resulting solution will increase. By the gradual addition of B to A the vapour pressure of the solution will at first become greater than p_1, will then reach a *maximum,* and, finally, by the addition of a great excess of B, will sink and approximate the value p_2.

III. The vapour of the first liquid is easily soluble in the second, but that of the second is soluble with difficulty in the first. Then by the addition of a slight quantity of B to A the vapour pressure of the resulting solution will be slightly smaller; by the addition of a small quantity of A to B it will be slightly greater than that of the pure solvent. Now if the vapour pressure p_1 of A is greater than p_2 of B, then by the successive additions of B to A the vapour pressure of the solution will fall, passing continuously without maximum or minimum from p_1 to p_2 (III[a].). But let the case be imagined where the first addition of B to A raises the vapour pressure, but that of A to B diminishes it; then by the gradual addition of B to A the vapour pressure will at first increase, will reach a maximum, will then decrease till it becomes smaller than p_2, will pass a minimum, and finally, with the great excess of B, will approach the vapour pressure p_2 (III[b].).

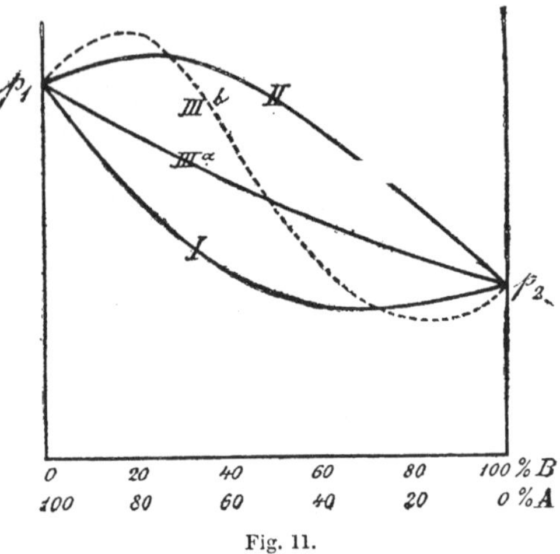

Fig. 11.

These various relations have been explained by the excellent theoretical and practical work of Konowalow;[1] the curve tracings (Fig. 11) will assist the general statement. The abscissæ are the varying proportions of the mixture in percentages of B; the ordinates are their varying vapour pressures. Case I. illustrates mixtures of formic acid and water from the measurements of Konowalow; case II., of water and propyl alcohol from the same source; case III[a]., from the same, experimentally proven for ethyl or methyl alcohol and water. The dotted line curve illustrating case III[b]., so far as I know, has not been realised.[2]

The Theory of Fractional Distillation.—We can now easily perceive what must be the relations in the isothermal evaporation of a solution, as we imagine it to be effected by the raising of a piston in a cylinder containing the solution. In general the vapour of the solution will have a different composition from that of the solution

[1] *Wied. Ann.*, **14.** 34 (1881).

[2] Ostwald has shown (*Lehrb. d. Allg. Chem.*, 2nd ed., II., **2.** 642 (1899) that this case is only possible for dissociating vapours; a very clear account of the evaporation of mixtures, including some new matter, is given (p. 617 ff.) in this work.

itself, and therefore as a result of the evaporation, the composition and, at the same time, the vapour pressure of the solution will vary. *Now the change in composition of the mixture must be such that its vapour pressure resulting from the evaporation, must be diminished.* If this were not the case, by raising the piston the pressure would increase somewhat; by lowering the piston it would diminish; so that a more stable equilibrium between the pressure on the piston from the vapour, and that exerted from without, is debarred. The evaporation proceeds in such a way that the more volatile ingredients evaporate first. Thus in case I., the mixture must remain behind with a minimal vapour pressure; in case II., one of the two solvents, according to the relative proportions of the original solution; in case III[a]., since we neglect the case III[b]., which is not as yet realised (and may be unrealisable ?), after a sufficiently long evaporation, that solvent with the smaller vapour pressure remains behind.

In practical laboratory work one does not as a rule boil and distil the mixture at constant temperature, but rather evaporates it usually at constant pressure, viz. that of the atmosphere. Since, as a rule, the vapour has a different composition from the remaining solution, we thus possess one of the most powerful and convenient methods of separating substances. Of course the separation will never be complete by one distillation, but may usually be carried to a high degree by interrupting the process and separating the parts which pass over at the successive temperatures ("fractional distillation"), and by suitable repetition of the operation. The law that a change of composition of the solution must result in decreasing the vapour pressure, results in this case that *the boiling-point must rise during the distillation.* We can now easily see that, in case I., one will finally obtain as the least volatile *residue*, the mixture with the minimal vapour pressure; and as the product of the sufficient repetition of the fractional distillation, the *distillate* will be *one* of the pure solvents; and that component will form the ultimate distillate, which was in excess in the mixture at its minimum of vapour pressure. In case II., on the other hand, one obtains as the final *distillate*, a *mixture* corresponding to the maximum of vapour pressure, and as the less volatile *residue*, that one of the pure solvents which was present in excess. Only in case III[a]. is it possible to effect a complete separation of the two ingredients of the mixture by repeating the operation sufficiently.

It is, moreover, possible, under certain definite conditions, which will be discussed later, to obtain *mixtures having constant boiling-points*: thus such mixtures, as for example the water solution of hydrochloric acid which distils at constant boiling-point, have occasionally been wrongly claimed as chemical compounds, as hydrates, since their chief characteristic was the property of being unchanged by distillation. Aside from the fact that this property of such a mixture is ascribed, so to speak, to the casual coincidence that the partial pressures of the two ingredients of the mixture stand in such relations that the vapour emitted has the same composition as the mixture itself, it must also be remembered that the relative proportions of the ingredients change with the pressure at which the distillation is conducted, which is not the case with genuine chemical compounds.

A quantitative theory of fractional distillation—not applicable to case III[a]. however—is given by Barrell, Thomas, and Young (*Phil. Mag.* (5), **37**. 8 (1894); it rests on an assumption justified by the experiments of F. D. Brown (*Trans. Chem. Soc.*, 1879, p. 550; 1880, pp. 49, 304; 1881, p. 517), that the ratio of the components passing off in vapour at any moment is proportional to the ratio of the components in the liquid mixture at that moment. Thus if x and y are the masses of the components in the liquid, the quantities dx and dy passing off as vapour satisfy the condition

$$\frac{dx}{dy} = c\frac{x}{y},$$

where c is a constant factor depending on the nature of the liquids. See also researches on the separation of ternary mixtures by distillation.

The Thermodynamic Treatment of Liquid Mixtures.—In the preparation of a mixture, a certain amount, Q, of heat will usually be developed or absorbed. The external work produced thereby, and which is measured by the product of the external pressure and the change of volume, has a negligible value, so that we can make the diminution of the internal energy U, as a result of the mingling, equal to the heat of mixture Q; this can be experimentally determined by means of a very simple calorimetric measurement; and thus

$$U = Q.$$

If there are x mol. of the second component to 1 mol. of the first, then Q is a function of x, *i.e.*

$$U = Q(x);$$

this can sometimes be expressed by a simple formula; thus the development of heat observed on mixing 1 mol. of H_2SO_4 with x mol. of water, was found by J. Thomsen[1] to be very accurately expressed by the formula,

$$Q(x) = \frac{17,860x}{x + 1{\cdot}8}.$$

The variation of the heat of mixture with the temperature, is given by the difference between the heat capacities of the components, and

[1] *Thermochem. Untersuchungen*, Bd. III. 34.

of the mixture: if K_0 denotes the heat capacity of the components before the mixture, and K that of the liquid resulting from their mixture, then we have

$$\frac{dU}{dT} = \frac{dQ}{dT} = K_0 - K.$$

This equation shows that only when

$$\frac{dQ}{dT} = 0, \quad \text{can } K = K_0;$$

i.e. it demands as a necessary and sufficient condition, that the specific heat of the mixture shall be calculated in an additive way from that of the components (p. 106).

Very much harder, but very much more interesting, is the determination of the diminution of *free energy* associated with the intermixture; to calculate this we must find some way to accomplish the intermixture isothermally and reversibly; if it is possible to find several methods of this sort, then all must give the same result, a law which will give particularly valuable results in the discussion of dilute solutions. Now as a matter of fact, it is possible to effect the operation of mixing two liquids isothermally and reversibly, in very many ways, viz. by *isothermal distillation* (Kirchhoff), by *electrical transference* (Helmholtz), by *osmose* (van't Hoff), by *selective solubility*, and the like; this operation of intermixture occurs spontaneously as soon as the liquids are brought into contact with each other, provided that they are mutually soluble. But since these methods for the most part first found practical application in the case of dilute solutions, which will be thoroughly treated in the following chapter, there will be given in this place only a short description of the method which is most universal and most important, viz. *mingling by means of isothermal distillation.*

We will take as the basis of our considerations, the mixture formed by the mingling of 1 mol. of the first component with x mols. of the second, and it is also assumed that the gas laws are applicable both to the vapours emitted by the mixture, and also to the vapour emitted by each component. Moreover, in case of necessity, one can easily generalise the formula for the case so that another equation of condition —an equation of dissociation, or the van der Waals' equation, for example—holds good for the case of the vapour.

Now let us imagine there to be brought into the mixture two semi-permeable partitions, one of which is permeable by the first, the other by the second ingredient. The vapour emitted by the mixture contains both ingredients; and let P be the pressure of the first ingredient of the saturated vapour standing in equilibrium with the liquid mixture, at the absolute temperature T; and let p be the pressure of the second ingredient; then, according to Dalton's law, $P + p$ is the maximal pressure of the mixture. Let the two ingredients, at the same temperature, in the pure state have the pressure P_0, and p_0, respectively.

Now let us imagine the following process to take place:—We will allow 1 g.-mol. of the first component, in the pure liquid state, to evaporate to saturated vapour; the work thus obtained will be—

$$p_0 v_0 = RT;$$

then let it expand till the pressure p_0 shall have fallen to p, whereby the work performed by the gas will be—

$$RT \ln \frac{p_0}{p};$$

finally, bring it into contact with the membrane which is permeable by it and compress it; in consequence of this, it passes through the membrane and is condensed in the mixture; if this condensation is performed under constant pressure p, we need to introduce the external work, $pv = RT$.

In order to realise the condition that the composition of the mixture, and hence that the value of p shall not change during this condensation, we cause x mol. of the second ingredient to evaporate at the same time; thus we obtain the work,

$$xP_0V_0 = xRT;$$

and if we let this expand whereby pressure P_0 sinks to P, there is performed the work,

$$xRT \ln \frac{P_0}{P},$$

and at the same time we condense it with 1 g.-mol. of the first ingredient by forcing it through the second semipermeable partition. The two vapours meet in the space enclosed by the two semipermeable membranes with the pressures P and p. Thus we have it always completely in our power to send through the partitions such quantities that there will condense a mixture having the composition just described; and if we do this, there will condense the 1 g.-mol. of the first ingredient under the constant pressure p, and the x g.-mol. of the second under the constant pressure P; as just stated, there is needed to condense the first, an introduction of external work RT, and for the second,

$$xPV = xRT.$$

The external work in the transference of 1 g.-mol. of the first component to the mixture, is

$$RT + RT \ln \frac{p_0}{p} - RT = RT \ln \frac{p_0}{p},$$

and for the transference of x g.-mol. of the second ingredient,

$$xRT + xRT \ln \frac{P_0}{P} - xRT = xRT \ln \frac{P_0}{P};$$

and the sum of the work accomplished by the system is

$$RT\left(\ln \frac{p_0}{p} + x \ln \frac{P_0}{P}\right).$$

Now, since the processes above described can all be completed reversibly and isothermally, the last expression refers to the maximal external work or the diminution of free energy A. This last expression, which may be suitably designated by A (x), thus becomes

$$A(x) = RT\left(\ln \frac{p_0}{p} + x \ln \frac{P_0}{P}\right) \quad . \quad . \quad . \quad (1)$$

and this gives

$$\frac{dA(x)}{dT} = R\left(\ln \frac{p_0}{p} + x \ln \frac{P_0}{P}\right) + RT\frac{d}{dT}\left(\ln \frac{p_0}{p} + x \ln \frac{P_0}{P}\right).$$

Introduced into the formula (p. 23),

$$A - U = T\frac{dA}{dT};$$

and remembering that $U = Q(x)$ on p. 114, we have

$$Q(x) = -RT^2\frac{d}{dT}\left(\ln \frac{p_0}{p} + x \ln \frac{P_0}{P}\right) \quad . \quad . \quad (2)$$

This formula affords the calculation of the heat of mixture of two liquids, from the temperature coefficients of their vapour pressures and that of their mixture, and the pressure of the vapour emitted by the mixture.

Moreover A(x) can be found by allowing x g.-mol. of the second ingredient to distil over into 1 g.-mol. of the first liquid ingredient; then the vapour pressure, π, of the second ingredient in the mixture rises from 0 to P, and the work obtained thereby is given by the integral,

$$A(x) = RT\int_0^x \ln \frac{P_0}{P} dx \quad . \quad . \quad . \quad (3)$$

in the evaluation of which, π must be known as a function of x.

From (1) and (3) we obtain

$$\ln \frac{p_0}{p} + x \ln \frac{P_0}{P} = \int_0^x \ln \frac{P_0}{P} dx.$$

Introducing this into (2), we obtain

$$Q(x) = -RT^2\frac{d}{dT}\int_0^x \ln\frac{P_0}{P}dx,$$

or, differentiated for x,

$$\frac{\partial Q(x)}{\partial x} = -RT^2\frac{\partial \ln\frac{P_0}{P}}{\partial T}.$$

This last equation, which one can easily derive directly by considering the isothermal distillation of one component into a large quantity of the mixture, was first derived by Kirchhoff,[1] and is a special case of the preceding generalisation, which, so far as I know, has not been developed before this; it was applied by him to the calculation of Regnault's vapour-pressure curve, measured for mixtures of water and sulphuric acid, the mixing heat of which is known (p. 114); the calculation gives a fairly satisfactory result.

The physical significance of $\frac{\partial Q}{\partial x}$ is given by the development of heat observed on the addition, for example, of 1 mol. of water to a very large quantity of a mixture having the composition of $H_2SO_4 + xH_2O$. Now, if one tries to calculate the quantity of heat from Regnault's vapour pressures measured for such a mixture, and from the vapour pressure of pure water, he obtains widely differing results.[2] This is due to the fact that small variations in the value of P—the measurement of which, by the way, is very difficult—change the results of the calculation very greatly; this is also the reason why hitherto the successful quantitative application of Kirchhoff's formula for the calculation of the heat of dilution, has failed, viz. on account of the difficulty of measuring the vapour pressures of mixtures with sufficient accuracy. But this result can be inferred from Tammann's[3] measurements—at least in a qualitative way—viz., that in the sense of Kirchhoff's formula, and in the case of solutions which develop heat on the addition of water (as H_2SO_4, $CaCl_2$, and the like),

$$\frac{\partial Q}{\partial x} > 0,$$

and that $\frac{P_0}{P}$, and accordingly $\frac{P_0 - P}{P_0}$ also, the so-called "*relative lowering of the vapour pressure*," decreases with increasing temperature; but that

[1] *Pogg. Ann.*, **104.** (1856); *Ges. Abh.* (Separate Papers), p. 492.
[2] Compare R. von Helmholtz, *Wied. Ann.*, **27.** 542 (1886).
[3] *Mem. d. Petersburger Akad.*, **35.** No. 9 (1887).

these values increase with increasing temperature, in the case of solutions exhibiting an absorption of heat on the addition of water.

The above formulæ show, as was first pointed out in the second edition of this book, that the composition of the vapour can be calculated, when the vapour pressure curves of the components (solid or liquid) are known, and the partial pressure of one component as a function of the composition of the liquid. A number of other deductions will be found in Duhem, "Solutions and Mixtures," *Trav. et Mém. des facultés de Lille*, Nos. 12 and 13 (1894); also M. Margules, *Wien. Ber.*, **104**. 1243 (1895) who calculated examples. See also Lehfeldt, *Phil. Mag.* (5), **40**. 397 (1895), and Dolezalek, *Z. S. phys. Chem.*, **26**. 321 (1898).

On differentiating (1) and (3) we get from (1)

$$\frac{dA}{dx} = RT\left(- \frac{d \ln p}{dx} - x\frac{d \ln P}{dx} - \ln P\right),$$

from (3)

$$\frac{dA}{dx} = - RT \ln P,$$

whence follows at once the differential equation given by Duhem (*loc. cit.*)—

$$\frac{d \ln p}{dx} + x\frac{d \ln P}{dx} = 0.$$

For applications of this equation see Gahl, *Z. S. phys. Chem.*, **33**. 178 (1900); and Zawidzki, *ibid.* **35**. 129 (1900).

The Critical Point of Mixtures.—In the warming of mixtures, there are observed certain characteristic phenomena which we have usually called "critical" (p. 67). But hitherto, of the critical data, in the case of mixtures, only the temperature has been measured for considerable ranges; from this the law[1] was observed that the critical temperature t, of a solution containing n per cent by weight of one liquid having the critical temperature t_1, and 100 – n per cent by weight of another liquid having the critical temperature t_2, can be calculated by the mixing rule,

$$t = \frac{nt_1 + (100 - n)t_2}{100}.$$

This formula can also be used conversely to calculate the critical temperature of one ingredient, provided that of the mixture and also that of the other ingredient be known.

G. C. Schmidt[2] has recently found the preceding rule very well

[1] Pawlewski, *B. B.*, **16**. 2633 (1883). [2] *Lieb. Ann.*, **266**. 266 (1891).

established for a number of mixtures, which places the *critical point on the list of the additive properties;* the variations between calculation and experiment never amount to more than 4°, and as these lie irregularly on both sides of the mean, it is possible that they may be explained as due, at least in part, to the uncertainty of observation.

Kucnen[1] has recently found that the critical temperatures of mixtures of ethane and nitrous oxide sometimes lie considerably below that of the components.

Isomorphous Mixtures.—The capacity of two crystallised substances for uniting to form a homogeneous mixed crystal, is observed quite rarely, although this property is not so exclusively limited to substances related chemically and crystallographically (*i.e.* substances possessing the same properties of symmetry and almost the same geometrical constants), as was formerly supposed; but since it has been found that pairs like ferric chloride and ammonium chloride, tetra-methyl-ammonium iodide and chrysoidine hydrochloride, etc., are able to form mixed crystals, and also substances in which one seeks in vain for any analogy of composition and constitution,[2] no doubt can be entertained that further investigation will considerably increase the list of such instances. (Compare the section on Isomorphism.)

It is undeniable that such solid mixtures are to be classified with liquid mixtures in many respects: thus we find liquids, as for example water and alcohol, which are miscible in all proportions, and also liquids of limited miscibility, as, for example, water and ether, which are able to dissolve each other only slightly; and there are also substances which can crystallise together in all proportions, as, for example, the alums, and also other substances where gaps exist in the series of mixtures, as, for example, the sulphate and the selenate of beryllium.

Very often two salts can crystallise together in molecular proportions when they have only a partial, and sometimes hardly any recognisable miscibility in each other; thus the *double salts* form a particular point in the gap exhibited by the series of mixtures. An analogy to this in many respects, is found in certain pairs of liquids which mutually dissolve each other only to a limited degree, and which at the same time are able to form a homogeneous molecular mixture in the shape of a chemical compound. Thus, for example, amylene, C_5H_{10}, and water, H_2O, are almost mutually insoluble towards each other; yet they are able to unite with each other in molecular proportions to form amyl alcohol, $C_5H_{10} + H_2O = C_5H_{12}O$, and thus this substance appears as a particular point in the gap exhibited by the series of mixtures of amylene and water. But, of course, the question

[1] *Phil. Mag.*, **40.** 173 (1895).
[2] See especially the work of O. Lehmann, *Molekular-physik.*, Bd. I. and II.

as to whether the kind of chemical union in the two cases is not essentially different,—this cannot be determined by this analogy.

Moreover, the property of water of uniting in molecular proportions with many substances, as *water of crystallisation*,—a property, by the way, not limited to water alone, as is shown by *alcohol of crystallisation, benzene of crystallisation*, etc.,—this may be referred to the list of the double salts just mentioned.

Of course, in the case of the crystallised mixtures, their relations are more various than in the case of liquid mixtures, since there the crystal form is a new factor of standard significance. Following a classification given by Retgers,[1] I give here a list of the different cases observed in the mutual miscibility of two crystalline substances.

1. The two substances form a complete series of mixtures, *i.e.*, it is possible to make mixed crystals of them in every desired proportion; pure isomorphism. Example: $ZnSO_4 \cdot 7H_2O$ and $MgSO_4 \cdot 7H_2O$.

2. On account of slight differences in the crystal angles or in the molecular volume, the two substances can intermix only to a limited degree; this case also is usually called pure isomorphism. Example: $BaCl_2 \cdot 2H_2O$ and $SrCl_2 \cdot 2H_2O$.

3. The different substances have forms more or less different, but mix with each other in considerable quantities; the gap in the series of mixtures is relatively small. Example: $NaClO_3$ regular and $AgClO_3$ quadratic.

4. Intermixture occurs only to a very slight extent, and can be recognised only by micro-chemical reactions on fragments free from inclusions; the gap in the series is relatively very large. Example: KNO_3 and $NaNO_3$.

Cases 3 and 4 are instances of *Iso-di-morphism*; this term designates the phenomenon where both of the two different substances occur in two crystal modifications which are isomorphous in pairs. Very often the two modifications are not both known in the free state, but one may be known in the form of an isomorphous mixture. Thus in such mixtures as are instanced in 3 and 4, that substance which is in excess, forces its crystal form in some degree upon the other substance.

5. There occurs an isolated double salt standing in the middle of the series, which crystallises differently from the end members, and does not mix with either of the simple salts; the gap between the end members is usually very great. The recognition of this case is of great importance, since by a superficial examination of the double salt, and especially if it resembled one of the end members, one would easily mistake it for an isomorphous mixture, and the phenomenon as one of pure isomorphism.

(Example: Calcite and magnesite which crystallise in molecular proportions, as dolomite.)

6. The two substances do not mix with each other appreciably.

[1] *Zeitschr. physik. Chem.*, 5, 461 (1890).

In order to decide the degree of miscibility by a practical case, according to Retgers, one should prepare a number of solutions holding each ingredient in greatly varying proportions, and by evaporation of the solvent obtain crystals which should be investigated regarding their behaviour. In order to obtain a product as homogeneous as possible, it is best to use a large quantity of solvent, and to study only the crystals which separate first. For the more the composition of the solution changes with the separation of the earlier portions, so much greater is the danger of the formation of heterogeneous products.

Hereafter by an "*isomorphous mixture*" will be understood a mixed crystal, the composition of which is capable of a constant change within certain limits; every member of a complete or incomplete series of mixtures we will also call an isomorphous mixture, but not the isolated double salt.

The Physical Properties of Mixed Crystals.—The analogies between a liquid and a solid mixture extend thus far, viz., that in both cases many properties of the mixture are additive, *i.e.* they can be calculated from the properties of the particular ingredients according to the so-called "partnership" calculation (p. 106).

This holds good especially for the volume of mixed crystals, which is frequently equal to the sum of the volumes of the ingredients in the free state; in other words, *in the intermixture of two crystals there is observed neither contraction nor dilation;* as an example there is given below a table of the specific volumes of mixed crystals of potassium and ammonium sulphates, by Retgers,[1] to whom we are indebted for these very exact measurements.

Composition.		Spec. Volume.		Difference.
$(NH_4)_2SO_4$.	K_2SO_4.	Obs.	Calc.	
0	100	0·3751	(0·3751)	...
5·45	94·55	0·3885	0·3855	+0·0030
8·33	91·67	0·3879	0·3906	−0·0027
15·03	84·97	0·4042	0·4037	+0·0005
18·45	81·55	0·4080	0·4098	−0·0018
20·55	79·45	0·4112	0·4138	−0·0026
26·47	73·53	0·4270	0·4250	+0·0020
29·30	70·70	0·4305	0·4307	−0·0002
42·67	57·33	0·4572	0·4556	+0·0016
65·35	34·65	0·4990	0·4988	+0·0002
83·37	16·63	0·5311	0·5324	−0·0013
100	0	0·5637	(0·5637)	...

[1] *Zeitschr. physik. Chem.*, **3**. 497 (1889); see also **3**. 289; **4**. 189; **5**. 436; **6**. 193 (1890); **8**. 6 (1892).

Although the law given above, as far as known, holds good and with great exactness for the case of isomorphous mixtures, yet a modification is necessary in the case of *iso-di-morphous mixtures.* Thus the sulphates of magnesium and iron, $MgSO_4 . 7H_2O$, and $FeSO_4 . 7H_2O$, the first being orthorhombic, the second being monoclinic, are able to form orthorhombic and monoclinic mixed crystals, and yet are miscible to only a limited degree, since mixed crystals containing more than 54 per cent, or less than 81·22 per cent of magnesium sulphate cannot be obtained, at least at ordinary temperatures; and, moreover, the mixed crystals having the magnesium sulphate in excess are orthorhombic, and those having the iron salt in excess are monoclinic. If one tries to calculate the specific volume of the mixed crystals from that of the two salts in the pure state (Fe 0·5269, Mg 0·5963), he meets one-sided deviations; but if one takes the specific volume of the iron salt in the orthorhombic crystals as 0·5333, and in the monoclinic crystals as 0·5269, and similarly the volume of the magnesium salt in the monoclinic crystals as 0·5914, and in the orthorhombic crystals as 0·5963, the deviations vanish almost completely. It is very probable that the two values assumed above represent respectively the specific volume of theoretical orthorhombic iron, and of the monoclinic magnesium salts which have not yet been produced in the free state. *Probably in general every salt in iso-di-morphous mixtures possesses a specific volume more or less different from that in the free state;* this is in harmony with the fact that di-morphous modifications in the free state have different specific gravities.

It follows from the observations of Dufet (1878), Fock (1880), Bodländer (1882), and others, that the optical properties of mixed crystals are emphatically additive; but of course the crystal structure is attended with certain complications.[1]

Fusion and Solidification of Mixtures.—Investigation of the laws regulating the solidification of liquid mixtures are of much practical importance, on account of the method of fractional crystallisation that is so often used. The theory of this phenomenon cannot be developed so far as that of fractional distillation, at present, and also experimental study is still much needed. Still the leading principles are well established and will be given here.

When a mixture freezes the solid deposited may be homogeneous, *i.e.* it may be a single component of the liquid, or an isomorphous mixture; or else more than one solid may separate at the same time; thus from an aqueous solution either ice, salt, or hydrate may crystallise out, or more than one of these may appear together. In this lies a complication that is absent in evaporation, for gases are always homogeneous.

We will take first the simple case that one component separates

[1] See especially F. Pockel's *Jahrb. f. Mineral,* **8.** 117 (1892).

out in the pure state; then we have the simple rule that the freezing-point of such a mixture is always lower than that of the separated substance in the pure state.

The proof of this follows from the consideration (p. 111) that the vapour pressure of a substance in a mixture is always less than in the pure state; the melting-point, being the intersection of the vapour pressure curves of solid and liquid (pp. 74 and 97), is accordingly always lowered.

Numerous examples of this rule are to be found. Salt solutions freeze (*i.e.* deposit ice) at a lower temperature than pure water; on the other hand, if they deposit salt, it is at a temperature below the melting-point of the salt, so that we may look on the crystallisation of a dissolved salt as freezing at a temperature very much lowered by addition of water. Phenol liquefies, when wet, much below its proper melting-point (42°); this too may be simply regarded as a lowering of melting-point due to mixture with water. Again the fusibility of alloys is a well known and technically important fact of the same order; iron rich in carbon melts at a lower temperature than when containing little; and the alloys of Rose, Wood, Lipowitz, etc., are striking instances of how greatly the components of a mixture can lower each other's melting-point. Similarly, mixtures of salts, of fatty acids, and so on, are noticeable for their low melting-point. The latter case was fully investigated by Heintz.[1]

It will appear in the following chapter that the lowering of the melting-point of dilute solutions follows very simple laws.

It is essentially different when an isomorphous mixture or a chemical compound in molecular proportions (hydrate, double salt, substance with alcohol of crystallisation, benzene of crystallisation, etc.) separates from a mixture; then the melting-point of the mixture may, according to circumstances, lie higher or lower than that of the pure substances. For example, the melting-point of naphthalene is raised by addition of β-naphthol, and accordingly pure naphthalene does not freeze out from the liquid, but an isomorphous mixture.[2] It is noteworthy that under these circumstances the freezing-point of a liquid, like the boiling-point, may be either raised or lowered. If the solid mixture separating out changes continuously in composition with change in that of the liquid, the freezing-point varies continuously from that of one component to that of the other. This case is rare; often there is a sudden change in the composition of the solid owing to replacement of one component by the other, or by a compound in molecular proportions (*e.g.* a new hydrate); for instances see the work of Vignon[3] and Miolati.[4]

[1] *Pogg. Ann.*, **92.** 588 (1854).
[2] Van't Hoff, *Z. S. f. phys. Chem.*, **5.** 388 (1890); Van Bijlert, *ibid.*, **8.** 343 (1891)
[3] *Bull. soc. chim.* (3), **7.** 387, 656 (1892).
[4] *Z. S. phys. Chem.*, **9.** 649 (1892).

In a few cases we find the phenomenon of continuous change in the solid deposit as the composition of the liquid is continuously changed; then the freezing-point curve is continuous, showing no sudden changes of slope. This happens in the case of mixtures of isomorphous and chemically similar materials. Küster[1] was the first to observe this case, and found that the melting-point could be calculated from the composition by the simple rule of mixtures. The agreement between observed and calculated melting-points is better when the composition is expressed in molecular percentage (mols. in 100 mols. of mixture) than in percentage by weight. It is necessary, in order that this rule be true, that the composition of the frozen part should be the same as that of the liquid; this can be tested by seeing whether the melting-point is independent of the quantity melted or not. Only if the temperature does not change during the fusion, is the rule true. When there is much difference in composition between the liquid and the frozen mass, the melting-points differ considerably from the values calculated by the (linear) rule. The regular behaviour is well shown by mixtures of hexachlor-α-keto-γ-R-pentene (C_5Cl_6O) and pentachlor-monobrom-α-keto-γ-R-pentene (C_5Cl_5BrO) (Küster).

Molecular percentage of C_5Cl_5BrO.	Freezing-point.	
	Obs.	Calc.
0·00	87·50	...
5·29	87·99	88·04
8·65	88·30	88·38
25·32	89·85	90·09
42·26	91·61	91·81
71·33	94·59	94·78
90·45	96·67	96·74
98·00	97·49	97·50
100·00	97·71	...

Often, however, more than one solid separates from the liquid; in this case it is not at present possible to give a rule for the position of the melting-point, but experience seems to show that it may lie above or below those of the components. In general the composition of the remaining liquid is altered by the separation; and whatever the composition of the solid that separates, the change must always be in such a direction that the melting-point of the remaining liquid is lowered. This is the analogue of the law that during distillation the temperature always rises (p. 113). If, then, a mixture be repeatedly fractionally crystallised, a liquid will eventually be reached with minimum freezing-point; if this is frozen, the solid separating must have the same com-

[1] *Z. S. phys. Chem.*, **5**. 601 (1890); **8**. 577 (1891).

position, otherwise we should obtain a liquid of still lower freezing-point. Conversely, the solid has a melting-point that is constant, *i.e.* independent of the amount melted. Such a substance was called by Guthrie[1] who discovered these relations, an *eutectic mixture*; it melts and freezes like a chemically simple substance, and is comparable with the mixture of constant boiling-point mentioned on p. 113; there are no grounds for assuming that it is a chemical compound, nor does experiment point in that direction.

Thus Guthrie found that a mixture of 46·86 per cent lead nitrate and 53·14 per cent potassium nitrate melts at 207°, and that by altering these proportions in either sense a higher melting-point is obtained. The following table gives the melting-point t of eutectic compounds of bismuth containing p per cent of the metal alloyed with bismuth:—

	p	t
Lead	44·42	122·7
Tin	53·30	133
Cadmium . . .	40·81	144
Zinc	7·15	248

In the same way, by repeated partial freezing eutectic mixtures of more than two components can be obtained, *e.g.* the mixture

Bi 47·75, Pb 18·39, Cd 13·31, Sn 20·00 per cent.

melts at 71°, the lowest melting-point known of any alloy except those containing alkali metals and the amalgams.

"Cryohydrates" are also examples of eutectic mixtures. When a salt solution is cooled it first deposits ice, and so becomes more concentrated; hence on cooling further a point must be reached when the salt in solution has become saturated; then a mechanical mixture of ice and salt will come down, and necessarily in the same proportions as they occur in the liquid. This temperature is at the intersection of the curve of saturation with the curve of lowering of freezing-point; the solution freezes completely and at constant temperature, as if it were a single substance. It was at one time supposed to be such, and described as a cryohydrate.

The temperature at which the solution freezes as a whole, to form a mechanical mixture, is also the lowest that can be obtained by mixing ice and salt. Thus, according to Guthrie[2] ice and NaCl give $-22°$, ice and NaI $-30°$; the cryohydric temperature is clearly lower, the more the salt lowers the freezing-point of water, and the more soluble it is. The temperature can be still further reduced by using

[1] *Phil. Mag.* (5), **17.** 462 (1884).
[2] *Ibid.* (4), **16.** 446; (5), **2.** 211; **6.** 35, 105.

more than one salt simultaneously; Mazotto[1] investigated the case of water, NH_4Cl and $NaNO_3$, and found the cryohydric point at $-31°·5$.

Apparently the solid that deposits from eutectic mixtures arrived at in this way consists always of a mechanical mixture of two or more substances; there are, however, cases, as mentioned above, in which a liquid mixture freezes to a single homogeneous substance (double salt, hydrate, etc.). If the composition of such a mixture be altered, then unlike an eutectic (*i.e.* easy melting) mixture, the freezing-point is lowered; therefore the mixture is one of maximum melting-point, and may be described as *dystectic*. Thus Roozeboom[2] found that a mixture of composition $FeCl_3 + 12H_2O$ freezes constantly at 37° to a solid hydrate; if the composition is varied, within certain limits, whether by addition of $FeCl_3$ or of water, a mixture of lower freezing-point is obtained.

From what has been said it may be seen that the different kinds of mixtures that we have distinguished must show characteristic differences when allowed to solidify gradually. Both eutectic and dystectic mixtures, like simple substances, solidify at one temperature. When the solid separating out changes in composition continuously with the change in the residual liquid, the fall of temperature, on cooling, will be made slower as soon as the separation begins, and the latent heat comes into play; this condition lasts until, after a longer or shorter interval of temperature, the whole has become solid. In cases in which this condition is not satisfied, but new substances appear, twice or more, in the solid deposit, the fall of temperature will be made slower each time, a phenomenon noticed in 1830 by Rudberg ("repeated melting-point"). Further, the course of the cooling may be influenced by phenomena of undercooling, and by subsequent gradual allotropic modification of solid deposits.[3]

Finally, this difference may be noted between the behaviour of a simple substance and a mechanical mixture. A simple substance always melts at the same temperature, and it is not possible to keep it solid even for a short time at a higher temperature. It is otherwise with a mixture; a finely powdered mixture of two metals can often be kept a long time at a temperature above the melting-point of the alloy that would result from it. Thus in a mixture, melting-point and freezing-point are not identical, as they are for a single substance.

On the other hand, even a rough mixture will sometimes liquefy at a temperature below the melting-point of either pure substance. Thus if a mixture of tin and lead be kept for some hours at 200° it liquefies, although the melting-point of tin is 230° and lead 325°; and a mixture of 1 part Cd, 1 part Sn, 2 parts P6, 4 parts Bi (Wood's metal), all the components of which melt above 200° can be liquefied by keeping for

[1] *Beibl.*, **15**. 323 (1891). [2] *Z. S. phys. Chem.*, **10**. 477 (1892).
[3] See Ostwald, *Lehrb. d. allg. Chem.*, 2nd ed., **1**. 1023 (1891).

some hours or days in a water-bath, under slight pressure (Hallock);[1] a mixture of sodium acetate and potassium nitrate melts when kept for some hours at 100°, though the melting-point of the components is above 300° (Spring).[2]

The Thermodynamics of Isomorphous Mixtures. — The observations which we have advanced above (p. 111) concerning liquid mixtures, from the standpoint of energy, can be applied directly to solid mixtures, in their essential principles. With the union of two crystals to form a new homogeneous mixed crystal, there is usually associated a certain diminution of the *total energy* U, which may be designated here as the heat of mixing. To be sure, as far as I know, experimental data on this subject have not yet been obtained; but without this, the heat of mixing could doubtless be ascertained easily and exactly from the difference between the heat of solution of the mixture, and that of the ingredients. Further, concerning the diminution of free energy associated with the mixing, we must find some method of making the mixed crystals reversibly and isothermally. Strictly speaking, if we could employ here the same methods which were successfully used in the case of liquid mixtures, viz. isothermal distillation or direct sublimation, we should obtain the same formulæ as in the preceding case. But, since the vapour pressure of crystals is far too small for us to hope to measure, we must devise some other way: there is such a desired method in *isothermal solution* and *crystallisation.* By distilling over into the two ingredients, sufficient water isothermally and reversibly, we can at the same time bring the crystals isothermally and reversibly into solution. Then we mix the two solutions. This is a process which can be conducted isothermally and reversibly, as we will see in the theory of dilute solutions (next chapter). Now, if we withdraw both ingredients from the resulting solution by the isothermal distillation of water, we can obtain the mixed crystal. If we know the conditions of solubility of the mixed crystal and its ingredients, as well as the vapour-pressure of their respective solutions, it is possible to calculate the maximal work obtained by mixing, as there is here no difficulty in applying the second law of thermodynamics. Of course, instead of water, we can select any other solvent desired, where we get the same maximal work. The condition for the identity of this work leads to certain conditions of solubility for the different solvents that have recently been tested and verified by E. Sommerfeldt.[3] It appears that the maximal work is of the same order of magnitude as in the case of liquid mixtures, and, therefore, there can be no doubt that the isomorphous mixtures arise by mutual molecular penetration, and not by a variable intercalation of very thin lamellæ, as is sometimes supposed.

[1] *Z. S. phys. Chem.*, **2.** 378 (1888). [2] *Ibid.* 536.
[3] *Neues Jahrb. f. Mineral*, 1900, vol. ii.

The analogy between liquid and isomorphous mixtures is also shown in this, viz. that in the same way that the maximal pressure, and the composition of the vapour emitted by the liquid mixture vary steadily with the relative quantities of the ingredients, so also do the concentration and the composition of the saturated solution of the mixed crystals vary steadily with the relative quantities of the ingredients. In the third book we will return again to this subject, which would take us at the present too far into the region of the *doctrine of affinity*.

Adsorption.—Charcoal shaken with an iodine solution or placed in an atmosphere of iodine vapour condenses appreciable amounts of iodine on its surface; this is known as "adsorption." The quantity adsorbed increases with the partial pressure (gas or osmotic) of the iodine. A definite state of equilibrium is set up; this was shown among others by Chappius,[1] who found that a definite amount of charcoal took up a quantity of carbon dioxide that depends only on the pressure of that gas. The adsorption of dissolved substances by solids has been studied especially by Bemmelen.[2] The thermodynamics of the subject will not be discussed here; but it is easy to see that the heat of adsorption may be calculated from the influence of temperature on equilibrium.

Lagergren[3] has lately put forward remarkable experiments and views on the phenomena of adsorption in aqueous solutions. Here it is probable that the most important influence is the formation of an aqueous layer on the surface of the adsorbing powder, and that this is in a highly compressed state owing to cohesive forces. The heat evolved by wetting insoluble powders would, according to this view, be due to compression of the adsorbed water.

If foreign substances are present in the water their solubility may be greater or less than in ordinary water; greater if the solubility increases with pressure, and *vice versâ*. Hence the dissolved substance will be concentrated or diluted in the water film, *i.e.* positive or negative adsorption will occur. Lagergren actually demonstrated negative adsorption, *e.g.* when sodium chloride solution is shaken with charcoal its concentration is raised. The data so far collected seem to support this view.

1 *Wied. Ann.*, **12.** 161 (1881).

2 *Z. S. phys. Chem.*, **18.** 331 (1895); *Z. S. anorg. Chem.*, **13.** 233 (1896); see also G. C. Schmidt, *Z. S. phys. Chem.*, **15.** 56 (1894); Georgewics u. Lowy, *Wien. Akad.*, **104.** (1895); Walker and Appleyard, *J. Chem. Soc.*, **69.** 1334 (1896).

3 *Bihang till. K. Sv. Vet. Akad. Handl.*, Band **24.** Afd. II. Nos. 4 and 5 (1898).

CHAPTER V

DILUTE SOLUTIONS

General Remarks.—One class of liquid and solid mixtures, which has attracted great attention in recent investigation, because their behaviour is very simple in many respects, is that of the *dilute solutions, i.e.* mixtures containing one ingredient greatly in excess of the others. This ingredient is called the "solvent"; the other, or others, the "dissolved substances."

On studying the relations of dilute solutions more closely, they explain themselves on several grounds. In the first place, the laws of thermodynamics can be applied here with especially good results, which are simple and obvious, and which by reflection clear up old principles, and throw light on new subjects. The method proposed by van't Hoff for the study of solutions, holds good, in a typical way, for the treatment of questions like this. Also, the study of dilute solutions attracts great practical interest from the consideration that most chemical processes which the analyst uses, and also those which claim attention in animal and plant physiology, take place in dilute water solutions. Finally, although most of the results here obtained were obtained independently of the molecular hypothesis, yet the study of the properties of dilute solutions has led to a development of our conceptions regarding molecules, which presents this subject in an entirely new phase.

The investigations on this subject had for a starting-point, the attempt to answer this question, viz., *What is the maximal work to be obtained by the addition of a pure solvent to a solution?* After methods of solving the problem were found, the experimental work gave general results of an unexpected nature, the very simplicity of which required a theoretical explanation; it lay close at hand, and consisted in such an extension of the rule of Avogadro as was previously unexpected.

Osmotic Pressure.—In order to answer the question—What is the maximal work capable of being obtained by mixing a dilute solution with a pure solvent?—we may propose the same way as that employed by us above (p. 114), where we met a similar problem concerning the mixing of two liquids; the same formulæ developed there can be transferred without further change to this case, if we cause the pure solvent to distil over isothermally into the solution. We will return to this a little later, but will next consider a simpler and plainer apparatus which permits the mixing and the separating to be conducted isothermally and reversibly. The fortunate use of this apparatus was the artifice, by means of which van't Hoff (1885) discovered the laws prevailing here almost by one stroke; this artifice, moreover, has already advantageously served us in a similar question (p. 102).

Let us imagine, for example, a sugar solution covered with pure water; as is well known, such a system at once suffers a change, since the sugar begins to wander from the lower to the upper part, *i.e.* from places having a higher to those having a lower concentration; this diffusion process, as this phenomenon is called, does not cease till the concentration has become the same in all parts of the solution. Now let us imagine the solution to be separated from the layer of water above it, by a so-called "semi-permeable" partition, *i.e.* such a partition as allows free passage to the water, but not to the dissolved sugar; it must happen, of course, that the sugar will exert a *pressure* on the partition, which opposes its endeavour to fill the whole solution. If the partition is movable in a cylinder, as shown in Fig. 12, in which L is the solution and W the layer of water above, and which are separated by the partition just described, then we have the desired apparatus to conduct the mixing isothermally and reversibly. When the piston is pressed downwards the sugar in L is compressed, and the water passes over from L to W; on the other hand, if the piston is raised, water passes from W to L, and the sugar solution is correspondingly diluted. But from the fact that the sugar will wander up into the pure water by means of diffusion, if we imagine the piston to be removed, it necessarily follows that there will be exerted a pressure on the piston in the direction of the arrow; and the greater this pressure is, so much greater, of course, is the work which the dissolved sugar can perform in its expansion, *i.e.* as it takes up new solvent material through the "semi-permeable" partition; the amount of this work can be calculated from the product of the pressure, by the volume through which the piston is displaced.

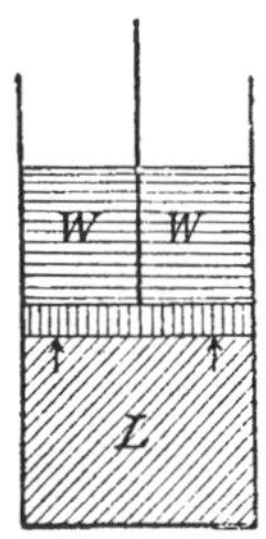

Fig. 12.

We will call the pressure measured by such an apparatus "*the osmotic pressure of solution.*" There is at once called to mind the analogy with the pressure exerted by a gas on its enclosing walls. If we imagine the solvent to be removed, and the space L to be filled with a gas, and the space W to be made vacuous, then we have obviously an entirely analogous experimental arrangement, since the ordinary gas pressure is acting, instead of the osmotic pressure. Moreover, the molecules of a dissolved substance, just the same as the molecules of a gas, have the tendency to occupy *the greatest possible space*; and just as the molecules of a gas can remove from each other to any desired distance, so the same is true of the tendency of the molecules of a dissolved substance, provided that we constantly add more pure solvent material.

The Direct Measurement of the Osmotic Pressure.—The question now arises, whether we are able to realise the conditions of the experiment just described, whether we can construct partitions having the desired properties. Now, as a matter of fact, such is really the case; the semi-permeable partitions do occur in nature ready formed, and can also be artificially prepared, at least in certain cases. According to the researches of M. Traube,[1] *the membranes of precipitation*, which are formed on the surface separating a solution of cupric sulphate from one of potassium ferro-cyanide, and which consist of cupric ferro-cyanide, do possess the property of being permeable for water, but not for many substances soluble in water, as cane sugar, for example. Therefore when Pfeffer [2] dipped, into a weak solution of potassium ferro-cyanide, a porous jar filled with a solution of sugar containing a little copper sulphate, and fitted with an upright tube, there was formed at once on the interior of the porous jar a precipitated membrane. By means of this, the outward passage of the sugar molecules through the cell wall is hindered, but not so the inward passage of the water. As a result there follows the pressure action on the semi-permeable membrane, as above described; but the membrane cannot yield since it is fastened to the resistant porous cell; therefore, according to the principle of action and reaction, there will be exerted on the solution an impulse which will tend to drive it away from the membrane. This impulse will cause the solution to rise in the upright tube from the simultaneous influx of the water; and if the upright tube be long enough, the opposing hydrostatic pressure will increase till it stops the further inflow of water. Of course after the equilibrium is established, this opposing hydrostatic pressure is equal to the osmotic pressure of the solution. But since the pressure amounts to several atmospheres, as we will see later, Pfeffer used, instead of an open manometer, a closed mercury manometer, and thus, besides having the advantage of a quicker adjustment, he also succeeded in avoiding the influx of large quantities of water and the associated changes of concentration which are difficult to control.

If one uses the cupric ferro-cyanide membrane, it must be noticed

[1] *Archiv f. Anatomie und Physiologie*, 1867, p. 87.
[2] *Osmotische Untersuchungen.* Leipzig, 1877.

that it is permeable for many soluble substances, such as saltpetre, hydrochloric acid, and many colouring substances, and therefore it does not satisfy the desired conditions for these substances. As would be expected, the osmotic pressure in such cases is too small. The circumstance that Adie,[1] in repeating the experiments of Pfeffer, did not give sufficient attention to this, although Pfeffer clearly recognised the significance of this point and clearly emphasised it, makes a large part of Adie's results useless, in the calculation of the theory of osmotic pressure. Some other precipitated membranes, such as zinc ferro-cyanide, tannic acid sizing, etc., likewise find application.

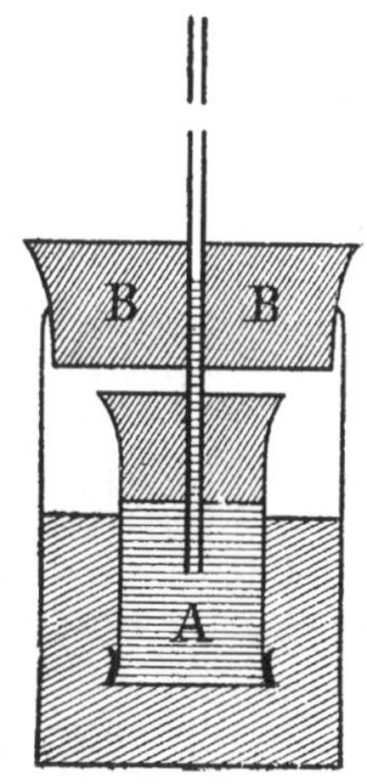

Fig. 13.

The efficiency of a semi-permeable membrane used by the author,[2] depends on the principle of *selective solubility;* thus, for example, ether can diffuse into water since it is partially soluble in water; but a substance dissolved in ether, and which is also insoluble in water, cannot so diffuse into water. In order to give stability to the separating membrane, just as Pfeffer arranged the membrane of cupric ferro-cyanide in a porous cell, so I arranged the water in contact with a plant or animal membrane. By means of the simple apparatus shown in Fig. 13, the efficiency of an osmotic cell can be very easily demonstrated, for lectures for example. A is a glass tube (for example, a piece cut from a test tube), with a bit of pig's bladder, moistened with lukewarm water, tied over the lower end; it is closed at the upper end with a well-fitting cork, provided with one hole bored through. After filling the cell A completely, with ether containing a considerable amount of benzene, the opening in the cork is closed with a narrow upright tube which fits tight; the cell is then dipped into a wider glass filled with ether. The cork B, which is not air-tight, serves both to hinder the evaporation of the outer ether, and also to hold the cell in place. Both the solution and the solvent must be previously saturated with water, in order that the solvent action of the ether may not injure the delicate partition membrane. Also, it is well to add a colouring substance which is insoluble in water, to the ether in the inner cell, so that the action of the experiment will be visible, and one can also be thus assured that the membrane is tight. After the cell has been left to itself for some time, one will notice the rising of the ether column from the osmotic pressure of the benzene; this usually amounts in an hour to more than a decimetre.

A very elegant method for the investigation of solutions having the same osmotic pressure, viz. the so-called *is-osmotic solutions,* in an

[1] *Jour. Chem. Soc.*, 1891, p. 344.
[2] Nernst, *Zeitschr. physik. Chem.*, **6**, 37 (1890).

optical [visual] way is described by Tammann,[1] who placed a drop of a solution of potassium ferro-cyanide in a solution of copper sulphate. The drop is at once surrounded with a membrane of precipitation; accordingly as the osmotic pressure is greater in the inner or the outer solution, the cell will expand or contract, with the inflow or outflow of water. Tammann observed the osmotic stream produced by the changes of concentration by means of a so-called *schlieren apparatus*.[2] When the *schlieren*[3] (*i.e.* the heterogeneous layers) disappear, the osmotic pressure is the same outside and inside. If one adds to both solutions something foreign to both substances producing the membrane, and which does not diffuse through the membrane, then the osmotic action remains in equilibrium. These phenomena can be beautifully used for demonstration. If one adds, by means of a capillary pipette, a drop of a strong solution of potassium ferro-cyanide to a moderately strong solution of copper sulphate, one can see with the naked eye that a *schliere*[4] (*i e.* thin layer) of concentrated solution of copper sulphate flows downwards from the cell; this is a proof that the inner solution continually extracts water from the outer one. This very instructive phenomenon can be shown to a large audience by the aid of a sciopticon.

Semi-permeable partitions and the associated action of the osmotic pressure play a very important rôle in the economy of living nature, and it should be emphasised that the investigations in plant physiology, and among others, those of Traube, of de Vries, and especially of Pfeffer, made possible the more detailed study of osmotic pressure; and this forms the basis of the modern theory of solutions. Thus the living protoplasmic layer, which entirely surrounds the surface of the cell-sap of plants, is very permeable for the solvent, water, but is an almost completely impermeable membrane[5] for the substances dissolved in the cell-sap, such as glucose, calcium, and potassium malates, etc., as well as some inorganic salts. If, therefore, some plant cells, taken for example from the leafy part of *Tradescantia discolor*, are sprinkled with a water solution having a greater osmotic pressure than that exerted by the substances dissolved in the cell-sap in the protoplasts, then the latter will contract, *i.e. plasmolysis* is said to occur: if, on the other hand, a smaller osmotic pressure exists in the outer solution, then the protoplasmic sheath expands as far as the cell wall permits. Thus, by means of microscopic observation, one can prepare solutions of selected substances which are *is-osmotic* (iso-tonic) with the cell-sap. Thus osmotic action cannot only be recognised in the red blood corpuscles,[6] and

[1] *Wied. Ann.*, **34.** 229 (1888).

[2] This term, for which I find no concise English equivalent, is in common use in Germany to denote a delicate apparatus of Topler used to detect small differences in the refractive power of the different layers ("schlieren") of heterogeneous media.—Tr.

[3] *Ibid.*

[4] *Ibid.*

[5] See especially the treatise written for those who are not professional botanists by de Vries, *Zeitschr. physik. Chem.*, **2.** 414 (1888).

[6] Hamburger, *Zeitschr. physik. Chem.*, **6.** 319 (1890). See also Löb, *ibid.*, **14.** 424 (1894); Koppe, *ibid.*, **16.** 261; **17.** 552 (1895); Hedin, **17.** 164; **21.** 272 (1896).

also very recently in bacteria cells,[1] and nerve cells, but these can also be applied to the study of osmotic solutions. Of course these methods are debarred when the dissolved substances either exert a specific poisonous action on the protoplasmic sheath of the cells in question, or else diffuse through them. It is not without interest to observe that the pressure in animal and plant cells under the most diverse conditions, amounts to from four to five atmospheres; and that sometimes it is even four times as great in those protoplasts which serve as the storehouse of dissolved substances held in reserve; examples of the latter are the cell contents of beets, and also the cells of bacteria. It is evident how these bacteria are supported in their destructive action by their abnormally high pressure.[2]

In concluding this sketch, I give the results obtained by Pfeffer, in measuring the osmotic pressure in a water solution of cane sugar, which was the substance most thoroughly investigated by him. The cupric ferro-cyanide membrane used by Pfeffer, satisfied most completely the conditions of *impermeability* for this substance.

OSMOTIC PRESSURE OF A ONE PER CENT WATER SOLUTION OF CANE SUGAR AT DIFFERENT TEMPERATURES

t	Pressure.		Diff.
	Obs.	Calc.	
6·8	0·664 Atm.	0·665 Atm.	+0·001
13·7	0·691 ,,	0·681 ,,	−0·010
14·2	0·671 ,,	0·682 ,,	+0·011
15·5	0·684 ,,	0·686 ,,	+0·002
22·0	0·721 ,,	0·701 ,,	−0·020
32·0	0·716 ,,	0·725 ,,	+0·009
36·0	0·746 ,,	0·735 ,,	−0·011

The figures in the third column are calculated from the formula

$$P = 0{\cdot}649(1 + 0{\cdot}00367t) \text{ atm.},$$

which is well suited to express the results of the investigation, as is shown by the smallness and irregularity of the differences given in the last column. The measurements arranged *for varying concentration*, can be also conversely calculated from the formula

$$P = n \, . \, 0{\cdot}649(1 + 0{\cdot}00367t),$$

in which n denotes the per cent strength of the solution, *i.e.* the number of g. sugar in 100 g. of the solution; thus the pressure of a 4 per cent solution, measured at 13·7°, amounted to 2·74 atm., while according

[1] Wladimiroff, *Zeitschr. physik. Chem.*, **7.** 524 (1891).
[2] See Tammann, *Zeitschr. physik. Chem.*, **8.** 685 (1891).

to the preceding formula it should be 2·73 atm. It is obvious that we have to do here with pressures of considerable magnitude.

Indirect Methods for the Measurement of Osmotic Pressure.—In most cases where one tries to measure osmotic pressures directly, he meets very great experimental difficulties, which depend upon the successful accomplishment of making a semi-permeable membrane of sufficient stability and osmotic activity. Fortunately the measurement in almost all cases can be easily and certainly obtained by *indirect methods.*

The indirect methods in general depend on the measurement of the expenditure of work which is necessary to separate the dissolved substance from the solvent. Now, according to what was said above, since the osmotic pressure is a direct measure of this expenditure of work, it is evident that if we know one we also know the other.

Of the many methods which can be employed for the reversible separation of solvent and dissolved substance, we will consider the following, viz. :—

1. Separation by evaporation.
2. Separation by selected solubility.
3. Separation by crystallisation.

But since each of these methods can be used in a double sense, *i.e.* we can remove either the *solvent*, or the *dissolved substance* from the solvent, we have the six following methods for the indirect measurement of osmotic pressure :—

A. The Separation of the Pure Solvent from the Solution

1. Separation by Evaporation.—We can see at once without further remark, that the partial pressure of the solvent above a solution, must always be less than that over the pure solvent, at the same temperature. Now, let us imagine a solution and a solvent at the same temperature, to be separated only by a semi-permeable membrane; then the solution will add to itself at the expense of the solvent. If we assume that the partial pressure of the saturated vapour of the solvent is greater over the solution, then there must necessarily occur a process of isothermal distillation, whereby some of the solvent would be forced back from the solution to the pure solvent. In that case we would have a perpetually automatic cyclic process, *i.e.* a *perpetuum mobile*, which would perform work at the expense of the heat of the environment, which is contrary to the second law of thermodynamics (p. 15).

In what follows we will limit ourselves to solutions of non-volatile substances, *i.e.* solutions the vapour pressure of which is equal to the partial pressure of the solvent, as described above. The diminution of pressure experienced by a solvent, on dissolving a small quantity of a foreign substance, is obtained as follows :[1]—

[1] See Gouy and Chaperon, *Ann. chim. phys.* [6], **13.** 124 (1888); and Arrhenius, *Zeitschr. physik. Chem.*, **3.** 115 (1889).

An upright tube for osmotic action holds the solution L (Fig. 14), and is dipped into the solvent, W, water for example. Water will enter through the semi-permeable membrane A, which closes the upright tube below, till the liquid has mounted to the height H above the outer level, corresponding to the osmotic pressure prevailing in the solution. The whole system must be closed air-tight, and must be maintained at the same temperature at all points, reckoned in absolute temperature T. Then between the vapour pressure p of the water, and the vapour pressure p′ of the solution, there must exist the relation that p′, increased by the vapour column acting on W of the height H, must be equal to p. This relation merely states that the system is in equilibrium. For if we assume that p′ is greater than that corresponding to this relation, then water would constantly distil over from the upright tube into W, to be again transported by osmotic pressure into L; in short, the water would constantly traverse a cyclic process, *i.e.* the system would represent a *perpetuum mobile*, which would be able to perform any desired amount of external work at the expense of the heat of the environment, and the existence of which would contradict the second law of thermodynamics.

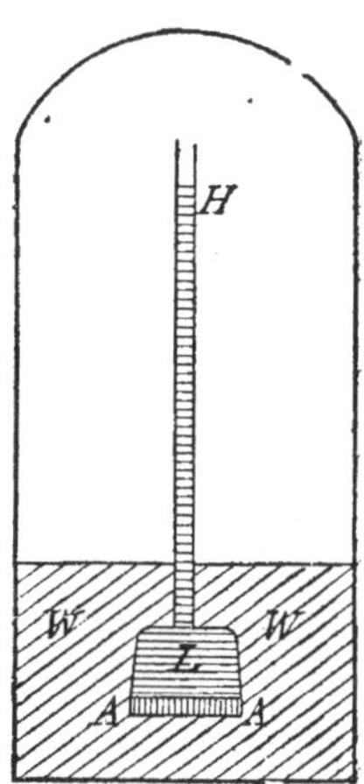

Fig. 14.

Also, conversely, if p′ were smaller than would correspond to the preceding relation, then water would distil over from W into the upright tube, would pass below from L into W, and again around in this reverse cycle. Equilibrium, therefore, can exist only when the excess of pressure of p over p′ is compensated by the hydrostatic pressure of a vapour column, equal in height to the difference between the levels.[1]

The pressure of this vapour column can be easily calculated. If we denote the molecular weight of the solvent by M, then from the gas formula, pv = 0·0821T, we obtain the specific weight of the vapour, compared with that of water as unity, and since there are M g. of the vapour in v liters, the weight of one c.c. in g. amounts to

$$\frac{M_0}{1000v} = \frac{M_0 p}{0{\cdot}0821T\ 1000}.$$

Instead of p we can introduce p′, since, according to the proviso that the solution is very dilute, their difference can be neglected. In the same way, considering that, although strictly speaking, the density of the vapour column varies along the upright tube, yet the hydrostatic pressure can be simply expressed by the formula

[1] It is of course noticed that the *hydrostatic* pressure of the liquid column H is a measure of the *osmotic* pressure; and that the pressure of a *vapour* column of height H is a measure of the *difference in the vapour pressure*. Tr.

$$\frac{HM_0p'}{0{\cdot}0821T\,.\,1000\,.\,76\sigma}\text{ atm.,}$$

where H is measured in cm., and σ denotes the specific gravity of mercury. The osmotic pressure of the solution corresponds to the pressure of the elevated column of liquid H. Therefore between H, and the osmotic pressure P expressed in atmospheres, there exists the relation

$$P=\frac{HS}{76\sigma},$$

where S denotes the specific gravity of the solution, or also that of the slightly differing solvent.

If we eliminate H in the expression for the hydrostatic pressure of the vapour column, by introducing

$$\frac{P76\sigma}{S}$$

for H, then we obtain the following expression for the relation of p and p′ (developed as above from the conditions of equilibrium), viz.,

$$p=p'+\frac{PM_0p'}{1000S\;0{\cdot}0821T},$$

and thus obtain the expression sought, for the osmotic pressure, viz.,

$$P=\frac{p-p'}{p'}\;\frac{0{\cdot}0821T\;1000S}{M_0}\text{atm.,} \qquad (1)$$

which was found by van't Hoff (1886),[1] by a method not differing in principle from the preceding.

The vapour pressure of a solution, containing 2·47 g. of ethyl benzoate to 100 g. of benzene, was found to be 742·60 mm. at 80°, while that of pure benzene at the same temperature amounted to 751·86 mm.;[2] the molecular weight M of benzene is 78; its specific weight at the preceding temperature is 0·8149; therefore the osmotic pressure of the aforesaid solution is calculated to be

$$P=\frac{9{\cdot}26}{742{\cdot}6}\;\frac{0{\cdot}0821\,.\,(273+80)\,.\,814{\cdot}9}{78}=3{\cdot}78\text{ atm.}$$

The simple yet obvious derivation given above for the fundamental relation between osmotic pressure and vapour tension is not sufficiently rigorous, because it assumes that $p-p'$ is only a small fraction (say the hundredth part) of p, a condition satisfied only in the case of very dilute solutions. We obtain a more exact formula in the following way.

By means of the apparatus figured on p. 131 (Fig. 12), we remove

[1] Van't Hoff, "Lois de l'équilibre chimique dans l'état dilué ou dissous." Stockholm, 1886. An abstract of this is to be found in *Zeitschr. phys. Chem.*, **1.** 481 (1887).

[2] Beckmann, *Zeitschr. phys. Chem.*, **6.** 439 (1890).

a small quantity of the solvent from the solution. The expenditure of work necessary for this amounts to Pdv, in which dv is the volume through which the piston is lowered. Let the quantity of solvent removed from the solution be dx g.-mol. Now we can also remove the same quantity of solvent from the solution by means of isothermal distillation, and denoting by p and p′ the respective vapour tensions of the pure solvent, and of the solution, at the temperature T, the expenditure of work necessary, according to the equations on p. 116, is

$$dxRT \ln \frac{p}{p'}.$$

Now since the work must be the same in these two isothermal and reversible ways, it follows that

$$P = \frac{dx}{dv} RT \ln \frac{p}{p'}, \quad . \quad . \quad . \quad . \quad (2)$$

Now if, on adding dx g.-mol. of the pure solvent to a solution, there does not occur either contraction or dilation, then dv is obviously equal to dx g.-mol., *i.e.* it follows that

$$dv = \frac{M_0}{S} dx \quad . \quad . \quad . \quad . \quad (3)$$

if M and S denote respectively the molecular weight, and the specific gravity of the solvent.

From (2) and (3) we find

$$P = \frac{S}{M_0} RT \ln \frac{p}{p'} \quad . \quad . \quad . \quad . \quad (4)$$

In order to obtain P in atmospheres, we must express the volume of 1 g.-mol. of the solvent in litres, and must write 0·0821 for R, when it follows that

$$P = \frac{0{\cdot}0821T \; 1000S}{M_0} \ln \frac{p}{p'} \text{ atm.} \quad . \quad . \quad (5)$$

For $\log \frac{p}{p'}$ in equation (5) we may write

$$\ln \left(1 + \frac{p - p'}{p}\right) = \frac{p - p'}{p'},$$

since $\frac{p - p'}{p'}$ is very small, cf. with 1. Equation (5) gives, with the former data for the pressure of ethyl benzoate in benzene

$$P = 3{\cdot}76 \text{ atms.},$$

while the less exact formula gives 3·78 atm.

It is recommended in practice to determine the *boiling-point* instead of the vapour pressure. Instead of the lowering of the vapour pressure $p - p'$, one can measure much more simply and accurately the *elevation of the boiling-point* produced by the addition of soluble substances. If the determination is conducted at the atmospheric pressure B, then the vapour pressure of the solution at its boiling-point will be $p' = B$. The vapour pressure p of the pure solvent at the temperature of the aforesaid B, is increased by the amount necessary to raise the temperature of the boiling-point of the solvent to that of the solution; this can be obtained from the vapour-pressure tables made by Regnault and others with satisfactory accuracy for the most available solvents.

For small elevations of the boiling-point t, we can calculate p, by means of the formula of Clausius on p. 62, viz.,

$$\frac{d \ln p}{dT} = \frac{\lambda}{RT^2},$$

which, when integrated, becomes

$$\ln p = -\frac{\lambda}{RT} + \text{const.}$$

Now, at the boiling-point T_0 of the pure solvent, we have

$$\ln B = -\frac{\lambda}{RT_0} + \text{const.},$$

and therefore

$$\ln \frac{p}{p'} = \ln \frac{p}{B} = \frac{\lambda}{R}\left[\frac{1}{T_0} - \frac{1}{T}\right].$$

This, introduced into formula (4), gives

$$P \frac{\lambda}{M_0} ST\left[\frac{1}{T_0} - \frac{1}{T}\right].$$

This, simplified, becomes

$$P = \frac{Slt}{T} \text{ or } P = \frac{1000S \,.\, l}{24{\cdot}25} \cdot \frac{t}{T} \text{ atm.} \quad . \quad . \quad (6)$$

where

$$l = \frac{\lambda}{M_0}$$

denotes the heat of evaporation of 1 g. of the solvent, reduced to litre-atmospheres by 24·25, the factor for 1 litre-atmosphere in g.-cal. (p. 12); and $t = T - T_0$ denotes the elevation of the boiling-point of the solution. As before S denotes the specific gravity of the solvent. The factor 1000 is due to expressing this volume of a mol. in litres. This formula does not depend on the condition that the vapour of the solvent must follow the laws of gases; but, on the other hand, it takes no account of the slight change of the heat of vaporisation with the temperature; yet it can be used for elevations of boiling-point of 5° – 10°, within 1 per cent of the correct value of the osmotic pressure prevailing at the temperature $T_0 + t$.

Thus the elevation experienced by the boiling-point of benzene, on the addition of 2·47 g. of ethyl benzoate to 100 g. of the solvent, is 0·403°. If, according to Schiff (p. 61), we put the heat of vaporisation of benzene as equal to 94·4 cal., it follows that

$$P = \frac{814{\cdot}9 \,.\, 94{\cdot}4}{24{\cdot}25} \cdot \frac{0{\cdot}403}{(273+80)} = 3{\cdot}62 \text{ atm.}$$

The difference, as compared with the value 3·76 found on p. 141, can be attributed to a slight uncertainty in determining the heat of vaporisation.

The osmotic pressure of *a water solution* at the boiling-point 100 + t°, may be estimated as equal to

$$P = \frac{0{\cdot}959 \,.\, 536{\cdot}4}{24{\cdot}25} \cdot \frac{t}{273+100},$$

or

$$P = 56{\cdot}8t \text{ atm.}$$

2. Separation by Selected Solubility.—The far-reaching analogy existing between the process of solution and that of evaporation, and to which repeated reference must be made, extends thus far, viz., that in the same way that the vapour pressure of a solvent A is depressed by the addition of a foreign substance B, so also the solubility of B in A is diminished by dissolving a third substance in B; and this happens according to the same laws. Thus, if we mix together two liquids, as ether and water, which are only slightly soluble in each other, and if we denote by L the solubility of pure ether in water at the temperature T, and by L′ the solubility in water of ether containing a foreign substance at the same temperature, then L′ must always be smaller than L; and we may derive the relation for the osmotic pressure P, prevailing for the foreign substance dissolved in the ether, as being

$$P = \frac{L - L'}{L'} \frac{0{\cdot}0819T \,.\, 1000S}{M_0} \text{ atm.} \quad . \quad . \quad (7)$$

where, as above, S and M denote respectively, the specific gravity and the molecular weight of the ether; this relation is completely analogous to the relation between the lowering of vapour pressure and osmotic pressure, equation (5).[1]

The proof of equation (7) is most easily arrived at from the observation that the solubility of ether in water must be the same as that of ether vapour. But as the latter, according to Henry's law, must have a solubility proportional to its partial pressure

$$L : L' = p : p'$$

equation (7) is a necessary consequence of this and (5).

[1] Nernst, *Zeitschr. physik. Chem.*, **6**. 16 (1890).

3. Separation by Crystallising (Freezing Out).—From the law that solutions of various substances in the same solvent are is-osmotic when they have the same vapour pressure, it follows directly that the same osmotic pressure must prevail in solutions of the solvent which have the same *freezing-point.* For inasmuch as the freezing-point is that temperature[1] at which the solid solvent (ice) and the solution are capable of existing together, it must also be the point at which the curves of the vapour pressures of the solution and of the solid solvent cut each other, *i.e., where both have the same vapour pressure.* For if we suppose, for example, that the vapour pressure of ice and that of the water solution were different at the freezing-point, then a process of isothermal distillation would commence which would not cease till equilibrium should have been established; but this could not happen till both the vapours of the solvent should have the same density. Solutions of equal freezing-point have therefore equal vapour pressures, and are isotonic.

These relations are shown in Fig. 15, in which, as in Fig. 4, the vapour pressure curves of solid and liquid are shown, but also—dotted—the vapour pressure curve of the solution. The latter must lie below the curve for the solvent, according to the foregoing argument; hence its intersection T, with the curve for the solid, must lie below T_0. $T_0 - T = t$ is the lowering of the freezing-point.

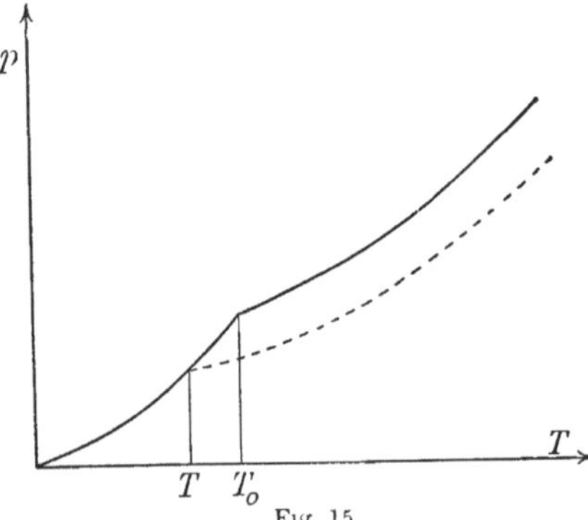

Fig. 15.

The relation existing between the lowering of the freezing-point and of the osmotic pressure of a solution, is found by combining the equation for the molecular heat of sublimation σ (p. 73), with that for the heat of vaporisation λ (p. 62), viz.,

$$\sigma = RT^2 \frac{d \ln p'}{dT}, \quad \text{and } \lambda = RT^2 \frac{d \ln p}{dT}; \quad . \quad . \quad (8)$$

in which p′ denotes the vapour pressure of the solidified solvent, equivalent to that of the solution according to the aforesaid law; and where p denotes the vapour pressure of the pure (undercooled) *liquid* solvent, both values being referred to the freezing-point of the solution: by integration we obtain

$$\ln p' = -\frac{\sigma}{RT} + C'; \quad \ln p = -\frac{\lambda}{RT} + C.$$

[1] It is assumed in all this argument that pure solvent crystallises out of the solution, as is usually the case. But see "Solid Solutions," p. 170.

Now, according to p. 74, at the freezing-point of the pure solvent, the temperature of which is T_0, the vapour pressures of the solid and of the liquid solvent are the same, *i.e.*, at T_0, we have $p' = p = p_0$, and therefore

$$\ln p_0 = -\frac{\sigma}{RT_0} + C'; \quad \ln p_0 = -\frac{\lambda}{RT_0} + C.$$

If we eliminate the integration constants C' and C respectively, by the subtraction of the equations standing under each other in the same column, we obtain

$$\ln \frac{p'}{p_0} = -\frac{\sigma}{R}\left[\frac{1}{T} - \frac{1}{T_0}\right]; \quad \ln \frac{p}{p_0} = -\frac{\lambda}{R}\left[\frac{1}{T} - \frac{1}{T_0}\right];$$

and by the subtraction of these last two equations we obtain

$$\ln \frac{p}{p'} = \frac{\sigma - \lambda}{R}\left[\frac{1}{T} - \frac{1}{T_0}\right]; \quad . \quad . \quad . \quad (9)$$

thus we obtain a formula which affords us the relation between the lowering of the vapour pressure and of the freezing-point of a solution, as well as *the molecular heat of fusion*, $\sigma - \lambda$, of the solution; this was found by Guldberg[1] as early as 1870; R here amounts to 1·99 if we express the heat of fusion in g.-cal.

In formula (4), p. 139, putting $\rho = \sigma - \lambda$ we have

$$P = \frac{\rho}{M_0} ST\left[\frac{1}{T} - \frac{1}{T_0}\right],$$

or simply as in equation (6), p. 140

$$P = \frac{Swt}{T_0} \quad \text{or} \quad P = \frac{1000\,S\,.\,w}{24{\cdot}25}\,\frac{t}{T_0}\ \text{atm.} \quad . \quad . \quad (10)$$

In this formula w denotes $\frac{\sigma - \lambda}{M}$, the heat of fusion of 1 g. of the solvent expressed in g.-cal.; T_0 is the fusing-point; S is its specific gravity; and t denotes $T_0 - T$, the lowering of the freezing-point.

Thus by the addition of 2·47 g. of ethyl benzoate to 100 g. of benzene, the freezing-point 5·5°, of the solvent is lowered about 0·840°: its heat of fusion amounts to 30·08 g.-cal., and its specific gravity at the freezing-point is 0·8875. The osmotic pressure of the solution is thus

$$P = \frac{887{\cdot}5\,.\,30{\cdot}08}{24{\cdot}25} \cdot \frac{0{\cdot}840}{273 + 5{\cdot}5} = 3{\cdot}324.$$

A similar value for water is

$$\frac{1000S\,.\,w}{24{\cdot}25\,T_0} = \frac{1000 \times 79{\cdot}6}{24{\cdot}25 \times 273} = 12{\cdot}03,$$

[1] *Compt. rend.*, **70**, 1349 (1870).

and therefore the pressure of a water solution freezing at $T_0 - t$, is $P = 12{\cdot}03\ t$ atm.

According to the very accordant measurements of Abegg[1]; Ponsot[2]; and Raoult[3]; the freezing-point of a 1 per cent sugar solution is about $-0{\cdot}0546°$. Its osmotic pressure is therefore 0·657 atmos., in good agreement with Pfeffer's direct measurement of 0·649.

Moreover, we observe from the preceding formula that the thousandth part of a degree of temperature, a quantity quite difficult to measure, corresponds to a pressure of 0·012 atm., *i.e.* about 9·1 mm. of mercury; this latter value could be determined to one part in a thousand, provided one had a semi-permeable membrane sufficiently sure and rapid in its action.

On account of the simplicity of determining the lowering of the freezing-point, in practice the method above described finds application almost always, in preference to the measurement of the osmotic pressure: its experimental treatment will be considered in the chapter on the determination of the molecular weight.

B. The Separation of the Dissolved Substance from the Solution

1. Separation of Evaporation.—We will consider two solutions of the same substance in the same solvent, of very nearly the same degree of concentration. Let P be the osmotic pressure of the substance dissolved in solution I.; let p be its vapour pressure over the solution; let V be the volume of the solution containing 1 g.-mol. of the substance dissolved; and let v be the volume assumed by the g.-mol. in the gaseous state at the pressure p. Let the corresponding magnitudes in solution II., be $P + dP$, $p + dp$, $V - dV$, and $v - dv$, respectively. We will now conduct the following cyclic process at the constant temperature T. We will first set free from solution I., 1 g.-mol of the dissolved substance, whereby the work PV will be performed *against* the osmotic pressure, and the work pv *by* the vapour pressure p. Then we will compress the g.-mol now existing in the gaseous state from v to $v - dv$, whereby there will be performed the work pdv. Next we will bring the g.-mol into solution II., whereby there must be applied the work $(p + dp)(v - dv)$ *against* the gas pressure, and the osmotic pressure *performs* the work $(P + dp)(V - dV)$. Finally, we bring the g.-mol from solution II. back again to solution I., whereby the osmotic pressure expends the work PdV, and then all is reduced to the original condition.

Now, since the sum of the work *expended on* the system, diminished by the work *performed by* the system, in a reversible, isothermal, cyclic process must be equal to zero, we have the equation,

$$PV - pv + pdv + (p + dp)(v - dv) - (P + dP)(V - dV) - PdV = 0,$$

[1] *Z. S. Phys. Chem.*, **20.** 221, 1896. [2] *Bull. Soc. Chim.* (3), **17.** 395, 1897.
[3] *Z. S. Phys. Chem.*, **27.** 617, 1898.

or simplifying it and neglecting the small magnitudes of the second order,

$$vdp = VdP \quad . \quad . \quad . \quad . \quad (11)$$

2. Separation by Selective Solubility.—We bring together two liquids which are only slightly soluble in each other, as, for example, carbon disulphide and water; we then dissolve a third substance, as iodine, which is divided between the two solvents. We will denote by P_1 and V_1, the osmotic pressure and the volume respectively, assumed by 1 g.-mol of the dissolved substance in one solvent, and by P_2 and V_2, the corresponding values in the second solvent. Then by means of a cyclic process completely analogous to that just described above, we can at once derive the relation

$$V_1dP_1 = V_2dP_2 \quad . \quad . \quad . \quad . \quad (12)$$

3. Separation by Crystallisation.—The method employed altogether most frequently in practice to separate a dissolved substance from its solution, consists in crystallising it out from its saturated solution, by means of a suitable change of temperature.

This process reminds us of the condensation of a substance from its saturated vapour, and we also easily see that the analogy between the processes of solution and of vaporisation is not merely external, but that it is really a deep-seated one. For if we vaporise a solid or a liquid body, its molecules are driven by a expansive force, called the "vapour pressure," into a space where they arrive under a certain pressure, viz., the pressure of the vapour saturated for the temperature in question. The conditions are similar when a solid goes into solution; in this case also the molecules are driven by a certain expansive force, called the *solution pressure*, into a space where they arrive under a definite pressure, viz., *the osmotic pressure* of the saturated solution.

We have developed on p. 62 a formula of Clausius, giving a simple relation between the change of vapour pressure with temperature, the increase of volume from vaporisation, and the heat of vaporisation. In a precisely similar way, by effecting the solution of a substance—instead of its vaporisation into a vacuum—and by making the process reversible by means of a semi-permeable partition, we can derive a relation between the change of "solution pressure" with the temperature, the increase of volume from solution, and the heat of solution of a solid substance: thus we obtain directly

$$L = T\frac{dP}{dT}(V - v')$$

Here L denotes the amount of *heat absorbed* when 1 g.-mol of a solid substance goes into solution, the osmotic pressure of the saturated solution being kept constant for the temperature T; v′ denotes the volume occupied by 1 g.-mol of the aforesaid substance before going into solution, and V the volume assumed by 1 g.-mol of the same substance in the saturated solution (van't Hoff).[1]

In case the substance going into solution is a liquid, there arises a complication from the solution of the solvent itself in this substance; but this can be overcome by easy calculation.

The Law of Osmotic Pressure.—The law of osmotic pressure naturally must consider these questions, viz., how is it conditioned (1) by the volume of the solution, *i.e.* the concentration? (2) how by he temperature? (3) how by the nature of the substance dissolved? (4) how by the nature of the solvent? The answer to these questions, which of course can be obtained only by experiment, *i.e.* by the direct or indirect measurement of the osmotic pressure in the most numerous and various conditions possible, is very simple, and has led to the following remarkable statement, viz. *the osmotic pressure is independent of the nature of the solvent, and in general obeys the laws of gases* (van't Hoff). The various proofs requisite for establishing this law will be given in the following sections.

Osmotic Pressure and Concentration.—Pfeffer found that the pressure of solutions of cane sugar was proportional to the concentration (p. 144); the rules, according to which the lowering of vapour tension (law of Wüllner), and the lowering of the freezing-point (law of Blagden), vary with the concentration of the dissolved substances, have long been known; these rules, in the light of the formulæ developed on pages 140 and 144, mean that the same holds good for the osmotic pressure. If we denote by c the number of g.-mol dissolved per litre, then

$$P = c \times \text{const.};$$

and we notice that

$$c = \frac{1}{V},$$

if V denotes the volume of the solution containing 1 g.-mol of the dissolved substance; then

$$PV = \text{constant};$$

i.e. the Boyle-Mariotte law holds good for the osmotic pressure.

Osmotic Pressure and Temperature.—Pfeffer's measurements of the pressure of solutions of cane sugar can be well represented by the formula

$$P = 0{\cdot}649(1 + 0{\cdot}00367t).$$

But the temperature coefficient 0·00367 is the same as in the case

[1] See also van Deventer and van der Stadt, *Zeitschr. phys. Chem.*, **9.** 43 (1891).

of gases, *i.e. the osmotic pressure is proportional to the absolute temperature.* This result will be established by a simple law which was discovered by Babo as early as 1848, and which has been verified repeatedly by experiment in recent times; viz. *that the relative lowering of the vapour pressure,* $\frac{p - p'}{p}$, *of a dilute solution, and also the quotient,* $\frac{p}{p'}$, *is independent of the temperature.* For if we observe in the equation

$$P = \frac{0{\cdot}0821\ T\ 1000\ S}{M_0} \ln \frac{p}{p'}$$

that S is inversely proportional to the volume V, of the quantity of solvent containing a definite amount of the dissolved substance, say 1 g.-mol, and if we also notice that according to Babo's law, the expression

$$\frac{0{\cdot}0821}{M_0} \ln \frac{p}{p'}$$

is independent of the temperature, then it also follows in this way that

$$PV = T \times \text{const.};$$

i.e. the law of Gay-Lussac holds good for the osmotic pressure.

This law will later be established more rigorously by thermodynamic considerations; but remark should here be made that from numerous coincident determinations of the elevation of the boiling-point and of the lowering of the freezing-point of dilute solutions, there follows this law, viz., that the osmotic pressure of solutions of similar dilutions at these two points referred to the same *spatial* concentration,[1] have the same ratio as the absolute temperatures of the boiling-point and melting-point.

Osmotic Pressure and the Heat of Dilution.—When one adds more solvent to a dilute solution, *there is no heat developed, neither is external work performed.* The total energy remains unchanged in this process. But if one conducts the operation in such a way that external work is obtained at the same time, then the solution must cool itself an equivalent amount, exactly as was the case with gases (p. 49). And if, on the other hand, one compresses the solution by the osmotic apparatus figured on p. 131 (Fig. 12), then the work applied will reappear as heat, according to the law of the conservation of energy. We may express this law also in this way, viz. that the content of energy of a dissolved substance is independent of the volume of the solution.

If we displace an osmotic piston through the volume v and apply the equation (*e*) on p. 23, viz.

[1] Such a reduction is necessary, since the expansion of the solution from heat increases the volume of the substance dissolved.

$$A - U = T\frac{dA}{dT},$$

when

$$U = 0,$$

then

$$A = T\frac{dA}{dT},$$

or integrated, we have

$$A = T \times \text{const.};$$

i.e. A is proportional to the absolute temperature. Now $A = Pv$, when v is the volume of 1 g.-mol.; then the osmotic pressure must also be proportional to the absolute temperature, conversely calculated for constant volume. We found this latter law established by experience. Now we see, moreover, since its validity is necessary and sufficient, that the *heat of dilution is equal to zero* (van't Hoff, 1885).

Quite generally, in the case of concentrated, and sometimes also in the case of dilute solutions, where the process of dilution is associated with the formation of new molecular complexes, or with the decomposition of those already occurring, the heat of dilution may have a positive or a negative value. In such cases, observation shows no ratio between the osmotic pressure and the absolute temperature (see section on ideal concentrated solutions).

The Osmotic Pressure and the Nature of the Substance dissolved.—In 1883, on the basis of a very extended series of observations, Raoult advanced the theorem that by dissolving equimolecular quantities of the most various substances in the same solvent, its freezing-point is lowered the same amount. Soon after, 1887, the same investigator saw that the same holds good for the lowering of the vapour pressure, and therefore of course for the raising of the boiling-point. But solutions having the same freezing-point and the same vapour pressure, have the same osmotic pressure; and thus the rules of Raoult can be condensed into the statement that *solutions having the same osmotic pressure can be obtained by dissolving equi-molecular quantities of the most various substances in the same solvent.*

Osmotic Pressure and Gas Pressure.—It has been shown in the preceding sections on the basis of many experimental results, that the osmotic pressure of dissolved substances depends on the volume and on the temperature, in the same way that the gas pressure does; further, that in both cases the amount of pressure is determined by the number of molecules, dissolved or gaseous, contained in unit volume; and, finally, that the analogy between the behaviour of dissolved and gasified substances is expressed in the theorem, that in both cases the content of energy at constant temperature is independent of the volume occupied. Now it is but a step to identify the osmotic pressure in its absolute amount, with the gas pressure which would be observed under similar relations.

From the observations conducted by Pfeffer on water solutions of cane sugar, and satisfying the formula (p. 135), viz.,

$$P = n\ 0{\cdot}649(1 + 0{\cdot}00366t)\ \text{atm.},$$

the pressure of a 1 per cent solution at 0° is estimated to be 0·649 atm. The volume of a 100 g.-solution at 0° is 99·7 c.c., and the volume, therefore, containing 1 g.-mol of cane sugar ($C_{12}H_{22}O_{11} = 342$), is $342 \times 99{\cdot}7$ c.c. = 34·1 litres. But the pressure in a volume v, containing 1 g.-mol of a gas, as calculated from the gas laws (p. 38), is

$$p = \frac{RT}{v} = \frac{0{\cdot}0821 \,.\, 273}{34{\cdot}1} = 0{\cdot}656\ \text{atm.}$$

This value coincides in a striking way with that found directly (viz. 0·649 atm.); *i.e. the osmotic pressure is exactly the same as the gas pressure which would be observed if the solvent were removed, and the dissolved substance were left filling the same space in the gaseous state at the same temperature.*

Thus the same equation of condition holds good for the dissolved cane sugar, as for a gas, viz.,

$$PV = RT = 0{\cdot}0821\ T\ \text{litre-atm.}, \qquad (13)$$

if P denotes the osmotic pressure of a solution containing 1 g.-mol of the substance dissolved in V litres, and measured in atmospheres, at the absolute temperature T.

It is proven by the numerous indirect measurements of the osmotic pressure, that this result has a universal value. We will next proceed to discuss a purely empirical law discovered by Raoult,[1] according to which *the relative lowering of vapour pressure experienced by a solvent on dissolving a foreign substance, is equal to the quotient obtained by dividing the number of dissolved molecules* n, *by the number of molecules* N, *of the solvent.* This law, especially well established for the case of dilute ethereal solutions, leads thus to the conclusion that

$$\frac{p - p'}{p'} = \frac{n}{N}. \qquad (14)$$

If we introduce this value for the *relative* vapour pressure, in the formula (1) on p. 138, we obtain

$$P = \frac{n}{N}\,\frac{0{\cdot}0821\ T \times 1000\ S}{M_0}.$$

But NM_0 denotes the number of grams of the solvent containing n-mol

[1] Raoult, *Zeitschr. phys. Chem.*, **2**. 353 (1888).

of the dissolved substance, and $\frac{NM_0}{1000S}$ denotes the volume of this in litres, and S denotes the sp. gr. of the solvent; and therefore, it follows that

$$\frac{NM_0}{n1000S} = V,$$

the volume of the solution in litres, containing 1 g.-mol.; and thus again we obtain the equation of condition

$$P = \frac{0{\cdot}0821T}{V}, \quad \text{or } PV = 0{\cdot}0821T \text{ litre-atm.}$$

The analogous law for the lowering of solubility has been proven by researches with various solvents,[1] viz. *the relative lowering of solubility is equal to the number of molecules of the dissolved substance, divided by the number of molecules of the solvent, i.e.*

$$\frac{L - L'}{L'} = \frac{n}{N}.$$

If we introduce this equation into the considerations given on p. 142, thus combining the lowering of the solubility with the osmotic pressure, we again obtain precisely as above, the equation

$$PV = RT = 0{\cdot}0821T.$$

Finally, this is proved by the measurements of the osmotic pressure conducted according to the method of freezing. In particular Blagden (1788), Rüdorff (1861), and Coppet (1871) studied the lowering of the freezing-point experienced by water from dissolving salts, but without arriving at any simple laws of general applicability. The reason of this, as will be shown later, is that on account of the electrolytic dissociation of salts in water, the relations are much more complicated, and, therefore, the laws are much more obscure than in the use of other dissolved or solvent materials. Therefore, as soon as Raoult turned his attention particularly to the study of the substances of organic chemistry, he discovered the validity of the following remarkable law, which is supported by very numerous observations; viz. *if one dissolves in any selected solvent, equi-molecular quantities of any selected substances, the freezing-point is lowered the same amount in all cases.*

If we denote the lowering of the freezing-point occasioned by the addition of m grams of the dissolved substance in 100 g. of the solvent, by t, then, by combining the proposition of Raoult with the law of Blagden (p. 146), we have

$$t = E\frac{m}{M}, \qquad (15)$$

[1] Nernst, *Zeitschr. phys. Chem.*, **6.** 19 (1890).

where M denotes the molecular weight of the dissolved substance; *i.e. the lowering of the freezing-point is proportional to the molecular content of dissolved substance.* The factor E is independent of the particular substance dissolved, but varies with the solvent used. Raoult called this factor E the "*molecular lowering of the freezing-point*" of the solvent in question. Its physical meaning is simply this, viz. that it furnishes the lowering of the freezing-point observed on dissolving 1 g.-mol of any selected substance in 100 g. of a solvent, provided that it is used only in such concentrations that the ratio between the molecular content and the freezing-point remains strictly constant, as in dilute solutions.

Compare Raoult's rule with equation (10), p. 143,

$$P = \frac{Swt}{T_0},$$

which gives the relation between osmotic pressure and lowering of freezing-point; multiply both sides by V, *i.e.* the volume of solution that contains 1 mol. of dissolved substance, then the gas equation (13) gives at the freezing-point T_0

$$PV = RT_0 = \frac{SVwt}{T_0}.$$

Now SV is the mass of solvent in which 1 mol. of dissolved substance is found; but as there are m grams, *i.e.* $\frac{m}{M}$ mols. for 100 g. solvent, we must have

$$SV = \frac{100M}{m}.$$

Hence

$$RT_0 = \frac{100Mwt}{mT_0},$$

or

$$t = \frac{RT_0^2}{100w} \cdot \frac{m}{M} \quad . \quad . \quad . \quad . \quad (16)$$

Equations (15) and (16) become identical when one puts

$$E = \frac{RT_0^2}{100w} \quad . \quad . \quad . \quad . \quad (17)$$

Thus Raoult's empirical molecular lowering of the freezing-point is calculated in equation (17), *from the gas constant* (R), *the absolute melting-point* (T_0), *and the heat of fusion* (w) *of the solvent* (van't Hoff, 1885).

If w is expressed in calories R must be in the same unit, *i.e.* is 1·991 (p. 48). For water we have

$$T_0 = 273; \quad w = 80·3 \text{ cal.}$$

Hence

$$E = \frac{1 \cdot 991 \times 273^2}{8030} = 18 \cdot 5,$$

whilst the latest determinations on very dilute solutions give (p. 144) as the most probable value

$$E = 18 \cdot 4.$$

The agreement would be precise if w has been underestimated and should be 80·7 instead of 80·3.

The following table gives for a number of solvents the molecular lowering of the freezing-point as observed and as calculated according to van't Hoff:—

Solvent.	E observed.[1]	E calc.	$T_0 - 273$.	w.[2]
Water	18·4	18·5	0°	80·3
Nitrogen dioxide	41	43–47	-10°	32–37
Formic acid	27·7	28·4	8·5°	55·6
Acetic acid	39	38·8	16·7°	43·2
Stearic acid	44	48	64°	47·6
Lauric acid	44	45·2	43·4°	43·7
Palmitic acid	44	55	55°	39·2
Capric acid	47	...	27°	...
Phenyl-propionic acid	88	...	48·5°	...
Stearin	51	...	55·6°	...
Ethal	62	...	46·9°	...
Ethylene-dibromide	118	119	7·9°	13
Chloral alcoholate	78	...	46·2°	...
Benzene	49	51	5·5°	30
Diphenyl	82	84	70·2°	28·5
Diphenyl-methane	67	...	26°	...
Naphthalene	71	69·4	80°	35·5
Phenol	74	76	39°	25
p-Mono-brom-phenol	107	...	63°	...
p-Cresol	74	73	34°	26
Thymol	83	85	48·2°	27·5
Anethol	62	...	20·1°	...
Benzo-phenone	95	96	48°	21·5
Urethane	50	50	48·7°	41
Urethylane	44	...	50°	...
Acetoxime	55	...	59·4°	...
Azo-benzene	82	82	66°	27·9
Nitro-benzene	70·7	69·5	5·3°	22·3
p-Toluidine	52	49	42·5°	39
Diphenyl-amine	88	88·8	54°	24
Naphthyl-amine	78	81·2	50·1°	25·6

[1] From the measurements of Raoult, *Ann. chim.*, (5) **28.** (6) **11.**; Beckmann, *Zeitschr. phys. Chem.*, **2.** 715; Eykman, *ib.* **3.** 113 and 203, **4.** 497; Ramsay, *ib.* **5.** 222.

[2] w, the heat of fusion, from the measurements of Berthelot, Pettersson, Eykman, Battelli, Bruner, and others. See especially Stillmann and Swain, *Z. S. phys. Chem.*, **29.** 705 (1899).

At first Raoult suspected that the relation between the molecular lowering of the freezing-point and the molecular weight of the various solvents was a simple one, represented by the equation $E = 0{\cdot}62M_0$. This was not supported by further observations. It was van't Hoff who first showed how the value of E could be calculated from the melting-point and the heat of fusion of the solvent.

The Law of Absorption of Henry and Dalton.—A simple law has long been known regarding the vapour pressure of a substance in solution, according to which there is a ratio between this vapour pressure and the concentration. It is usually formulated thus: *gases dissolve in any selected solvent in the direct ratio of their pressure* (Henry's Law of Absorption, 1803). So far as the law was proven, as was done chiefly for the solubility of the permanent gases, he established a genuine law of nature. The law, of course, holds good in a similar way for dissolved liquids, and thus, to take an example, the partial pressure of alcohol in the vapour over its dilute water solution, is proportional to its concentration in water.

In the sense of Henry's law there must exist between the vapour pressure p, and the osmotic pressure P, of the dissolved substance, a ratio which can be expressed by the equation

$$\frac{dp}{p} = \frac{dP}{P},$$

if we compare these with the equation developed on p. 145, viz.,

$$vdp = VdP,$$

it follows that

$$pv = PV,$$

i.e. we again obtain

$$PV = RT.$$

Thus for all gases or vapours which dissolve in any selected solvent proportionally to their pressure, i.e. *for those which obey Henry's law of absorption, their osmotic pressure is equal to the corresponding gas pressure.*[1] From the—so far as is known—high accuracy of the law of absorption it may be concluded that the osmotic pressure follows the laws of gases with equal accuracy. The exactness of the law of absorption offers the simplest, and also the most exact experimental proof, that the dissolved substance exerts the same pressure on a semi-permeable partition that it would exert on an ordinary partition, were it a gas at the same temperature and concentration.

As Dalton discovered in 1807, each gas dissolves in a gas mixture in accordance with its own partial pressure; and this can only mean that each of the dissolved gases exerts the same pressure that it would

[1] Van't Hoff *l.c.*

if it were alone; *i.e.* the same simple law of summation, as found by Dalton, holds good for the osmotic pressure of a solution of several substances, as for gases; and therefore the total lowering of the freezing-point occasioned by two dissolved substances in the same solution is equal to the sum of the lowerings which each would occasion were it alone; the same holds good also for the lowering of the vapour pressure, and of the solubility, provided, of course, that no chemical action occurs which would effect a change in the number of the molecules.

The Nature of the Solvent.—The question as to how the nature of the solvent influences the osmotic pressure of the dissolved substance, is at once settled by the fact, that, inasmuch as it is identical with the gas pressure, there is no dependence at all between the osmotic pressure and the nature of the solvent.

A direct proof is afforded by investigation of the partition of a substance between two solvents; in analogy to Henry's law it is found that the concentrations in the two solvents are proportional, just as in the partition between gas-space and solvent; thus, for example, if one mixes together carbon disulphide, water, and iodine, then the ratio of the concentration of the iodine in the carbon disulphide, to that in the water, at a temperature of 15°, is 410, regardless of the quantity of iodine used.

If we introduce this relation into the equation given on p. 145, viz. $V_1 dP_1 = V_2 dP_2$, then by a method exactly similar to that given in the preceding section, we obtain the result

$$P_1 V_1 = P_2 V_2;$$

i.e. under the same spatial concentration the osmotic pressure is the same in the two solutions. We must conclude, therefore, that if we could dissolve the same quantity of iodine in a litre of water as in a litre of carbon disulphide, we should obtain the same osmotic pressure in the two solutions.[1]

It is of course assumed in this that the dissolved substance has the same molecular weight in the two solvents; otherwise the osmotic pressure of equally concentrated solution would be different (*e.g.* if double molecules are formed in one solvent—as by acetic acid in benzene, whilst in another, such as water, the molecular weight is normal). If the molecular weight is not the same, the ratio of partition is not constant (Book III. chap. iii.).

The earliest experiments on partition between two solvents are due to Berthelot and Jungfleisch (*Ann. chim. phys.* (4), **26.** 396, 1872); as they investigated many substances that did not satisfy the above condition, Berthelot concluded that the partition coefficients vary with concentration even in dilute solution. It was shown by the present

[1] Nernst, *Zeitschr. physik. Chem.*, **8.** 16 (1891).

writer in 1891 that constancy of the partition coefficient is associated with similarity of molecular condition of the dissolved substance in the two solvents (*Z. S. phys. Chem.*, **8.** 110, 1891).

Since for a given concentration, osmotic pressure is independent of the nature of the solvent, it follows that it is not altered by compression of the solvent, provided that no change of molecular state is produced. This result may be easily proved thermodynamically.

Molecular State of Bodies in Solution.—The results of the preceding sections may be summarised as an extensive material for proving the law of osmotic pressure: the law, that is, that the osmotic pressure is equal to the gas pressure which could be observed manometrically if the solvent were removed and the dissolved substance left in gaseous form occupying the same space.

From this follows a practical application of the methods for measuring osmotic pressures that is of great importance. If the osmotic pressure of a substance—not too concentrated—in any solvent is measured, the corresponding gas pressure is known, and all the data are at hand for determining the vapour density, and so by Avogadro's law the molecular weight. We have thus methods of finding the molecular weight of substances whose vapour density it is difficult or impossible to measure, because they only evaporate at high temperatures or not at all without dissociation. Moreover, it is to the advantage of these methods, that with a proper choice of solvent the osmotic pressure is usually more convenient to measure in this way than the vapour density.

It must be observed, however, that determination of molecular weight by osmotic pressure is purely empirical. It is only a result of experience that the osmotic pressure is equal to the pressure that the dissolved substance would exert as a gas in the same space if the solvent were removed. It is, moreover, indifferent whether the experiment can be realised or not, *i.e.* whether the dissolved substance is capable of existing as a gas under the given conditions; it is true that we can reckon the molecular weight of any chemically well-defined material by means of Avogadro's law from the gas pressure, or indirectly by the gas pressure corresponding to the observed osmotic pressure. But the calculation has no other basis than the equality that is often observed between gas and osmotic pressure; no assumption can be made beforehand as to the molecular state of the substance in solution.

The remarkable relation between osmotic pressure and molecular weight demands, however, a theoretical explanation; and the explanation evidently must turn on the molecular state of the substance in solution. Hypothesis must of course come in here, because everything about the molecular state is hypothetical, and rests on the hypothetical assumption of a discrete arrangement of matter in space. The choice met with in establishing a suitable hypothesis is not often more easily made, however, than in this case.

The knowledge that dissolved substances follow the laws of gases leads by analogy to the conclusion that the molecular state of a dissolved substance is like that of a gas; or in other words, that Avogadro's law applies to the former. Thus we arrive at the following hypothesis:—

Is-osmotic solutions contain the same number of molecules of dissolved substance in a given volume, at a given temperature, and the number is the same as in an equal volume of a perfect gas at the same temperature and pressure (van't Hoff).

The extraordinary importance of this extension of Avogadro's law is obvious; as to its trustworthiness, this is rendered the more probable by the simplicity of its deduction by analogical reasoning. But more important than the probability of a hypothesis is its fruitfulness, for its existence is only justified by leading to new knowledge confirmed by experience. We shall therefore accept van't Hoff's extension of Avogadro's law and use it, especially in Book II., for further conclusions.

It is often stated that *experiment* shows the osmotic pressure to follow the laws of gases, namely, those of Boyle, Gay-Lussac, and Avogadro; this is an error in principle and must be especially guarded against. Only the two first are the expression of experimental facts. Avogadro's rule is equally hypothetical for solutions and gases, and one can hardly see how this can be altered by experimental facts.

Osmotic Pressure and Hydro-diffusion.—Reference has already been made regarding the well-known phenomenon that substances in solution, when left to themselves, wander from places of a higher to those of a lower concentration, and it should be emphasised here anew that this leads to the conclusion that external work can be developed by diluting a solution. The osmotic pressure, which we have come to regard as a most convenient help in calculation to express the value of this work numerically, must also be regarded as exerting a great influence in the phenomenon just described; this is generally known as "*hydro-diffusion*," or simply as "*diffusion*"; it plays a very important *rôle* in many processes of sub-organic structures in nature, but especially in plant and animal organisms.

This was first discovered in its general signification by Parrot (1815), but was first applied by Graham[1] in the thorough study of the subject, considering especially water solutions. It was shown that the coefficient of diffusion varies with the nature of the substance dissolved, and in all cases increases strongly with increasing temperature. Later investigations showed that a simple law may be formulated for the process of diffusion, which is completely analogous to that advanced by Fourier for the conduction of heat. This states that the mechanical

[1] *Lieb. Ann.*, **77.** 56 and 129 (1851); **80.** 197 (1851).

force which drives the dissolved substances from places of higher to those of a lower temperature, and also that the velocity with which the dissolved substance wanders in the solvent, are proportional to *the degree of concentration.* It is this fundamental law which makes possible a complete mathematical description of the process of diffusion, as was first suspected by Berthollet,[1] but was later and independently restated by Fick,[2] who subjected it to a thorough theoretical and experimental proof. In the sense of the preceding law, we may express by the following equation, the quantity of salt dS, which passes in the time dz, through a diffusion cylinder having the cross-section q, and when c is the concentration of the whole cross-section at the point x, and c + dc at the point x + dx; thus

$$dS = -Dq\frac{dc}{dx}dz;$$

D denotes a constant peculiar to the particular substance dissolved, and called the "*diffusion coefficient.*" Fick's law has nothing to say concerning the nature of the mechanical force: it is simply the "nature of the case." Moreover, later and thorough investigations of this have led to the result[3] that it can claim to hold good only approximately, since the coefficient of diffusion in general varies more or less with the concentration.

The author[4] has sought to develop the theory of the phenomena of diffusion, on the basis of the modern theory of solution. Thus, for example, we will consider the diffusion of cane sugar in water: if we pour a layer of pure water over a solution of cane sugar, the dissolved sugar at once begins to pass from points of higher to those of lower concentration, and this process does not cease till the differences of concentration are completely equalised. This obviously concerns the action of the same expansive force which we have learned to call the osmotic pressure; the nature of process is completely analogous to the equalisation of differences in density, developed by any cause whatever, in gases; and, moreover, under corresponding conditions the active forces are of the same magnitude. But, on the other hand, equalisation of density comes about very quickly in gases, while the substances dissolved in a liquid move very slowly and lazily. The reason for this is to be found in the fact that gas molecules in their movement meet with only very slight resistant friction, which, however, in the case of liquids is enormous.

The efficiency of the law of Fick is shown by the fact, that *the mechanical force occasioned by differences of pressure, is proportional to the "head" of the concentration.* But since we can also calculate the absolute

[1] *Essai de statique chimique.* Paris, 1803. Part I. chap. iv.
[2] *Pogg. Ann.*, **94.** 59 (1855).
[3] See especially Scheffer, *Zeitschr. phys. Chem.*, **2.** 390 (1888).
[4] Nernst, *Zeitschr. phys. Chem.*, **2.** 613 (1888).

value of the mechanical force, from the law of osmotic pressure, and since we can measure directly the velocity of diffusion, it is possible to calculate on the absolute scale the resistant friction experienced by the dissolved substance in its movement through the solvent. By carrying out the corresponding calculation for the resistant friction K, we obtain the formula

$$K = \frac{1{\cdot}99}{D} \times 10^9(1 + 0{\cdot}00367\ t)\,;$$

where D denotes the coefficient of diffusion measured for the temperature t. Thus, for example, K for cane sugar at 18° is calculated to be $4{\cdot}7 \times 10^9$ kg. in weight; *i.e.* it requires this enormous force to drive 1 g.-mol of cane sugar (= 342 g.) through the solvent (water) with a velocity of 1 cm. per second: this enormous value finds its explanation in *the smallness of the molecules.*

Osmosis by Isothermal Distillation

A very interesting experiment made by Magnus (1827) has recently been brought into notice again by Askenasy,[1] who carried it out in the following form. A glass tube is widened to a funnel at one end and closed there by a layer of gypsum; it is filled full of water and placed with the gypsum plate uppermost and its lower end dipping in a dish of mercury. The water used may conveniently be saturated with gypsum to avoid solution of the plate. The water evaporates slowly through the layer of gypsum—the process can be hastened by leading dry air over the plate. As the evaporation proceeds the mercury rises in the tube and may reach a height considerably exceeding that of the barometer at the time.

In a recorded experiment the mercury rose in fifteen hours to 89·3 cms., the barometer being at 75·3; it then touched the gypsum layer and so closed the experiment. The diameter of the tube was 3·3 mm. and evaporation took place in free air. Usually the end of the experiment was caused by formation of an air bubble under the gypsum, which consequently dried; then the mercury gradually fell again.

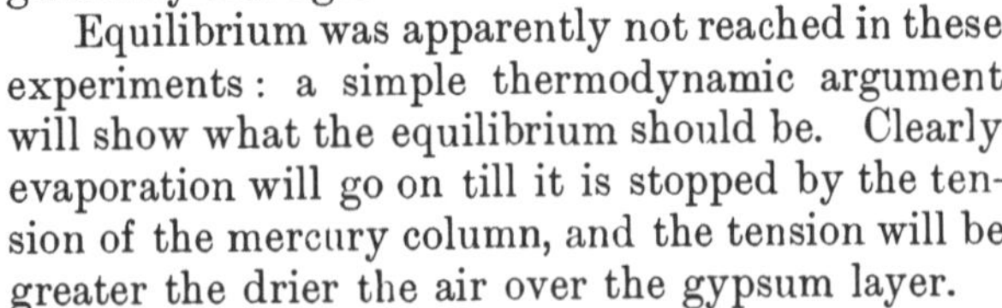

Fig. 16.

Equilibrium was apparently not reached in these experiments: a simple thermodynamic argument will show what the equilibrium should be. Clearly evaporation will go on till it is stopped by the tension of the mercury column, and the tension will be greater the drier the air over the gypsum layer.

For definiteness consider the following arrangement (Fig. 16):—

[1] E. Askenasy, "Beiträge zur Erklärung des Saftsteigens," *Verhandl. des naturh. med. Vereins zu Heidelberg*, **5.** (1896).

At a is a thin layer of pure water, under it mercury, over it a thin partition which has the property of being impermeable for gaseous, but permeable for liquid water. This property is possessed by Askenasy's moist gypsum through the pores of which water passes easily through, it is air-tight even for moderately high pressures.[1] Let the pressure above the gypsum be p′, which may be that of a solution bb; the vapour pressure of the pure solvent being p. To determine equilibrium apply the law of p, that at constant temperature variations in the neighbourhood of equilibrium must be reversible. If dx mols of water are distilled from the funnel to the solution, the system loses free energy to the extent $dxRT \ln \frac{p}{p'}$; at the same time mercury is raised through the height H and so free (potential) energy is stored up to the extent $H\frac{gM\sigma}{S}dx$, where g is the acceleration of gravity, M the molecular weight of the solvent, S its density, and that of mercury, so that $\frac{M\sigma}{S}dx$ is the mass of mercury lifted. Since the process is reversible, *i.e.* involves no loss in power to do work of the system (free energy), these quantities of work must be equal, or

$$dxRT \ln \frac{p}{p'} = H\frac{gM\sigma}{S}dx,$$

or the pressure P of the mercury column is

$$P = H\sigma g = \frac{S}{M}RT \ln \frac{p}{p'},$$

in other words, it is simply equal to the osmotic pressure of the solution in bb (p. 139).[2]

Conversely, measurements of the rise in the tube—which, it is seen is quite large—would be a means of determining osmotic pressure, or lowering of vapour pressure.

Osmotic Pressure in Mixture.—If for 1 mol of dissolved substance there are ν mols of a simple solvent then according to p. 149, we have

$$\frac{1}{\nu} = \ln \frac{p}{p'}, \text{ or } 1 = \nu \ln \frac{p}{p'} \quad . \quad . \quad . \quad (1)$$

where p is the vapour pressure of the solvent, p′ that of the solution.

If a substance occurs in a mixture of several solvents, a formula may be found on the very reasonable assumption that the osmotic

[1] The same property would be possessed by any solid partition that can dissolve water.

[2] An elementary proof analogous to that of p. 138, is given by Reinganum, *Wied. Ann.*, **59**. 764 (1896).

pressure of a substance (in normal molecular state) being independent of the solvent, follows the gaseous laws in mixed solvents also. It may be shown thermodynamically that if there be ν_1, ν_2 mols. of the solvents containing 1 mol. of dissolved substances, p_1, p_2 their vapour pressures, p'_1, p'_2 the same after addition of the dissolved substance[1]

$$1 = \nu_1 \ln \frac{p_1}{p_1'} + \nu_2 \ln \frac{p_2}{p_2'} \quad . \quad . \quad . \quad . \quad . \quad . \quad . \qquad (2)$$

Formula (3) which is a generalisation of the Raoult-van't Hoff formula (1) was satisfactorily verified by Roloff,[2] so one may conclude that the osmotic pressure follows the gas laws in mixtures also.

It is interesting to note that Roloff found some of the terms in the summand (2) may be negative. Thus when KCl is dissolved in a mixture of water and acetic acid the vapour pressure of the acid is raised; $p_2' > p_2$, and consequently the corresponding logarithm is negative.

It is to be observed that so far the considerations of this chapter relate only to very dilute solutions; we will now deal with the problem of calculating the osmotic pressure of concentrated solutions.

Osmotic Pressure at High Concentrations

The osmotic pressure reaches considerable amounts even for moderate concentrations; a solution of 1 mol (*e.g.* 46 grams of alcohol) in a litre gives 22·4 atmospheres at the freezing-point, according to the laws of gases (p. 38). As there is no prospect of making semi-permeable partitions capable of standing this or higher pressures, we are driven to indirect means of measurement for the osmotic pressure of concentrated solutions.

The calculation of this quantity gives the work needed to separate a mixture—as a concentrated solution may be called—a problem already considered in a general way on p. 114. The following calculations form therefore a special application of the principles developed there, but with the aid of the experience gained of a special class of mixtures, viz. dilute solutions.

It was found on p. 139, equation (4), by means of isothermal distillation, that

$$P = \frac{S}{M_0} RT \ln \frac{p}{p'},$$

or

$$P \frac{M_0}{S} = RT \ln \frac{p}{p'},$$

in the latter form the left-hand side stands for the osmotic work, the right-

[1] Nernst, *Z. S. phys. Chem.*, **11.** 1 (1893). [2] *Ibid.*, 17.

hand that done in isothermal distillation ; equality of the two follows, in all cases, from the second law as expressed in equation (*e*) of p. 23.

Now the expression for the work in isothermal distillation holds good for any range of concentration, provided only the vapour of the solvent follows the laws of gases and the difference between the specific volumes of the solution and solvent can be neglected by comparison with that of the saturated vapour. The latter condition is practically always satisfied, and if the former is not, a correction can easily be applied to the calculation if the characteristic equation of the vapour is known.

The osmotic work, on the other hand, is equal to the osmotic pressure multiplied by the volume through which the semi-permeable piston must be lowered to press out 1 mol of solvent; this volume however is not necessarily equal to that of the expelled solvent, as is the case in dilute solutions. That is only the case when on adding a small amount of solvent to the mixture the volume of the latter is increased by the volume of the solvent added, *i.e.* if the mixture occurs without contraction or expansion. Still this assumption is usually permissible even for strong solutions.

E.g. if 2 grams of water are added to 100 g. of a 50 per cent sugar solution we get a $\frac{50}{1 \cdot 02} = 49 \cdot 02$ per cent solution; the density of the former at 17·5° is 1·2320 S, of the latter 1·2275 S where S is the density of water at the same temperature. The change of volume is therefore

$$\frac{100 + 2}{1 \cdot 2275\ \mathrm{S}} - \frac{100}{1 \cdot 2329\ \mathrm{S}} = \frac{1 \cdot 987}{\mathrm{S}},$$

whilst $\frac{2}{\mathrm{S}}$ is what it would be if there were no contraction. Calculation of a number of such cases shows that equation (3), p. 139,

$$dv = \frac{M_0}{S} dx,$$

may be applied even to 20 to 30 per cent solutions with an error of less than 1 per cent. Moreover, the circumstance that a compressed solution (inside the osmotic cell) is mixed with an uncompressed solvent, matters little, for the compressibility of liquids is always minute. In what follows we shall, to avoid complication, assume the liquid to be incompressible, though that is of course not the case.

If Π be the osmotic work for a solution of any concentration, we have

$$\Pi = P\frac{M}{S}(1 + \epsilon) \quad . \quad . \quad . \quad . \quad (1)$$

in which ϵ is the relative increase in volume due to mixture of a mol of solvent with a large quantity of solution (under pressure P)—consequently a mostly negligible amount. Further

$$\Pi = RT \ln \frac{p}{p'} \quad . \quad . \quad . \quad . \quad (2)$$

with the restrictions already mentioned, and which are of no practical consequence.

According to the second law (equation *e*, p. 23)

$$\Pi - q = T\frac{d\Pi}{dT} \quad . \quad . \quad . \quad . \quad (3)$$

in which q is the heat evolved on adding a mol of solvent to a large quantity of solution. From (2) and (3) follows the relation already given on p. 118,

$$q = -RT^2 \frac{\partial \ln \frac{p}{p'}}{\partial T} \quad . \quad . \quad . \quad (4)$$

The meaning of q as $\frac{\partial Q(x)}{\partial x}$ has been referred to. If $q = 0$ $\ln \frac{p}{p'}$ is independent of T, so that according to (2) Π is proportional to T even for concentrated solutions.

The osmotic work may also be calculated from the freezing-point of strong solutions without much difficulty.

To integrate equation 8, p. 142, more exactly, set, according to p. 142,

$$\lambda = \lambda_0 - (C_1 - C_p)T \quad . \quad . \quad . \quad (5)$$

and the heat of sublimation

$$\sigma = \sigma_0 - (C_2 - C_p)T \quad . \quad . \quad . \quad (6)$$

where C_1, C_2 are the molecular heats of the liquid and solid solvent. The integrals become

$$\ln p = -\frac{\lambda_0}{RT} - \frac{C_1 - C_p}{R} \ln T + K_1 \quad . \quad . \quad (7)$$

$$\ln p' = -\frac{\sigma_0}{RT} - \frac{C_2 - C_p}{R} \ln T + K_2 \quad . \quad . \quad (8)$$

To determine the integration constants K_1, K_2 we have (p. 142) that at T_0

$$p' = p = p_0,$$

and therefore

$$\ln p_0 = -\frac{\lambda_0}{RT_0} - \frac{C_1 - C_p}{R} \ln T_0 + K_1 \quad . \quad . \quad (9)$$

$$\ln p_0' = -\frac{\sigma_0}{RT_0} - \frac{C_2 - C_P}{R}\ln T_0 + K_2 \quad . \quad . \quad (10)$$

(7) - (8) - (9) + (10) gives

$$\ln\frac{p}{p'} = \frac{\sigma_0 - \lambda_0}{R}\left[\frac{1}{T} - \frac{1}{T_0}\right] + \frac{C_1 - C_2}{R}\ln\frac{T_0}{T} \quad . \quad (11)$$

If we put $t = T_0 - T$ and develop in series

$$\ln\frac{T_0}{T} = \ln\left(1 + \frac{t}{T}\right) = \frac{t}{T} - \frac{t^2}{2T^2} + \frac{t^3}{3T^3} \dots$$

retaining only three terms (11) becomes

$$\ln\frac{p}{p'} = \frac{t}{R}\left[\frac{\sigma_0 - \lambda_0 + (C_1 - C_2)T_0}{T_0 T} - \frac{C_1 - C_2}{2}\frac{t}{T^2} + \frac{C_1 - C_2}{3}\frac{t^2}{T^3}\right] \quad (12)$$

Now

$$\sigma_0 - \lambda_0 + (C_1 - C_2)T_0 = \rho,$$

where ρ is the heat of fusion at the melting-point of the solution; hence (12) becomes

$$\ln\frac{p}{p'} = \frac{t}{R}\left[\frac{\rho}{T_0 T} - \frac{C_1 - C_2}{2}\frac{t}{T^2} + \frac{C_1 - C_2}{3}\frac{t^2}{T^3}\right] \quad . \quad (13)$$

and the osmotic work at temperature T, the freezing-point of the solution is

$$\Pi = RT\ln\frac{p}{p'} = t\left[\frac{\rho}{T_0} - \frac{C_1 - C_2}{2}\frac{t}{T} + \frac{C_1 - C_2}{3}\frac{t^2}{T^2}\right] \quad . \quad (14)$$

If we require the osmotic work, not at the variable freezing-point of the solution, but always at the same temperature, say the freezing-point of the pure solvent T_0, we must integrate equation (4) and get

$$\ln\frac{p}{p'} = \frac{q}{RT} + \text{const.} \quad . \quad . \quad . \quad . \quad (15)$$

for as we are only concerned with a small correction, q may be regarded as constant with sufficient accuracy.

Then for T_0 (15) becomes

$$\left(\ln\frac{p}{p'}\right)_{T_0} = \frac{q}{RT_0} + \text{const.} \quad . \quad . \quad . \quad . \quad (16)$$

(13) - (15) + (16) gives

$$\left(\ln\frac{p}{p'}\right)_{T_0} = \frac{t}{R}\left[\frac{\rho - q}{T_0 T} - \frac{C_1 - C_2}{2}\frac{t}{T^2} + \frac{C_1 - C_2}{3}\frac{t^2}{T^3}\right];$$

and the osmotic work at temperature T_0

$$\Pi_0 = RT_0 \left(\ln \frac{p}{p'}\right)_{T_0} = t\left[\frac{\rho - q}{T} - \frac{C_1 - C_2}{2}\frac{T_0 t}{T^2} + \frac{C_1 - C_2}{3}\frac{T_0 t^2}{T^3}\right] \quad (17)$$

This formula was obtained by Dieterici (*Wied. Ann.*, **52**. 263, 1894) who also showed by examples that (2) and (17) give values for the osmotic work in satisfactory agreement. The formula was also given independently by Th. Ewan (*Z. S. Phys. Chem.*, **14**. 409, 1894) who in integrating (4) paid attention to the variation of q with temperature; the correction term thus obtained however—involving the specific heats of solution and solvent—is almost always negligible. To show the usefulness of equation (17) we will take the case of some potassium chloride solutions, whose vapour pressures (Dieterici, *Wied. Ann.*, **42**. 513, 1891; **50**. 47, 1893) and freezing-points (Roloff, *Z. S. Phys. Chem.*, **18**. 572, 1895) have been accurately measured. In the following table m signifies the number of grams to 100 grams solvent, t the lowering of freezing-point, q the heat of dilution, p′ the vapour pressure of the solution at 0°; p that of pure water at the same temperature is 4·620 mm. :—

m	t	g	p′	Π_0	
				calc.$_1$	calc.$_2$
0	0	0	4·620	...	...
3·72	1·667	− 1·63	4·546	8·80	8·75
7·45	3·284	− 5·96	4·472	17·55	17·67
14·90	6·53	− 19·5	4·326	35·18	35·71
22·35	9·69	− 34·3	4·190	52·64	53·12

Under calc.$_1$ are given the values derived from equation (17), assuming that

$$T_0 = 273 \qquad \rho = 18 \times 80{\cdot}3 = 1445 \qquad C_2 - C_1 = 18 \times 0{\cdot}475 = 8{\cdot}55.$$

Under calc.$_2$ are the values derived from equation (2) assuming that

$$R = 1{\cdot}991.$$

In both cases we obtain the osmotic work in ordinary calories; the agreement is very satisfactory and would be made still better by taking the heat of fusion of ice—which unfortunately is still uncertain—as 80·7 instead of 80·3 (see p. 152).

Vapour Pressure of Concentrated Solutions.—The thermodynamic treatment of concentrated aqueous solutions is simple—as Dolezalek[1] lately showed—if we make use of the rule, that seems to be pretty generally true, that the logarithm of the vapour pressure p

[1] *Verh. d. Deutsch. phys. Ges.*, **5**. 4 (1903).

is approximately a linear function of the number x of salt mols per mol of water, *i.e.*

$$-\frac{d \ln p}{dx} = a \quad . \quad . \quad . \quad . \quad (1)$$

where a is a characteristic constant for the dissolved substance.

Now (p. 119)

$$\frac{d \ln p}{dx} + x\frac{d \ln P}{dx} = 0 \quad . \quad . \quad . \quad (2)$$

where p is the vapour pressure of water and P that of the dissolved substance. From this and (1) we have

$$x\frac{d \ln P}{dx} = a,$$

or integrated,

$$\ln P = a \ln x + \text{const.}$$

Thus the vapour pressure of HCl over concentrated solutions of hydrochloric acid is satisfactorily given over a wide range by the expression

$$\log_{10} P = 7{\cdot}9 \log_{10} x + 6{\cdot}6421$$

as is shown by the following table :—

x	P	
	beob.	ber.
0·288	277	230
0·257	112	95
0·226	31·5	34
0·198	11·2	12
0·173	4·1	4·1
0·146	0·06	1·1
0·122	0·52	0·6

For the work required to remove 1 mol of dissolved substance from solution I. to solution II. we have (p. 117)

$$A = RT \ln \frac{P_1}{P_2} = a\, RT \ln \frac{x_1}{x_2} \quad . \quad . \quad (3)$$

for the work required to remove 1 mol of water from I. to II. we have from (1)

$$A' = RT \ln \frac{p_1}{p_2} = a\, RT\,(x_2 - x_1) \quad . \quad . \quad (4)$$

Equation (1) is only approximate. The simplicity of equations

(3) and (4) is noticeable, for they contain only a constant characteristic of the dissolved substance. Dolezalek gives the following table :—

	a		a		a
H_2SO_4 . .	8·4	KJ . . .	2·5	$NaNO_3$ . .	1·6
HCl . . .	7·9	KBr . . .	1·8	LiBr . . .	5·8
H_3PO_4 . .	3·0	KCl . . .	1·8	LiCl . . .	4·6
NaOH . .	7·4	KNO_3 . .	1·2	NH_4Br . .	2·1
KOH . .	3·9	NaJ . . .	4·1	NH_4Cl . .	1·7
K_2CO_3 . .	7·1	NaBr . .	3·5	$ZnCl_2$ . .	8·4
KF . . .	4·8	NaCl. . .	2·8	$C_3H_8O_3$ . .	1·01

Ideal Concentrated Solutions.—It is natural to compare the change in total with that in free energy, or in other words, the heat of reaction with the osmotic work, in concentrated solutions.

If a mol of water be taken from solution I. to II. then the heat evolved is

$$\left(\frac{\partial Q(n)}{\partial n}\right)_{n=n_2} - \left(\frac{\partial Q(n)}{\partial n}\right)_{n=n_1} \quad . \quad . \quad . \quad (1)$$

if there are n_1 mols. of water to one of salt in solution I. and n_2 in II.

The osmotic work according to equation (2), p. 165, is

$$\Pi = RT \ln \frac{p_1}{p_2} \quad . \quad . \quad . \quad . \quad (2)$$

where p_1, p_2 are the vapour pressures of the two solutions.

Comparison shows that there are concentrated solutions for which these two quantities differ very little from each other, and as we shall see, such solutions have a remarkably simple behaviour, it is convenient to describe them as "ideal concentrated solutions."[1]

For example, take solutions of sulphuric acid containing 29·2 mols and 4·76 mols of H_2O for one of H_2SO_4. The heat of dilution can be found from the equation of p. 114.

$$Q(n) = \frac{17{,}860n}{n + 1{\cdot}8}, \quad \frac{\partial Q(n)}{\partial n} = \frac{32{,}150}{(n + 1{\cdot}8)^2}$$

Hence for (1) we get

$$747{\cdot}0 - 33{\cdot}5 = 713{\cdot}5 \text{ cal.}$$

and for (2)

$$\Pi = 1{\cdot}991 \times 273 \ln \frac{4{\cdot}284}{1{\cdot}206} = 690{\cdot}5 \text{ cal.},$$

taking the vapour pressures (at 0°) from Dieterici;[2] these numbers agree within the limits of error of measurement.

[1] W. Nernst, *Wied. Ann.*, **53.** 57 (1894). [2] *Wied. Ann.*, **50.** 47 (1893)

In the following table the calculation is made for a number of solutions :—

n.	p.	$1252 \log_{10}\frac{p_1}{p_2}$	$\frac{\partial Q(n)}{\partial n}$	$\left(\frac{\partial Q(n)}{\partial n}\right)_{n=n_2} - \left(\frac{\partial Q(n)}{\partial n}\right)_{n=n_1}$
∞	4·620		0	
		10·1		3·96
91·6	4·535		3·96	
		31·0		29·5
29·2	4·284		33·48	
		85·2		85·1
14·66	3·664		118·6	
		117·6		116·9
9·93	2·952		235·5	
		307·1		308·5
5·89	1·679		544·0	
		180·6		202·0
4·76	1·206		746·0	
		1084·0		982·0
2·51	0·164		1728·0	

In the third column is the osmotic work, in the last the corresponding heat of dilution, both quantities relating to the pair of solutions between which they are placed. The comparison shows that for $n = 91{\cdot}6$ even the region of ideal concentrated solutions is reached; the agreement between that point and $n = 2{\cdot}51$ is striking—the deviations are so irregular that they probably are due for the greater part to error of measurement.[1]

Just as in ideal dilute solutions the lowering of vapour pressure and change of boiling- and freezing-points can be calculated from the osmotic pressure, and consequently also from the rise in boiling-point or fall in freezing-point, so here in ideal concentrated solutions the same things can be deduced from the heat of dilution. The relations are simple in both limiting cases because in the fundamental equation

$$A - U = T\frac{dA}{dT}$$

one of the terms vanishes—in ideal dilute solutions U in ideal concentrated solutions $\frac{dA}{dT}$.

If now the identical equations

$$\frac{dA}{dT} = 0 \text{ and } A = U$$

are to hold not merely at a singular point but over a considerable range of temperature we must also have

$$\frac{dU}{dT} = 0,$$

[1] The new numbers of Dieterici agree a good deal better than the earlier ones of Regnault used in the above calculation.

i.e. the heat of dilution of an ideal concentrated solution is independent of temperature.

Thus

$$H_2SO_4 . 4H_2O + 5H_2O = H_2SO_4 . 9H_2O$$

gives 2559 cal.; the thermal capacity before mixture is according to Thomsen

$H_2SO_4 + 4H_2O$	92·7
$5H_2O$	90
	182·7 cal.

and after

$H_2SO_4 . 9H_2O$	182·0 cal.

The difference (0·7 cal.) is according to p. 8 the change of heat of dilution per degree of temperature; so the heat of dilution 2550 cals. varies only 3 parts in 10,000 per degree.

The Extent of the Domain of the Laws of Solutions.—The question now arises whether the laws of osmotic pressure, which have been developed in part directly, and in part have been developed on the basis of thermodynamic considerations, hold good for all cases strictly and without exception. Although both the direct and the indirect measurements of osmotic pressure need improvement in point of exactness, yet it may be assumed with considerable probability that the law *holds good only approximately*, and that it constantly loses its exactness with the increasing concentration of the substance dissolved. The analogy with the laws of gases holds good also in this case, for it is noticeable that, according to experiments up to date, the osmotic pressure of concentrated solutions, like the pressure of strongly compressed gases, increases more rapidly than it should according to the Boyle-Mariotte law.

Solid Solutions.—We have already seen (p. 120) that the aggregates produced by mutual molecular interpenetration, existing in the solid state, and concisely designated "isomorphous" mixtures, can be compared with liquid mixtures in many respects. The supposition is thus at once forced upon us, that the behaviour of a solid mixture, like that of a liquid mixture, would be very simple in the case where one of its components is present in great excess, *i.e.* where we are considering a "*dilute solid solution.*"

Experiment shows that this supposition is fulfilled. At least van't Hoff,[1] to whom we are indebted not only for the theory of solid solutions, but also, on the whole, for its introduction into science, has made it appear in the highest degree probable that we may venture to

[1] *Zeitschr. phys. Chem.*, **5.** 322 (1890).

speak of the *osmotic pressure of substances existing in solid solution*, as being analogous to that of the liquid solution, and as obeying the same laws.

To be sure there is no opportunity to measure this directly by means of a semi-permeable partition, since the chance of its realisation in a solidified system is as good as impossible; but it is to be hoped that we may obtain measurements of these enormous pressures by indirect methods. Indeed, the property of substances dissolved in solidified systems, of spreading through the solid *by diffusion*, certainly speaks clearly and beyond all doubt in favour of the existence of an inherent expansive force, which can be regarded as comparable to the osmotic pressure.

Various facts point to diffusion in solids. Hydrogen dissolved in platinum and palladium gradually spreads through the whole metal, as has long been known. Bellati and Lussana[1] observed a clear instance in the easy passage of nascent hydrogen through iron at ordinary temperatures, with various experimental arrangements. *E.g.*, a barometer was closed at the top by an iron plate, and by cementing a glass ring on this, an electrolytic cell was made in which hydrogen was generated at the iron plate; the mercury at once fell by diffusion of the hydrogen through the plate into the barometric vacuum. Carbon finds its way into hot iron, and can even pass through a porcelain crucible. Further, Roberts-Austen[2] observed that gold diffuses through lead at 251° (75° below the melting-point of the latter), and even to a perceptible extent at atmospheric temperature in the course of several years. According to experiments by Spring (1886), solid barium sulphate and sodium-carbonate react to the extent of reaching a state of equilibrium, which appears hardly possible without molecular interpenetration.

Again the fact that many substances conduct electrolytically indicates the possibility of diffusion in solids, for, as we shall see later, ionic transport and diffusion are intimately connected.

It must, however, be remarked that there are other cases that do not show the slightest trace of diffusion in the solid state. Thus in petrography sharply bounded portions are sometimes found in a homogeneous crystal that show different colouring to the rest; here no appreciable equalisation by diffusion has taken place in thousands of years, although the colouring matter must be regarded as in a state of solid solution. If osmotic pressure can be assumed in such cases the dissolved molecules must suffer quite extraordinary resistance to displacement in their solvent.

When one brings a liquid solution to the freezing-point, as is well known, the solvent separates as a rule in the pure state; the Raoult-van't Hoff formulæ hold good for this case. But in some cases there

[1] *Atti R. 1st Veneto* (7), **1**. 1173 (1890); abstract *Z. S. phys. Chem.*, **7**. 229 (1891).
[2] *Proc. Roy. Soc. Lond.*, **67**. 101 (1900).

are observed considerably smaller lowerings of the freezing-point, than those calculated from the molecular proportions of the dissolved substance, according to the preceding formulæ; usually this is explained by polymerisation of the dissolved substance. But there are cases in which such an explanation is highly improbable or even quite inadmissible, and then it appears that a mixture of solid solvent and dissolved substances crystallises out in place of the pure solvent—the hypothesis originally put forward by van't Hoff.

It is easy to see that in such cases the lowering of the freezing-point must be too small. According to the considerations of p. 124 the freezing-point of a mixture must sink continuously, only becoming constant when the part crystallising out has the same composition as that remaining liquid. Thus if the solid separating out contains more of the dissolved substances than the residual liquid the latter is diluted by freezing, *i.e.* the freezing-point must, in this case, rise with increasing concentration; if, on the other hand, the frozen-out solid is less rich in the dissolved material than the residual liquid, the freezing-point must fall with increase of concentration; but clearly not so fast as when the pure solvent crystallises out, *i.e.* the lowering of freezing-point is lessened.

If, after van't Hoff, we apply the notion of osmotic pressure to solid solutions, the vapour pressure of the solid solvent must be reduced by the taking up of dissolved substance. If the vapour pressures of solid and liquid solvent suffer equal change the freezing-point remains unaltered, since this point is characterised by equality of vapour pressure in the two states of aggregation. The freezing-point must rise if the vapour pressure of the solid is more affected than that of the liquid, fall in the contrary case. These relations can be easily followed by means of Fig. 15, p. 142, if the vapour pressure of the solid solution be introduced by a line lying below and parallel to the curve of vapour pressure of the pure solid solvent.

It is found in fact that some abnormally small freezing-point depressions are produced by co-crystallisation of dissolved material. Thus, according to van Bijlert,[1] a solid solution freezes out of solutions of thiophene in benzene; its composition is, according to Beckmann,[2] practically 0·42 times the concentration in the liquid, for all strength of solution; there is thus a constant ratio of partition of thiophene between the liquid and solid solvent. Solution of antimony in tin and β-naphthol in naphthalene raises the melting-point, and van Bijlert found that the crystals contain a larger percentage of the dissolved substance than the liquid, in accordance with the above theory. Beckmann and Stock[3] found that iodine which gives abnormally small depressions in benzene is shared between solid and liquid solvent, and in an approximately constant ratio, like thiophene.

[1] *Z. S. phys. Chem.*, **8.** 343 (1891). [2] *Ibid.*, **22.** 609 (1897).
[3] *Ibid.*, **17.** 107 (1895).

Ferrantini and Garelli[1] have made extensive investi ions on co-crystallisation of dissolved substances, especially of cyc ic-organic compounds.

Respecting the measurement of the osmotic pressure P, of solid solutions, we must, of course, consider the accomplishment of the reversible separation of the solvent and the dissolved substance. If dA is the minimum work to be expended and that necessary for the removal of a volume dv, of the pure solvent from the solution, then there exists the relation

$$Pdv = dA.$$

If a gas dissolves in a solid in proportion to its pressure, then it may be concluded by analogy, with what is stated on p. 155, that its osmotic pressure is equal to the corresponding gas pressure. If a substance distributes itself in constant ratio between a liquid and a solid solvent, then, according to what is stated on p. 155, for equal spatial concentration the osmotic pressure must be the same; therefore it must be inferred from the measurements of van Bijlert, that this is at least approximately true for the solutions of thiophene in solid and liquid benzene.

According to the considerations stated on p. 128, we have a method capable of universal applicability for effecting, in a reversible way, the separation of the components of a solid mixture.

Further expression (7) of p. 141 may be at once applied to the relative lowering of solubility which is experienced in any solvent by a solid to which another soluble in it has been added.

It should be added that Küster (*Z. S. phys. Chem.*, **17**. 367, 1895) maintains the very reasonable view that a distinction should be made between solid solutions and isomorphous mixtures. Only in the former is diffusion of a dissolved substance possible, according to this view, while in the latter the molecules of the added material form part of the crystalline structure and are held in fixed positions of equilibrium by the forces that produce orientation of the crystal molecules. Again an opinion of Bodländer should be noted, that in the formation of solid solutions, *e.g.* separation of iodine with crystallising benzene, absorption phenomena play an important part (see for this and the theory of solid solutions generally Bodländer, *Neues Jahrbuch fur Mineralogie*, Beilage-Band, **12**. 25, 1898).

[1] *Z. S. phys. Chem.*, **13**. 7 (1894), communicated by Ciamician; see also Garelli, *Gazz. chim.*, **23**. 354; **24**. 229 (1894).

BOOK II

ATOM AND MOLECULE

CHAPTER I

ATOMIC THEORY

Combining Weight and Atomic Weight.—The question whether a well-defined chemical substance is an element or a compound of several different elements, and in the latter case to what extent each element is contained in unit weight of the substance, is a problem of a purely experimental nature: this problem, according to the case, can be answered with great certainty and exactness by means of an equipment for analytical chemical methods, without recourse to theoretical speculation. The elementary analysis of a compound is included among the commonest operations of a chemical laboratory, and a description of the purely chemical methods of such an investigation does not lie in the province of this treatise.[1]

The question as to the relative number of atoms in the molecule of a compound is, however, entirely different. To answer this question, besides knowing the combining weights, which may be obtained directly from experiment, we must also know the *relative weights of the atoms* which make up the compound considered: this can be learned only by the aid of theoretical speculation, and even then not with absolute certainty, but only with more or less probability. From the principles of the atomic theory stated on p. 30, it follows that the atomic weights and the combining weights are related to each other in the simple ratios of rational numbers; but the values of these numerical ratios remain undetermined. But, if one obtains the same results by very different methods, the probability is very great that

[1] A critical summary of the combining weight determinations thus far made will be found in Ostwald's *Lehrbuch der Allgemeinen Chemie*, Leipzig, 1891, vol. i. p. 18 and fol. [See also a condensed statement in Walker's translation of Ostwald's smaller *Outlines of General Chemistry*, Book I. chap. iii.—Tr.]

theoretical considerations will lead to safe conclusions. The question of the relative weights of the atoms has been established with such certainty that its general correctness is at present no longer called in question; it is very instructive to study the different methods pursued, by means of which different investigators, without wandering very much from the direct path, have finally attained the same desired goal.

If one wishes to explain facts obtained by experiment, by means of an hypothesis, as here, for instance, where we call in the atomic hypothesis to aid us in explaining the laws of definite and multiple proportions, it should be remembered that that method of explanation is to be chosen which is the *simplest.* This can afterwards be expanded, if later discoveries force us to adopt more complicated conceptions. Thus Dalton (1808)[1] proceeded in this way when he arranged the first table of the atomic weights.[2] In the case of those compounds which consist of only two elements, it is the simplest to assume that the same number of atoms of the two elements have united to form the compound. Thus it was [provisionally] assumed that carbonic oxide contained the same number of atoms of oxygen as of carbon, and that water contained the same number of atoms of oxygen as of hydrogen, etc. In this way Dalton tried to learn the relative atomic weights of the most important elements, and in the same way, advancing to compounds consisting of more than two elements, he sought to determine the number of atoms, and thus to obtain a consistent system of the atomic weights.

But this method was by no means free from arbitrary assumption; for the same logic that regarded carbonic oxide as composed of an equal number of atoms of oxygen and carbon, would regard carbon dioxide as containing twice as many atoms of oxygen as of carbon; and, on the other hand, if Dalton had regarded carbon dioxide as containing the same number of atoms of carbon and oxygen, then carbonic oxide must contain twice as many atoms of carbon as of oxygen. Thus there was an opportunity for choice in selecting the atomic weight. That the choice in the preceding case was the right one was purely a matter of chance. We need to consult other experimental facts and their meaning, on the basis of a more extended development of the atomic hypothesis, in order to obtain a system of the atomic weights which shall be free from arbitrariness.

In judging of the fruitfulness of atomistic conceptions it is of interest to note that Dalton did not invent the atomic theory as a subsequent explanation of the laws of constant and multiple proportions, as was formerly thought; on the contrary, he was led by considerations of molecular theory to the discovery of the fundamental laws of chemistry (see Roscoe and Harden, *Dalton's Atomic Theory*).

[1] Ostwald's *Klassiker*, No. 3. Leipzig, 1889.

[2] The first table was arranged in 1804. See Thomson, *History of Chemistry*, vol. ii. p. 289.—TR.

The Rule of Avogadro.—The necessary experimental facts were found in Gay-Lussac's laws [combination by volume], according to which the volumes of those gases which combine with each other stand in simple ratios to each other, and also the volumes of the resulting compounds, if gaseous, show such simple ratios. The theoretical interpretation of this, in the light of the atomic hypothesis, led to the law which was first introduced by the hypothesis of Avogadro (1811); this states that all the different gases, both simple and compound, contain the same number of molecules in the same unit of space (p. 40). After a method was found for determining the relative weight of the *molecules* by measuring the vapour densities, by applying the principle of the simplest explanation, it was easily possible to obtain constant determinations of the *atomic weights* of all elements, at least of all having a sufficient number of volatile compounds. If the molecules of a compound are really produced by the union of a small number of the atoms of each particular element, then it may be safely assumed that among a large number of molecules, whose molecular weights can be determined from their vapour density, some molecules will occur containing only *one atom* of the particular element in a molecule. Thus we arrive at the conclusion *that the smallest relative quantity of an element which can enter into the molecule of a compound, or the smallest quantity representing the difference between the combining weights of the element in different compounds, corresponds to the atomic weight.* But even if only a few compounds of an element are investigated there is hardly any doubt as to the atomic weight; thus if the quantity of that element present in a mol of the first compound is a, in the second $3a$, in the third $4a$, then a may be taken as the atomic weight. In practice, as other tests are not wanting, a few vapour-pressure measurements suffice, and the atomic weight is taken as that weight which, multiplied by whole numbers, as small as possible, gives the amount present in each molecule. To be sure in this way, strictly speaking, one can arrive only at the upper limit of the atomic weight; but with the investigation of a large number of compounds, the probability that one is not dealing with a multiple of the atomic weight, but with the desired value itself, becomes very great. Thus it follows that in a gram molecule of the numerous chlorine compounds, there is contained at least either 35·4 g. of chlorine, or an *exact multiple* of this, and similarly in the case of many other elements.

After determining the relative atomic weights, the vapour density determination of an element shows the number of atoms contained in its molecule. The fact that the molecule consists of one atom, in the case of only a few elements, where the atomic and molecular weights are identical with each other, such not being the case for all the elements, this occasions only passing doubts; for a broader development of the doctrine of valence shows at once that similar, as well as dissimilar, atoms can unite firmly with each other by means of chemical forces.

The Law of Dulong and Petit.—Another series of experimental facts, the theoretical significance of which certainly holds good, is the relation between the atomic weight and the specific heat of the respective element in the solid state; this was discovered by Dulong and Petit in 1818. The product of the atomic weight by the specific heat is called the *atomic heat;* by this is to be understood the amount of heat expressed in g.-cal., which must be added to 1 gram atom of an element in order to raise its temperature 1°. The following is the simple statement of the law, viz. :—*The atomic heat of elements in the solid state of aggregation is approximately constant, and amounts to about* 6·4.

This law is not strictly exact, for some of the elements having atomic weights smaller than 35, have atomic heats considerably removed from the mean. H. F. Weber in 1875 showed, in the case of the most marked examples, viz. boron (2·6), carbon (2 to 2·8, according to the modification), and silicon (about 4), that the atomic heats of these elements increase strongly with increasing temperature, and approach the values necessitated by the Dulong-Petit law.[1] Beryllium, which is also a marked exception (its atomic heat being 3·71), according to the measurements of Nilson and Pettersson (1880), shows a strong increase of the atomic heat with rising temperature. The law finds its most exact application in the case of the metals, where it holds good for even those having a low atomic weight, as lithium, magnesium,[2] etc. Thus it takes about the same amount of heat (6·6 cal.) to raise the temperature of 1 g. atom of lithium (7·03), 1°, that is required to raise 1 g. atom of uranium (239), 1°. This clearly shows that we have here to do with a very remarkable law.[3]

Accordingly in this, by measuring the specific heat of a new element, we have a simple and a perfectly sure method for the determination of its atomic weight, provided that special attention is given to certain points; viz. firstly, the specific heat must be measured at different temperatures in order to make sure that it does not vary too much with the temperature; secondly, the determination must not be made too close to the melting-point; and, finally, the atomic weight of the element in question must not be too small.

Thus recently [4] an investigation of the specific heat of the element germanium, discovered by Winkler, gave the atomic heat of 5·6, which speaks in favour of the atomic weight 72·3, assumed for the element.

It is very remarkable, and also of great assistance in determining the atomic weight, that the *constancy of the atomic heat also holds good for*

[1] Moisson and Gautier found the same result for boron (*Ann. chim. phys.*, (7), **7.** 568, 1896).

[2] See especially Waterman, "Specific Heat of Metal," *Phys. Rev.*, **4.** 161, 1896.

[3] For a very clear summary of this subject in English, see M. M. P. Muir, *Prin. of Chem.* 2nd edit., pp. 49-67, Cambridge (1889); Remsen, *Theoret. Chem.* 4th edit., pp. 65-75; and Lothar Meyer, *Mod. Theories of Chem.*, trans. by Bedson and Williams, pp. 63-95, London and New York (1888).—Tr.

[4] Nilson and Pettersson, *Zeitschr. physik. Chem.*, **1.** 31 (1878).

compounds existing in the solid state. By means of the painstaking work of F. Neumann (1831), of Regnault (1840), and especially of Kopp,[1] who in his classic work on the specific heat of solid salts added a great deal to the subject, the principle has been formulated in a remarkably broad generalisation. According to this, the specific heat of solid substances is expressly an *additive property: the molecular heat of a solid compound* (i.e. *the product of the specific heat and the molecular weight*) *is equal to the sum of the atomic heats of the elements contained.*

The atomic heats have the following values, viz. for C, 1·8; H, 2·3; B, 2·7; Be, 3·7; Si, 3·8; O, 4·0; P, 5·4; S, 5·4; Ge, 5·5;[2] and for the other elements approximately 6·4.

Thus, for example, the specific heat of solid water, *i.e.* ice, amounts to 0·474, and the molecular heat therefore is $18 \times 0·474 = 8·5$; while the molecular heat calculated from the figures given above for the formula H_2O, is

$$\overline{2 \times 2·3} + 4 = 8·6.$$

The specific heat of calcium carbonate, $CaCO_3$, is 0·203, corresponding to a molecular heat of 20·4, while that calculated according to Kopp's law is

$$6·4 + 1·8 + \overline{3 \times 4} = 20·2.$$

Conversely, if one calculates the specific heat by dividing by the molecular weight, he obtains 0·201 instead of 0·203; and similar coincidences are found in hundreds of other cases investigated, and, although small errors are not wanting, yet they do not exceed the errors of observation.

Thus we find that the atomic heats of the elements, as calculated from the specific heats of their compounds, coincide with the atomic heats of the elements in the free state, in so far as these can be studied in the solid state. It may therefore be concluded, with great certainty, that chlorine in the solid state has the specific heat

$$\frac{6·4}{35·5} = 0·180,$$

and thus obeys the Dulong-Petit law. Thus, in general, it is possible to calculate the atomic heats from the specific heats of solid compounds.

The following facts, viz. that the heat capacity of solid substances remains the same before and after chemical union; that the increase of the kinetic and potential energy, experienced by a rise of 1° in temperature, is the same for the atoms of a free element as for those in chemical union with other elements; and, further, that this increase of energy—law of Dulong and Petit—is nearly the same for

[1] *Lieb. Ann.*, Supplement, **3.** 1 and 289 (1864). [2] Nilson and Pettersson, *l.c.*

the atoms of most elements;—all these will undoubtedly be of fundamental significance in developing a future theory of the solid state of aggregation.

Kopp's law is of value in atomic weight determinations because by it the atomic heat of elements can be deduced from the specific heat of their solid compounds.

It will be useful to illustrate this point by an example. Analysis of corrosive sublimate shows it to contain 100 gr. mercury to 35 gr. (1 gr. atom) of chlorine. The specific heat according to Regnault is 0·069. The molecular heat according to possible formulæ would be—

Molecular weight.	Molecular Heat Observations.	Calc.
$HgCl = 100 + 35$	$135 \times 0{\cdot}069 = 9{\cdot}3$	12·8
$HgCl_2 = 200 + 70$	$270 \times 0{\cdot}069 = 18{\cdot}6$	19·2
$HgCl_3 = 300 + 105$	$405 \times 0{\cdot}069 = 28{\cdot}0$	25·6

Only the triatomic formula $HgCl_2$ gives a molecular heat (according to Kopp's law) in accordance with the facts: the atomic weight of mercury must therefore be 200. Such a calculation does not give any information as to molecular weight, for if the formulæ were $(HgCl_2)_n$ both observed and calculated molecular heats would be n times as great, and the agreement would still hold.

The specific heat of liquids is of a more complicated character, or at any rate no simple results have so far been arrived at. That of gases, which is sometimes of high importance in atomic weight determinations, will be dealt with in the next chapter. We will only remark here that the specific heat of monatomic gases is 3·0, *i.e.* about half that of solid atoms according to Dulong and Petit's law.

Isomorphism.—The relation, discovered by Mitscherlich (1820), between atomic weight and isomorphism, affords another independent way of determining the atomic weight. This, to be sure, alone is insufficient, but as an accessory it is of the greatest importance, and has repeatedly furnished practical results.

The following points may be cited as the most important characteristics of isomorphism:—

1. This term primarily denotes *identity of crystal form*, which must show complete coincidence of the properties of symmetry, and approximate coincidence of geometrical constants.

2. The property of forming *mixed crystals* in any selected proportion, at least within certain limits.

3. The property of *mutual overgrowth*, *i.e.* a crystal of one substance (as a nucleus) increases in size in a solution of the other substance.

Ostwald (*Z. S. phys. Chem.*, **22.** 330, 1897) has proposed as a test of isomorphism the ability of a crystal to remove supersaturation of a solution and so act as nucleus for its crystallisation. This noteworthy criterion seems to be a combination of 2 and 3; but it will not be further considered here, as experimental tests of it are wanting, and that is precisely what is needed for any application in the study of isomorphism—at present an entirely empirical region.

There are a great number of examples known where different elements are exchanged in proportion to their atomic weights, and thus far no clear example has been found where the exchange implies any inconsistency of the accepted atomic weights.[1]

The following table contains the isomorphous series of elements, according to Arzruni.[2] The elements (or radicals) in a series can appear isomorphically in their analogous compounds, often in the ratio of their atomic weights (without change of crystalline form, and with little change in the geometric constants); the elements separated by a semicolon show isomorphism only in a few compounds. Regularly crystallising compounds are but little suited for showing isomorphism even when they have the same habitus.

Isomorphic Series

I. H(?), K, Rb, Cs, NH_4, Tl; Na, Li; Ag.

Example:—

Tl_2SO_4	rhombic	a : b : c = 0·5539 : 1 : 0·7319
$(NH_4)_2SO_4$	,,	,, 0·5643 : 1 : 0·7310
Rb_2SO_4	,,	,, 0·5723 : 1 : 0·7522
K_2SO_4	,,	,, 0·5727 : 1 : 0·7464
Cs_2SO_4	,,	,, 0·5805 : 1 : 0·7400
$KHSO_4$	,,	,, 0·5806 : 1 : 0·7489
NH_4HSO_4	,,	,, 0·6126 : 1 : 0·7436
Ag_2SO_4	,,	,, 0·5713 : 1 : 1·2382
Na_2SO_4	,,	,, 0·5914 : 1 : 1·2492

II. Be, Zn, Cd, Mg, Mn, Fe, Os, Ru, Ni, Pd, Co, Pt, Cu, Ca.
Example: $CoPtCl_6 . 6H_2O$, $FePtCl_6 . 6H_2O$, etc.

III. Ca, Sr, Ba, Pb.
Example: $CaCO_3$, $SrCO_3$, $BaCO_3$, $PbCO_3$.

IV. La, Ce, Di, Y, Er.
Example: $Di_2(SO_4)_3 . 8H_2O$, etc.

V. Al, Fe, Cr, Co, Mn, Ir, Rh, Ga, In, Ti.
Example: Cr_2O_3, Al_2O_3, Fe_2O_3, Ti_2O_3.

1 It should be noticed that the exchange of one element for another occurs not only in accordance with the ratio of their accepted atomic weights, but also the substituting element must assume the *valence* of the element substituted; thus in the thallium alums, thallium has the valence of a monad, imitating the monad valence of K, Na, etc.; in short, it is a *thallous*, not a *thallic* compound.—Tr.

2 "Relations between Crystalline Form and Chemical Composition," Braunschweig, 1893. In vol. i. of Graham Otto, *Lehrbuch d. Chemie*.

VI. Cu, Hg, Pb, Ag, Au.

VII. Si, Ti, Ge, Zr, Sn, Pb, Th, Mo, Mn, U, Ru, Rh, Ir, Os, Pd, Pt, Te.

Example: K_2PtCl_4, K_2PdCl_4.

VIII. N, P, V, As, Sb, Bi.

Example: As_2S_3, Sb_2S_3, Bi_2S_3.

IX. Nb, Ta.

X. S, Se, Cr, Mn, Mo, Wo; Te(?), As, Sb.

XI. F, Cl, Br, I; Mn : CN.

The elements B, Sc, C, O cannot be classified.

Usually isomorphism appears the more readily the more complex the compound; obviously because, as Kopp (1863) remarked, the dissimilar influence of the element on the crystalline form is overpowered by the influence of the remaining components. Thus the usual potassium and sodium salts are not isomorphous, but such complicated compounds as the alums $K_2Al_2(SO_4)_4 . 24H_2O$ and $Na_2Al_2(SO_4)_4 . 24H_2O$ are.

Mitscherlich discovered isomorphism from the four following salts crystallising in the tetragonal system, viz. H_2KPO_4, H_2KAsO_4, $H_2(NH_4)PO_4$, $H_2(NH_4)AsO_4$, and also from corundum and hematite.

Some of the best known series of isomorphous substances are the alums; $CaCO_3$ (as aragonite), $BaCO_3$, $SrCO_3$, and $PbCO_3$ (orthorhombic); $BaSO_4$, $SrSO_4$, $PbSO_4$ (in the same forms); $MgSO_4 . 7H_2O$, $ZnSO_4 . 7H_2O$, $NiSO_4 . 7H_2O$ (orthorhombic hemihedral). A complete classification has been given by H. Topsoë.[1]

It should also be noticed that elements and radicals may be mutually isomorphous, as for instance K and NH_4.

There is no doubt that isomorphism expresses a very remarkable law, but on further consideration, and especially by a careful examination of the abundant observations[2] made in this department, certain misgivings arise.

The condition of "approximate coincidence" of geometrical constants at once brings up the question: Where shall we draw the line between identity and difference of crystal form? Or if one seeks the decisive criterion for legitimate isomorphism in the capacity for mixed crystals wherever this is possible, he at once meets the same difficulty that we have already seen in the mutual miscibility of solid substances (p. 120), viz. of finding all conceivable gradations. Or finally, regarding the third mark of isomorphism, viz. the property of overgrowth, it has been observed that substances which do not possess the slightest

[1] *Tidskrift f. Fysik og Chemi.*, **8.** 5, 193, 321 (1869), and **9.** 225 (1870); see also the monograph on "Isomorphism," by Arzruni referred to above.

[2] It is especially recommended to read the work of Retgers already cited. A *résumé* may be found in the *Jahrbuch für Min.* (1891), **1.** 132, and in the *Chem. Centralbl.* (1891-92).

chemical or crystallographic analogy [as far as known], exhibit this property; and therefore, on this ground, its value as a criterion of isomorphism has been criticised by Retgers.[1]

Therefore, no criterion will be found in the future to decide in all cases, with an unequivocal yea or nay, whether isomorphism is present or not; but investigation should primarily be directed to the *degree of isomorphism*, since the discussion of such vague and idle questions as whether isomorphism is shown in any given case or not, is not a subject for deliberate investigation. The only question at issue is: What are the properties suitable for adoption as standard for the *degree of isomorphism?*

Of all the properties of a crystal, its *form* is very important and very striking to the eye; but that is no reason for conceding to it a significance, in precedence above the density, elasticity, optical constants, etc. That the first relations studied were those between crystal form and chemical composition, was obviously because this property obtruded itself upon the observer; as thus discovered it was an undoubted acquisition, even though it could not always be retained as a rule without an exception; but to elevate that discovery to the grade of a guiding principle, would amount to voluntarily assuming the fetters of a historical accident.

The analogy between two crystals is shown more clearly in their *miscibility* than in their crystal form, and thus the statement appears to be completely established, as Retgers so strongly emphasised in his investigations, that though a more or less complete similarity of form is not without interest, yet the whole problem appears to centre about the question of miscibility. The various grades of miscibility, according to Retgers, may be represented as follows:—

1. Miscibility in all degrees, a rather rare case. The physical properties—as well as the crystal form—vary gradually and evenly, in the series of mixtures, and are decidedly additive.

2. Limited miscibility, without the formation of double salts; the physical properties of the mixed crystals are emphatically additive, *i.e.* they can be calculated from those of the two substances when pure.

3. Limited miscibility, without double salt formation; the physical properties of the mixed crystals are additive here also, but in calculating the properties of the series of mixtures, it is necessary to ascribe properties to the end member on one side of a gap, which are different from those of the end member on the other side, and moreover properties different from those which actually occur, *e.g. another crystal form. Sometimes there is a labile (unstable) form for one end member which conducts itself in the free state as though it were produced by intermixture with the mixed crystal, the properties of which have been transported beyond the gap by extrapolation; the so-called iso-di-morphism or iso-poly-morphism.*

[1] *Zeitschr. physik. Chem.*, **5.** 460 (1890.)

4. Limited miscibility, with double salt formation, such that it indicates an important *chemical contrast;* the properties of the double salt are more or less different from those calculated from its end members.

5. The last degree; no marked miscibility, either with or without double salt formation.

Thus in the series of mixtures of ammonium and potassium sulphates, the specific volumes of the mixed crystals, which can be produced in every desired proportion, can be calculated quite accurately from the specific volumes of the end members, as was shown on p. 123; this pair of salts would be included under the first class just enumerated. The sulphates of iron and magnesium, which form an iso-di-morphous mixture, are doubtless to be referred to class 3.

It is obvious that a sharp line can be drawn at least between classes 1 and 2; the question as to the existence of a gap in the series of mixtures, can be decided definitely by experiment, provided we possess a solvent common for both of the crystals to be mixed. On this basis one should attempt to obtain miscibility in all proportions, as the decisive criterion of isomorphism. But this would amount to emphasising isomorphism as one particular phenomenon, and very unsuited for the purpose, because many crystals showing only a limited miscibility can, in all probability, be made miscible in all proportions by changing the temperature; just as many liquids, as water and phenol, which show limited miscibility at certain temperatures, at others dissolve each other in all proportions.

The study of mixtures is doubtless to be preferred to the narrow study of the crystal form, as shown by many of the cases observed by Retgers. "Thus if we were shown a prism of KNO_3, a rhombohedron of $NaNO_3$, a tabular crystal of $KClO_3$, and a cube of $NaClO_3$, the chemical composition of all four of these being unknown, no one would suppose, from this exhibit, that they were substances chemically analogous. But that this is the case is shown beyond a doubt by their crystallising together."

On the other hand certain cases suggest caution, because the property of forming mixed crystals does not belong exclusively to substances which are chemically analogous; this ammonium chloride, which, according to all experience hitherto, has shown unvarying properties, can unite, to a certain extent, with substances which are entirely different chemically;[1] solid benzene and iodine can crystallise together (p. 171).

In conclusion, let me define the present state of crystallography as it appears to the writer, in the following statements:—

The property of forming solid molecular mixtures is a very common property of solid substances; but in general, in most cases,

[1] Lehmann, *Zeitschr. f. Krystall.*, **8.** 438 (1883); Retgers, *Zeitschr. physik. Chem.*, **9.** 385 (1892).

mixed crystals can be prepared only up to certain limits, which limits are the points of mutual saturation. Every solid substance may contain traces, at least, of another substance, and thus there is formed a solidified solution (p. 169), even though very dilute. When the solid substance is a metal and the dissolved substance is a non-metal, or *vice versa*, the concentration is excessively small, and though it may be much greater in the preceding case, which deals with the solution of a solid salt in a salt, usually escapes notice. *The degree of miscibility grows with increasing chemical relationship*, so that the property of forming mixed crystals, within broad limits or in all proportions, finally appears in the case of substances which are completely analogous chemically. But in the case of complete miscibility, all the properties of the mixed crystal, including the crystal *form*, must be a regular function of the composition, exactly as in the cases of liquids and gases; but since a gradual, regular coincidence of crystalline form is capable of being experimentally realised, only when the two pure crystallising substances have *an original similarity* of form, it therefore follows, as a special case of the aforesaid more general rule, that the rule of Mitscherlich is true, viz. that chemically analogous substances usually have similar crystal forms.

If the series of mixtures shows a gap, the crystal forms of the end members may be very different from each other, since the chemical analogy may be very remote; but since each crystal on the one side of the gap is forced to adapt itself to the other crystal respectively, the very fact of a great extension of the series of mixtures indicates that each crystal shows this tendency; as a matter of fact, it is often observed that one crystal, even when in a perfectly pure state, may assume a labile form of the other (iso-di-morphism or iso-poly-morphism).

The rule that *miscibility varies directly with increasing chemical relationship*, is hidden in buried foundations, so to speak; it is a rule moreover which, in the case of liquids, is expressed in the statement that closely related substances dissolve each other in all proportions. Now, other things being equal, since a solid mixture is formed with greater difficulty than a liquid one, and especially so since the crystal form here exerts a limiting influence, much greater demands are made upon chemical analogy in the case of solids, than in the case of liquids.

The Periodic System of the Elements.[1]—In addition to the facts already described, by means of which the tables of the atomic weights can be completed with a large degree of confidence, there now appears a strong support in the shape of the so-called "natural" or "periodic" system of the elements; this unites certain relations between the atomic weights and the physical properties which have

[1] In the preparation of this section use was made of Lothar Meyer's *Grundzüge der theoretischen Chemie*. Leipzig, 1890.

been long suspected, into a complete, well-rounded system, and broadens our knowledge in many respects. But the fact that most of these regularities would disappear only by a radical change of the atomic weights, makes it appear that no attempt of that kind will be made within the immediate future.

As early as 1829 Doebereiner called attention to the fact that there are certain triads of elements which show close analogies in their physical properties and certain regularities in their atomic weights. The following table shows series of such similar triad groups, the atomic weights of which exhibit *differences which are fairly constant*:—

	Atomic Weights.	Diff.
Lithium	7·03	
		16·02
Sodium	23·05	
		16·1
Potassium . . .	39·15	
Calcium . . .	40·10	
		47·5
Strontium . . .	87·6	
		49·8
Barium	137·4	
Sulphur	32·06	
		47·14
Selenium . . .	79·2	
		48·4
Tellurium . . .	127·5	
Chlorine . . .	35·45	
		44·51
Bromine . . .	79·96	
		46·89
Iodine	126·85	

On the other hand, in the following table we find triad groups of related elements having *only slightly different atomic weights*:—

Iron	55·9
Cobalt	59·0
Nickel	58·7
Ruthenium	101·7
Rhodium	103·0
Palladium	106·5
Osmium	191·0
Iridium	193·0
Platinum	194·8

Attempts were not wanting to advance farther along the line of these regularities. But the thorough generalisation and consequent utilisation of this did not come till there appeared a simple classification of the elements in the works of Mendelejeff and of Lothar Meyer (1869); these two by slightly different ways arrived at the same conclusion, viz., that the properties of the chemical elements are periodic functions of their atomic weights; this will appear in detail from the following arrangement:—

H = 1.

0.	I.	II.	III.	IV.	V.	VI.	VII.	VIII.		
He 4	Li 7·03	Be 9·1	B 11	C 12·00	N 14·04	O 16·00	F 19			
Ne 20	Na 23·05	Mg 24·36	Al 27·1	Si 28·4	P 31·0	S 32·06	Cl 35·45			
A 39·9	K 39·15	Ca 40·1	Sc 44·1	Ti 48·1	V 51·2	Cr 52·1	Mn 55·0	Fe 55·9	Co 59·0	Ni 58·7
…	Cu 63·6	Zn 65·7	Ga 70	Ge 72·5	As 75	Se 79·2	Br 79·96			
Kr 81·8	Rb 85·4	Sr 87·6	Y 89	Zr 90·6	Nb 94	Mo 96·0		Ru 101·7	Rh 103·0	Pd 106
…	Ag 107·93	Cd 112·4	In 114	Sn 119·0	Sb 120·2	Te 127·6	I 126·85			
X 128	Cs 133	Ba 137·4	La 138·9	Ce 140		…	…			
…			Yb 173		Ta 183	W 184	…	Os 191	Ir 193·0	Pt 194·8
…	Au 197·2	Hg 200·0	Tl 04·	Pb 206·	Bi 208·	…	…			
…				Th 232·5		U 238·5	…			

In column 0 are the new elements discovered in the atmosphere (the so-called "noble" gases); they were of course unknown to Meyer and Mendelejeff, and have only recently been introduced into the periodic system by Ramsay. We shall return to this column later; in what follows here, it will be left out of consideration.

In each of the first two periods (horizontal rows) are seven elements, the respective members of which in the same column are very similar. In the four following rows we find some similarity of the elements in the same column; but we find the greatest similarity by omitting one, and comparing with each other K with Rb, Cu with Ag, Zn with Cd, Br with I, etc. The seventh and eighth rows are very incomplete; but perhaps elements as yet not much investigated will be found to follow in after Ce. The elements of the ninth row, as far as known, accord well with the corresponding members of the sixth.

The elements in the eighth column, and including the iron and platinum groups, occupy an exceptional position: the atomic weights of each group of three stand nearer together than most of the elements in the horizontal rows, and a similar statement may be made of their [physical] properties. The three triad groups together play the part each of an element, but at the same time of elements which do not fit well into the preceding scheme. This arrangement shows very clearly, as Mendelejeff emphasised, the relation between their respective chemical values, as compared with oxygen, and their atomic weights, as the following table of the oxides instead of the elements still further illustrates:—

I.	II.	III.	IV.	V.	VI.	VII.	VIII.
Li_2O	Be_2O_2	B_2O_3	C_2O_4	N_2O_5			
Na_2O	Mg_2O_2	Al_2O_3	Si_2O_4	P_2O_5	S_2O_6	Cl_2O_7	
K_2O	Ca_2O_2	Sc_2O_3	Ti_2O_4	V_2O_5	Cr_2O_6	Mn_2O_7	
Cu_2O	Zn_2O_2	Ga_2O_3	Ge_2O_4	As_2O_5	Se_2O_6	Br_2O_7	
Rb_2O	Sr_2O_2	Y_2O_3	Zr_2O_4	$Nb_2{}^*O_5$	Mo_2O_6	...	R_2O_8
Ag_2O	Cd_2O_2	In_2O_3	Sn_2O_4	Sb_2O_5	Te_2O_6	I_2O_7	
Cs_2O	Ba_2O_2	La_2O_3	Ce_2O_4	$Ta_2{}^\dagger O_5$	$W_2{}^\dagger O_6$	...	Os_2O_8
Au_2O	Hg_2O_2	Tl_2O_3	Pb_2O_4	Bi_2O_5	$U_2{}^\dagger O_6$		

But in spite of this regularity we must take warning not to follow it blindly; for, as is well known, many elements form other oxides than will fit into the preceding table, and also some elements which do not fit into this table, are omitted.

As a matter of fact, it cannot be denied that the arrangement given above does show certain contradictions which still wait for explanation. Thus there is no great chemical analogy clearly expressed for the copper, silver, and gold row, and the elements of the eighth column have not a fortunate location. Therefore, many modifications of Mendelejeff's arrangements have been suggested, as in recent times by Walker (*Chem. News*, **63**. 251, 1891), and J. Thomsen (*Z. S. anorg. Chem.*, **9**. 190, 283, 1895). The latter lays stress chiefly on the electropositive or negative character of the elements, and brings this out more clearly in his arrangement; Walker accentuates the distinction between metals and metalloids by the following scheme:—

Groups.			Even Series. 4	6	8	10	12
I.			K	Rb	Cs		
II.			Ca	Sr	Ba		
III.			Sc	Y	La	Yb	
IV.			Ti	Zr	Ce	...	Th
V.			V	Nb	Di	Ta	
VI.			Cr	Mo	...	W	U
VII.			Mn Fe	Ru	...	Os	
VIII.			Co Ni	Rh Pd	... ...	Ir Pt	
I.	Li	Na	Cu	Ag	...	Au	
II.	Be	Mg	Zn	Cd	...	Hg	
III.	B	Al	Ga	In	...	Tl	
IV.	C	Si	Ge	Sn	...	Pb	
V.	N	P	As	Sb	...	Bi	
VI.	O	S	Se	Te			
VII.	F	Cl	Br	I			
	1	3	5	7	9	11	
	Uneven Series.						

In this way oxygen comes with sulphur, fluorine with the halogens, etc.; also lithium is near the closely-related magnesium, beryllium is near aluminium, boron near silicon, etc. The non-metals are sharply divided from the metals: it is possible to draw a line separating these two groups. To be sure there are certain irregularities in this classification, as the placing of lithium and sodium beside the metals copper, silver, and gold, instead of with potassium, rubidium, and cæsium, to which they are obviously more closely related.

The elements discovered by Ramsay—helium, argon, etc.—are

completely neutral in chemical behaviour; Ramsay therefore arranges them in a new column, as having the valency zero. The atomic weights (only approximately known) harmonise with this view, as may be seen from the table, p. 184; actually these elements form a transition between the strongly positive and strongly negative univalent elements, being themselves neutral on account of their chemical indifference; the following table given by Ramsay illustrates these points:—

H 1	He 4	Li 7	Be 9
F 19	Ne 20	Na 23	Mg 24
Cl 35	A 40	K 39	Ca 40
Br 80	Kr 82	Rb 85	Sr 88
I 127	X 128	Cs 133	Ba 137

Remembering that hydrogen according to its physical properties is throughout of a metalloid character, and also in many compounds, such as the hydrocarbons, stands much nearer chemically to the halogens than to the univalent metals, its position at the head of the univalent metalloids is not without reason. A certain number, at any rate, of the elements are satisfactorily arranged, then, in the above order (see Ramsay, *Ber. deutsch. chem. Ges.*, **13.** 311 (1898), and *Modern Chemistry*, London, 1900, p. 50).

Staigmüller's (*Z. S. phys. Chem.*, **39.** 243, 1902) similar arrangement of the elements is given on p. 185; the noble gases are introduced in it in the same way as in Ramsay's. Various difficulties of the old arrangement are thus avoided; the elements in the last vertical column occur naturally; the metalloids are separated from the metals by the heavy line; moreover, C, B, Si, may very well be looked upon as metals, and in this way the grouping gains in simplicity of outline. By a slight rearrangement, already made in this table, the analogy is clearly brought out between the three triads, Fe, Ru, Os; Co, Rh, Ir; Ni, Pd, Pt (see the remarkable observations of H. Biltz, *Ber. chem. Ges.*, **35.** 562, 1902).

It should be mentioned that Newlands, as far back as 1864, attempted to arrange the elements systematically according to their atomic weights; for further history of the periodic system see K. Seubert, *Z. S. anorg. Chem.*, **9.** 334 (1895). On the didactic value of the periodic system, see Lothar Meyer (*Ber. deutsch. chem. Ges.*, **26.** 1230 (1893).

It is very remarkable that the atomic weights of a number of elements, especially those of low atomic weight, approximate closely to whole numbers; a hypothesis put forward on the strength of this by Prout (1815), that they are exact multiples of the atomic weight of

He 4	Li 7	Be 9·1	B 11	C 12											N 14	O 16	F 19
Ne 20	Na 23	Mg 24·4	Al 27·1	Si 28·4											P 31·0	S 32·1	Cl 35·5
A 40	K 39·1	Ca 40·1	Sc 44·1	Ti 48·1	V 51·2	Cr 52·1	Mn 55·0	Fe 56	Ni 58·7	Co 59	Cu 63·6	Zn 65·4	Ga 70	Ge 72	As 75	Se 79·2	Br 80·0
Kr 82	Rb 85·4	Sr 87·6	Y 89	Zr 90·6	Nb 94	Mo 96·0		Ru 102	Rh 103	Pd 106	Ag 107·9	Cd 112	In 114	Sn 119	Sb 120	Te 127	J 126·9
X 128	Cs 133	Ba 137·4	La 139	Ce 140													
			Yb 173		Ta 183	W 184		Os 191	Ir 193	Pt 195	Au 197·2	Hg 200·3	Tl 204·1	Pb 206·9	Bi 208		
				Th 232		U 238·5											

hydrogen is certainly false; but the fact remains that many elements nearly satisfy the hypothesis in a way that cannot possibly be accidental.

Rydberg (1886) undertook to follow out these unmistakable regularities, and in a recent very thorough study (*Z. S. anorg. Chem.*, **14**. 66, 1897) has brought to light a whole series of striking rules that will doubtless be of meaning in the future development of the periodic system. The atomic weights P can be put in the form P = N + D, where N is a whole number and D is a quantity always small compared with N and for the elements of low atomic weight, small compared with unity. If M be a whole number—Rydberg calls it the ordinal number of the element—elements of odd M have odd valency and odd values of N = 2M + 1; elements of even M have even valency and even values of N = 2M. The values of D, which can be pretty accurately determined in view of the above rules, form a pronounced periodic function of the atomic weight. From this follows the suggestion that in studies on the periodic system either N or perhaps better M should be taken as the independent variable instead of the atomic weight. Adopting this principle the difficulty disappears that is due to the newly determined atomic weight of tellurium (127·6 instead of 125). According to the views hitherto adopted, tellurium (127·6) would come after iodine—in a quite hopeless position; according to Rydberg, it should stay in its natural place with M = 60, N = 120, D = 7·6, the last value being more satisfactory than D = 5, as compared with the values of D for neighbouring elements. For hydrogen (with odd valency) M = 0, which well characterises its exceptional position. Neon and argon also come satisfactorily into the system if their atomic weights are taken as 20 and 40. Where the curve showing the relation of any property E to M passes through a maximum or minimum, the value of E of course varies little with M, and as such maxima and minima occur in about the same place for the most varied properties, elements arise with little difference in character ("twin elements," according to R. Lorenz, who has worked out a number of rules for these, *Z. S. anorg. Chem.*, **12**. 329; **14**. 103, 1897).

For further observations on regularity in atomic weights see Jul. Thomsen (*Bull. Acad. Danemark*, Dec. 14, 1894), and M. Töpler (*Ges. Isis Dresden*, 1896, Abh. 4).

Physical Properties of the Elements.—Some physical properties also stand in more or less definite relation to the atomic weights. A superficial consideration shows at once that the metallic elements (except the last series) are collected in the outer vertical columns only, while the metalloids are in the middle. The atomic volume, *i.e.* volume of 1 gram-atom of the element in the solid state, also comes clearly in its relation to atomic weight, as appears from Fig. 17.

The higher periods, which are very incomplete, are not included, as regular change of atomic volume cannot be recognised in them; the maximum is for the alkali metals, the minimum for the group C, Al, Ni, Ru, Os. Melting-points behave in the same way, as may be seen from the following table.

Melting-point of Elements on the Absolute Scale (from $-273°$)—(*n.g.* not melted, *s.h.* very high, *s.n.* very low, *ub.* over, *u.* under, *h.a.* higher than, *n.a.* lower than, *r.* red, *f.* colourless).

MELTING-POINTS OF THE ELEMENTS IN ABSOLUTE TEMPERATURE (*i.e.* FROM $-273°$).

I.	II.	III.		IV.	V.	VI.	VII.	VIII.		
H 13										
Li 453	Be ub. 1270	B s.h.		C n.g.	N 59	O 38	F s.n.			
Na 369	Mg 1070	Al 1000		Si s.h.	P r. 528 f. 317	S 388	Cl 171			
K 335	Ca h.a. Sr	Sc ?		Ti n.g.	V n.g.	Cr üb. 2270	Mn 2170	Fe 1977	Co 2070	Ni 1870
Cu 1355	Zn 691		Ga 303	...	As üb. 773	Se 490	Br 266			
Rb 311	Sr h.a. Ba	Y ?		Zr h.a. Si	Nb n.g.	Mo s.h.	...	Ru 2070	Rh 2270	Pd 1973
Ag 1241	Cd 591		In 449	Sn 503	Sb 700	Te 800	I 387			
Cs 299	Ba 1123	La üb. 710	Ce u. 1273	...	...	...	...	...	...	...
...	...	...		...	Ta n.g.	W s.h.	...	Os 2770	Ir üb. 2500	Pt 2050
Au 1345	Hg 233		Tl 563	Pb 597	Bi 542	...	...			
...	...	...		Th ?	...	U s.h.	...			

If the atomic weights are taken as abscissæ and melting-points as ordinates of a curve, the regularity comes out more clearly; the curve is wave-shaped, with maxima in the fourth or fifth columns. The curve of atomic volumes in Fig. 17 has a similar course, and comparison of the two shows that all the gaseous and easily fusible (liquid below red heat) are on the rising branches and maximal points of the volume curve; while the elements that are melted with difficulty, or are not fusible at all with the means now available, are on the descending branches and at the minima. Comparing the elements in each vertical series, *i.e.* each natural family, the melting-point is usually found to rise with increasing atomic weight; but the alkalis Li, Na, K, Rb, the group Zn, Cd, Hg, and probably the alkaline earths, Be, Mg, Ca, Sr, behave in the opposite manner. Indeed, all the rules based on the periodic system are of this approximate and uncertain character.

Of other physical properties that have a more or less marked periodic character we may note crystalline form (see the Isomorphous Series (p. 179) which has clear relations to the vertical columns of the periodic system (see also G. Winck, *Z. S. phys. Chem.*, **19**. 193 (1896), Ortloff, *ibid.*, 201), extensibility, thermal expansion, conductivity for heat and electricity, heat of formation of oxides and chlorides, magnetic and diamagnetic properties, refraction equivalents (see, on these properties, Lothar Meyer, *Modern Theories of Chemistry*, 1883, p. 144 ff.), "Hardness of the Free Elements" (Rydberg, *Z. S. phys. Chem.*, **33**. 353, 1900), "Change of Volume on Fusion" (M. Töpler, *Wied. Ann.*, **53**. 343, 1894), "Viscosity of Salts in Aqueous Solution" (Jul. Wagner, *Z. S. phys. Chem.*, **5**. 49, 1890), "Colour of Ions" (Carey Lea, *Sill. Am. Journ.* [3], **49**. 357, 1895), "Ionic Mobility" (Bredig, *Z. S. phys. Chem.*, **13**. 289, 1894), etc. The relations between emission of light and atomic weight are dealt with in the last section of this chapter.

Attempts to find quantitative relations in this region have so far proved almost without result. It is, however, possible that so far properties have always been compared under arbitrary conditions of temperature and pressure, and that exact laws are not to be expected thus; if, too, the elements have much more power of existing in several modifications than we know at present, this would introduce another factor of arbitrariness as the comparison is made between the modifications that happen to be known to us. Thus we find the most marked regularity in the solid state, which offers the safest basis of comparison, as the density of solids varies but little with temperature, pressure, or even with change of molecular condition.

Significance of the Periodic System in the Table of Atomic Weights.—At first the establishment of the periodic system was regarded as a discovery of the highest importance, and far-reaching conclusions as to the unity of matter were expected from it; more recently a disparaging view has been taken of these noticeable, but unfruitful, regularities. One now often finds them underrated, which is intelligible enough, since this region, which especially needs scientific tact for its development, has become the playground of dilettante speculations, and has fallen into much discredit. It is all the more

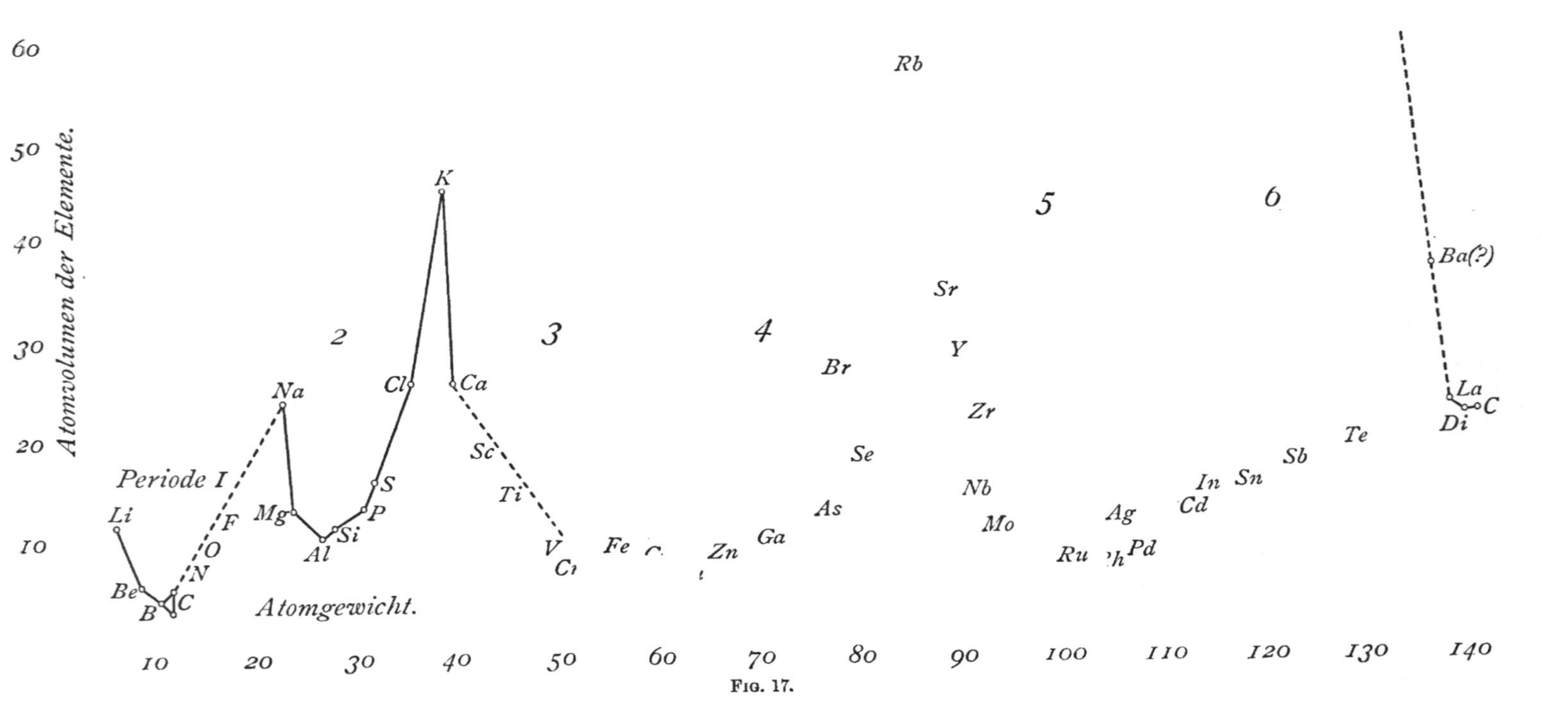

Atomvolumen der Elemente.
60
50
40
30
20
10
Periode I
2
3
4
5
6
Li
Be
B
C
N
O
F
Na
Mg
Al
Si
P
S
Cl
K
Ca
Sc
Ti
V
Fe
Zn
Ga
As
Se
Br
Rb
Sr
Y
Zr
Nb
Mo
Ru
Pd
Ag
Cd
In
Sn
Sb
Te
Ba(?)
La
C
Di
Atomgewicht.
10
20
30
40
50
60
70
80
90
100
110
120
130
140

Fig. 17.

satisfactory an intelligent and thorough investigation of this subject of fundamental importance in theoretical chemistry has begun anew.

The periodic system is of especial importance in settling atomic weights. Although the rules so far known do not possess—any of them—completely convincing validity, between them they give a striking proof that the choice of atomic weights underlying them is a happy one, and constitute a valuable indication in studying new or little known elements. It is usually possible to find the place of a new element in the periodic system by the analogies it shows to better known ones. Thus it was long doubtful whether beryllium had atomic weight $9{\cdot}08 = 2 \times 4{\cdot}54$ or $13{\cdot}62 = 3 \times 4{\cdot}54$; but only the former value fitted without strain into the periodic system, and since Nilson and Pettersson measured the vapour pressure of beryllium chloride and showed it to have, most probably, the formula $BeCl_2$, the lower atomic weight has been considered alone admissible. On the other hand, obvious gaps in the system lead to searches after new elements; scandium, gallium, and germanium, discovered since the establishment of the periodic system, have fitted in in this way. In such cases the chemical behaviour and some of the physical properties of still unknown elements can be predicted—Mendelejef did so for the metals just mentioned; this must be regarded as a second practical consequence of knowing the relations between the atomic weights of the elements and their properties. The periodic system, as may be seen from this, is of the greatest service to chemists as a mnemonic help in dealing with the huge mass of data in their science.

Besides the reasons mentioned above, there are other grounds for accepting the usual atomic weights as correct. The numerous recent measurements of molecular weight in solution have never led to contradiction with those values; again dissociation has been shown to occur in many of the polyatomic gases, but never in those regarded as monatomic by the usual molecular theory.

The Spectra of the Elements.—At the end of this chapter we are forced to this question, viz. in what way may we hope best to obtain a deeper insight into the nature and more intimate behaviour of these atoms, the relative magnitude of which we have already learned to determine by a number of distinct and safe methods. As far as we can judge at present, of all the physical properties, the consideration of the *spectra of the elements* offers altogether the best way to attack this problem.

An attempt to consider the phenomena of spectrum analysis, with even partial thoroughness, of course is beyond the limits of our task, which is to consider only those investigations which have served as a broader standpoint for the theoretical treatment of chemistry. But although spectrum analysis has been a great practical help in chemistry, and especially by its aid in the discovery of new elements, yet the hopes entertained of a theoretical explanation remain thus far unfulfilled. For this reason this section will give only a short summary concerning the emission spectra of the elements. Concerning the absorption spectra of organic compounds, another important class of spectrum phenomena, some reference will be made in a later chapter.

Irrespective of the so-called *phenomena of "luminescence,"* by which, after a proposal made by E. Wiedemann,[1] there is understood the property of certain substances of becoming luminous at very low temperatures, from external conditions, as illumination (photo-luminescence), electric discharge (electro-luminescence of gases), chemical processes (chemico-luminescence), crystallisation (crystallo-luminescence), gentle heating (thermo-luminescence), etc.; the normal development of light, *i.e.* light produced by the *heat movement of molecules*, and which we will consider exclusively in this section, is due to the *high temperature* of the substance emitting the light. According to Draper's law, all solid and liquid substances radiate light of shorter wave-lengths with increasing temperature, and at about 525° they all begin to glow with a dark red light, then become bright red, and finally "white hot." If one observes this gradual increase in glow with a spectroscope, there appears at first the red end of the spectrum, which gradually stretches over to the violet end with increasing temperature. Also all solid and liquid substances give *continuous spectra*; but the erbium and didymium earths give continuous spectra crossed by certain bright lines.

The glowing gases conduct themselves in a different, but a characteristic, way. These give out only, or at least mostly, rays of a definite wave-length; and therefore they give spectra, which consist of single bright bands sharply separated. According to the number and breadth of these bright bands, they are called *band-spectra* or simple *line-spectra*. Emission spectrum analysis, the introduction of which into science was the immortal work of Bunsen and Kirchhoff, and whose first-fruit was the discovery of several new elements, is based on this fact, viz. that under certain conditions of temperature and pressure every gas gives a definite, and, to a large degree, a characteristic spectrum. Therefore, in order to examine a substance by spectrum analysis,[2] it is necessary to bring it into the state of a glowing gas. For easily volatile substances, such as the salts of the alkalies, the flame of a Bunsen burner is sufficient; for substances volatile with great difficulty, one uses, according to the circumstances, the oxy-hydrogen blow-pipe, the electric spark, or the electric arc. As is well known, the production of monochromatic light, which is necessary for so many optical investigations, depends upon the selected rays emitted by incandescent gases. It

[1] *Wied. Ann.*, **34.** 449 (1888).

[2] For spectral analysis, see especially the review by Kayser in Winkelmann's *Handbuch der Physik*, Breslau, 1894, vol. ii. pp. 390-450. Of later work may be mentioned the extensive collection of references by Landauer, *Spektralanalyse*, Braunschweig, 1896.

should, however, also be pointed out that gases can give continuous spectra. Thus, according to Frankland, hydrogen, *e.g.* when burnt in oxygen under a pressure of twenty atmospheres, gives a continuous spectrum.

It cannot be doubted that each molecular species, whether representing isolated atoms or chemical compounds, has its own characteristic spectrum; but the question as to which particular molecular species corresponds to an observed spectrum, can be very rarely answered with certainty. The difficulties which are met consist especially in this, viz. that it is very difficult to discriminate in regard to the molecular condition of an incandescent gas; that this is usually very much complicated at high temperatures where the capacity of the substance for reaction is apparently very much increased, and certainly is totally changed; and that *the spectrum observed where a compound or even an element is volatilised, consists of the superimposed spectra belonging to different molecular species.* The dissociation phenomena, which, from all that is known, occur more often at high temperatures than under ordinary conditions, play an especially important *rôle.* It is certain in many cases that the volatilised substance, by reaction with the gases of the flame or with atmospheric air, forms new molecular species, and although only very slight quantities of these may be present, yet such is the great sensitiveness of the spectroscope, that these may be sufficient to produce clearly recognisable lines. We have good reason for assuming that *single atoms give line spectra, and atomic complexes give band spectra.*

Regularities in the Distribution of the Spectral Lines of the Elements.—Inasmuch as there is scarcely any reason to doubt that there is the most intimate dependence of emission spectra upon the configuration and state of vibration of the molecule and atom of a luminous substance, it requires only a glance at the laws (by which, on the one hand, the lines of the same substance arrange themselves in the spectrum, and by which, on the other hand, the arrangement varies with the substance), to hope that very soon, perhaps, we shall have further disclosures concerning the behaviour and the condition of movement of the atom. Although at present we are far removed from a thorough knowledge of the laws concerning the subject, yet already a beginning has been made which is worthy of attention, and which spurs us on to a continued search for the goal.

The most important result thus far obtained is in the calculation of the lines of the *first hydrogen spectrum* so-called, which is obtained by a Geissler's tube, for example, in which the pressure must not be too small. As Balmer[1] discovered, the wave-length λ, of each line can be calculated very accurately from the simple formula

$$\lambda^{-1} = A(1 - 4m^{-2}), \quad \text{or } \lambda = \frac{1}{A} \cdot \frac{m^2}{m^2 - 4},$$

[1] *Wied. Ann.*, **25.** 18 (1885).

in which λ^{-1} denotes the vibration number; the successive numbers 3, 4, 5, etc., are to be substituted for m. By substituting in this formula the value 3647·20 of the constant $\frac{1}{A}$, for the calculated values corresponding to m, the following wave-lengths are obtained:—

The Hydrogen Spectrum

Line.	m	Calculated.	Observed.	Difference.
Hα	3	6564·96	6564·97	0·0
Hβ	4	4862·93	4862·93	0·0
Hγ	5	4341·90	4342·00	+0·1
Hδ	6	4103·10	4103·11	0·0
Hε	7	3971·40	3971·40	0·0
Hζ	8	3890·30	3890·30	0·0
Hη	9	3836·70	3836·80	+0·1
Hθ	10	3799·20	3799·20	0·0
Hι	11	3771·90	3771·90	0·0
Hκ	12	3751·40	3751·30	−0·1
Hλ	13	3735·60	3735·30	−0·3
Hμ	14	3723·20	3722·80	−0·4
Hν	15	3713·20	3712·90	−0·3

The wave-lengths in the "observed" column are taken from the latest measurements by Ames,[1] and are expressed in ten-millionths of a millimetre; the correspondence between the wave-lengths observed, and those calculated from Balmer's formula, is very remarkable.

The preceding formula does not apply at all to the so-called *second hydrogen spectrum;* this consists of very numerous fine lines; it is suspected that the two spectra belong to two different molecular conditions of hydrogen. It is possible that the discovery of further regular cases of this sort may be rendered difficult from the fact, that the production of the spectra of other elements is due to the superimposition of spectra belonging to several different molecular conditions, and that their separation does not take place automatically, as in the case of hydrogen, where we obtain the first spectrum by itself.

Concerning the relations which should exist between the spectra of allied elements, it has been shown by Lecoq de Boisbaudran, that the spectra of the alkali metals are displaced towards the red end of the spectrum, with increasing atomic weight. Recently Kayser and Runge[2] volatilised a number of metals by means of the electric arc,

[1] *Phil. Mag.* (5), **30.** (1890); see also Cornu, *Journ. de phys.* [2], **5.** 341.

[2] *Wied. Ann.*, **41.** 302 (1890), 43,385 (1891), more completely in *Abh. d. Berl. Akad.* 1890, 1891, 1892; collection of results on p. 200 of the section of Winkelmann's hand-

and photographed their spectra by means of a Rowland concave grating; the wave-lengths of the lines were exactly determined to within a hundred-millionth of a millimetre. They showed that the lines may be fairly well calculated by means of the formula,

$$\lambda^{-1} = A - Bm^{-2} - Cm^{-4};$$

this formula is a generalisation of Balmer's formula; yet although it has two more constants than the other, the observations of the metals are not attended with the same degree of success as was the case with hydrogen. Moreover, it is not possible to obtain the spectrum of a metal by a formula like this; but it must be resolved into a number of series in order to calculate the values of the constants A, B, C.

Now it happens that for the elements of Mendelejeff's first group, the series (of which each element has several), consists not of lines, but of pairs of lines. Kayser and Runge distinguish the following series:—

1. The Principal Series; their pairs are the strongest lines of the spectrum, and are easily reversible, *i.e.* they appear dark when the vapour has a sufficient density; their vibration difference diminishes with increasing "m."

2. The First Subordinate Series; strong but very hazy pairs of lines; they have a constant vibration difference.

3. The Second Subordinate Series; weaker pairs of lines, but better defined, with a constant vibration difference, as in 2.

The first pair of the principal series has also this same vibration difference: this difference is the most important spectroscopic constant of the element in question, and, as will appear in a moment, it seems to stand in a definite relation to the respective atomic weight. The principal series appears to be found only in the case of the alkali metals; the spectra of all other metals seem to consist only of secondary series; no series are found in the spectrum of barium; the spectrum of lithium seems to consist of only simple lines.

The elements that belong to the same vertical column of the periodic system show an obvious similarity in the construction of their spectra; this is established at any rate for the first two groups and a part of the third; in the others it is as yet little known. These three groups can each be divided into two parts that show special chemical analogy; thus we have I. Li, Na, K, Rb, Cs; II. Cu, Ag; III. Mg, Ca, Sr; IV. Zn, Cd, Hg; V. Al, In, Tl. The relations inside each of these groups are particularly close, as may be seen from the following table. Within each group the spectrum is displaced towards the red by increase in atomic weight; but in passing from one group to the next, strongly towards the blue. The rules that

book quoted above. Rydberg put forward similar views independently of Kayser and Runge, see *Svenska Akad. Handl.*, **23.** (1889-1891); *Wied. Ann.*, **50.** 629 (1893); *Astrophysical Journal*, **6.** 233 (1897).

hold are best seen from the numbers in the accompanying table, which give the constants A, B, C (multiplied by 10^8) for the first line of each pair or triplet in the subsidiary series :—

	Allied Series.			Allied Series.			ν	$\frac{\nu}{a^2}$
	A.	B.	C.	A.	B.	C.		
Li	28,587	109,625	1,847	28,667	122,391	231,700	...	...
Na	24,475	110,065	4,148	24,549	120,726	197,891	17	325
K	21,991	114,450	111,146	22,021	119,363	62,506	57	381
Rb	20,939	121,193	134,616	...	...	...	234	322
Cs	19,743	122,869	305,824	...	...	...	545	309
Cu	31,592	131,150	1,085,060	31,592	124,809	440,582	249	622
Ag	30,712	130,621	1,093,823	30,696	123,788	394,303	921	794
Mg	39,796	130,398	1,432,090	39,837	125,471	518,781	41	713
Ca	33,919	123,547	961,696	34,041	120,398	346,067	102	638
Sr	31,031	122,328	837,473	...	...	...	394	517
Zn	42,945	131,641	1,236,125	42,955	126,919	532,850	386	918
Cd	40,755	128,635	1,289,619	40,797	126,146	555,137	1159	929
Hg	40,159	127,484	1,252,695	40,218	126,361	613,268	4633	1161
Al	48,308	156,662	2,505,331	48,245	127,527	687,819	112	1534
In	44,515	139,308	1,311,032	44,535	126,766	643,584	2213	1721
Tl	41,542	132,293	1,265,223	41,506	122,617	790,683	7795	1879

These numbers show that A falls off with increase of atomic weight within each series—A being the frequency for $m = \infty$ or limiting frequency; *each series, therefore, is pushed towards the red end by increasing atomic weight.*

B and C are nearly the same for the lines of a pair or triplet belonging to the same series. The B values vary but little—especially those of the second series.

The constant difference between pairs (or the first two lines of triplets) in both series is given under the heading ν; in the last column this quantity is divided by the square of the atomic weight a. Within each of the five groups this quotient is roughly constant, *i.e.* the width of the pairs and triplets, measured in frequency of vibration, is approximately proportional to the square of the atomic weight within each group.

Series of subsidiary character have also been found in the spectra of oxygen and sulphur. The lines of the series are triplets.

There are likewise found in *the band spectra* of the metalloids certain regularities which are partially related to the preceding. Thus Deslandres[1] found that the distribution of the bands can be

[1] *C. R.*, **104.** 972 (1887); **106.** 842 (1888); **110.** 748 (1890); **112.** 661 (1891); *Ann. chim. Phys.* [6], **14.** 5 (1888).

expressed by the formula

$$\lambda^{-1} = A + Bm^2;$$

when, again, "m" denotes the vibration number, and A and B constants which are characteristic for the respective series of the element in question; λ denotes the wave-length of some characteristic line of the band which serves as its representative: thus the distribution of lines within any one band is given by the formula

$$\lambda^{-1} = a + bm^2,$$

where again "a" and "b" are constants. By making $m = 0$, we obtain the fundamental vibration of the band, *i.e.* the line which, as stated above, was chosen by Deslandres to represent the band. This last rule requires that each of the bands belonging to the same series of an element, shall show a manifest similarity as regards the number of their maxima and minima of brightness, and also as to their relative distances and arrangement; while the former rule requires just what was found to appear in the case of the line-spectra of the metals.

As is obvious from what has preceded, the following law holds good, viz. *the farther we advance toward the violet (more refrangible) end of the spectrum, the closer do the lines approach each other*, so that the line-spectrum prevails almost exclusively at the ultra-violet end.

According to Riecke,[1] the difference between line and band spectra is that in the former the frequencies in the series form a simple progression, like those of a string or an organ pipe; the waves are therefore due to atomic or molecular vibrations similar to those of linear systems; while the vibrations of molecules producing band spectra are determined by three independent conditions, and are thus analogous to the vibrations of bodies possessing extension in space.

[1] *Lehrbuch d. Physik*, 2 edit. Bd. 1, pp. 518 and 520 (1903).

CHAPTER II

THE KINETIC THEORY OF THE MOLECULE

General Observations.—As we have thus, in the preceding chapter, considered the properties of the atoms, which are to be regarded as the building stones which by their union produce molecules, we will in this and the following chapter seek to frame a conception of the structures which these building-stones form. It cannot be denied that we depart from simple experiment at the outset: if it were possible to construct a system of theoretical chemistry without speculations of that sort, but under the safe guidance of thermodynamics, even then much important knowledge would escape us. "It is certainly an advantage that the laws of thermodynamics contain truths which cannot be shaken, since they do not rest upon any hypothesis of the constitution of substances. But if one should refrain from a thorough investigation into the being of substances, from a fear that he must thereby abandon the region of indestructible truths, he would wilfully ignore a path leading to new truths."[1]

There are two very different ways which lead to conceptions regarding the world of the molecule.

Thus, on the one hand, we may seek to derive many properties of molecules, and at the same time the properties of the substances formed from them, by means of purely mechanical principles: this procedure primarily would have a purely deductive nature; but with the further development of this method, since it serves to bring the properties of the substance into an increasingly intimate connection with the properties of the molecule, here also one must work inductively, as is invariably true in science. We regard it as the greatest result of this method of investigation, that it makes the nature of heat intelligible by means of simple kinetic conceptions.

On the other hand, by studying the innumerable carbon compounds, one is led to thorough conceptions regarding the arrangement of the atoms in the molecule. The conjectures which at first were

[1] Van der Waals, *Kontinuität*, etc., p. 119. Leipzig, 1881.

vague, begin to assume tangible form, and they support the work of the investigators so effectively that the fear that chemists would branch out into hypotheses almost too daring, must be regarded as groundless in view of the undeniable results to which these hypotheses have led: in this way arose *structural chemistry* and its consequence, stereo-chemistry.

In this chapter we will consider the molecular theory in its physical aspect, and on account of the limited space in this work, only the more important subject will be noted. It should be noticed, as a matter of historical interest, that as early as 1740 Daniel Bernoulli entertained conceptions regarding the behaviour of gases which were essentially identical with the views universally entertained at present; but it was not till 1845 that J. J. Waterson presented before the London Royal Society a paper containing a very happy and complete development of the kinetic theory of gases. Unfortunately the paper at that time was not printed, and its publication was accomplished only lately by Lord Rayleigh,[1] who found it in the archives. Thus it happened that Kroenig in 1856, and Clausius in 1857, developed essentially the same views independently. After the fundamental ideas were thus clearly established, Clausius, Maxwell, Boltzmann, van der Waals, and others, participated in the further development of the theory.

For the literature see O. E. Meyer, *Kinetic Theory of Gases*, in which weight is laid on simplicity of presentation and experimental proof; and especially Boltzmann, *Gas Theory*, i. and ii. (Leipzig, 1895-1898), which aims rather at the most exact working out of the fundamental hypotheses, also the remarkable work of Clausius, *Mechanical Theory of Heat*, iii. (Braunschweig, 1889-1891), which unfortunately was never finished.

The Kinetic Theory of Gases.—This theory represents the first successful attempt to explain a number of the material properties of substance, by means of assumptions concerning the nature of the molecule which are characterised by simplicity and clearness; the theory starts out by assuming that the molecule of a gas is indeed very small, but yet has *a certain finite extension*, so that the space occupied by the molecule itself, or by its sphere of activity, is very small as compared with the volume occupied by the gas as a whole. The molecules are separated by distances which are very *large* in comparison with their own size, and therefore exert no marked influence on each other. Only when they come very close to one another do *colliding* forces appear, and these at once cause them to separate; or, in other words, the molecules conduct themselves on collision like *absolutely elastic* bodies.

Let us consider the path of any selected molecule; this will

[1] *Phil. Trans.*, **183.** 1 (1892).

continue to move with a certain uniform velocity, since it is not subject to the action of any force, until it collides with another molecule; after rebounding from this, it will, as a rule, move with a uniform velocity, but changed in direction and magnitude, till there follows another collision, etc. The path of a molecule will be a zigzag, the component parts of which will range through fluctuating velocities which vary around a certain mean value. Thus the length of the free path, *i.e.* the path traversed without colliding with any other molecule, the so-called "free path," is constantly subject to change; but when the external conditions are maintained constant, it varies about a definite mean value, viz. the "mean free path."

The *mean kinetic energy of the motion* of translation of any particular molecule amounts to $\frac{m}{2}u^2$, if "m" denotes its mass, and "u" its mean velocity. But besides the energy of the motion, the molecule has a certain internal energy which is given by the kinetic energy both of the rotatory motion of the molecule, and also of the vibration of the atoms composing the molecule. Regarding the internal energy of a molecule, we must assume that it is by no means constant, but that it also, in the course of time, varies about a certain mean value.

That pressure, under which an enclosed volume of gas stands, and which it conversely exerts on the walls of the enclosing vessel, can be at once calculated from this assumption without knowing anything more; it is occasioned by the blows of the molecules on the walls of the vessel as they strike and bound back from it; and it is obvious that, other things being equal, the number of these blows is proportional to the quantity of the molecules in unit volume, *i.e.* to the density of the gas. But this only amounts to saying that the pressure exerted by an enclosed quantity of gas must be inversely proportional to the space occupied by it, as is explained by the Boyle-Mariotte law (p. 38).

In order to calculate the pressure quantitatively, let us suppose a cube, the content of which represents unit volume, and the sides of which therefore represent unit area, to contain any selected quantity of a simple gas; let the mass of one molecule of the gas be "m," and the number of molecules be N; then the density of the gas will be represented by the equation

$$mN = \rho.$$

The mean square-velocity u^2 of the molecule determines the pressure exerted, as is easily seen; for both the force and the frequency of the blows of the gas molecules depend upon u. A molecule striking in a direction normal to a cube face, will be thrown back in a reverse direction, but with the same velocity, and has therefore the same degree of inverse momentum; the momentum therefore of the wall experiences the change

$$2mu.$$

Now if we denote the number of molecules striking the cube wall in unit time by ν, then the pressure exerted on unit of surface amounts to

$$p = 2mu\nu.$$

In order to calculate ν, let us imagine the irregular movement of the molecules to be arranged in order, for an instant; and during this time, let them all move in the same direction, and that at right angles to one of the sides of the cube, and with the mean velocity u; then during this moment the number of molecules which will bound back from the wall, provided the external conditions remain constant, will be

$$Nu,$$

and it is obvious that during this time, this wall only will be subjected to the total pressure exerted by the gas. But as a matter of fact, this total pressure distributes itself between the six cube faces in the movements actually taking place; then in reality we will obtain[1]

$$\nu = \frac{Nu}{6},$$

and the pressure sought therefore will be

$$p = \frac{1}{3} Nmu^2 = \frac{1}{3} \rho u^2.$$

From this equation there can be calculated the mean velocity u for all sorts of gases. For example, let us consider a quantity of hydrogen enclosed in 1 c.c. at 0° and under atmospheric pressure; the weight of this amounts to 0·00008988 g. and the pressure exerted by this upon 1 sq. cm. amounts to 1033·3 g. weight, or 1033·3 × 980·6 absolute units (980·6 = gravity acceleration).

Thus the value of u is found to be

$$u_0 = \sqrt{\frac{3 \times 1033{\cdot}3 \times 980{\cdot}6}{0{\cdot}00008988}} = 183{,}900 \frac{\text{cm}}{\text{sec}}.$$

According to this the hydrogen molecules move with a velocity which averages about the enormous mean value of 1·84 kilometers per second. Since, according to Avogadro's rule, ρ is proportional to the

[1] This conclusion can really be based upon more satisfactory reasoning (p. 207 of the above quoted work of Boltzmann).

molecular weight M, we can find the velocity for other gases by the formula

$$u_0 = 183{,}900\sqrt{\frac{2{\cdot}016}{M}}\frac{\text{cm.}}{\text{sec.}} = \frac{261{,}100}{\sqrt{M}}.$$

We observe that u is independent of the pressure; but u increases with increasing temperature; and since p is proportional to the absolute temperature, then $\frac{m}{2}u^2$, *the mean kinetic energy of the translatory motion, is proportional to the absolute temperature; and conversely the mean kinetic energy of the molecules of a gas is a measure of the temperature.*

Now this definition of temperature has led to entirely new conceptions of the nature of heat. The content of heat of a body, no matter what the state of aggregation, in the sense of these kinetic considerations, is represented by the sum total of the kinetic energy of its molecules. The kinetic energy consists of *the energy of progression* of the molecules (or more correctly, of their centres of gravity), together with *their internal energy;* this latter consists of the kinetic energy of the rotatory movements of all the molecules, and also of the movements performed by the atoms in the molecular groups. Apparently the former variety [the molecular movement] is proportional to the absolute temperature, not only in the case of gases, but also universally. According to this, at absolute zero, $-273°$, all movements of the molecule would cease, and matter would lapse into the death of heat.

The irregular movement of the molecule, which constitutes the heat content of a substance, primarily is not different from the orderly movement of a body, *i.e.* a movement whereby all the molecules move through space in the same direction, and with the same velocity, when the body moves as a whole. But from the practical point of view, it makes a great difference whether the problem given to the experimenter is to determine the work expended in the kinetic energy of the regular, or of the irregular movements of matter. Let us imagine kinetic energy to be introduced into any body in two forms: first, in the form of orderly motion, whereby it attains a definite velocity either of progression or of rotation, as a whole; second, in the form of irregular motion (heat), whereby the particular molecules of the body attain velocities of progression and rotation which vary in degree and direction,—in other words, the temperature of the body is raised a certain amount depending on its specific heat. There is no difficulty in the first case in ascertaining the amount of energy in the substance, as by applying it to the accomplishment of a certain amount of work, or in causing it to warm another substance, or even itself. But to apply the total energy of irregular motions to the performance of any kind of work, to change it completely into energy of progressive motion of another body, this is at present an insoluble problem for our human powers of experimentation. The problem could be easily solved by a being small enough to consider energy of motion of molecules in the same way that we consider the motions of large bodies, but not by beings like us to whose senses the molecules are imperceptible. However, if we cool the original body and collect the heat introduced, by suitable intermediate substances, we can, at least in part, convert this into energy of progressive motion. In this way, by the aid of kinetic considerations, we get a clear conception of the mutual convertibility of heat and external work; and as the first law of thermodynamics, *i.e.* the indestructibility of energy and the equivalence of heat and work, follows directly from the definition of heat as kinetic energy, which is obtained from kinetic principles, so we are also led to the second law of thermodynamics,[1] *i.e.* the limited convertibility of heat into external work, by the essential difference, which is noted according to our present condition in the art of experimentation, between the realisation of the kinetic energy of regular (macroscopic) and of irregular (molecular) motion.

The Rule of Avogadro.—Let us compare two different gases at the same temperature and pressure; and let us denote by N_1, m_1, and u_1, the number of molecules in unit volume, their mass, and their mean velocity respectively for one gas, and similar values for the other gas by N_2, m_2, and u_2 respectively; then the common pressure p of both gases, by the formula (p. 203), must amount to

$$p=\frac{1}{3}N_1m_1u_1^2=\frac{1}{3}N_2m_2u_2^2 \quad . \quad . \quad . \quad (1)$$

Now, experiment shows that by mixing different gases there occurs no change of pressure or temperature; the different molecular species preserve their kinetic energy unchanged after the mixture. We might make the very improbable assumption that one kind of molecule gains as much kinetic energy as the other loses, on mixture; but even this assumption is contradicted by experience, for each gas in a mixture exerts (*e.g.* on a semipermeable partition) the same pressure as if the others were not there, and that would not be possible if the addition of another gas affected its kinetic energy. Hence in the mixture the two kinds of molecules must have mean kinetic energy $\frac{1}{2}m_1u_1^2$ or $\frac{1}{2}m_2u_2^2$ respectively. But as the two gases are in thermal equilibrium they must have the same mean kinetic energy, else, according to the laws of collision of elastic spheres an interchange of energy would occur.

[1] Boltzmann, the work quoted above, p. 207. To him we owe also the deduction of the second law of thermodynamics from kinetic concepts (1886).

Therefore we find that

$$\frac{1}{2}m_1u_1^2 = \frac{1}{2}m_2u_2^2 \quad . \quad . \quad . \quad . \quad . \quad (2)$$

and accordingly

$$N_1 = N_2 \quad . \quad . \quad . \quad . \quad . \quad (3)$$

That is, unit volumes of all the different gases, at the same temperature and pressure, contain the same number of molecules; or the molecular weights of gases have the same ratio to each other as their densities. Thus the law of Avogadro follows from the standpoint of the kinetic theory, a result of fundamental importance for physical chemistry, since its conclusions are largely based on Avogadro's law.

The Ratio of the Specific Heats.—As already pointed out, the heat content of 1 g.-mol. of a gas having the molecular weight M, consists of the energy of the progressive motion of the molecule $\left(=\frac{M}{2}u^2\right)$, plus the energy of the internal movements of the molecule. If we denote the increase of the internal energy per degree of temperature by E, then the specific molecular heat at constant volume C_v (see p. 49) will be

$$JC_v = \frac{1}{2}M\frac{u^2}{T} + E;$$

and the specific molecular heat at constant pressure will be

$$JC_p = \frac{1}{2}M\frac{u^2}{T} + E + \frac{1}{3}M\frac{u^2}{T};$$

in which

$$\frac{1}{3}M\frac{u^2}{T} = \frac{pv}{T}$$

refers to the external work performed; J is the mechanical equivalent of heat; the ratio of the two specific heats is given by the equation

$$\frac{C_p}{C_v} = k = \frac{\frac{5}{6}M\frac{u^2}{T} + E}{\frac{1}{2}M\frac{u^2}{T} + E}$$

Inasmuch as E, from necessity, has a positive value, k must always be less than $\frac{5}{3} = 1{\cdot}667$; and k approaches this limiting value only when E is very small. If, on the other hand, E becomes very great, then k approaches the value 1. These anticipations are most perfectly established by experiment, as the following table shows :—

	k		k
Mercury . .	1·666	Chloroform .	1·10
Oxygen . .	1·404	Methyl ether .	1·113
Nitrogen . .	1·410	Ethyl ether .	1·029
Ammonia . .	1·30		

In only one case, viz. that of mercury, in which k was measured by Kundt and Warburg[1] by the "dust figure" method (p. 51), does the ratio of the specific heats reach its upper limit; but mercury is a monatomic gas; and it was to be expected on *a priori* grounds that, in this case, the internal kinetic energy would have an infinitesimal value as compared with the energy of progression. This brilliant coincidence between anticipation and experiment is one of the most beautiful results of the kinetic theory of gases.

In case of the other gases studied, k is always smaller than 1·667, and in general it falls below this upper limit, *i.e.* the internal energy is larger as compared with the external energy, just in proportion as the number of atoms in the molecule increases; and this is in accord with the theoretical explanation, that as a molecule becomes more complicated, a greater fraction of the heat introduced will be used in increasing the kinetic energy of the atoms in the molecule: thus in the case of ethyl ether, the relative difference between the specific heats is very small in comparison with their absolute magnitudes.

The molecular heat at constant volume may thus be calculated for monatomic gases, for which $E = 0$, in absolute measure; it is $C_v = \frac{M}{2}\frac{u^2}{T}$. But $u^2 = \frac{(261{,}100)^2 T}{273 \times M}$, therefore

$$C_v = \frac{(261{,}100)^2}{2 \times 273} \text{ absolute units}$$

$$= \frac{(261{,}100)^2}{2 \times 273 \times 41 \cdot 77 \times 10^6} = 2 \cdot 990 \text{ cal.}$$

Direct measurement of $C_p/C_v = 1 \cdot 666$ for mercury vapour together with the relation $C_p - C_v = 1 \cdot 991$ (p. 49), gives $C_v = 2 \cdot 989$. The remarkable agreement of these numbers is, of course, only another version of the test of theory already given. The monatomic gases do not fall under Le Chatelier's rule (p. 46); their molecular heat at constant pressure does not converge to 6·5 at the absolute zero, but remains at $2 \cdot 99 + 1 \cdot 99 = 4 \cdot 98$ for all temperatures. But by a slight change in the value of d the rule might be put in the form $C_p = 4 \cdot 98 + \alpha T$; it would then have a very simple kinetic meaning, that with fall of temperature the energy of internal movement lessens relatively to that of translation, and vanishes at the absolute

[1] *Pogg. Ann.*, **157** 353 (1876).

zero; this would point—in harmony with kinetic views on dissociation—to a shrinking of the molecules to massive points with no energy of rotation at the absolute zero. Lord Rayleigh, it is well known, found the ratio of the specific heats 1·667 for the newly discovered gaseous elements, helium, argon, etc., from which, by analogy with mercury, they were concluded to be monatomic. This is in accord with their position in the periodic system (p. 187).

The Mean Length of the Free Path.—The kinetic theory of the gaseous state has in a similar way led to a clearer conception concerning a number of other properties of gases, and in particular it has thrown light upon the *properties of diffusion, of internal friction, and of the conduction of heat*. All of these properties can be explained by the backward and forward motion of the gas molecules; these motions cause the impinging layers of gases of different composition to mingle with each other (ordinary diffusion); they also bring about the equalisation of the different velocities of the various gas molecules (internal friction), and thus effect the exchange of kinetic energy (conduction of heat). These three properties appear to be very closely related; we may define the first as the diffusion of matter, the second as the diffusion of momentum, and the third as the diffusion of kinetic energy (heat).[1]

Each one of these three processes has a very close dependence upon the "mean free path," L, of the molecule, which, according to Clausius, is calculated to be

$$L = \frac{\lambda^3}{\frac{4}{3}\pi s^2},$$

where we denote by λ the mean distance of the supposedly spherical molecules; by λ^3 the cube which in cross-section contains a molecule; and by s, the distance of the centres of gravity of two molecules on collision, *i.e.* when at the nearest approach that they can make to each other. The free path is inversely proportional to the number of molecules in unit volume, *i.e.* it is inversely proportional to the density of the gas. In deriving this formula, it is presupposed that s is small compared with L, and also that the same (mean) velocity of progressive motion is to be ascribed to the molecules. If one dispenses with this last assumption, then in the preceding formula, according to Maxwell, in establishing his law of distribution, there is obtained $\sqrt{2}$ instead of $\frac{4}{3}$; (*i.e.* 1·41 instead of 1·33).

The first determination of a mean free path was made by Maxwell

[1] See Maxwell, *Theory of Heat*, chap. xxii.

(1860), who obtained for the value of η, *i.e.* the internal friction of a gas, the equation

$$\eta = \frac{L\rho \frac{12}{13} u}{\pi},$$

where ρ denotes the density, and $\frac{12}{13}u$ denotes the mean velocity calculated from the law of distribution. Later, the mean free path was calculated from the conductivity of heat (Maxwell and Clausius), and from the diffusion of gases (Maxwell and Stefan), and in all cases the values, found by these different methods, coincided with each other, at least approximately.

Thus O. E. Meyer[1] calculated these values at 20°, and 760 mm. pressure, for the following substances—

	L
Hydrogen	0·000185 mm.
Methane	0·000085 ,,
Carbon monoxide	0·000098 ,,
Carbon dioxide	0·000068 ,,
Ammonia	0·000074 ,,

The Maxwell law of distribution is this: out of N molecules of a gas, the number whose velocity lies between v and v + dv at any moment is

$$4\sqrt{\frac{27}{8}}\,\frac{Nv^2}{u^2}\,e^{-\frac{3v^2}{2u^2}}dv;$$

where u is the molecular velocity already calculated (p. 204), u, according to its deduction, expresses the velocity corresponding to the mean kinetic energy of the molecules. The chief fault of this law is that its complexity makes it almost impossible to carry out extensive numerical calculations based on kinetic concepts.

Thus the theory of neither thermal conduction nor diffusion can be regarded as complete. The only certainty is that the coefficient of heat conduction K takes in the kinetic theory a form

$$K = \kappa\eta c_v,$$

where κ is a numerical factor (apparently depending on the ratio of the specific heats), and c_v is the specific heat of the gas. This relation is proved by Schleiermacher (*Wied. Ann.*, **36**. 1889, p. 346).

The following remarkable result is due to Maxwell. L being inversely proportional to the density of the gas, the viscosity η, and consequently the thermal conductivity K, is independent of the density. Unlikely as this appears at first sight it is fully confirmed

[1] *Kinetische Theorie der Gase*, Breslau, 1877, p. 142.

by experience. No one who grasps the order of these thoughtful conclusions and their exact experimental confirmation, would decide to give up the kinetic theory until some other self-contained theory of the phenomena appears.

For two gases of equal molecular weight and equal viscosity the coefficient of diffusion is found to be

$$D = \frac{\eta}{\rho}$$

e.g. CO_2 and NO_2 gave 0·089, whilst the calculated value from $\eta = 0·000160$ and $\rho = 0·00195$ is 0·082. The general theoretical treatment of gaseous diffusion involves great difficulties (see the above-mentioned work of Boltzmann quoted on p. 201).

The Behaviour of Gases at Higher Pressures.—When we reduce a gas to a great density by the application of great pressure, as we approach the liquid state, the gases repudiate the gas laws entirely (as was shown on p. 54), and we meet a difficult problem in attempting to account for this modification of gases. This was studied with extraordinary success by van der Waals;[1] his theoretical explanation of the deviations, shown by strongly compressed gases, from the Boyle-Mariotte law, have given us some wonderful glimpses into the nature of the liquid state.

The accepted view of the kinetic gas theory serves as the guiding idea. In ascribing the pressure exerted by the gas on the walls, to the bombardment of the molecules in their backward and forward movement, two assumptions were made; viz. firstly, that the whole space of the enclosing vessels is utilised by the moving molecules, or, in other words, the volume actually occupied by the molecules is very small in comparison with the whole volume; and, secondly, that the molecules exert no very great reciprocal action on each other. These two assumptions, of course, would be realised at a great attenuation of a gas, and they would become less exact the nearer the molecules were brought together. Thus it is necessary to introduce the influence of these two factors into the gas equation

$$pv = RT.$$

We must, under all circumstances, hold fast to the corollary that the kinetic energy of the translatory motion of the molecule is proportional to the absolute temperature, and is independent of the particular nature of the molecule in question. Then the influence of a spatial extension of the molecules will have this effect, viz. that, as a result of the consequent contraction of their "play space," they will strike

[1] *Kontinuität der gasförmigen und flüssigen Zustandes.* German trans. by F. Roth. Leipzig, 1881.

the wall so much the oftener. As a result of this circumstance, the pressure exerted will be *greater* than that calculated from the gas formula, and, indeed, the pressure will be increased in the same ratio that the mean free path is shortened from the spatial extension of the molecules. In this way van der Waals found that, as a result of the increase of the volume of the molecule, the pressure would apparently increase in the ratio $\frac{v}{v-b}$; where "b," the so-called "volume correction," *is fourfold as large as the volume of the molecules.* Apparently "b" diminishes somewhat when the molecules are very close to each other; but the important question of the rate of decrease needs further study.

Moreover, there are attracting and repelling forces active between the molecules as they approach each other in compression. From the experimental fact, discovered by Joule and Thomson (1854), that strongly compressed gases are noticeably cooled by expansion which takes place without doing work against external pressure, it is to be inferred that the expansion performs work against the action of molecular forces, *i.e. the molecules attract each other.* Therefore we must ascribe a certain *cohesion* to gases also, and this will be the more noticeable the greater their density. Regarding the mode of action of this molecular attraction, many facts lead in common to the conception that *these forces are active only when the molecules are very near each other*, and that they quickly vanish on their separation from each other. Thus the explanation of the fact that the gas molecules do *not* agglomerate into one mass in spite of their attraction, and of their being separated by only empty space, this we must ascribe to the heat motion which resists a diminution of volume, and acts like a repulsive force. Thus likewise the moon, though attracted to the earth, does not fall upon it, because its centrifugal force, resulting from its circular motion, resists the attraction, and exactly compensates it, at least in finite time.

A molecule in the interior of a gas mass, experiences, on the whole, no attractive force, because the molecules are distributed around it in homogeneous density, while a molecule existing in the surface is attracted inwards. This attraction acts against the momentum of the blow with which the molecules strike the wall: *thus from the molecular attraction there results a diminution of the pressure acting outwards.* In view of our ignorance of the law, according to which the molecular attraction decreases with separation, the following may be stated regarding its dependence upon the density of the gas in question.

If we consider a part of the surface, then the force attracting it inwards will be proportional to the number of molecules in the interior, *i.e.* to the density of the gas; but, on the other hand, this force is also proportional to the number of molecules existing in the part of the surface considered, and this also increases with the density;

hence the *attraction sought varies directly with the square of the density, or inversely with the square of the volume of the gas mass.* If we denote by p_0 the pressure of the gas, referred to its density and the kinetic energy of its molecules, and by p the effective pressure of the gas mass as actually measured by the manometer, then we have—

$$p_0 - p = K = \frac{a}{v^2},$$

where "a" denotes a constant which refers to the calculated molecular attraction of the gas, and where K denotes the molecular pressure.

The Equation of van der Waals.—If we introduce into the gas equation (p. 41), in place of the volume occupied by the gas mass, the volume corrected for the space actually occupied by the molecules, and instead of the pressure actually exerted by the gas mass, the pressure which would exist if there were no molecular attraction, then the equation assumes the form

$$\left(p + \frac{a}{v^2}\right)(v - b) = RT.$$

This is van der Waals' *characteristic equation*, which also applies to the liquid state, as will appear in considering the kinetic theory of liquids.

Now the formula given above with three constants, shows, in a remarkable way, the dependence of any given gas mass upon the pressure, volume, and temperature. If we consider the case where a gas is compressed at constant temperature, experiment shows that Boyle's law holds good for large volumes; and, as a matter of fact, for large values of the volume v, both the corrections are infinitesimally small. As we pass over to conditions of greater pressure, gases in general are compressed more easily than they should be, to correspond with Boyle's law; this is explained on the supposition that in compression the molecules are drawn more closely together by their attractive force, which thus tends to aid the action of external pressure. At very great compression, on the other hand, gases resist a diminution of volume more strongly than they should to correspond with Boyle's law; the reason for this is, on the one hand, that, with a slight change of volume, the quantity $\frac{a}{v^2}$ increases very slowly; and, on the other hand, the "volume correction," which acts in the sense just considered, begins to be an important factor, and becomes more so, the nearer "v" approaches "b" by compression.

Amagat found (p. 54), in a series of gases (viz. nitrogen, methane, ethylene, and carbon dioxide), that the product pv, instead of remaining constant as it should, according to Boyle's law, at first began to decrease, and afterwards increased very strongly: this is very well explained by the formula of van der Waals.

A picture of the degree of quantitative coincidence is shown in the following table, in which are given the values of pv as calculated by Baynes [1] for ethylene according to the formula

$$\left(p + \frac{0{\cdot}00786}{v^2}\right)(v - 0{\cdot}0024) = 0{\cdot}0037(272{\cdot}5 + t),$$

and the corresponding values of pv observed by Amagat:—

TABLE FOR ETHYLENE

p	1000 pv		p	1000 pv	
	Obs.	Calc.		Obs.	Calc.
31·58	914	895	133·26	520	520
45·80	781	782	176·01	643	642
59·38	522	624	233·58	807	805
72·86	416	387	282·21	941	940
84·16	399	392	329·14	1067	1067
94·53	413	413	398·71	1248	1254
110·47	454	456			

The pressures are estimated in atmospheres; the measurements are referred to t° = 20°.

But hydrogen, which Regnault called a "more than perfect gas" (*gaz plus que parfait*), shows at the very first that, at ordinary temperatures,[2] it resists compression more than corresponds to Boyle's law, at least in the region thus far covered by experiment; thus if $p = 2{\cdot}21$ and $p_1 = 4{\cdot}431$ m. of mercury pressure, we obtain

$$\frac{pv}{p_1 v_1} = 0{\cdot}9986.$$

It follows from this that, in the region studied, this gas shows the influence of the "volume correction" as preponderating over that of the molecular attraction. If one assumes "a" to be infinitesimally small, then from the preceding figures "b" is calculated to be 0·00065. It must therefore be inferred from van der Waals' interpretation of the volume correction, that at 0°, and under 1 m. pressure of mercury, the molecules of hydrogen, as a matter of fact,

[1] van der Waals, *l.c.* 101.

[2] Wroblewski found that at very low temperatures even hydrogen began to show evidence of a diminution of the value of pv.—*Wiener Monatshefte*, **9**. 1067 (1888).

occupy only $\frac{0{\cdot}00065}{4} = 0{\cdot}00016$ of the apparent volume. We must therefore conclude that, even under the greatest pressures possible, hydrogen could not be compressed so as to occupy less space than $\frac{1}{7000}$th of what it occupies at 0° and under 1 m. pressure.[1]

On the other hand, entirely in harmony with the supposition that the molecular attraction in the case of hydrogen is very small, is the observation made by Joule and Thomson (1854), that this gas, hydrogen, was the only one of all those investigated that showed no cooling on expanding without the performance of external work.

Finally, it must be emphasised that the van der Waals' formula may claim to hold good *only when the gas experiences no change in its molecular condition as its volume changes;* for the theory advanced above provides that the molecules, even under the highest degrees of condensation, must remain single individuals, and not unite to form larger groups. It cannot be stated *a priori* whether or not this occurs in particular cases, but the accuracy of the formula makes this very probable. Deviations of this sort from Boyle's law, which are shown by certain gases, are to be ascribed to the polymerisation of the molecules in case of an increase, or to dissociation in case of a decrease, of pressure; such deviations, which usually belong to a different order of magnitude, cannot of course be explained by van der Waals' equation of condition. The explanation of this will be found in the laws of dissociation, in the third and fourth books.

If we consider the dependence of the pressure, at constant volume, upon the temperature, then by applying the equation of condition to two temperatures, T_1 and T_2, there is obtained

$$\left(p_1 + \frac{a}{v^2}\right)(v - b) = RT_1,$$

and

$$\left(p_2 + \frac{a}{v^2}\right)(v - b) = RT_2;$$

and by subtraction there is obtained

$$\frac{p_2 - p_1}{T_2 - T_1} = \frac{R}{v - b}.$$

Since there are no quantities on the right side of the equation which change with the temperature, the increase of the pressure for highly compressed gases is proportional to the increase of the temperature; and, indeed, the pressure increases more rapidly than in the case of the

[1] This of course assumes that the molecules cannot suffer any *internal* condensation; the figures given above would place 0·824+ as the superior limit of the sp. gr. of condensed hydrogen.—TR.

ideal gases, *i.e.* more than $\frac{1}{T}$ per degree, since instead of $\frac{R}{v}$, as in the case of ideal gases, we have here $\frac{R}{v-b}$. This also, substantiated as it is by experiment, proves that neither a nor b change very much with the temperature. The proviso is also, of course, made here that there is no change of molecular condition associated with the temperature.

It was noticed in the last edition of this book that van der Waals' equation can be reduced to apply to perfect gases, and used to obtain an exact value of the gas constant R (see also Leduc, *C. R.*, **124**. 285, 1897). This has been done meanwhile by Guye and Friedrich (*Arch. scien. phys. nat.*, **9**. 505, 1900) as follows. For $p = 1$ atm. and $T = 273$, van der Waals equation becomes

$$\left(1 + \frac{a}{v_0^2}\right)(v_0 - b) = R \, . \, 273,$$

where v_0 is the volume of a mol. of gas under normal conditions, whilst the equation

$$p_0 V_0 = RT$$

gives the volume V_0 of an ideal gas under normal conditions. Hence

$$V_0 = v_0\left(1 + \frac{a}{v_0^2}\right)\left(1 - \frac{b}{v_0}\right),$$

or, accurately enough,

$$V_0 = v_0\left(1 - \frac{b}{v_0} + \frac{a}{v_0^2}\right).$$

The correcting factor in the bracket differs by one per mille. or less from unity for the so-called permanent gases; it raises the mean value (p. 41) for H_2, O_2, CO, N_2 from 22·42 to 22·43.[1]

The Kinetic Theory of Liquids.—The views advanced by van der Waals to explain the behaviour of gases under great pressure, lead to some very remarkable conclusions respecting the liquid condition. It has been already inferred from the critical phenomena which make it possible to transform the two states into each other without interruption (p. 66), that the molecular condition in the two is not very different; and, in fact, the following considerations lead also to this same conclusion.

According to the original hypothesis advanced for the case of ideal gases, the mean kinetic energy of the translatory motion of the molecule (in distinction from the kinetic energy contained in their

[1] Guye and Friedrich find 22·41, but they take for the molecular weights of H_2 and N_2 2·015 and 28·01 instead of 2·016 and 28·08.

atoms), is proportional to their absolute temperature, but independent of their nature; this hypothesis can now be also applied to gases of any desired density; we will next consider the consequences resulting from assuming that it also holds good *for liquids.*

In the light of this hypothesis, we come at once to the view that on account of the great velocity of the molecular movement (p. 203), and the very close approximation brought about by the condensation of the gas, the molecules must collide with each other very often, and therefore must exist under a very great partial pressure. But there results from this very active movement a tendency of the individual molecules to separate from each other; this tendency manifests itself unmistakably, not only in the vapour pressure, but also in the resulting property of liquids of filling a space completely with its molecules by way of evaporation; but it seems very small in comparison with the enormous expansive force of gases compressed gradually into liquids. Associated with the recognition of this, the further question at once arises, viz. in that case what hinders the liquid molecules, on account of their active movement, from separating from each other explosively? or, in other words, how is the enormous partial pressure held in check?

As an answer to this question this suggestion arises, viz. the attractive force between the molecules which it was found necessary to introduce in explaining the behaviour of gases at high pressures. It was shown (p. 212) that this force of attraction vanished [*i.e.* was counterbalanced] in the case of molecules existing in the interior, and manifested itself only in the case of molecules at or near the bounding surface, inasmuch as the force gave a resultant perpendicular to the surface. Now the resultant directly counteracts the expansive force resulting from the heat motion of the molecules, and thus appears suited to hold this in equilibrium. In general a molecule coming from within, to the free surface, will be held back by the molecular attraction, and so will be retained within the liquid mass. Only those molecules which by chance reach the free surface with a very great velocity, will be able to free themselves from the control of the molecular forces, and thus *to evaporate.* If there is a free space above a liquid this will always become filled with its molecules, but it must be noticed that the pressure of the evaporated molecules in the gaseous state cannot pass a certain maximal limit. On the other hand, any of those molecules in the gaseous state, which approach too near the surface of the liquid, will at once be drawn in by the molecular attraction; and thus there results a continual exchange between the liquid and gaseous parts of the system.

Obviously the pressure of the gaseous molecules can go only so far that the number of molecules striking upon and retained by unit area of the surface in unit time, is the same as the number of molecules passing from the liquid to the gaseous part of the system through the bounding surface; it is also easily seen that this maximal pressure must be independent of the relative amounts of the gas and liquid, and therefore this pressure corresponds in every respect to the saturation pressure of the liquid (Clausius, 1857).

Only those molecules can evaporate from the liquid which have a kinetic energy greater than the mean, since only these can overcome the molecular attraction. Also, the mean kinetic energy of the liquid molecules must be *diminished* by evaporation; that is to say, the evaporation is attended with an absorption of heat, which of course is in accord with experiment.

Moreover, the phenomena exhibited by the surface tension of liquids (p. 57) are to be ascribed to molecular attraction. In order to bring a molecule from the interior part into the free surface, work must be expended against the attractive force: it follows directly from this that a force must be overcome in order to produce a free surface, namely, the surface tension, and that the free surface of a liquid tends to reduce itself to a minimum.

The way to obtain quantitative results from the preceding considerations is obvious. The following formula of van der Waals, viz.—

$$\left(p + \frac{a}{v^2}\right)(v - b) = RT,$$

holds good both for a simple homogeneous gas and for a simple homogeneous liquid. In both cases a refers to the molecular attraction; b refers here to the correction to be made for the total volume of the liquid, resulting from the contracted "play space" of the molecules. Now the constants a and b can be determined from the behaviour of gases under high pressure, and thus the theory leads to the surprising result, that from the behaviour of the gas as a gas, we can quantitatively derive its behaviour when it shall have been condensed into a homogeneous liquid.

Let us test the requirements of the preceding law by some special practical example. Thus van der Waals, from the compressibility of gaseous carbon dioxide, calculated the value of a to be 0·00874, and b, 0·0023, when the atmospheric pressure is the unit of pressure, and the unit of volume is the volume containing 1 g. of a gas at 0° and under 1 atm. pressure. We then have $\frac{1}{M}$ mol of gas and

$$\left(p + \frac{0{\cdot}00874}{v^2}\right)(v - 0{\cdot}0023) = \frac{R}{M}T.$$

If we make p and v both equal to one, then T will equal 273, and there follows for $\frac{R}{M}$ the equation

$$273\frac{R}{M} = (1 + 0{\cdot}00874)(1 - 0{\cdot}0023) = 1{\cdot}00646,$$

and we thus obtain for the equation of condition of carbonic acid,

$$\left(p + \frac{0{\cdot}00874}{v^2}\right)(v - 0{\cdot}0023) = 1{\cdot}00646\ \frac{273 + t}{273},$$

where t denotes the temperature on the ordinary Celsius scale. This formula represents very satisfactorily the observations made by Regnault and Andrews on the compression of gaseous carbon dioxide. Let us also see whether the formula represents the behaviour of liquid carbon dioxide.

For this purpose we will calculate a series of isothermals for different temperatures, putting the above equation into the more convenient form

$$p = \frac{1{\cdot}00646}{v - 0{\cdot}0023} \cdot \frac{273 + t}{273} - \frac{0{\cdot}00874}{v^2},$$

and substituting various values of v, whilst t is kept the same. Thus for t = − 1·8 we get

v	p	v	p
0·1	9·37	0·008	38·8
0·05	17·47	0·005	20·9
0·015	39·9	0·004	42·0
0·01	42·6	0·003	457·0

Here p at first increases with decrease of v, reaches a maximum (for about v = 0·01), decreases, and finally increases rapidly.

For this purpose we will plot the curves, corresponding to the definite temperatures t, on a system of co-ordinates where the abscissæ represent volumes, and the ordinates represent the corresponding pressures. These curves are the so-called "isotherms," and are shown in Fig. 18. On studying these we are impressed by the fact that above 32·5° only one pressure corresponds to one volume, *i.e.* the latter is determined unequivocally by the former. But below this temperature, within the limits of a pressure interval (which is marked by a heavy curve for the isotherm of 13·1°), *for the same particular pressure there correspond three different volumes.* At first glance this would appear to be absurd. But we know that at these vapour pressures, and only at these pressures, the same substance is capable of occupying two different volumes, viz. the one as a homogeneous gas, and the other as a homogeneous liquid; but what is indicated by the third? Of course to suggest the volume of the substance in the solid state is out of the question. The van der Waals' formula does not consider this.

The matter will become clearer by considering this in the light of experiment. The behaviour of gaseous and liquid carbon dioxide was studied very exactly by Andrews for those temperatures the isotherms of which are plotted. Let us consider, for instance, the isotherm corresponding to 13·1°. Andrews found, by beginning with small pressures and large volumes, that the gaseous carbon dioxide could be condensed to volume v_0 and pressure p_0, corresponding exactly to the

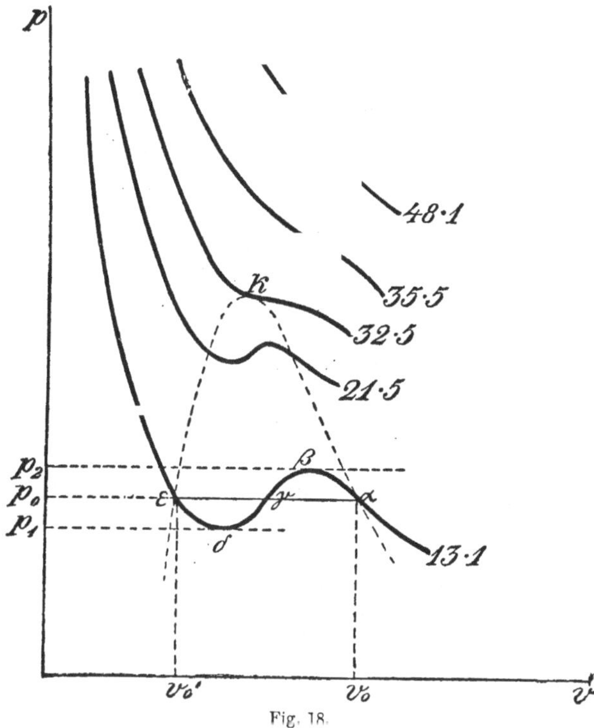

Fig. 18.

path of the curve. The constants a and b, as in the van der Waals' formula, were so determined as to make the compressibility of the gas coincide as well as possible with the calculation. That these constants may, as a matter of fact, be so chosen has been previously shown (p. 213). But when the pressure was made greater than p_0, the diminution of volume did not correspond to the pressure of the advancing curve $\alpha\ \beta\ \gamma\ \delta\ \epsilon$, but there occurred a partial liquefaction, corresponding to the vapour pressure of carbon dioxide at 13·1°. The pressure remained constant until the volume of the saturated vapour had fallen from v_0 to the volume of the liquid v_0', *i.e.* until the whole of the substance was condensed.

After this any farther diminution of volume was attended by an increase of pressure, and a very great increase too, as must be the case from the magnitudes of the coefficients of compressions of liquids. Beyond ϵ, Andrews found the path as given by the rest of the curve; beyond ϵ, the figures of the formula go hand in hand with those of observation, and the rapid ascent of this part of the curve, as compared with that part before α, is due to the fact that liquid carbon dioxide is much less compressible than the gaseous form.

The formula rejects only the part of the curve from α to ϵ. Instead of passing from the first point to the second by the serpentine path $\alpha\ \beta\ \gamma\ \delta\ \epsilon$, investigation shows that the substance passes from the condition at α to that at ϵ by the direct line. The formula does not hold good here. The substance from α to ϵ is not homogeneous; it is part gas and part liquid. The formula is applicable to both gaseous and liquid forms, but it insists that each shall be homogeneous. The process of gradual liquefaction is such that the adequacy of the formula must be temporarily interrupted, and it is interrupted.

The question now arises whether it is possible to realise the portion of the curve $\alpha\ \beta\ \gamma\ \delta\ \epsilon$, *i.e.* whether it is possible to change a gas into a liquid isothermally and continuously. A glance serves to make this appear improbable; for in the part of the curve $\beta\ \gamma\ \delta$, an increase of pressure would be attended with an increase of volume, and conversely a decrease of pressure with a decrease of volume; thus the substance would appear to be in an unstable form, the realisation of which is impossible.

But cannot the portions of the curve α-β and δ-ϵ at least be realised? These portions represent respectively the conditions of supersaturated vapour and of over-heated[1] liquid. In the first case, the pressure is in fact greater than p_0, and the volume less than v_0; and there can be no doubt that the conditions of a vapour after saturation represent a continuance of those before saturation. Thus the fact that the advance of a sound wave in a saturated vapour gives no evidence of interruption, shows conclusively that the vapour conducts itself in a normal way in the compression associated with supersaturation. On the other hand, the part of the curve ϵ-δ expresses the capacity of a liquid for existing under a lower pressure than that corresponding to the vapour pressure of the respective temperature, a condition which would be unstable, as shown by many observations.

The further study of the curves plotted in Fig. 18 shows also that the three volumes (at which carbon dioxide can exist at the temperature in question, and designated by the points $\alpha\ \gamma\ \epsilon$, for the temperature 13·1°), approach each other with increasing temperature, and finally at the isotherm corresponding to 32·5° coincide in the point k. If we connect the points corresponding to α and ϵ, for the other isotherms, (*i.e.* the points where liquefaction begins and ends with increasing pressure), we obtain the dotted siphoidal (*berg-förmige*) curve shown in

[1] How "over-heated" when the curves are *isotherms*? Rather it is a "super-attenuated" liquid, so to speak.—Tr.

the figure. The isotherm of 32·5 is tangential to this at the point k. The serpentine curve $\alpha\,\beta\,\gamma\,\delta\,\epsilon$, here is crowded together into a point; the physical interpretation of this obviously is that the specific volume of the liquid carbon dioxide is the same as that of the gaseous carbon dioxide condensed to this vapour pressure. Here, and here only, is it possible to convert the gas into a liquid continuously and isothermally, and the reverse; k corresponds to the critical point of carbon dioxide (p. 66).

We thus arrive at the conclusion, viz. that by the aid of the constants a and b, of van der Waals' formula, all the critical data can be obtained. For this purpose we need only to plot a few isotherms until we arrive at the clearly marked k, where the curve portions α, β, γ, δ, ϵ are crowded together in one point of inflection.

We can obtain the same result more simply and easily by the analytical discussion of the van der Waals' formula,

$$\left(p + \frac{a}{v^2}\right)(v - b) = RT\,;$$

or solved for v :

$$v^3 - \left\{ b + \frac{RT}{p} \right\} v^2 + \frac{a}{p} v - \frac{ab}{p} = 0.$$

The equation is of the third degree for v. Then if the three roots are x_1, x_2, and x_3, we will have, as is well known,

$$(v - x_1)(v - x_2)(v - x_3) = 0.$$

The roots may be real and imaginary. Of course only the first have any physical significance. Since the product of the three $(v - x)$ values is real, then either two or none of the roots can be imaginary, because it is only by the product of two imaginary quantities that a real quantity can result. Therefore for one value of p at a given temperature, there are either one or three values of v. This is seen at once from an inspection of Fig. 18 on p. 219. Thus at 32·5°, in general, for one value of p there is only one value of v. For lower temperatures, as at 13·1°, for example, in general the case is similar, and it is only in the interval from p_1 to p_2, that there may be three values of v for one of p.

The critical point k sought is that where the three real roots become equal to each other, *i.e.* if ϕ_0 denotes the critical volume, it must follow that

$$x_1 = x_2 = x_3 = \phi_0\,;$$

and it follows also that

$$(v - \phi_0)^3 = v^3 - \left\{ b + \frac{R\theta_0}{\pi_0} \right\} v^2 + \frac{a}{\pi_0} v - \frac{ab}{\pi_0}.$$

In this equation π_0 denotes the critical pressure, and θ_0 the critical temperature. And since the coefficients of the different powers of v must be equal, we obtain the following equations :—

$$\phi_0{}^3 = \frac{ab}{\pi_0},$$

$$3\phi_0{}^2 = \frac{a}{\pi_0},$$

$$3\phi_0 = b + \frac{R\theta_0}{\pi_0}.$$

These, when simplified, give

$$\phi_0 = 3b\,; \quad \pi_0 = \frac{a}{27b^2}\,; \quad \theta_0 = \frac{8}{27}\frac{a}{bR}.$$

Thus the constants a and b in the equation given above determine the critical volume ϕ_0, the critical pressure π_0, and the critical temperature θ_0, and in this way are found the co-ordinates and the point k of the curve tracing, and also the particular isotherm in which k falls.

The critical temperature of carbon dioxide was calculated from the values for "a" and "b," assumed above, to be 273° + 32·5° ; while Andrews, by direct observation, found 273° + 30·9°. The critical pressure was calculated to be 61 atm., while Andrews found 70 atm. by observation ; and the theoretical critical volume was 0·0069, while Andrews observed 0·0066.[1]

Of course conversely, "a" and "b" can be calculated from the critical data, a fact of great importance in practical work. Thus the fact that the critical data can be calculated so approximately by the deviations of gases from Boyle's law, harmonises with the remarkable fruitfulness of van der Waals' theory, as already mentioned.

By calculating the molecular volume $\frac{b}{4}$ from the critical data, there is obtained this simple result, viz. that at their respective boiling-points and at atmospheric pressure, *the molecules of the most various liquids, such as water, ether, carbon disulphide, benzene, chlor-ethane, ethyl acetate, sulphur dioxide,* etc., *occupy a space very nearly* 0·3*ths of the total apparent volume.* (See Chap. V., Molecular Volumes.)

Finally, let us calculate the superficial molecular tension K, which resists the expansive force resulting from the heat motion of the molecules ; and from the preceding, we have

$$K = \frac{a}{v^2}.$$

[1] Guye and Friedrich give a valuable collection of numerical data on critical pressures and temperatures (*Arch. sci. phys. nat.*, **9**. 505 (1900) ; reference in *Z.S. phys. chem.*, **37**. 380, 1901).

Liquid carbon dioxide at 21·5° takes up about 0·003 of the volume occupied by the substance as a gas at 0° under atmospheric pressure. From this K is calculated to be 970 atm., and this value indicates the enormous forces of pressure which are met here. These pressure forces have thus far eluded a direct determination.

The Reduced Equation of Condition.—The combination of the general equation of condition, viz.—

$$\left(p + \frac{a}{v^2}\right)(v - b) = RT,$$

with relations derived as above, between the critical data of a substance on the one hand, and the constants "a" and "b" on the other hand, has led to a very simple result. By introducing into the equation of condition the following values derived on p. 222, viz.—

$$a = 3\pi_0\phi_0^2, \quad \text{and}$$

$$b = \frac{\phi_0}{3}, \quad \text{and}$$

$$R = \frac{8}{3}\frac{\pi_0\phi_0}{\theta_0};$$

then instead of the constants a, b, and R, there appear the critical data, thus

$$\left(p + \frac{3\pi_0\phi_0^2}{v^2}\right)\left(v - \frac{\phi_0}{3}\right) = \frac{8}{3}\pi_0\phi_0\frac{T}{\theta_0}.$$

By dividing the left and right side of the equation by $\frac{\pi_0\phi_0}{3}$, we obtain

$$\left(\frac{p}{\pi_0} + \frac{3\phi_0^2}{v^2}\right)\left(3\frac{v}{\phi_0} - 1\right) = 8\frac{T}{\theta_0}.$$

Now if we make, as follows :—

$$\pi = \frac{p}{\pi_0}; \quad \phi = \frac{v}{\phi_0}; \quad \theta = \frac{T}{\theta_0},$$

we obtain

$$\left(\pi + \frac{3}{\phi^2}\right)\left(3\phi - 1\right) = 8\theta.$$

That is, by expressing the pressure, volume, and temperature respectively in fractions of the critical pressure, volume, and temperature, the equation of condition assumes the same form for all substances.

If one plots the values of π and ϕ for a definite value of θ in a system of co-ordinates, where the abscissæ represent values of ϕ, and the ordinates values of π, he will obtain isotherms similar in form to those plotted in Fig. 18 (p. 219), and which will hold good for all substances. Thus, for example, for $\theta = 1$, when $\pi = 1$ and $\phi = 1$, this isotherm passes through the critical point.

The pressure divided by the critical pressure, we will call the *reduced pressure*, in accordance with van der Waals; and also the respective quotients of the volume by the critical volume, and of the temperature by the critical temperature, we will call the *reduced volume* and the *reduced temperature*. Such reduced pressures, volumes, and temperatures, when identical are better called *corresponding* for short, and we may speak of two substances whose pressures, volumes, and temperatures "correspond" in the sense of what precedes, as being in a *corresponding condition* (*ubereinstimmenden Zustanden*).

It is not at all easy to form a conception of the *boldness* of this equation, which claims to express the general behaviour of all homogeneous liquid and gaseous substances, as regards their changes in pressure, temperature, and volume. Of course those cases are excepted where changes result in a chemical reaction, such as polymerisation [condensation], or dissociation. Therefore it will be useful to follow out some application of this equation in order, on the one hand, that we may add to its examples, and, on the other hand, to effect its proof from experiment.

Application of the Theory of Corresponding States.—If we solve the following equation:—

$$\left(\pi + \frac{3}{\phi^2}\right)(3\phi - 1) = 8\theta,$$

for ϕ, we will obtain

$$\phi = \mathrm{f}(\pi, \theta),$$

in which the function f is the same for all substances. By raising the temperature from θ_1 to θ_2, the expansion produced at constant pressure π will amount to

$$\phi_2 - \phi_1 = \mathrm{f}(\pi, \theta_2) - \mathrm{f}(\pi, \theta_1).$$

Dividing both sides of the equation by

$$\phi_1 = \mathrm{f}(\pi, \theta_1),$$

and observing that

$$\phi = \frac{\mathrm{v}}{\phi_0},$$

we obtain

$$\frac{\mathrm{v}_2 - \mathrm{v}_1}{\mathrm{v}_1} = \frac{\mathrm{f}(\pi, \theta_2) - \mathrm{f}(\pi, \theta_1)}{\mathrm{f}(\pi, \theta_1)}.$$

Here v_1 and v_2 denote the specific volumes of the liquid according to the ordinary standards, and therefore the fractional expansion, experienced by the substance on raising the temperature, is equal to the quotient given, viz.—

$$\frac{v_2 - v_1}{v_1}.$$

The right side of the equation is independent of the special nature of the substance, and therefore the left must be so also; *i.e. the fractional expansion, experienced by the most diverse liquids or gases, will be the same when they are heated, at a constant corresponding pressure, from one corresponding temperature to another.*

This relation gives the calculation of the specific volume of any selected liquid at all temperatures, if its critical temperature, and also if its specific volume at some one temperature are known, inasmuch as we can compare it with any other liquid which is already well investigated. Such a liquid, for example, is fluor-benzene, the specific volumes of which have been measured even up to its critical point ($\theta_0 = 560°$).

Thus the specific volume of ethyl ether at 10° above the freezing-point of ice is 1·3794, and suppose that we wish to calculate its value at 33·8°. For this purpose, from the values of θ_1 and θ_2, viz.—

$$\theta_1 = \frac{273 + 10}{467 \cdot 4} = 0 \cdot 6055,$$

and

$$\theta_2 = \frac{273 + 33 \cdot 8°}{467 \cdot 4} = 0 \cdot 6564,$$

let us calculate the two absolute temperatures, expressed in fractions of the critical temperature (467·4°). At the temperatures θ_1 and θ_2, *i.e.*

$$T_1 = 560 \ \theta_1 = 339 \cdot 1,$$

and

$$T_2 = 560 \ \theta_2 = 367 \cdot 6,$$

the specific volumes of fluor-benzene, in absolute numbers, are 1·0339 and 1·0741 (p. 231), and the relative increase in volume of fluor-benzene, in consequence of a rise in temperature from 339·1° to 367·6° (abs.), amounts to

$$\frac{1 \cdot 0741 - 1 \cdot 0339}{1 \cdot 0339} = 0 \cdot 0389.$$

Now this is the increase of the volume of ether sought, and therefore its specific volume at 33·8° amounts to

$$1 \cdot 3794 \cdot 1 \cdot 0389 = 1 \cdot 4331,$$

which is very close to the value as determined by experiment, 1·4351.

All the preceding specific volumes are measured at atmospheric pressure, which, strictly speaking, does not give the same reduced or corresponding pressure; since the critical pressure of fluor-benzene is about 20 per cent greater than that of ether, then in the calculation the specific volume of the former must be measured at 1·2 atmospheric pressure; but this is insignificant in view of the very slight compressibility of liquids. In general, in this sort of calculation of liquid volumes, *the atmospheric pressure can be regarded as one which is corresponding for all liquids.*

In a precisely similar way we derive the result that the per cent diminution of volume experienced by the most diverse substances, whether liquid or gaseous, is the same when one corresponding pressure is raised to another, the corresponding temperature remaining constant.

The coefficient of compression (*i.e.* the diminution of volume of one c.c. resulting from raising the external pressure one atmosphere) for ether amounts to 0·00011 at 0°. Now, according to the preceding law, the coefficient of compression for all liquids, at corresponding temperatures, must be inversely proportional to the critical pressure. Thus, the compression coefficient for chloroform, for instance, is calculated to be

$$0{\cdot}00011\frac{36}{55} = 0{\cdot}000072,$$

55 and 36 atm. being the critical pressures for chloroform and for ether respectively. This refers to the value of chloroform at a temperature of about 40°, corresponding with that of ether at a temperature of 0°, since the absolute critical temperature of ether, increased by about one-seventh of its amount, gives that of chloroform. Observations at this temperature gave about 0·000076.

Amagat has given (*C. R.*, **123**. 30, 83, 1896) a particularly clear test of the law of corresponding states. If isothermals of two substances are drawn in such a way as to take for units of volume and pressure the critical values, the two series of curves must fit, *i.e.* in combination they must look as if they belonged to one substance: no two such reduced isothermals can cut, however close they may lie. Without knowing the critical values, this can be tested by drawing the isothermals in the usual manner (not reduced) and seeing if it is possible to alter the scale so as to bring the isothermals to form a series without intersection. Amagat accomplished the change of scale by projecting one diagram on the other with (nearly) parallel light; the diagram to be projected was at the same time turned round one or other axis in order to alter the relative proportions of ordinates and abscissæ; by a combination of the two changes, any derived change of scale could be arrived at.

C. Raveau (*J. de phys.* [3], **6**. 432, 1897) reached the same object by an even simpler and more ingenious means. He plotted the logarithms of the volumes and pressures as co-ordinates; now the unit of pressure or volume can be altered by merely adding a constant quantity to the logarithm; it must therefore be possible to make the two diagrams fit by merely displacing one relatively to the other.

Amagat and Raveau both find a remarkably close fit between the diagrams of various substances [C_2H_4, CO_2, $(C_2H_5)_2O$, air], *i.e.* a striking confirmation of the theory of corresponding states.

To test further how far the special form of the function $f(\pi, \phi, \theta) = 0$, given by van der Waals, is applicable, Raveau drew a diagram of isothermals, using the logarithms of p and v from that equation. It appeared that this diagram would not fit those for carbon dioxide and acetylene without intersections extending to somewhat widely separated isothermals. This shows—in harmony with the preceding—that van der Waals' formula is only an approximation to the truth; it naturally deviates more from the truth the smaller v, and consequently the greater the correction terms $\frac{a}{v^2}$ and $\frac{b}{v}$.

It was remarked by Meslin (*C. R.*, **116**. 135, 1893) that the theorem of corresponding states must be true for any characteristic equation that contains only as many constants as there are determining quantities (*i.e.* three-volume temperature and pressure), and which includes the critical point. For it is always possible to put the equation with three constants a, b, R,

$$f(p, v, T, a, b, R) = 0$$

into the form

$$f(p, v, T, \pi_0, \phi_0, \delta_0) = 0,$$

as was done with van der Waals' equation on p. 224.

Since this equation must hold independently of the units adopted, and on the other hand it is not possible, without a fresh condition, to express one of the three determining quantities in terms of the other two (p. 5), it follows that the equation must have the form

$$f\left(\frac{p}{\pi}, \frac{v}{\phi}, \frac{T}{\theta}\right) = 0;$$

hence the equation can contain nothing characteristic of the particular substance considered except the critical constants, *i.e.* can only contain numerical constants. For the conditions that such an equation must satisfy see Brillouin (*J. de phys.* [3], **2**. 113, 1893).

The Coexistence of Liquid and Vapour.—As has been repeatedly emphasised, both the original and the reduced equations of condition apply only to *homogeneous* substances, whether liquid or gaseous. They cease to hold true as soon as there occurs a partial evaporation or partial condensation of the liquid in question, and, therefore, the equation has nothing to say regarding the substance as soon as it has lost its homogeneity through evaporation or condensation; for such magnitudes as vapour pressure, boiling-point, volume of the saturated vapour, or of the liquid, these lie beyond the region of its applicability.

If we consider an isotherm, such, for instance, as that plotted by van der Waals' equation for 13·1° in Fig. 18 (p. 219), we will seek in vain for any marked point which indicates the beginning of liquefaction; but we know this much from what precedes, viz. that for a vapour pressure p_0 there correspond three points of the curve [of volumes], viz. α, γ, and ϵ. But, nevertheless, it is possible to develop a simple law in a thermodynamic way, for the position of the straight line α-ϵ. Thus we can imagine a gas mass to be carried from α to ϵ by the path α-β-γ-δ-ϵ, and back again by the straight path ϵ-α to the original point. The sum of the work performed *by* the system and *upon* the system in this reversible and isothermal cyclic process must be equal to zero (p. 20). But since the former refers to the area bounded by α-β-γ and the straight line α-γ, and the latter to the area bounded by γ-δ-ϵ and the straight line γ-ϵ, *then these two areas must be equal to each other*, and the line α-ϵ must be chosen so as to satisfy this condition (Maxwell, 1875; Clausius, 1880). To be sure, we cannot regard the proof on this basis as very stringent since it is impossible to realise this cyclic process, but, nevertheless, the law appears very probable from this consideration. It should be again emphasised here that there has entered into the problem an element which is foreign to what was originally a *purely kinetic* theory.

The equality of the area embraced within the three straight lines αv_0, $v_0 v_0'$, $v_0'\epsilon$, and the curve α-β-γ-δ-ϵ, with the area embraced within the same three lines and the straight line α-ϵ, gives the following new relation, which is independent of the nature of the substance, viz.—

$$F(\pi, \theta, \phi_1, \phi_2) = 0,$$

in which π denotes the reduced vapour pressure, θ the reduced boiling-point, and ϕ_1 and ϕ_2 the reduced volumes of the liquid and the saturated vapour. This equation also, like that developed on p. 224, expresses a nature law of unusually wide application, since it requires that the nature of the function $F(\pi, \theta, \phi_1, \phi_2)$ shall be *the same for all substances.*

Although we can make no use of the special nature of this function, yet we will briefly show its universal importance. The equality of the two areas just described requires that

$$\int_{v_0'}^{v_0} p\,dv = p_0(v_0 - v_0'),$$

or, by substituting in accordance with the equation of condition,

$$p = \frac{RT}{(v - b)} - \frac{a}{v^2},$$

we have, on integrating,

$$RT \ln\frac{v_0 - b}{v_0' - b} + \frac{a}{v_0} - \frac{a}{v_0'} = p_0(v_0 - v_0').$$

By dividing both sides by $\pi_0\phi_0$, and by substituting the values of "a" and "b" (p. 223) by the critical data, and thus introducing the reduced pressure, volume, and temperature, the preceding equation becomes

$$\left(\pi + \frac{3}{\phi_1\phi_2}\right)(\phi_2 - \phi_1) = \frac{8}{3}\theta \ln\frac{3\phi_2 - 1}{3\phi_1 - 1},$$

in which

$$\pi = \frac{p_0}{\pi_0}, \quad \phi_1 = \frac{v_0'}{\phi_0}, \quad \phi_2 = \frac{v_0}{\phi_0}, \quad \text{and } \theta = \frac{T}{\theta_0},$$

and denote respectively the corresponding values of the reduced condition, and by which the nature of the universal function mentioned above is explained.

Moreover, aside from establishing the equation,

$$F(\pi, \theta, \phi_1, \phi_2) = 0 \quad . \quad . \quad . \quad . \quad (1)$$

the universal equation of condition (p. 224) when applied to the case of a saturated vapour, and again to a vapour in equilibrium with a liquid, gives the two following new relations, viz.—

$$\left(\pi + \frac{3}{\phi_1^2}\right)(3\phi_1 - 1) = 8\theta \quad . \quad . \quad . \quad (2)$$

$$\left(\pi + \frac{3}{\phi_2^2}\right)(3\phi_2 - 1) = 8\theta \quad . \quad . \quad . \quad (3)$$

By eliminating π and ϕ_2 from equations (1) and (3), then π and ϕ_1, and finally ϕ_1 and ϕ_2, we obtain the three following equations, viz.—

$$\phi_1 = f_1(\theta),$$
$$\phi_2 = f_2(\theta),$$
$$\pi = f_3(\theta),$$

in which again the three functions $f_1(\theta)$, $f_2(\theta)$, *and* $f_3(\theta)$, *are independent of the nature of the substance in question.*

If we estimate the temperature in fractions of the critical temperature, then the specific volumes of the saturated vapours of all substances form a *constant temperature function*, provided that we estimate the volumes in fractions of the critical volumes, and that this holds good for both the volumes of liquids, and also for their vapour pressures.

We may formulate this law thus: at equally reduced boiling-points, the respective quotients, of the specific volume of the saturated vapour by the critical volume, of the specific volume of the liquid by the critical volume, and of the vapour pressure by the critical pressure, are identical for all substances.

Of course, according to this law, the reduced specific volumes of the vapour and liquid of the most diverse substances must be the same *when they are compared with each other at equal fractions of the critical pressure.*

The Demonstration by Young.—These laws have been recently subjected to a thorough test by S. Young.[1] Unfortunately, space does not permit us to give completely the material brought together in his work, which furnishes an example of a problem subjected to investigation both by calculation and by experiment; and it also deserves the more attention because it has such a fundamental and universal significance, as is illustrated by few problems in the whole extent of physics or chemistry.

The method proposed by Young consisted in the comparison of the specific volumes of different substances in the liquid state and the state of saturated vapour, as well as in the comparison of their vapour pressures, with the corresponding values afforded by a suitable "normal" substance in the "corresponding states." Fluor-benzene may be recommended as an example which has been very well studied. Below are given its complete data as an important basis for future calculations:—

[1] *Phil. Mag.* [5], **33.** 153 (1892).

[Table

FLUOR-BENZENE

Mol. Wt. = 95·8.

T	p	Mv_0	Mv_0'	Obs. vapour density / Calc. ,, ,,
272·27	20	...	91·47	...
289·3	50	...	93·20	
303·9	100	...	94·92	
320·25	200	...	96·80	
338·75	400	...	99·05	...
358·1	760	...	101·59	...
367·3	1,000	22,000	102·90	1·037
382·0	1,500	15,000	105·10	1·056
393·25	2,000	11,400	107·00	1·073
410·4	3,000	7,680	110·03	1·107
423·8	4,000	5,785	112·64	1·138
434·85	5,000	4,634	114·98	1·166
444·25	6,000	3,857	117·06	1·193
452·8	7,000	3,298	119·14	1·217
460·4	8,000	2,871	121·19	1·247
473·6	10,000	2,265	125·04	1·300
484·95	12,000	1,862	128·80	1·349
499·7	15,000	1,447	134·64	1·431
519·7	20,000	1,009	145·08	1·600
536·0	25,000	733	158·40	1·818
544·5	28,000	601	169·35	2·011
550·0	30,000	516	179·40	2·206
555·0	32,000	440	193·0	2·450
559·55	33,912	270·4	270·4	3·79

The vapour pressure p is expressed in millimetres, the molecular volumes of the saturated vapour Mv_0 and of the liquid Mv_0' in c.c. The critical data of the substances investigated, measured partly by Young alone, and partly together with Ramsay, are given in the following table:[1]—

CRITICAL DATA

Substance.	Formula.	Mol. wt.	θ_0	π_0	ϕ_0
Fluor-benzene . .	C_6H_5F	95·8	559·55	33,912	2·822
Chlor-benzene . .	C_6H_5Cl	112·2	633·00	33,912	2·731
Brom-benzene . .	C_6H_5Br	156·6	670·00	33,912	2·059
Iodo-benzene . .	C_6H_5I	203·4	721·00	33,912	1·713
Benzene . . .	C_6H_6	77·84	561·50	36,395	3·293
Carbon tetrachloride	CCl_4	153·45	556·15	34,180	1·799
Stannic chloride .	$SnCl_4$	259·3	591·70	28,080	1·347
Ether . . .	$(C_2H_5)_2O$	73·84	467·40	27,060	3·801
Methyl alcohol .	CH_3OH	31·93	513·00	59,760	3·697
Ethyl alcohol . .	C_2H_5OH	45·90	516·10	47,850	3·636
Propyl alcohol .	C_3H_7OH	59·87	536·70	38,120	3·634
Acetic acid . .	CH_3CO_2H	59·86	594·60	43,400	2·846

[1] *Phil. Mag.* [5], **34**, 505 (1892).

The critical volumes are not observed directly, but are extrapolated by the rule of Cailletet and Mathias;[1] according to this, the arithmetical mean between the densities of the liquid and of the saturated vapour diminishes linearly with the temperature. By the extrapolation of these mean values to the critical temperature we obtain the critical data.

The table on the following page contains an extract from the calculations of Young. The first three horizontal columns contain the comparison of the respective substances (given in column I.) with fluor-benzene at the corresponding temperatures, which are given under the horizontal column marked θ; the absolute temperatures of fluor-benzene are found in the horizontal column marked T; the second column (marked II.) contains the molecular volume of the saturated vapour of the substance divided by that of fluor-benzene; the third column (marked III.) contains the molecular volume of the substance in a liquid state divided by that of fluor-benzene; the fourth column (marked IV.) gives the vapour pressure of the substance divided by that of fluor-benzene; the sixth column (marked VI.) contains the boiling-points of the substance at the reduced pressure π, divided by that of fluor-benzene, referred to the same reduced pressure, while under p are given the actual vapour pressures of fluor-benzene.

[1] *C. R.*, **102.** 1202 (1886).

[TABLE

TABLE OF COMPARISON.

T C_6H_5F	θ	C_6H_5F	C_6H_5Cl	C_6H_5Br	C_6H_5I	C_6H_6	CCl_4	$SnCl_4$	$(C_2H_5)_2O$	CH_3OH	C_2H_5OH	C_3H_7OH	CH_3COOH	I.
367·3	0·656	1·000	1·118	1·174	1·272	0·842	0·828	1·336	1·101	1·244	2·076	2·678	0·894	II.
410·4	0·733	,,	1·150	1·201	1·293	0·885	0·895	1·316	1·083	0·945	1·414	1·714	0·818	
460·4	0·822	,,	1·143	1·189	...	0·923	0·938	1·300	1·054	0·729	1·028	1·216	0·745	
519·7	0·928	,,	...	..	...	0·926	0·977	1·262	1·030	0·572	0·763	0·943	0·681	
559·55	1·000	,,	...	...	...	0·940	...	...	...	...	...	...	0·631	
272·25	0·486	1·000	1·1247	1·1802	1·2761	0·9457	..	1·2657	...	...	...	...	0·6215	III.
358·1	0·639	,,	1·1238	1·1805	1·2782	0·9487	1·0273	1·2744	1·0297	0·4136	0·5971	0·7748	0·6198	
410·4	0·733	,,	1·1236	1·1793	1·2777	0·9489	1·0280	1·2775	1·0279	0·4087	0·5895	0·7661	0·6191	
460·4	0·822	,,	1·1249	1·1824	...	0·9481	1·0257	1·2797	1·0302	0·4048	0·5861	0·7633	0·6171	
519·7	0·928	,,	...	...	...	0·9499	1·0281	1·2913	1·0357	0·4047	0·5859	0·7711	0·6204	
559·55	1·000	,,	...	...	...	0·9400	...	...	...	...	...	...	0·6310	
272·25	0·486	1·000	0·998	0·987	0·985	1·338	1·460	0·706	...	...	...	...	0·476	IV.
358·1	0·639	,,	1·005	0·998	0·998	1·200	1·206	0·778	0·740	0·678	0·409	0·323	0·708	
410·4	0·733	,,	0·993	0·988	0·997	1·144	1·117	0·805	0·761	0·966	0·681	0·573	0·859	
460·4	0·822	,,	1·000	1·007	...	1·114	1·066	0·821	0·785	1·259	0·962	0·810	1·008	
519·7	0·928	,,	1·007	...	...	1·088	1·029	0·828	0·804	1·583	1·253	1·028	1·190	
559·55	1·000	,,	...	...	...	1·073	1·008	0·828	0·798	1·762	1·411	1·124	1·280	
p C_6H_5F.	π													V.
20	0·0006	1·000	1·1315	1·1985	1·2902	0·9890	0·9699	1·0679	...	1·0127	1·0494	1·0997	1·1278	VI.
760	0·0023	,,	1·1307	1·1980	1·2889	0·9932	0·9774	1·0633	0·8411	0·9859	1·0061	1·0432	1·1151	
3,000	0·0885	,,	1·1321	1·1988	1·2890	0·9963	0·9826	1·0607	0·8394	0·9667	0·9801	1·0117	1·1045	
8,000	0·2360	,,	1·1315	1·1965	...	0·9987	0·9864	1·0585	0·8370	0·9485	0·9577	0·9904	1·0905	
20,000	0·5900	,,	1·1300	...	...	1·0015	0·9910	1·0574	0·8343	0·9260	0·9347	0·9690	1·0718	
33,912	1·0000	,,	...	...	...	1·0035	0·9939	1·0575	0·8353	0·9168	0·9223	0·9592	1·0626	

Thus, for example, the molecular volume of the saturated vapour of alcohol at the absolute temperature T(= 338·75), is 45700 c.c., therefore, since the critical temperature of alcohol is 516·1 [see preceding table of critical data], the reduced temperature is

$$\theta = \frac{338 \cdot 75}{516 \cdot 1} = 0 \cdot 656 ;$$

the corresponding temperature of fluor-benzene amounts to

$$T = 559 \cdot 55 \times \theta = 367 \cdot 3^\circ .$$

In the table on p. 231 the molecular volume of the saturated vapour of fluor-benzene at this temperature is found to be 22,000, and thus the quotient is

$$\frac{45,700}{22,000} = 2 \cdot 076 ;$$

and this last number, as a matter of fact, is found in column II. (p. 233) under C_2H_5OH, corresponding to the value $\theta = 0 \cdot 656$.

It appears from a study of the preceding table, that the figures for the three halogen derivatives of benzene meet the demands of the theory as the comparison requires they should at *corresponding temperatures*, as shown by the constancy of the figures in the vertical rows; but that the numbers for benzene, carbon tetrachloride, stannic chloride, and ether, given in the vertical rows of each particular column, are not constant as they should be, inasmuch as regular deviations can be recognised; and that the deviations are much greater still in the case of acetic acid and the three alcohols; and further, that the requirements of the theory are well satisfied only in the case of *the molecular volume in the liquid state* (see column III.)

The comparison of the boiling-points at *corresponding pressures*, again, gives very good results, except in the case of the substances last mentioned.

Moreover, the fact that there is less variation in comparing the quotients of the boiling-points at *corresponding pressures*, than in comparing the quotients of the vapour pressure at *corresponding temperatures*, is easily explained as follows: since the pressures diminish in the ratio of 16,000 : 1, when the reduced temperature θ sinks from 1 to 0·5°, the inexactness of the theory will be rendered more apparent by those quotients which are developed from the pressures, than by those developed from the temperatures.

By means of the equations on p. 223, the critical volume ϕ_0 is estimated to be

$$\phi_0 = \frac{3}{8} \frac{R\theta_0}{\pi_0},$$

if the gaseous laws held up to the critical point it would be $\frac{R\theta_0}{\pi_0}$.

Thus we find that the actual critical vapour density should for every substance equal $\frac{8}{3}$ of the theoretical, *i.e.* the density according to Avogadro's law.

This law holds good to this extent as actually found by Young;[1] for omitting the alcohols and acetic acid from the values given (p. 222) for the critical volume, the critical vapour density of all the substances investigated was several times as great as the theoretical; thus it was not $\frac{8}{3}(=2{\cdot}67)$, but about 3·8 times as great as the theoretical. *In general the statement holds good that the ratio of the observed to the calculated vapour densities is the same at corresponding pressures (but not quite so exact at corresponding temperatures).* These quotients are calculated for fluor-benzene on p. 231.

Young and Thomas[2] have studied, further, the behaviour of a series of esters. The reduced temperatures and volumes for these, as well as for the substances previously investigated, are given in the following table, for reduced pressure 0·08846 :—

Substance.	Reduced Co-ordinates (p. 229).		
	θ	ϕ_1	ϕ_2
Methyl formate	0·7348	0·4007	29·4
Ethyl formate	0·7385	0·4004	29·6
Methyl acetate	0·7445	0·3992	30·2
Propyl formate	0·7430	0·4010	29·4
Ethyl acetate	0·7504	0·4006	30·3
Methyl propionate	0·7485	0·4006	29·6
Propyl acetate	0·7541	0·3994	29·75
Ethyl propionate	0·7540	0·3996	30·05
Methyl butyrate	0·7522	0·4005	29·5
Methyl isobutyrate	0·7502	0·4021	29·2
Methyl alcohol	0·7734	0·3949	34·2
Ethyl alcohol	0·7794	0·4047	32·1
Propyl alcohol	0·7736	0·4028	31·1
Acetic acid	0·7624	0·4106	25·5
Ethyl ether	0·7371	0·4044	28·2
Fluorbenzene	0·7334	0·4067	28·4
Chlorobenzene	0·7345	0·4046	28·65
Bromobenzene	0·7343	0·4041	28·4
Iodobenzene	0·7337	0·4042	28·45
Benzene	0·7282	0·4053	28·2
Carbon tetrachloride	0·7251	0·4072	27·4
Tin chloride	0·7357	0·4021	28·1

Within each group the theorem of corresponding states holds

[1] *Phil. Mag.* [5], **34**, 507 (1892). [2] *Trans. Chem. Soc.*, 1893, p. 1191.

remarkably for the volume of the liquid ϕ_1, and fairly for the vapour volume ϕ_2, and temperature θ.

A comparison of the reduced co-ordinates for isopentane with those of benzene turned out quite in accord with the theorem.[1]

The special form of the characteristic equation developed on p. 229 for coexistence of liquid and vapour does not agree at all with the observations,[2] a further proof that van der Waals' special equation for matter in a highly compressed state is only true qualitatively.

The Heat Content of Compressed Gases and of Liquids.—We will now finally consider the answer to certain questions regarding the heat content of *gaseous and liquid substances* in the light of the kinetic theory of van der Waals. We will omit the question regarding the amount of work performed against the external pressure, for this can be ascertained in all cases exactly and easily in calorific equivalents, in the case both of gases and of liquids; on the one hand, the heat introduced will increase the kinetic energy of the molecules, and on the other hand, it will perform a certain amount of work against the molecular forces, because in consequence of the associated thermal expansion, the molecules remove farther apart. But now since the molecular forces increase greatly with condensation, they will be much greater in liquids than in gases, where they are inconsiderable at ordinary condensation; thus we explain the universal phenomenon that the specific heat of a liquid is always greater than that of its vapour (p. 59). If a liquid is kept at constant volume during the warming, then, provided that the theory of van der Waals is rigorously applicable to it, its specific heat must be exactly the same as that of its vapour, provided, of course, that this latter is taken at constant volume; as far as I know, no experiments have been made in this direction.

The same conclusion holds good for compressed gases; it was, in fact, found that up to pressures of 6000 to 7000 atm. the specific heat at constant volume was independent of the volume (p. 45). As is well known, water does not change its volume very considerably by being heated in the vicinity of 4°, in spite of the fact that the specific heat is much greater in the liquid condition than in the gaseous condition; we have here an exception which can be no more explained by van der Waals' theory than can all the anomalous densities of this substance.

On the other hand, it seems to me that the following result speaks emphatically in favour of the preceding conclusion. From the equation developed on p. 55, viz.—

$$\frac{\partial C_v}{\partial v} = T\frac{\partial p}{\partial T^2}$$

[1] Young, *Proc. Phys. Soc. Lond.*, 1894-95, p. 602.
[2] Riecke, *Gött. Nachr.*, 1894, No. 2.

when

$$\frac{\partial C_v}{\partial v} = 0, \text{ it follows that } p = AT - B,$$

if we denote by A and B, two integration constants. As a matter of fact, Ramsay and Young[1] found that the pressure of a liquid or gaseous substance, when kept at constant volume, varied linearly with the temperature; or, in other words, that the *isochores* (*i.e.* curves which are obtained from varying pressure at constant volume) are straight lines.

Moreover, it is possible to derive this last result directly from the equation of condition (p. 212), by writing it in the form—

$$p = \frac{R}{v-b}T - \frac{a}{v^2};$$

thus we have

$$A = \frac{R}{v-b}; \quad \text{and } B = \frac{a}{v^2}.$$

As was shown (on p. 48), the work performed in compressing an ideal gas appears in the interior as heat. This law, like the other gas laws, becomes inexact with a high degree of compression, and thus the interesting question arises, as to how we are to conceive of these deviations in the light of the theory of van der Waals.

The answer is very simple. On the approach of the molecules towards each other by compression, their mutual attraction performs a certain amount of work which appears in the form of heat; the conditions, therefore, require that a greater amount of heat should appear from compression than what corresponds exactly to the work applied. The molecular pressure,

$$K = \frac{a}{v^2},$$

may be regarded as a measure of the force of attraction. The heat developed by this force, by compressing the volume from v_1 to v_2, amounts to

$$\int_{v_1}^{v_2} K dv = \int_{v_1}^{v_2} \frac{a}{v^2} dv = a\left(\frac{1}{v_1} - \frac{1}{v_2}\right).$$

If a gas expands without performing external work, as, for example, when it flows into a vacuum, then the aforesaid quantity of heat must be absorbed. As a matter of fact, van der Waals[2] verified the preceding formula quantitatively by calculating the experiments of Joule and Thomson with air and carbon dioxide.

[1] *Zeitschr. physik. Chem.*, **1.** 433 (1887); **3.** 49, 63 (1889). [2] *l.c.* 114 and foll.

Moreover, the theory affords many explanations regarding the heat of evaporation. For, in the first place, it is at once obvious that the heat of evaporation must become zero at the critical point, because the distinction between a liquid and its saturated vapour ceases at that point. Mathias[1] found experimental proof for this corollary from measurements conducted with carbon dioxide and nitrous oxide. This necessarily follows from the formula of Clausius; for at the critical point, when

$$v_0 = v_0', \quad \text{therefore } l = 0.$$

It now becomes possible to go a step farther and calculate the *heat of vaporisation* from the molecular forces.[2] If a gas is condensed, the molecular forces perform the work calculated above, which must reappear in the form of heat. By calculating the latter for 1 g.-mol (according to p. 58), we obtain

$$\lambda - p(v_0 - v_0') = a\left(\frac{1}{v_0'} - \frac{1}{v_0}\right) \quad . \quad . \quad . \quad (1)$$

By comparison with the equation deduced on p. 229.

$$RT \ln \frac{v_0 - b}{v_0' - b} + \frac{a}{v_0} - \frac{a}{v_0'} = p(v_0 - v_0') \quad . \quad . \quad . \quad (2)$$

we get

$$\lambda = RT \ln \frac{v_0 - b}{v_0' - b} \quad . \quad . \quad . \quad (3)$$

From the critical data (according to p. 223),

$$a = \frac{27}{64}\frac{R^2\theta_0^2}{\pi_0};$$

now on considering the heat of evaporation at the boiling-point, it follows that v_0 must be very large compared with v_0', and therefore similarly $\frac{1}{v_0'}$ as compared with $\frac{1}{v_0}$; and we thus find as the value of the molecular heat of evaporation

$$\lambda = \frac{27}{64}\frac{R^2\theta_0^2}{\pi_0 v_0'} + RT.$$

For benzene we have

$$\theta_0 = 561^\circ; \quad \pi_0 = 42 \text{ atm.};$$
$$v_0' = 0{\cdot}096 \text{ lit.}; \quad T = 273 + 80:$$

now by writing $R = 0{\cdot}0821$, we obtain λ in litre-atm. (p. 48): then by reducing to g.-cal., by dividing by 24·25, we obtain $\lambda = 5413$,

[1] *Ann. chim. phys.* [6], **21.** 69 (1890).
[2] Bakker, *Z. S. phys. Chem.*, **18.** 519 (1895).

instead of 7200 g.-cal. Thus the demands of the theory are met by only a very remote approximation; but perhaps the deviation in this case may be ascribed to the partial formation of complex molecules from condensation, which would greatly increase the heat of evaporation. Nevertheless, it is not without interest to know that it is possible to calculate the heat of evaporation, at least approximately, so easily from the critical data; and also that the formula is in harmony with experiment in showing a gradual diminution [for the value of λ], till the critical point is reached, when λ becomes zero.

Similarly is found for

Ether	$\lambda=4600$	instead of	6640	at the boiling-point.
Carbon tetrachloride	$\lambda=5250$	,,	7100	,, ,,
Tin tetrachloride	$\lambda=6270$	,,	7900	,, ,,
Ethyl alcohol	$\lambda=5640$	,,	10500	at 0°.
Ethyl acetate	$\lambda=5500$	,,	9000	

In all cases experiment gives a higher value than the theory.

Equation (3) is obtained by combination of (2) derived entirely from the kinetic theory with (1), which is purely thermodynamic; it can also be deduced kinetically as Kammerlingh Onnes showed (*Verh. d. Ak. Wiss.*, Amsterdam, 1881; *Arch. Néerl.*, **30**. 101 (1897); see also W. Voigt, *Gött. Nachr.*, 1897, Heft 3). This problem, obviously associated with that of the kinetic explanation of the second law of thermodynamics, is fully treated by Boltzmann (*Gas Theory*, ii. p. 167 ff.).

Critique of the Results.—By means of the numerical material given above, it is shown that the theorems of van der Waals do not find a stringent proof from experiment, but that there are certain undoubted deviations between theory and fact.

Now the question arises whether these deviations are of such a sort that the whole theory must be given up, or is it possible to account for them by some change in developing the theory. The verdict cannot be doubted for a moment; for the results of the theory are so undeniable, and the region of the phenomena which it claims to control is so extensive, that it would be a very profitable problem to follow up these deviations, and to prove carefully what is necessary, *i.e.* whether a re-shaping or an extension of the theory is required. The study of the changes which must be made in the gas laws in order to adapt them better to the facts, was what led van der Waals to his wonderful discoveries. Is it not probable that the desire to invest the laws of van der Waals with yet greater accuracy would be rewarded with surprising and unexpected results?

The results so far arrived at with regard to van der Waals' theory may be summarised as follows: the equation

$$\left(p+\frac{a}{v^2}\right)(v-b)=RT$$

only holds exactly for not too strongly compressed gases; it quite fails near the critical point. The theory of corresponding states, which is not bound up with any special characteristic equation, is of wider applicability; in especial it allows of calculating accurately the volume and thermal expansion of liquids. The theorem is much further from the truth as regards numerical calculations on the coexistence of liquid and vapour (volume and pressure of saturated vapour). It is natural to conclude that in this the deviation is due to a polymerisation of the molecules, which may be quite small; it is therefore quite possible that the theorem of corresponding states may belong to the class of exact natural laws. Kirstine Meyer (*Z. S. phys. Chem.*, **32**. 1, 1900) adopts the same view in a numerical discussion of the theory.

It is of course a question whether the characteristic equation can be put in a final form; the changes necessary in van der Waals' formula have lately been critically investigated, at my suggestion, by Reinganum (*Theory and Deduction of a Characteristic Equation*, Diss., Göttingen, 1899). The descriptions are especially great in reckoning heats of expansion: Reinganum shows that the attraction constant a is decidedly a function of the volume, and to a certain extent of the temperature. Now, as the attractive forces exercise an influence on the volume constant b by accelerating molecular collisions, b must also be a function of the volume and temperature. For greater density van der Waals' method of treatment becomes more exact. In this way Reinganum arrives at an equation that is partly deductive and partly based empirically on the observations of Young.

$$\left(p + \frac{a}{v^2}e^{\frac{\epsilon}{T}}\right) \cdot \frac{\left(v - \beta e^{\frac{\gamma}{T}}\right)^4}{v^3} = RT,$$

in which β is the volume occupied by the molecules, and ϵ and γ are given by the expressions

$$\epsilon = \frac{0{\cdot}0345a\left[(v - 2\beta)^3 + 12{\cdot}2\frac{\beta^4}{v}\right]}{\beta R v^3}$$

$$\gamma = \frac{0{\cdot}0726a\left[(v - 2\beta)^3 + 3{\cdot}34\frac{\beta^4}{v}\right]}{\beta R v^3}.$$

Reinganum shows by the method of Raveau (p. 226) that his equation does agree strikingly with the facts.

For other modifications of the original equation of van der Waals, see van der Waals, *Z. S. phys. Chem.*, **38**. 257, 1901.

The Kinetic Theory of the Solid State of Aggregation.— The attempt to press forward into the constitution of solid substances by means of molecular considerations has as yet only begun. The assumption, that the kinetic energy of the translatory motion is proportional to the absolute temperature, and independent of the nature of the molecule, may possibly be introduced here, as was the case with gases and liquids; but here the belief is at once forced upon us that the heat motion of the molecules of solid substances consists of an oscillation within very sharply-defined limits, in which they are restrained by the forces of cohesion, and that no single molecule can continue to change its place as is the case with gases and liquids. In fact, the peculiarities of the solid state become intelligible in this way.

Thus heat introduced may become effective in three ways, viz.—

1. It may increase the kinetic energy of the molecule, or of the atoms within the molecule.

2. It may perform work against the forces of cohesion between the molecules.

3. It may perform work against the external pressure.

On account of the slight change of volume occasioned by heating solid bodies, in general the third mode, and the second also, perhaps, may be neglected. Nothing further can be noted here [1] regarding the attempt to press forward on the basis either of the preceding assumption, or of other assumptions.

The process of crystallisation is especially noteworthy from the point of view of molecular theory. We have to imagine that to form a crystal (*e.g.* in an undercooled fused mass or a supersaturated solution) a considerable number of molecules have to meet in one spot, and in an appropriate constellation; most probably the smallest crystalline individual capable of existence consists of very many molecules. Thus the probability of formation of crystals may be very small; experience shows, in fact, that it may take a very long time before crystals appear in an undercooled fused mass, and that crystalline precipitates possess a number of centres of crystallisation (nuclei) that is excessively small compared with the number of molecules concerned.

The circumstance that discrete centres of crystallisation can arise in apparently homogeneous melts or solutions is a striking instance of the appropriateness of the molecular theory; for this alone makes it intelligible that even in the interior of an apparently homogeneous fluid, local differences should occur due to special constellations of molecular movements, and that these should be very sparsely scattered in comparison with molecular distances.

According to G. Tammann (*Z. S. phys. Chem.*, **25**. 441, 1898) the points at which crystallisation begins in a melt are hardly 1000 per minute per cubic millimetre under the most favourable circumstances. With increase of undercooling the number of these nuclei at first increases, then diminishes, to practically zero.

[1] See H. A. Lorentz, *Wied. Ann.*, **12**. 134 (1881); van der Waals, *Zeitschr. physik. Chem.*, **5**. 133 (1890).

The molecular explanation of this remarkable fact seems to me this. The greater the undercooling, the smaller the smallest possible crystalline individual must be, and so the greater the probability of the molecular constellation required to form it. But on the other hand, the molecular movements fall off in vigour with fall of temperature, so that the occurrence of any particular constellation is rendered more difficult and at the absolute zero, impossible. These two factors in combination would give the results observed by Tammann. See especially Küster (*Z. S. anorg. Chem.*, **33**. 363 (1903).

The Kinetic Theory of Mixtures.—In all our theoretical considerations regarding molecules thus far, we have always had in mind a substance which should be simple from a chemical standpoint, *i.e.* one consisting of nothing but molecules *of one kind.* The question now arises, viz. how the preceding considerations will be modified when we study a mixture.

And here we must distinguish between two cases, viz. the different kinds of molecules may react chemically upon each other, or they may hold themselves indifferent towards each other.

It must be noticed, in the first case, that the condition of chemical equilibrium prevailing in the mixture will, in general, vary with the pressure, volume, and temperature; such a mixture, for example, is the vapour of acetic acid, which consists partly of doubled molecules and partly of normal ones, the ratio of which changes with the external conditions. We will consider such mixtures more thoroughly in the part devoted to "the doctrine of affinity." In the following we consider only mixtures which consist of molecules which are chemically indifferent.

Now, the van der Waals' equation of condition,

$$\left(p + \frac{a}{v^2}\right)(v - b) = RT,$$

can, of course, be applied directly to a homogeneous mixture of this last sort, whether it be a liquid or a gas. In fact, it is as fully applicable to strongly-compressed atmospheric air as to a simple gas, only the meaning of the constants "a" and "b" is somewhat different from that which they have in the case of pure substances, and their magnitudes also change with the relative proportions of the ingredients in the mixture.

Van der Waals (*Z. S. phys. Chem.*, **5**. 134, 1890) gives the following equations to calculate the constants a_x and b_x of a mixture containing $1 - x$ mols. of one component to x of the other.

$$a_x = a_1(1 - x)^2 + 2a_{12}x(1 - x) + a_2x^2$$
$$b_x = b_1(1 - x)^2 + 2b_{12}x(1 - x) + b_2x^2.$$

Here a_1 and b_1 are constants of the first component, a_2 and b_2 of the second; a_{12} is the attraction constant between the two, and b_{12} the volume constant for the combination. Also, for b_{12},

$$\sqrt[3]{b_{12}} = \tfrac{1}{2}\{\sqrt[3]{b_1} + \sqrt[3]{b_2}\}$$

holds.

According to this the properties of a mixture can be deduced from those of its components by the help of one new constant a_{12} only.

The treatment of the question regarding the coexistence of liquid and vapour is entirely different in the case of mixtures, since in general, in distinction from simple substances, the composition of the liquid and vapour differ. As shown on p. 228, this question has not yet been treated for the case of a simple substance from a pure kinetic standpoint. Much less, therefore, would this be the case for mixtures; and this is also true of van der Waals' quoted work, although entitled *The Molecular Theory of Bodies consisting of Two Different Substances;* this is emphatically a thermodynamic study, but its description does not belong to this chapter. It should, however, be observed that, in addition to the investigation of the coexistence of liquid and vapour of mixtures, there also comes up the question regarding the coexistence of liquids of different composition (mutually saturated solutions).[1]

The Kinetic Theory of Solutions.—We learned, in the last chapter of the first book, a fact drawn from experiment, viz. that the laws of gases are applicable also to dissolved substances, and that there is a very far-reaching analogy between the condition of matter in dilute solution, and in the gaseous state. Now it is but a step to *transfer this analogy also to the molecular condition;* and the conception is at once suggested that the osmotic pressure of a solution, like gas pressure, has a *kinetic nature, i.e.* that it results from the bombardment of the molecules of the dissolved substance against the semi-permeable partition. Boltzmann,[2] Riecke,[3] and Lorentz[4] have made attempts of this sort to develop the laws of dilute solutions immediately from the play of molecular forces and of molecular movements, without having recourse to the aid of thermodynamics.

Each of these investigators assumed that the *mean kinetic energy of the translatory motion of the molecules in solution was as great as in the case of the molecules of a gas at the same temperature as the solution.* This assumption appears the more probable, because the theory of van der Waals has been already applied to liquids with good results.

According to Boltzmann, in the calculation of the osmotic pressure,

[1] See the important new work of van Kuenen (especially *Z. S. phys. Chem.*, **24**. 667, 1897; **41**. 43, 1903) and the second part of v. d. Waals, *Continuität*, etc. (Leipzig, 1900).

[2] *Zeitschr. physik. Chem.*, **6**. 474 (1890); **7**. 88 (1891).

[3] *Ibid.*, **6**. 564.

[4] *Ibid.*, **7**. 36.

one must take into consideration not only the mutually opposed action between the wall and the molecules bounding against it, as was done in explaining gas pressure, but also the mutual action between the dissolved molecules and those of the solvent. But it is a difficult problem to state in just what way this occurs, and to make such a supposition as shall be explanatory and shall harmonise with the facts, and the satisfactory solution of this problem remains for the future. Boltzmann assumed, on the one hand, the distance between the centres of two neighbouring salt molecules, to be very great as compared with the distance between the centres of two neighbouring molecules of the liquid solvent; and, on the other hand, the space changed by the presence of the molecule of the dissolved salt, to be small in comparison with the total space occupied by the liquid solvent; and in this way he obtained this result, viz. *that the osmotic pressure exerted by the dissolved substance on a semi-permeable membrane is equal to the corresponding gas pressure.*

Lorentz assumed that the dissolved substance is subjected to an attractive force proceeding from the solvent, which is equal to, and opposite to, the force acting upon the quantity of the liquid displaced. Riecke proceeds upon the supposition that the number of collisions between the molecules of the dissolved substance is infinitesimally small as compared with the number of collisions between the molecules of the solvent. Each, on the basis of his proviso, arrived at the same conclusion as Boltzmann.

Boltzmann and Riecke both showed in this way, respecting the diffusion of dissolved substances, and of electrolytes especially, that the same results are obtained which I have already developed independently from the special kinetic conceptions preceding (p. 157). Thus Riecke succeeded in deriving the *mean free path* of a dissolved substance from its diffusion velocity: the following is the value, in which D denotes the coefficient of diffusion (p. 158), and "u" denotes the mean calculated velocity of the molecule, viz. :—

$$L = \frac{\cdot 3D}{8.64 \times 10^4 . u}.$$

In this way he found, for instance, the mean free paths, at 8° to 10°, as follows :—

Substance.	D	Mean Free Path, "L"
Urea	0·81	$0·68 \times 10^{-8}$ mm.
Chloral hydrate . .	0·55	0·94 ,,
Mannite	0·38	0·68 ,,
Cane sugar . . .	0·31	0·77 ,,

These mean free paths are small compared with the dimensions of the molecule (Chap. XII.)

The explanation recently offered by J. H. Poynting (*Phil. Mag.*, **42**. 289, 1896) of relative lowering of vapour pressure as due to association of the dissolved molecules with those of the solvent is an arbitrary *ad hoc* hypothesis. There is no proof that this hypothesis can explain anything but the particular fact it was brought forward to explain, *e.g.* electrolytic conductivity. The hypothesis of electrolytic dissociation gives deductions in harmony with the facts of conduction, whereas the association hypothesis contradicts them.

Finally, I will refer to one more point which seems to have some importance. It was shown in the case of many liquids, for example ether, at the temperature of a dwelling-room, that the space actually occupied by the molecules is about 0·3ths of the space apparently occupied. If we dissolve the ether in any selected solvent, then the space at the disposal of the oscillating molecules is certainly much smaller than the volume of the solution; therefore it would be expected that this "volume correction" would exercise as correspondingly great an influence on the osmotic pressure as on the pressure of a strongly compressed gas; and it is inconceivable that it should nevertheless be so nearly equal to the gas pressure. Yet a more careful examination shows that *this correction may be neglected in the methods hitherto applied for measuring the partial pressures of the dissolved substances.* The methods thus far used always give the osmotic work P, *i.e.* the work required to remove a unit volume of the pure solvent from the solution, and this value of the pressure (being multiplied by the unit of volume) turns out to be equal to the gas pressure. But if we assume that the pressure of the dissolved substance must be increased on account of the volume correction, in the ratio of $1 : 1 - \beta$, when β denotes the amount of the diminution of the space at the disposal of the movement of the dissolved molecules, then this pressure P will become

$$\frac{P}{1-\beta};$$

therefore the osmotic work required to remove unit volume of the solvent would be calculated as the product of the pressure and the volume, *i.e.*

$$\frac{P}{1-\beta} \times (1-\beta) = P,$$

since the volume through which the pressure is to be forced back is also diminished in the ratio $1 - \beta : 1$, and for the same reasons. In the methods thus far employed for measuring the osmotic pressure, the volume correction is thus wholly neglected. It would be very interesting to find a method for measuring directly the actual pressure of the dissolved substance, viz.,

$$\frac{P}{1-\beta}.$$

On osmotic pressure of strong solutions in terms of the molecular theory see Abegg (*Z. S. phys. Chem.*, **15**. 209, 1894).

CHAPTER III

THE DETERMINATION OF THE MOLECULAR WEIGHT

The Molecular Weight of Gaseous Substances.—Formerly gases were the only substances the molecular weights of which could be directly determined, and it was only very recently that the possibility was suggested of measuring the same values for substances in dilute solution. The theoretical reasons which underlie these methods have been already thoroughly treated in discussing the theory of the gaseous state, and of solutions. The chief aim of this chapter will be devoted to the experimental realisation of these methods.

Since, according to Avogadro's rule (p. 205), under similar conditions of temperature and pressure, different gases contain the same number of molecules in a litre, therefore the densities of any two gases are in the ratio of their molecular weights. It is customary to refer the density, D, of a gas *to atmospheric air at the same temperature and pressure.* The density of air at 0° and atmospheric pressure in latitude 45° is 0·0012932, compared with water at 4°; hence the density of the gas under the same conditions is

$$0{\cdot}0012932D.$$

Since 1 mol of an ideal gas under normal conditions occupies 22,400 c.c., the density is

$$\frac{M}{22,400}$$

if M be the molecular weight.

Hence

$$M = 0{\cdot}0012932 \, . \, 22,400D = 28{\cdot}994D,$$

or nearly enough

$$M = 29{\cdot}0D.$$

All the practical methods for the determination of the density of a gas amount to this, viz. one determines—

1. The mass of a gas filling a measured volume at a known temperature and pressure.

Or, 2. The volume filled by a weighed amount of a vaporised substance.

Or, 3. The pressure exerted by the evaporation of a known amount of the substance in a measured volume at a given temperature.

Moreover, in practical work in the laboratory, it is not so necessary that a vapour density method shall be capable of measurement with a high degree of accuracy, as that it shall be certain and simple. For since the percentage composition is furnished by the analysis, a determination of the vapour density within 1 per cent is amply sufficient to determine *which one* of the different possible molecular weights in question is to be selected as the correct one.

Regnault's Method.—For the determination of the density of a substance which is already gaseous under the ordinary conditions of temperature and pressure, Regnault used two glass bulbs of nearly the same size, which were hung from the arms of a sensitive balance. One of the bulbs was first exhausted, and then filled with the gas to be investigated at a known pressure. The second bulb merely served to avoid the very considerable correction due to the buoyancy of the air.[1]

As is well known, Regnault in this way was able to measure the density of the permanent gases with satisfactory accuracy (p. 41). But for use in the chemical laboratory, when liquid and solid substances are to be investigated, other methods are employed, of which the following are the most important.

Dumas's Method (1827).—A light glass bulb, holding about 0·25 l., is drawn out at the opening to a long thin point, and then weighed. After partially exhausting it, by previous warning, about 1 g. of the liquid substance is caused to be drawn up into it. It is then placed in a heating bath, which has a measured constant temperature which must be above the boiling-point of the substance to be volatilised. After the contents have been completely volatilised, the point of the opening is sealed by melting the glass with a blow-pipe. The bulb, when cooled and cleaned, is then weighed again. Then it is filled with water by breaking off the sealed point under water. The experiment is regarded as satisfactory, only when it is filled completely with water, showing that only traces of air remain in the bulb. Its volume is obtained by weighing it filled, including the point broken off; an approximate weight suffices here.

Let m denote the weight of the bulb filled with air, m′ that when filled with the vapour, and finally M that when filled with water.

[1] This method has been used recently by Rayleigh, Crafts, Leduc, Morley, etc., and worked out more thoroughly; see Morley, *Z. S. phys. Chem.*, **17.** 87 (1895); **20.** 68 and 242 (1896).

Let t and b be the temperature and barometric pressure at the moment of sealing the opening, and t′ and b′ the corresponding values at the first weighing. Let λ be the weight of 1 c.c. of air at pressure b′ and temperature t′. Then will the density D be equal to

$$D=\left(\frac{m'-m}{M-m}\frac{1}{\lambda}+1\right)\frac{b'}{b}\frac{1+0{\cdot}00367t}{1+0{\cdot}00367t'}.$$

Since 1 c.c. of atmospheric air at 0° and 760 mm. weighs 0·001293 g., therefore λ amounts to

$$\lambda=\frac{0{\cdot}001293}{1+0{\cdot}00367t'}\frac{b'}{760}.$$

The Gay-Lussac-Hofmann Method (1868).—A measured quantity, m, of the liquid in question is enclosed in a tiny bottle with a glass stopper. This is then introduced into an upright barometer and allowed to rise into the vacuum above the mercury. By means of a steam jacket, heated with the steam of a suitable liquid, the barometer is warmed to such a temperature that the contents of the tiny bottle are vaporised. From the volume v, finally occupied by the vaporised substance at the temperature t, the density D is found to be

$$D=\frac{m}{v}\frac{1}{\lambda};$$

where again the value of λ, the weight of 1 c.c. of air, at the temperature and pressure of the vaporised substance, is obtained from the formula,

$$\lambda=\frac{0{\cdot}001293}{1+0{\cdot}00367t}\frac{b-h-e}{760}.$$

Here the pressure, at which the vapour in the barometer tube stands, is equal to the external barometric pressure, b, at the time of the experiment, reduced to zero, minus the height, h, of the mercury column over which the vapour stands, reduced to zero, and minus also the vapour tension of mercury e, at the temperature t.

Victor Meyer's Method by the Displacement of Air (1878).—All the methods thus far described are far surpassed in simplicity and certainty by this method. It shares with Hofmann's method the advantage of being able to use a very small quantity of a substance.

A long "pear-shaped" vessel, A in Fig. 19, is heated to, and kept at, *a constant temperature*, which must be above the boiling-point of the substance to be investigated, either by means of a heating bath,[1] or in any other suitable manner. It is not necessary to know this temperature in order to calculate the vapour density. The pear tube

[1] For the description of a simple bath, see V. Meyer, *Ber. deutsch. chem. Ges.*, **19.** 1861 (1886).

extends into a long tube of smaller diameter. This is closed at its upper end, near which are two side tubes. One of these leads to a small delivery tube or to a gas burette. The other carries a glass rod which is provided at the opening with a tightly-fitting rubber tubing. This serves to drop the substance. It is somewhat enlarged in Fig. 19. On the small glass rod rests the substance to be used, enclosed within a tiny glass flask if it is a liquid, or if a solid, in the form of a little rod, obtained by sucking the molten substance up into a glass tube.

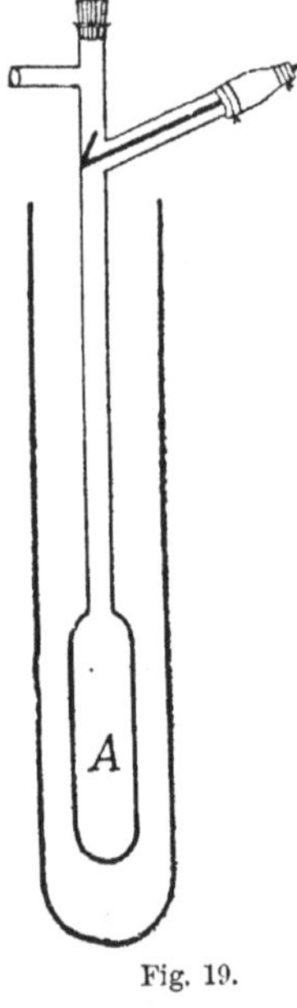

Fig. 19.

After the temperature has become nearly constant, as will be shown by the cessation of bubbles from the delivery tube, the substance is dropped by slipping back the glass rod, which is drawn back into position again by the elastic rubber. The substance falls to the bottom of the pear-shaped flask, which is provided with a cushion of asbestos, spiral wire, mercury, or the like, to avoid its being broken. The substance at once is vaporised. The disengaged vapour displaces the air above it, blocking its way; and if this occurs before there is any marked diffusion or condensation into the colder part of the apparatus, then the vaporisation occurs quickly enough. The volume of air displaced is measured in the eudiometer, or in the gas burette; and this represents the volume which would be occupied by the substance at the same conditions of temperature and pressure. Thus the density D is

$$D = \frac{m}{b}\frac{760}{0{\cdot}001293}\frac{1 + 0{\cdot}004t}{v} = 587{,}800\frac{m}{bv}(1 + 0{\cdot}004t).$$

Here m denotes the quantity of the substance used in grams, b the pressure in millimetres, and t the temperature at which the displaced volume v, of air (measured in c.c.) is observed. The factor 0·004 is taken instead of the usual coefficient of expansion 0·00367, in order to take account of the moisture of the air.

One may also drop the substance into the pear-shaped flask, by opening the stopper and closing as quickly as possible. A heating-jacket of glass around the "pear" may be used when the temperatures do not run too high.

It must also be observed that the partial pressure at which the vaporised substance stands is *neither definite nor constant;* but that from the bottom, where the substance is only slightly intermixed with the air present, and therefore where the pressure is almost equal to the atmospheric pressure, the partial pressure gradually sinks to zero at the top, and thus that it diminishes at all intermediate points with increasing diffusion. But this variation of the partial pressure, resulting from dilution with the obstructing air, may be disregarded in the measurement (*i.e.* in the measurement of the displaced air), *provided that the disengaged vapour conducts itself as an ideal gas and obeys Boyle's law.* Only in such cases will the quantity of displaced air be as great as though no intermixture had occurred through diffusion.

But the behaviour is entirely different when the substance which is to be studied exhibits dissociation (see Chap. VI.), and when it dissociates more and more as the partial pressure diminishes. In that case, the quantity of air displaced will be greater the more quickly the mingling occurs, and the results obtained will no longer bear a simple interpretation, because one does not know how far the mixture had gone at the time of the measurement, and therefore what the relative pressures may be under which the vaporised substance was studied.

The way in which the air is displaced may also indicate in a qualitative way, whether the vaporised substance is conducting itself properly, or whether dissociation is taking place; *in the latter case, the method is unsuitable for exact measurements, and will not give simple results.* (See Book III. Chap. II., "Influence of Indifferent Gases.")

The Determination of the Vapour Density at Very High Temperatures.—At high temperatures, Hofmann's method is debarred on account of the great vapour tension of mercury, and also for other reasons. The method of Dumas is here quite difficult to use, since we are debarred from using a glass bulb at high temperatures (above 650°), and if a porcelain flask is used according to the modification of Deville and Troost, its point must be sealed with the oxyhydrogen blow-pipe. On the other hand, the method of the displacement of air has been used recently with very good results, even at very high temperatures (up to 2000°).

For the material of the heating bath, there may be used boiling sulphur (444°), or boiling phosphorus pentasulphide (518°), or boiling stannous chloride (606°), and for still higher temperatures a charcoal furnace, or a Perrot's gas furnace which is fed with an air blast. As shown by the researches of Nilson and Pettersson (1889), and also by those of Biltz and V. Meyer (1889), it is possible in this way to reach and measure high temperatures which were sufficiently constant during the investigation; these temperatures were easily regulated by controlling the supply of gas, and mounted as high as 1730°.

Using water-gas in a Perrot[1] furnace Biltz[2] attained a temperature of 1900°; suitable materials for a vessel to use at this temperature are not known however.

[1] For a description of the Perrot furnace see V. Meyer and C. Meyer, *Ber. d. chem. Ges.*, **12**. 1112 (1879).

[2] *Z. S. phys. Chem.*, **19**. 385 (1896).

The pear-shaped vessels and the extension tubes were protected against injury, by making them of platinum free from soldering with any other metal, or of porcelain glazed, both inside and outside. The latter has the advantage, as it can be subjected directly to the gas flames; while the former, on account of its ready permeability by gases at high temperatures, must be protected from direct contact with the flame by a surrounding mantle of porcelain. Inasmuch as the porcelain "pears" begin to soften at 1700°, it is recommended that they be wrapped in thick platinum foil in order to increase their resistance to heat.

The dropping apparatus, and the tube delivering to the gas burette, were as before made of glass, and provided with rubber fittings at the ends of extension tubes which reached far enough out of the furnace; they were protected from being heated, by a screen at the side of the furnace.

In general the measurement differs in no detail from that conducted at lower temperatures, except that the substance cannot be vaporised in air on account of the great chemical activity of oxygen at high temperatures; therefore, before the experiment, the apparatus must be filled with some neutral gas, as nitrogen or carbon dioxide.

Although, as previously emphasised, it is not necessary to know the temperature at which the substance is vaporised, yet it is nevertheless of great advantage in the case of substances which change their molecular weight with rising temperature. As a matter of fact, Nilson and Pettersson, V. Meyer, and others, without making the apparatus very complicated, succeeded in making vapour density determinations, combined with quite reliable determinations of the temperature.

This experiment[1] consisted simply in using the "pear" as an *air thermometer* by heating it from the initial temperature (as 0° or the temperature of the room), and measuring the volume of the air displaced; from this the final temperature was calculated from the well-known expansion coefficient of gases.

Finally, it is necessary to introduce a correction due to the expansion of the pear with the temperature; this is given by the coefficient of volume expansion of the material of which it consists Now it must be noticed that only the "pear" itself is exposed to the temperature which it is desired to measure, and this temperature diminishes along the extension tube down to the temperature of the room. Inasmuch as this correction introduces much difficulty into the calculation, it is determined directly by means of a *compensator;* this is a companion tube closed below, of the same material as the extension tube, of the same form, and placed parallel to it and as near it as possible.

[1] Nilson and Pettersson, *J. pr. Ch.* [2], **33.** 1 (1886); Biltz and Meyer, *Zeitschr. physik. Chem.*, **4.** 249 (1889).

By subtracting the amount of gas driven out of the compensator, from that driven out of the vapour density apparatus, one obtains the amount actually driven out of the "pear" by the elevation in temperature, which, reduced to 760 mm. and 0°, amounts to v. Further, if V denotes the content of the pear (= the total content minus the content of the compensator), if α denotes the coefficient of expansion of gases (0·00367), and if γ denotes the cubic coefficient of expansion of the material of the pear (0·0000108 for porcelain and 0·000027 for platinum), then the final temperature is given by the simple formula:

$$t = \frac{v}{V(\alpha - \gamma) - v\alpha}.$$

At very high temperatures, when only about $\frac{1}{7}$th of the original air remains in the apparatus, the method loses its accuracy, because here with a very great increase in temperature there is associated the expulsion of only a very slight quantity of air.

Another method, used at the same time by Crafts and V. Meyer,[1] works much better; this requires a more complicated apparatus, and consists in driving out the air or nitrogen contained in the apparatus at atmospheric temperature, and again at the temperature of the experiment by carbon dioxide or by hydrochloric acid; measuring the former in a gas burette, while the displacing gas, in a quick stream, is absorbed by a solution of caustic potash, or by water in a gas burette. The displacing gas is introduced into the apparatus through a narrow tube which is fused into the lower part of the pear, and runs parallel with the extension tube; and similarly the introduction of the gas into the compensator was effected by means of a tube sealed into it, and running parallel with the extension tube. The use of such an apparatus was rendered impossible in the case of porcelain pears by reason of technical difficulties, but pears of glass and platinum proved well suited for this purpose.

If V denotes the volume of the pear at 0° and 760° mm., and v that of the heating jacket containing dry nitrogen referred to 0° and 760 mm., then the temperature of the experiment (as before), is

$$t = \frac{V - v}{v\alpha - V\gamma}.$$

Of course V and v are regarded as corrected by the subtraction of the corresponding quantity of air driven out of the compensator.

The author[2] has recently succeeded both in simplifying the procedure described above and in extending the range of temperature

[1] Crafts and Fr. Meier, *C. R.*, **90.** 606 (1880); V. Meyer and Züblin, *B. B.*, **13.** 2021 (1880). See also Langer and V. Meyer, *Pyrochemische Untersuchungen*, Brunswick, 1885; Mensching and V. Meyer, *Zeitschr. physik. Chem.*, **1.** 145 (1887).

[2] W. Nernst, *Z. S. f. Elektroch.*, 1903, p. 622.

available to 2000°. A small iridium vessel (made by Heraeus of Hanau) containing about 3 c.c. was heated by an electrically heated iridium tube: the small amount of gas driven out was measured by the movement of a mercury drop in a calibrated glass tube. The substance to be gasified was weighed—usually a fraction of a milligram—on a "microbalance," constructed like a letter balance, the pan being fastened at right angles to horizontally stretched quartz fibre. The temperature was measured by the emission of light from the bottom of the iridium vessel; this was compared with a photometric, electrolytic glower (of a Nernst lamp). With this apparatus fairly accurate measurements could be taken easily and rapidly.

The Determination of the Vapour Density at Diminished Pressure.—An important method of vaporising substances consists not only in raising the temperature, but also in diminishing the pressure; and indeed the latter alone is applicable when an elevation of temperature is attended by a decomposition of the substance which makes the determination of its vapour density impossible. The method of Hofmann is the only one of those described above, which permits the evaporation of the substance in a vacuum, and which therefore gives a determination at any desired diminished pressure. By combining the bulb used in the method of Dumas, with a water-pump and a manometer,[1] the pressure can be reduced very low; but it is obvious that, on weighing the very small residue remaining in the bulb, the method loses in accuracy very considerably.

Finally, it has been recently shown by V. Meyer,[2] as appears from his experience in vapour density determination, that it is possible to vaporise the substance in question at temperatures from 20° to 40° below its boiling-point *by taking pains to vaporise it quickly.* In order to accomplish this it is necessary to spread the substance out quickly on the bottom of the pear. This is accomplished by dropping solid substances in the form of tiny rods; and liquids enclosed in little flasks of Wood's metal, which melt immediately on arriving at the bottom of the pear. Also the use of hydrogen, as the packing gas, was found to be of advantage in increasing the speed of vaporisation, on account of its ready diffusibility [mobility].

Finally, a number of particular methods have been described for the determination of vapour densities at diminished pressures.

Malfatti and Schoop,[3] and Eykmann[4] as well as Bleier and Cohn[5] also in a modified way, measure the increase of pressure occasioned, in a space of known volume and which is almost completely exhausted, by the vaporisation of a known quantity of the substance in question.

[1] Habermann, *Lieb. Ann.*, **187.** 341 (1877).

[2] Demuth and Meyer, *B. B.*, **23.** 311 (1890); Krause and V. Meyer, *Zeitschr. physik. Chem.*, **6.** 5 (1890).

[3] *Zeitschr. physik. Chem.*, **1.** 159 (1887).

[4] *B. B.*, **22.** 2754 (1889).

[5] *Monatshefte f. Chem.*, **20.** 909 (1900).

Schall[1] compares the increase of pressure, resulting from such evaporation as that just described, with the increase of pressure from the admission of a known quantity of air into the exhausted apparatus, or from the escape of carbon dioxide from a known quantity of soda; in this method it is not necessary to know the volume of the apparatus. An apparatus has been described by Lunge and Neuberg,[2] which consists of a combination of V. Meyer's apparatus with a Lunge gas burette, and allows a very elegant control of the pressure under which the substance is vaporised.

The gas baroscope of Bodländer's[3] arrangement using the method of displacement, but measuring the increase in pressure at constant volume; a special advantage of this is that it avoids reduction to normal volume.

Calculation of Atomic Weight from Gas Density.—Atomic weights can be calculated from the molecular weights determined according to Avogadro's law (p. 175); usually this has only been done in an approximate manner, the exact value being subsequently arrived at by analysis.

Now, according to all probability Avogadro's law is strictly true for ideal gases; hence if the measurements on ordinary gases and vapours can be reduced to the ideal state, we have a purely physical method for exact determination of atomic weights from gas densities.

This method has been carried out with remarkable results by D. Berthelot.[4]

Let v_0 be the actual volume of a mol of gas under normal conditions, V_0 that which a mol of ideal gas would occupy under the same circumstances, then (p. 215)

$$V_0 = v_0\left(1 - \frac{b}{v_0} + \frac{a}{v_0^2}\right).$$

Hence to reduce the measured density of a gas to the ideal condition and obtain the exact molecular weight it must be multiplied by

$$\frac{v_0}{V_0} = \left(1 + \frac{b}{v_0} - \frac{a}{v_0^2}\right).$$

The simplest way to get the expression in brackets is by measurements of compressibility. As we are only concerned with gases in a condition approaching the ideal, the quantities $\frac{b}{v}$ and $\frac{a}{v^2}$ in van der Waals' formula are small compared with unity; we may therefore write with sufficient approximation

[1] *B.B.*, **22.** 140; **23.** 919, 1701 (1892).
[2] *Ibid.*, **24.** 724 (1891).
[3] *Ibid.*, **27.** 2263 (1894).
[4] *J. de Phys.* (3), **8.** 263 (1899).

$$pv = RT\left[1 + p\left(\frac{b}{RT} - \frac{a}{(RT)^2}\right)\right],$$

i.e. pv must, for constant temperature, be a linear function of the pressure p. If, then, pv be measured for various pressures and extrapolated linearly to p = 0, we shall have accomplished the reduction to the ideal state. D. Berthelot found the following molecular weights :—

	H_2	N_2	CO	O_2	CO_2
d	0·062865	0·87508	0·87495	1·0000	1·38324
A	−0·064	+0·038	0·046	0·076	0·674%
M	2·0145	28·013	28·007	32·000	44·000

	N_2O	HCl	C_2H_2	SO_2
d	1·38450	1·14836	0·81938	2·04835
A	+0·761	0·790	0·840	2·358%
M	44·000	36·486	26·020	64·050

Under A are given the corrections in per cent, calculated as described above; the observed densities must be reduced by these amounts to yield correct molecular weights. The atomic weights deduced from the latter are given in line (I.) below, while in (II.) are the atomic weights from the table, p. 34.

	O	H	C	N	S	Cl
(I.)	16·000	1·0075	12·004	14·005	32·050	35·479
(II.)	16·000	1·008	12·00	14·04	32·06	35·45

The remarkable agreement is evidence that Avogadro's law is true, in the limit, with great, or perhaps complete accuracy.

The Determination of the Molecular Weight from the Osmotic Pressure of the Dissolved Substance.—It has been already stated that it is possible to determine the molecular weight of any desired substance in any selected solvent, provided the concentration is not too great; this depends upon van't Hoff's generalisation of *Avogadro's* rule; according to this, *the osmotic pressure of a substance in solution, like the pressure of a gas, is independent of the nature of the molecules, but simply proportional to their number, and this is identical with the corresponding gas pressure* (see p. 155). Thus, if there are c grams of a substance dissolved in 1 litre of any selected solvent, and if the pressure at the temperature t, exerted by this on a partition (which is permeable for the solvent, but not for the dissolved substance), amounts to p atmospheres, then the molecular weight of the dissolved substance is calculated to be

$$M = 22{\cdot}42(1 + 0{\cdot}00367t)\frac{c}{p};$$

for, as shown by Regnault's determinations, 1 g.-mol of any selected gas, when enclosed in the space of 1 litre, at 0°, exerts a pressure of 22·42 atm., and at t° a pressure of 22·42 (1 + 0·00367t); and the osmotic pressure is equal to this, multiplied by the relative number of molecules, viz. $\frac{c}{M}$: 1, from which we derive the preceding formula.

But when one comes to the experimental determination of the osmotic pressure, he meets with great difficulties: hence osmotic measurements for molecular weight determination have been used only occasionally, as once by de Vries,[1] who determined the molecular weight of raffinose by comparing a solution of it with the known osmotic pressure of a plant-cell, by the plasmolytic method.

On the other hand, we possess several methods for the indirect measurement of the osmotic pressure, and which give simple and exact results; we are indebted to Raoult especially for their discovery. Since the lowering of the freezing-point, of the vapour pressure, and of the solubility, experienced by *a solvent* on adding a foreign substance, are all proportional to the osmotic pressure of the substance added, by measuring these depressions we obtain directly the molecular weight of the dissolved substance.

The Depression of the Freezing-Point.—If the addition of m grams of a substance to 100 g. of a solvent lowers the freezing-point of the latter t degrees, then the molecular weight of the dissolved substance is calculated from the formula

$$M = E\frac{m}{t};$$

here E, the "molecular depression of the freezing-point" of the particular solvent used, is obtained from the heat of fusion w, of 1 g. of the solvent, expressed in g.-cal. and from its fusing-point T in absolute temperature; thus

$$E = R\frac{T_0^2}{100w} = 0·02\frac{T^2}{w}.$$

The molecular depressions of the freezing-point for the solvents thus far investigated are given in the table on p. 153. The accuracy of the formula requires that the *pure* solvent shall freeze out of the solution, and not a mixture of the solvent with the dissolved substance (p. 170).

A great variety of pieces of apparatus which, however, are essentially the same in principle have been described to measure the depression of the freezing-point.[2] Of these we will describe the

[1] *C. R.*, **106.** 751 (1888).

[2] Raoult, *Ann. chim. phys.* [6], **2.** 93 (1884); **8.** (1886); *Zeitschr. physik. Chem.*, **9.** 343 (1892); Hollemann, *B. B.*, **21.** 860 (1888); Auwers, *ib.*, 701; Eykmann, *Zeitschr. physik. Chem.*, **4.** 497 (1889); Fabinyi, *ib.*, **2.** 964 (1888); Klobukow, *ib.*, **4.** 10 (1889).

apparatus constructed by Beckmann,[1] which was generally adopted in a very short time, and which enables one to measure the lowering of the freezing-point very simply and accurately.

The vessel A, Fig. 20, which is designed to hold the solvent, consists of a thick-walled test-tube, having a side tube. After pouring in the solvent, about 15 or 20 g. being weighed out or measured out with a pipette, there is introduced a stirrer made of thick platinum wire, and then the thermometer D, fitted with a cork. In order to hold the test-tube in place, and also to protect it with an air-jacket, it is fitted with a cork ring into another wider test-tube B. The whole is placed in a glass jar C filled with the cooling mixture, the temperature of which should be from 2° to 5° below the freezing-point of the solvent.

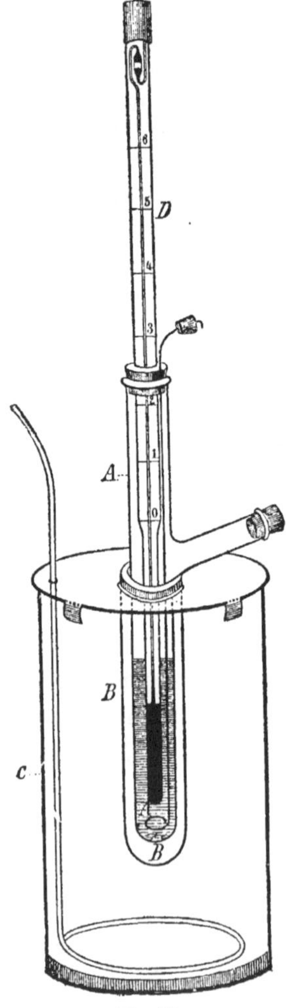

Fig. 20.

A measurement is conducted in the following way: First, the test-tube containing the requisite quantity of the pure solvent is dipped into the cooling mixture, and at the same time is constantly stirred till the whole is a little undercooled, and there occurs a separation of finely-divided ice, attended by a sudden rise in the thermometer to the freezing-point of the solvent. In this way, when the test-tube is protected by the air-jacket, which prevents its heat from being given off too suddenly, we obtain an exact measurement of the freezing-point of the solvent itself. We then add, through the side tube, a weighed amount of the substance to be investigated, and determine the freezing-point of the solution. The resulting depression of the freezing-point is obtained by subtraction.

Although we are now in possession of all the data requisite for the calculation of the molecular weight, yet it is recommended that a series of observations be made by adding successive portions of the substance, in order to determine whether the molecular weight is independent of the concentration, or whether the degree of dissociation varies with the concentration.

[1] *Zeitschr. physik. Chem.*, **2.** 638 (1888). See especially G. Fuchs, *Anleitung zu Molekulargewichtsbestimmungen*, Leipzig (1895).

In investigating solutions of greater concentration, the amount of ice separated must be as little as possible, in order to avoid the errors consequent upon too great a change of concentration from the freezing out of too large a portion of the solvent. This can be accomplished easily with a little practice. The side tube serves to introduce the solid substances. To introduce liquid substances one may use as a wash bottle the Sprengel-Ostwald pycnometer. More recently Beckmann has described a simple apparatus which enables one to use solvents which are strongly hygroscopic.[1]

Inasmuch as the measurement of the freezing-point can be accomplished with a little practice, within some thousandths of a degree, it is recommended to use a thermometer divided to read directly to hundredths of a degree. In order to use such a thermometer over wide ranges of temperature ($-6°$ to $+60°$), Beckmann devised a thermometer where the capillary above passed [through an enlargement of the bore] into a bent mercury reservoir (see Fig. 20). According as the thermometer is to be used with solvents having a higher or lower freezing-point, more or less of the mercury is passed from the capillary into the lower part of the reservoir by warming and carefully tapping the thermometer. The value of the scale of the thermometer practically remains unchanged, since here one is merely concerned in measuring differences of temperature. [Hence such thermometers are called *differential.*]

A method, given by the author, for measuring the freezing-point of solutions, especially at high concentrations, has been worked out by M. Roloff.[2] The principle of the method is to determine the composition of the solution that is in equilibrium with the frozen solvent at a given temperature. This may also be looked upon as a measurement of the solubility of the solvent in the given solution. In working out the method the first difficulty met with was to maintain constant low temperatures. Cryohydrates formed by mechanical mixture of snow and salts are not finely enough divided to give really constant temperatures. If, however, the cryohydrate be formed by freezing saturated salt solutions, they keep their cryohydric temperatures exactly till the freezing is complete. By the use of appropriate salts all temperatures from $0°$ to $-30°$ can be obtained.

Freezing-point of very Dilute Solutions.—It is important to make sufficiently exact measurements on very dilute solutions, both on account of the theoretical importance of such solutions and practically to obtain the molecular weight of slightly soluble substances. Early observers obtained somewhat erroneous values for the aqueous solutions with which, in the first place, they concerned

[1] *Zeitschr. physik. Chem.*, **7.** 323 (1891); **22.** 616 (1897).
[2] *Ibid.*, **18.** 572 (1895).

themselves; but recently results of value seem to have been arrived at.

It has been found that the influence of the cooling-bath and of the heat generated by stirring produce errors of some thousandths of a degree or more; this is of no consequence in careful measurements of molecular weight of substances at moderate concentrations, but must—and can—be avoided in dealing with very dilute solutions.

The following is a brief account of the theory of the establishment of equilibrium on freezing:[1]—Given a large mass of freezing liquid: let the true freezing-point be T_0, *i.e.* the temperature at which the separated solid and residual liquid are in equilibrium; let the temperature at time x be t. Then t will approach T_0 by the melting or freezing of some of the solvent, and the accompanying absorption or evolution of heat, according as t is above or below T_0. The outside temperatures exert no influence, since we have assumed a large mass of liquid.

According to experience on the solution of solids (Book III. Chap. V.), the rate of solution of the solid solvent may be taken as proportional, other circumstances being equal, to its difference in temperature from the point of equilibrium; but as the heat absorbed is proportional to the mass dissolved, the rate of change of temperature at any moment may be put proportional to the difference between T_0 and t. Thus

$$dt = K(T_0 - t)dz \quad . \quad . \quad . \quad . \quad (1)$$

We need not trouble as to the meaning of K except that it is proportional to the total area of the solid solvent and its heat of fusion. On integrating we get

$$K(z_2 - z_1) = \log \text{nat} \frac{T_0 - t_1}{T_0 - t_2} \quad . \quad . \quad . \quad . \quad (2)$$

where t_1, t_2 are the temperatures at times z_1, z_2.

Actually the mass of liquid is limited, so that there will be an interchange of heat with the surroundings by radiation, etc., as well as a production of heat by stirring. If we call t_0 the temperature that the solution would tend towards if no freezing took place (the "convergence temperature"), the course of the temperature, if there were no separation of solid, would be given by Newton's law as

$$dt = k(t_0 - t)dz \quad . \quad . \quad . \quad . \quad . \quad (3)$$

with the integral

$$k(z_2 - z_1) = \log \text{nat} \frac{t_0 - t_1}{t_2 - t_0} \quad . \quad . \quad . \quad . \quad . \quad (4)$$

Again the physical meaning of k is unimportant; we need only

[1] Nernst and Abegg, *Z. S. phys. Chem.*, **15.** 681 (1894).

remark that k is smaller the greater the ratio of thermal capacity of the solution to surface exposed.

The actual course of the thermometer in a finite mass of solution is found by adding (1) and (3).

$$dt = [K(T_0 - t) + k(t_0 - t)]dz \quad . \quad . \quad . \quad (5)$$

which gives, on integration,

$$(K + k)(z_2 - z_1) = \log \text{nat} \frac{KT_0 + kt_0 - (K + k)t_1}{KT_0 + kt_0 - (K + k)t_2} \quad . \quad (6)$$

The final temperature t′, which may conveniently be called the apparent freezing-point, is given by

$$\frac{dt}{dz} = K(T_0 - t') + k(t_0 - t') = 0,$$

or

$$t' = T_0 - \frac{k}{K}(t' - t_0) \quad . \quad . \quad . \quad . \quad (7)$$

The thermometer settles therefore, not at T_0, but at the somewhat different temperature t′; the latter is the closer to the former, the nearer the convergence temperature t_0 is to the freezing-point T_0, and also the larger K by comparison with k. The true and apparent freezing-points are only identical when

$$t_0 = t',$$

or

$$K = \infty.$$

Hence, to obtain correct values, either the temperature of the freezing mixture must be carefully regulated, so that the convergence temperature practically coincides with the point to which the thermometer settles, or else the correction $\frac{k}{K}(t' - t_0)$ must be obtained by special experiments (see the work of Abegg and Raoult already quoted on p. 144). It should be observed that for electrolytes, as for water, K is usually so large that the correction is negligible (though K of course varies according to the way the ice separates). It is otherwise with non-electrolytes, such as sugar; for these Raoult,[1] Jones,[2] and others found erroneous values, which were only corrected by means of experiments conducted in accordance with the above theory of Abegg and the present writer. Only recently have Raoult[1] and Loomis[3] taken up the same standpoint as ourselves. Hausrath[4] has succeeded, by refinements in temperature measurement, in measuring the freezing-point of solutions to 0·00001°.

The Lowering of the Vapour Pressure.—The law which

[1] *Z. S. phys. Chem.*, **9.** 343 (1892).
[2] *Ibid.*, **11.** 529 ; **12.** 623 (1893).
[3] *Ibid.*, **32.** 584 (1900).
[4] *Wied. Ann.*, **9.** 522 (1902).

was developed theoretically by van't Hoff, and experimentally by Raoult, states that the lowering of the vapour pressure experienced by a solvent on adding a non-volatile substance, is equal to the number of molecules of the dissolved substance divided by the number of molecules of the solvent; this leads at once to the determination of the molecular weight. Let p be the vapour pressure of the pure solvent at a selected temperature, and p′ that of a solution containing m grams of a foreign substance dissolved in 100 g. of the solvent; then, according to the preceding law, we shall have

$$\frac{p - p'}{p'} = \frac{mM_0}{M100},$$

where M denotes the molecular weight of the dissolved substance, and M_0 that of the solvent, as ascertained from a determination of the vapour density. Then we shall have, in terms of those quantities only which are directly accessible,

$$M = M_0 \frac{mp'}{100(p - p_1)}.$$

But very great experimental difficulties are found to interfere with the practical application of this formula. For many reasons the exact measurement of the vapour pressure of a solution is not a simple problem. Also, since the difference in the preceding formula, between any two vapour pressures to be measured, amounts to only a small percentage, it is necessary that this measurement should be very exact. Moreover, the selection of a method [1] for measuring simply and accurately the difference between the vapour pressure of a solvent and of the solution has not hitherto been attended with good results. It is only recently that Beckmann [2] accomplished this in another way, viz. instead of determining the lowering of the vapour pressure after the method of Raoult, *he determined the corresponding elevation of the boiling-point.*

It follows directly from the equations, developed on p. 140, that

$$M = E\frac{m}{t},$$

where m has the same meaning as above, and t denotes the observed elevation of the boiling-point. E, the "molecular elevation of the boiling-point" [for the particular solvent], is calculated as follows,

[1] Raoult, *Ann. chim. phys.* [6], **20.** (1890); Will and Bredig, *B. B.*, **22.** 1084 (1888); Beckmann, *Zeitschr. physik. Chem.*, **4.** 532 (1889).

[2] *Zeitschr. physik. Chem.*, **4.** 532 (1889); **6.** 437 (1890); also Raoult, *C. R.*, **87** 167 (1878).

from the heat of vaporisation l, of 1 g. of the solvent at its boiling-point T, in the absolute scale; thus

$$E = \frac{0{\cdot}02T^2}{l}.$$

The following table shows, E, calculation and experiment, giving results in agreement for the solvents thus far investigated:—

	E	T		E	T
Benzene	26·7	273 + 80	Ethyl alcohol	11·5	273 + 78
Chloroform	36·6	273 + 61	Ethyl acetate	25·1	273 + 75
Ethylene bromide	63·2	273 + 132	Ethyl ether	21·2	273 + 35
Carbon disulphide	23·7	273 + 46	Acetone	16·7	273 + 56
Acetic acid	25·3	273 + 118	Aniline	32·2	273 + 182
Phenol	30·4	273 + 132	Water	5·2	273 + 100

The measurement of the elevation of the boiling-point can be conducted very accurately by means of the apparatus shown in Fig. 21.[1] The three-neck flask A serves to heat the liquid; the bottom is pierced with a short thick piece of platinum wire, set in with cement glass (*Einschmelzglas*), and is also half-filled with glass beads. A [differential] thermometer, provided with a mercury reservoir as in the freezing apparatus, and thus made serviceable for temperatures from 30° to 120°, is fitted in one opening. An inverted condenser is fastened into the middle opening (*b*); this condenser has an opening at *d*, which serves to give free passage for the steam. On account of its ready condensation, the apparatus of Soxhlet is used, and this is protected from the moisture of the air by a chloride of calcium tube. A third opening, C, serves to introduce the substance to be dissolved.

In conducting the operation, the flask is first partly filled with a definite quantity of the solvent, which is either weighed or measured with a pipette. It is then heated by a gas lamp, which can be suitably regulated, and is protected against too sudden heating by an asbestos jacket. The flame is so regulated that a drop shall fall from the condenser B every ten or twenty seconds. The upper part of the apparatus is protected against the heat by two pieces of asbestos paper, one of which is cut out with a circular opening for a support. The heat is largely transmitted through the platinum wire which is fused into the bottom of the flask, and which touches the lower asbestos paper. As a result of this, bubbles of vapour form at its upper end when the liquid boils, and on account of the glass beads present, the bubbles must pass up through the liquid in serpentine courses, thus having sufficient time to bring themselves into temperature equilibrium

[1] Beckmann, *Zeitschr. physik. Chem.*, **4.** 543 (1889); see also the short monograph by Fuchs and Beckmann, *ibid.*, **40.** 129 (1902); **44.** 161 (1903).

corresponding to the external pressure, and to the concentration. After the temperature has been kept constant within a few thousandths of a degree, the substance to be dissolved is introduced in a weighed

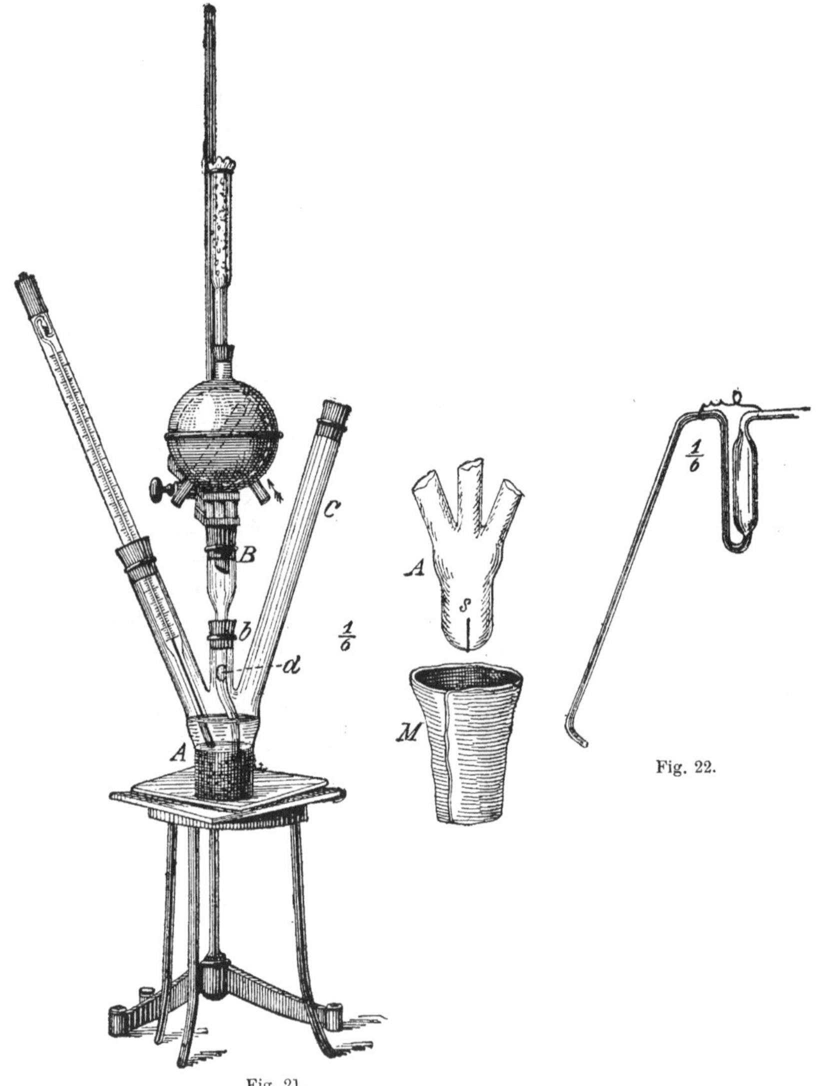

Fig. 21.

Fig. 22.

quantity, through the opening C, solid substances being in the shape of little sticks or lozenges, and liquids being introduced by means of the pipette shown in Fig. 22. As was done in the case of the freezing-point method, so here also it is recommended to make several determinations with increasing concentration.

Beckmann[1] has since described a modified form of his boiling-point apparatus, by the use of which it is not only possible to employ liquids of a higher boiling-point, but it is also so shaped that a very little of the solvent and of the dissolved substance will give very exact results. The flask A, containing the liquid, is shaped (Fig. 23) like the flask in the freezing-point apparatus figured on p. 258. It is a side-necked test-tube, 2·5 cm. wide, is provided with a thick platinum wire sealed into the bottom, and is filled to a depth of 3-4 cm. with glass beads. The thermometer is fitted in by means of a cork. The boiling-flask is surrounded by a steam jacket of glass, B, containing about 20 c.c. of the solvent, and shown in a special figure. Between the boiling-flask and the vapour jacket is a roll of asbestos paper, and both are provided with return condensers, K_1 and K_2, which can be replaced in the case of volatile solvents by small Liebig condensers. Bunsen burners serve for heating, and, as shown in the figure, these are placed at the side of the vapour jacket. Where the flames strike there are pieces of wire gauze, and also crescent-shaped plates of asbestos paper, which are arched above by a little asbestos saucer *d*; also the rings h_1 and h_2, protect the boiling-flask from the direct flame; *ss* are little chimneys of asbestos paper for the flame gases. If necessary, the boiling-flask itself is also heated directly by a small flame which does not touch it, but, as a rule, the heat introduced from the outer vapour jacket is sufficient to bring the inner liquid to complete boiling. The right condition of boiling will be shown, as in the other form of apparatus, by the constant reading of the thermometer.

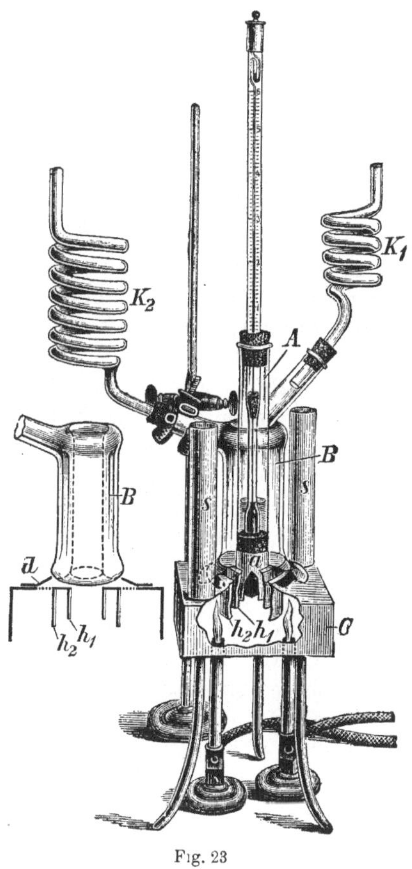

Fig. 23

The Investigation of Volatile Substances.—The use of the

[1] *Zeitschr. physik. Chem.*, **8.** 223 (1891); see also Beckmann, Fuchs, and Gernhardt, *ibid.*, **18.** 473 (1895).

forms of apparatus just described implies that the substance dissolved is not lost by volatilisation. Experience shows that this condition is met when its boiling-point lies about 130° above that of the solvent.

If the dissolved substance begins to volatilise in a noticeable way, then, according to the statements on p. 154, it has the same molecular weight in the state of a vapour and in the solution, if the partial pressure of its vapour is proportional to its concentration in the solution; or, in other words, if the vapour of the dissolved substance follows Henry's law of absorption. If the dissolved substance has a molecular weight different from that in the gaseous condition, then there occur the most glaring deviations from a simple ratio. We will consider these laws further in the third book, Chapter III.

As already shown by the writer,[1] there is no difficulty, on the basis of these considerations, in extending the theory of the boiling-point apparatus to the case where the dissolved substance has its own vapour pressure. I will here merely indicate briefly that the boiling apparatus, in this case also, is able to afford information regarding the molecular condition of the dissolved substance. Provided the dissolved substance is sufficiently volatile, the Henry's law will, of course, lead to changes of the boiling-point proportional to the concentration, which in this case (see also p. 105) may consist in an elevation or a lowering, according to circumstances. If this is so, the substance has the same molecular condition in the state of vapour as in solution. If the latter condition is not fulfilled, there is not even approximate proportionality between the changes of boiling-point and concentration.

Of course it is immaterial for the freezing-point method whether the dissolved substance is volatile or not.

The Lowering of the Solubility.—A third method, recently added, to the two so-called Raoult-van't Hoff methods, by the author, has both a theoretical and an experimental basis (see pp. 141 and 148). According to this, *the relative lowering of solubility experienced by a solvent, as ether, in another solvent, as water, on the addition of a third substance [which is soluble only in the first solvent], is equal to the number of dissolved molecules of this third substance divided by the number of molecules of the [first] solvent.* Thus, let L denote the solubility of the [first] pure solvent in the second solvent, and L′ its solubility when there are 100 g. of the first solvent for m grams of the [third] substance; then the molecular weight of the [third] dissolved substance can be calculated as accurately as from the lowering of the vapour pressure (p. 260), and with only such magnitudes as can be determined directly, as follows:—

$$M = M_0 \frac{mL'}{100(L - L')}.$$

[1] Nernst, *Zeitschr. physik. Chem.*, **8.** 16 (1891); Beckmann, *ibid.*, **17.** 110 (1895).

Here M_0 denotes the molecular weight which the first solvent has when dissolved in the second.

There are several chemical and physical methods which can be used to determine the solubility, and it is not necessary to determine the absolute values, but only the ratio of the solubility before and after adding the substance. When using ether and water, one can advantageously employ Beckmann's freezing apparatus with 20 c.c. of ether and 20 c.c. of water, to determine the change of solubility.[1] Then the freezing-point of water is at that temperature which corresponds to the depressed freezing-point ($-3\cdot85°$), occasioned by its saturation with ether. Now, if a third substance is added to the ether, according to the aforesaid theorem, the solubility of ether in the water will diminish in proportion to the molecular content of the third substance that it [*i.e.* the ether] absorbs, and thus the freezing-point of the water will *rise.* This elevation of the freezing-point can be exactly determined, and, as numerous experiments have shown, the method is capable of exactly the same degree of accuracy, as in the case with the Raoult-van't Hoff method. As the non-volatility of the substance in question was presupposed in the boiling-point method, so here we must assume that the [third] substance is insoluble in water [or in the second solvent used, whatever it may be].

The above method has been worked out further by F. W. Küster,[2] who used phenol as one solvent and saturated common salt solution as the other, measuring the solubility of the phenol by titration; also by Tolloczko,[3] who used ether and a relatively large quantity of water, and measured the change in volume of the latter due to solution of foreign bodies, directly; the latter method leaves nothing to be desired in simplicity, especially for strong solutions.

The Distribution of a Substance between Two Solvents.—A very simple and exact method for comparing the molecular condition of a substance in two solvents, which are only partially soluble in each other, is to *determine the dependence of the relative distribution upon the concentration.* If dissolved substance has the same molecular condition in the two solvents the distribution is independent of the concentration (p. 155). If the molecular condition is not the same, then the coefficient of distribution will vary with the concentration in a very pronounced way. This subject will be considered again in the third book.

The Rôle of the Solvent.—All the methods thus far described

[1] *Zeitschr. physik. Chem.*, **6.** 573 (1890).
[2] *Ber. deutsch. chem. Ges.*, **27.** 324 and 328, 1894.
[3] *Z. S. phys. Chem.*, **20.** 389 (1896).

for the determination of the molecular weights of substances in solution are based on the same principle; this consists in the measurement of the osmotic pressure and its evaluation, in the sense of Avogadro's rule as generalised by van't Hoff. Therefore all these methods thus far used, though apparently very diverse, lead to the same results, when the investigation concerns the same substances in the same solvents. But numerous instances are known where the aforesaid substances when dissolved in different solvents show a *different* molecular condition. Thus acetic acid when dissolved in benzene, at a sufficient degree of concentration, consists almost entirely of molecules having the formula $(CH_3CO_2H)_2 = 120$; when dissolved in ether, of $CH_3CO_2H = 60$; and when dissolved in water, as will be seen later, it is split electrolytically into the ions, $CH_3\overset{-}{COO}$ and $\overset{+}{H}$. In the gaseous state, according to the external conditions, we find acetic acid to consist more or less entirely of the "normal molecules, CH_3CO_2H.

This result, of course, speaks neither for nor against the correctness of the methods, neither is it at all surprising. The molecular condition of a vaporised or of a dissolved substance varies not only with the external conditions of temperature and pressure, as shown in numerous instances, but also in the case of dissolved substances it will vary with the nature of the solvent employed.

The question now arises whether the difference between the molecular condition of a substance in the gaseous state, and of the same substance in solution, is to be ascribed to *a chemical action* by the solvent. This question is of surpassing interest, but it is impossible to answer it at present. For it must be emphasised that any compound of the molecules of the solvent with the molecules of the dissolved substance, at slight degrees of concentration, *would have no effect in changing the osmotic pressure of the latter*, and therefore would not appear in the figures of the lowering of the freezing-point, and the like.

We do not know at present, and we have no good reason to show whether the possibility that the "dissociating force" of water, in its capacity as a solvent, may cause it to form compounds with the dissolved substance in the way of hydration, is well grounded or not.[1] In general this holds good, though it is often overlooked, viz. that the osmotic pressure makes no disclosures regarding any compounds of the molecules of the solvent with those of the dissolved substance. (See further under the "Doctrine of Affinity.")

We must be forewarned also against another very common error concerning the aforesaid osmotic method; it has been supposed that it afforded some information regarding the molecular condition of the

[1] Brühl, *Z. S. phys. Chem.*, **18.** 514 (1895); **27.** 319 (1898).

solvent itself in the liquid state. Thus Raoult and Recoura,[1] instead of obtaining the formula (as on p. 262),

$$M = 60 \frac{mp'}{100(p - p')},$$

for the lowering of the vapour pressure of acetic acid, where 60 denotes the [normal] molecular weight,—actually obtained

$$M = 1{\cdot}61 \,.\, 60 \frac{mp'}{100(p - p')};$$

from which they concluded the molecular weight of liquid ascetic acid to be

$$M_0 = 1{\cdot}61 \,.\, 60 = 97.$$

This conclusion is not well grounded; the results of this investigator can be more simply explained, by supposing that the density of the saturated vapour of acetic acid at the temperature of the experiment (118°) was 1·61 times as great as the theoretical value, 2·08; and therefore that their calculated molecular weight (97) should be introduced into the formula. At the same time, these experiments show that the osmotic pressure of dissolved substances is normal, *i.e.* that it obeys the gas laws, even when the vapour of the solvent itself is abnormal.

The relation that seems to exist between dissociating power of a solvent and its dielectric constant is discussed in Chapter VII. of this book.

The Molecular Weight of Strongly-Compressed Gases and of Liquids.—We are not at present in the possession of any method which is universally applicable to strongly-compressed gases and to liquids. The application of the rule of Avogadro (that equal volumes, at the same temperature and pressure, contain the same number of molecules) to substances like these is not only destitute of good support, but it is directly improbable. It has been already shown, on p. 239, that the further development of the theory of van der Waals will lead, perhaps, to the solution of this question; but at present this much may be safely asserted: viz. that those substances which meet the requirements of the theory (as for example the halogen derivatives of benzene, p. 231), have approximately the same molecular condition in the state of strongly-compressed gases and of liquids, as in the state of attenuated gases or of dilute solutions; and, moreover, it is but a step to assume that when a substance exhibits great deviations from the theory of corresponding states,

[1] *C. R.*, **110.** 402; *Zeitschr. physik. Chem.*, **5.** 423 (1890).

this is probably due to a change in the molecular condition resulting from condensation or rarefaction.

It is probable that a comparison of the *pressure* and of the *molecular volume* of the saturated vapour of a substance, with the corresponding values of fluor-benzene at corresponding states, could be used very advantageously to decide this question; but the comparison with the molecular volume of the *liquid* could not be used, for this probably would be only very slightly affected by a polymerisation of the molecule.

A series of important criteria from this point of view have been suggested by Guye:[1]—(1) The critical density for most substances is 3·8 times that reckoned from the laws of gases; in cases in which this ratio is noticeably greater (*e.g.* 4·0), polymerisation is to be concluded. (2) The same when the law of Cailletet and Mathias does not hold—that is, when the arithmetic mean between the densities of the liquid and saturated vapour is not a linear function of the temperature, *e.g.* alcohol shows a noticeable departure from this law. (3) When the latent heat of evaporation reaches a maximum on rise of temperature, the liquid is more polymerised than the saturated vapour. (4) When the vapour-pressure curve of a liquid cuts that of another liquid, which, according to the above criteria, is not polymerised in the liquid state, the first is to be regarded as polymerised.

A very promising method of determining molecular weight of liquids is due to Eötvös,[2] who found the following rule confirmed by a series of experiments, calling γ the surface tension of a liquid expressed in dynes, v its molecular volume (that is, the volume occupied by one mol in the liquid state), the relation holds that

$$\gamma v^{\frac{2}{3}} = \mathrm{k}(\mathrm{T} - \mathrm{T}_0) \qquad (1)$$

where T_0 is a temperature not very different from the critical, and k is a constant independent of the nature of the liquid. This relation was later tested experimentally by Ramsay and Shields[3] in a very thorough manner, and found in good agreement with the observations. Conclusion (1) was put in the form

$$\gamma v^{\frac{2}{3}} = \mathrm{k}(\tau - \mathrm{d}) \qquad (2)$$

where τ is reckoned downwards from the critical temperature, and d is usually about 6; conclusion (2) only holds accurately when τ is greater than 35°; not, therefore, in the immediate neighbourhood of the critical point.

Imagine that we have a mol of the liquid in question in spherical form, its radius is $\sqrt[3]{v}$, its surface is proportional to $v^{\frac{2}{3}}$, and accord-

[1] *Archives des Sciences Phys. et Nat. de Genève*, **31.** 38 (1894).
[2] *Wied. Ann.*, **27.** 452 (1886).
[3] *Zeitschr. physik. Chem.*, **12.** 433 (1893).

ingly $\gamma v^{\frac{2}{3}}$ is a quantity proportional to the *molecular surface energy of the sphere.* Conclusions (1) and (2) therefore state *that the temperature coefficient of the molecular surface energy* (except in the immediate neighbourhood of the critical point) *is independent of the special nature of the liquid.*

Putting the molecular surface energy at $\gamma v^{\frac{2}{3}}$, then the temperature coefficient k is 2·27 according to Eötvös, and 2·12 according to Ramsay and Shields.

Eötvös's law may be expressed in this manner: *to form the surface of a spherical mol* of a liquid the work required varies with the temperature in the same manner for all liquids. This is the case with the production of a mol of gas under constant pressure according to Avogadro's law; in the latter case, according to the laws of gases, the work required is simply proportional to the absolute temperature.

The following liquids behave normally—that is, give a temperature coefficient of the molecular surface energy in the neighbourhood of 2·12:—

	k.
Benzene	2·17
Chloro-benzene	2·08
Tetrachloride of carbon	2·11
Ethyl ether	2·17
Bi-sulphide of carbon	2·02
Benzaldehyde	2·16
Nitrobenzene	2·09
Aniline	2·05
Pyridine	2·23
Phosphorus trichloride	2·10

and a large number of esters.[1]

Anomalous values of K (lower and varying with the temperature) are given by the alcohols (1·0 – 1·6), organic acids (·8 – 1·6), aceton (1·8), propionitril (1·5), nitro-ethane (1·7), methyl-urethene (1·6), valeroxime (1·7), water (0·9 – 1·2). To obtain the normal value of the temperature coefficient for these liquids their molecular weight must be raised—that is, association of molecules must be assumed. But it does not appear possible in the present state of our knowledge to conclude satisfactorily *the degree of association* from the divergence from the normal value. The method of calculation adopted by Ramsay, in particular, is subject to certain objections.

There is, at the present time, no satisfactory theoretical application of Eötvös's law; see the remarks on this subject by the author,[2] and especially the very complete study of van der Waals.[3] It may be concluded from the results at present obtained: (1) When a substance has the same molecular weight in the gaseous and liquid states, k is about 2·12, that molecular weight being used to calculate the molecular

[1] Ramsay and Aston, *Zeitschr. physik. Chem.*, **15.** 98 (1894); Guye and Baud, *ibid.*, **42.** 379 (1903). Cf. also the work by Schenk mentioned on p. 98.

[2] *Jahrb. der Chemie*, **3.** 18 (1893). [3] *Zeitschr. physik. Chem.*, **13.** 713 (1894).

surface energy; (2) k is found to be smaller than 2·12 when the molecular weight in the liquid state is smaller than in the gaseous. It is, however, an open question whether the molecular weight in the liquid state alone determines k, or whether also that in the vapour. Possibly this may be tested empirically by the study of the surface tension of mixtures. In this way a rule for deriving the average molecular weight of a liquid from the observed value of k might be deduced: in fact, Pekár[1] recently found that in some mixtures at least the temperature coefficient of the molecular surface energy determines the average molecular weight.

A third method that can be used to study the molecular condition of a liquid is given in Chapter V. of this book, in the section on rules of boiling-point. Linebarger[2] has given an interesting study of the vapour pressure of mixtures, and uses the partial pressure of the components to decide the question whether association exists. Finally, it may be remarked that all the methods here quoted agree remarkably in showing that the majority of liquids studied are not polymerised, and that certain classes of bodies, such as acids, alcohols, and, in particular, water, form complexes (mostly double molecules) in the liquid state, and that the last-mentioned substances show a tendency to do the same in the gaseous and dissolved conditions.

Guye[3] has investigated the application of the critical density to decide the question of molecular condition in the critical state, but the measurements of this quantity so far obtained are too uncertain to form the basis of a further step in this direction.

Molecular Weight of Solids.—There is no method known at the present time which leads to a knowledge of the molecular weight of solids; indeed, our molecular conceptions on the nature of the solid state are very vague (p. 240). The case is somewhat more hopeful with *dilute solid solutions*, and perhaps Avogadro's law can be applied to them (p. 170).

The results obtained in the last-mentioned way are not free from objection; see Nernst,[4] Küster,[5] Hoitsema,[6] Würfel.[7]

[1] *Zeitschr. physik. Chem.*, **39.** 433 (1902).
[2] *Amer. Chem. Journ.*, **17.** 615, 690 (1895).
[3] *Compt. rend.*, **112.** 1257 (1891); more fully, Thèse, Paris, 1891.
[4] *Zeitschr. physik. Chem.*, **9.** 137 (1892).
[5] *Ibid.*, **13.** 445 (1894); **17.** 357 (1895).
[6] *Ibid.*, **17.** 1 (1895).
[7] *Dissertation*, Marburg, 1896.

CHAPTER IV

THE CONSTITUTION OF THE MOLECULE

Allotropy and Isomerism.—The properties of all substances vary with the external circumstances under which they are studied. Those external conditions which exert the greatest influence upon the physical and chemical conduct of substances are the *temperature* and the *pressure*; also, according to circumstances, the properties are modified in one way or another by *magnetisation*, *electrification*, *illumination*, etc. In describing the behaviour of a chemically simple substance, it must always be stated what the external conditions are under which it is studied.

Further, under all circumstances the properties of any two substances having a different chemical composition, are different: if only one atom in the molecule be replaced by another, even then there is a difference in both the physical and the chemical behaviour of the compound; but this change is a variable quantity. Those atoms which can replace each other in the molecular group without occasioning a deep-seated change in the whole behaviour of the compound, are said to be related chemically. A number of such related groups of elements have been already pointed out in the vertical columns of the periodic system (p. 184). Although the change in the properties of a compound, when one atom is replaced by one of another element of similar behaviour, may be very slight, yet it is definite in all cases: *two substances which behave alike in all their properties must have the same composition.*

But the converse of this is by no means true; viz. that two substances which conduct themselves differently under the same external conditions have a different composition. This is neither true of elements nor of compounds, as has been already seen in the capacity of substances to assume different states of aggregation under the same external conditions. It is also known that many elements exist in the solid state, in different modifications, called "allotropic forms." Phosphorus is an element known to us in two modifications, called the yellow and the red; and these varieties are so different under the same external conditions, both in their physical and in their chemical behaviour, that one might easily believe that he had two essentially different substances. Carbon occurs in nature in the forms of the diamond, of graphite, and of so-called amorphous carbon. Sulphur appears in crystalline forms of the orthorhombic and the monoclinic system, according to the method of formation, etc. etc.

We do not certainly know the reason for the difference between the allotropic forms of solid elements, but the probability, in the light of the atomic hypothesis, is very strong that the atoms unite either with a different number in the molecule, or with a different mode of union. To be sure, certain proof for this supposition is wanting at present, since hitherto it has been impossible to obtain a glimpse into the molecular constitution of solids.

More often we find the case where a chemical compound can occur in different modifications, not only in the solid state, but in all the states of aggregation. Such compounds are called *isomeric.*[1] The molecular hypothesis has done much good work in explaining these cases of isomerism, by showing how to obtain new examples of isomerism; and, on the other hand, it has repeatedly been found to be a fact in the history of theoretical chemistry, that the attempt to explain isomerism in its new relations, led to a bold extension of the molecular hypothesis, which, in turn, reacted in new requirements on experimental research.

These are some of the points in the *history of the knowledge of isomerism*, which is, at the same time, the history of constitutional theories: in 1823 Liebig remarked that the silver fulminate analysed by him had the same constitution as the silver cyanate discovered by Wöhler in 1822; in 1825 Faraday found that benzene, discovered by him, agreed in composition with acetylene; and in 1828 Wöhler succeeded in directly converting one isomer into another, namely, ammonium cyanate into urea. The number of cases of isomerism grew very rapidly; in 1832 Berzelius discovered the isomerism of racemic and tartaric acids, and very many instances were discovered in organic chemistry, especially after the theoretical explanation had been given. The most striking advance towards the development of stereochemistry of carbon was the investigation of Wislicenus on isomerism in lactic acids (1871), and that of fumaric and maleic acids in 1887, whilst the first experimental material discovered for stereochemistry of nitrogen was in the isomerism of benzyl-dioxim discovered in 1832 by Goldschmidt, and in 1888 by v. Meyer and Auwers.

The Constitution of the Molecule.—The question may now be asked at once, whether or not the differences in the properties of isomeric compounds are based upon a difference in the size of the molecule; *i.e.* whether the atoms unite to form the different molecules

[1] Oxygen and ozone form an example of isomerism amongst the elements.

in the same relative proportions, but not in the same number. Experience shows that this circumstance often explains some material differences, but by no means can it explain all exhibitions of isomerism. For, on the one hand, there are isomeric substances having the same per cent composition and different molecular weights, as acetylene C_2H_2, and benzene C_6H_6 (*isomerism* in the broader sense of *polymerism*); and, on the other hand, there are isomeric substances with physical and chemical properties clearly distinct, and these are found especially in the carbon compounds, which have both the same per cent composition and the same molecular weight (*isomerism* in the special sense of *metamerism*).

The existence of metameric compounds at once gives us a chance to frame more definite conceptions regarding the mode of union of the atoms in the molecule; it at once excludes the supposition that the atoms can unite in a chemical molecule in all conceivable positions with reference to each other, like the molecules in a homogeneous liquid mixture. For otherwise, just as by bringing together definite quantities of different substances, there can result only one physical mixture of definite properties, so also by the union of a certain number of atoms of different elements into a molecule there could, in that case, result only one chemical compound, having the same properties; and thus the formation of metameric compounds would be impossible.

Therefore it must be assumed that certain forces are exerted between the atoms in the molecule, which determine the relative positions of the atoms; also that the relative positions of the atoms can vary with the mode of union of the atoms with each other.

The differences, shown in the physical and chemical behaviour of metameric compounds, must thus be ascribed to differences in the arrangement of the atoms in the molecule, or as is said, to differences in the constitution of the molecule.

It cannot be denied that in attempting to develop the atomistic science from this standpoint, and to frame more definite conceptions regarding the arrangement of the atoms in the molecule, we go into a region of a purely hypothetical nature—a region which can only be reached by a leap of a bold phantasy. The usual requirement customary in the careful study of nature, of abstaining from such an attempt, does not appear to be justified on these grounds; for such a requirement, on the one hand, would amount to a refusal to obtain many obvious conceptions regarding many important phenomena, which neither the experimental nor the theoretical student can explain in any other way; and, on the other hand, it would not harmonise with the fundamental principle of the method of natural science which commands us to follow out, to the ultimate, such a practical and fruitful hypothesis as the atomic hypothesis is well known to be.

The Chemical Forces.—At present scarcely anything definite is known regarding either the nature of the forces which bind the atoms together in the molecule and which hinder them from flying apart in consequence of the heat motion, or regarding their laws of action; but there are many reasons which lead us to suppose that these forces, like the forces in the explanation of capillarity and related phenomena, act only in the immediate neighbourhood of the atoms, and diminish in strength very rapidly as the distance from the atoms increases.

Further, in order to explain the various capacities for reaction of the elements, and the various degrees of stability with which the atoms are linked together, we must assume that the mutual action of the atoms varies greatly with their nature; and in order to explain the fact that atoms of the same sort unite to form molecules, we must assume that chemical forces are active between the atoms of the same element, and that these vary very much with their nature.

The answer to the question, how these forces vary with the reacting elements, is rendered exceedingly difficult from the fact that in the great majority of cases we have to explain, not a single exchange, but the sum total of several exchanges. Thus in the formation of hydriodic acid,

$$H_2 + I_2 = 2HI,$$

we find that the reaction is determined not only by the affinity between the hydrogen and the iodine, but also before a single atom of th reacting elements can enter into the exchange, the bonds holding th two atoms in the molecules H_2 and I_2 respectively, must be loosened When the reaction takes place from left to right in the sense of th preceding equation, as happens at temperatures not too high, then i works against the affinities holding two hydrogen atoms, and two iodin atoms together, and tends to come to rest with the affinity between hydrogen and iodine predominant. But when the reaction takes plac from right to left, as happens at high temperatures, then it work against the affinity between different kinds of atoms, and tends t come to rest with the affinity between atoms of the same sor predominant. And the case is similar in almost all reactions whic are carefully studied, so that the case is very rare where the course o the reaction can have but one issue depending solely on the chemica forces. The affinity certainly changes with the external conditions o temperature and pressure in all cases, although it may be very differen in a qualitative way; but we usually are entirely ignorant as to the whereabouts of the cause. Thus, in the preceding case, we canno state why the affinity changes between the like and the unlike, with the temperature; we can merely conclude from the course of th reaction, that at lower temperatures the affinity between unlike atom prevails, and at higher temperatures that between like atoms.

In order to obtain a deeper insight into the manner of the action of the forces of affinity, it is, of course, necessary at first to direct ou attention to those reactions where the simplest conditions find expression. This is found in those cases where a complex molecule breaks up into simpler ones (dissociation), or conversely where several simpler molecules condense into a more complicated one (addition). In such a case the chemical change occurs either against, or with, only *one* affinity. The simplest case is where two elementary atoms unite to form a single molecule, or conversely where a diatomic molecule of an element breaks up into its atoms; as is the case, for example, in the dissociation of iodine vapour; thus

$$I_2 = I + I.$$

The study of the conditions under which these reactions come to a pause will give us some light regarding the affinity which is acting between the elementary atoms in question. We may perhaps hope that, by the thorough study of some such simple case, some light may be thrown on these questions, the answer to which has hitherto seemed so remote.

The Doctrine of Valence.—Without framing any definite conception of the nature of chemical affinity, it is possible to consider the mode of union of the atoms in the molecule, according to a certain scheme, which not only gives us much general information on the classification of chemical compounds, but which also serves to make their capacities for reaction intelligible in many respects, and also renders it easy to remember them.

Much observation has taught us that often certain elements and radicals may replace each other without occasioning a deep-seated change in the properties, and especially in the reaction capacity of the molecule. Thus in many instances it is possible to replace a hydrogen atom in a molecule, by an atom of F, Cl, Br, I, Li, Na, K, etc., or by certain radicals as NH_2, NH_4, CH_3, C_2H_5, C_6H_5, always to be sure with some associated change in the physical and chemical behaviour of the compound, but never entirely destroying the clearly expressed similarity between the new and the original compound.[1] The experience based on an enormous amount of observation is embraced in the statement that *such elements or radicals are chemically equivalent.*

Some other similarly chemically equivalent elements are O, Mg, Zn, Ca, Sr, Ba, etc.; these, as a rule, easily replace each other, and in such a way as not to change radically the whole habit of the compound.[2] A number of such groups of elements are given in the vertical columns of the table on p. 184.

A further observation has been made, that elements which are not

[1] This statement must be taken "with a grain of salt"; as a rule, it may be said that the least change is produced when acidiferous atoms or residues replace acidiferous ones, and basiferous replace basiferous; and the greatest change when basiferous ones replace acidiferous ones, or the reverse.—TR.

[2] See last footnote.—TR.

of the same equivalent group can replace each other, but not in such a way that an atom of one sort can replace an atom of another sort in the molecule; but the replacement happens in this wise, viz. that in the place of a certain number of atoms of one group, there are substituted a different number of atoms of another group. Thus two atoms of H, Li, Na, F, Cl, etc., can usually replace one atom of O, Mg, etc.

In this way it is possible to compare the *chemical value or valence* of the elements belonging to the different groups, and to make a quantitative determination of this. Since no element is known where more than one of its atoms are required to take the place of an atom of hydrogen, the valence of this latter element is assumed as standard; and, accordingly, hydrogen and the related elements are said to be *univalent;* then oxygen and the related elements are *bivalent*, phosphorus is *trivalent*, carbon, silicon, etc., are *quadrivalent*, and the like.

It is usually assumed, in order to explain these relations, that the chemical force of the elementary atoms does not act equally in all directions in space, which must be a proviso in considering the attraction of a gravitating mass-point, or in considering the mutual attraction of the molecules of a liquid; but it is rather to be assumed that the *affinity is entirely, or at least chiefly, active in certain directions; according to this view, the number of these directions or rays corresponds to the chemical value* (*valence*) *of the atoms.* Thus the chemical force of hydrogen acts in one direction, that of oxygen in two, that of carbon in four, etc. The union of the atoms in the molecule is to be conceived of in this wise; viz. the outgoing line of force of one atom coincides with one from another atom, or in other words, the valences of the different atoms *mutually and oppositely satisfy each other.*

By such conceptions as these it is possible to obtain a general view, if not of all the possible compounds, at any rate of those which are formed most readily, and which are characterised by great stability.

Thus two univalent elements, as hydrogen and chlorine, or two bivalent elements, as calcium and oxygen, will unite in the sense of what has preceded, and, in harmony with experiment, in very stable forms in such compounds as the following, viz. :—

H—Cl and Ca=O.

The uniting dashes, as commonly used, represent the force-lines which anastomose into each other. Similarly, without further remark, we can explain the ready formation of such molecules as

```
                      H
   /H                 |
N<—H    or    H—C—H,
   \H                 |
                      H
```

and also molecules composed of the same kind of atoms, as

H—H, O=O,

and the like; for it is to be expected that the same scheme will be used to explain the constitution of molecules composed of like, as of those of unlike, atoms. Regarding the saturation capacity of the valences, it must be supposed that each valence is satisfied by another valence, whether it be from the same kind of element or from a different kind, although there do occur very pronounced quantitative distinctions in the complete saturation of the linking of the atoms.

The Dualistic and the Unitary View.—It must be admitted that, according to previous experiments, the power of saturation of valences is almost unlimited, that as all observable matter produces mutual attraction regardless of what its character may be, so also two valences can unite their lines of force under certain circumstances whatever the atoms may be from which they arise. On the other hand, the intensity of these is in the highest degree dependent on the nature of the two atoms and also on the number and character of the other atoms present in the molecule. It is, therefore, remarkable that on the whole the atoms and atom complexes comparable to them in behaviour (radicals) may be placed in two groups, between which a polar distinction is recognisable. Whilst the atoms and radicals belonging to either group are more or less indifferent to one another, the members of one group show marked affinity towards those of the other group.

The existence of a polaric contrast in their chemical action is undoubted, and will be explained more clearly in the subject of electrolysis, where it will be shown how the representatives of the first class (positive) "wander" to the cathode, and those of the second class (negative) to the anode. It was the discovery of these which led Davy, and especially Berzelius (1810), to the advancement of the electrochemical theory; this regarded the undeniable dualism in the mutual affinity, as the guiding star of chemical investigation, and explained the polaric contrast, as resembling that between the positive and negative electrostatic charges.

But very soon it appeared that it was impossible to carry out this supposition; for, aside from the faulty consideration of the physical side of this question implied in the theory, and which shows the premises to be inadmissible, there are well-known chemical processes which occur in distinct contradiction to an exclusively dualistic conception.

For on that basis, how can one explain the mutual and very active capacity for union shown by two atoms of the same sort, and illustrated in the molecules of many elements, as H_2, O_2, Cl_2, etc.? Or how explain the abnormal behaviour of carbon, which can unite firmly with such characteristically positive and negative elements as hydrogen and chlorine.

Instead of drawing the conclusion that, in chemical action, aside from the polaric forces (analogous to electric attraction and repulsion), there is some other force acting simply (like the Newtonian attraction of ponderable matter) which must be taken into account, the tendency has been for a long time, and also is at present, to develop a one-sided *unitary* theory in opposition to the one-sided *dualistic* theory of Berzelius; this is easily explained by the historical consideration that the dualistic view was unsuited to the treatment of the carbon compounds which form the subject of the flourishing department of organic chemistry. It is possible in the immediate future, as the phenomena of electrolysis are exciting an awakened interest, that this dualistic theory will receive justice, and that a deeper meaning will be found in the "positive" and "negative" behaviour of many atoms and radicals, though the terms "*positive*" and "*negative*" are intolerable to many modern chemists.

The Changeableness of Chemical Valence.—The great variety of chemical transformations are explained by the fact that the properties of the chemical forces vary with external conditions of temperature and pressure, with the presence of other substances, and finally and especially with the nature of the atoms concerned in the exchange. It may be regarded as the chief problem of theoretical chemistry, to express the mode of the variation of the affinity, by designated factors of measure and number. How far this problem has been solved will be considered under the subject of the "Doctrine of Affinity"; we will here only anticipate what is most important from the standpoint of the doctrine of valence.

Both the number of valences acting from an atom in a molecular group, and the intensity with which each one preserves the integrity of the union, vary within certain limits. There has hardly ever been any doubt regarding the latter [*i.e.* the variation of the intensity of the valences]; but the attempt to maintain the doctrine of a constant valence was very general till recently; for to explain the cases in question, it was urged, on the one hand, that certain valences remained unsatisfied, when an atom used less valences than were allotted to it; and, on the other hand, the conception of "molecular compounds" was advanced to account for the possibility of the existence of those molecular complexes where the number of effective valences seemed too small to explain their constitution.

We cannot discuss here how far in this method of investigation mere words helped on the faulty conceptions of the nature of valence; but, at all events, the fact remains that all chemical compounds cannot be arranged under the structure plans of the accepted chemical rubric, and also that nothing else primarily remains but to give careful attention to any change in the chemical valences, since this implies no especial encroachment on the significance of valence in systematic chemistry.

At all events, in the case of many compounds, where at first glance the number of active valences appeared to be too small, it later appeared possible to devise a constitution formula which harmonised well with the doctrine of valence; this is shown in the structure-plans of the so-called "unsaturated compounds," which maintain the four-valence theory of carbon. The simple device which here brought about the desired result consisted in making the well-known assumptions, that several valences may be mutually satisfied between any two carbon atoms, and that the carbon atoms also can form *rings*. This assumption has been accorded great triumph both in theory and in experiment.

But such results as these, for which we are indebted to the application of the doctrine of constant valence, must not therefore blind our eyes to the fact that there are many things which appear, at least temporarily, to be unexplained by this same view. A brilliant example of this is shown in a recent discovery of Nilson and Pettersson,[1] in which they proved that three chlorides of indium can undoubtedly exist in the gaseous state, viz.

$$\mathrm{In{-}Cl}, \quad \mathrm{In}\begin{smallmatrix}\diagup \mathrm{Cl} \\ \diagdown \mathrm{Cl}\end{smallmatrix}, \quad \mathrm{In}\begin{smallmatrix}\diagup \mathrm{Cl} \\ -\mathrm{Cl} \\ \diagdown \mathrm{Cl}\end{smallmatrix}$$

even among the carbon compounds change of valency is to be found, although it is rare. Carbon monoxide,

$$C{=}O,$$

has long been assumed to contain a bivalent carbon atom; recently Gomberg[2] realised the first case of obviously trivalent carbon in triphenylmethane $(C_6H_5)_3C$. In this respect porphyrexide prepared by Piloty[3] possesses particular interest, as nitrogen in it appears tetravalent instead of either tri- or penta-valent as usual.

As a matter of fact, the considerations of the kinetic theory of gases lead to the result that we should hardly expect to find such simple conduct as is implied in the doctrine of constant valence. The so-called stability of chemical compounds, accordingly, appears to be the resultant of two opposing forces, one of which, the chemical force proper, tends to hold the atoms in the molecular group, while the other, controlling the atoms by heat motion, strives to break the molecule apart. It is probable that the latter force changes, and probably increases with the temperature. Regarding the former [*i.e.* the chemical force] we know nothing, and nothing can be stated regarding its variation with the temperature. The more the chemical force preponderates, the more stable will the compound be.

Here we meet with relations which are comparable with those considered in the kinetic theory of liquids (p. 215). The vapour

[1] *Zeitschr. physik. Chem.*, **2.** 669 (1888).

[2] *Ber. deutsch. chem. Ges.*, **33.** 3150 (1900).

[3] *Ibid.*, **36.** 1283 (1903).

tension of a liquid (which results, on the one hand, from the concurrence of the forces which control the heat motion of the molecules, and which results, on the other hand, from the action of those forces which exert an attraction between the molecules), can be regarded as a standard of the capacity of the liquid to be vaporised. So in the same way, the capacity of compounds for dissociation, and for chemical reaction also, can be regarded as conditioned by the concurrence of analogous forces. Thus the well-known dependence of the stability of molecular groups upon the external conditions of temperature and pressure is at once explained on the basis of these considerations. The laws which prevail here will be thoroughly considered in detail in the section on the Doctrine of Affinity.

Thus we can easily account in this way for the numerous instances observed of the variation of the chemical valence, by supposing that the heat motion of the atom within the molecule may cause some particular lines of force, called valences, to vanish.

Molecular Compounds.[1]—More difficult to explain than those cases called the occurrence of unsaturated valences, or, in other words, where elementary atoms in many compounds employ less valences than the normal number, according to the valence doctrine, is the formation of those well-characterised chemical compounds where more valences appear to be active than can be ascribed to the atoms in question in any of their other relations.

Water, and also many salts, must be classified under the compounds which are completely saturated, in which no more free valences are available for use in uniting with other atoms; yet, in spite of this, the salts containing water of crystallisation are obviously chemical compounds which are built up in accordance with the rules of multiple proportion. The phenomenon of water of crystallisation is analogous to that where many salts crystallise together as double salts in stoichiometric proportions. The property of not being volatile without decomposition is by no means characteristic of these compounds, but is held in common by many others which obey the rules of the valence doctrine most perfectly.

If one clings to the notion of the trivalence of phosphorus, there is no clear reason for the existence of the molecule PCl_5 in the gaseous state. Methyl ether is able to unite with a molecule of hydrochloric acid, which is in striking contradiction to the well-developed valence theory of organic compounds.

If we regard oxygen as regularly divalent, then the existence of the molecules $(H_2O)_2$ is unexplained, and yet we cannot hesitate to assume the existence of molecules of this size both in water vapour and also in solutions, if we believe that Avogadro's rule also holds good even when applied to solutions.

[1] See A. Naumann, *Die Molekülverbindungen*, Heidelberg, 1872.

Molecular complexes of this sort, occasioned by the union of saturated compounds, are called "*molecular compounds.*" It is assumed on the basis of the valence doctrine, that the components of this aggregate retain to a certain extent their individuality in the new complex, and that the binding force does not inhere in force lines passing between the single atoms, but that the union centres in the attraction of the original molecules acting as units.

This view at least accounts for the fact that the so-called "molecular compounds," in general, are characterised by a lower degree of stability than that shown by chemical compounds. On the other hand, it should be remembered that, in spite of all attempts in this direction, no characteristic distinction has been found, either in physical or in chemical behaviour, between the ordinary compounds and the molecular compounds; *and therefore, strictly speaking, from the phenomena exhibited, at present no other conclusion can be drawn except that chemical compounds do undoubtedly exist which cannot be included in the structure scheme which is based on the doctrine of a constant valence.*

A remarkable research of A. Werner's towards the systematics of molecular compounds will be dealt with in Chapter VII.

The Compounds of Carbon.—The valence doctrine has been hitherto applied almost exclusively, but with great and undoubted results, to the carbon compounds, or the so-called organic compounds; and thus it has been already done, or it is already possible, to construct for every actual compound, whose reactions are known to a fairly satisfactory degree, a structure scheme which represents the quintessence of its chemical reactions. The reason of this is to be found in the fact that the theory of valence here found, in the abundance of material and the great variety of behaviour, ample opportunity for an experimental "baptism by fire"; but the great development of organic structural chemistry is to be explained, as far as we know, from the fact that the behaviour of organic compounds is actually much simpler than that of other groups, and also that carbon surpasses other elements in its more regular behaviour, and in the great variety of its compounds.

The quadrivalence of carbon is the basis of organic structural chemistry; after the preparatory work of Frankland, this was distinctly emphasised by Couper and Kekulé contemporaneously (1858); its fruitfulness was recognised and proven by the latter in particular.

If one imagines the hydrogen atoms as in the formula

$$\begin{array}{c} H \\ | \\ H—C—H \\ | \\ H \end{array}$$

to be substituted by other atoms or radicals successively, or two hydrogen atoms to be replaced by one divalent atom or radical, and so on, it is possible to represent structure formulæ for the whole host of carbon compounds; and these formulæ cannot only be regarded as possible, but they also will disclose to those who are familiar with the language, much regarding the reaction capacities and the physical properties of the compounds in question.

The Peculiarities of the Carbon Compounds.—The very fact that there is an "organic chemistry" gives rise to the question regarding the peculiarities which so characterise the compounds of this branch of chemistry, that they receive a special method of treatment, a treatment which is distinct from the general domain of chemistry, not only on the part of the teacher, but also on the part of the investigator. We see here, in the fact that the materials of organic chemistry not only claim the interest of the animal and plant physiologist in particular, but also in that they are of value for the physician and the technologist,—a circumstance which, though perhaps external and casual, has nevertheless a most impressive significance. But let us rather inquire after the causes which make the organic compounds, in their physical and chemical behaviour, actually appear so different in many respects from other compounds.

It is doubtless carbon itself which stamps its character on the "chemistry of the carbon compounds." We must consider carefully to what extent this element occupies a special position. The following items are from a statement of van't Hoff.[1]

1. The quadrivalence of carbon necessitates an enormous number of derivatives of a carbon compound.

2. The capacity of carbon atoms uniting with each other, and in very many ways, allows the possibility of the greatest variety of combination.

3. The position of carbon, standing as it does between positive and negative elements, invests it with a peculiar capacity for uniting with the most different elements, as hydrogen, nitrogen, oxygen, chlorine, etc. To this is due the property of adapting itself alternately to the processes of oxidation and reduction, which have such significance in animal and plant life. If we consider the first horizontal row of the periodic system (p. 184), viz.

Li Be B C N O F,

we find carbon well balanced between the affinity extremes; for the elements to the right are emphatically negative; and those to the left are emphatically positive in character.[2] Moreover, the temperature

[1] *Ansichten über die org. Chemie*, I. p. 34 ff.; II. p. 240 ff. Braunschweig (Brunswick), 1881.

[2] The statement that Be and B are positive elements must be taken with "a grain of salt," in view of their well-known passive acidiferous properties.—Tr.

clearly exerts an influence, for at high degrees of heat the affinity of carbon for oxygen grows, and it is emphatically *positive* itself;[1] possibly a lowering of temperature might act in the inverse way.

4. According to the style and manner of the saturation of three of the carbon valences, the fourth will have a decided positive or negative character, or something intermediate; thus the free valence of the following group is usually negative,

$$\begin{matrix} H_2 \\ \diagdown \\ \quad C— ; \\ \diagup \\ NO_2 \end{matrix}$$

but that of the methyl group,

$$H_3 \equiv C—,$$

is most nearly comparable to hydrogen in its reaction value, and is distinctly positive; finally, the cyanogen group,

$$N \equiv C—, \text{ [or } C \equiv N—, \text{ Tr.]},$$

may be at times positive or negative in reaction.

5. Another characteristic property is the *inertia* of the carbon compounds, and the *slowness of reaction* associated therewith, which characterises organic chemistry where carbon compounds come into play, and which, therefore, is found exhibited in the life activity of plants and animals. Thus it is probable that the existence of the compound

$$Zn \begin{matrix} \diagup CH_3 \\ \diagdown CH_3 \end{matrix}$$

its analogous hydrogen compound being unknown, viz.

$$Zn \begin{matrix} \diagup H \\ \diagdown H \end{matrix}$$

is to be explained, not by the greater affinity of the methyl groups for zinc as compared with hydrogen, but rather in that the former compound decomposes much more slowly than the latter. Further, the fact is well known that many carbon derivatives are more stable than the mother substances; thus methyl sulphonic acid $CH_3 \cdot SO_2 \cdot OH$ is much more stable than sulphurous acid $H \cdot SO_2 \cdot OH$; also the esters are well known in the case of the unstable ortho-carbonic acid, etc.

[1] Yet the unsaturated carbon compounds, like propargyl alcohol and phenol, are clearly acidiferous, showing that, atom for atom, carbon is more negative than hydrogen is positive; but this may be merely a relative affair.—Tr.

This same inertia renders it possible to build on to molecules, the arrangement of which is very artificial, so to speak, and therefore unnatural; by a more or less energetic blow, there occurs a transformation to a more stable form, and a closer union of the atoms. Usually there is a great quantity of energy set free in this, and the rearrangement leads to an explosion, which in turn leads to the dissolution of the molecule; in this way we find the explanation for the large number of explosive compounds produced in organic chemistry.

The Methods for the Determination of the Constitution.— The great development of organic chemistry and the extraordinary experimental results, for which we are indebted to the most deliberate application of the conception of the "constitution of the molecule," prove, in a most striking way, how fortunate the application of this idea has been. In what follows I have tried to represent, in brief, according to van't Hoff,[1] the principles which guide the organic chemist in representing the structure plans.

1. The *composition* of the compound, which is known to be pure, must first be ascertained from an *analysis*, and from a molecular weight determination by some one of the methods described in Chapter III. Next, in the attempt to trace the *method of union* of the atoms in the molecule, attention must be given to the *valence* of the atoms entering into the compound in question; thus carbon has a valence of four, nitrogen of three or five, oxygen of two, hydrogen and the halogens of one, etc.; this gives a foothold, so that the number of conceivable formulæ is seen to be very limited, and the more limited according as the number of valences of each element is more constant.

2. *The method of preparation* of the unknown compounds from those of known constitution, or conversely the transformation of the former into the latter, gives a still better grasp of the subject. It may be assumed in many cases that the new compounds have a constitution which is related to that of the original, and this is the more probable when the change and the retransformation occur easily, and when the changes of energy associated with the decomposition are very slight. The peculiar *inertia* of the carbon compounds, which gives the chemistry of this element its peculiar character (p. 284), in certain cases justifies the proviso that *the number of old valences discharged, or of new ones attached*, is reduced to a minimum. This method of determining the constitution is altogether the most reliable, and the one most commonly employed. On account of the readiness with which the valences in inorganic compounds are accustomed to change their *rôles*, this method is almost entirely limited to the carbon compounds.

3. On the basis of a very extended experience, showing that the reaction capacity of certain atomic groups, as OH, CO, C_6H_5, NH_2,

[1] *Ansichten über die org. Chem.*, Brunswick, 1881.

etc., usually remains unchanged, irrespective of the composition of the rest of the molecule, one may make certain inferences regarding the existence of these groups in the molecule from the reaction capacity of the compound. This is the *Principle of Analogous Reactions.*

4. A very elegant principle consists in the investigation of the *number of isomeric derivatives*. Thus, by ascertaining how many new compounds can result by the same sort of substitution, as of a hydrogen atom by a chlorine atom, one may safely infer whether the variously substituted atoms exert the same functions in the molecular group or not. Thus the existence of only one phenyl-chloride, $C_6H_5 . Cl$, led to the recognition of the identical union of the hydrogen atoms in benzene. Similarly, the discovery of three hydroxy-benzoic acids necessitated the distinction between the ortho, para, and meta positions, which is of fundamental significance in representing the structure of benzene. See also the discussion on p. 292 regarding the number of methylene chlorides.

5. From the degree of readiness with which a cleavage product is formed from a compound, a probable reference may be drawn as to the contiguity of the components of the separated products, in the molecule, of the original compound. Thus, other things being equal, the nearer together that the hydroxyl group and the hydrogen of the detached water molecule are in the original compound, so much the more readily can an anhydride be found. It will be seen later what important service has been given to stereochemistry by this *principle of intermolecular action.*

6. Sometimes certain elements or radicals, in the molecule, exert a mutual influence on their aptitude for reaction.[1] This will be manifested more clearly, as the reacting atoms are nearer to each other. From this aptness of reaction of particular atoms or radicals in a compound of unknown constitution, one may make some inference as to their respective distances from each other, and thus get a starting-point for the representation of their structural formulæ. Thus Ostwald[2] has systematically studied, and also used with good success, the mutual influence exerted by various atoms and radicals on the reaction aptitude of the acid hydrogen atoms in organic acids, resulting from the relative location of these atoms and radicals in the respective molecular groups. Methods of this sort have become very important in determining *the spatial arrangement of atoms*, and they are the more important inasmuch as they furnish the quantitative determination of the reaction aptitude of many radicals.

7. All isomeric compounds are distinguished more or less by their properties, such as the melting-point, the boiling-point, the density, the refractive index, etc.; and the constitution is no less a standard

[1] For a very interesting summary of this line of argument see Remsen, *Theoret. Chem.*, 4th edit., chap. xix.—TR.

[2] *Zeitschr. physik. Chem.*, **3**. 170, 241, 360 (1889).

factor than the composition in determining the respective physical conduct. Thus, when one has found a connection between the constitution and the physical properties of a large number of compounds which are known to be related in their structure, then conversely from the *physical properties* of an unknown compound, he can safely draw some inference regarding its constitution. Usually such relations as these, as for instance the connection between the constitution and the refractive index, are of a purely empirical nature, and the safety with which they may be employed in any given case will depend simply on the number of the analogous cases. But again, as is the case in the relation between the constitution and optical activity, one sometimes obtains a deeper insight into their nature, and then he may infer with a large degree of confidence. The material thus far discovered in this direction will be found grouped in the following chapter.

Benzene Theory.—By means of the methods described, it is possible to ascertain with great certainty the constitution of a legion of carbon compounds, the number of which increases every day. These constitutional formulæ so obtained have much significance in the eyes of those who would regard them as established, entirely independently of all molecular and theoretical speculations; for these formulæ represent to a large degree, and in a very condensed form, their "facies" in many empirical relations, and he who is skilled in reading this symbolism, learns very much regarding the nature of the compound, from what is expressed through the formula.

Thus, for instance, in considering the constitutional formula of phenol, thus

```
        H
        |
        O
        |
        C
      // \
  H—C     C—H
     |     ||
  H—C     C—H
      \\ /
        C
        |
        H
```

at the first glance it is apparent that one hydrogen atom will react differently from all the others, since it can be substituted quite readily by a positive radical; also that in substituting any one of the other hydrogen atoms by a univalent atom or radical, it is possible to obtain any one of three isomers, according to the location of the substitution; also that a dissolution of the ring union would be accompanied by a fundamental change in the molecular structure, etc., etc.

The three isomers that can be formed by replacement of a hydrogen atom directly combined with carbon in phenol and analogous compounds are distinguished as ortho- meta- and para-compounds, but, strictly speaking, the number of cases of isomerism should be greater; for example, a difference should exist between substitution of the two atoms next the hydroxyl group in phenol. The fact that this case of isomerism has not been realised has led to many explanations and to the suggestion of many modified forms of Kekulé's benzene formula.[1] Quite recently a modified view of the nature of a double compound has been developed which is worth inserting here.

It has often been remarked that the expression "double bond" and the symbol for it C = C corresponds but little to the actual behaviour and does not lend itself to calculation of the characteristic peculiarities of the double bond, its instability and its tendency to addition products. J. Thiele[2] has offered an extension of the concept of a double bond which certainly brings in a new hypothetical element, but is well suited to bring together a number of observed facts. He believes that, in the formation of a compound with double bond, only *partial saturation of the two valencies occurs.* Thiele found the strongest support for this belief in the behaviour of the so-called conjugate double compounds. When a hydrogen or bromine molecule is added to the atomic complex C = CH − CH = C, the result is not the elimination of one double bond by formation of a group like CBr − CHBr − CH = C, but both disappear, and another double bond appears in the middle instead; thus we get CBr − CH = CH − CBr.

This is difficult to understand according to accepted views. Thiele assumes that the affinities are not completely used up in the formation of a double bond, but that a residue of affinity or partial valency remains in each atom. The two neighbouring partial valencies saturate each other mutually, whilst the extreme one remains unsaturated and capable of forming addition products. He writes the following symbol for it:—

The addend is attached to the carbon atom with the free partial valencies, and hence the product BrC − CH = CH − CBr. Other phenomena of addition that have been noticed in the aldehydes, quinone, benzyls, and so forth, are explained in a similar manner.

An especially good application of these views is to be found in

[1] See especially the critical study by Marckwald, *Benzoltheorie*, Stuttgart, 1897. F. Enke.

[2] *Lieb. Ann.*, **306**. 87 (1889).

the benzene problem. Thiele's modification of the old formula is explained by the symbol

```
        H
      ⁄C⟍
    HC    CH
    ||    |)
    HC    CH
      ⟍C⁄
        H
```

or, if all the bonds are regarded as equivalent, by

```
        H
      ⁄C⟍
    HC    CH
    ||    ||
    HC    CH
      ⟍C⁄
        H
```

Since benzene, according to this formula, contains six inactive bonds, it is to be regarded as a saturated compound in harmony with its chemical behaviour.

The objection made against Kekulé's formula disappears in the case of Thiele's, on account of its complete symmetry.

But, at the same time, experience has shown that the constitutional formulæ are not adequate to the complete description of the compounds represented. For on the one hand it appears that the anticipated compounds cannot in all cases be represented by the constitutional formulæ; thus the two isomers, methyl cyanide and methyl isocyanide (the nitril and the isonitril), are known,

$$N{\equiv}C{-}CH_3, \quad \text{and} \quad C{\equiv}N{-}CH_3,$$

but only one hydrocyanic acid is known, although there should be *two*, corresponding to the formulæ

$$N{\equiv}C{-}H, \quad \text{and} \quad C{\equiv}N{-}H.$$

In this case the constitutional formulæ cover too much ground, since fewer compounds are known than were to be expected. Yet this proves nothing against the formulæ, because it is entirely possible that the desired isomer may exist, but that the proper method of preparation has not as yet been found, nor the conditions suitable for its existence. Also the assumption is not improbable that both molecules may exist in the case of hydrocyanic acid, and that these two forms are converted into each other so readily, that the *one* acid may be able to react in the sense of *both* structural formulæ. (For further details see Book III., chapter on Chemical Kinetics, Tautomerism.)

On the other hand, it would be an entirely different case *if more isomers were known than were required by the theory, or than could be accounted for by it.*

The classical case is that of lactic acid. This was shown by the investigations of Wislicenus [1] to have several distinct forms of really different properties, all of which yet had one and the same formula, viz.,

$$CH_3CH(OH)CO_2H.$$

Being greatly disturbed by this observation, van't Hoff [2] proposed this question (1877), viz., What change or extension must be introduced into the theory of structural chemistry in order to adapt it to all the observed compounds? The following abstract gives the essentials of the order of thought by which van't Hoff was led to the creation of stereochemistry. Le Bel developed these views simultaneously.

The Stereochemistry of Carbon.—The fundamental assumption on which all the following considerations are based is this: viz. *that the four valencies of carbon are like each other in every particular.* The correctness of this is shown by the negative proof that not nearly so many isomers are known as would be known, if one or several of the valencies of carbon were different. Thus, for example, only one methyl chloride is known; but there would be several if the mode of union with the chlorine atoms varied with their location.

The question regarding the number of mono-substitution products has been treated very systematically by L. Henry.[3] We will briefly indicate the course pursued by this investigator, on account of its great importance as the basis of stereochemistry.

If we assume that the four valencies of carbon are different, then we must write the formula of methane thus, $CH_IH_{II}H_{III}H_{IV}$; here the Roman numerals denote that the respective hydrogen atoms are united to the carbon each in a different way; then there would be four monosubstitution products, according to the particular hydrogen atom substituted. Now, let the univalent radical A replace H_I, producing the compound $CAH_{II}H_{III}H_{IV}$. We will now replace A by the radical B, producing the compound $CBH_{II}H_{III}H_{IV}$; and again in this latter compound introduce A, which, let us say, takes the place of

[1] *Lieb. Ann.*, **156.** 3; **157.** 302 (1871).

[2] *Dix années dans l'histoire d'une théorie*, Rotterdam, 1877. The arrangement of atoms in space, 2nd edition (Braunschweig, 1894). See also A. Hantzsch, *Grundriss der Stereochemie* (Breslau, 1893).

[3] *Bull. Acad. Belg.* (3), **12.** No. 12 (1886); **15.** 333 (1888).

H_{II}, producing $CBAH_{III}H_{IV}$; finally, B is replaced by hydrogen, producing $CH_IAH_{III}H_{IV}$. Now if the valencies I and II are different, then $CAH_{II}H_{III}H_{IV}$ and $CHAH_{III}H_{IV}$ should have different properties [*as is not the case*]; and in the same way all the other valencies of carbon were investigated. In this way Henry successively prepared all of the four nitro-methanes, *and always obtained the same substance* [*showing that all the valencies are alike*].

Let us next advance the question as to the directions in which the four valencies radiate from the carbon atom. Now it follows from the likeness of the four valencies, that they must be symmetrically distributed in space, and there are only two conceivable modes of such arrangement, viz.: the four valencies must either lie in a plane, intersecting each other at angles of 90°, or else they must be symmetrically distributed in space as the four axes of an equilateral tetrahedron.[1] Here again the isomeric forms must decide between these two possibilities. Thus, in the former case [the arrangement in a plane], by replacing two hydrogen atoms by chlorine, we should obtain the two following isomeric methylene chlorides [dichlor-methanes], thus,

```
     H                      H
     |                      |
Cl—C—Cl,       and      Cl—C—H ;
     |                      |
     H                      Cl
```

these should be distinguished by the circumstance that in one the chlorine atoms are opposite each other, while in the other they are beside each other. On the other hand, if the four valencies are arranged tetrahedrally, two chlorine atoms when introduced into the molecule would always lie beside each other, and the conditions would allow of only one methylene chloride. Now, as a matter of fact, only one methylene chloride is known. [*Hence the tetrahedron formula is probably the correct one.*]

The case is similar when two hydrogen atoms in methane are replaced by two different radicals, or when three hydrogen atoms are replaced by two radicals of one kind and one of another. In all cases, unless we assume, as is very improbable, that there are isomers which have not yet been observed, *we are forced to regard the tetrahedron arrangement of the valencies in space as correct.*

1 These axes are not the crystallographic axes (say of the 1st system), but the axes converging from the solid angles of a tetrahedron to the centre; their distribution in space may be most readily pictured thus; *any two* of the four axes being taken as a pair, the remaining two form another pair; the two axes of any pair intersect each other at an angle of 120° in their plane like a very obtuse V; the planes of the two pairs cut each other at an angle of 90°, but so that the two obtuse Vs shall have their apices pointing in opposite directions.—Tr.

Optical Isomerism.—The only possible case of isomerism of the substitution products of methane, in the sense of the tetrahedron proviso, is found when three of the hydrogen atoms are replaced by different radicals, or expressed in more general terms, *where the four carbon valencies are replaced by four different radicals [or atoms.]*

By denoting the four different atoms or radicals which satisfy a carbon atom, by the letters *a*, *b*, *c*, and *d*, we obtain such formulæ types as those shown in Fig. 24.

These formulæ, in all probability, should correspond to two different

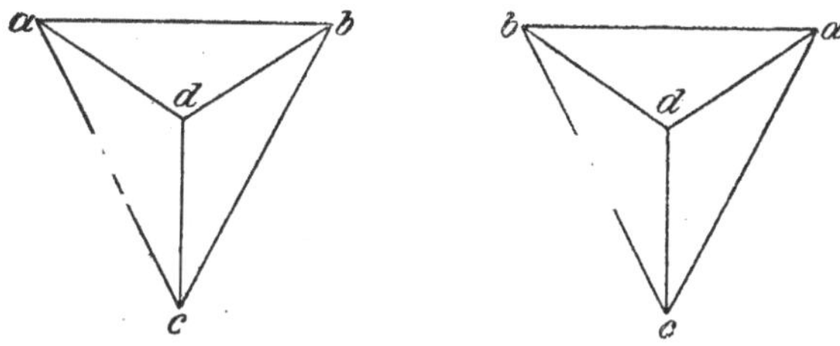

Fig. 24.

compounds, because they cannot be made to coincide by superposition; the difference between them is comparable to that between the right and left hand, or between a real object and its reflected image.

A single glance at Fig. 24 shows that such isomers as these two must represent the slightest conceivable difference between their respective physical and chemical properties. The degree of separation between the atoms is the same in both cases; thus the distance between *a* and *b* is the same. The sole difference consists in this, viz., that if we consider any selected angle as that marked by the radical *d*, then in the left figure the course *a*, *b*, *c* is directed like the movement of the hands of a watch, while the corresponding order *a*, *b*, *c* in the right figure is in the inverse direction. Inasmuch as these two types, figured above, have no planes of symmetry, such a carbon atom, the valencies of which are satisfied by four different [atoms or] radicals, is said to be *unsymmetrical.*

In 1874 Le Bel and van't Hoff[1] independently and contemporaneously suggested that the right- and left-handed optically active isomers correspond to these two types. As a matter of fact, the physical and chemical properties are exactly the same, and only differ in their property of rotating the plane of polarisation equally strongly but in opposite directions, which means that the respective substances have a [fundamental] difference in their molecular structure. This kind of isomerism is accordingly called "*optical*";

[1] See references on p. 291.

the subject of optical rotation will be further considered in the next chapter.

This kind of isomerism usually exhibits itself in the solid state by occasioning the crystallisation of the two substances in two respectively opposite enantiomorphic forms (forms turned back upon themselves).

The two isomers shown in Fig. 24 obviously have the same constitution, since the conditions are alike in every respect; they differ only in the spatial arrangement of the particular groups in the molecule, or they are said to have a different "*configuration.*"

Geometrical Isomerism.—Another variety of isomerism, which is clearly inexplicable by means of the ordinary structure formulæ, is met in the case of compounds which have a "double linkage" between two carbon atoms. And here again it was due especially to the experimental work of Wislicenus[1] that there was recognised a distinction between isomers of this sort, and again it was to van't Hoff that their theoretical explanation was due.

When two valencies of two different carbon atoms mutually satisfy each other, or, in the language of stereochemistry, when two solid angles of one tetrahedron are joined to two solid angles of another tetrahedron, then the four free valencies lie in one plane;[2] thus, if the four free valencies become satisfied by the four radicals R_1, R_2, R_3 and R_4, so that one carbon atom holds the first two, and the other the last two, then there is produced the compound

$$\begin{matrix} R_1 & & & & R_3 \\ & \diagdown & & \diagup & \\ & & C{=}C & & \\ & \diagup & & \diagdown & \\ R_2 & & & & R_4 \end{matrix}$$

Now, according to van't Hoff's view, two compounds of this sort would be looked for, in one of which R_1 and R_3, in the other R_1 and R_4, would lie on the same side of the double tetrahedron.

A similar case of isomerism would also be expected when the two free valencies of each of the doubly linked carbon atoms become united to a like pair of radicals, as in the compound

$$\begin{matrix} R_1 & & & & R_1 \\ & \diagdown & & \diagup & \\ & & C{=}C & & \\ & \diagup & & \diagdown & \\ R_2 & & & & R_2. \end{matrix}$$

---

1 *Abh. d. kgl. sächs. Akad.*, 1887.

2 This plane is at right angles to the plane of the edge in which lie the saturated valence solid angles of the two tetrahedra.—TR.

This case is illustrated in the following diagram of the isomerism between fumaric and maleïc acids, as in Fig. 25.

This case of isomerism is also very frequently satisfactorily expressed by the following constitutional formulæ, where the spatial relations are illustrated :—

Maleïc Acid.	Fumaric Acid.
H—C—COOH	H—C—COOH,
‖	‖
H—C—COOH	HOOC—C—H.

It would not be expected that this variety of isomerism would exhibit optical activity, because the four radicals satisfying the four

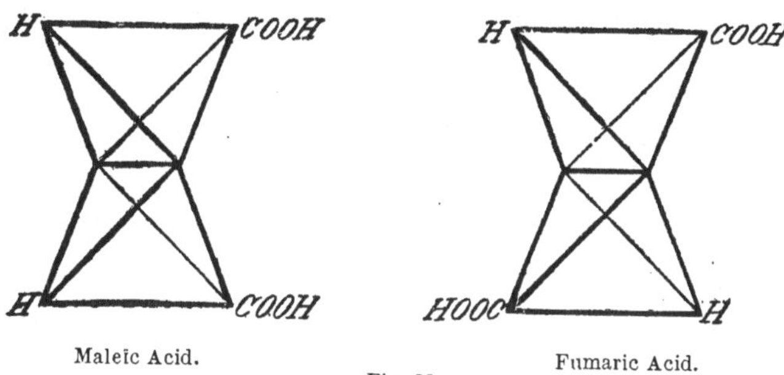

Maleïc Acid. Fumaric Acid.

Fig. 25.

[external] valencies of the double tetrahedron all lie in the same plane. Also, since the relative degree of separation of the radicals from each other, as shown in Fig. 25, is not the same, the two isomers will behave differently as regards aptitude for reaction, boiling-point, melting-point, solubility, etc. As to the question regarding the relative contiguity of the respective groups, in any such case of which Fig. 25 is a type, this can be decided only by means of the principle of intermolecular reaction (p. 287).

The theory suggests no new case of isomerism in the formation of a triple bond, or so-called "acetylene bond," but as in the case of di-methylene (ethylene), so in the poly-methylenes, cases of geometric isomerism must occur. Considering the simplest carbon ring, tri-methylene

$$\begin{array}{c} CH_2 \\ \diagup \quad \diagdown \\ CH_2—CH_2 \end{array}$$

, it may be seen that in the arrangement of the atoms in space, geometrical isomerism must be possible by the entry of two substituting groups, according as they take a "cis" position on the same side of the plane of the ring, or a "trans" position on both sides of the plane. In the latter case two isomers are possible, forming non-congruent images; we get, therefore, altogether three isomers which can be understood by the following models :—

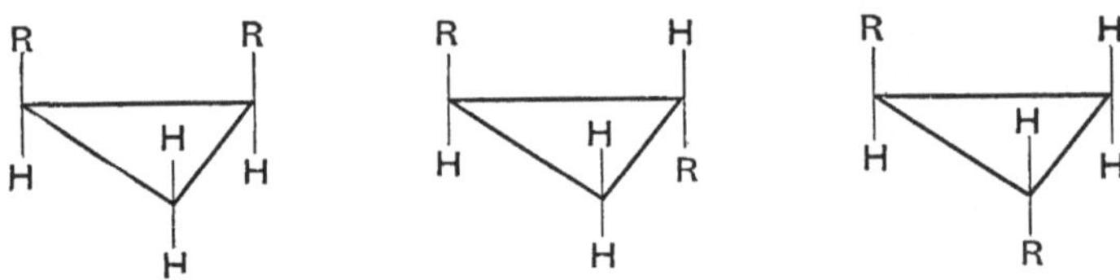

Actually two different tri-methylene di-carbon acids have been observed,[1] one of which can apparently be split into opposed active components.

Cases of geometrical isomerism are also known in derivatives of larger rings; the investigations of v. Baeyer's[2] on the hydrogenised phthalic acids have been largely productive of material confirming this theory. Whilst no case of isomerism is known in the singly substituted derivatives of hexa-methylene two hexa-hydro-teraphthalic acids are known corresponding to the formula

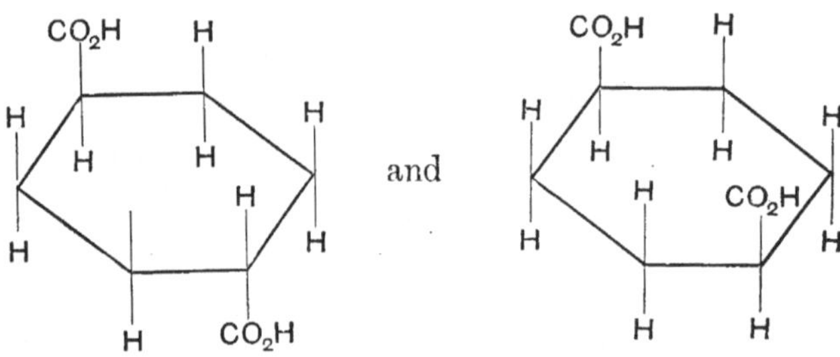

The first is analogous to fumaric, the second to maleic acid.

The Stereochemistry of Nitrogen.—There has recently been begun, in addition to the stereochemistry of carbon, a stereochemistry of nitrogen, which has already led to considerable results. According to experiments thus far made, there are two groups of stereo-isomeric nitrogen compounds, which are completely analogous to the two groups of stereo-isomeric carbon compounds described above.

Firstly, Le Bel[3] has recently prepared a compound which contains, instead of the asymmetric carbon atom, the group NX, where X denotes a univalent radical, the four free valencies being saturated by four different radicals. The compound produced by Le Bel was isobutyl-propyl-ethyl-methyl-ammonium chloride, which was obtained

[1] Buchner, *Ber. deutsch. chem. Ges.*, **23.** 702 (1890).
[2] *Lieb. Ann.*, **245.** 103 (1888); **251.** 258 (1889); **258.** 1, 145 (1890).
[3] Le Bel *C. R.*, **112.** 724 (1891); **129.** 548 (1899).

in an optically active form by means of a fungus (*Pilz*). It has the formula

$$\begin{array}{c} CH_3 \\ | \\ C_4H_9—NCl—C_2H_5 \\ | \\ C_3H_7. \end{array}$$

The activity remained even when the chlorine was replaced by the acetyl radical. The preceding compound seems to be completely analogous to the optical isomers of carbon.

Secondly, certain compounds were prepared before that of Le Bel, and were claimed to be geometrical isomers by Hantzsch and Werner.[1] These compounds can be regarded as analogous to the corresponding carbon isomers, by supposing the group CR to be replaced by *trivalent nitrogen.* Now, since this latter is very often possible, according to experiments, the inference is a direct one that the three nitrogen valencies do not lie in a plane, but that they occupy relative positions which are similar in direction, at least approximately so, to the three free valencies in the group CR. Therefore it follows that when *two nitrogen valencies are bound to one carbon atom, there must result a case of geometrical isomerism which is completely analogous to that described on p. 295, and which may be expressed by spatial formulæ written as follows:*

$$\begin{array}{c} R_1—C—R_2 \\ \| \\ R_3—N \end{array} \quad \text{and} \quad \begin{array}{c} R_1—C—R_2 \\ \| \\ N—R_3 \end{array}$$

Numerous examples of this interesting variety of isomerism have been found among the asymmetric oximes, *i.e.* compounds where the hydroxyl group plays the part of the radical R_3.

The question as to which of the isomers corresponds to which of the configurations given above, can be answered by means of the principle of intermolecular reaction between groups which are spatially contiguous. Thus one of the aldoximes,

$$\begin{array}{c} H \\ R \end{array}\!\!>C=N—OH,$$

produced by the reaction between aldehydes and hydroxyl-amine, decomposes with difficulty, but the other readily into the nitril R—C≡N; therefore the inference is drawn directly that in the one, the H and the OH, expelled as water, are nearer neighbours than in

[1] See references on p. 291.

the other; and therefore they are given the respective formulæ, viz.—

In an analogous way the three isomeric benz-aldoximes may be expressed by the respective formulæ, viz.:—

CHAPTER V

PHYSICAL PROPERTIES AND MOLECULAR STRUCTURE

General Observations.—According to the views advanced in the preceding chapter regarding the structure theory of molecules, there are three circumstances which are of fundamental significance in determining the physical properties of any compound. These are—

1. The chemical composition.
2. The constitution, *i.e.* the mode of linking of the atoms.
3. The configuration, *i.e.* the spatial arrangement of the atoms.

A change in any one of these factors occasions a more or less extended change in the properties of the compound.

The knowledge of this at once suggests the problem of determining the relations which exist between the physical and chemical properties of a compound, and the *structure of the molecule*; this latter expression denotes the sum and substance of the three factors mentioned above. The complete solution of this problem will enable us to infer in every respect the behaviour of a substance from its structural formula; and further, to predict the conditions for the existence of, and the properties of, compounds not yet prepared; and this would mean the attainment of a goal, to reach which has been the chief aim of all chemical investigation.

Even at present, structural chemistry, at least among organic compounds, has already progressed so far that the structural formula of a substance enables us to draw many inferences regarding its aptitude for reaction, as has been repeatedly mentioned in the preceding chapter. But as yet the science is wanting in the formulation of sharp and precise laws, so that many conclusions which are drawn are the insight of a certain "chemical instinct," rather than the necessary deductions from clearly perceived chemical laws. We would not undervalue this "chemical touch," the fortunate possessor of which doubtless must enjoy an unusual scientific intuition, which is sharpened by long experience; but this perhaps might be replaced by something better, were more effort directed towards translating into the language of natural science, or at least reducing to some degree of regularity, that experience by which this instinct is developed.

In this state of the case, in the "doctrine of affinity" we can only incidentally consider the question regarding the relation between the reaction-aptitude and the structure of a compound. In this chapter will be given only the other side of the question, which is certainly a growing one, viz. concerning the regularities which have been found thus far, regarding the connection of the *physical properties* with the structure of the molecule of a substance. Naturally the question will be almost entirely limited to the consideration of carbon compounds, because thus far it is only here that we find well-grounded notions regarding the arrangement of the atoms in the molecule. As was true in the case of the salt solutions, so here also it should be noticed, that this subject occupies a special position, and therefore a special chapter will be devoted to the physical relations.

But at the same time, in this chapter will be described the most important *physical methods* which are used by the chemist in his investigations; and not only those which are used at present, but also those which will be used to a much greater degree in the future.

Specific Volume and Molecular Volume.—By the *specific volume* of a substance there is meant the volume, expressed in c.c., occupied by a gram; the reciprocal of this specific volume, *i.e.* the weight of a unit volume, is the so-called *specific gravity*. It is of much advantage, for purely algebraic reasons, to derive the relations between the density and the composition of substances from the specific volume.

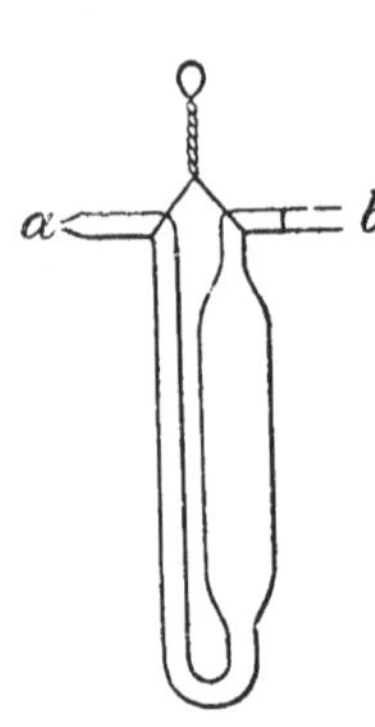

Fig. 26.

The essentials of the determination of the specific volumes of gases, and their relations to the respective molecular weights, have been already described in the chapter on the determination of the molecular weight (p. 247); the simple result was obtained that the molecular volumes of the most different gases are the same, under the same external conditions. We will consider here only those substances which exist in the *liquid* and *solid* states.

The *specific volume of liquids* can be easily and exactly determined, either by measuring the resistance caused by the immersion of a solid body of known volume (the araömeter, Mohr's balance, etc.), or by weighing the liquid contained in a vessel of known volume (the pycnometer). The first method is quicker and more easy to execute, but the second is much more exact, and therefore to be preferred for all scientific purposes. Of the many forms of pycnometer which have been described, that shown in Fig. 26, which was designed by Sprengel and modified by Ostwald, is at once the simplest and the most convenient. It consists of a small bent pipette, provided on one side with a capillary opening at *a*, and on the other side with a narrow tube *b* with a mark; it is filled by dipping the capillary opening *a* into the liquid in question, and sucking through a rubber tube which is attached to *b*; it is then hung in a bath which has the temperature of the determination, the liquid meniscus being adjusted to the mark in *b*, either by sucking out the excess of the liquid from the capillary opening at *a*, by means of a bit of filter paper, or by adding what liquid is wanting by a drop on a glass rod applied at *a*, when it will be sucked in by the capillary force.

Let p, p_1, and p_2 be the weights respectively of the pycnometer empty, when full of a liquid (as water) of known density s, and when filled with the liquid which is to be studied; and if Δ is the weight of the air displaced, and which is approximately equal to $(p_2 - p_1)\ 0{\cdot}0012$, then the specific gravity sought is

$$S = s\frac{p_2 - p - \Delta}{p_1 - p - \Delta}.$$

F. Kohlrausch and Hallwachs[1] have lately perfected the areometric method to such an extent that changes in the specific gravity of aqueous solutions of one-millionth can be distinguished with certainty, a notable result for the investigation of very dilute solutions. Smooth cocoon silk was used for the suspension, and as immersed body a glass bulb of 133 grammes weight and 129 c.cm. content.

While by this method the density of a liquid can be determined without any difficulty, yet the methods used to determine the *specific gravities of solid bodies* are not quite satisfactory in accuracy, for the reason that only small pieces of the substance can be used; and this is true of the methods which depend upon the measurement by means of the buoyancy in a liquid of known density, or by weighing it in a liquid pycnometer, or by the application of gas laws (volumenometer), or by its swimming free in a liquid mixture, the density of which is already determined. The "swimming method" usually gives the best results when small pieces only can be used. This method has at the same time this advantage, which should not be undervalued, viz., that from the way the substance floats in the liquid, by observing whether a part of it floats at the top, and whether another part sinks to the bottom, one can infer very much regarding the purity of the substance, and thus one can raise the grade of purity.[2]

A liquid, in which the substance shall swim free, without either rising or sinking, can be made by a mixture of methylene iodide, CH_2I_2, which has a sp. gr. of 3·3, with such lighter hydrocarbons as toluene, xylene, etc. The specific gravity of the solid substance is

[1] *Wied. Ann.*, **53.** 14 (1894); see also F. Kohlrausch, *Wied. Ann.*, **56.** 185 (1895).
[2] Retgers, *Zeitschr. physik. Chem.*, **3.** 289 (1889).

then equal to that of the liquid in which it swims free, and that of the liquid can be determined by a suitable method.

The question regarding the dependence of the specific volume upon the composition and the constitution of compounds has thus far, as is the case with many other properties, been successfully studied *only for carbon compounds.* It is well known, as shown by Kopp (1855), that it is possible to calculate the volume occupied by 1 g.-mol of a liquid organic substance at its boiling-point from its composition; this volume is equal to the product of the molecular weight and the specific volume, and therefore is suitably called the "*molecular volume*"; it is calculated in the following way.

If a molecule of thé substance in question contains m atoms of carbon, n_1 atoms of "carbonyl oxygen" (*i.e.* oxygen with both its valencies united to the same carbon atom), n_2 atoms of oxygen which has its two valencies distributed between two carbon atoms or between two atoms of other elements, o atoms of hydrogen, p atoms of chlorine, q atoms of bromine, r atoms of iodine, and s atoms of sulphur; then its molecular volume at the boiling-point, amounts to

$$\begin{aligned} \text{M.V.} = {} & 11{\cdot}0\text{m} + 12{\cdot}2\text{n}_1 + 7{\cdot}8\text{n}_2 + 5{\cdot}5\text{o} + \\ & 22{\cdot}8\text{p} + 27{\cdot}8\text{q} + 37{\cdot}5\text{r} + 22{\cdot}6\text{s}. \end{aligned}$$

This formula is by no means strictly exact, as deviations amounting to several per cent are not uncommon; but nevertheless it should be particularly noticed that by means of a few empirical constants it is possible to calculate, at least approximately, the molecular volume; and thus by dividing this value by the molecular weight, it is possible to obtain the specific gravity of numerous compounds which are made up of the aforenamed elements. Thus the measurement of the molecular volume of acetone at its boiling-point gave as the result 77·5; and from its formula $CO\begin{smallmatrix}CH_3\\CH_3\end{smallmatrix}$ it was estimated as follows:—

3 atoms of carbon	33·0
1 atom of carbonyl oxygen . . .	12·2
6 atoms of hydrogen	33·0
Sum . . .	78·2

The presence, in the molecule, of carbon atoms with doubled linkage increases the molecular volume proportionally.[1] Reference should be made to the thorough epitome of this material, which has been compiled by Horstmann.[2] He has shown from éxtended researches that, in general, simple additive relations really exist with some very considerable deviations, and has ascribed these deviations to peculiarities in the constitution.

The specific gravities of liquid chlorine and bromine at the boiling-

[1] Horstmann, *B. B.*, **20.** 766 (1887).

[2] Graham Otto, *Lehrbuch der Chemie*, 3rd edit., Braunschweig, 1893.

points amount respectively to 1·56 and 2·96;[1] *the atomic volumes deduced from these figures are* 22·7 *and* 26·9 *respectively, or nearly the same as the figures,* 22·8 *and* 27·8, *which Kopp calculated from the organic compounds.*

Moreover, it cannot be doubted that we must regard $\frac{1}{5 \cdot 5} = 0 \cdot 18$ as the specific gravity of boiling hydrogen.

Concerning the question as to how far Kopp was justified *in selecting the boiling-point as the point for comparison,* we will obtain information from the following. In the light of the theory of van der Waals, the molecular volume at the critical point should be three-eighths of that calculated from the laws of ideal gases. Experience vindicates this law (p. 235) to this extent, viz. that the critical volume of many substances amounts to the same fraction, viz. $\frac{1}{3 \cdot 8}$.

Further, from the equation given on p. 223, viz.,

$$\phi_0 = 3b,$$

it follows that the molecules of all substances at their critical points occupy the same fraction of the critical volume; *this latter therefore may be regarded as a measure of the space actually occupied by the molecules.* But now the volume of liquids in corresponding states is the same fraction of the critical volume. *Therefore the molecular volumes measured in corresponding states,* i.e. *at equal reduced pressures and temperatures, may be regarded as a measure of the space actually occupied by the molecules;* and this is almost the same thing as regarding the comparison temperature as a reduced temperature, or making it equal to the same fraction of the critical temperature. Less value is to be attached to a comparison with the critical pressure, as already shown on p. 226, because the volume of a liquid would be subject to only very slight change if it were measured at a pressure of one-half, and then at two atmospheres; and no greater variations of pressure are concerned here.

But now Guldberg[2] called attention to the fact that, for the most various substances, the boiling-point, on the absolute scale, is about two-thirds of the critical temperature; i.e. *for the most various substances, the boiling-point at the atmospheric pressure is itself an identical reduced temperature.* The fact that the identity may not be strictly fulfilled, is to be ascribed to the fact, which is not considered enough, viz. that the volume of a liquid is only slightly affected by changes of 5° or 10°; and only such variations as these are found.

We thus obtain the result that not only is the boiling-point a

[1] Dammer's *Handbuch der anorg. Chem.,* I. pp. 474 and 520 (1892).

[2] *Zeitschr. physik. Chem.,* **5.** 374 (1890).

suitable temperature for comparison, but also that the molecular volumes so determined, for the most different substances, represent very nearly the same multiple of the space occupied by the molecules. Now, inasmuch as we must regard Kopp's molecular volumes as a measure of the volumes actually occupied by the molecules, therefore *we must conclude that the volume of the molecule can be calculated additively from the volumes of the atoms.* This result will also be established later in an entirely different way.

The Density of Solid Bodies.—The volume relations of solid compounds have as yet been only slightly investigated. The molecular volume is obviously additive in certain cases, as of salts of analogous constitution, as shown by the following table :—

I.	Diff.	II.	Diff.	Diff. I.—II.
KCl = 37·4		NaCl = 27·1		10·3
	6·9		6·7	
KBr = 44·3		NaBr = 33·8		10·5
	9·7		9·7	
KI = 54·0		NaI = 43·5		10·5

We find constant differences both between the corresponding potassium and sodium salts, and also between the corresponding chlorides, bromides, and iodides, to be a necessary condition for the calculation of the molecular volume in summation from the atomic volumes, by the choice of suitable constants. Schröder, who has done good service by investigating these relations, stated, in 1877, in harmony with the preceding, that in the homologous series of the silver salts of the fatty acids, the molecular volume increases quite constantly, about 15·3, for every additional CH_2 group.

The question as to what is the proper temperature for comparison is here of less importance on account of the slight change of density of solid substances from a change in temperature. The relation between the atomic weight and the density of the elements in the solid state has been already treated in the description of the periodic system (p. 192).

The Refractive Power.—The optical coefficient of refraction for homogeneous gases, liquids, or solids can be measured very accurately by enclosing them in a hollow prism with a prismatic cavity, and then measuring, by means of a spectrometer, the deviation of a ray of monochromatic light. The method of total reflection, which is employed in the refractometers devised by Kohlrausch, Abbe, and others, can be used much more easily, and this method has recently been applied by Pulfrich in the construction of a very convenient apparatus for the special study of liquids. Inasmuch as we are practically concerned only with the determination of the optical behaviour of liquids, in the exigencies of the chemical laboratory, and inasmuch also as the Pulfrich [1] refractometer combines ease of manipulation with an accuracy which is more than sufficient for most purposes, we will give here a description of this apparatus.

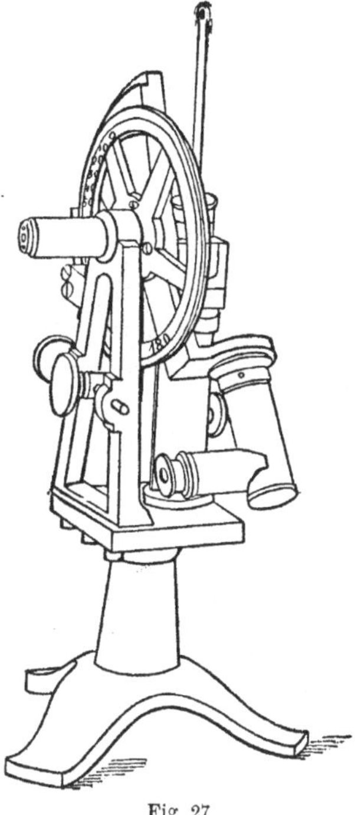
Fig. 27.

The monochromatic light given out by a Bunsen burner, fed with salt for instance, and concentrated by passing through a [collimating] lens which is adjusted to the apparatus (but not shown in the cut, Fig. 27), is allowed to fall on the horizontal surface of a prism which must be provided with a refractive angle of 90°. The top of the prism is adjusted to receive the glass cylinder, which is to hold the liquid to be studied in the following way :—The lower edge of the glass cylinder rests a little lower than the round flat upper face which fits into it, and thus the cemented edge does not interfere with the passage of the light. Then it is necessary to moisten the upper surface of the prism with only a few drops of the liquid to be investigated, an advantage which is not to be undervalued in chemical work. The prism, with its smoothly ground hypothenuse surface, rests on a bevelled block of brass, which in turn rests on a wooden support. This rests on a heavy tripod base. By means of this it is possible to obtain very certain manipulation, so that after removing the prism it can be returned again to its original position, and a readjustment is not required even after long use. The normal positions of the two surfaces of the prisms are determined by means of an adjusting apparatus which is fastened, with the wooden support, to the brass block on which the prism rests, parallel with the plane of rotation of the divided circle. The observer may convince himself of the correct adjustment most easily, by observing the refractive index of pure water at the atmospheric temperature, which is known, and deducing from it the small correction which must be introduced into all readings.

The measurement consists simply in moving the telescope which is attached to the alidade of the divided circle, until the cross wires coincide with the sharply defined limit between the light and the dark portions of the field, and thus obtaining the exit angle i of the

[1] *Zeitschr. f. Instrumentenkunde*, **8.** 17 (1888); *Zeitschr. phys. Chem.*, **18.** 294 (1895).

light ray. Then, as follows from the simple application of the law of refraction, we have for the desired value of the refractive index of the liquid in question—

$$n = \sqrt{N^2 - \sin^2 i},$$

where N denotes the respective value of the index of refraction of the glass prism, for the sort of light used. A table furnished with the instrument gives, without further calculation, the value of the refractive index for the angle i for sodium light. Whenever it is necessary to identify an unknown substance, this determination should never be neglected, inasmuch as scarcely any other property of liquids can be determined so easily and exactly. The apparatus may also be used for analytical purposes, as in the study of solutions.

Recently, Le Blane[1] has suggested a very simple method for measuring very exactly the refractive index of solid bodies which are optically isotropic by means of this same instrument. A decigram or so of the finely powdered substance is shaken into the glass cylinder of the refractometer, and this is then covered with a mixture of two liquids, for example, brom-naphthalene and acetone, between the refractive indices of which that of the substance in question must lie: if that of the latter is very far removed from that of the mixture, the field of the telescope will remain almost uniformly dark; then more of either brom-naphthalene or acetone is added, when a light field appears after the addition of one or the other liquid. If the desired point is already nearly obtained, so that two parts of the field are somewhat different in brightness, then one can easily judge whether the liquid of the greater, or of the lesser refractive index must be added. Thus, if the refractive index of the liquid mixture is greater than that of the powder, there appears a light band at the limit between the light and the dark parts; this band is due to the total reflection of the incident light by the solid substance; it is wanting in the opposite case. *It is possible, by the further addition of the required liquid, to so nearly equalise the coefficients of the solid substance and of the mixture, that the limit between the light and dark half of the field will be as sharp as in the case of a single pure liquid.* When this sharp limit shall have been found, the refractive index of the solid substance can at once be obtained from the table furnished with the apparatus.

This method is also applicable to the measurement of the refractive index of the ordinary ray of optically uniaxial bodies; but not, at least not in this form, to the measurement of the extraordinary rays, nor to the investigation of substances which are optically poly-axial.

The Molecular Refraction of Organic Compounds.—The refractive coefficient of a substance changes with its temperature, and also especially with its state of aggregation. If, as in the preceding

[1] *Zeitschr. physik. Chem.*, **10.** 433 (1892).

case, the question only concerns a clear statement of the relations between the optical and the chemical behaviour, then the only way in which we may hope for results is *to find some function of the refractive index which, when freed from other influences, shall be conditioned essentially by the chemical nature of the substance.*

The specific refractivity, expressed by the index minus 1, and divided by the density, thus

$$\frac{n-1}{d}$$

partially satisfies the preceding requirements, as shown by Gladstone and Dale,[1] and especially by Landolt.[2] In fact, in most cases, the specific refractivity is only slightly dependent upon the temperature; moreover, each particular substance in a mixture preserves its own specific refractivity nearly unchanged, as already shown on p. 107. But the preceding expression does not apply to changes in the state of aggregation, and the specific refractivity of a liquid, in general, is considerably different from that of its vapour.

This condition of being independent of changes in the state of aggregation is very well satisfied by a formula which was suggested at the same time (1880) by Lorenz in Copenhagen and by Lorentz in Leyden; it was developed from the following considerations.[3]

The Clausius-Mossotti theory of dielectrics proceeds upon the assumption that the molecules, assumed to be spherical, are electrical conductors, and that to this is due the weakening action experienced by the mutual attraction or repulsion of two electrically charged points, on interposing a dielectric. Let u denote a fraction, assumed to be very small, of the total volume which is actually occupied by the molecules; then, according to this view, the dielectric constant[4] is calculated to be

$$K = \frac{1+2u}{1-u};$$

from which it follows that

$$u = \frac{K-1}{K+2}.$$

Now according to the electromagnetic theory of light,

$$K = N^2,$$

if N denotes the refractive index for light of very long wave-length; by introducing this, and thus dividing by d, we obtain

$$\frac{u}{d} = \frac{N^2-1}{N^2+2} \cdot \frac{1}{d} = R;$$

[1] *Phil. Trans.*, 1858, p. 8; 1863, p. 523.
[2] *Pogg. Ann.*, **123.** 595 (1864).
[3] Lorentz, *Wied. Ann.*, **9.** 641; Lorentz, *ibid.*, **11.** 70 (1880).
[4] Clausius, *Ges. Abh.* (Separate Papers), II., 135 (1867).

and this expression must be independent of the temperature, the pressure, and of changes in the state of aggregation, because, according to the definition, $\frac{u}{d}$ denotes the real specific volume of the molecule, *i.e.* the volume actually occupied by the molecules of 1 gram of the substance. R *denotes the specific refraction for light of infinitely long wave-length.*

In order to determine the value of N, for want of anything better, use is made of Cauchy's formula for dispersion, viz.—

$$n = A + \frac{B}{\lambda^2} + \frac{C}{\lambda^4} + \ldots;$$

here $\lambda = \infty$, and therefore

$$N = A.$$

This extrapolation will of course give less exact results, the more the light is dispersed by the substance in question; but, as a rule, a satisfactory approximation is obtained by making N equal to the refractive index of red light.

Now, as a matter of fact, experiment leads, in a very remarkable way, to the result that the specific refraction is a very characteristic quantity for the given substance, irrespective of the conditions under which it may be studied; and this holds true, not only when calculated from the so-called "refractive index of light of infinitely long wave-length," but also when calculated from the definite refractive index of any selected kind of light in the visible spectrum. As already shown on p. 108, the rule of mixtures gives better results by calculating with the n^2 formula than with the n formula; and all experiment speaks in favour of the view, viz. that the former decidedly ought to be preferred in studying the connection between the chemical composition and the optical behaviour.

Both expressions are very satisfactory in the relatively small changes of the refractive indices, due to the expansion by heat, as is shown by the following table of values for water from the determinations of Rühlmann, referred to the D-line:—

t	$\frac{n-1}{d}$	$\frac{n^2-1}{n^2+2} \cdot \frac{1}{d}$
0°	0·3338	0·2061
10°	0·3338	0·2061
20°	0·3336	0·2061
90°	0·3321	0·2059
100°	0·3323	0·2061

A comparison of the n^2 expression with the old formula shows that it alone is independent of the state of aggregation, as is illustrated by the following example,[1] in which again the refractive index n is referred to the sodium line.

	$\frac{n-1}{d}$			$\frac{n^2-1}{n^2+2} \cdot \frac{1}{d}$		
	Vapour.	Liquid.	Diff.	Vapour.	Liquid.	Diff.
Water . . .	0·3101	0·3338	- 0·0237	0·2068	0·2061	+0·0007
Carbon-disulphide .	0·4347	0·4977	- 0·0630	0·2898	0·2805	+0·0093
Chloroform . .	0·2694	0·3000	- 0·0306	0·1796	0·1790	+0·0006

The temperature of the liquid was 10°, of the vapour 100°.

The product of the specific refraction R and the molecular weight M is called the *molecular refraction*; thus

$$MR = \frac{n^2 - 1}{n^2 + 2} \cdot \frac{M}{d}.$$

A regularity of conduct is shown in the case of the molecular refraction, which is similar to that in the case of molecular volume; but the influence of the constitution on the optical behaviour is manifested much more decidedly than in the relations between the density and the nature of the substance. *The molecular refraction of a compound can be calculated by the summation of the atomic refractions*; but the atomic refraction is fairly constant, *i.e.* independent of the other elements occurring in the molecule, only in the case of the univalent elements; it varies considerably for oxygen and carbon according to the mode of union.

It is to J. W. Brühl[2] that we are chiefly indebted since 1880, together with Landolt, for the development of the science of the molecular refractions, in that he has worked up and calculated a great number of observations. The following table contains the principal atomic refractions recently recalculated by Brühl. The figures which refer to sodium light are taken from the calculations of Conrady.[3]

[1] Brühl, *Zeitschr. physik. Chem.*, **7.** 4 (1891).
[2] *Zeitschr. physik. Chem.*, **7.** 140 (1891). See also the voluminous investigations of Kannonikoff, *J. pr. Chem.* [2], **31.** 339 (1885).
[3] *Zeitschr. physik. Chem.*, **3.** 210 (1889).

[TABLE

TABLE OF SOME OF THE ATOMIC REFRACTIONS.

	For the Red H-line.	For the Na-line.	For the Blue H-line.	Atomic Dispersion Blue-Red.
Simply bound carbon	2·365	2·501	2·404	0·039
Hydrogen	1·103	1·051	1·139	0·036
Hydroxyl oxygen	1·506	1·521	1·525	0·019
Ether oxygen	1·655	1·683	1·667	0·012
Carbonyl oxygen	2·328	2·287	2·414	0·086
Nitrogen united simply and only to carbon	2·76	...	2·95	0·19
Chlorine	6·014	5·998	6·190	0·176
Bromine	8·863	8·927	9·211	0·348
Iodine	13·808	14·12	14·582	0·774
Ethylene union	1·836	1·707	1·859	0·23
Acetylene union	2·22	...	2·41	0·19

Atomic refractions of a great number of other elements have been determined mostly from their organic compounds by the use of the above values. They have generally been found to show considerable variation according to the nature of the chemical combination into which they enter, so that the atomic refractions have no wide range of applicability for the other elements. The variation increases naturally with the variety of the modes of combination; it is therefore greatest for multivalent elements. On the other hand, this constitutive variability of the atomic refraction, if only its dependence on constitution is thoroughly studied, serves as a valuable help in determining the constitution, as was shown especially for the nitrogen compounds by the extensive investigations of Brühl.[1]

Oxygen itself in some states of combination shows an atomic refraction different to that given above, as appears from the investigations of Nasini, Carrara, Anderlini, and others.[2]

The refractive power of the enol- and keto- forms are markedly different in a number of solvents.[3]

The use of these figures may be most easily explained by means of an example; we will calculate the molecular refraction of benzene, C_6H_6, referred to the figures for the red H-line.

$$\begin{array}{ll} \text{6 carbon atoms} & = 6 \times 2{\cdot}365 = 14{\cdot}190 \\ \text{6 hydrogen atoms} & = 6 \times 1{\cdot}103 = 6{\cdot}618 \\ \text{3 double unions} & = 3 \times 1{\cdot}836 = 5{\cdot}508 \\ \hline \text{Sum} \quad . & \quad MR = 26{\cdot}32 \end{array}$$

[1] *Zeitschr. physik. Chem.*, **16.** 193, 226, 497, 512; **22.** 373; **25.** 577 (1895-1898).
[2] *Gazz. chim. ital.*, **24.** I. (1894); **25.** II. (1895); *Zeitschr. physik. Chem.*, **17.** 539 (1895).
[3] Brühl, *Zeitschr. physik. Chem.*, **30.** 61 (1899).

Observation with the same kind of light and at 20° gave $n = 1{\cdot}4967$, $d = 0{\cdot}8799$, $M = 78$, and therefore

$$MR = \frac{n^2 - 1}{n^2 + 2}\frac{M}{d} = 25{\cdot}93,$$

which coincides fairly well with the value as calculated above.

The molecular refraction calculated for acetone $CO(CH_3)_2$, and with the Na-light, gives 3 carbon atoms + 6 hydrogen atoms + 1 carbonyl oxygen = 16·10; while observation gives 16·09.

Small variations of the expression $\frac{n^2 - 1}{n^2 + 2}$ from proportionality with the density on change of temperature have been found which, however, exceed errors of observation. Eykman[1] suggested from a large and varied collection of observations the expression $\frac{n^2 - 1}{n + 0{\cdot}4}$ as expressing this proportionality more completely, but, as this has no theoretical foundation, it can only be considered as an interpolation formula.

The figures given in the last column of the preceding table (p. 310), and called the atomic dispersions, are the differences of the respective values, referred to the blue and the red hydrogen lines. Brühl finds, by a discussion of the material thus far observed, that the *molecular dispersion* of a compound for a definite substance, liquid or gaseous, is independent of the temperature, and fairly independent of the state of aggregation; this *molecular dispersion*, in a manner analogous to that for the molecular refraction, may be defined by the formula

$$\left(\frac{n_\gamma^2 - 1}{n_\gamma^2 + 2} - \frac{n_\alpha^2 - 1}{n_\alpha^2 + 2}\right)\frac{M}{d},$$

where n_α and n_γ are the respective refractive indices for the H_α and the H_γ lines. Thus the molecular dispersion, like the molecular refraction, becomes of value as a specific manifestation of the material nature and composition of chemical substances.

As far as the available data extend, it is possible in many cases to calculate the molecular dispersion from the sum of the atomic dispersions; yet in this case it appears that the constitutive influence appears more often and in a more decided way than in the case of the molecular refraction. A comparison of atomic refraction and atomic dispersion shows that there is no simple connection between refraction and dispersion. The atomic refraction of carbon is about twice as great as that of hydrogen, but their atomic dispersions are about the same. The atomic refraction of bromine is once and a half as large as that of chlorine; while its atomic dispersion is twice as great. The iodine atom refracts twice as strongly as the chlorine

[1] *Rec. Trav. chim. Pays-Bas*, **14.** 185; **15.** 52 (1895 and 1896).

atom; but it disperses four times as strongly, etc., etc. It is interesting to know that the influence of multiple unions of atoms in the molecule is manifested more clearly in the case of dispersion than in the case of refraction.

An observation of Brühl[1] is very important, to the effect that the atomic refractions of hydrogen and chlorine, as calculated from their behaviour in their compounds, coincide respectively with the powers of refraction of these gases in the free state; the former, according to the table on p. 310, are respectively 1·05 and 6·00 for Na-light; the latter are found by observation to be 1·05 and 5·78 respectively. On the other hand, the atomic refraction of free oxygen amounts to 2·05, which is considerably greater than the value for hydroxyl oxygen, viz. 1·52.

Since the mixture formula on p. 105 usually holds, solids can be investigated and their molecular refraction determined in a suitable solvent (that is, a chemically indifferent one). It is not improbable that, if these substances were studied in dilute solution, more remarkable results might be obtained than by the investigation of pure liquids. For in the latter the degree of polymerisation introduces complications as to which the constitutional formula gives us no indication. To perform such experiments it would be advisable to use an optical differential method which would yield directly the difference of refractive index between solution and solvent.

The following may be said in brief, regarding the *theoretical foundation* of the regularities which obtain in molecular refraction. According to the results given on p. 307, the molecular refraction is a measure of the space actually occupied by the molecules; it had been previously (p. 303) inferred that the molecular volume is made up additively, or at least approximately so, from the volumes of the atoms; and the same must also be true for the molecular refraction, as in fact was found to be the case.

The molecular volumes and the molecular refractions thus appear to be closely related values, because they are both proportional to the space occupied by the molecules.

Now, as a matter of fact, it cannot be denied that there is a certain parallelism between the relations of the molecular volume on the one hand and of the molecular refraction on the other, regarding the molecular structure. Both properties are peculiarly additive. Also the influence of constitution is clearly the same in both cases, in that carbonyl oxygen has a greater atomic volume and a greater atomic refraction than that possessed by hydroxyl oxygen, or by ester oxygen. Also, both the molecular volume and the molecular refraction are increased by the presence of a doubled carbon union.

The attempt to express the molecular refraction of a *gaseous compound* as the sum of the molecular and atomic refractions of its

[1] *Zeitschr. physik. Chem.*, **7.** 1 (1891).

components, employing the n^2 formula, has given the following results :[1]—

Substance.	Molecular Refraction.	
	Observed.	Calculated.
Hydrochloric acid	6·70	$H + Cl =$ 6·83
Carbon dioxide	13·32	$CO + 2Cl =$ 16·59
Hydrocyanic acid	6·63	$H + CN =$ 7·21
Water	3·82	$2H + O =$ 4·14
Ammonia	5·63	$3H + N =$ 5·36
Carbon dioxide	6·71	$CO + O =$ 7·08
Ethyl chloride	16·35	$HCl + C_2H_4 =$ 17·49
Nitric acid	4·46	$N + O =$ 4·25
Nitrus acid	7·58	$2N + O =$ 6·45

The refractive values of the components used were obtained from the direct measurements which Dulong, Mascart, Jamin, and others have made on the free gases, and are as follows :—

H	O	N	Cl	CO	CN	C_2H_4
1·5	2·05	2·20	5·78	5·03	6·16	10·66

The differences between experiment and calculation are less than the errors of observation; at the same time the molecular refraction is not a purely additive property, but constitutive influences make themselves felt. Thus oxygen in the free state has undoubtedly a different atomic refraction than in the carbon compounds, and in the latter there is a difference according as the oxygen has both its valencies saturated by the same or by two different carbon atoms. The atomic refraction of elements such as sulphur or nitrogen is still more variable.

Dielectric Constants.—The mutual electrostatic action of two electrically charged bodies varies with the nature of the medium in which they are placed; if in a vacuum they attract with a force k, then in another medium the force amounts to $\frac{k}{D}$, in which D, the dielectric constant of the medium in question, is always greater than one, though in gases only slightly greater. Electrostatics teaches that the mutual action of two bodies kept at a constant potential difference is proportional to the dielectric constant of the medium, that, further, if c is the capacity of a condenser in a vacuum (or, what is practically the same thing, air), it becomes cD when the condenser is placed in a medium of dielectric constant D. It follows from the theory of electric oscillations that the velocity of propagation of electric waves in wires is inversely proportional to the square root of the dielectric constant of the surrounding insulator.

[1] Brühl, *Zeitschr. physik. Chem.*, **7.** 140 (1891).

From these facts arise a number of experimental methods for determining the dielectric constant; we may mention the following:—

1. *The electrometer method* consists in observing the throw of the needle of a suitably constructed quadrant electrometer, first in air, then in the liquid in question (Silow, 1875). Cohn and Arons[1] used alternating currents instead of continuous currents to charge the electrometer. With this method, slightly conducting liquids may be investigated.

2. *The condenser method* consists in comparing by any suitable method the capacity of condensers, first in air, and then when filled with the liquid to be studied. Measurement by means of a telephone and Wheatstone's bridge is to be recommended; the chief difficulty which arises from want of insulation of the substance under investigation can be eliminated by the use of variable resistances placed in parallel to the condensers to be compared. The method thus becomes suitable for the investigation of slightly conducting liquids, and is simpler and more exact in practice than the electrometer method.[2]

3. *Measurement of the length of stationary electric waves.*—Some arrangement is used for producing stationary electric waves and for measuring their length, which is directly proportional to the velocity of propagation and, therefore, according to what precedes, inversely to the square root of the dielectric constant of the medium. A convenient apparatus has recently been described by P. Drude,[3] and used for numerous experiments. By working with short waves this method allows of studying relatively good conductors.

Dielectric action can be simply explained, as was remarked on p. 307, by assuming conducting particles surrounded by an insulator (luminiferous ether); the greater the volume of these conducting particles the greater the dielectric constant of the medium in question.

The following table contains the dielectric constants of a number of liquids at 18°; the numbers show that the dielectric constant varies extraordinarily from one substance to another, and so is highly suited for characterising the substances:—

Dielectric Constants of some Liquids at 18°

Liquid		Liquid	
Benzene	2·29	Ethyl alcohol	26·1
Xylene	2·35	Propyl alcohol	22
Carbon disulphide	2·58	Isobutyl alcohol	19
Ether	4·35	Amyl alcohol	16
Aniline	7·28	Ortho-nitro-toluene	28
Chloroform	5·0	Nitro-benzene	36
Ethyl chloride	11	Water	80
Ethyl bromide	4·8	Ethyl acetate	6·3
Methyl alcohol	32		

---

[1] *Wied. Ann.*, **33.** 13 (1888). For the description of simple and suitable electrometers see F. Smale, *Wied. Ann.*, **57.** 215 (1896).

[2] Nernst, *Zeitschr. physik. Chem.*, **14.** 622 (1893).

[3] *Zeitschr. physik. Chem.*, **23.** 267 (1897).

If in the expression (p. 307)

$$R = \frac{N^2 - 1}{N^2 + 2} \cdot \frac{1}{d},$$

N^2 be replaced by D,

$$R = \frac{D - 1}{D + 2} \cdot \frac{1}{d}$$

means *the specific refraction for very long waves*, since the measurements of dielectric constants are always made with wave lengths considerably greater than those of visible light. It may be expected, therefore, that the regularities observed in optically measured refraction would become much more obvious if there were used, instead of the optical refractive index N, the corresponding value of the electric refractive index ($= \sqrt{D}$) which is free from the influence of dispersion (that is, refers to very long waves). This anticipation does not turn out to be true, as was shown by the extensive investigations of Landolt and Jahn.[1] The expression for the electric refraction is not even approximately constant for the same substance with varying temperature and state of aggregation, nor can the specific refraction of mixtures be calculated with any certainty from the mixture formula.[2]

The explanation of this behaviour lies in the fact that the refractive index extrapolated from Cauchy's formula for very long waves rarely agrees with the square root of the dielectric constant. To take the crassest example, the refractive index of water extrapolated for long waves is 1·3, the square root of its dielectric constant is 9; in other words, there is a region of very strong and, in fact, anomalous dispersion in the infra-red spectrum. Dispersion phenomena in the infra-red region of the spectrum have so far been but slightly studied, we have, therefore, no means of tracing the complete dispersion curve; for the long heat waves investigated are still very much shorter than the shortest electric waves which have been produced and measured, so that a large and certainly highly interesting region remains inaccessible to experiment; but the recent investigations of Drude (*l.c.*, p. 314) bring out clearly that dispersion is mainly associated with presence of *hydroxyl groups* in the molecule, and that the observed phenomena of anomalous dispersion accompanied by the appearance of absorption bands in the infra-red part of the spectrum is in accordance with Helmholtz's theory.

The circumstance that mixtures of pyridine or betaine with water show much more marked anomalous dispersion than these substances in the pure state is explained by Bredig[3] as due to formation of hydrates in solution after the analogy of $NH_4(OH)$.

[1] *Zeitschr. physik. Chem.*, **10.** 289 (1892).

[2] See also, amongst others, the researches of F. Ratz, *Zeitschr. physik. Chem.*, **19.** 94 (1896); Linebarger, *ibid.*, **20.** 131 (1896); J. Philip, *ibid.*, **24.** 18 (1897).

[3] *Zeitschr. f. Elektrochem.*, **7.** 767 (1901).

From the point of view of the above theory of Clausius and Mossotti these facts may mean that the molecules consist of a well-conducting core, or several such, and a less conducting sheath; then the optical refraction would be the measure of the volume of the well-conducting core, whilst the electric refraction measured for very long waves would give the volume of the core plus the badly conducting sheath. The additive behaviour of the optical molecular refraction would show that the volume of the core is practically equal to the sum of the volumes of its components, whilst the volume of the surrounding sheath would be greatly influenced by constitution. Of course this assumption is, at the present time, very hypothetical, still we may draw certain conclusions on the molecular constitution of matter from the explanation of anomalous dispersion in the region of long waves.

Magnetic Rotation of the Plane of Polarisation.—In 1846 Faraday discovered that transparent substances, when placed in a magnetic field, acquire the property of rotating the plane of polarisation in the direction of the lines of force; and also that the observed rotation is proportional to the thickness of the layer, and to the intensity of the magnetic field. The sense of this rotation is the same for most substances, as for all organic substances for instance; it is in the same direction, as seen by the observer, in which the magnetising current circles; but its magnitude is dependent upon the nature of the particular substance.

The problem of explaining the relation between the degree of rotation and the chemical nature of the substance has been recently (since 1882) attacked in a very thorough way by W. H. Perkin. As a measure of the coefficient of the magnetic rotation of substances, Perkin took the observed angle of rotation in a definite intensity of the magnetic field, divided by the density of the substance and by the rotation angle of a layer of water of the same thickness in the same field; this value he named the *specific rotation;* and the product of this with the molecular weight of the substance divided by the molecular weight of water, he called the *molecular rotation.* Here we meet with relations which are similar to the dependence of the molecular refraction upon the composition and the constitution respectively. The molecular rotation of organic compounds can usually be calculated very approximately, by summation of the *atomic rotations,* the values of which must be suitably chosen; and here also, in the case of the multivalent elements, the atomic rotation varies very distinctly with the mode of union.[1]

Magnetism.—The molecular magnetism, *i.e.* the specific magnetism referred to that of water as unity, and multiplied by the molecular weight, has recently been investigated for a number of organic com-

[1] For further information see Ostwald, *Lehrbuch der allg. Chem.*, 2nd edit., I. p. 499 (1891). [Also see Walker's translation of Ostwald's *Outlines*, p. 105.—Tr.]

pounds by Henrichsen,[1] who used the torsion method of G. Wiedemann. All of the substances studied were *diamagnetic.* As the molecular magnetism could be calculated by summation of suitably chosen *atomic magnetisms*, it proves that this property is also emphatically *additive.* The presence of double carbon-unions in the molecule seems to increase the diamagnetism.

According to recent investigations by G. Jäger and St. Meyer,[2] the atomic magnetism of the paramagnetic elements, nickel, cobalt, iron, manganese in equivalent solutions of their compounds stands in the simple ratio 2 : 4 : 5 : 6, and chromium appears to lie between nickel and cobalt.

It is remarkable that the magnetic susceptibility appears to be independent not merely of the anion, but of the valency of the cation; for solution of ferrous and ferric salts containing equal amounts of iron are equally susceptible.

The Heat of Combustion.—By the heat of combustion of a substance there is meant the quantity of heat produced by the complete oxidation of 1 g. of the substance. It makes a slight difference whether the combustion is conducted at constant *pressure*, or at constant *volume*, the former value being a trifle smaller, viz. by the amount of the heat value of the external work performed by the combustion. In the case of the hydrocarbons, *e.g.*, this difference is usually less than 0·5 of a per cent of the total value. The details of this, as well as the methods of the determination of the heat of combustion, will not be given till the chapter on "Thermochemistry." We will give in this place only a statement of the regular relations which were ascertained by a study of organic compounds.

The *hydrocarbons* in particular have been carefully studied; according to J. Thomsen,[3] the following may be theoretically prefaced regarding these values. The process of combustion may be regarded as occurring in two stages, viz.:—

1. The decomposition of the molecule into the individual atoms.
2. The combination of the individual atoms with oxygen.

The quantity of heat produced in the combustion of a hydrocarbon is equal to the *heat developed* by the union of the isolated atoms with oxygen, minus the *heat absorption* which would be observed in the dissociation of the hydrocarbons into separate carbon atoms and separate hydrogen atoms. It is a matter of indifference whether the process actually occurs in this way; because according to the law of the conservation of energy, the quantity of heat developed must be the same, whatever may be the order in which it occurs.

[1] *Wied. Ann.*, **34.** 180 (1888).
[2] *Ibid.*, **63.** 83 (1897).
[3] *Thermochem. Untersuchungen*, Bd. IV., Liepzig, 1886; *Zeitschr. physik. Chem.*, **1.** 369 (1887).

Now, if we make the assumption that the same quantity of heat absorption S is always required in the separation of one hydrogen, wherever the separation occurs; and that there are required the same amounts of heat, viz. U, V, and W, in the separation of single, double, or triple carbon unions respectively, wherever the separations occur; then in the dissociation of a carbon compound C_aH_{2b}, the amount of heat absorption A_1, is,

$$A_1 = 2bS + xU + yV + zW,$$

where x, y, and z denote respectively the number of single, double, or triple unions.

Now, since there are in the molecule 4a carbon valencies, 2b of which are satisfied by hydrogen, then 4a − 2b carbon valencies must be satisfied *in pairs;* and therefore, and because every single union employs two valencies, every double union four valencies, and every triple union six valencies, it follows that

$$4a - 2b = 2x + 4y + 6z.$$

Therefore

$$x = 2a - b - 2y - 3z,$$

and therefore for the heat absorption A_1, we find

$$A_1 = 2bS + (2a - b - 2y - 3z)U + yV + zW.$$

Now, if P denotes the heat of combustion of an isolated carbon atom, and Q that of an isolated hydrogen atom, then the heat developed A_2 in the combustion of the isolated atoms amounts to

$$A_2 = aP + 2bQ.$$

And thus for the heat of combustion at constant volume, we have

$$A_2 - A_1 = aP + 2bQ - 2bS - (2a - b - 2y - 3z)U - yV - zW.$$

In order to obtain the heat of combustion at constant pressure, it must be observed that the combustion of the gaseous molecule $C\,H_{2b}$ requires $a + \frac{b}{2}$ molecules of oxygen, and that there are produced a molecules of gaseous carbon dioxide, and b molecules of liquid water; there results, therefore, a diminution of the volume of $1 - \frac{b}{2}$ mol, which corresponds to

$$0{\cdot}580 - 0{\cdot}290b \text{ Cal.}$$

Now, if we add these corrections, which are usually very small, to

$A_2 - A_1$, we obtain *as the heat of combustion* $\mathfrak{B}(C_aH_{2b})$ *of the hydrocarbon* (C_aH_{2b}), *at constant pressure,*

$$\mathfrak{B}(C_aH_{2b}') = aA + bB + yC + zD + 0{\cdot}580,$$

where

$$A = P - 2U, \qquad C = 2U - V,$$
$$B = 2Q + U - 2S - 0{\cdot}290, \qquad D = 3U - W.$$

As a matter of fact, this formula allows a very exact calculation of the heat of combustion of the *hydrocarbons of the fat series*, expressed in large calories, if we assume the following values, viz. :—

$$A = 106{\cdot}17, \qquad C = 15{\cdot}465,$$
$$B = 52{\cdot}53, \qquad D = 43{\cdot}922.$$

Thus, for example, by the calculation of the heat of combustion of di-allyl, we find

$$\mathfrak{B}(C_6H_{10}) = 6A + 5B + 2C + 0{\cdot}58 = 931{\cdot}2 \text{ Cal.},$$

while Thomsen determined it experimentally as 932·8 Cal. Of course it is not possible to ascertain the particular values of P, Q, S, U, V, and W from the empirically determined values of A, B, C, and D.

The preceding formula is not applicable to closed chain hydrocarbons, such as trimethylene or benzene. Thus the heat of combustion of the latter substance, C_6H_6, in the gaseous state, on the assumption of nine single unions, according to the preceding formula, would give

$$\begin{aligned} 6A &= 637{\cdot}02 \\ +3B &= 157{\cdot}59 \\ +0{\cdot}580 &= 0{\cdot}58 \\ \hline \text{Sum} &= 795{\cdot}19 \end{aligned}$$

while by assuming three doubled unions in benzene, we would have

$$795{\cdot}19 + 3 \times 15{\cdot}465 = 841{\cdot}585.$$

Both of these values differ too much from that obtained by direct measurement, viz. 787·5, which, however, is quite near the value calculated on the supposition of nine single carbon unions.

It happened by chance that Thomsen, in using his universal burner, obtained heats of combustion which were a little too high, and so his old values for the heat of combustion of gaseous benzene, when corrected, coincide quite well with the value calculated on the assumption of nine single unions. This circumstance, viz. that the values as first determined by him seemed for the time to speak against the Kekulé formula, gave opportunity for much controversy.

Thus it resulted, particularly by the work of Berthelot in Paris, and Stohmann in Leipsic, both of whom worked with the calorimetric bomb, that more extended material gave a broader basis for the theoretical development of the relations between the heat of combustion and the chemical constitution. Then a renewed discussion of the observations led Thomsen[1] to the conclusion *that the benzene union has a heat value different from that of the acetylene union.*

In order to obtain satisfactory material to prove this view, Thomsen then calculated the heat of combustion of a number of aromatic hydrocarbons, which were measured *in the solid state.*

The assumption that the same formula, of course with changed constants, would hold good here, was justified by this circumstance, viz. that the heat of sublimation, which obviously is the excess of the heat of combustion in the gaseous state over that in the solid state, in all cases has a regular connection with the constitution, and that, at all events, *this value is only a very small fraction of the heat of combustion.*

In fact, it was shown that the heat of combustion of benzene, naphthalene, anthracene, phenanthrene, and chrysene, all being taken *in the crystallised state,* can be calculated from the formula

$$\mathfrak{V}(C_aH_{2b}) = 104{\cdot}3b + 49{\cdot}09m + 105{\cdot}47n,$$

where m denotes the number of single hydrocarbon unions, and n the number of double unions. The formula harmonises much better with the results of experiment by ascribing to benzene three double unions, to naphthalene four, to anthracene, phenanthrene, and chrysene each six. It is also possible to calculate the heat of combustion of the phenyl methanes, by ascribing a heat value of 723·7 to the radical $C_6H_5 - C$. The preceding formula is also in harmony with other determinations; thus the heat of combustion of methane in the solid state is calculated to be

$$\mathfrak{V}(CH_4) = 2 \times 104{\cdot}3 = 208{\cdot}6,$$

a result a little smaller than that for gaseous methane (211·9), as it should be.

A comparison of the numerical values obtained by the new formula, with those given on p. 319, which were obtained for gaseous hydrocarbons of the open-chain variety, leads to this conclusion; viz. *that the heat of formation of an ethylene union is considerably smaller than that of the formation of a benzene union.* And this result is entirely in harmony with the well-known fact that the formation of addition products is attended with much greater difficulty in the case of aromatic nuclei than in the case of the olefines; or, in other words, that the double benzene unions are much harder to dislocate than the olefine unions. This difference is all the more remarkable because the compounds have the same optical behaviour (p. 310).

In other organic compounds[2] the heats of combustion show a

[1] *Zeitschr. physik. Chem.*, **7.** 55 (1891).
[2] Stohmann, *Zeitschr. physik. Chem.*, **6.** 334 (1890).

prevailingly additive character. But the unmistakable influence of constitutional differences (for instance, in isomeric acids) may be distinguished with some certainty as the difference in the value of the heats of combustion, and can be compared with the much more remarkable variations of other physical properties, specially the dissociation constant.[1] In the sense of Berthelot's principle (Book IV. Chap. V.) the less stable isomeric form has usually, but not always, the greater content of energy, and consequently the greater heat of combustion; but as, according to Baeyer's[2] stereochemical theory, double and treble bonds between carbon atoms are more unstable the more their valencies vary in direction from the tetrahedric arrangement, *i.e.* the greater the tension inside the molecule, so usually the heat of combustion increases with the magnitude of the tension, as was found by Stohmann[3] in general, and especially for the poly-methylenes $(CH_2)_n$.

The heats of combustion of the hydrogenised benzenes measured by Stohmann[4] are very characteristic of the thermal behaviour of the aromatic compounds. He found the heat of combustion under constant pressure—

		Diff.
Benzene	779·8	
		68·2
Di-hydro-benzene	848·0	
		44·0
Tetra-hydro-benzene	892·0	
		41·2
Hexa-hydro-benzene	933·2	
		58·0
Hexene	991·2	

Similar differences were found for the terephthalic acids and their products.[5] The transformation of benzene into its first reduction product is accompanied by a much greater absorption of heat than the transformation of already hydrogenised products to the next higher product of reduction. The entire thermal behaviour of benzene and its derivatives agrees well with Thiele's assumption (p. 289), according to which the saturation in aromatic compounds is much more complete than was formerly supposed.

Regularities in the Boiling-Points of Organic Compounds.—The problem of calculating the boiling-points of organic compounds from their composition and constitution, and with the same exactness that is possible in the case of the molecular refraction, has not been solved as yet; and, indeed, this problem, which is interesting both in a practical and a theoretical way, presents no little difficulty, for the reason that the boiling-point, doubtless, depends to a large degree

[1] Stohmann and Schmidt, *Zeitschr. physik. Chem.*, **21.** 314 ref.; *Journ. pr. Chem.*, **53.** 345 (1896).

[2] *Ber. deutsch. chem. Ges.*, **18.** 2278 (1885).

[3] *Journ. pr. Chem.*, **45.** 305, 475; **46.** 530 (1892).

[4] *Sitzungsber. der sachs. Akad.*, 1893, 477.

[5] Stohmann and Kleber, *Journ. f. prakt. Chemie*, **43.** 1 (1891).

upon the constitution, and that certainly makes the problem all the more fascinating.

Now, investigation has disclosed a number of instances of regular behaviour[1] which are sufficiently remarkable to deserve a more thorough treatment, although it has not been possible, as yet, to group them from any general standpoint. These regularities, for the most part, consider the *change of boiling-point* experienced by an organic compound from substitution.

The Substitution of CH_3.—In the homologous series of the normal alcohols, acids, esters, normal nitriles, and ketones, the boiling-point increases about 19°-21° regularly from series to series; in the case of the aldehydes it is about 26°-27°. In the glycolic acid esters there is a normal rise of about 20° when the methyl group is in the alcohol residue, as contrasted with a rise of about 10° when it is in the acid residue. Regarding the aromatic compounds, there is a normal difference of 21° in pyridine and its derivatives; in some of the amine bases, as aniline, toluidine, and piperidine, a difference of 10°-11°.

On the other hand, in the case of the entrance of the CH_3 group into the normal hydrocarbons, or further into the benzene nucleus, or into the side chains, there is always a rise in the boiling-point, to be sure, but it shows no decided regularity.

The Substitution of Cl, Br, and I.—The introduction of the first chlorine atom into a methyl group occasions a rise in the boiling-point of about 60°; the actions of the second and third Cl atoms are much weaker, as shown in the example of the chlor-acetic acids; thus

		Diff.
CH_3CO_2H boils at	118°	
		67°
CH_2ClCO_2H ,,	185°	
		19°
$CHCl_2CO_2H$,,	194°	
		3°
CCl_3CO_2H ,,	195°-200°	

The replacement of Cl by Br makes the rise about 24° greater throughout; by I about 50° greater.

The Substitution of H by OH.—This in general occasions a rise of about 100°; the phenoles and the corresponding amines have the same boiling-points; thus here the effect of the substitution of NH_2 and of OH is the same.

A very complete and valuable study on the action of substituting negative radicles has been made by L. Henry,[2] who established that

[1] These have been compiled by W. Marckwald, *Dissertation*, Berlin, 1888; see also Fehling's *Handwörterbuch*, art. "Siedepunkt" ("boiling-point"), (1893), to which also reference should be made for the bibliography; and Nernst and Hesse, *Sieden und Schmelzpunkt*, Braunschweig, 1893.

[2] *Bulletin de l'Académie belgique* [3], **15.** Nos. 1 and 2 (1888).

the accumulation of negative radicals, especially oxygen, at one point of the molecule of an organic compound causes great increase of volatility, that the influence is greatest when the substituting negative radicals are attached to a single carbon atom, whilst the action is still noticeable if the substitution takes place in two carbon atoms directly combined.

When two compounds unite with each other *with elimination of water*, the boiling-point of the resulting product can be roughly calculated by adding the boiling-points of the two components, and then subtracting 100°-120°. Thus

Acetic acid boils at	118°
Ethyl alcohol ,, ...	78·5°
	Sum = 196·5°
Acetic ether boils at	77·0°
	Diff. = 119·5°

A Comparison of Isomeric Compounds shows that in the fat series the normal compound has the highest boiling-point.[1] The more the chain of carbon atoms assumes a "branching" type, or the more "spherical" the molecule becomes, so much the more does the volatility increase. Pentane, C_5H_{12}, may serve to illustrate this—

	$CH_3(CH_2)_3CH_3$;	$(CH_3)_2CHCH_2CH_3$;	$(CH_3)_4C$
B.-P.	38°	30°	9·5°

In general a change of $CH_3(CH_2)_2$ into $(CH_3)_2CH$ corresponds to a fall in boiling-point of about 7°.

Further, in the case of isomers containing oxygen, the boiling-point will be the lower the nearer the oxygen atom approaches the centre of the molecule: a change from a primary to a secondary alcohol corresponds to a fall in the boiling-point of about 19°; thus

$CH_3(CH_2)_2CH_2OH$ boils at . . .	116·8°	Diff. = 17·8°
$CH_3(CH_2)CH(OH)CH_3$,, . . .	99·0°	

Moreover, in the case of the isomeric halogen substitution products, that one will have the lower boiling-point where the chlorine seems to be more central in the chain of atoms; thus

$CH_3CH_2CH_2Cl$ boils at	46·5°
$CH_3CHClCH_3$,, ...	37°

Of the *isomers of the benzene derivatives*, in general the *ortho compounds* boil at a higher temperature than the *meta compounds*, and these in turn about the same as, or higher than, the *para* compounds.

Regarding the influence of *the double union* in hydrocarbons, there is no general rule; but it is found that the hydrocarbons corresponding to the formula C_nH_{2n+2}, and those corresponding to C_nH_{2n}, have the same boiling-point, so that in this case the double union

[1] B. Tollens, *B. B.*, **2.** 83 (1869).

seems to have the same influence as two hydrogen atoms. But in other cases a corresponding change in molecular structure, *e.g.* carbon union instead of two hydrogen atoms, is associated with a decided change in the boiling-point; thus compounds of the type

$$\begin{matrix} C_6H_5 \searrow \\ \qquad R \\ C_6H_5 \nearrow \end{matrix}$$

boil about 40°-41° lower than those of the type

$$\begin{matrix} C_6H_4 \searrow \\ | \qquad R, \\ C_6H_4 \nearrow \end{matrix}$$

where R represents any selected divalent radical, as O, CH_2, S, NH, etc.

When an *acetylene compound* is produced by the more extended elimination of hydrogen, the boiling-point rises: thus the propargyl compounds boil about 19·5° higher than the corresponding propyl compounds.

Increase of molecular weight due to substitution causes a rise in the boiling-point, as was shown by Earp[1] in the replacing of oxygen by sulphur; at the same time another action is distinguishable, viz. that increase of symmetry in the molecule lowers the boiling-point. The two influences may help or oppose one another; the passage of H_2S into CH_3SH both increases the molecular weight and lowers the symmetry so that a change from − 61° to + 21° in the boiling-point is observed. In the passage of CH_3SH into $(CH_3)_2S$ increase of molecular weight causes a rise, gain of symmetry a fall, so that only a relatively small rise from 21° to 41° results. By exception the substitution of methyl in the place of the hydrogen atom of a hydroxyl group lowers the boiling-point; when therefore methyl alcohol is converted into methyl ether, the substitution and production of a symmetrical molecule act in the same sense, and a great fall of the boiling-point is produced (from 67° to − 23°).

The attempt of Vernon[2] to apply regularities of boiling-point to determine molecular condition of a liquid is noticeable as he starts from the very plausible assumption that the deviations which individual substances show from general rules have their ground in polymerisation of the liquid molecules. Thus usually doubling the molecular weight raises the boiling-point about 100° (ethylene boils at − 105, butylene at − 5, octylene at 126, the hydrocarbon $C_{16}H_{32}$ at 274); now in the series hydroiodic acid, hydrobromic acid, hydrochloric acid, the boiling-point falls from − 25, − 73, to − 100. We may therefore expect for hydrofluoric acid about − 120, whereas its boiling-point is actually + 19·4°, that is, 140° too high. Vernon concludes therefore for a molecular formula of hydrofluoric acid lying between H_2F_2 and H_4F_4.

[1] *Phil. Mag.* [5], **35.** 458 (1893). [2] *Chem. News*, **64.** 54 (1891).

Similarly water, by analogy with sulphuretted hydrogen, should boil at $-100°$, as the boiling-point lies $200°$ higher the formula $(H_2O)_4$ is probable. By similar considerations Vernon found the molecular formulæ S_{12}, SO_2, $SOCl_2$, SO_3, $(H_2SO_4)_4$, $(SeO_2)_4$. It is especially noticeable that the hydroxyl compounds, according to these observations, are in general strongly polymerised; for the thio-compounds which contain SH instead of OH boil 30° or more below the corresponding hydroxyl compounds, whereas in general replacing of oxygen by sulphur causes a rise of 40° or 50° in the boiling-point; further, many ethers boil at a lower temperature than the corresponding alcohols (for instance, methyl ether at $-23°$, methyl alcohol at $+66°$), whereas otherwise replacing of oxygen by an alkyl raises the boiling-point (see pp. 269-272).

The Boiling-Point Curves.—It is obvious that the comparison of the boiling-points of the different compounds, which has been summarised in the preceding sketch, can lay no claim to any particular scientific value, because the "boiling-point" so-called represents only one casual point in the boiling-point curve (p. 65), and also because any regularities observed between the boiling-points at atmospheric pressure would vary more or less when compared with points selected at another pressure.

If we denote by T_0' the absolute boiling-point at atmospheric pressure of a liquid which is selected as "normal," and by T_0 its boiling-point at any selected pressure, then, according to Ramsay and Young (p. 64), the following formula expresses the boiling temperature T of another liquid, viz.,

$$T = T_0A[1 + B(T_0 - T_0')].$$

It would, doubtless, be more rational to investigate the question, how the constants A and B depend upon the nature of the substances compared with each other, instead of the simple comparison of the boiling-points of different substances at normal pressure; then instead of isolated boiling-points, one would consider the whole boiling curve.

No matter how well the preceding formula may accord with experiments at vapour pressures which are not too high, it is, nevertheless, purely empirical, and does not include the critical point. The theory of the boiling-point curves and of the vapour pressure curves of van der Waals (p. 227) has a decided advantage in this respect, viz. that it presents in itself a well-rounded development of views which include a consideration of the critical phenomena; and no doubt would exist as to the way to be followed, if it were satisfactorily vindicated by the results of experience, and if the critical data in particular should actually make a more certain calculation of the boiling-point curves possible. But, as already shown, this is not the case (p. 234); for the aforesaid theory is capable of improvement in many respects.

For this reason it would perhaps be profitable if some one should first use the formula, given by Ramsay and Young, for all that it is worth, and push on still further in the direction of the relations between the boiling-point curve and the chemical composition.[1]

The Critical Data.—On the other hand, there can be no doubt that the critical data have great significance for very well-defined chemical substance, and that the problem of tracing out their relationship to the constitution is a very important one. Unfortunately the experimental data thus far accumulated are neither very extensive nor very exact, as there is considerable difference between the results of different observers. For this reason, the results up to date are very small, and we will limit ourselves to the description of a rule discovered independently by Guye and Heilborn.

In keeping with experiment (p. 234), and according to van der Waals, the critical volume, which is determined experimentally with great difficulty, may be made proportional to the critical temperature divided by the critical pressure, *i.e.* proportional to the so-called "*critical coefficient*" k,

$$k = \frac{\theta_0}{\pi_0}.$$

But now since, according to the reasoning on p. 301, the critical molecular volume is an additive property, then the critical coefficient must be so also (Heilborn). On the other hand, the critical coefficient, like the molecular refraction (p. 309), is a measure of the space actually occupied by the molecules; and since the latter is decidedly an additive property, therefore the former is also (Guye).[2]

Now as a matter of fact it is found that it is possible to make

$$k = \frac{\theta_0}{\pi_0} = 1{\cdot}8MR,$$

if the critical pressure is estimated in atmospheres, and the refraction R refers to infinitely long wave-lengths (p. 307). This rule is only a rough approximation, inasmuch as the numerical factor in the preceding equation varies between the limits of 1·6 and 2·2; therefore in calculation one may use the red rays, the refraction of which is only slightly different, in comparison. And conversely, it is possible to calculate the critical coefficient from the atomic refraction given on p. 309, with about the same degree of approximation, *i.e.* about 10-20 per cent.

Heat of Evaporation.—According to p. 60, the molecular weight of evaporation is

[1] An extensive collection of data has been made by G. W. A. Kahlbaum, *Studien über Dampfspannkraftsmessungen*, 1893, Basel, Benno Schwabe.

[2] *Ann. chim. phys.* [6], **21**. 206 (1890); Thesis, Paris, 1892.

$$\lambda = RT^2\frac{d \ln p}{dT} \quad . \quad . \quad . \quad . \quad . \qquad (1)$$

or

$$\frac{\lambda}{T} = R\frac{d \ln p}{d \ln T} \quad . \quad . \quad . \quad . \quad . \qquad (2)$$

According to van der Waals's theory (see p. 227), the differential coefficient on the right must be the same for all substances in corresponding states; but as, according to p. 303, the boiling-point is approximately a corresponding temperature for all substances, it follows that *the quotient of the molecular weight of evaporation by the absolute boiling-point must be approximately constant* (Trouton's law). The following table [1] shows the degree of exactness of the rule.

Substance.	Heat of Evaporation.	T − 273	$\frac{\lambda}{T}$
Benzene	94·4	80·2	20·65
Toluene	86·8	110·8	20·61
m-Xylol	82·8	138·5	21·03
Pyridine	101·4	117	20·62
Water	536·6	100·0	25·64
Alcohol	216·5	78·2	28·09
Acetic acid	89·8	118·5	13·74
Methyl formiate	110·1	31·8	21·45
Ethyl formiate	94·4	54·3	21·13
Methyl acetate	97·0	57·1	21·53
Propyl formiate	90·2	80·9	22·38
Ethyl acetate	88·1	77·15	21·93
Methyl propionate	89·0	79·7	21·99
Propyl acetate	83·2	101·25	22·45
Ethyl propionate	81·8	99·2	22·22
Methyl butyrate	79·7	102·7	21·43
Methyl iso-butyrate	75·0	92·3	20·74

The substances which have normal molecular weight in the liquid state give quotients that vary only a little from the normal value 21; substances which have normal vapour density but are associated as liquids, such as water and alcohol, give noticeably higher values (25·64 and 28·09). Acetic acid finally which forms double molecules in the gaseous state has a small value for the quotient (13·74); but if the heat of evaporation for this substance is reduced to that for normal molecules by the addition of 80 (the heat required for dissociation of the saturated vapour in the normal molecules), we get the quotient 26·28, indicating a considerable degree of association in the liquid state.[2]

[1] According to the measurements of Marshall and Ramsay, *Phil. Mag.* [5], **41**. 38 (1896), and W. Longinine, *Arch. sc. phys. nat.*, Genève, **9**. 5 (1900).

[2] Batschinski (*Zeitschr. phys. Chem.*, **43**. 369, 1903) gives a deduction of Trouton's rule from Clausius's characteristic equation.

The Melting-Point.—In determining the melting-point experimentally, one will best accomplish it by surrounding the thermometer with the finely pulverised substance and heating it to incipient fusion. On account of the latent heat the thermometer remains constant for some time, when it is carefully stirred till all is melted at a fixed temperature, which can be exactly measured, and which is the melting-point. We can in this way obtain very exact measurements, especially by using large quantities of the particular substances, and by fixing the zero-point of the thermometer scale very accurately by observing the melting-point of water according to this method.

Another method, which allows of very accurate determination, and which also makes it possible to use considerably smaller quantities of the substance, say 10-20 g., depends upon the phenomenon of *under-cooling* or *over-melting*. For although a solid substance at its melting-point assumes the liquid form under all circumstances, yet there is usually a marked delay in the reverse process, and it requires some external stimulus to induce the liquid to assume the solid state which corresponds to its temperature. If one brings this about by stirring with a glass rod, or more certainly still by introducing a bit of the substance in the solid state, the freezing of the under-cooled substance occurs; and then, as a result of the latent heat, the temperature rises to the freezing-point. This pause in the fall of the temperature may be determined with great precision and certainty by this method (see also p. 258).

The methods of melting and of freezing with large quantities of the substance give results which coincide to the hundredth part of a degree. These methods, according to a recent publication by Landolt,[1] are the *only* ones which give certain results. Nevertheless, it is the common practice to renounce all chance of extreme accuracy, and either by working with substances which are not completely pure, or with quantities of the substance too small to use with the preceding method; usually the determination is made by introducing the substance into a capillary tube which is fastened to a thermometer bulb, and the two together are dipped into a bath of water, oil, paraffin, or sulphuric acid. The temperature shown by the thermometer at the moment when the substance (which is opaque in the solid state) begins to become transparent is noted as the melting-point. The moment of the change of colour, as a rule, can be sharply noted in a clear bath and with favourable light; but sometimes, when the substance exhibits an appearance of translucency before it really melts, the error of observation may amount to several degrees. In such a case the moment when the melted substance begins to flow down into a deeper part of the tube may be taken as the sign of incipient fusion. This

[1] *Zeitschr. physik. Chem.*, **4.** 349 (1888); see also R. v. Schneider, *ibid.*, **22.** 225 (1897).

point may be made more visible to the eye by a simple apparatus suggested by Piccard.[1]

Mention was made on p. 189 of the regularity in the melting-point of the elements. A rule announced by v. Baeyer[2] for organic compounds has general interest. According to this, the even members of homologous series are clearly different from the odd members—

Series of the Fatty Acids

		M.-P.
Normal $C_2H_4O_2$	. . . . .	+16·7°
$C_3H_6O_2$	. . Does not solidify at	−21°
$C_4H_8O_2$	. . . . .	− 0°
$C_5H_{10}O_2$	. . Does not solidify at	−16°
$C_6H_{12}O_2$	. . . . .	− 2°
$C_7H_{14}O_2$	. . . . .	−10·5°
$C_8H_{16}O_2$	. . . . .	+16°
$C_9H_{18}O_2$	. . . . .	+12°
$C_{10}H_{20}O_2$	. . . . .	+30°
$C_{16}H_{32}O_2$	. . . . .	+62°
$C_{17}H_{34}O_2$	. . . . .	+59·9°
$C_{18}H_{36}O_2$	. . . . .	+69·2°

Succinic Acid Series

Normal Succinic acid	$C_4H_6O_4$	180°
Normal Pyrotartaric acid	$C_5H_8O_4$	115°
Adipic ,,	$C_6H_{10}O_4$	148°
α-Pimelic ,,	$C_7H_{12}O_4$	103°
Suberic ,,	$C_8H_{14}O_4$	140°
Azelaic ,,	$C_9H_{16}O_4$	106°
Sebacic ,,	$C_{10}H_{18}O_4$	127°
Brassylic ,,	$C_{11}H_{20}O_4$	108°

In both series, without exception, the member with an uneven number of carbon atoms has a lower melting-point than the preceding member with one less carbon atom. In the succinic acid series, with increasing molecular weight, the melting-point of the acids having an uneven number of carbon atoms falls, and that of acids having an even number of carbon atoms rises, so that the two sets of numbers together seem to have a common mean variation.

The following may be stated respecting other irregularities :—

Bromine compounds appear to have a higher melting-point than the corresponding chlorine compounds; and the nitro-compounds higher than the corresponding bromine and chlorine compounds.[3]

In the pyrotartaric acid series the melting-point is higher the more the structure of the acid deviates from the normal type, *i.e.* the more side chains are built on to the normal. This may be formulated in the statement that in the pyrotartaric acids the melting-point rises with the number of methyl groups.[4]

[1] *B. B.*, **8.** 687. See also art. "Schmelzpunkt" ("melting-point") in Fehling's *Handwörterbuch* (1890).
[2] *B. B.*, **10.** 1286 (1877).
[3] Petersen, *B. B.*, **7.** 59 (1874).
[4] Markownikoff, *Lieb. Ann.*, **182.** 340.

Of the isomers of the aromatic series, the para derivatives seem to have the highest boiling-points, but there is an exception to this rule in the case of the amido derivatives of the substituted benzene sulphonic acids,[1] and also in the case of the isomers of the toluene sulphonic amides, anilides, and toluidides.

Among the amides of the halogen derivatives of the benzene sulphonic acids, in the para series, the melting-point rises about 20°, when a halogen is substituted by the next heavier halogen. Thus the chlorine compound melts at 143°-144°, the bromine compound at 160°-166°, and the iodine compound at 183°. Also the melting-points of the halides of any particular series appear to rise from the fluorine to the iodine compounds, but not regularly.[2] For a series of further rules see the monograph quoted on p. 322. Another point which is of much consequence for this question must not be overlooked; it gives to melting-points occasionally, perhaps more often than is supposed, a certain arbitrary character. Many solids have the faculty of existing in several modifications, possessing different physical properties, and in particular often very different melting-points (p. 96). This is especially common among organic compounds, and it is not impossible that allotropy (polymorphism) is a general phenomenon, *i.e.* that every solid substance can exist in several modifications corresponding with different conditions of pressure and temperature. If this view is correct, then when comparing melting-points the question must first be settled *what modifications are comparable*, and it is quite possible that many exceptions to the rules of melting-point so far discovered are only apparent, and would be avoided by the discovery of new modifications.

Although in the case of the other physical properties, the theory of their relation to the constitution of organic compounds in particular has been developed much further than in the case of the melting-point, yet, on the other hand, there is no other physical constant so well suited to identify a well-defined chemical substance. The melting-point has a pre-eminent place among all the physical properties, as is shown by the fact of its easy and accurate determination, by its great sensitiveness to foreign adulteration, and also by the fortunate circumstance that these foreign adulterations almost always act in the same way, viz. to lower it (p. 142), and finally, it varies considerably, even with very slight changes in the composition.

Internal Friction.—According to the gas theory, p. 209, the internal friction in the case of *gases* and *vapours* is directly proportional to the path of the molecule, and this latter is inversely proportional to the cross-section of the molecule; therefore by measuring the internal friction of gases, one may obtain a measure of the space occupied by

[1] Beilstein, *Handbuch*, **1.** 60 (1886).
[2] Lenz, *B. B.*, **12.** 582 (1879).

the molecule. As a matter of fact, L. Meyer[1] found a proportionality, which was at least approximate, between the volumes of molecules determined in this way and Kopp's values; and we have already seen that the latter (p. 303) may be regarded as a measure of the space actually occupied by the molecules. It would be very interesting to have a comparison between the values as obtained by internal friction, and those from the molecular refraction, since these latter values, according to p. 312, are also a measure of the volumes of molecules.

The internal friction of liquids has been the subject of much investigation, but without affording any regularities of a general nature.[2] It is noticeable that the peculiar influence on other physical properties of the mode of combination of oxygen appears also in the viscosity.[3]

The Natural Rotation of the Plane of Polarisation.—While the property of being optically active under the influence of magnetism is a universal property (p. 316), only certain substances are able of themselves to rotate the plane of polarisation. The natural rotation, like the magnetic, is proportional to the thickness of the section used, and varies with the temperature, and also with the wave-length of the light used.

The rotation may be measured by the polarisation apparatus or by the polari-strobometer; this in its simplest form, as devised by Mitscherlich, consists of two Nicol's prisms, which, when placed respectively before and behind the stage, produce darkness; the stage is for the reception of the substance. The angle of rotation is read directly from the divided circle of the ocular, or else the rotation is compensated by a quartz wedge which can be interposed to any desired thickness (Soleil). Finally, a number of changes have been proposed in order to obtain still greater accuracy; these have been applied in the so-called "*half-shadow*" apparatus, which has been well received in both scientific and practical work. The fundamental principle of this is that the field of view of the polarising apparatus is filled, not with one bundle of light rays, but with two; their planes of vibration form a certain angle with each other, and by turning the analysing Nicol, the two halves of the field of view are brought to the same degree of brightness.[4]

The natural circular polarisation was discovered in quartz by Arago in 1811, and was afterwards observed in many other crystals.

The natural circular polarisation of liquids, or of amorphous solids, was first observed by Biot in 1815 in sugar solutions, and although it

[1] *Wied. Ann.*, **7.** 497 (1879); **13.** 1 (1881); **16.** 394 (1882). See also Steudel, *ibid.*, **16.** 369 (1882).

[2] Ostwald, *Allg. Chem.*, 2nd edit. **1.** 550 (1891).

[3] Thorpe and Rodger, *Philosophical Transactions*, London, 1894 and 1896; *Zeitschr. physik. Chem.*, **14.** 361; **20.** 621.

[4] For a more detailed description, see Landolt, *Optisches Drehungsvermögen org. Substanzen*, 2nd edit., Braunschweig, 1879.

has been found since then in very many substances, yet these are *all carbon compounds.*

The rotation property of crystallised substances depends apparently upon their molecular arrangement (crystal structure); while that of organic compounds depends upon a constitution which inheres in the molecule. A fact among others, in favour of this latter view, is that Biot (1819) found that those organic substances, which were active in the liquid state, were also capable of rotating the plane of polarisation in the gaseous state; this was more thoroughly studied by Gernez (1864) for the vapour of oil of turpentine. We will consider here only the property of rotation of the carbon compounds.

The property of rotation of organic substances has primarily been of great importance as an aid in analysis. This can be determined not only directly for active substances, as, for instance, sugar in water solution, but also under certain circumstances, as shown by Landolt,[1] indirectly, in that the influence of inactive substances on active substances may be measured by the polari-strobometer. An active substance is characterised by its *specific coefficient of rotation* $[\alpha]$; α denotes the rotation angle for a definite kind of light (sodium light for example), measured at a definite temperature (20° for example), and expressed in degrees of the divided circle; l denotes the length of the section used, expressed in decimeters; c denotes the number of grams contained in 1 c.c. of the liquid used (whether of a solution or a pure substance); then we have

$$[\alpha]\frac{20^\circ}{D} = \frac{\alpha}{lc}.$$

Moreover, the specific coefficient of rotation of the substance in solution in general varies with the nature of the solvent, and also with the concentration; it therefore will be advisable to give a statement of this phase of the rotation property. The product of the specific coefficient of the rotation and the molecular weight is called the *molecular coefficient of rotation.*

The connection with constitution is shown in no other physical property so clearly as it is in this of optical rotation. As already stated on p. 293, if we omit the asymmetrical nitrogen compounds, p. 296, which are as yet only very little known, *only those compounds are active which have one or more unsymmetrical carbon atoms, i.e.* carbon atoms, the four valencies of which, in the sense of structural organic chemistry, are satisfied by four different atoms or radicals. But, on the other hand, the existence of such carbon atoms by no means conversely necessitates optical activity; this is explained by the following fundamental laws regarding the polari-strobometric behaviour of

[1] *B. B.*, **21.** 191 (1888).

organic compounds, which in turn are an immediate consequence of the views developed on p. 293.

1. Only those compounds are active in an amorphous state, whether solid, liquid, or gaseous, which possess one or more unsymmetrical carbon atoms in the molecule.

2. There corresponds to every optically active substance a twin, which polarises light with the same degree of strength but in the opposite direction; if the compound has only one *unsymmetrical* carbon atom, then the two twins have the relative rotations, $+A$ and $-A$. But if several unsymmetrical carbon atoms exist in the molecule, two for instance, then, if we denote by A and B the rotations produced by each of these respective carbon atoms, there will result the following combinations, viz. :—

$$A+B; \quad -A-B;$$
$$A-B; \quad -A+B.$$

The two in the first horizontal row are isomeric twins, since they have the same degree of rotation but in contrasted directions. If the compound has n unsymmetrical carbon atoms, then the number of optical isomers is 2n, two of which are always twins.

3. But, conversely, it is not necessary that all compounds containing asymmetric carbon atoms must be optically active; for these may be—

(*a*) *Compounds having a very small coefficient of rotation*, so that it baffles measurement. The quantitative relations between the strength of the rotation and the nature of the atoms or radicals satisfying the four valencies of the carbon atoms have not been determined as yet; so that nothing certain can be stated at present regarding the absolute degree of the rotation.

(*b*) *Compounds exhibiting internal compensation*; if a molecule contains an even number of unsymmetrical carbon atoms, then it may happen that the particular action is exactly neutralised, and that the molecule as a whole is inactive, in case the carbon atoms are alike in pairs. This is the case with tartaric acid,

$$\begin{array}{l} CHOH—CO_2H \\ | \\ CHOH—CO_2H, \end{array}$$

which has two similar unsymmetrical carbon atoms; the possible isomers are, of course, as follows :—

$+A+A$	$+A-A$	$-A-A$
Dextro tart. acid;	Inactive tart. acid;	Lævo tart. acid.

Of course $+A-A$ and $-A+A$ correspond to identical substances, which are optically inactive in consequence of internal compensation.

(c) *Compounds consisting of an equi-molecular union of the right and left varieties.* Thus this is the case with racemic acid, which is an equi-molecular union of right and left tartaric acid.

Such mixtures as the latter are obtained in the *synthetic production* of compounds with unsymmetrical carbon atoms from inactive substances. This follows necessarily from the fact, viz. that in the synthesis of such a compound and in the replacement of one c in the compound $Cabe_2$ by d, on account of the complete likeness in value between the places occupied by the two c valencies, two kinds of substitution can always occur.

The Decomposition of a Mixture of Optical Isomers.[1]—It remains as a problem for the experimenter to separate the equi-molecular mixture of the dextro and laevo compound. There is no particular difficulty in the synthetical preparation of such a mixture above the other syntheses of organic chemistry; but certain special methods of treatment are requisite for the separation. The customary methods applicable to the separation of physical mixtures, and which depend upon incidental differences in the melting-points, the vapour pressures, or the solubilities, all fail here completely, because, according to p. 294, both components have the same melting-point, the same vapour pressures, and the same solubility. There are three methods applicable in this case, for all of which we are indebted to Pasteur. A short sketch of these follows.

1. By allowing an active substance, + B, to react on the mixture, the components of which are designated as + A and − A, we obtain thereby a mixture of B + A and B − A. Now, these last compounds are no longer twins, but will in general exhibit greater or smaller differences in melting-point, vapour pressure, and solubility, therefore they can be separated by the usual methods. By decomposing the separated compounds, the isomers + A and − A are obtained in the pure state, and the desired separation of the original inactive mixture is accomplished.

In this way Pasteur used cinchonicine to neutralise a solution of racemic acid, which is an equi-molecular mixture of dextro and laevo tartaric acids; the respective salts were separated from each other by crystallisation, and then by decomposing these cinchonicine salts the dextro and lævo tartaric acids were obtained in the free state, thus the velocity of reaction in the saponification of an optically active alcohol by a racemic mixture of an active acid is noticeably different for the two components of the mixture.[2]

[1] For details see Van't Hoff, *Lagerung der Atome im Raume*, Braunschweig, 1894, and especially in the extensive work of Landolt referred to on p. 331.

[2] Marckwald and McKenzie, *Ber. deutsch. chem. Ges.*, **32.** 2130 (1899).

2. Certain organisms affect the two isomers differently. Thus Pasteur observed that the vegetation of *penicillum* in a dilute solution of ammonium racemate destroyed the dextro acid, while the lævo acid remained.

Since the organisms in question, or the enzymes contained in them, possess themselves an asymmetrical structure, this method is essentially the same as the first.

3. Sometimes the dextro and lævo compounds can be separated from each other in contrasted hemihedral crystals, which can be picked out from each other. Pasteur observed this in the crystallisation of sodium-ammonium racemate. But usually the two isomeric salts crystallise out as a compound. This is the case with the tartaric acids which, in an equi-molecular solution, unite in crystallising out as racemic acid. Of course in such a case the method is debarred.

To produce an optically active compound from its antipodes the following method can sometimes be used. On rise of temperature an optically active compound is commonly converted into the racemic mixture, which then, by any suitable process of decomposition, will give both the dextro and lævo compound

Quantitative Relations of the Rotatory Power].—The fact that optical activity is conditioned by the presence of an unsymmetrical carbon atom suggests another question, viz. What are the relations between the nature of the radicals which saturate the four valencies of the unsymmetrical carbon atom on the one hand, and the degree and direction of the rotation on the other hand? Although not very much is certainly known at present, yet a few general remarks will be given regarding the kind of relation which should be looked for here.

We will denote by e the numerical value of the property of a radical which is influential in occasioning the polarisation; then the power of rotation $[\alpha]$ will be determined by the combination of the values e_1, e_2, e_3, and e_4, which this property assumes for each one of the four respective radicals of the unsymmetrical carbon atoms; and the mathematical expression which may be calculated as the power of rotation from these values must obviously fulfil the following conditions, viz. :—

1. It must become equal to zero when two or more of the four values of e are the same, for in the latter case the asymmetry, and, therefore, of course, the rotation, will disappear.

2. It must remain unchanged in degree, but in the opposite direction, when two values of e exchange places with each other. Such an exchange means nothing more than that a dextro-isomer is changed into a lævo-isomer, or the reverse.

By a few trials it is found that an expression of the following form is suited to the conditions outlined above, viz. :—

$$(e_1 - e_2)(e_1 - e_3)(e_1 - e_4)(e_2 - e_3)(e_2 - e_4)(e_3 - e_4),$$

or

$$\ln\frac{e_1}{e_2}\ln\frac{e_1}{e_3}\ln\frac{e_1}{e_4}\ln\frac{e_2}{e_3}\ln\frac{e_2}{e_4}\ln\frac{e_3}{e_4}.$$

Of course many such expressions as the preceding may be found, but these two are the simplest. They are, moreover, fundamentally identical, for it is only necessary to use log e in the second expression, *i.e.* to regard this property as a measure of the power of rotation, in order to obtain the first expression. It is also probable that there exists a property of the atom or radical, which is related to the molecular rotatory power according to the following law, viz. :—

$$M[\alpha] = (e_1 - e_2)(e_1 - e_3)(e_1 - e_4)(e_2 - e_3)(e_2 - e_4)(e_3 - e_4).$$

Of course it cannot be determined on *a priori* grounds, whether or not the value of e may depend both on the nature of the particular radical and also on the nature of the other three radicals which constitute the unsymmetrical carbon atom group, *i.e.* whether or not there exists a mutual influence between the radicals. If such were the case the problem would be much complicated, because then the calculation would have to take account of the mutual influence of each of the values of e, though probably this mutual influence would not be very great.

It was suspected by Guye[1] that the product of the mass of the radical, by its distance from the central point [centre of gravity], of the unsymmetrical carbon atom, or, since the latter is approximately constant, that the *mass of the radical* is a measure of the degree of the rotatory power. The molecular rotation accordingly would be

$$M[\alpha] = (m_1 - m_2)(m_1 - m_3)(m_1 - m_4)(m_2 - m_3)(m_2 - m_4)(m_3 - m_4),$$

where the values of m are either proportional to the weights of the four radicals, or at least are greater the greater the weights become. But this has not been confirmed by experience.

The Absorption of Light.—If monochromatic light, *i.e.* light having a definite wave-length, falls at right angles upon an absorbent layer which has the thickness d, then a part of the light will be used in warming the substance permeated, and thus will be lost by absorption. The fundamental law of absorption states that the intensity of the emerging light J′, is proportional to that of the immerging light J, and that there exists the relation

$$J' = J(1 - \gamma)^d;$$

here γ denotes a numerical factor which is characteristic of the absorbing

[1] *C. R.*, **110.** 744 (1890); thoroughly in a Thesis, Paris (1891); see also the work of Landolt quoted on p. 331.

substance; this *absorption coefficient* γ varies with the wave-length of light. In order to detect this variation, it is best to illuminate the absorbing substance with white light, and then decompose the emerging rays by the aid of a spectral apparatus (see also p. 193).

Regarding the construction of the spectral apparatus, the student is referred to the handbooks of physics.[1] It should be noted that the apparatus usually made for laboratory use depends upon the dispersive power of transparent substances, glass in particular, and therefore *give refraction spectra*; it is only quite recently that scientific investigation has commonly begun to use spectra produced by *diffraction gratings*. Inasmuch as this latter method (reflection gratings) gives regular and strong dispersion, and is also entirely free from the absorption phenomena of glass, it now commonly finds almost universal application, as it thus has a great advantage, especially in studying the infra-red and the ultra-violet parts of the spectrum. *Photography* is employed to study the ultra-violet rays, and it is also possible to render the ultra-violet rays visible by *fluorescence*. In the study of the infra-red part of the spectrum, use is commonly made of heat effects caused by these rays on the thermopile, bolometer, radiometer, etc.

The following kinds of absorption are recognised:—

1. Absorption which steadily increases or decreases with the wave-length of light: *uni-lateral absorption*. As a rule the absorption increases with decreasing wave-length, *i.e.* the violet end of the spectrum is absorbed more strongly than the red end.

2. Absorption which exhibits a minimum in the spectrum, with regular increase on both sides of this minimum: *bi-lateral absorption*.

3. Absorption exhibiting shaded maxima; the spectrum appears to be crossed by *absorption bands*.

4. Absorption exhibiting sharply defined maxima; the spectrum appears to be crossed by *absorption lines*.

This latter variety, which is altogether the most characteristic, is met in incandescent gases.

Inasmuch as the absorption capacity of substances varies with the nature of the substance to such an extraordinary degree and in such a variety of ways, and beyond all other properties, therefore both light absorption and light emission would appear to be primarily suited to give us some information regarding molecular structure. But the results thus far obtained in this line do not at all justify the anticipations. We know of innumerable examples where very slight changes in molecular structure are associated with deep-seated changes in the capacity for absorption. But we cannot infer a single generalisation regarding the laws according to which this happens. Thus, for example, if we break up the unstable variety of nitrogen peroxide, N_2O_4, by raising the temperature into the simpler molecule NO_2, we

[1] See also the monographs on spectral analysis by Kayser, Berlin, 1883, new edition being prepared, and by H. W. Vogel, Berlin, 1886, and the literature quoted on p. 194.

obtain a very slight change in the chemical behaviour;[1] but the gas changes from a slight yellow to a dark reddish-brown; this gas, when interposed in the path of a bundle of light rays, gives a spectrum crossed by innumerable absorption lines. Also the absorption may vary strongly with unchanged molecular weights. Thus iodine, when dissolved in carbon disulphide, gives a violet colour; in ether, a reddish-brown; yet in both cases the iodine was shown, by the boiling-point method, to have the molecule I_2. Of course it is not impossible that the different colours are occasioned by reaction with the solvent, *i.e.* by forming compounds consisting of $I_2 + n$ molecule of the solvent. Moreover, the cases are not rare where, as in this case, the absorption varies more or less strongly with the nature of the solvent. Usually, but not always, the absorption bands advance towards the red, according as the refractive power of the solvent increases.

Up to the present the most thorough study of absorption in solution has been done in the case of organic compounds; here there were found certain remarkable regularities regarding the influence of substitution on the position of the absorption bands.[2]

According to this, the introduction of hydroxyl, methyl, hydroxymethyl, carboxyl, phenyl, and the halogens occasioned a displacement towards the red; the introduction of the nitro and amido groups, and also the addition of hydrogen, occasioned a displacement towards the violet; these rules hold good almost without exception. For reasons which will be given in the following paragraph, the former group is designated as *batho-chromic*, and the latter as *hypso-chromic*. In the case of all chemically related groups, the displacements increase with increasing *molecular weight* of the radical introduced: thus it is greater in the case of iodine than of bromine, and again greater in the case of phenyl than of methyl. Of the solvents used, concentrated sulphuric acid gave the sharpest colour reactions.

The Theory of Colouring Substances.—Inasmuch as the colour of a substance depends, as is well known, upon its selective absorption, therefore, as has been shown by M. Schütze,[3] the regularities observed regarding the absorption of light give opportunity for some interesting observations upon organic colouring matters. The mixture of all the colours of the sun's spectrum appears to be white, because for every colour there is an equally strong *complementary* colour; and these two, when combined, produce in the eye the sensation of whiteness. Therefore, when any colour of the spectrum is extinguished by absorp-

[1] This must be taken *cum grano salis*; moreover, the polymerisation of NO_2 into N_2O_4 is attended with considerable evolution of heat.—Tr.

[2] G. Krüss and Oeconomides, *B. B.*, **16.** 2051 (1883); Krüss, *B. B.*, **18.** 1426 (1885); *Zeitschr. physik. Chem.*, **2.** 312 (1888); E. Koch, *Wied. Ann.*, **32.** 167 (1887); E. Vogel, *ibid.*, **43.** 449 (1891); M. Schütze, *Zeitschr. physik. Chem.*, **9.** 109 (1892); G. Grebe, *ibid.*, **10.** 674 (1892); Hartley and Dobbie, *Trans. Chem. Soc.*, 1899, p. 640.

[3] *Zeitschr. physik. Chem.*, **9.** 109 (1892).

tion, the substance appears to have the colour of the complementary emerging rays. The paired complementary colours are as follows :—

Violet—Greenish-yellow.
Indigo—Yellow.
Cyanide blue—Orange.
Bluish-green—Red.
Green—Purple.

Many colourless substances produce absorption bands in the violet. These bands are displaced towards the less refrangible [red] end of the spectrum by the introduction of batho-chrome groups ; by the entrance of bands into the visible part of the spectrum, *violet* will be first absorbed, when the substance will have a *greenish* hue. As the bands retreat regularly back towards the red, the colour of the substance will pass successively into yellow, orange, red, purple, and finally back up into violet, when the absorption bands pass from green into greenish-yellow. Continued displacement in the same direction will cause the appearance of the colours, indigo, cyanide blue, bluish-green, and when the bands pass into the infra-red spectrum, the body will again be colourless.

This series of absorption colours, which is the simplest possible, is very rarely seen, because usually, before the one set of bands has passed down through the spectrum, another set appears in the violet end of the spectrum, and of course there occur complications between these two sets.

The introduction of "hypso-chromic" groups works in a way contrary to that of the batho-chromic groups ; the former produce an elevation of the "tone colour," the latter a depression, whence their respective names. Inasmuch as the hypso-chrome groups are of exceptional occurrence, the following law may be stated, viz. :—*the simplest colouring substances are in the greenish-yellow and yellow, and with increasing molecular weight the colour passes into orange, red, violet, blue, and green.* In fact, this law was empirically discovered in 1879 by Nietzki ; it is by no means perfect, because disturbances arise from various sources ; thus the very groups introduced to increase the molecular weight may have a specific hypso-chrome action in displacing the bands towards the violet ; or else complications may arise from the existence of several bands in the region of the visible spectrum.

One remark of Schütze is very interesting, viz. that in the case of analogous elements an increase of the atomic weight is attended by a deepening of the colour ; a beautiful illustration of this is found in the series including colourless fluorine,[1] yellowish-green chlorine, reddish bromine, and violet iodine.

Now experience shows that the colour of many organic colouring substances is due to the presence of certain specific groups in the

[1] Fluorine has been recently found by Moissan to have a light greenish-yellow colour similar to, but lighter than, that of chlorine.—TR.

molecule; thus the colour of the azo derivatives is due to the presence of the azo group; moreover, the supposition is an immediate one, that these same groups are responsible for the absorption of light in the molecule, and that the change of colour is due to the effect of the influence exerted by the group which is introduced on the group producing the colour.

O. N. Witt,[1] who first developed this view, called those groups which produce the colours *chromophores*. Inasmuch as their influence becomes greater the nearer spatially the group introduced is placed to the chromophore, this theory suggests an interesting method for determining the value of the influence of relative distance of the groups in the molecule, by means of spectroscopic investigation. In fact, Schütze, in his work to which repeated reference has been made, by the spectroscopic study of a number of azo colouring substances, showed that the distances of the atoms as given by the structure formulæ, on the whole, corresponded to the distances evaluated in the way mentioned above.

Interesting direct applications of the above theory are to be found in Wallach,[2] who recognised the combinations

$$C=CH.CO \quad \text{and} \quad C=CHCOCH=C$$

as chromophores.

Fluorescence.—According to the investigations of C. Liebermann and R. Meyer, the somewhat singular property of fluorescence is associated with the existence of certain atomic groups in the molecule, the *fluorophores*; such groups are chiefly rings of six members, mostly heterocyclic, such as pyrine, azine, oxazine, and thioazine rings, and further atomic rings contained in anthracine and acridine. As in the case of colour the nature of the surrounding atomic groups influences the action of the fluorophores.

Crystal Form.—The connection between crystal form and molecular structure has not been very thoroughly studied as yet, and the results thus far obtained are limited to some rules given by Groth[3] regarding the change in the ratio of the axes occasioned by the substitution of certain radicals. This phenomenon in mineralogy is called *morphotropy*.

It has been already stated that optical isomerism is shown in the solid state by the contrasted right and left hemihedral (or tetartohedral) crystal forms (pp. 80 and 334). This is doubtless due, like the optical rotatory power, to an asymmetric structure of the molecule. But such a development of planes may also occur when the structure of the molecule is not unsymmetrical; in such cases it is highly probable that the molecules unite in an unsymmetrical way to build up the crystal.

[1] *B. B.*, **9.** 522 (1876).

[2] *Gött. Nachr.*, 1896, Heft 4.

[3] *Pogg. Ann.*, **141.** 31 (1876); see also Hintze, *Zeitschr. Kryst.*, **12.** 165 (1887); and Retgers, *Zeitschr. physik. Chem.*, **6.** 193 (1890).

Hemihedrism may, like the rotary power of crystallised substances (p. 332), be produced in two entirely different ways, one of which, the chemical, refers to the method of *the arrangement of the atoms in the molecule;* the other, the physical, to *the arrangement of the molecules in the crystal.* Both conditions allow of spatial isomerism; but by passing into the liquid or gaseous state, the latter [physical] variety of isomerism is destroyed, while the former [chemical] persists.

Retgers gives some interesting statistics with regard to a law proposed by Buys-Ballot in 1846 that the simple *chemical* compounds take for preference a regular or hexagonal shape, that is, crystallise in the *simpler forms.*[1]

Of 40 monatomic substances (elements) there are—

	%
regular	50
quadratic	5
hexagonal	35
rhombic	5
monoclinic	5
triclinic	0

Thus: regular and hexagonal 85 %, all other systems 15 %.

Of 67 diatomic substances there are—

	%
regular	68·5
quadratic	4·5
hexagonal	19·5
rhombic	3·0
monoclinic	4·5
triclinic	0·0

Thus: regular and hexagonal 88 %, all other systems 12 %.

Of 63 triatomic substances there are—

	%
regular	42·0
quadratic	19·0
hexagonal	11·0
rhombic	23·5
monoclinic	3·0
triclinic	0·0

Thus: regular and hexagonal 53 %, all other systems 47 %.

Of 20 tetratomic substances there are—

	%
regular	5
quadratic	5
hexagonal	35
rhombic	50
monoclinic	5
triclinic	0

Thus: regular and hexagonal 40 %, all other systems 60 %.

Of 50 pentatomic substances there are—

	%
regular	12
quadratic	
hexagonal	3
rhombic	3
monoclinic	
triclinic	6

Thus: regular and hexagonal 50 %, all other systems 50 %.

Of 673 polyatomic inorganic compounds there are—

	%
regular	5·8
quadratic	7·0
hexagonal	14·6
rhombic	27·3
monoclinic	37·3
triclinic	8·0

Thus: regular and hexagonal 20·4 %, all other systems 79·6 %.

Of 585 organic substances there are—

	%		%
regular	2·5	rhombic	33·0
quadratic	5·0	monoclinic	47·5
hexagonal	4·0	triclinic	7·0

Thus: regular and hexagonal 6·5 %, all other systems 93·5 %.

[1] *Zeitschr. physik. Chem.*, **6.** 193 (1890).

The Systemisation of the Physical Properties.—We will give in the concluding paragraph of this chapter only a summary of the rather heterogeneous material found here; for unfortunately space does not allow more than a short sketch of a subject which is broad and which deserves generous treatment.

A large number of physical properties have been shown to be clearly additive, *i.e.* the value of the property in question can be calculated as though the compound were such a mixture of its elements that they experience no change in their properties; thus *we can calculate the properties of a compound from the properties of its components in exactly the same way as was done for many properties of the physical mixture* (p. 105). This, as already mentioned in a preceding chapter (p. 178), is more clearly shown by the specific heat of solid salts than is the rule elsewhere; but there are many properties of organic compounds which are more or less clearly additive, as the volume, refraction, magnetism, heat of combustion, etc. Several of these have this in common, viz. *that their numerical values are a measure of the space actually occupied by the molecule.* Thus we infer this [additional] law from these various sources, viz. *that these magnitudes may very often be calculated as though they were approximately equal to the sum of the volumes actually occupied by the atoms.*

In some cases, conversely, the numerical values deduced for the properties of the elements from their compounds coincide with the values actually shown by the elements in the free state; thus this is the case for the specific heat of solid elements for the atomic volume and the atomic refraction of chlorine, but not for the atomic refraction of oxygen. We have seen previously, in the case of isomorphic mixtures (p. 123), that in some cases the specific volume of a salt in a mixed crystal was the same as in the free state, but in other cases very different.

Now the properties of compounds are no more *strictly* additive than those of physical mixtures. For with compounds the deviations from simple additive relations are, as a rule, much more emphatic. This is not to be wondered at; for the mutual influence of the properties is much smaller in a *simple mixture of the molecules,* whereby there results a physical mixture, than when the atoms unite to form a *chemical compound.*

The kind of influence of the atom in a compound is primarily dependent upon the mode of its union, *i.e.* upon the constitution and configuration of the compound. Those properties which are *clearly* traceable to mutual influence of this sort, and which generally require very exact measurements, are called "*constitutive,*" following the example of Ostwald,[1] who has done such great service in the systemisation of the physical properties. An excellent example of a strongly constitutive property is the absorption of light. Similar examples are

[1] Ostwald, *All. Chem.*, 2nd edit., 1. 1121 (1891).

found in the optical activity, the melting-point, etc. Moreover, our knowledge as to how far the influence of constitution is shown in certain cases varies with the different properties, but in general it is developed to only a slight degree.

Moreover, it does not seem to the writer to be certainly established that the deviations from simple additive relations are to be ascribed to the influence of constitution alone. It is quite possible that the *molecule as a whole* may exert an influence upon these deviations, as is evidenced by many experiments with physical mixtures; for, in addition to the forces which are active between the atoms, there are also forces active between the molecules, but the action of the latter may be eliminated by studying substances in the gaseous state.

A third species of properties depends, neither upon the nature of the atom in the molecule, nor upon the mode of the union of the atoms, but only upon the sum total of the molecule. Properties of this kind, which are called "*colligative*" by Ostwald, we have considered in the chapter on the Determination of the Molecular Weight. Upon these properties are based the methods for ascertaining the relative weights of molecules.[1]

[1] Additive properties give no foothold in determining the molecular weights, because, as shown in the specific heats of solid salts, they are not affected by a change in the size of the molecule; but constitutive properties sometimes do render accessory service in determining the molecular weights.

CHAPTER VI

THE DISSOCIATION OF GASES

Abnormal Vapour Densities.—In a preceding chapter (p. 247) we have considered the methods which enable us to determine the molecular weights of gaseous and of dissolved substances; in this and the two following chapters a discussion will be given regarding some conclusions which are necessitated by the experimental results obtained by the methods mentioned above. These conclusions deal with the *molecular condition of gaseous and of dissolved substances.*

In some cases, which, to be sure, are not very numerous, one at once finds values for the molecular weights which are in glaring contradiction to those chemical formulæ which are probable from all analogies. Thus the vapour of ammonium chloride has a density about one-half as large as it should have if calculated from the formula NH_4Cl; that of ammonium carbamate has a value about one-third of what it should be if calculated from the formula $NH_2CO_2NH_4$; acetic acid, on the other hand, at lower temperatures, has a density considerably greater than that corresponding to the formula CH_3CO_2H.

The inference was close at hand, in spite of their unexpected behaviour, that Avogadro's rule still holds good, and that the *abnormal vapour densities* were to be explained by *abnormal molecular conditions.* Thus, at almost the same time, Cannizzaro (1857), Kopp (1858), and Kekulé (1858) expressed the opinion that the low vapour densities must be ascribed to a more or less complete decomposition; in fact, on the basis of this, Kopp succeeded in showing in many cases that the decrease of vapour density followed in accordance with the increase in the number of molecules resulting from the decomposition. Thus ammonium chloride has about only a little more than half its theoretically normal vapour density, because it is almost entirely decomposed into $NH_3 + HCl$; and similarly ammonium carbamate only about a third of its vapour density, because it is decomposed into $2NH_3 + CO_2$, etc. Similarly, on the basis of this view, it must be concluded that acetic acid [at first] is partially polymerised, and that in addition to the simple molecules, more complicated ones occur in greater quantity.

The scientific world were perfectly conscious of the importance of this question. It was attacked by the most various means, and, in short, by the collection of most of the abundant experimental material, it was decided, most convincingly and undeniably, that the rule of Avogadro holds good even in the case of gases exhibiting abnormal vapour densities. Thus, in detail, it was not only shown that the gaseous decomposition products of the substances in question could be recognised by their special physical and chemical behaviour, but also that these ingredients could be at least partially separated from each other into the free state by means of diffusion.

Dissociation.—The more or less complete dissociation of a molecule into its components or *dissociation products,* we will call *dissociation,* following the usage of St. Claire Deville,[1] to whom we are greatly indebted for his investigations regarding the definition of those chemical processes which are of fundamental importance in these phenomena.

The thorough treatment of dissociation phenomena belongs to the department of the so-called "reversible" chemical reactions, and the theoretical derivation of the laws of dissociation forms a special case, and a very simple one, of the universal laws of reaction. These will be considered under the subjects of the *doctrine of affinity,* and also of *thermochemistry.* Yet, nevertheless, in order to understand what follows, it is necessary to introduce some remarks on the dissociation of gases.

The Extent of Dissociation.—Dissociation is a chemical reaction, and consists in separating a complex molecule into simpler ingredients, thus leading to a *continual increase* in the number of molecules. The pressure exerted on the walls of a vessel by a definite quantity of a gas, other things being equal, will be greater the greater the number of new molecular species produced from the original molecules, and the further the decomposition of the latter proceeds; and also the density of the original gas must diminish in the same proportion, if it is not maintained at constant volume, but at constant pressure.

Let us denote by δ the vapour density observed when no dissociation occurs, and by Δ that actually observed; then Δ must always be smaller than δ. Let the number of molecules into which the original molecules become decomposed be n; thus in the dissociation of iodine vapour,

$$I_2 = I + I.$$

[1] "Sur la dissociation ou la décomposition spontanée des corps sous l'influence de la chaleur." *Compt. rend.*, **45.** 857 (1857). The term "thermolysis," suggested by Fr. Mohr, has not come into extensive use.

Here n = 2; and in the case of ammonium carbamate, viz.,

$$NH_4CO \, . \, ONH_2 = 2NH_3 + CO_2,$$

n = 3. If the decomposition were complete, Δ would be the nth part of δ. In general Δ lies between δ and δ/n.

Let the dissociated part of the gas mass, or the extent of the dissociation amount to α; then the undissociated part is $1 - \alpha$; *i.e.* of 100 molecules, 100α are dissociated and produce $100n\alpha$ new molecules, while $100(1 - \alpha)$ molecules remain undecomposed. Thereby the number of molecules is increased by dissociation, from

$$100 \text{ to } 100n\alpha + 100(1 - \alpha) = 100[1 + (n - 1)\alpha],$$

and the vapour density diminishes in the same proportion, viz.,

$$\frac{1}{1 + (n - 1)\alpha} = \frac{\Delta}{\delta}.$$

The extent of the dissociation α is calculated, therefore, to be

$$\alpha = \frac{\delta - \Delta}{(n - 1)\Delta}.$$

When $\Delta = \delta/n$, then the extent of the dissociation $\alpha = 1$, and we have complete dissociation; but when $\Delta = \delta$, then $\alpha = 0$, and no dissociation has occurred.

According to experiment, the extent of the dissociation varies with the temperature and the pressure. It increases with increasing temperature, and diminishes with increasing pressure. This is expressed by the statement that by lowering the temperature and increasing the pressure, the values of the vapour densities Δ and δ are brought nearer together.

The Physical Conduct of Dissociated Gases—Effusion.—If a gas which exists in a state of dissociation is allowed to escape (effuse) through a narrow opening into a vacuum, or into a space filled with an indifferent gas, then the velocity of the effusion will decrease as the density or the molecular weight of the gas increases. Moreover, there occurs a partial separation of the products of dissociation, as the effusate will contain an excess of the lighter molecules, and the residue an excess of the heavier molecules.

On this phenomenon is based the method of research of Pebal[1] in proving that ammonium chloride owes its diminished density to an extended dissociation into ammonia and hydrochloric acid. This method may be used in demonstration at the lecture desk, in the following simplified form given by Strauss.[2] In a combustion tube of about

[1] *Lieb. Ann.*, **123.** 199 (1862).
[2] Exner's *Répert. d. Phys.*, **21.** 501 (1884).

10-12 mm. diameter, near the middle there is fastened an asbestos plug about 5 mm. thick. A piece of ammonium chloride is then placed at one side of the plug in the middle of the tube, and, in addition, a piece of litmus paper in each division of the tube. A blue piece is placed on the same side of the asbestos plug with the ammonium chloride, and a red piece on the other side. On heating the piece of ammonium chloride by a Bunsen burner, there occurs a change of colour in the inner ends of the litmus paper in both parts of the tube. This proves that the effusate reacts alkaline, while the residue reacts acid. The experiment should be interrupted in season, as the reaction is reversed after a little.

Colour.—The physical properties of a mixture of any two gases are intermediate between those of the particular components, while the properties of a compound of two gases exhibit in many respects a deep-seated change, as shown for example in the absorption of light. Now, dissociation must cause the physical properties of gases to approach those of their components, and thus it is usually possible to draw some inference from the change in the physical properties of gases relative to the progress of the dissociation. In this way Deville observed that the colourless vapour of phosphorus pentachloride became distinctly green at higher temperatures. This is explained by a dissociation in the sense of the following equation, viz. :—

$$PCl_5 = PCl_3 + Cl_2,$$

which is likewise in harmony with the vapour density determinations.

An experiment which can be easily carried out consists in heating a flask filled with nitrogen dioxide. This will become almost colourless by cooling, and then, on gentle heating, it will appear to be filled with a dark-brown vapour, which is again decolorised by cooling. In this case Salet[1] succeeded in showing, in a quantitative way, that the colour varied with the change in vapour density, and that it was entirely explained on the supposition that nitrogen dioxide consists of a mixture of a colourless compound N_2O_4, and a brownish-red gas NO_2, and that the latter molecular species increases with increasing temperature at the expense of the former.

Specific Heat.—The specific heat of a gas, existing in a state of dissociation, is abnormally great. This is due to the fact that the heat introduced is used not only in raising the temperature, but also in effecting dissociation, which latter is attended with considerable absorption of heat. We are indebted to Berthelot and Ogier for measurements of this sort. These measurements were made with nitrogen dioxide[2] and acetic acid vapour.[3] The molecular heat of

[1] *Compt. rend.*, **67.** 488 (1868).
[2] *Ibid.*, **94.** 916 ; *Bull. Soc. chim.*, **37.** 434 (1882).
[3] *Bull. Soc. chim.*, **38.** 60 (1882).

nitrogen dioxide, referred to $NO_2 = 46$ g. and at constant pressure at 0°, is about 95·1°; at 100° it is only about 39·1° greater, while at 157° it falls to 7·1°. At this latter temperature, where the dissociation is almost complete, we find a value of the molecular heat which would naturally be expected from the figures on p. 46.

Gaseous acetic acid conducts itself in a similar way. In this case the investigators mentioned above found the following values for the molecular heat C_p at the respective temperatures:—

$t° =$ 129°	160°	200°	240°	280°
$C_p =$ 90·1	76·2	57·0	38·2	28·5 g.-cal.

Inasmuch as the degree of dissociation varies not only with the temperature, but also with the pressure, it is to be expected that the specific heat of those gases, which are already in a state of dissociation, would vary considerably when compared with those gases which either are not at all dissociated, or only partly so, according to the pressure.

Thermal Conductivity.—The kinetic theory allows of an interesting application of the thermal conductivity of gases in a state of dissociation. As we have already seen (p. 208), heat is carried through an ideal gas in the direction of the fall of temperature by equalisation of the mean kinetic energy of the molecules; in gases in a state of dissociation a new factor enters. At the higher temperature the dissociation is more considerable than at the lower; the result is that when, in consequence of the irregular heat movements, undissociated molecules pass into the hotter part of the gas, they partly decompose there; conversely, when the products of dissociation come into the colder parts they partially recombine to form non-dissociated gas. But since dissociation is accompanied by absorption of heat and reformation of undissociated molecules by generation of heat, the above described process furthers the equalisation of temperature between the two parts of the gas, that is, it *considerably increases the thermal conductivity.*

Actually Magnanini and Malagnini found that the thermal[1] conductivity of nitrogen dioxide in the dissociating state is more than three times as great as when the gas is completely dissociated. This appears to be a remarkable confirmation of the kinetic conception of matter.

The quantity of heat transported in the form of heat of dissociation in a partially dissociated gas can be calculated if the amount of

[1] *Nuovo Cim.*, **6.** 352 (1897); later Magnanini and Zunino, *Mem. Acad. Modena* (3), **2.** (1899). R. Goldschmidt has shown (*Sur les rapports entre la dissociation et la conductibilité thermique des gaz*; Thèse, Brüssel, 1902) that, in determining the thermal conductivity of gases, we possess a means of detecting dissociation at very high temperatures where measurement is hardly possible otherwise.

dissociated and non-dissociated molecules which pass through an area in the gas at right angles to the temperature gradient be known; the quantity of heat in question is then proportional to the excess of undissociated molecules which travel in a unit of time against the temperature gradient over the quantity that travels in the opposite direction.

The Condition of Dissociation.—The results given in this chapter show that the state of an ideal gas is realised only when there is no dissociation at all, or else when it is complete. In the first case we have a simple gas; in the latter case a gas mixture, the properties of which are a make-up of the properties of the components. The gas laws cease to hold good only in the case of *partly dissociated* gases, for here changes in the temperature and the pressure are associated with changes in the molecular condition, and the gases vary enormously in their behaviour, not only as regards their compressibility and their expansion by heat, but also as regards all their other physical properties.

This passage from the normal behaviour of a simple gas to another normal behaviour of a gas mixture is called passing to the *state of dissociation.* The laws of the state of dissociation will be considered in the section devoted to the *doctrine of affinity.* Here we will only refer to the results of pure experiment. These show that those vapour densities which seemed to contradict Avogadro's rule most decidedly are associated with very remarkable deviations in all their physical behaviour, and these deviations which at first seemed to contradict Avogadro's rule really argue strongly in its favour.

CHAPTER VII

ELECTROLYTIC DISSOCIATION

Dissociation in Solutions.—As already stated on p. 268, the abnormal vapour densities are completely analogous to the abnormal values of the osmotic pressure of dilute solutions observed in many cases. It is an immediate inference to explain these abnormal osmotic pressures, in the sense of van't Hoff's generalised form of Avogadro's rule, in a similar way, viz. as being due to abnormal molecular conditions. This assumption almost amounts to a certainty from this fact, viz. that almost all the instances of dissociation which have been met with thus far in the case of gases are also observed if the particular gas is investigated, in a suitable solvent, at about the same temperature. *Very often the abnormal vapour density of a substance is also found to correspond to an abnormal value of its osmotic pressure.*

Let us suppose a molecule, existing in solution, to decompose into n new and smaller molecular species. These may consist of one or of several atoms, and may be similar or dissimilar. Let t_0 denote the lowering of the freezing-point, or a value which is proportional to the osmotic pressure, as, for example, the lowering of the vapour pressure, the raising of the boiling-point, etc. Let t_0 be such that it can be calculated from the molecular proportions of the substance in question, on the assumption that the dissolved substance is not at all dissociated; and in accordance with this, let nt_0 be the lowering of the freezing-point corresponding to complete dissociation. Then the value of t, as observed, will lie between these limiting values. An increase in the number of molecules, in the proportion of $1 : 1 + (n - 1)\alpha$, will correspond to the degree of dissociation α. Therefore it follows that

$$1 + (n - 1)\alpha = \frac{t}{t_0};$$

and therefore

$$\alpha = \frac{t - t_0}{(n - 1)t_0}.$$

The numerous experimental investigations, for which we are

particularly indebted not only to Raoult, but also to Beckmann[1] and Eykmann,[2] have led to this result; viz. that it is usual to find substances in solution not existing as simple molecules, but in a state of dissociation. Thus the organic acids in benzene solution, as in the gaseous state, consist of double molecules, but these separate into single ones on further dilution. Chloral hydrate in acetic acid, as in the gaseous state, decomposes partially into chloral and water, as is shown from the following table:—

m	t	α
0·266	0·095	0·52
1·179	0·385	0·38
2·447	0·755	0·31
4·900	1·450	0·25

Here m denotes the number of grams of chloral hydrate dissolved in 100 g. of acetic acid. The lowering of the freezing-point t_0, which would be observed in the normal behaviour of chloral hydrate, is calculated from the formula given on p. 151, viz.—

$$t_0 = 39\frac{m}{165·5}.$$

Here 39 denotes the molecular depression of acetic acid, and 165·5 the molecular weight of the undissociated chloral hydrate. The degree of dissociation, given in the third column of the preceding table, is calculated from the formula

$$\alpha = \frac{t - t_0}{t_0}.$$

It is obvious that, with increasing concentration, this becomes smaller; and that, conversely, with decreasing concentration, it converges towards a value of one. At a high degree of concentration the behaviour of the solution would approach that of an ideal one, provided that certain discrepancies did not occur, independently of dissociation, at high degrees of concentration. At a high degree of dilution there would be a mixture of the two substances, water and chloral, which would perfectly obey the laws of ideal solutions, while the state [or process] of dissociation would be intermediate [between these extremes].

[1] *Zeitschr. physik. Chem.*, **2**. 715 (1888); **6**. 437 (1890).
[2] *Ibid.*, **4**. 497 (1889).

Water Solutions.—Reference has already been made to the fact that abnormal values of the osmotic pressure are very common in the ease of water solutions. All the methods of measuring the osmotic pressure lead, with quantitative coincidence, to this result; viz. that it is much greater for *acids*, *bases*, and *salts* in water solution than it should be if calculated from the molecular weights of the respective substances, either in the gaseous state or in other solvents than water. If we assume that the Avogadro-van't Hoff law holds good here, we are forced to the unavoidable conclusion that these substances, when dissolved in water, are in a peculiar molecular condition, and *must be dissociated to a greater or less degree.*

But one of the first difficulties here consists in answering this question, viz., What are the products of the dissociation? Such a substance as hydrochloric acid, for instance, undoubtedly has, as a gas, a molecular weight corresponding to the formula HCl. But inasmuch as the lowering of the freezing-point, in water solution, is almost twice as great as that which should correspond to that molecular formula, the original molecule, when dissolved, must be separated into two new ones, *i.e.* HCl must be dissociated into H and Cl. Both of these dissociation products are such that we cannot recognise them in any other way. At ordinary temperatures hydrogen and chlorine are known to us as H_2 and Cl_2 respectively: if we boil a water solution of hydrochloric acid, we obtain HCl to be sure, but no free hydrogen or free chlorine. So an assumption like that mentioned above appears, at first glance, very improbable; and there would be some difficulty in understanding it if the same assumption were not shown to be necessarily true from an entirely different aspect, and if we did not have, at the same time, more detailed explanations regarding the nature of the dissociation products of salts, acids, and bases, in water solution.

In the case of salts one would at once think of a separation into the acid and the base, and thus explain their great osmotic pressure: in fact, we will consider later this so-called "*hydrolytic dissociation*"; but it is not possible to consider it such a common phenomenon as the abnormal pressures are, because salts can be separated, by diffusion, into the respective acids and bases in only a very few cases. For, disregarding for the moment many other reasons which contradict such a method of explanation, and to which we will return later, this assumption fails utterly in the case of acids and bases, which, as well as salts, show abnormal osmotic pressures. Therefore we are forced to seek another explanation.

Now it happens that those substances, and only those substances, which can conduct the galvanic current in water solution, i.e. the electrolytes, are the ones which exhibit osmotic pressures very much greater than those which are calculated for their concentration and their molecular weight in the gaseous state; and, moreover, if the same substances, when dissolved in other solvents, are unable to conduct electricity in any marked degree, then they also lose their abnormal behaviour. Thus we are forced to the view that, if dissociation actually takes place, it is intimately connected with the conduction of electricity by electrolytes.

Electrolytic Conduction.—Let us try to frame a mental image of the conduction of the electric current. It is a well-known fact that, in contrast to the metallic conduction of electricity, it is associated [in electrolytes] with a transportation of matter; and, moreover, that the passage of the galvanic stream from the metallic electrode into the solution is, according to circumstances, either associated with the solution of the metal, or else with the separation of the substance in the solution and its deposition on the electrode.

Thus if we introduce gaseous hydrochloric acid between the platinum electrodes of a galvanic battery, no appreciable transference of electricity occurs; neither is there any transference of electricity if we introduce *very pure water* between the poles; but the electricity passes readily if the poles are introduced into water containing hydrochloric acid. Thus, it is fair to suppose that the hydrochloric acid, which is dissolved in water, is in a molecular condition *different* from that in the gaseous state; because in the one case it can conduct electricity, but not in the other. The following picture is both clear and simple in attempting to account for the definite conception entertained.

When a current passes through the solution, free chlorine is separated at the anode, where the current enters the solution; and free hydrogen at the cathode, where it leaves the solution: one component of the electrolyte "wanders" in one direction, the other in the other direction. This is explained most simply by supposing that the electrolytes consist of different parts which are oppositely polarised, *i.e.* of molecules charged respectively as electro-positive and electro-negative. These parts, after the usage introduced into science by Faraday, are called "*ions.*" According to this view, in the solution, the galvanic current consists in the passage of the positively charged ions, the "*cations,*" in one direction, and in the passage of the negatively charged ions, the "*anions,*" in the other direction. The passage of positive electricity from the [one] electrode *into* the solution is accordingly associated with a separation of anions; and the passage of positive electricity *out* of the solution into the [other] electrode is associated with a separation of cations.

The capacity of dissolved substances to conduct the electric current thus assumes, as a proviso, a polarised cleavage, a decomposition into positively and negatively charged molecules, which we will call "*electrolytic dissociation.*"[1] Of course it is not necessary for the

[1] The expression "*ionisation*" sometimes used should rather be reserved for the formation of ions in gases produced by the action of Röntgen rays and the like (see

decomposition to be complete, for in addition to the electrolytically dissociated molecules, there may also exist in the solution undecomposed molecules, which are electrically neutral.

Only the former [*i.e.* the dissociated molecules] are instrumental in current conduction, and it is apparent, other things being equal, that a solution will conduct electricity more readily in proportion as its resistance is less, *i.e.* the larger the fraction of the electrolytically dissociated molecules in the solution. Cane sugar, when dissolved in water, does not conduct electricity noticeably; therefore it must consist entirely, or at least almost entirely, of uncleaved molecules which are electrically neutral. On the other hand, hydrochloric acid is a good conductor of electricity; and therefore the gas which has

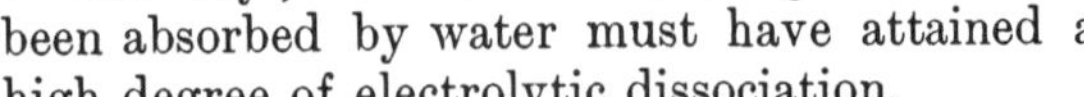

been absorbed by water must have attained a high degree of electrolytic dissociation.

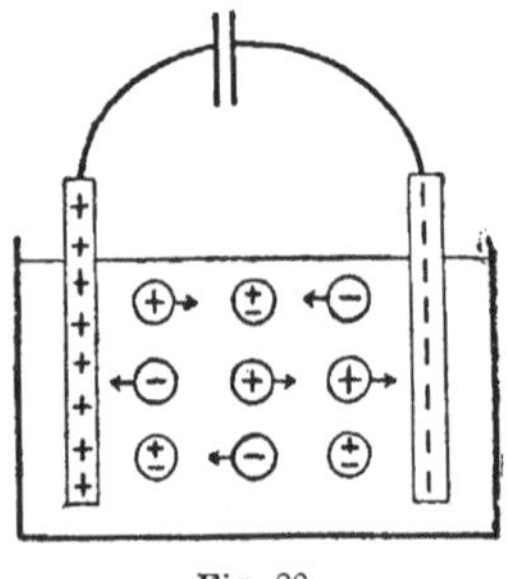

Fig. 28.

On the basis of these preceding remarks, we will now attempt to draw a picture of the mechanism of electrolytic conduction. In order to conform to the preceding supposition, let there be a water solution of hydrochloric acid between two platinum plates (Fig. 28). These two platinum electrodes are in connection with a source of electricity, as for example, with a galvanic battery. The first result of this is that the platinum plate, which is connected with the positively conducting pole, becomes charged with positive electricity, and the other becomes charged with negative electricity.

As a result of these charges, the free electricities of the electrodes will exert an electrostatic attraction or repulsion on the free ions in the solution, which are charged with free electricity; the positive ions will be attracted by the negative electricity of the cathode, and will be repelled by the positive electricity of the anode; thus a force will be developed, acting in the direction of the arrows. The reverse of this will happen in the case of the negative ions; and thus there will be developed a force in the opposite direction. No force of this sort will be exerted upon the electrically neutral molecules.

As a result of this electrical attraction and repulsion, there results a displacement of the free ions in the solution, whereby the positive ions wander from the anode to the cathode, and the negative ions in

Chap. IX. of this book). [Referring to Professor Nernst's criticism of *ionisation*, we would suggest that Faraday used *ion* before Arrhenius made his discovery; that in all deference to both of these great discoverers, *ionisation* is more concise than the synonymous phrase "electrolytic-dissociation"; that in English, *ion* naturally suggests the words *ionise*, *ionic*, *ionisation*, *ionised*, *ionisable*, etc., etc.; and finally, that it is neither for author nor translator to dictate, but to wait for the common-sense use of intelligent users to decide which expression is preferable. It is possible that both terms may be convenient.—Tr.]

the opposite direction. *This wandering [or "migration"] of the ions represents to us that which we call an electric current in an electrolyte.*

The Free Ions.—It now remains to determine the material nature of the ions and of the electrically neutral molecule. In the electrolysis of hydrochloric acid, free hydrogen is separated at the cathode and free chlorine at the anode: therefore the former [the hydrogen] must be positively charged, the latter [the chlorine] negatively. Thus we arrive at the result that hydrochloric acid, when dissolved in water, must be dissociated into positively charged hydrogen ions and negatively charged chlorine ions. Of course we must regard the undissociated hydrochloric acid molecules as being electrically neutral. A similar inference must of course be drawn for all other electrolytes.

We may learn something about the molecular dimensions of the ions, and about the degree of electrolytic dissociation, by the aid of principles developed previously.

The osmotic pressure in a dilute solution of hydrochloric acid, as shown by the quantitative coincidence of results obtained in different ways, is almost twice as great as that corresponding to its gaseous molecular weight; therefore two new molecules must result from the solution of one in water. This means nothing else than that the electrolytic dissociation is *almost complete;* thus—

$$HCl = \overset{+}{H} + \overset{-}{Cl}.$$

The ions of hydrochloric acid are thus monatomic.

Now since the two kinds of electricity are produced in the equivalent quantities by this cleavage, then the positive charge of the hydrogen must be exactly as great as the negative charge of the chlorine, *i.e. an atom of hydrogen and an atom of chlorine are electrically equivalent.* An important conclusion follows from this. If a definite quantity of electricity passes through our electrolytic cell, then electrically equivalent quantities of the ions must separate out on the two electrodes; for otherwise there would occur an enormous accumulation of free electricity in the circuit, which is impossible. Thus hydrogen and chlorine must be set free, at the electrodes, in equivalent electrical quantities, or what amounts to the same thing, *in equivalent chemical quantities.*

Now if we put into the same electric circuit a cell filled with hydrobromic acid, then on the cathodes of the two electrolytic cells the same quantities of hydrogen will be set free; and on the respective anodes there will be set free equivalent quantities of chlorine and bromine. And in general it may be said that *chemically equivalent quantities of the most various ions are set free from the most various solutions by the same quantity of electric current.*

As is well known, this statement is most completely substantiated by experience; it is simply the fundamental law of electrolysis, advanced by Faraday and demonstrated most exactly by experiment.

Those ions which are charged with the same equivalent quantities of electricity as the hydrogen ion, the chlorine ion, etc., are called *univalent;* those ions which are charged respectively with a twofold, or a threefold, etc., quantity of electricity are called *divalent, trivalent,* etc.

Thus sulphuric acid, for instance, in dilute solution, exhibits an osmotic pressure which is *three times* as great as that which should correspond to its concentration, and to the molecular formula H_2SO_4: this means simply that, under these circumstances, three new molecules have been produced from one; and that an almost complete dissociation has occurred, in the sense of the following equation, viz.:—

$$H_2SO_4 = \overset{+}{H} + \overset{+}{H} + \overset{=}{SO_4}.$$

The negative ion, SO_4, must be as heavily charged as the two positive ions together, *i.e.* with twice as great a quantity of electricity as a hydrogen ion; this is indicated by a double dash ($\overset{=}{SO_4}$).

Accordingly we call the SO_4 a *divalent* ion, and sulphuric acid a *di-basic* acid; and in general we call an acid which produces n hydrogen atoms, by complete dissociation, *n-basic.*

The radical hydroxyl (hydroxyl ion) OH, which has a negative charge, is a univalent ion; it is produced by the electrolytic dissociation of water, something which happens to only a very slight degree under ordinary circumstances, thus—

$$H_2O = \overset{+}{H} + \overset{-}{OH};$$

and also in the dissociation of bases, as sodium hydroxide, *e.g.* thus—

$$NaOH = \overset{+}{Na} + \overset{-}{OH}.$$

Those bases, the molecules of which, like the preceding, give only one hydroxyl ion on ionisation as above are called *monacid;* those bases which, like barium hydroxide, produce two hydroxyl ions, are called *di-acid,* etc., thus

$$Ba(OH)_2 = \overset{++}{Ba} + \overset{-}{OH} + \overset{-}{OH}.$$

It is very remarkable that similar ions may have very different charges, even in the same solution; thus the iron ion, produced by the dissociation of ferrous chloride, is *divalent,* thus—

$$FeCl_2 = \overset{++}{Fe} + \overset{-}{Cl} + \overset{-}{Cl},$$

while the iron ion, produced by the dissociation of ferric chloride, is *trivalent*, thus—

$$FeCl_3 = \overset{+++}{Fe} + \overset{-}{Cl} + \overset{-}{Cl} + \overset{-}{Cl}.$$

Thus we see that the *electric value of an element is by no means a constant property*, but may vary by leaps in certain cases. It is possible that a thorough investigation of this change in the electrical value of certain elements might prove the starting-point for a deeper insight into the reason for the change in valence, and into the nature of valence.[1]

In the manner just described, it is usually possible to answer the question regarding the nature and value of the ions quite certainly, by combining the results of the determination of the molecular weight and of the conductivity.

Moreover, the *chemical behaviour* of solutions usually yields important information on this point (see Chap. IV. of the third book).

The Determination of the Degree of Electrolytic Dissociation.—It is not only important to answer the question regarding the nature of the ions, but also another question, viz. *how far the decomposition of the electrically neutral molecules in the solution has proceeded.* Therefore a knowledge of the degree of dissociation of an electrolyte is of great significance, inasmuch as many other properties, besides the conductivity and the osmotic pressure, depend upon the extent of the electrolytic dissociation; and especially is this true of the *share of the dissolved substances in chemical reactions:* a thorough reference will be made to this point under the subject of the *doctrine of affinity.*

The degree of the dissociation, or the value of the coefficient of dissociation, by which is meant the ratio of the dissociated molecules to the whole number of molecules, can be directly obtained in two independent ways: viz. from the osmotic pressure and from the conductivity.

The first method is of course treated in exactly the same way as the determination of the ordinary (*i.e.* the non-electrolytic) dissociation in solution (p. 350). Thus if P_0 denotes the osmotic pressure, as calculated from the gas laws without reference to dissociation, and if P denotes that actually observed, then it will follow that

$$1 + (n - 1)\alpha = \frac{P}{P_0}, \quad \text{and } \alpha = \frac{P - P_0}{(n - 1)P_0},$$

where α denotes the degree of electrolytic dissociation. Instead of the ratio of the respective osmotic pressures, we may, of course, use the relative depressions of the freezing-point, or of the vapour pressure.

[1] In this connection, see the paragraph on "The Changeableness of Chemical Valence" (p. 280).—TR.

Another way [to find the value α] is given by the determination of the electrical conductivity; this can be easily and exactly done by the method proposed by F. Kohlrausch.[1] A full description of this method, so commonly used in the laboratory, need hardly be given here. It need only be mentioned that this method depends upon the use of Wheatstone's bridge; and that instead of a constant current, an alternating current produced by an induction coil, is used,—whereby the disturbing effect of polarisation is eliminated—and that instead of a galvanometer, as a current indicator, use is made of a telephone which is sensitive to the alternating current.

As unity of conductivity is now adopted, that of a body of which a column 1 cm. long and 1 cm.^{2} in cross-section has a resistance of 1 ohm. If, therefore, a body of this size possesses a resistance w, it has a conductivity κ :—

$$\kappa = \frac{1}{\mathrm{w}}.$$

The conductivity of an electrolyte divided by its concentration (= number of g-equivalents in cm.^{3}) is called the molecular conductivity Λ—

$$\Lambda = \frac{\kappa}{\eta}.$$

Formerly the conductivity of mercury at 0° was adopted as the unit of conductivity; in the above measure it is 10,630, hence the conductivity in terms of mercury

$$\mathrm{k} = \frac{\kappa}{10{,}630}.$$

Similarly the molecular conductivity λ was formerly defined as

$$\lambda = \frac{\mathrm{k}}{\mathrm{c}},$$

where $\mathrm{c} = 1000\eta$, that is, expresses the normality.

Now, according to the statements on p. 354, the conductivity is proportional to the number of free ions, *i.e.* it is proportional to the product $\alpha\eta$, and therefore it is directly proportional to the degree of dissociation, α itself. That is, we may put

$$\Lambda = \mathrm{K}\alpha,$$

while K is a proportional factor. For very great dilution the electrolytic dissociation becomes complete, that is, $\alpha = 1$; if the molecular

[1] This method is found in F. Kohlrausch, *Lehrb. d. prakt. Physik.*, 9 Aufl., S. 409; see further Ostwald, *Physiko-chem. Messungen*, Leipzig, 1893, p. 265; and especially Kohlrausch and Holborn, *Leitvermögen der Elektrolyte*, Leipzig, 1898.

conductivity measured for sufficiently great dilution be written $\Lambda\infty$, we simply get

$$\Lambda\infty = K \quad \text{and } \alpha = \frac{\Lambda}{\Lambda\infty}.$$

The following table gives the molecular conductivities obtained by Kohlrausch [1] for various dilutions of *potassium chloride* at 18° :—

KCl.

$c=1000\eta$	Λ	α	$c=1000\eta$	Λ	α
1·0	98·2	0·748	0·005	124·6	0·950
0·5	102·3	0·780	0·001	127·6	0·973
0·1	111·9	0·853	0·0005	128·3	0·978
0·03	118·3	0·902	0·0001	129·5	0·987
0·01	122·5	0·934	$\frac{1}{\infty}$	131·2	1·000

$\Lambda\infty = 131{\cdot}2$.

It has been already remarked that the change of the conductivity with the concentration from 0·1 normal downwards is very nearly the same for all salts made up by combination of univalent radicals, *i.e. that in equivalent solutions these salts experience the same high dissociation.* Therefore the value of α in the preceding table can be used to calculate the degree of dissociation of almost all these salts, *e.g.* NaCl, $LiNO_3$, $CH_3CO_2NH_4$, etc.

The combination of these two methods for the determination of α gives

$$\alpha = \frac{A}{\Lambda\infty} = \frac{P - P_0}{(n-1)P_0}.$$

This was first clearly developed by Arrhenius [2] in his work on the hypothesis of electrolytic dissociation; this law unites the abnormal values of the osmotic pressure of electrolytes with the changes of conductivity with varying concentration. This law has been well established not only by the preliminary calculation of the available material by Arrhenius, but also by its subsequent more exact proof.[3] The following table will serve as an illustration; here are given the values of the factor $1 + (n-1)\alpha$, *i.e.* the ratio in which the number of molecules is increased by dissociation; under column I. are given the results according to the plasmolytic method (p. 134); under II. the results according to the freezing-point method; and under III. the results according to the method of conductivity.

[1] *Wied. Ann.*, **26.** 161 (1885). [2] *Zeitschr. physik. Chem.*, **1.** 631 (1887).

[3] Arrhenius, *Zeitschr. physik. Chem.*, **2.** 491 (1888); Van't Hoff and Reicher. *ibid.*, **3.** 198 (1889).

Substance.	n	c	1+(n−1)α		
			I.	II.	III.
Cane sugar . .	0	0·3	1·00	1·08	1·00
CH_3CO_2H . .	2	0·33	...	1·04	1·01
KCl . . .	2	0·14	1·81	1·93	1·86
LiCl . . .	2	0·13	1·92	1·94	1·84
$MgSO_4$ . . .	2	0·38	1·25	1·20	1·35
CaN_2O_6. . .	3	0·18	2·48	2·47	2·46
$SrCl_2$ . . .	3	0·18	2·69	2·52	2·51
K_4FeCy_6 . .	5	0·356	3·09	...	3·07

Hittorf's Transportation Values and Kohlrausch's Law of the Independent Migration (Wandering) of Ions.—On turning back to consider the mechanism of electric conduction, the question at once arises as to the velocity with which the ions are transported through the solution, at a given difference of potential between the electrodes, and with given dimensions of the electrolytic cell. The magnitude of the force acting on the ions, and depending upon the charges of the electrodes, other things being equal, is of course equally great for all univalent ions; and, moreover, the pull exerted upon the positive ions in the direction of the current is just exactly as strong as the pull exerted upon the negative ions in the opposite direction; and of course the force which drives n-valent ions is n-fold as great.

But the *friction resistance* of the different ions will vary with their varying nature. It may be predicted, with a large degree of probability, that this friction resistance will be very great; for, if we notice how slowly a fine precipitate in water sinks to the bottom, and that it requires more time the finer the precipitate is—then it follows that such extraordinarily fine particles as the ions are would require an enormous force to transport them through the solution with any noticeable velocity. *On this assumption, which will be justified later, we may regard the ions as massive points with very great friction* (p. 13), *and therefore set their velocity proportional to the forces acting on them.* But as the intensity of current is proportional to the velocity at which the ions travel in the solvent, it is proportional to the electromotive force, that is, *Ohm's Law* holds.

It appears from section 7, p. 14, that force and velocity are not proportional at the first moment, but on account of the very large friction this initial stage is not accessible to experimental proof (see Cohn).[1]

The term *resistance friction* (*Reibungs-widerstand*) will be understood as meaning that force—expressed *e.g.* in kilogram weights—which will

[1] *Wied. Ann.*, **38.** 217 (1889).

be required to drive 1 g.-ion, *i.e.* the molecular weight of the ion expressed in grams, against the solvent, with a velocity of 1 cm. per sec.; and the term *mobility* (*Beweglichkeit*) or *velocity of transport* will mean the reciprocal of this force, *i.e.* the velocity with which 1 g.-ion will be transported under the influence of a pull of 1, *e.g.* 1 kilogram weight.

Now, for the sake of simplicity, let us consider a binary electrolyte, *i.e.* one composed of two univalent ions, and let us suppose that the quantity of electricity, E, is sent through the electrolytic cell. Then the transference of ions occurs so that the positive ones are carried "in the direction of the current" through the solvent, and the negative ones in the opposite direction. If we take a cross-section through the electrolytic cell, at any selected place, at right angles to the current, then the electricity conveyed by the positive ions in the direction of the current, plus that conveyed by the negative atoms in the opposite direction, is equal to E.

Now if we denote the mobility of the positive and negative ions by U and V, then their respective velocities will be in the same ratio as their mobilities; since they are, at every instant of the passage of electricity, under the influence of the *same forces*. Moreover, the transport of a quantity of electricity, E, occurs in this way: viz. in the direction of the current, there "wander"

$$E\frac{U}{U+V} \text{ positive ions,}$$

and in the opposite direction

$$E\frac{V}{U+V} \text{ negative ions.}$$

Now there separate out on the cathode, E equivalents of + electricity, in the form of positive ions; and these disappear from the solution, and are precipitated, as gas or metal, on the electrode (of course in an electrically neutral state): $E\frac{U}{U+V}$ ions have thus passed by the "ionic migration." Moreover, at the close of the experiment, there will be found a diminution of concentration in the solution in the neighbourhood of the cathode, corresponding to the removal of $E\frac{V}{U+V}$ cations; or, according to what has preceded, to the removal of the same quantity of negative ions by the wandering [in the opposite direction], and therefore corresponding to $E\frac{V}{U+V}$ equivalents of the electrolyte. In a precisely analogous way it follows that the liquid in the neighbourhood of the anode must be attenuated by the loss of $E\frac{U}{U+V}$ equivalents of electrolyte.

Now it is usually the case that the separated quantities of anion and cation respectively react, in a secondary way, either upon the solution or upon the metal of the electrode. Thus by the electrolysis of potassium chloride, the metallic potassium, which is deposited upon the cathode by the passage of the quantity of electricity E through the cell, does not remain there as such; but, as is well known, it reacts upon the solvent, forming potassium hydroxide and hydrogen. If we electrolyse silver nitrate between silver electrodes, there does not result an accumulation of the free radical NO_3, equivalent to E, at the anode, but this reacts with the metal of the electrode forming the equivalent quantity of silver nitrate. Therefore, the variations in concentration in the neighbourhood of the electrodes are changed in a corresponding way, to be sure, but in such a way that it can be easily calculated in each particular case.

For the thorough study of these changes in concentration produced by the wandering of the ions, and also for their theoretical interpretation in the sense given above, we are indebted to Hittorf.[1] His work is of fundamental importance, both in framing a conception of electrolysis, and also in the theory of solutions. Following the usage of Hittorf, we will designate the following ratios, which can be determined directly by experiment, as the *transport values* of the *cations* and *anions* respectively, viz.,

$$n = \frac{U}{U+V} \quad \text{and} \quad 1 - n = \frac{V}{U+V}.$$

In practice it is often possible to decide upon the nature of the ions by means of measurements of migration, especially in complex molecules.

The discovery of the relation between the conductivity and the transport number was made by Friedrich Kohlrausch:[2] we will proceed to derive it by the consideration of electrolytic dissociation.

Kohlrausch's Law of the Independent Wandering of Ions.—The conductivity of a solution of a binary electrolyte is greater in accordance as it contains more free ions, and according as these have a greater mobility; and since the particular quantities of electricity transported by the anion and cation are directly proportional to their mobility, therefore the conductivity must be proportional to their sum; that is, the specific conductivity κ of a solution, which contains c g.-mol. of the electrolyte in a litre, must be

$$\kappa = \alpha\eta F(U + V),$$

[1] *Pogg. Ann.*, **89.** 177; **98.** 1; **103.** 1; **106.** 337 (1853-59); see also Ostwald's *Classiker*, Nos. 21 and 23.

[2] *Wied. Ann.*, **6.** 1 (1879); **26.** 161 (1885).

where α denotes the degree of dissociation at the respective concentration. F is a proportional factor, depending on the system of units chosen. By making

$$u = FU \quad \text{and } v = FV,$$

and by introducing the molecular conductivity Λ for κ/η, we obtain

$$\kappa = \alpha\eta(u + v).$$

At very great dilution $\alpha = 1$; then by noticing that u and v are proportional to the magnitudes U and V introduced above, we obtain the three equations,

$$\Lambda\infty = u + v; \quad u = n\Lambda\infty; \quad v = (1 - n)\Lambda\infty.$$

These express Kohlrausch's law of the *independent wandering of the ions.* They state primarily *that the molecular conductivity of a binary electrolyte is equal to the sum of the conductivities of the two ions,* i.e. *that it is an additive property;* further, that there exists between the conductivity and the transport values an intimate relation, such, that if the transport value of one electrolyte is known, the transport values of the other electrolytes can be calculated from the respective conductivities.

These rules hold good for those electrolytes which are completely dissociated, but they likewise apply to the comparison of electrolytes existing in the same state of dissociation, as one can easily convince himself. In the way that Kohlrausch developed his law, it is impossible to distinguish between the conducting molecules (*i.e.* the free ions), and the inactive molecules of an electrolyte; and Kohlrausch found the proof for his law, both in the conductivities as measured by himself and also in the transport values obtained earlier by Hittorf. He found good coincidence, in all cases, between calculation and experiment, when he compared electrolytes in the same degree of dissociation, but otherwise he found decided deviations. As is obvious from the deduction given above, Arrhenius showed that the sufficiency of Kohlrausch's law must necessarily fail in comparing dissociated electrolytes which have been reduced to different degrees of ionisation [*i.e.* dissociation].

Recently an experimental investigation [1] has been conducted which is suited for showing this. This work concerns the conductivity and transport values of dilute silver salts; and it shows the strict sufficiency of Kohlrausch's relation between the conductivity and the transport values, provided that these two values refer to electrolytes which are completely dissociated.

The following table gives the molecular conductivity for a number of ions: the values are referred to a temperature of 18°:—

[1] M. Loeb and W. Nernst, *Zeitschr. physik. Chem.*, **2.** 948 (1888).

	K	NH_4	Na	Li	Ag	H		
u =	65·3	64·2	44·4	35·5	55·7	318		
	Cl	Br	I	NO_3	ClO_3	CO_2H	$C_2H_3O_2$	OH
v =	65·9	66·7	66·7	60·8	56·5	45	33·7	174

The values increase 1·5 to 2·7 per cent per degree of temperature; the temperature coefficient is smaller the greater the mobility, so that with increasing temperature the differing mobilities of different ions are more nearly equalised. Thus from the preceding table the molecular conductivity of potassium chloride, *e.g.*, is 131·2 at infinite dilution; and its transport value is calculated to be 0·50; while the respective values obtained by experiment are 131 and 0·51.

The great practical value of this law consists in this, viz. that *we may calculate quite certainly the limiting value of the molecular conductivity of those electrolytes at infinite dilution, in the cases of which we cannot reach the limit by experiment.* Thus we can obtain the limiting value of the potassium chloride quite certainly by extrapolation (p. 359); but this is not the case with ammonia, as is shown by the following figures :—

Concentration . . .	= 1·0	0·1	0·01	0·001	g. equivalents per litre.
Mol. conductivity of NH_3 .	= 0·89	3·3	9·6	28	

The dilution of ammonia, as shown by the first measurements, falls too far short of complete dissociation to enable us to reach its constant molecular conductivity. But from Kohlrausch's law we can calculate, with full certainty, the limiting value of the molecular conductivity of ammonia (ammonium hydroxide $= NH_4 + OH$) at infinite dilution, by means of the mobility of the two ions, thus—

$$\Lambda\infty = 64{\cdot}2 + 174 = 238{\cdot}2.$$

Thus the conception that those salts which conduct the galvanic current when they are dissolved in water are more or less dissociated into their ions, is seen to be supported in every respect, both by the phenomena and by the laws of electrolysis; moreover, the law of Avogadro, which has been applied by van't Hoff to solutions, can be also shown to meet the requirements fully in the case of electrolytes.

The Characteristics of Electrolytic Dissociation.—Electrolytic dissociation differs from ordinary dissociation chiefly in this, viz. that in the former the dissociation products are electrically charged. but not in the latter. The introduction of the idea of electrical dissociation into science marks a new era in our molecular conceptions, since it has revealed to us an entirely new species of molecule, viz. the *electrically charged ion.* According to circumstances, the same substance may decompose either into electrolytic or into non-electrolytic molecules. Thus by dissolving ammonium chloride in a good deal of water it dissociates electrolytically, and almost completely in the sense of the equation,

$$NH_4Cl = \overset{+}{N}H_4 + \overset{-}{Cl},$$

while by volatilisation, at sufficiently low pressures, it decomposes quite completely in the sense of the following equation, into non-electric molecules, thus

$$NH_4Cl = NH_3 + HCl.$$

These two processes are distinctly different, and no connection between them, however probable, is known at present.

The products of ordinary dissociation may be mixed together in any selected proportion; but in electrolytic dissociation there are in the system the same number of positive and of negative ions, so that they neutralise each other electrically.

In ordinary dissociation no other expenditure of work is requisite in order to separate the products than that usually required to effect the separation of a mixture (p. 101); but in the case of electrolytic dissociation, in addition to this work, there is required the expenditure of the much greater work necessary to overcome *the electric attraction of the oppositely charged ions.* Such a separation of this kind is observed to only a very inconsiderable degree by electrostatic influence on electrolytes;[1] here the free electricities collect on the surfaces in the form of ions; but no one has ever detected a weighable quantity of "free ions" in this sense, which are unaccompanied by the opposite kind. This could not be very easily produced, because the accumulation of free ions of the same kind is opposed by the enormous electrical forces resisting it.

The Diffusion of Electrolytes.—The preceding observations explain at once why we cannot conduct an experiment like that described on p. 346 [effusion] for the separation of the ions. If we allow a dissociated gas to diffuse, through a small opening for instance, there will occur a partial separation of the components, since the more mobile components will outstrip the sluggish ones; but it is quite otherwise in the case of electrolytic dissociation.

Let us imagine two solutions of hydrochloric acid, having a different concentration, to be brought into contact with each other; and let these be sufficiently diluted, in order that we may disregard the undissociated molecules of HCl. Then on every molecule existing in the solution (145), on the positively charged hydrogen and on the negatively charged chlorine ions, there will be exerted the same force due to the gradient of the osmotic pressure; this will tend to move them from places of a higher to those of a lower concentration. But

[1] Ostwald and Nernst, *Zeitschr. physik. Chem.*, **3**. 120 (1889).

we know from the conductivities, that a hydrogen ion has a greater mobility than a chlorine ion (p. 364); then the former will, to a corresponding extent, outstrip the latter, and thus there will occur *a partial separation of the two ions.*

But this can only happen to a limited degree. For as soon as one dilute solution contains an excess of H-ions, and the other an excess of Cl-ions, then the one will become positively charged, and the other negatively. As a result of these electric charges, there will arise an electrostatic force; this will drive the H-ions from places of a lower to those of a higher concentration, and the Cl-ions from those of a higher to those of a lower concentration. Thus the diffusion of the H-ions will be retarded, and that of the Cl-ions will be accelerated; and the equalised condition will obviously be that where both ions diffuse with the same velocity. A separation of the ions will occur only during the first instant, and, on account of the great electrostatic capacity of the ions, only to an inappreciable degree.

Thus the decomposition products of electrolytic dissociation cannot be separated from each other, like the products of ordinary dissociation, by diffusion, to any marked degree. On the contrary, this can easily be done by removing the electrostatic charges which result from diffusion, *i.e.* the solution can be *electrolysed.*

Thus, let a diffusion cylinder, having the height x, be filled with a highly dissociated solution of a binary electrolyte, having the concentration η, in the total cross-section q, and let the osmotic pressure of each particular ion be p; then at the position $x + dx$ these values will become respectively $\eta - d\eta$ and $p - dp$; the volume of the section thus increased is qdx, and it contains a quantity of electrolyte (of hydrochloric acid, *e.g.*) equal to $\eta q dx$ g.-mol. Let this be acted upon by the force qdp; this will give per g.-mol.

$$\frac{1}{\eta}\frac{dp}{dx}.$$

Now the resistance friction, which must be overcome by the two ions in their movement, amounts respectively (according to p. 360) for the cation to

$$\frac{1}{U}$$

and for the anion to

$$\frac{1}{V}$$

The quantity of each ion which will migrate in unit of time dz, through a cross-section of the diffusion cylinder, if acted upon only by the forces due to the osmotic pressure, can be obtained by the product of the cross-section × the concentration × the force per g.-mol × the mobility × the time; it amounts respectively to

$$- Uq\frac{dp}{dx}dz, \quad \text{and} \quad - Vq\frac{dp}{dx}dz.$$

But, as a matter of fact, the electrostatic forces described above come into play; these equalise the velocity of the two kinds of ions; let their electrostatic potential amount to P. Then the electrostatic attraction or repulsion, exerted per g.-ion, will amount respectively to

$$-\frac{dP}{dx}, \quad \text{and} \quad +\frac{dP}{dx};$$

and the quantities of each ion which will migrate through the cross-section, from the influence of this force alone, calculated by means of the same products as above, are respectively

$$- Uq\eta\frac{dP}{dx}dz, \quad \text{and} \quad + Vq\eta\frac{dP}{dx}dz.$$

Now, as a matter of fact, the two forces act together; and the quantity of salt which will migrate through the cross-section, in the time dz, is

$$dS = - Uqdz\left(\frac{dp}{dx} + \eta\frac{dP}{dx}\right) = - Vqdz\left(\frac{dp}{dx} - \eta\frac{dP}{dx}\right);$$

or, after the elimination of $\frac{dP}{dx}$,

$$dS = - \frac{2UV}{U + V}q\frac{dp}{dx}dz.$$

Now, according to the law of the osmotic pressure (p. 148),

$$p = \eta RT,$$

where c, according to definition, is equal to the reciprocal of the volume of the solution which contains 1 g.-mol. By introducing this, we obtain

$$dS = - \frac{2UV}{U + V}RTq\frac{d\eta}{dx}dz.$$

A comparison of this formula with that on p. 157 shows that

$$D = \frac{2UV}{U + V}RT \quad . \quad . \quad . \quad . \quad . \quad (1)$$

and this denotes *the diffusion coefficient of the electrolyte.*

Regarding U and V, we know already (p. 363) that they are proportional to the ion mobilities u and v; thus

$$u = FU \quad \text{and} \quad v = FV \quad . \quad . \quad . \quad . \quad (2)$$

The value of the proportional factor, F, depends upon the choice of the unit of measurement; its absolute value can be calculated from the equation derived by Kohlrausch on p. 363, where $\alpha = 1$, in the case of fully dissociated electrolytes, viz.,

$$\kappa = \eta F(U + V).$$

If we express all these magnitudes on the same system of measurement, then F, of course, will $= 1$. We may suitably choose the C.G.S. system.[1]

In order to obtain the conductivity k, on this scale, we must multiply it by 10^9 because an Ohm $= 10^9$ C.G.S. Further, the unit of *ion concentration*, of course, is that which holds the quantity of electricity ± 1 bound to the ions contained in 1 c.c. If the unit quantity of electricity is held by ν-equivalents of a positive ion, then a c.c. contains

$$\frac{\eta}{\nu}, \quad \text{and 1 c.c. contains } \frac{c}{1000\ \nu}$$

units of + electricity; and of course the same quantity of − electricity. We thus find that

$$\kappa 10^{-9} = (U + V)\frac{\eta}{\nu};$$

or, after introducing the molecular conductivity (p. 363),

$$U + V = \Lambda\nu \,.\, 10^{-9} \quad . \quad . \quad . \quad . \quad (3)$$

Now 96,540 Coulomb (Ampère seconds) (= 9654 C.G.S. units) deposit 1 mol of a univalent ion (Book IV. Chap. VI.); hence we have for univalent ions

$$\nu = \frac{1}{9654}$$

further

$$U + V = \frac{\Lambda}{9654} \,.\, 10^{-9} = 1{\cdot}036\Lambda \,.\, 10^{-13},$$

or, according to p. 363,

$$U = 1{\cdot}036u \,.\, 10^{-13}, \quad V = 1{\cdot}036v \,.\, 10^{-13} \quad . \quad . \quad (4)$$

If we think of an electrolytic cell (p. 354) as connected with the

[1] See the Text-books of Physics, *e.g.* Kohlrausch, *Prakt. Physik.*, Leipzig, 1892, Appendix. [In English, see, *e.g.* Daniel's *Physics*, 2nd edit., p. 13.—Tr.]

poles of a Daniel's element which has an electromotive force of 1·11 volt; and if we separate the electrodes 1·11 cm., then we have a potential gradient of 1 volt per cm., provided we disregard the internal resistance of the element and the polarisation of the electrodes; if the electrolytic cell is filled with dilute hydrochloric acid, then the values of u and v, calculated according to p. 364, are respectively

$$U' = 0·00329, \quad \text{and } V' = 0·00068 \frac{\text{cm.}}{\text{sec.}}$$

It thus becomes apparent that the velocities controlled by the current are very small in comparison with the velocities of the molecules moving back and forth; now these latter were introduced (p. 243) to explain the osmotic pressure, and they constitute the measure of the temperature in the sense of the kinetic theory. The increase in the kinetic energy of the ions occasioned by the current is extremely slight; and the work which the current does is, practically, completely used up in overcoming the friction, *i.e. in developing Joule's heat.*

The migration in the case of coloured ions, such as the anion of potassium permanganate, can be made visible by a method suggested in 1887 by O. Lodge. Using a U-shaped apparatus[1] the migration may not only be made visible, but the absolute velocities may be measured with fair accuracy in a lecture.

Now if we introduce into equation (1), developed for the diffusion coefficients, the values of U and V, from equations (3) and (4), we obtain

$$D = \frac{2uv}{u+v} \cdot \nu RT \cdot 10^{-9} \quad . \quad . \quad . \quad . \quad (5)$$

Now, it remains to express the gas constant R, in the same system of measurement, *i.e.* in its evaluation, to select that unit of volume in c.c., and that unit quantity of ions which shall hold the unit quantity of electricity 1. Now the pressure in a space containing this quantity of ions per c.c., according to p. 41, at T = 273, is

$$22{,}420\ \nu \text{ atm.} = 22{,}420 \times 981{,}000 \times 1·033\ \nu \text{ abs. units,}$$

and, therefore, in the C.G.S. system (for univalent ions $\nu = \frac{1}{9654}$),

$$RT = 2·351 \frac{T}{273} \times 10^6 \quad . \quad . \quad . \quad . \quad (6)$$

Now, by introducing this into equation (5), we obtain the diffusion coefficients on the absolute scale, *i.e.* the quantity of salt which will migrate per sec. through a cross-section of 1 sq. cm., where the gradient of concentration per centimetre is 1. It is a matter of

[1] See *Zeitschr. f. Elektr.*, **3**, 308 (1897).

indifference what scale we take to express the quantity of salt, provided only that we measure the concentration (*i.e.* the quantity of salt × cm.$^{-3}$) on the same scale. It must be noticed, however, that the numerical values usually given are based on the day as unit of time, and thus the factor 8·64 × 10^4, the number of seconds in a day, must be introduced in order to express the diffusion coefficient D in the conventional system of measurement; then

$$D = \frac{2uv}{u+v}\nu \,.\, 2 \cdot 351 \,.\, \frac{T}{273} \times 10^6 \times 10^{-9} \,.\, 8 \cdot 64 \times 10^4 \quad . \qquad (7)$$

or, simplified and calculated for T = 273 + 18,

$$D = 0 \cdot 04485 \frac{uv}{u+v}[1 + 0 \cdot 0034(t - 18)] \quad . \quad . \qquad (8)$$

The temperature coefficient, as calculated theoretically for 18°, from that of the mobility of the ions, and from that of the osmotic pressure, is in complete accord with observation, viz. 0·024 for bases and acids, and 0·026 for neutral salts. The following table contains the observations of Scheffer calculated for 18°, and the theoretical values calculated from the ion mobilities u and v, according to equation (8):—

Substance.	D obs.	D theor.	Substance.	D obs.	D theor.
HCl	2·30	2·45	$NaNO_3$	1·03	1·15
HNO_3	2·22	2·29	$NaCO_2H$	0·95	1·00
KOH	1·85	2·13	$NaCH_3CO_2$	0·78	0·86
NaOH	1·40	1·58	NH_4Cl	1·33	1·46
NaCl	1·11	1·19	KNO_3	1·30	1·41

W. Oeholm has made a thorough experimental study of the accuracy of the diffusion theory; the numbers he obtained[1] for very dilute solutions (about 0·01 normal) at 18° are given in the following table:—

Substance.	D obs.	D theor.	Substance.	D obs.	D theor.
HCl	2·32	2·45	KCl	1·46	1·46
KOH	1·90	2·13	KI	1·46	1·47
NaOH	1·43	1·58	LiCl	1·00	0·994
NaCl	1·17	1·19	CH_3CO_2H	(0·930)	1·37

Since acetic acid at this concentration is only dissociated to a small extent, the observed value 0·93 is considerably less than that calculated for complete dissociation 1·37. Otherwise the agreement found,

[1] "Diffusion von Elektrolyten," *Akademische Abhandlung*, Helsingfors, 1902.

especially in the recent measurements of Oeholm, is surprising, when one thinks that the calculated value involves diffusion coefficients and laws of gases and electrical measurements, and that parts of physics are thus brought in relation between which a connection was hardly imagined previously. The fact that not only in acetic acid but in the other substances the calculated values are usually a little larger than the observations may be attributed to incomplete dissociation, since the undissociated molecules diffuse more slowly than free ions.

The theory of diffusion in mixtures of salts can also be completely developed; observation again supports the results of the theory, as was shown by me in 1888, and subsequently in a series of researches by Arrhenius.[1]

If we measure the molecular conductivity, the mobility of ions, the diffusion coefficient, and the gas constant, by one system of measurement, consistently applied, it will be found that

$$\Lambda = U + V,$$

and according to equation (5)

$$\frac{1}{D} = \left(\frac{1}{U} + \frac{1}{V}\right)\frac{1}{RT}.$$

That is, that while the molecular conductivity is simply equal to the sum of the ion mobilities, the reciprocal of the diffusion coefficient is proportional to the sum of the reciprocals of the mobilities, these being also included under the additive properties.

The Friction of the Ions.—The friction of the ions can be calculated on the absolute scale from the mobilities found on p. 369. Now, since the velocities assumed by the ions, when unit force acts on ν-equivalents, according to equation (4), are respectively,

$$U = 1{\cdot}036u \times 10^{-13}, \quad \text{and } V - 1{\cdot}036v \times 10^{-13};$$

then, per ν-equivalents, there will be respectively the forces

$$\frac{1}{1{\cdot}036u \times 10^{-13}}, \quad \text{and } \frac{1}{1{\cdot}036v \times 10^{-13}} \text{ abs. units};$$

or, per g.-ion respectively, the forces

$$K = \frac{1}{1{\cdot}036u \times 10^{-13} . 981000\nu} = \frac{0{\cdot}950}{u}, \quad \text{and } \frac{0{\cdot}950}{v} \times 10^{+11} \text{ kg.} \qquad (9)$$

required to cause the ions to move with velocity 1 cm./sec. (Kohlrausch).

We may apply the same formula to a non-electrolyte or to a salt,

[1] *Zeitschr. physik. Chem.*, **10.** 51 (1892); see also the interesting studies of U. Behn, *Wied. Ann.*, **61.** (54 1897); and Abegg and Bose, *Zeitschr. physik. Chem.*, **30.** 545 (1899).

by making u = v, and then calculating its mobility according to equation (8); then we obtain

$$u = \frac{D}{0{\cdot}02242[1 + 0{\cdot}0034(t - 18)]},$$

and by considering equation (9), we obtain

$$K = \frac{2{\cdot}13}{D}[1 + 0{\cdot}0034(t - 18)] \times 10^9 \text{ kg.} \quad . \quad (10)$$

a formula which was already referred to.

Now we will use equations (9) and (10) in order to calculate the resistance friction for an ion and for an electrically neutral molecule of any related molecular structure. Thus the conductivity v of the ion of caproic acid, viz. $H_3C—CH_2—CH_2—CH_2—CH_2—C\begin{smallmatrix}\nearrow O \\ \searrow OH\end{smallmatrix}$, at 10°, amounts to 18; and, therefore, according to equation (9), the force in kilograms required to give it a velocity of 1 $\frac{\text{cm.}}{\text{sec.}}$, is

$$K = 5{\cdot}3 \times 10^9.$$

Moreover, from the diffusion constant 0·38 of mannite, $C_6H_{14}O_6$, measured at the same temperature, according to equation (10), we calculate this force to be

$$K = 5{\cdot}5 \times 10^9.$$

The fact that these resistance frictions are so nearly coincident, in spite of the very different ways in which they are derived, is a new proof that the views developed in the preceding pages are based on a sure foundation.

Further comparisons are given in the following table :—[1]

Anion or Kation of	Molecular weight	*v* or u	*K* 10⁻⁹	Dissolving substance	Molecular weight	*D*	*K* 10⁻⁹	Temperature
Acetic acid .	59	45	2·1	Acetic acid .	60	·90	2·4	18
Tartaric acid .	149	30	3·2					
Racemic acid .	149	30	3·2	Tartaric acid	150	·50	4·2	18
Isobutyl-Sulphuric acid .	153	29	3·3	Racemic acid	150	·52	4·1	18
Ammonium Ion	18	64	1·5	Ammonia .	17	1·64	1·3	19

Other Solvents.—The faculty of breaking up dissolved substances into ions and so yielding good conducting electrolytes, which is most marked in the case of water, is found also in other solvents, but mostly only to a small extent and hardly ever to an extent comparable with water.

It should be noticed that conductivity depends not only on the

[1] Euler, *Wied. Ann. Jabelband*, p. 273 (1897).

number of ions, that is on the degree of dissociation, but also upon their mobility. According to the preceding pages, the velocity of diffusion is a measure of this. This is remarkable because in weakly dissociated solvents we can neither determine $\Lambda\infty$ directly nor obtain α from freezing-point determinations on account of its smallness; hence to find α and $\Lambda\infty = u + v$ separately in the equation

$$\kappa = \alpha\eta\ (u + v)$$

one is thrown back upon measurements of diffusion. The measurements made so far show that ionic velocities vary from one solvent to another. Thus Kawalki found the mobilities of a large number of ions in alcohol to be 0·34 times as great as in water.[1] Yet they remain at least of the same order of magnitude, so that the conductivity of dissolved substances yields at least a rough indication of the dissociating power of the solvent.

We may form the following conclusions on the power of solvents to break up dissolved substances into ions. Clearly the electrostatic attraction of oppositely charged ions plays an important *rôle* in electrolytic dissociation; that force naturally tends to recombine the ions into electrically neutral molecules.[2] We must therefore assume that other actions whose nature is unknown to us, perhaps the kinetic energy of the components of the molecule, cause separation, and that dissociative equilibrium has its origin in the concurrence of these two opposed tendencies. If the electrostatic force is weakened the electrolytic dissociation should increase; now electrostatics shows that two oppositely charged bodies attract each other more weakly the higher the dissociation constant of the medium in which they are placed. It follows, therefore, *ceteris paribus*, that electrolytic dissociation will be greater the higher the dielectric constant of the medium. This view is supported by the following table:—

Medium.	Dielectric constants.	Electrolytic Dissociation.
Gaseous Space .	1·0	Indistinguishable at normal temperatures.
Benzene . .	2·3	Extremely small but measurable conductivity indicates a trace of dissociation.
Ether . .	4·1	Noticeable conductivity of the dissolved electrolyte.
Alcohol . .	25	Moderately strong dissociation.
Formic acid .	62	Strong dissociation of dissolved salts.
Water . .	80	Very strong dissociation.

The parallelism between electrostatic dissociation of dissolved substances and the dielectric constant of the solvent is put beyond

[1] *Wied. Ann.*, **52**. 300 (1894).
[2] Nernst, *Gott. Nachr.*, Nov. 12, 1893; *Zeitschr. physik. Chem.*, **13**. 531 (1894).

doubt by this and a number of other examples; but absolute proportionality is not to be expected, for other influences have to be taken into account, especially as we are unable to say how far the forces tending to separation of the ions vary with the nature of the medium.

The behaviour of formic acid as a solvent studied by Zanninovitch-Tessarin [1] is especially noticeable. Salts, such as NaCl, KBr, etc., are hardly less dissociated in this solvent than in water, as measurements of conductivity and freezing-point both show; hydrochloric acid, on the other hand, hardly conducts at all, and consists for the most part of double molecules; tri-chlor-acetic acid also, which is extensively dissociated in water, has almost its normal molecular weight in formic acid. This shows, as was to be expected, that besides the undoubted influence of the dielectric constant other specific influences make themselves felt. Apparently association of the ions with molecules of the solvent is important in this respect. Methyl alcohol as a solvent has been thoroughly investigated by Carrara [2] and by Zelinsky and Krapiwin,[3] acetone by St. v. Laszczynski [4] and Carrara; [5] for other literature in this interesting field see the *Jahrbuch der Elektrochemie* from 1895 (Halle, W. Knapp).

It is worth mentioning that solvents with considerable dissociating power are also those which have a tendency to polymerisation in the liquid state.[6] This favours the assumption that electrolytic dissociation is helped by association of the free ions with the molecules of the solvent, an assumption which is all the more plausible, as the ions have a strong tendency to form molecular compounds, as will appear in the following section.

The fact noted by F. Kohlrausch [7] that we know of no good electrolytically conducting liquids at usual temperatures, that consequently electrolytic conduction is first seen in mixtures, supports this view.

Werner's Theory of Molecular Compounds.—Whilst in dealing with the carbon compounds the unitary conception is the most helpful; in dealing with the majority of the inorganic compounds, especially the salts, acids, and bases, Berzelius's dualistic hypothesis is more suitable, the hypothesis which, as we have seen in this chapter, can be referred to decomposition into negative and positive ions. Werner has shown that the investigation of ionic dissociation gives valuable insight into the constitution of so-called molecular compounds, especially the metal ammonia derivatives.

[1] *Zeitschr. physik. Chem.*, **21.** 35 (1896).
[2] *Ibid.*, **19.** 699 (1896).
[3] *Ibid.*, **21.** 35 (1896).
[4] *Zeitschr. f Elektrochemie*, **2.** 55, 214 (1895).
[5] *Gaz. chim.*, **27.** [1]. 207 (1897).
[6] Nernst, *Zeitschr. physik. Chem.*, **14.** 624 (1894); Dutoit and Aston, *Compt. rend.*, **125.** 240 (1897).
[7] *Pogg. Ann.*, **159.** 270 (1875).

The following observations serve as starting-point. Amongst the metallic ammonia compounds of the trivalent metals, cobalt, chromium, and rhodium, four series are known with 3, 4, 5, and 6 molecules of ammonia. Thus considering the chlorides, we have as the member richest in ammonia the luteo salts $Me(NH_3)_6Cl_3$, where Me is one of the metals mentioned. In these salts, as in other metallic chlorides, the chlorine atoms can be separated as ions; this is shown by the conductivity and by chemical behaviour, for instance, immediate reaction with silver nitrate, decomposition by sulphuric acid with formation of hydrochloric acid, etc. When these compounds lose a molecule of ammonia and turn into the purpureo salts $Me(NH_3)_5Cl_3$, the remarkable phenomenon appears that *one of the chlorine atoms can no longer be separated as ion.* This can be rendered symbolically by the formula

$$\left(\begin{matrix}Me(NH_3)_5\\Cl\end{matrix}\right)Cl_2,$$

which is to express that the complex within the bracket does not dissociate on solution. Coming to the next member of the series we find that as each molecule of ammonia disappears the chlorine atom becomes incapable of acting as ion. Thus we have the series

$$\underset{\text{Luteo salts.}}{\left(Me(NH_3)_6\right)Cl_3} \quad \underset{\text{Purpureo salts.}}{\left(\begin{matrix}Me(NH_3)_5\\Cl\end{matrix}\right)Cl_2} \quad \underset{\text{Praseo salts.}}{\left(\begin{matrix}Me(NH_3)_4\\Cl_2\end{matrix}\right)Cl} \quad \underset{\text{Hexamine salts.}}{Me\left(\begin{matrix}(NH_3)_3\\Cl_3\end{matrix}\right).}$$

The last member of the series is not dissociated.

The series of the cations of these salts is

$$\overset{+++}{Me(NH_3)_6} \qquad \overset{++}{Me\begin{matrix}(NH_3)_5\\Cl\end{matrix}} \qquad \overset{+}{Me\begin{matrix}(NH_3)_4\\Cl_2\end{matrix}} \qquad Me\begin{matrix}(NH_3)_3\\Cl_3\end{matrix}.$$

Thus we here find the neutral ammonia molecule replaced by univalent anions, so that, naturally, the negative charge of the anions is neutralised by the positive charge of the complex. Hence for each atom of chlorine that enters it the complex loses one charge and finally becomes electrically neutral. If an anion be introduced in the place of an ammonia molecule in the neutral hexamine salt, we obtain the anion

$$\overset{-}{Me\begin{matrix}(NH_3)_2\\(NO_2)_4\end{matrix}}$$

which is capable of forming salt with potassium.

These relations are excellently shown by the platin-ammonia compounds investigated by Werner;[1] the following series of radicals has been found:—

[1] *Zeitschr. anorg. Chemie*, **3.** 267 (1893); **8.** 153 (1895); Werner and Miolati; *Zeitschr. f. phys. Chem.*, **12.** 35 (1893)

$$\overset{++++}{Pt(NH_3)_6},\quad \overset{+++}{Pt(NH_3)_5Cl},\quad \overset{++}{Pt(NH_3)_4Cl_2},\quad \overset{+}{Pt(NH_3)_3Cl_3},\quad Pt(N$$

$$\overset{-}{Pt(NH_3)Cl_5},\quad \overset{--}{PtCl_6}.$$

The original platin-ammonia ion with a quadruple charge l charge by substitution of chlorions, becomes electrically neutı finally even negative. The following curve gives the o molecular conductivity of the chlorides and potassium salts above radicals :—

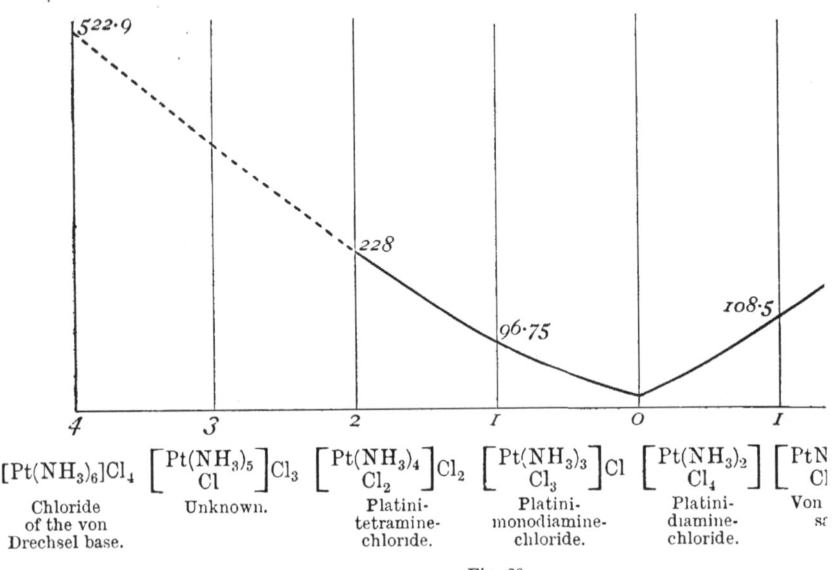

Fig. 29.

Similarly a series can be obtained by the substitution of cl in compounds of divalent platinum :—

$$\left(Pt(NH_3)_4\right)Cl_2 \quad \left(Pt\begin{matrix}(NH_3)_3\\Cl\end{matrix}\right)Cl \quad \left(Pt\begin{matrix}(NH_3)_2\\Cl_2\end{matrix}\right) \quad \left(Pt\begin{matrix}NH_3\\Cl_3\end{matrix}\right)K \quad \left(Pt\right.$$

The position of ammonia in the examples here given taken by other atomic groups, such as water. Thus we kn roseo compounds $\left(Me\begin{matrix}(NH_3)_5\\H_2O\end{matrix}\right)Cl_3$ and the tetramine-roseo com $\left(Me\begin{matrix}(NH_3)_4\\(H_2O)_2\end{matrix}\right)Cl_3$. In these the radical remains trivalent becaı substituent is neutral. By introducing this water molecule we hexahydrates, and this is actually the form in which the heavy most commonly appear. When a hydrate contains more molecules of water, for example the vitriols, this is attribu Werner to a combination of water with the anion.

Werner and Gubser[1] have given an interesting case of "hydrate isomerism" in the chromium chlorides. It has long been known that, like other chromium salts, these exist in two modifications, a green and a violet. The solid salts both have the composition $CrCl_3 . 6H_2O$. In solution the green form changes into the violet, whilst the conductivity rises about threefold. In the violet solution the three chlorine atoms are present as ions, it has therefore the constitution $(Cr(H_2O)_6)Cl_3$; in the green solution only one chlorine atom is present as ion, the constitution is therefore $\left(Cr{(H_2O)_4 \atop Cl_2}\right)Cl + 2H_2O$. That we have here actually two molecules of water combined in a different way to the others, is shown by the fact that the green salt can be transformed into the tetra-hydrate without losing any of its characteristic properties. It may be seen from these examples that the negative atom can be combined in two different ways: one so that it is easily separated as an ion, the other so that it combines with the metallic atom to form a radical; the latter is the case when it takes the place of ammonia or water. Werner believes that in the latter case the radical is directly attached to the metallic atom and exists in the inner sphere of the latter. Ions which contain other kinds of molecules added in this way are called complex ions. The number of radicals directly combined is a characteristic constant of each metal known as the co-ordination number; it is to be distinguished from the valency.

A very remarkable research of Abegg and Bodländer[2] shows how the affinity of the elements for the electron regulates the formation of complex compounds; the less firmly the ion holds its electric charge the more readily it forms "molecular compounds" in general. This hypothesis cannot, however, be followed out with exactness, since we do not know how to give precise determination to the electro-affinity.

Ions with Positive and Negative Charge.—A peculiar kind of ion must be formed, as Küster[3] showed first, when an electrically neutral molecule gives off simultaneously a positive and a negative ion (either uni- or multi-valent); in that case an electrically neutral molecule remains, but one which is capable of acting as an ion, since it carries free electric charges; Küster gave such the name *zwitter-ion.* An example is to be found in methyl orange and substances which possess the formula

$$N(CH_3)_2 - R - SO_3H.$$

There are in solution both a base (substituted ammonia) and an acid (sulphonic acid), therefore must form the ion

$$\overset{+}{N}H(CH_3)_2 - R - \overset{-}{S}O_3.$$

[1] *Ber. deutsch. chem. Ges.*, **34.** 1579 (1901).
[2] *Zeitschr. anorg. Chem.*, **20.** 453 (1899).
[3] *Ibid.*, **13.** 136; see also Winkelblech, *Zeitschr. physik. Chem.*, **36.** 546 (1901).

Evidently this conception of a *zwitter-ion* is partly identical with that of an "internal salt"; the new theory therefore admits the possibility of two isomers, according as the charges remain free or neutralise each other. Nothing is known experimentally as to the existence of these two isomers; they would correspond to the difference, for example, between dissociated and non-dissociated acetic acid.

It would evidently be of much importance to find a means of testing for *zwitter-ion*; since the entire ion suffers no force in an electric field, such ions do not, like the ordinary kind, contribute to the conductivity. But the electric field must exert a moment of rotation on their free charges, and hence must have a directive action on them, but on this subject nothing is known at present. It is possible that *zwitter-ions* are much commoner than has hitherto been assumed, for example, the carbon in carbonic oxide would be tetravalent if the latter substance had the constitution

$$\overset{+}{\underset{-}{C}} = O,$$

but this view can neither be confirmed nor disputed at the present time.

Incandescent Gases.—Arrhenius,[1] on the basis of his own investigations, has recently developed the view that the conductivity of incandescent gases, which had been repeatedly studied, is an electrolytic phenomenon, and, therefore, that it argues the existence of ions. The flame of a Bunsen burner was fed by means of an atomiser (*Zerstäuber*) with a mixture of air and little drops of different salt solutions, and thus was brought to a degree of saturation which was fairly constant and capable of approximate determination. Of course, this was proportional to the concentration of the solutions used, and was determined on an absolute scale by photometric measurement. Inasmuch as it was possible to measure the intensity of a galvanic current which was sent through, by means of two platinum electrodes adjusted in the flame, it was thus possible to measure the conductivity of the incandescent salt vapours. The flames which were free from salt vapours did not conduct in a marked degree.

In the case of the akali salts the conductivity was nearly proportional to the square root of the quantity present. It was independent of the nature of the negative ingredient of the salt. Thus, for example, all potassium salts having the same degree of concentration conducted equally well. The conductivity of the salts of the alkali metals increased with the atomic weights. Mixtures of the vapours of potassium and sodium salts showed a conductivity similar to that which can be calculated for the mixture of two electrolytes in water solution.

Acids and ammonium salts did not conduct in a marked way.

[1] *Weid. Ann.*, **42.** 18 (1891).

The salts of the heavy metals showed only traces of conductivity. The behaviour of the akaline earths was much more complicated.

According to Arrhenius, the reaction which leads to the formation of free ions, and which consists of the action of the water vapour of the flame on the akali salt, is probably as shown in the following scheme, viz. :—

$$KCl + H_2O = HCl + KOH.$$

The potassium hydroxide formed then breaks up slightly into ions, as follows, viz. :—

$$KOH = \overset{+}{K} + \overset{-}{OH}.$$

Since direct observation shows that the resulting acid does not conduct in the flame, there follows immediately from this view a result obtained experimentally, but which is, at first sight, very surprising, viz. that *salts having the same positive ion, when in the same degree of concentration, conduct equally well.* Moreover, a number of other phenomena are explained by the view given above, as, particularly, the fact that the conductivity increases in proportion to the square root of the concentration.

Historical Observations.—The idea of electrolytic dissociation, which has exerted such a fruitful and regenerating effect upon various departments of physics and chemistry, like other discoveries of this sort, had its historical anticipation. As early as the year 1857, Clausius,[1] from the laws of electrolytic conductivity, developed the view, that in conductors of that sort there must be traces of free ions leading, at least, an ephemeral existence.

H. v. Helmholtz,[2] in 1880, from a similar starting-point, arrived at similar conclusions.

"*Since the complete equilibrium of electricity is produced in the interior of electrolytic liquids, as well as in metallic conductors, by the weakest distribution of the electric forces of attraction, therefore it must be assumed that no other (chemical) force is opposed to the free movement of the positively and negatively charged ions, save the forces of their electrical attraction and repulsion.*"

In this law Helmholtz has clearly formulated the fundamental idea of the theory of electrolytic dissociation. But the idea remained latent till very recently, because no significant inferences were drawn from it. And, therefore, Arrhenius must be regarded as the particular father of this theory (1887); for on the one hand he showed how to determine the nature and the number of the ions; and on the other hand, by a series of applications, he clearly showed the fruitfulness of the theory.[3]

[1] *Pogg. Ann.*, **101.** 338 (1857).
[2] *Wied. Ann.*, **11.** 737 (1880).
[3] *Zeitschr. physik. Chem.*, from 1887 on; also **9.** 330 (1892).

CHAPTER VIII

THE PHYSICAL PROPERTIES OF SALT SOLUTIONS

The Necessity of the Additive Relation.—A solution which is sufficiently dilute contains the electrolytes in a state of complete dissociation. The dissolved substance has now become a mixture of different molecules, *i.e.* of its ions. Now it is a common experience that, in a dilute solution, or in a gas mixture (p. 100), the properties of each particular component remain unchanged, and, moreover, if the components are known, the properties of the mixture can be predicted. From this there follows the fundamental law that—

The properties of a salt solution are composed, additively, of the properties of the free ions.

Inasmuch as this simple law has been at times subject to much misconception, particular warning should be given against certain illegitimate applications.

The law presupposes *complete* dissociation of the electrolytes. Nothing can be said *a priori* regarding the behaviour in cases where the dissociation is incomplete.

The law may hold good in cases where the dissociation is not complete; and this would obviously be the case where the properties in question are not changed by the union of the ions to form electrically neutral molecules. To what extent this may happen in any given case depends expressly upon this, viz. whether the property in question is decidedly additive or not.

It is unreasonable to reject this law because it occasionally holds (*e.g.* with regard to absorption of light) even in cases in which it is not obviously necessary that it should.

The Density of Salt Solutions.—If we add to water a salt which is completely dissociated electrolytically when in solution, then the change in volume so occasioned will be brought additively by intermixture of the particular ions, and therefore this will be an additive property of the ions. A similar thing may occur in the case of salts which are only slightly dissociated, as, for instance, when the electrolytic dissociation is associated with no change of volume. But here this is not a necessary consequence. It has not been possible as yet to state *what part of the change of volume is due to each particular ion.*

Let us denote by s the density of a solution containing in m grams of water 1 g.-mol of a salt having a mol wt. of M; and by s_0, the density of pure water at the same temperature; then the change of volume occasioned by the solution of that quantity of salt will be

$$\Delta v = \frac{M + m}{s} - \frac{m}{s_0}.$$

The following table contains some values of the change of volume,[1] as calculated from dilute solutions:—

I.	Diff.	II.	Diff.	Diff. I.-II.
KCl = 26·7		NaCl = 17·7		9·0
	8·4		9·0	
KBr = 35·1		NaBr = 26·7		8·4
	10·3		9·4	
KI = 45·4		NaI = 36·1		9·3

These values are uncertain within the limit of one unit. But within this limit of error, the preceding law holds good, as shown by the constancy of the differences between Cl and Br, Br and I, and Na and K.

The change in volume, Δv, is, as a rule, no longer additive, when weakly dissociated salts are concerned; thus

KAc = 50·6	NaAc = 40·0	HAc = 51·1
KCl = 26·7	NaCl = 17·7	HCl = 17·4
Diff. = 23·9	= 22·3	= 33·7

The differences refer to differences in the increase of volume occasioned by Ac (the acetic acid radical) and Cl. They are equal when salts are compared which exhibit the same degree of dissociation; but they have an entirely different value when the highly dissociated hydrochloric acid is compared with the slightly dissociated acetic acid. Acetic acid also is highly dissociated at great dilution; then Δv would diminish to about 40·5.

It is very noticeable that the values of Δv for electrolytically dissociating substances are quite exceptionally small, in fact, they are smaller than the molecular volume in the solid state, and in some instances (for example, sodium carbonate, magnesium sulphate, zinc sulphate) are even negative. It seems, therefore, that with increasing separation of ions Δv has in general a marked tendency to fall off; this may be seen from the exact measurements by Kohlrausch and Hallwachs on p. 301. The effects hitherto observed may be most easily explained

[1] J. Traube, *Zeitschr. anorg. Chem.*, **3.** 1 (1892).

by supposing *the solvent, water, suffers a strong contraction on account of the presence of free ions.*

It is of interest to know that such a contraction might be expected from the electric charges of the ions; every electric liquid must suffer a contraction if, as usually happens, the dielectric constant increases with pressure. It is therefore natural to explain the observed contraction of water as due to the electrostatic field of the ions in a dissolved electrolyte.[1] Carrara[2] has shown the existence of such "*electro-striction*" in non-aqueous solutions.

If the change of volume, as defined above, is an additive property, the same thing of course will not be strictly true regarding changes in the specific gravity, and for algebraic reasons. But this is approximately true, because the changes of density amount to only a small part of the total density. In fact, it is found that the densities of normal solutions of Na and K compounds, *e.g.* show a constant difference when combined with the same negative component. And furthermore, as the changes in density are nearly proportional to the amount of the content, then the difference must be proportional to [the difference in] the content.

These regularities can be used to calculate, from a few data which are empirically determined, the density of any selected solution of a salt having the ions A and B.[3]

Let us denote by D_μ the density of an μ-fold normal solution of ammonium chloride, which is chosen as the starting-point, and let a and b represent the differences between the densities of a normal ACl solution and of a normal NH_4B solution respectively; then the density of an μ-fold normal AB solution is calculated from the formula

$$d = D_\mu + \mu(a + b).$$

The constants a and b, which are to be empirically determined, are called the *moduli* of the respective ions. In the following table there are given the densities of solutions of ammonium chloride, and also the moduli of a number of ions.[4] All the values refer to a temperature of 15°.

μ	Densities of NH_4Cl Solutions.
1	1·0157
2	1·0308
3	1·0451
4	1·0587
5	1·0728

1 Drude and Nernst, *Zeitschr. physik. Chem.*, **15.** 79 (1894).
2 Carrara and Levi, *Gaz. chim. ital.*, **30.** II. 197 (1900).
3 Valson, *C. R.*, **73.** 441; **77.** 806.
4 Bender, *Wied. Ann.*, **20.** 560 (1883).

MODULI IN $\frac{1}{10000}$ UNITS

$NH_4 = 0$; $K = 289$; $Na = 238$; $Li = 78$; $\frac{1}{2}Ba = 735$; $\frac{1}{2}Sr = 500$; $\frac{1}{2}Ca = 280$; $\frac{1}{2}Mg = 210$; $\frac{1}{2}Mn = 356$; $\frac{1}{2}Zn = 410$; $\frac{1}{2}Cd = 606$; $\frac{1}{2}Pb = 1087$; $\frac{1}{2}Cu = 434$; $\frac{1}{2}Ag = 1061$.

$Cl = 0$; $Br = 373$; $I = 733$; $NO_3 = 163$; $\frac{1}{2}SO_4 = 206$; $C_2H_3O_2 = -15$.

Thus the density of a threefold normal solution of $Sr(NO_3)_2$, which contains 1·5 g.-mol per litre, is calculated to be

$$d = 1{\cdot}0451 + 0{\cdot}0003(500 + 163) = 1{\cdot}2440 ;$$

while a direct determination gave the result 1·2422. Sometimes the differences are considerably greater.

The Refractive Power of Salt Solutions.—For the reason that the change of volume, on adding a salt to a solvent, must be necessarily an additive property of the ions the same would be expected of the change in the specific refractive power. In fact, it is found to be true that *the molecular refractive power of a salt, in water solution, is an additive property*, provided that this is completely, or at least nearly, decomposed into the free ions; but this may (not necessarily must) cease to hold good when the electrolytic dissociation is slight. Gladstone[1] has stated that the specific molecular refractive powers of similar K and Na compounds show quite constant differences. Le Blane[2] has shown most conclusively that the same is true of the differences between the specific refractive powers of acids and of their Na salts, in so far as the former are highly dissociated; but not so when the dissociation is slight, *i.e. there is a marked difference between the atomic refraction of combined hydrogen, and that of hydrogen which is electrolytically separated.*

No way has yet been found to determine the specific refraction of particular ions; nor any method to determine the amount by which each ion changes the specific volume of water (compare p. 381).

It would be simpler, but less rational, to calculate from the refractive index instead of from the specific refractive power. The change experienced by the refractive index of pure water, on the addition of a salt, is, at least approximately, made up additively of the changes occasioned by the particular atoms; and, therefore, by the use of suitably chosen numerical values or moduli of the ions, it is possible to calculate the refractive coefficients of salt solutions in exactly the same way that was shown to be possible for the specific gravities

[1] Gladstone, *Phil. Trans.*, 1868; Kanonikoff, *J. pr. Chem.* [2], **31.** 339 (1885).
[2] *Zeitschr. physik. Chem.*, **4.** 553 (1889).

(p. 383). Thus let us denote by N_μ the refractive index o of KCl containing μ equivalents per litre; then the refra n of a solution of a salt of this content, and when the mo ions are a and b, may be calculated from the formula

$$n = N_\mu + \mu (a + b).$$

Bender,[1] using the following refractive coefficients for K at t° = 18°, as a starting-point, found the following moduli:

μ	$H\alpha$	D	$H\beta$	$H\gamma$
1	1·3409	1·3428	1·3472	1·3505
2	1·3498	1·3518	1·3565	1·3600
3	1·3583	1·3603	1·3651	1·3689

MODULI IN $\frac{1}{10000}$ UNITS

	$H\alpha$	D	$H\beta$	$H\gamma$
K	0	0	0	0
Na	2	2	2	2
½Cd	38	...	40	41
Cl	0	0	0	0
Br	37	38	41	43
I	111	114	123	131

Thus the refractive coefficient for a twofold normal Na and the line $H\alpha$, is calculated to be

$$n = 1{\cdot}3498 + 0{\cdot}0002(2 + 37) = 1{\cdot}3576;$$

while experiment gave 1·3578. The coincidence usually is as this.

Light Absorption and Colour.—The theory of dissociation requires that *the absorption of an electrolyte, which dissociated, should be made up additively from the absorption of the negative ingredients*; and therefore that the colour of a dilu tion depends upon the colour of the free ions. Some ar favour of this view are the well-known facts that, when in concentration, all chromates are yellow, all copper salts ar

[1] *Wied. Ann.*, **39.** 89 (1890); see also Valson, *Jahresber. f. Chem.*, 18

all salts made up of nothing but colourless ions—as Cl, Br, I, NO_3, SO_4, etc., and as K, Na, Ba, Ca, NH_4, etc.—in water solution are colourless. Moreover, this has been recently subjected to a quantitative proof. Ostwald,[1] in particular, has investigated the absorption spectra of a number of salts of permanganic acid, of fluorescein, of eosin, of rosolic acid, of aniline violet, etc., and fixed them photographically,

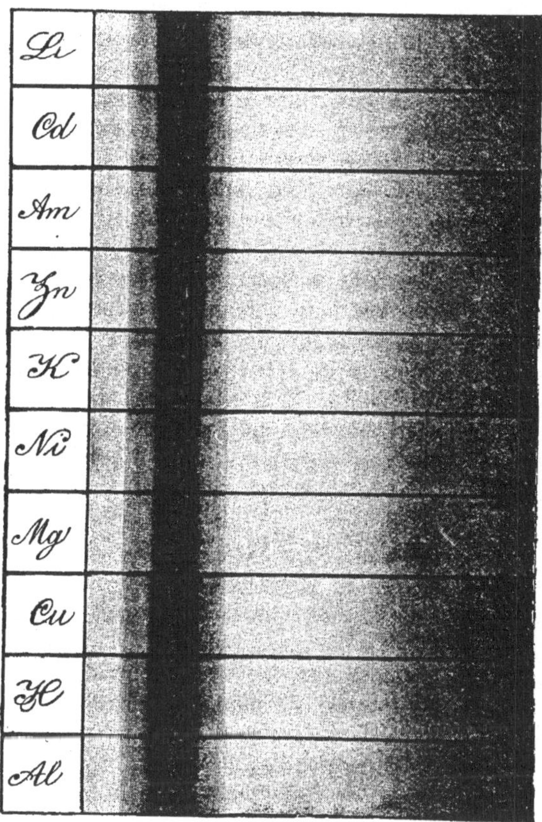

Fig. 30.

thus obtaining actual proof, undistorted by any subjective influence. The paper on the subject included some accompanying illustrations, made up of the photographed spectra, which were compared with each other as exactly as possible. These represented a number of dilute solutions of salts containing one coloured ion, and contributed, in a striking way, to demonstrate the question at issue regarding the

[1] *Abh. d. kgl. sächs. Akad.*, **18.** 281; *Zeitschr. physik. Chem.*, **9.** 579 (1892); also Ewan, *Phil. Mag.* [5], **33.** 317 (1892).

dissociation theory; *it was found that the spectra of dilute solutions of salts containing the same coloured ion are identical.*

The accompanying cut (Fig. 30) shows the absorption spectra of a number of salts of permanganic acid, which were studied in equivalent solutions, *i.e.* in solutions containing the same number of coloured ions; they all show absolutely identical bands in the yellow and green. If we remember that the absorption of light is usually (p. 336) changed considerably by very slight changes in the structure of the molecule, it would appear very strange not to seek its simple explanation in the dissociation theory.

General rules for the behaviour of the spectrum, when the ions unite to form molecules which are electrically neutral, have not been found as yet; thus the change is quite small when a solution of copper sulphate, *e.g.* is diluted, but this change in the colour is clearly recognisable when one adds water to a solution of copper chloride; in the latter case, the originally green solution gradually assumes the blue colour of the copper ions, a clear and simple experiment.

Of course the same ion may change its colour by changing the value of its electrical equivalent [*i.e.* its valence]; thus the divalent iron ion is coloured green, as in $FeSO_4$; but the trivalent iron ion is coloured yellow, as in $FeCl_3$. The spectrum also changes when a coloured ion unites with another substance to form a new ion; thus the negative ions of the hydroferro- and of the hydroferri-cyanic acids [H_4FeCu_6 and H_3FeCu_6] have colours different from that of the free iron ions.

The Natural Power of Rotation.—*Those salts which contain one optically active ion must, when in equivalent solutions, and when in a state of complete dissociation, have the same power of rotation.* This law was first established from certain measurements of Landolt,[1] but the results of Oudemans,[2] are still more striking; he developed the law in a purely empirical way, and formulated it as follows. Alkaloids, in their salts, are independent of the acid with which they are combined; and optically active acids are independent of the nature of the metal with which they are combined: these show the same rotation, at the same degree of concentration; thus the molecular rotation of the following salts of quinic acid, as measured in about $\frac{1}{7}$th normal solution, were nearly identical, viz.:—

$$K = 48{\cdot}8 \; ; \; Na = 48{\cdot}9 \; ; \; NH_4 = 47{\cdot}9 \; ; \; Ba = 46{\cdot}6 \; ; \; Sr = 48{\cdot}7 \; ; \; Mg = 47{\cdot}8.$$

The *power of magnetic rotation* is also an *additive* property, as is inferred from the investigations of Jahn.[3] It should also be mentioned

[1] *B. B.*, **6.** 1073 (1873); W. Hartmann, *ibid.*, **21.** 221 (1882).
[2] *Beibl.*, **9.** 635 (1885).
[3] *Wied. Ann.*, **43.** 280 (1891).

here, in brief, that, according to the investigations of G. Wiedemann,[1] the atomic magnetism of dissolved salts of magnetic metals is independent of the nature of the acid, see p. 316, and the phenomenon of *fluorescence* can be shown by ions as well as by electrically neutral molecules.[2]

Ionic Mobility.—Ostwald, Walden, and especially Bredig,[3] investigated the relation between the chemical nature of a number of anions and cations and their mobilities (see Chap. VII.) as deduced from the limiting values of the molecular conductivities. The relations so far known are chiefly the following:—

The mobility of *elementary* ions is a marked *periodic function of their atomic weight*, and *rises* in each series of similar elements when the atomic weight increases until the latter reaches the value 35.

Compound ions travel equally fast when they are *isomeric* and of *equal degree of substitution* with respect to the dissociating group; thus we find equal mobility for

the anions of ortho-toluic acid $CH_3 . C_6H_4 . COOH$ (29·9) and toluic acid $C_6H_5 . CH_2 . COOH$ (29·8), butyric acid $CH_3CH_2CH_2COOH$ (30·7) and isobutyric acid $(CH_3)_2CHCOOH$ (30·9);

the cations of propyl amine $CH_3CH_2CH_2NH_3OH$ (40·1) and isopropylamine $(CH_3)_2CHNH_3OH$ (40·0);

on the contrary, when the degree of substitution is not the same the mobilities are unequal as with the

cations of propyl amine $C_3H_7NH_3OH$ (40·1) and trimethyl amine $(CH_3)_3NHOH$ (47·0).

The higher the degree of substitution of isomeric ions the faster they travel.

The substitution of the same group in different ions alters the mobility in the *same sense*; mobility in general falls off with increase in the number of atoms, but less so when the number of atoms is already large, so that the mobility tends with increasing complexity of the ion asymptotically to the minimum value of 20 (the fastest ions are $\overset{+}{H} = 325$ and $\overset{-}{OH} = 167$). Numerical relations have not been established on account of numerous imperfectly known constitutive influences.

Systematic Résumé of the Properties of Ions.—In the study of any casual physical property of a salt solution, the first attempt made is to give it an additive form. This is usually done simply

[1] See art. "Magnetismus," in Ladenburg's *Handwörterbuch*, **7.** (1889).

[2] Compare the interesting study of Buckingham, *Zeitschr. physik. Chem.*, **14.** 129 (1894).

[3] *Zeitschr. physik. Chem.*, **13.** 242 (1894).

by means of the so-called "uni-lateral" [or "one-sided"] properties (p. 109), *i.e.* those where the dissolved substances alone are efficient, but not the solvent. Thus the conductivity, the colour of a substance when dissolved in a colourless solvent, etc., are "uni-lateral" properties; but not the density, the refractive power, the internal friction, and the like, because here the solvent contributes to the values measured. In the latter case this problem must first be solved, viz. the determination of the very interesting action exerted by the *dissolved substance.* We have, in the preceding sections, several illustrations of the way that this is done in special cases.

When this shall have been finally accomplished, viz. the bringing of the property in question into an additive form, as by combining it with another property, according to circumstances (as, *e.g.* by combining the refraction of light with the density), then there remains the further question, viz. *How far does each particular ion contribute to the observed value of the property?*

In certain instances, only one ion is active, as in the case of salts, where a coloured or an optically active radical is combined with a colourless or optically inactive radical; then in such a case the preceding question answers itself. In other cases, the attempt to answer the question has not met with good success, as shown, *e.g.* in the increase of the volume of water on dissolving a salt; on the other hand, the attempted explanation was successful in the conductivity, and in the capacity of diffusion, and also in the case of those properties regarding the nature of which we have some clear conceptions.

Not until these preliminary questions are solved, can we undertake with safety the investigation of the problem *as to how far the properties considered are dependent upon the nature of the ion.* But, in general, it may be predicted that the property will have a more or less decidedly *constitutive nature;* therefore the point of view attained in the paragraph on the *résumé* of the properties of electrically neutral molecules (p. 342) finds here its full value.

Moreover, in addition to the composition and constitution of the ion, it appears that sometimes another factor in the case of ions is of considerable importance, viz. the *magnitude of the electric charge, i.e.* the same ion may conduct itself with varying electrical valencies. This question regarding the influence of a change in the electric charge on the physical properties, although interesting in many respects, has not yet been subjected to a systematic examination.

Finally, *regarding the electrolytes which are not completely dissociated,* this question arises, viz. what part of the observed value of the respective properties of the dissolved substances is due to the electrically neutral components, and what part is due to those which are dissociated. This question decides itself in the conductivity, since the alone is active. For those cases where the problem ve *a method for determining the degree of electrolytic dis-* most important result of the two preceding chapters the theorem: *that the ions show all the properties ecules as well as certain new ones due to the electrical*

CHAPTER IX

THE ATOMISTIC THEORY OF ELECTRICITY

General.—In the two foregoing chapters we have been concerned with the properties of free ions, that is with positively or negatively charged molecules; in the third book we shall see that the theory of free ions is of the greatest importance in understanding numerous chemical processes.

We are thus naturally led to the question of the nature of electricity itself, a problem which certainly belongs more to the region of pure physics, but is of such extreme importance for the theoretical chemist that some mention of it appears to be in place here, the more so as a chemical conception of electricity has come forward in recent times.

It is not superfluous to warn the reader against a misunderstanding of the recent development of electricity which was associated with the names of Maxwell and Hertz: a misunderstanding that is somewhat widespread, and which makes the following theory difficult or impossible to understand. It is generally known that recent physics has been occupied with electrical vibrations; it is clearly under this influence that a notion has arisen that the so-called fluid theory of electricity regarding the latter as a corporal agent is superseded, and now one often finds the entirely needless supposition that electricity is a state of vibration. The electromagnetic theory of light has given a conclusive proof that the phenomenon of light which has long been known to consist of wave motion is essential by an electrical phenomenon, or in other words, that there is no essential difference between optical vibrations and electrical. Hence optics has become a special chapter in the theory of electricity, just as has long been the case with magnetism. But the question as to the nature of electricity remains the same as before.

This may be made clear by an example. Physics has shown that the sensations of tone are to be attributed to vibrations of the air, hence acoustics becomes a special branch of hydrodynamics, viz. the theory of vibrations of gaseous substances. If any one were to say on the strength of acoustics that air was a state of vibration, the absurdity of this expression would be obvious; yet the same false conclusion has been drawn recently with regard to electricity. Purely chemical investigations have shown us the nature of air, and the development of acoustics has hardly even contributed a detail towards it. Similarly we may look for knowledge of electricity, and perhaps also of the luminiferous ether, to investigations that have a great similarity to the methods used in chemistry.

Ion and Electron.—The consideration of electrolytic phenomena has shown that the quantity of free electricity which the ions are capable of taking up is as invariable a quantity as that of the atoms of the chemical elements; the simplest explanation of this behaviour is that put forward by Helmholtz that electricity itself has an atomistic structure, that therefore we must assume the existence of *positive and negative elementary particles.*

Just as the law of constant and multiple proportions was unintelligible without an atomistic conception of matter, so the existence of ions—that is, of atoms or radicals which are always capable of taking up a defined quantity of free electricity, or twice or three times that quantity—would be quite unintelligible if electricity were a continuum. But if we attribute an atomic structure to electricity which we know is as indestructible as matter, and can be regarded as an elementary substance like the chemical elements, it is clear that each ion can only combine with a whole number of electrical elementary particles.

Thus we have to assume two new univalent elements whose atoms exert no, or very little, Newtonian attraction on the other elements but act on each other, similar atoms repelling each other, and dissimilar attracting (Coulomb's law). We shall use the signs $\oplus$ and $\ominus$ as symbols for these elements. The chemical compounds of these two univalent elements with the others are to be considered ions; Faraday's law is then nothing else than the law of constant and multiple proportions, applied to the compound atoms of ordinary elements with the positive and negative electrons.

The compound $H\oplus(=\overset{+}{H})$ arises by substitution of a positive electron in a molecule of hydrochloric acid (HCl) in the place of the chlorine; similarly, if a negative electron be substituted for the hydrogen, we obtain the compound $Cl\ominus(=\bar{Cl})$; double substitution of negative electrons in the molecule of sulphuric acid H_2SO_4 in the place of hydrogen gives the compound $SO_4<^{\ominus}_{\ominus}(=\overline{\overline{SO_4}})$, and so on. This formation of free ions comes into the scheme of valence (p. 277). The dualism which is found in the valence theory, and which leads to a distinction into positive and negative elements, is to be explained by the faculty of some elements (such, for instance, as strongly positive hydrogen) to combine with positive electrons, with others (such, for example, as strongly negative chlorine) to combine with negative electrons.

The ions behave as saturated chemical compounds in the sense of the valence theory, as may be seen from their behaviour towards molecular compounds; as we saw on p. 375 that ions are capable of replacing ammonia molecules in the platin-ammonia compounds.

The relation between positive and negative electrons recalls that between optical isomeric twins (p. 293). It is a question of much importance whether a compound of the positive and negative electrons ($\oplus\ominus$ = neutron, an electrically neutral massless molecule) really exists; we shall assume that neutrons are everywhere present like the luminiferous ether, and may regard the space filled by these molecules as weightless, non-conducting, but electrically polarisable, that is, as possessing the properties which optics assumes for the luminiferous ether.

Free Electrons.—The chemical theory of electricity put forward above raises the question whether the elementary atoms of electricity can be isolated, or whether they are only found in chemical combination with ordinary atoms or radicals.

The following consideration [1] makes it probable that free electrons can exist at least momentarily. If we consider the chemical equilibrium (see Book III. Chap. I.)

$$\overline{Cl} + Br = Cl + \overline{Br}$$

which exists kinetically in aqueous solution, we must suppose that an exchange of negative electrons occurs between the bromine and chlorine atoms, and hence we are driven to assume that the electrical atoms must be capable of independent existence apart from matter at least momentarily. The question as to the amount of free electrons in that or similar cases has not been answered so far.

It has been found possible by two quite different means to isolate at least negative electrons, that is, to get them free from original matter. Closer investigation of the cathode rays discovered by Hittorf has led to the knowledge (Wiechert, 1897) that they consist of negative electrons projected with great velocity. By studying the action which electrostatic and electromagnetic fields of force exert on the cathode rays, it has been found possible to measure the velocity of these electrical particles with certainty; according to W. Kaufmann's measurements [2] for discharge potentials of 3000 to 14,000 volts, it amounts from 0·31 to 0·68 . 10^{10} cm. per second, that is, from $\frac{1}{10}$ to only $\frac{1}{4}$ of the velocity of light.

Phenomena very similar to this are shown as cathode rays in

[1] Nernst, *Ber. deutsch. chem. Ges.*, **30**. 1563 (1897).

[2] See also Riecke, *Lehrbuch der Physik*, 2nd edit., vol. ii. p. 348 ff.

strongly evacuated Geissler tubes, and we meet them in an entirely unexpected manner in the remarkable rays discovered a few years ago by Becquerel. Becquerel found, first, that metallic uranium and its compounds can give off continuously rays which pierce comparatively thick masses of wood, glass, and even metals, which are photographically active, and make air slightly conducting. The active substance is not uranium itself, but is formed of small quantities of other elements of which radium has so far been isolated with certainty, and is considered a homologue of barium, with an atomic weight between 230-250.

The compounds of radium possess the faculty of giving off Becquerel rays with surprising strength, so that these preparations, which are due in the first instance to M. et Mme. Curie, have been more precisely investigated. The noteworthy result has been obtained in this case also that the emission consisted usually of cathode rays, that is, of free negative electrons. Their velocity appears to lie between 2·4 to $2{\cdot}8 \,.\, 10^{10}$, that is, only a trifle less than the velocity of light $(3 \,.\, 10^{10})$.

Their high velocity compared with that of the cathode rays explains the difference between their properties and those of cathode rays, especially their much greater penetrative power.

It has not so far been found possible to prove the existence of free positive electrons in any similar way. Probably the positive electron has much greater affinity for ordinary atoms and radicals than the negative, so that its isolation is much more difficult.[1] There appears to me at the present time to be no reason to doubt that positive electrons might be isolated.

For numerous details, especially on the experimental side of this remarkable region of investigation, see the exposition by E. Riecke, *Lehrbuch der Physik*, 2nd ed. vol. ii. p. 339.

The fact that negative electrons can be projected by electrostatic forces in the cathode rays, and by chemical forces in the Becquerel rays, and that they travel with constant velocity (in a vacuum) shows that, like ordinary matter, they possess a certain inertia or mass. This mass can be calculated from the magnetic or electrostatic deviation of the rays, and has been found in the cathode rays to be $\frac{1}{2000}$ of the mass of a hydrogen atom. But this mass is considerably greater at higher velocities. Apparently it is not a mass in the ordinary sense, but an inertia of electromagnetic character, due to the action of the moving electrons on the luminiferous ether.[2]

It should be mentioned that the theory given by Lorentz of the Zeemann phenomenon (see Riecke, *l.c.* p. 447) is based on the vibrations of free negative electrons round positively charged metallic

[1] See Nernst, *Bedeutung elektrischer Methoden und Theorien für die Chemie*, Göttingen, O. Vandenhoek u. Ruprecht, 1901, p. 25.

[2] See H. A. Lorentz, *Physikal. Zeitschr.*, **2.** 78 (1900); Abraham, *Gött. Nachr.*, Heft 1, 1902, and W. Kaufmann, *ibid.*, Heft 3, 1903.

atoms, and that, in this case also, the mass of the negative electrons is about the same as in the cathode rays.

Action of Electrons on Gases.—In the previous section we saw that rays of electrons are characterised by their penetrative power, by their electrostatic and electromagnetic deviation, by their action on photographic plates, and by the phosphorescence they produce. All these actions therefore are due to the enormous velocity of the electrons; it is therefore of great importance that we possess indications of slowly moving or motionless electrons in the action of electrons on gases.

It has been observed in fact that electrons immediately form ions, that is, electrically charged molecules in a gas; it is true that the ions are of quite a different kind to those known in the theory of electrolytic dissociation. To distinguish these from the electrolytic ions, since they are produced only in gas, we shall describe them as *gas ions*.

The numerous observed facts on this subject may be simply shown and described by means of the following conceptions.[1] With free electrons in a mass of gas, for example air, each electron attaches to itself a certain number of molecules of the gas and so forms a *gas ion*; if the velocity of the electron is very great, it may produce a relatively large number of positive and negative gas ions.

The formation of gas ions may be most simply conceived as due to the dielectric attraction of the free electrons of the gas molecules, just as on p. 382 we were obliged to assume that an electrolytic ion surrounded by a sheath of water molecules can cause a contraction of the solvent. The complex thus produced must go into thermal equilibrium with the other molecules, that is, it must assume an amount of kinetic energy corresponding to the temperature of the gas.

If the free electron possesses a very high velocity, it may break up a large number of neutrons in the course of gradually losing its excess of kinetic energy, that is, it may form new positive and negative electrons, each of which, according to its kinetic energy, will either form a gas ion at once, or will break up fresh neutrons.

As with electrolytic ions in a solvent the gas ions impart a certain conductivity of the gas. Cathode rays which penetrate into a gas through a thin metallic window from a Geissler tube, or Becquerel rays, projected by a radio-active preparation brought into the mass of gas, give to the gas a noticeable conductivity, since the negative electrons thus entering with high velocity produce a considerable number of positive and negative gas ions.

This conductivity, unlike that of ordinary molecules, is not stable

[1] See the lengthy exposition of J. Stark, *Elektrizität in Gasen*, Leipzig, 1902, by Ambrosius Barth; a short but extraordinarily clear and precise summary by Langevin, *Recherches sur les gaz ionisés*, Thèse, Paris, 1902; *Ann. chim. phys.* [7], **28.** 289 and 433 (1903).

or lasting, re-combination of the gas ions, or, more correctly, the combination of the electrons contained in a positive gas ion with the electron contained in a negative gas ion causes it to disappear in a short but measurable time. Here, also, then is an important difference between the electrolytic and the gas ions ; in the former equilibrium is produced in an unmeasurably short time, in the second case quickly, but not so fast as to prevent any measurement.

Many measurements have shown that the law of Guldberg and Waage (Book III. Chap. I.) is applicable to the recombination of the gas ions ; if η is the concentration of positive, and η' of positive ions, we have

$$-\frac{d\eta}{dt} = K\eta\eta';$$

usually but not always η can be put $=\eta'$. Apparently the gas molecules with which the free electrons are loaded cause the re-combination of the two kinds of gas ions to be measurably slow, though probably the reaction

$$\oplus + \ominus = \oplus\ominus$$

which takes place in a vacuum, that is, in the luminiferous ether, occurs with a prodigious velocity, so that ether cannot be made conducting.

It is of very great importance that the chemical nature of the gas is of secondary consequence to the degree of ionisation ;[1] according to the views put forward here, it is primarily the dissociation of neutrons which happens, gas molecules play only a small part in forming a sheath round the free electrons in consequence of dielectric forces. According to the customary view that the ionisation consists primarily in breaking up the gas molecule into a positive ion and a free negative electron it might be expected, contrary to the observations, that the chemical nature of the gas would be of as much importance as in electrolytic dissociation.

Other Methods of Formation of Gas Ions.[2]—Gases can be made conducting by various ways other than by rays of negative electrons. Apparently the reaction

$$\oplus\ominus = \oplus + \ominus$$

is always produced by a sufficiently high potential gradient, such, for example, as that at a charged point, when gas molecules are present which can attach themselves to the free electrons and so delay their

[1] See Stark, *l.c.* p. 74. For distinction from the stable electrolytic dissociation of solutions, salts, and also gases (p. 378), the transitory conductivity of gases by means of gas ions may be described as ionisation.

[2] Our knowledge of gas ions is mainly due to the valuable investigations of J. J. Thomson and his students (Rutherford, Zeleny, McClelland, Wilson, Townsend, and others) in Cambridge, 189-71900.

re-combination. At any rate gases can be made decidedly conducting by means of the so-called point discharge.

Röntgen rays have a very strong ionising influence on gases; it is not unlikely that the violently impulsive ether waves, such as Röntgen rays are supposed to be, produce potential differences capable of decomposing neutrons.

Whether the ionic content of flame gases is due to spontaneous formation of gas ions by high temperature is doubtful; it must be observed in passing that at high temperatures gases possess a certain amount of stable conductivity depending upon the chemical nature of the substance, and due to ordinary electrolytic dissociation (p. 378).

Mobility of the Gas Ions.—The number of gas ions present in the gas at the highest degree of ionisation hitherto produced is very small compared with the whole number of molecules in the gas, less than one gas ion in a billion gas molecules.[1] Hence the gas ions, like an extremely dilute electrolytic solution, must have a mobility independent of their concentration.

The conductivity K of an ionised gas is therefore

$$K = k\eta + k'\eta',$$

where η and η' are the concentrations, k and k′ the mobilities of the positive and negative gas ions.

The values of k and k′ have been obtained by widely differing methods; without going into experimental details we will give here the most reliable values given by Zeleny.[2]

	k	k′	$\frac{k'}{k}$	Temp.
Dry air	1·36	1·87	1·37	13·5
Damp air	1·37	1·51	1·10	14
Dry oxygen	1·36	1·80	1·32	17
Damp oxygen	1·29	1·52	1·18	16
Dry carbon dioxide	0·76	0·81	1·07	17·5
Damp carbon dioxide	0·82	0·75	0·92	17
Dry hydrogen	6·70	7·95	1·19	20
Damp hydrogen	5·30	5·60	1·05	20

The numbers are the velocities in cm. per second which the gas ions assume with a potential gradient of 1 volt per cm.

On comparison with the mobility of the hydrogen ion (p. 369) we see that the positive ions in dry air move some 400 times faster under the same potential gradient; evidently the friction of the ions in air

[1] Langevin, *l.c.* p. 322.

[2] *Phil. Trans.*, **195.** 193 (1901).

must be very much smaller than in water; the positive gas ions in hydrogen travel 2000 times faster than the hydrogen ions in water.

In spite of these relatively great mobilities the conductivity due to ionisation in gas is very small because the content in gas ions, as remarked above, is less than 10^{-12} of the gas, or since the latter at atmospheric pressure contains only $\frac{1}{22}$ of a mol per litre the normality only reaches about $0{\cdot}45 \cdot 10^{-13}$. Even in hydrogen, in which we have seen the mobility is 2000 times that of the hydrogen ion, the conductivity is only of the same order of magnitude as that of an acid solution of 10^{-10} normal. Consequently the conductivities of the ionised gases must usually be determined by electrostatic measurement, but even the small amount of electricity required to charge an electrometer usually causes noticeable changes of concentration in the gas ions. Consequently the methods of investigation of the conductivity of ionised gases differ considerably from these for the investigation of electrolytic conductivity. They cannot be discussed here however.

The numbers given above for the mobility refer to gases at atmospheric pressure; according to the kinetic theory (p. 308), the mobility of a gas ion must be inversely proportional to the number of collisions it makes with the gas molecules; in other words, it must increase in inverse proportion to the pressure of the gas.

This conclusion is confirmed by measurements made at pressures of 0·1 to 0·2 atmospheres; for lower pressures than this the mobility increases much faster than inversely proportionally to the pressure; this is the case especially with the negative gas ions.[1]

This result clearly means that at low pressures the number of gas molecules which surround the electron diminishes, and that hence the gas ion is more mobile. The circumstance that this diminution is more noticeable in the negative than in the positive gas ions confirms the conclusion drawn on other grounds that the positive electrons hold ordinary matter more firmly than negative, so too the invariably greater mobility of the negative ion as compared with the positive (see the ratio $\frac{k'}{k}$ in the table, p. 396) shows that the number of gas molecules attached to a positive electron is greater than to a negative.

The great influence that water vapour exercises on the mobility, and also, as may be seen from the table in the following section, on the diffusivity, is clearly to be attributed to the greater number of water molecules which attach themselves to electrons, perhaps on account of their greater dielectric constant (see also the next section).

Diffusion of Gas Ions.—The theory of the diffusion of electrolytes in dilute solution given by the author is applicable without change to

[1] See Langevin, *l.c.* p. 514.

the diffusion of gas ions. The diffusion coefficient is therefore, according to p. 370 (7),

$$D = \frac{2UV}{U+V} RT \quad . \quad . \quad . \quad . \quad (1)$$

Here, if instead of volts we introduce absolute units,

$$U = k10^{-8}, \quad V = k' . 10^{-8}.$$

We have found for the gas constant of univalent ions, p. 369 (6),

$$RT = 2{\cdot}351 \frac{T}{273} . 10^6,$$

and since an equivalent of an n-valent ion possesses one nth of the osmotic pressure we have for n-valent ions

$$RT = \frac{2{\cdot}351}{n} \frac{T}{273} . 10^6,$$

and thus finally

$$D = \frac{2kk'}{k+k'} \frac{0{\cdot}02351}{n} \frac{T}{273} \quad . \quad . \quad . \quad . \quad (2)$$

Townsend[1] has given the following values of the diffusion constant at 15° (cm.^{2} sec.^{1}) :—

	Obs.	Cal.
Air dry	D=0·0347	0·0391
Air damp	D=0·0335	0·0356
Oxygen dry	D=0·0323	0·0384
Oxygen damp	D=0·0323	0·0386
Carbon dioxide dry . . .	D=0·0245	0·0194
Carbon dioxide damp . .	D=0·025	0·0194
Hydrogen dry	D=0·156	0·1805
Hydrogen damp	D=0·135	0·1345

According to p. 210, the diffusion coefficient, for example, of CO_2 into NO_2 is 0·089, and since these two gases have almost the same free path and molecular weight, the diffusion of carbon dioxide molecules in carbon dioxide (Maxwell's "*Diffusion into itself*") must be equally great. Now we saw, p. 372, that an electrolytic ion has about the same mobility as an ordinary molecule, we may therefore reasonably expect that the mobility, and hence the diffusivity of a gas ion would be the same as that of an ordinary molecule if the two are of the same magnitude. We find, however, that in carbon dioxide—

[1] *Physikalische Zeitschrift*, **1.** 313 (1900) ; more fully, *Phil. Trans.*, **193.** 129 (1899).

and other gases behave similarly—the gas ion travels only about one-third as fast as the carbon dioxide molecules (0·0245 as compared with 0·089); hence the assumption is made very plausible that a greater number of molecules is grouped round the electron in a gas ion (Langevin, *l.c.* p. 332, assumes that there are 7).

Carrying out for gas ions calculations analogous to that on p. 370, we find agreement between the observed and calculated diffusion coefficients when

$$n = 1 ;$$

this indicates *that a gas ion is electrically univalent, that is, it is only a single positive or negative electron.*

We saw on p. 365 that the diffusion of an electrolyte produces potential differences when the mobility of the positive and negative ions is different. The corresponding phenomenon for gas ions has been demonstrated by Zeleny,[1] who stated that a metallic conductor dipping into an ionised gas is negatively charged because the more mobile negative gas ions diffused to it faster than the positive.

Condensation by Gas Ions.—When air saturated with water vapour is expanded it cools, and therefore becomes supersaturated with water, which, when there is dust in the air, condenses to form a cloud; when the air is free from dust it remains supersaturated; but when the supersaturation is sufficiently great, gas ions are capable of acting as centres of condensation, so that we have an important new means of detecting them.

J. J. Thomson,[2] who, with his students, cleared up this phenomenon by his brilliant researches, showed that when the ratio of expansion is greater than 1·33, both the positive and negative gas ions act as nuclei for the condensation of cloud drops; if it lies between 1·29 and 1·33 all the negative but only some of the positive ions act as nuclei. When the ratio of expansion is between 1·27 and 1·29 only the negative ions act as nuclei; and below 1·27 these cease to be effective.

The action of the gas ions just described is probably, like the attachment of gas molecules to the electrons, referred to dielectric forces; in the last chapter of this book we shall return to this phenomenon, which is of the greatest importance for atomistics. It is especially remarkable that the negative ions produce condensation more strongly than the positive.

Positive and Negative Elements.—The electron theory allows us to form a clear picture of the relation between the dualistic and unitary conceptions. The different elements and radicals have different chemical affinities towards the positive and negative electrons;

[1] *Phil. Mag.*, **46.** 134 (1898).
[2] *Ibid.* [5], **46.** 528 (1898); [6], **5.** 346 (1903).

those which have a strong tendency to combine with the positive electrons form positive groups of elements ; similarly negative elements are characterised by their affinity for negative electrons. Besides this the different elements exercise a chemical attraction on each other which is not of a polar character. Accordingly, two atoms of an element may form a stable chemical compound without any action on the part of the electrons ; the stability with which the two hydrogen or two nitrogen atoms unite with one another in the molecule is an example of this. The same is true of many of the compounds of the metalloids with each other, such as iodine chloride, phosphorus sulphide, and so on. So too the metals make numerous compounds with each other, in which we have no reason to suppose that the electrons take part. Carbon especially, which forms a transition between the well-marked positive and negative elements, reacts with both groups of elements, and since electrons seem here to be out of the question, the possibility of a purely unitary conception of the carbon compounds is noticeable.

But so soon as a positive and a negative element react with one another we have separation of ions, that is, the additional separation of a massless electrically neutral molecule is associated with this chemical process ; it is remarkable that this process involves a far deeper change in the ordinary behaviour than one in which the electrons do not take part ; whilst the compounds of the metals with each other retain their metallic character, and similarly, compounds of metalloids recall the properties of their components, when a metal reacts with a metalloid something entirely new and peculiar is produced. Substances like sodium chloride show the greatest possible difference of character from that of their components, and in the formation of such compounds clearly the chemical forces acting are the most intense.

It is of course possible that in the non-polar reactions electrical forces are in the background, just as it has often been hoped to refer the Newtonian attraction to electrical phenomena, and as has been accomplished with optics. That, however, is a matter of the future ; at present it is necessary to distinguish the forces of polar character from unitary actions.

The views put forward here admit of supposing that an element or radical can react with a positive or negative electron, without a simultaneous reaction between another element of the opposite character combining with the electron which is set free. If this happens the free electrons should possess a certain dissociation pressure in analogy with that of ordinary chemical processes ; this would cause differences in the kinetic energy of the electrons given out ; possibly the Becquerel rays are due to some such chemical process.

In any case the recent investigations in the region of the electron theory have an importance for purely chemical questions that can hardly be overrated.

CHAPTER X

THE METALLIC STATE

General.—Matter is found in a singular condition as metal, and although transition stages are known, there is on the whole a sharp distinction between this and the non-metallic state.

The noticeably distinct position of the metallic substances may be seen in the first place in their facility for combining to form homogeneous mixtures with one another, but not with non-metallic substances; no single non-metallic solvent is known for any metal apart from obvious cases of chemical combination, nor are any cases of isomorphous mixtures of a metallic and a non-metallic substance known.[1]

An equally important criterion for the metallic state is the power of conducting electric currents "*metallically,*" that is, unlike electrolytic conduction, without any transport of matter; metallic and electrolytic conductors behave so differently that they are distinguished as conductors of the first and second class.

The opacity of metals is closely related to their electric conductivity; there are no known non-metallic substances which are so opaque in thin films as the metals. According to the electromagnetic theory of light, the opacity of the metallic state is due to the absence of material transport associated with electricity, the conduction in the metal is therefore practically free from inertia in contradistinction to electrolytic conduction; the metals behave as good conductors even towards electric oscillations as rapid as the waves of light.

It is noteworthy that metallic conductivity is usually at ordinary temperatures of a much higher order of magnitude than the highest known conductivity of electrolytes, and it seems from numerous observations that metallic conduction would assume extraordinary values in the neighbourhood of the absolute zero of temperature at which electrolytic conduction would almost vanish; and on the other

[1] The fact that hydrogen and the other gases, for which many non-metallic solvents are known, are also occluded by many metals, may perhaps be regarded as an exception, but occlusion by metals is probably merely a phenomenon of adsorption, p. 129.

hand, at very high temperatures, whilst electrolytes gain considerably in conductivity, metallic conduction tends to cease.

Certain partly conducting materials, such as carbon, silicon, tellurium, and similar elements lie on the boundary between the metallic and non-metallic states.[1] Metallic conduction in chemical compounds is rare and not well marked.[2]

The characteristics of the metallic state vanish in gases; vapours of metals mix with all other gases and show no trace of metallic conductivity.

Metallic Solutions.[3]—The general behaviour of liquid metallic mixtures is quite analogous with that of mixtures of non-metallic substances. Metals are known, for example lead and tin, which mix in all proportions in the liquid state and behave in this respect therefore like water and alcohol; others, such as zinc and lead, form two separate layers which only dissolve each other to a small extent, and recall the behaviour of water and ether. Moreover, this limited solubility is much the more frequent case.

When a liquid mixture freezes, usually one of the components separates out in the pure state first, and in the same way the components of a metallic mixture can usually be separated by fractional distillation. Consequently the thermodynamic theory of metallic mixtures follows the same lines as that for non-metallic substances, and in particular, molecular weight determinations in dilute solution can be made on metals as on other substances. The most important result that has been obtained in this respect is that the metals in dilute solution are for the most part monatomic; this has also been shown by vapour pressure determinations for mercury and cadmium vapour; it is of course possible that the dissolved metal forms a compound with the solvent, but this can only contain a single atom of the dissolved metal.

Molecular weight determinations in metallic solutions have been made according to the Raoult van't Hoff method of Ramsay[4] and Tammann,[5] and recently in a very complete manner by Heycock and Neville;[6] the most important results of their investigations have been collected by Bodländer in the publication referred to.[3] They found[7] that *tin* is deposited in the pure state from alloys with sodium, aluminium, indium, copper, zinc, silver, cadmium, gold, mercury, thallium, lead, and bismuth; that the molecular weight of these metals

[1] See also the arrangement of the elements, p. 189.

[2] See also F. Streintz, *Leitvermögen gepresster Pulver*, Stuttgart, 1903, F. Enke.

[3] See also the article "Chemische Natur der Metall-legierungen," by F. Forster, *Naturw. Rundschau*, **9.** Nos. 36-41 (1892); and "Konstitution einiger Legierungen," by G. Bodlander, *Berg und Huttenmännische Zeitung*, 1897, Nos. 34 and 39.

[4] *Zeitschr physik. Chem.*, **3.** 359 (1889).

[5] *Ibid.*, **3.** 441 (1889).

[6] *Journ. of Chem. Soc.*, since 1888.

[7] *Chem. Zentralblatt*, **1889,** Bd. I. p. 666.

in the liquid mass is the same as their atomic weight; that, therefore, the molecules consist of single atoms like those of vapour of mercury. Aluminium and indium form double atoms in tin. *Antimony* alone raises the melting-point of tin. *Sodium*[1] deposits in the pure state from alloys containing gold, thallium, mercury, and lead, and these metals occur in sodium in the form of single atoms. Cadmium, potassium, lithium, and zinc lower the melting-point of sodium less than the same number of atoms of the former metals, they therefore probably form double atoms as well as single. *Bismuth*[2] also separates in the pure state from alloys with most metals, and the following metals form monatomic molecules in fused bismuth: lead, thallium, mercury, tin, palladium, platinum, gold, cadmium, sodium, and silver. Zinc, copper, and arsenic form larger molecules, di- or tri-atomic; antimony raises the melting-point of bismuth and must therefore be deposited in the form of a mixture or compound. *Cadmium* forms compounds with silver, since the latter metal raises its melting-point; otherwise cadmium separates in the pure state. Antimony, platinum, bismuth, tin, sodium, lead, and thallium exist in these alloys in the state of monatomic molecules. Copper, zinc, mercury, palladium, potassium, gold, and arsenic form polyatomic molecules in fused cadmium. *Lead* seems always to separate in the pure state from its fused alloys. Whilst gold, palladium, silver, platinum, and copper form mainly monatomic molecules in fused lead, the molecules of the other metals in this solvent are larger. Antimony, cadmium, mercury, and bismuth in particular occur as double atoms in fused lead. If gold and cadmium are simultaneously dissolved in *tin*, the compound AuCd is formed in solution,[3] and crystallises out as such if the concentration is sufficiently great. The same alloy has been found in later investigations,[4] when gold and cadmium are dissolved in tin, thallium, bismuth, or lead. Silver forms with cadmium in solution in tin, lead, or thallium, the compound Ag_2Cd, but in fused bismuth Ag_4Cd. Gold combines with aluminium when both are dissolved in tin to form the crystalline compound Al_2Au, which is insoluble in tin. By adding aluminium to tin containing gold the whole of the gold is precipitated in this form, so that pure tin is left behind. Pure *thallium* separates from a solution of gold, silver, or platinum in thallium;[5] the dissolved metals, namely, gold, silver, and platinum, form monatomic molecules in thallium. Tin, bismuth, thallium, lead, antimony, and magnesium are monatomic in *zinc*, and all these solutions on freezing deposit zinc in the pure state.[6]

The dissolved metals, like ordinary dissolved materials, show the phenomenon of *diffusion* as might be expected, since they obey the laws

[1] *Chem. Zentralblatt*, **1889**, Bd. II. p. 1043.
[2] *Ibid.*, **1891**, Bd. I. p. 129.
[3] *Ibid.*, **1892**, Bd. I. p. 153.
[4] *Ibid.*, **1894**, Bd. I. p. 410.
[5] *Ibid.*, **1894**, Bd. I. p. 266.
[6] *Journ. Chem. Soc.*; **71.** 383 (1897).

of osmotic pressure. G. Meyer[1] has shown that the diffusion constants of metals dissolved in mercury are of the same order of magnitude as those of the salts in water, and that this is also true of the frictional resistance. Metallic powder can be welded at high pressure, as has been shown by Spring[2] in a series of interesting researches, so that even solid metals are capable of fusing into each other (see also p. 170).

Metallic Alloys.—Investigation of liquid metallic mixtures meets with considerable difficulty on account of their opacity, so that the experimenter usually finds a difficulty in distinguishing between a homogeneous liquid, an emulsion, or a solution filled with minute crystals. The investigation of these mixtures in the solid state is still more difficult, for here it is almost impossible to say whether the substance is amorphous or crystallised. The behaviour of the zinc-cadmium mixtures[3] is simple; the melting-point of the alloy falls with increasing content of cadmium from 418·5°, that of pure zinc, to 264·5°, the melting-point of an alloy with 75 atoms per cent of zinc. On the other hand, the addition of zinc lowers the melting-point of cadmium from 320·5° to 264·5° when an alloy of composition 75 atoms per cent of zinc is reached. This alloy, which apparently freezes in a homogeneous state at the last-named temperature, is completely analogous to the so called cryohydrates (p. 126), and the alloy of the composition stated formed on freezing is a chemical mixture of zinc and cadmium crystals. More frequently, however, compounds are deposited from such mixtures, and often of more than one composition, so that metals show a strong tendency to form in the solid state compounds analogous to salts with water of crystallisation, double salts, and the like. Many metallic alloys are undoubtedly deposited as an intimate mechanical mixture of such compounds of the pure metals. This may even be made visible by the study of etched figures produced by acting on a polished surface with nitric or similar acid.

The view put forward, especially by Bodländer, that most alloys have a structure like that of granite, more or less fine, would be more easily demonstrated if it were possible to prepare thin transparent sections of alloys as of granite, in which the crystals of the different substances might be seen by the microscope lying side by side. This would not seem possible on account of the opacity of an alloy even in the thinnest layers. However, Heycock and Neville[4] succeeded in accomplishing this problem with the aid of the invisible rays discovered by Röntgen. They took Röntgen photographs of thin plates of an alloy of sodium and gold. A fern-like growth of easily penetrable crystals

[1] *Wied. Ann.*, **61.** 225 (1897).
[2] *Zeitschr. physik. Chem.*, **15.** 65 (1894).
[3] Heycock and Neville, *Journ. Chem. Soc.*, **71.** 383 (1897).
[4] *Proceed. Chem. Soc.*, **1896-97,** 105.

of the shape of salammoniac crystals was found embedded in a less penetrable mass. The penetrable crystals consisted of pure sodium which crystallises in the regular system. The ground mass was formed of a eutectic mixture of small gold and sodium crystals, and on account of the gold, it is almost opaque to the Röntgen rays. Plates containing a very large proportion of gold show the reverse phenomenon of opaque crystals of pure gold in a more transparent mass containing sodium. Plates of pure sodium show no crystallisation, because the whole mass crystallises together, so that the individual crystals are unable to grow. In the alloy, on the other hand, at the beginning of the crystallisation single crystals are separated from each other by the less easily frozen eutectic mixture, and therefore have the power of developing. It is to be hoped that this method will be applied to the structure of alloys whose components show smaller differences in penetrability. E. Maey [1] has shown that the specific volume is an important means of investigating the chemical compounds that exist in alloys.

It seems to be definitely shown in many cases by recent experiments *that the faculty of forming isomorphous mixtures is very rare in metals;* this is the more plausible, since the formation of isomorphous mixtures between wide limits or in all proportions is almost exclusively met with in chemically analogous materials. The chemical distinction between two elements is almost always too great to allow the formation of isomorphous mixtures. Even in the *liquid* state, as we have seen, metals do not often mix in all proportions.

Electric Conductivity of Alloys.—Most physical properties of metallic mixtures have been but little investigated on account of the great experimental difficulties; it may be remarked that the specific volume of copper gold, antimony bismuth, and other alloys is an additive property; copper tin, silver gold, tin gold show a contraction, antimony tin, zinc cadmium, cadmium lead a marked dilatation. The specific heat of solid alloys, according to Regnault,[2] is decidedly additive; but it must be remembered that even chemical compounds in the solid state show the same capacity for heat as the solids from which they are produced; in other cases it is easily seen that the properties are considerably modified by mixture; this of course suggested the view that mixtures of metals form chemical compounds; this view is supported by the fact that the heat of mixture, for example, in mixing fused copper and zinc to form brass, and in the preparation of sodium amalgam, reaches an unusual magnitude, and even gives rise to phenomena of incandescence.

The study of the *electric conductivity of alloys* yields important conclusions on this point, and C. Liebenow [3] has put forward valuable

[1] *Zeitschr. physik. Chem.*, **38.** 292 (1901). [2] *Ann. chim. phys.* (2), **73.** (1840).
[3] *Zeitschr. für Elektrochemie*, 1897, Bd. IV. p. 201. It appears separately as: "*Der elektrische Widerstand der Metalle*," Knapp, Halle, 1898.

theoretical considerations on it. For want of space we can here only indicate the outlines of his theory. The conductivity of an alloy must in general be smaller than that reckoned for the conductivity of the pure metal according to the rule of mixtures, *because thermoelectric forces arise which oppose the electric current.*

Let us suppose a conductor, for simplicity, to consist of thin alternate layers of the two metals which form the alloy, then at the surface of contact Peltier effects will be produced (alternate heating and cooling), that is, the points of contact will be alternately heated and cooled, so that the conductor becomes a thermopile whose electromotive force must oppose the current. Since, as is easily seen, this opposing force must be proportional to the strength of current, its effect appears simply as an increase in the resistance of the alloy. The same effect must be found when the two metals are not arranged in plates, as was assumed above, but in the form of very small particles, as actually occurs.

We get for the specific resistance of an alloy the formula

$$C_0(1 + \gamma t) = A_0(1 + \alpha t) + B_0(1 + \beta t);$$

here A_0 is the true resistance at 0°, B_0 is the apparent resistance due to the thermoelectric effect, or what is physically the same, the opposing electromotive force per unit strength of the current; C_0 is the observed resistance, α, β, γ are the temperature coefficients.

It is found that the temperature coefficient of most pure metals is 0·004 (about the same as the expansion coefficient of gases), or, in other words, the resistance of most pure metals increases approximately in proportion to the absolute temperature, as was pointed out by Clausius (1858). As to the thermoelectric effect, it may be assumed in the first place that it is little affected by temperature; since the experimental results support this assumption we may put

$$\beta = 0$$

as the first approximation.

Then from these assumptions a series of interesting conclusions may be drawn:—

1. In general (when B_0 possesses a considerable value) the specific resistance of an alloy will be much greater than that of its components, but the temperature coefficient much smaller than that of the pure metals.
2. For metals which are thermoelectrically indifferent B_0 is small, therefore the specific resistance can be calculated from that of the components; the temperature coefficient for such alloys should be the same as that of the pure metals.
3. If the two metals in mixing form a compound, and if the proportions are taken so as to produce it, B_0 must vanish and the temperature coefficient must be the same as that of the

pure metals; but if either of the components is present in excess, B_0 must reach a noticeable value, and therefore the temperature coefficient must fall.

These conclusions were thoroughly tested by Liebenow so far as the experimental material would allow, and were found correct to a surprising degree; there is no doubt that this indicates a means of investigation of the constitution of alloys which will lead to the most important results. (See here the original treatment by the same author of the constitution of mercury.)[1]

The Theory of Metallic Conduction.—It is hardly doubtful from what precedes that the theory of metallic conduction should lead to a deeper knowledge of the metallic state, just as the theory of electrolytic conduction has led to a knowledge of the peculiar constitution of the electrolytes.

Only the beginnings of this line of progress exist at the present time. The first assumption is obviously that electric particles are present in the metals and travel in the direction of the current, but these, unlike the ions, appear to be massless. The remarkable conductivity of many metals leads to the supposition that the number of such particles is very great; this view is supported by the fact that matter in the metallic state has a very great density, *i.e.* that the numerous electric particles cause very great *electrostriction* (p. 382). The existence of massless electric atoms is rendered probable by the considerations on p. 391; according to these, we must regard the metals as solvents in which the dissociation

$$\oplus\ominus = \oplus + \ominus$$

reaches considerable magnitude.

If it is assumed that the free electrons and metals follow the principles of the kinetic theory of gases, and especially that their kinetic energy can be considered as in equilibrium with that of ordinary molecules, the ratio of thermal and electric conduction in the metals can be calculated in good agreement with experience (see Drude).[2]

[1] *Zeitschr. für Elektrochemie,* **4.** 515 (1898).

[2] *Ann. d. Phys.,* **1.** 566 (1900); Reinganum, *ibid.,* **2.** 398 (1900).

CHAPTER XI

COLLOIDAL SOLUTIONS

Colloids and Crystalloids.—From his examination of hydro-diffusion, Graham,[1] came to this conclusion, viz. that substances existing in solution fall into two groups, the physical and chemical behaviour of which are characteristically different. These differences appeared to Graham so pronounced that he designated them as "two different worlds of matter," the "crystalloidal condition" of matter being in the sharpest contrast to the "colloidal condition."

The difference between them is shown most clearly in their solutions. Up to this point we have considered only solutions of crystalloids; let us now turn to the examination of solutions of *colloids*.

The most important examples of the latter are silicic acid, aluminium hydroxide, ferric hydroxide, and many other metallic hydroxides; also starch-flour, dextrin, gum, caramel, tannin, white of egg, glue, etc. etc.; also colloidal solutions of certain elements, as silver, selenium, etc., have been recently prepared.

The distinctions consist in this, viz. that crystalloidal substances are characterised by the facility with which they assume the crystallised condition; while the colloids, which were named by Graham from their representative *glue* [κόλλα], are incapable of crystallising, or at least do so very slowly. Still more important are the differences between the conduct of these two classes when in a state of solution. It will be shown shortly *how pronounced is the difference between crystalloid and colloid solutions regarding their osmotic pressures.*

As has been mentioned already, Graham was led to his discovery of the peculiar nature of colloidal solutions by the study of their diffusion velocity; the diffusion of colloids, in general, resembles that of crystalloids, but is relatively a much slower operation. This is shown by the following time periods found by Graham to be required to effect the same degree of diffusion at 10°.

[1] *Lieb. Ann.*, **121.** 1 (1862).

Crystalloids.		Colloids.	
Hydrochloric acid	1	White of egg	49
Sodium chloride	2·33	Caramel	98
Sugar	7		
Magnesium sulphate	7	...	

Osmotic Pressure.—The question was formerly raised whether a colloidal solution was to be regarded as a simple emulsion, or as a true solution. But, at present, when we can detect a "head" of pressure directly from the phenomena of diffusion, we must in all probability suppose that there is no essential difference between solutions of colloids and of crystalloids.[1] But the extreme slowness of the diffusion of colloids speaks decidedly regarding two things, viz. on the one hand, the driving force must be very small, *i.e.* the osmotic pressure is very small; and, on the other hand, the resistant friction experienced by the molecules in their passage through the water must be enormous; both these conditions are explained by the assumption *that the colloids possess exceedingly high molecular weights.*

Now, as a matter of fact, the experiments conducted with colloid solutions show very small values for the osmotic pressure. Thus Pfeffer,[2] by the use of a membrane of cupric ferrocyanide, obtained the following values for gum-arabic at the respective concentrations, and at a temperature of 15°; the corresponding values for cane sugar are appended for comparison:—

Con.	Gum-Arabic.	Cane Sugar.	Mol. Wt. of Gum-Arabic.
1 %	6·9 cm. Hg	51·8 cm. Hg	2570
6 %	25·9 ,, ,,	310·8 ,, ,,	4110
14 %	70·0 ,, ,,	725·2 ,, ,,	3540
18 %	119·2 ,, ,,	932·4 ,, ,,	2680

The molecular weights of the gum-arabic given in the last column are obtained by the well-known rules (p. 256), from the mol. wt. (342) of cane sugar, multiplied by the ratio of the corresponding osmotic pressure.

In the following table are given some molecular weights, calculated in a similar way, from the osmotic pressures observed by Pfeffer:—

[1] Suspensions are to be distinguished from solutions in the fact that they exercise no measurable osmotic pressure, and also that they do not diffuse; this criterion has been overlooked in many recent researches.

[2] *Osmotische Untersuchungen*, Leipzig, 1877.

Substance.	Membrane.	Con.	Temp.	Pressure.	Mol. Wt.
Dextrin .	Cupric ferrocyanide	1 %	16°	16·6 cm. Hg	1080
Conglutin .	Calcium phosphate	2 %	16°	3·8 ,, ,,	9500
Glue .	Cupric ferrocyanide	6 %	23·3°	23·7 ,, ,,	4900
Glue . .	Parchment	6 %	23·3°	21·3 ,, ,,	5200

Recently Linebarger[1] has determined the osmotic pressure of colloidal tungstic acid, using parchment paper as the semi-permeable membrane, as it is impermeable for the dissolved colloid. A solution, containing 24·67 g. of H_2WO_4 per litre at 17°, showed a pressure of 25·2 cm. of Hg, which gives a mol. wt. of 1700; while that from the formula $(H_2WO_4)_7$ requires 1750, a fairly good coincidence; another solution containing 10 g. per litre gave a mol. wt. of 1720.

The Freezing-Point and the Vapour Pressure.—In harmony with the small osmotic pressure [of colloidal solutions] stands the observation, that *the freezing- and boiling-points of colloidal solutions are only slightly different from the respective values of pure water.* Tammann[2] showed that the vapour pressure of water is only slightly depressed by the addition of considerable quantities of gelatin and gum. The following table of figures, given by Sabanejeff and Alexandrow,[3] gives an illustration of the slight depression of the freezing-point of colloidal solutions:—

Quantity of White of Egg in 100 g. of water.	Depression of the Freezing-Point.	Mol. Wt.
14·5	0·020°	14,000
26·1	0·037°	13,000
44·5	0·060°	14,000

Of course such results as these are open to the objection that the observed depression of the freezing, small as it is, may be accounted for on the supposition that the colloid is impure, and therefore that the mol. wts. given in the third column are to be regarded *only as the lower limit.* But at all events it is an evidence that the remarkable magnitude of the mol. wt. of substances occurring in colloidal solution is one of their chief characteristics, and therefore the question

[1] Silliman's *Journal* [*Am. Jour. Sci.*] [3], **43.** 218 (1892).

[2] *Mem. d. Petersb. Akad.*, **35.** No. 9 (1887).

[3] *Journ. russ. Ges.*, **2.** 7-19 (1891); ref. in *Zeitschr. physik. Chem.*, **9.** 88 (1892).

proposed by Graham[1] ought to be discussed, viz. "whether the complex nature of the molecule is not the basis of the colloidal condition." Although colloidal solutions may represent the transition state intermediate between emulsions and true solutions, yet they doubtless resemble the latter much more than they resemble a coarse mechanical suspension.

The Molecular Weight of Colloids.—The osmotic method described above is particularly suitable for the further investigation of colloidal solutions, and especially for the more exact determination of their molecular weights. For while exact measurements of the lowering of the vapour pressure and of the freezing-point are open to question, because of the extreme difficulty in purifying the colloids from foreign salts, yet these impurities may be disregarded in the measurement of the osmotic pressure, because these salts can easily penetrate the parchment paper, and thus do not affect the observed pressure action. Moreover, the differences of the freezing- or boiling-points here, although very small, some hundredths of a degree, correspond to pressures which can be easily measured by the manometer.

The same method of investigation, when applied to a mixture of several colloids, would lead to some conclusions respecting *the mutual action of colloid molecules.* Of course the very dark question regarding the relative capacity of the colloid molecules for reaction claims especial interest.

In the following table are given some figures collated from the results by the freezing-point method :—

Substance.	Mol. Wt.	Observer.
Inulin	2200	Brown and Morris.[2]
Maltodextrin	965	
Starch	ca. 25000	
Gum	,, 1800	Gladstone and Hibbert.[3]
Caramel	,, 1700	
Ferric hydroxide	,, 6000	
Tungstic acid	,, 800	Sabanejeff.[4]
Glycogen	1625	
Silicic acid	at least 49000	
Tannin	1100	

Although the results may be uncertain on account of the smallness of the depressions observed, yet at the same time they undoubtedly show that colloids do have a very large mol. wt.

[1] *l.c.* p. 71.
[2] *Chem. Zentralbl.*, **2.** 122 (1889).
[3] *Ibid.*, 189 ; *Phil. Mag.*, **28.** 38 (1889).
[4] *Ibid.*, **1.** 10 (1891).

Moreover, it appears that other solvents besides water may be utilised to prepare colloidal solutions; thus caoutchouc dissolved in benzene shows a mol. wt. of 6500 (Gladstone and Hibbert); and tannin in acetic acid a mol. wt. of 1100 (Sabanejeff). Further, it should be mentioned that Graham explained the peculiar inactivity of colloid substances as a result of their high mol. wt.; also, from the fact that arabic acid, *e.g.* (*Gummi-säure*), could be neutralised by very small quantities of calcium hydroxide or potassium hydroxide, he concluded that its mol. wt. must be much greater than $C_{12}H_{11}O_{11}$, the formula commonly given.

Sabanejeff[1] attempted to found *a classification of the soluble colloids* from the magnitude of the molecular weights. He called those having a mol. wt. greater than 30,000 the *higher or typical colloids.* To this group belong starch, silicic acid, and ferric hydroxide, and probably also antimonic sulphide, cupric sulphide, and potassium, ferric, and silver tartrates. The group of the *lower colloids,* or those having a mol. wt. of less than 30,000, includes tungstic acid, molybdic acid, arabic acid, tannin, glycogen, inulin, dextrin, albumin, etc.

By allowing a solution of the lower colloids to solidify completely by refrigeration, then, on remelting this, a colourless solution is obtained, while by this same process the colloids of the first species [the higher or typical] become completely insoluble, or nearly so;[2] in other words, solutions of the "typical colloids" are supersaturated.

Gelatination.—Many colloidal solutions are able to coagulate or gelatinise, either on the addition of foreign substances or sometimes spontaneously. What the nature of this process is, and whether it is comparable to the process of crystallisation, cannot be stated at present. As a result of his investigations, Graham concluded that the addition of a foreign substance only tends to accelerate a process which is spontaneous; and, in fact, it appears probable that those colloidal solutions which are made to coagulate or gelatinise by the addition of foreign substances, must be regarded as supersaturated, and, therefore, in a state of labile (unstable) equilibrium. Other colloidal solutions, as those of gelatin, agar-agar, etc., *e.g.* will solidify only below certain definite temperatures, and will become liquid again only when raised to the same temperature. Here of course we have to do with a condition of equilibrium between the gelatinous and the liquid states, and this is a parallel to that between the salt and the saturated solution.

The simplest conception of the nature of gelatinised solutions is that a solid substance is separated which has a "network" structure, the interstices of which are filled with water, which is held fast by capillary forces. The greater part of this water which is held by

[1] *Jour. russ. Ges.*, **1.** 80 (1891); ref. in *Zeitschr. physik. Chem.*, **9.** 89 (1892).

[2] See Ljubavin, *Chem. Zentralbl.*, **1.** 515 (1890).

"capillary affinity,"[1] is removed by evaporation, but where the interstices of the network are very fine, the water is given off only by drying thoroughly.

If a dried jelly is brought into contact with water, it begins to swell. This process apparently indicates an inward suction of the water into the capillaries, resulting in a large increase of the volume. That the force which drives the water into the colloid substances is very great is obvious from the fact that blocks of granite are split by the swelling of wood. The work performed by the capillary and adhesion forces occasions the development of heat, although this is not very considerable. This was observed by E. Wiedemann and Lüdeking,[2] but was regarded by them as "heat of hydration."

By the addition of plenty of water, some colloids, as gelatine, glue, etc., can be brought back to solution; but others, as silicic acid, cannot be. As stated by both the observers mentioned above, the solution is attended *by an absorption of heat.* This is in harmony with the fact that the heat of solution of solids is as a rule negative. The coagulation of colloidal silicic acid is accordingly attended by the development of a corresponding degree of heat.

A gelatinised solution presents several peculiarities,—an intermediate thing between a solid and a liquid, its enormous internal friction relates it to solids. It also has clearly defined elasticities of deformation which clearly distinguish it from true pulpy substances. Yet it preserves many properties of liquid solutions; thus it can dissolve crystalloids and allows them to diffuse freely. Thus, according to the older researches of Graham, who worked with solutions of gelatine, and especially according to the more recent work of Voigtländer,[3] who worked with agar-agar solutions, it appears that crystalloids diffuse through gelatinised solutions *with velocities which are remarkably near to those with which they diffuse through pure water.*

In harmony with this, the electric conductivity of salts, which, as shown by the principles already established, is closely connected with the capacity of diffusion, is only slightly changed by the gelatinisation of the solution. A similar thing seems to hold good for the reaction capacity, as is inferred from an observation made by Reformatzky.[4] According to this, the velocity with which methyl acetate is decomposed by acids in a solution gelatinised by agar-agar, is within 1 per cent of the velocity in pure water.

By means of these properties we can explain some of the applications which have been made of gelatinised solutions, and which possess great interest.

[1] Graham, *l.c.* p. 72, and interesting researches by Linebarger on velocity of coagulation, *J. Amer. Chem. Soc.*, **20.** 375 (1898).

[2] *Wied. Ann.*, **25.** 145 (1885).

[3] *Zeitschr. physik. Chem.*, **3.** 316 (1889).

[4] *Ibid.*, **7.** 34 (1891).

Thus the sensitive layer of photographic plates which were *wet* by the old methods, by the new methods are made of a *dried* gelatinised solution or emulsion of the sensitive substance, so that their practical use is rendered much easier than was formerly supposed to be possible.

When galvanic elements are to be transported, the electrolytic liquid with which they are provided is gelatinised. This led to the construction of the so-called "dry cell."

The introduction of "culture gelatine" into bacteriology by Robert Koch marked a new epoch in this science.

Finally, it should be noted that the technology of explosives has experienced an extended regeneration by the use of gelatinised solutions of pyroxylin (smokeless powder), and especially so since Alfred Nobel gelatinised the pyroxylin with "nitroglycerin," and thus used the solvent itself as an explosive.

It is remarkable that *a gelatinised solution is unable to absorb another colloid*, because it retards its diffusion almost completely; for this reason it makes a semi-permeable membrane, which gives free passage to water and its dissolved crystalloids, but not to colloids dissolved in water.

This fact, which was known by Graham, and which he rightly regarded as very important, explains why a clearly defined osmose is to be observed only when an animal membrane, parchment, or unsized (*planirtes*) paper is used with colloid solutions. Membranes of this sort have the property of colloid solutions, viz., of offering not a very much greater resistance to crystalloids than to pure water, but they hinder colloids almost completely. Before the membranes, which are semi-permeable for many crystalloids, and which are described on p. 132, were known, the tendency was to ascribe a very strong osmotic activity to colloidal solutions, and it was not until Pfeffer's experiments that the contrary was known to be the case.

It cannot be certainly stated at present why a gelatinised solution is impermeable to a dissolved colloid. Perhaps it may be reasonable to suppose that the meshes of the network are too fine to allow the colloid molecules to diffuse, but are wide enough to allow the smaller crystalloid molecules to diffuse.

Dialysis.—The property of gelatinised solutions just described is very important for the preparation of pure colloid solutions, because it makes the separation of colloids and crystalloids possible. Thus let a piece of parchment paper or of animal membrane be fastened to the lower part of a frame, and let this be filled with the solution to be purified, and let it be dipped into pure water; then, after a sufficient lapse of time, the crystalloids in the solution diffuse through into the water, especially if the water bathing the frame be replaced from time to time with fresh water; the colloid substances remain behind in the frame [the dialyser]. Thus to prepare colloidal silicic acid, *e.g.* we a solution of sodium silicate by hydrochloric acid, and then o the "dialyser," for so the discoverer, Graham, called the aratus, the method being called "dialysis." After several dium chloride formed, and the excess of hydrochloric acid, ay, and there remains a solution of silicic acid which is olutely pure. As a result of the osmotic action of the dis- oid during the process, the solution [of the colloid] becomes h water.

ent literature on the subject of the colloids see Lottermoser, *loide, Sammlung chem. techn. Vorträge*, Heft 6; and *Theory of* ipzig, 1903 (Deuticke); and especially the work by Bredig, *erments*, Leipzig, 1901.

CHAPTER XII

THE ABSOLUTE SIZE OF MOLECULES

The Superior Limit.—We have treated thus far of the properties of molecules, of their kinetic conditions, of the various forces exerted by them, etc. In this concluding chapter we will discuss briefly the problem of *their absolute dimensions*, a problem which, to be sure, is usually treated more for its fascination than for any decided results afforded.

The assumption that matter is not infinitely divisible requires that these dimensions [of the molecule], no matter how small, must yet be finite.

Many experiences of daily life speak in favour of the view that the divisibility of matter extends to a high degree. Infinitesimal quantities of strong perfumes can impart their odour to whole rooms; they have entirely filled them with their molecules. Infinitesimal quantities of strong colouring substances can distinctly colour large quantities of water. One part of fluoresceïn, when dissolved in more than a hundred million parts of water, suffices to impart to it a distinct fluorescence. Faraday prepared gold-foil, the thickness of which, at the most, was 0·5 millionths of a mm. Röntgen[1] succeeded in preparing layers of oil on water, having about the same thickness.

The attempt has been made repeatedly from determinations of this sort to find the lower limit of the size of the molecule, or at least the limit of the *sphere of activity of the molecular forces*, *i.e.* the distances within which the mutual attraction of the molecules begins to have a "noticeable value." As is obvious, there is no precise definition of this sphere of activity, for it leaves the question open regarding the boundary between the "noticeable" (*merklich*) and the "unnoticeable" action; nor will this be possible until we know something certain regarding the law of the variation of the molecular forces, with their distances apart.[2] Nevertheless, by means of the capillary forces, an

[1] *Wied. Ann.*, **41.** 321 (1890).

[2] See, *e.g.* Bohl, *Wied. Ann.*, **36.** 334 (1889); Galitzine, *Zeitschr. physik. Chem.*, **4.** 417 (1889).

estimation of these magnitudes can be attempted; in this way van der Waals found the order of the magnitude to be about 0·2 millionths of a mm.

If it were possible to make lamellæ of a homogeneous liquid sufficiently thin, then the otherwise constant surface-tension (p. 57) would begin to diminish as soon as the thickness of the lamellæ should become smaller than the sphere of activity of the molecular forces; because, from this point, the resultant of the forces, which are directed towards the interior of the liquid, would begin to diminish. The phenomenon of the dark spots[1] in soap-bubble films has been employed for this purpose; but even disregarding the fact that such a heterogeneous and ill-defined substance, chemically, as "soap solution," must at the outset be discarded for investigations like these, it is stated by Drude[2] that the surface-tension of these dark parts is not noticeably different from that of the other parts; this observer determined their thickness to be 17×10^{-6} mm.; while Reinold and Rücker[3] found it to be 12×10^{-6} mm. A. Overbeck[4] conducted an interesting research, in which he determined the slightest thickness with which a platinum plate must be covered by another metal, in order that this layer may receive an electric charge; but this method cannot give a safe determination of the sphere of activity of the molecular forces, because the observations argue for the formation of an alloy (a solid solution, p. 169), on the surface of the platinum, to explain them.

A very interesting and promising method of detecting small particles whose refractive index differs sufficiently from that of the medium has been discovered by Siedentopf and Zsigmondy;[5] by strong side illumination they succeeded in producing a diffraction image of the particles which are made self-luminous by the illumination; this can be done even when their dimensions are small compared with the wave-length of light. The method was found to be applicable to particles of about $6 \cdot 10^{-6}$ mm., so that direct observation has already approached molecular dimensions. The authors applied their method to ruby gold glasses; they succeeded in showing that the gold is distributed in small particles in these glasses. Although no doubt the gold particles (colloidal molecules?) thus made visible consist of many single gold atoms, it is none the less of the highest interest that by refining the observation the apparent continuum can be resolved into discrete points.

In the following paragraphs I have collated the methods, by means of which, and guided chiefly by the sound doctrine of the molecular

[1] See J. P. Cooke's *New Chemistry*, p. 29 *et seq.*—Tr.
[2] *Wied. Ann.*, **43.** 158 (1891).
[3] *Phil. Trans.* (1881), 447; (1883), 645.
[4] *Wied. Ann.*, **31.** 337 (1887).
[5] *Ann. d. Phys.* [4], **10.** 1 (1903).

kinetic theory, an estimation of these molecular dimensions has been obtained. Caution in selecting the material to be described here is doubly necessary; for perhaps nowhere else in the natural sciences does fantasy so overshadow the critical faculty. We are not yet in the possession of a method[1] for measuring the molecular dimensions with the same degree of accuracy which is possible in the measurement of the wave-length of light, but fair approximations have already been reached.

The Space occupied by the Molecules.—According to the present view, the molecules do not completely fill the space which is occupied by the body as a whole, but are separated from each other by intervening spaces. We have already repeatedly studied properties which are to be regarded as measures of the volume actually occupied by the molecules (p. 307), or which at least are related more or less closely in this way. We will now proceed to see how far *the absolute value* of this magnitude is determined.

The first method to determine this is derived from van der Waals' theory, according to which the constant "b" of the equation of condition is equal to fourfold the volume of the molecule; now its value can be calculated, either by means of the deviation of gases from the law of the ideal gas condition, or from the relation that it is threefold as large as the critical volume; or finally, from the equation (p. 223), viz.

$$b = \frac{1}{8} \frac{\theta_0}{273\pi_0};$$

if we calculate the critical pressure π_0 in atmospheres, then our unit of volume will denote the space occupied by the gas under normal conditions of temperature and pressure, and the equation

$$x = \frac{b}{4} = \frac{1}{32} \frac{\theta_0}{273\pi_0} \quad . \quad . \quad . \quad \text{I.}$$

denotes the fraction of the volume actually occupied by the molecules.

The three methods given above to determine the value of b give only approximately coincident results, on account of the inexactness of the theory, to which reference has often been made; but these deviations are of no importance in view of the object here considered. We find abundant data for calculations of this sort.[2] Equation I. is naturally not applicable to substances that are strongly dissociated (p. 271).

[1] For a summary of the different attempts in this line, see v. Hemmelmayer, *Bericht der Lesehalle der deutschen Studenten in Prag.*, 1892. See also W. Thomson, *Exner's Rép.*, **21.** 182, 1885. [See also W. Thomson's (Lord Kelvin's) published works in English.—TR.]

[2] *Kontinuität*, etc., p. 136.

The Clausius-Mossotti theory of dielectra (p. 316) gives another method; according to this, when D denotes the dielectric constant, the value of the space filled is

$$x = \frac{D - 1}{D + 2} \quad . \quad . \quad . \quad . \quad \text{II.}$$

The following table, taken from a work of Lebedew,[1] contains some results calculated by these two independent methods. The dielectric constant D is obtained from the equation

$$\frac{D - 1}{D + 2} \cdot \frac{1}{d} = \text{const.},$$

which is well established, both by theory and by experiment; here d denotes the density reduced to the gas condition at 0°, and to a pressure of 1 atm. The table is based on the measurements of Boltzmann (1875), Klemencic (1885), and Lebedew (1891).

Substance.	x		$\frac{M}{22420\,x}$
	Acc. to I.	Acc. to II.	
Nitrous oxide . .	0·00048	0·00038	3·7
Carbon dioxide . .	0·00050	0·00033	4·0
Ethylene . . .	0·00056	0·00054	2·4
Carbon disulphide . .	0·00082	0·00097	4·3
Benzene . . .	0·00128	0·00123	2·7

Although the deviations between the two columns of figures are far greater than any error due to observation, yet it is very remarkable that the orders of magnitude coincide so well. Therefore it need not appear questionable if those values should be used to develop further conclusions.

The Density of the Molecules.—The specific gravity w of the molecules may be calculated directly from the value of x. 22,420 c.c. of a gas at 0°, and a pressure of 1 atm., contain 1 g.-mol. M; therefore the density of the molecule, referred to water at 4°, amounts to

$$w = \frac{M}{22{,}420x}.$$

In the third column of the preceding table are given the respective values of this quantity, on the basis of the value of x, according to van der Waals' figures.

[1] *Wied. Ann.*, **44.** 288 (1891).

Values of the same order of magnitude are obtained by calculating w from the dielectric constant; thus

$$w = d\frac{D+2}{D-1}.$$

The specific refraction (p. 307) also gives a measure of the specific volume $\frac{1}{w}$ of the molecule, which is at least approximate.

The Size of Molecules.—By combining these units, obtained as just described, with the formula of Clausius and Maxwell (p. 208), viz.

$$L = \frac{\lambda^3}{\sqrt{2}\pi s^2}, \quad \text{or } \frac{1}{\lambda^3} = \frac{1}{\sqrt{2}L\pi s^2} \quad . \quad . \quad . \quad (1)$$

we are led a step further. The mean wave-length L has been determined, with at least an approximate coincidence, by three different methods, so that we are certain of the approximate value. Then λ^3, a cube, the cross-section of which contains a molecule of the gas, may be denoted by the reciprocal value of the number of molecules N in unit volume, so that for the sum Q, of the cross-section of the molecules, we may write

$$Q = \frac{N\pi s^2}{4} = \frac{1}{4\sqrt{2}L}. \quad . \quad . \quad . \quad (2)$$

Now if we assume that the molecules are *spherical*, it obviously results that the following equation, viz.

$$x = N\frac{s^3\pi}{6} = \frac{2}{3}sQ \quad . \quad . \quad . \quad . \quad (3)$$

represents the volume of the molecules; and therefore the molecular diameter would be expressed by the following formula, containing only magnitudes which can be determined, viz.

$$s = 6\sqrt{2}xL = 8{\cdot}5xL.$$

Thus for carbon dioxide (p. 209) we find

$$s = 8{\cdot}5 \times 0{\cdot}000068 \times 0{\cdot}00050 = 0{\cdot}29 \text{ millionths mm.},$$

and similar values for other substances.

The Number and the Weight of Molecules.—The number of molecules contained in 1 cubic mm. of a gas at 0° and at atmospheric pressure is obtained directly from the formula

$$N = \frac{1}{\sqrt{2}L\pi s^2} = \frac{1}{72\sqrt{2}L^3x^2\pi} = \frac{1}{320L^3x^2}.$$

According to Avogadro's law this must be the same for all gases, *i.e.* L^3x^2 must be a constant.

Again, if we use the mean free paths derived from the internal friction (p. 209), we obtain—

	$L\times 10^6$	$x\times 10^5$	$L^3x^2\times 10^{20}$
Nitrous oxide . .	68	48	7·2
Carbon dioxide . .	68	50	7·8
Ethylene . . .	58	56	6·1
		Mean	$7\cdot 0\times 10^{20}$

Therefore it follows that the number of molecules N contained in 1 cubic mm. of a gas at 0° and at atmospheric pressure is

$$N = \frac{1}{320\times 7\cdot 0}10^{20} = 4\cdot 5\times 10^{16}.$$

Now 1 cubic mm. of hydrogen weighs 0·000090 mg.; therefore an atom weighs

$$\frac{0\cdot 000090}{2N} = 1\cdot 00\times 10^{-21}\text{ mg.},$$

and a molecule of any gas of mol. weight M has the weight expressed by the formula

$$y = M\times 10^{-21}\text{ mg.}$$

These calculations were first given by van der Waals, though Loschmidt still earlier (1865) had tried to obtain an estimate of these values.

The colloidal molecule, for which M may be over 10,000, may weigh about 10^{-17} mg. If we dissolve 1 mg. of white of egg in a litre of water, there will be about 10^{11} molecules in each cubic mm.

The Electric Charge of an Ion.—One mg. of hydrogen, when in the ion condition, carries about 9·65 absolute units of electricity and therefore an atom has $9\cdot 65\times 10^{-21}$. This number in general represents the magnitude of the electric charge of univalent ions. The n-valent ions of course have an n-fold charge.

If a particular ion moves with a velocity of 1 $\frac{\text{cm.}}{\text{sec.}}$, then this represents a galvanic current of an intensity of $9\cdot 6\times 10^{-20}$ of an ampere. Now, by the aid of a very sensitive galvanometer, one can detect a current of about 10^{-11} of an ampere, *i.e.* what would correspond to the movement of about 100 millions of ions, with the velocity mentioned above.

Determination of the Absolute Charge of an Ion by J. J. Thomson.—In the earlier editions of this book it was suggested that the magnitude of the charge of single ions might perhaps be determined, and hence the absolute magnitude of the molecule.

This hope has meanwhile been most completely realised by the work of J. J. Thomson on gas ions, referred to in Chapter IX. of this book.

We saw on p. 399 that a gas ion can serve as the nucleus for cloud formation, and that with sufficient under-cooling all the gas ions act in this way.

Now the mass of condensed water can be reckoned from the degree of under-cooling; the size of the single drops may be determined from the rate at which they sink, so that the number of drops formed is known. But as the degree of ionisation produced, for example by means of Röntgen rays, is also known, we have a means of determining the amount of electricity contained in each drop, and this is identical with the charge of a single ion.

In this way Thomson found (*loc. cit.* p. 399) $11{\cdot}3 \times 10^{-21}$ absolute (magnetic) units as the charge of a single ion, whilst we calculated above by an entirely different method $9{\cdot}65 \times 10^{-21}$, a most striking agreement. The number of gas molecules in a cubic millimetre of a gas at 0° atmospheric pressure is $N = 3{\cdot}6 \times 10^{16}$, whilst we found above $4{\cdot}5 \times 10^{16}$.

Planck in his electromagnetic theory of radiation arrives by a method that is perhaps somewhat hypothetical at the present time at $N = 5{\cdot}0 . 10^{16}$, which is equally in good agreement with the above numbers.[1]

In any case we may regard the molecular dimensions as established with a remarkable degree of certainty, so that the atomistic conception begins to lose its hypothetical character.

[1] *Ann. d. Phys.*, **4.** 553 (1901).

BOOK III

THE TRANSFORMATION OF MATTER
(THE DOCTRINE OF AFFINITY, I.)

CHAPTER I

THE LAW OF CHEMICAL MASS-ACTION

The Aim of the Doctrine of Affinity.—As the final goal of the doctrine of affinity, this problem is proposed, viz., to ascribe those forces which are efficient in the transformation of matter to well-investigated physical transformations. The question of the nature of the forces which come into play in the chemical union or decomposition of substances was agitated long before a scientific chemistry existed. As long ago as the time of the Grecian philosophers, the "love and hate" of the atoms were spoken of as the causes of the changes of matter; and regarding our knowledge of the nature of chemical forces, not much further advance has been made even at the present time. We retain anthropomorphic views like the ancients, changing the names only when we seek the cause of chemical changes in the changing *affinity* of the atoms.

To be sure, attempts to form definite conceptions [regarding these affinities] have never been wanting. All gradations of opinion are found, from the crude notion of a Borelli or a Lemery, who regarded the tendency of the atoms to unite firmly with each other as being due to their hook-shaped structure (and we employ the same view at present when we speak of "the linking of the atoms in the molecule"), to the well-conceived achievements of a Newton, a Bergman, or a Berthollet, who saw in the chemical process a phenomenon of attraction which was comparable with the fall of a stone to the earth.

Incidentally it appears that a deeper insight into the nature of the chemical forces may be gained by identifying them with the attractions of the different electricities; for chemistry has completely emancipated herself from the authority of a Berzelius, whose theory, instead of leading to further discoveries, was only suited to obscure the impartial estimation of the facts. It cannot be too strongly asserted that there is no discovery of any physical interchange of bodies with each other which cannot be used by the speculative brain in the explanation of the chemical forces also; but up to the present the results are not at all commensurate with the ingenuity displayed. It cannot be emphasised enough that we are as yet very far from reaching the goal, viz. the explanation of chemical decompositions by the play of well-defined and well-investigated physical forces.

In view of this undeniable fact, we must ask ourselves the question, whether this problem has been well chosen, and also whether it should not be seasonably discarded. And, in fact, it appears that the strenuous endeavours made to reach the goal at present do not promise very much; and it seems as though the stepping-stones, from which some one will pluck the ripe fruit, are not at present built firmly enough nor high enough.

Nowhere does a master mind show itself more clearly than in the wise restraint which imposes upon the investigator the choice of the mark to be aimed at; and nowhere is danger more imminent than when valuable ability for work is squandered in attacking useless problems—problems which offer difficulties which are almost insuperable at present, but which may perhaps be easily overcome in a short time by the use of results gained from sources which [previously] seemed entirely foreign to the subject.

The history of chemistry offers a striking example of this. As long as the alchemists attempted to convert the worthless metals into gold, their endeavours remained fruitless; scientific chemistry did not develop until it began to consider questions which had been previously regarded as insignificant.

The most immediate aim requires that we must deliberately ignore the question regarding the *nature* of the force which is instrumental in chemical decompositions, and must fix the eye upon *its mode of action*, and especially as regards its dependence upon the external conditions, such as the mass-ratio, the temperature, and the pressure. And here brilliant results will undoubtedly be obtained. Similarly the laws which control the pressure exerted by gases were *discovered* before they were *explained*, by ascribing them to the collisions of the gas molecules. For, supposing that it had happened that some genius had gained an insight into the kinetic gas theory a little before the gas laws themselves were discovered, even then, as a matter of fact, the way leading to the kinetic conception of the gaseous state had to be levelled down by much painstaking endeavour.

We are as yet very far distant from having a clear conception of the course of a chemical combination, but we possess some fundamental laws which control it. The investigation of such laws should not be regarded as less worthy than the aim designated above; for, ultimately, that standpoint will receive vindicated recognition which recognises the *experimental* discovery of the laws of gases as being on an equality with their *theoretical* establishment.

We may compare the present condition of the doctrine of affinity in many respects with that of theoretical astronomy. The latter, on the basis of Newton's law, attained a development which at that time was not reached by the other sciences. This law, according to which two mass-points attract each other directly as the product of their masses, and inversely as the square of their distance, merely describes the mode of action of attraction; it does not explain its nature, which is unexplained, even at present. The question which anticipated the important discovery of this law was not *why* does a stone fall to the earth, but *how* does it fall.

Similarly, a great result in chemical mechanics was obtained when the question was asked, not *why*, but *how*, do acids invert water solutions of cane-sugar. And although such a simple and universally efficient law as Newton's law has not been formulated as yet for the complexities of chemical phenomena,—where, in strong contrast to astronomy, the specialised nature of matter is expressed most clearly,—yet, nevertheless, theoretical chemistry does possess a number of universal nature laws, which make a formal description of the course of a chemical reaction possible, just as theoretical astronomy predicts the paths of the heavenly bodies.

The demonstration and application of these laws will form the second part of this description of theoretical chemistry.

The Condition of Chemical Equilibrium.—When we bring together a number of substances which are capable of chemical reaction,—and thus constitute *a chemical system*, as we will call it,—a reaction occurs which will terminate after the lapse of a sufficient interval of time; then we say that the system reaches a state of *chemical equilibrium.*

In general, the state of equilibrium is associated with definite conditions of temperature and pressure; and a change in these latter values implies a change in the state of equilibrium. Moreover, it requires great precaution before one can regard a system as actually having arrived at a state of equilibrium. For the observation that no material change can be observed, even after the lapse of quite a long time, is by no means always sufficient.

Thus, as many observations have shown, a mixture of hydrogen and oxygen may be preserved for years in sealed glass bulbs, without the formation of a noticeable amount of water. Yet, in spite of this, the two gases are by no means in equilibrium; but, on the other hand, we have many reasons for asserting that, at ordinary temperature, the reaction goes on spontaneously, but far too slowly to be recognised in such a lapse of time as is practically feasible; and also that the actual equilibrium would not be reached until all of the gases should have united almost but not quite completely to form water; and therefore that the completion of this, at ordinary temperatures, requires an enormously long lapse of time.

Reversible Reactions.—As an expression of the equalisation of a reaction we will, in what follows, use this as a universal symbol, viz.

$$n_1A_1 + n_2A_2 + \ . \ . \ . \ = n_1'A_1' + n_2'A_2' + \ . \ . \ .$$

This states that n_1 molecules of the chemically simple substance A_1, and n_2 molecules of the substance A_2, etc., unite to form n_1' molecules of the substance A_1', and n_2' molecules of the substance A_2', etc. The substances A_1, A_2 . . ., A_1' A_2', . . ., may be present in any casual number and quantity, and may exist in any of the states of aggregation. We say that the substances present, in certain proportions, in reference to the preceding reaction are in chemical equilibrium when they can remain together for an indefinitely long time in these same proportions, without the occurrence of a decomposition, in the sense of the preceding reaction, either in one direction or the other.

We call a reaction which progresses in the sense of the preceding scheme, *reversible* when, in the sense of the reaction equation, it goes on from left to right, with any arbitrary excess of some (not all) of the reaction products A_1', A_2', etc., if we start out with any arbitrary quantity of the substances A_1, A_2, etc.; and also if the reaction occurs spontaneously, in the opposite sense, with some arbitrary excess of the reaction products A_1, A_2, etc., if we start out with arbitrary quantities of A_1', A_2', etc.; so the final state of equilibrium reached will be the same in both cases if we start out with equivalent quantities.

An excellent illustration of a reversible reaction is the production of an ester, which occurs spontaneously according to the equation

$$\underset{\text{Alcohol}}{C_2H_5OH} + \underset{\text{Acetic acid}}{CH_3CO_2H} \rightleftarrows \underset{\text{Ethyl acetate}}{C_2H_5\text{-}CH_3CO_2} + \underset{\text{Water}}{H_2O}$$

One molecule of alcohol and one molecule of acetic acid unite to form one molecule of the ester (ethyl acetate in this case) and one molecule of water, or the reverse. If we bring together alcohol and acetic acid, a reaction occurs, in the sense of the equation, from left to right. On the other hand, if we mix a molecule of the ester with a molecule of water, a reaction occurs, in the sense of the equation, from right to left. In neither case is the course of the reaction complete, *i.e.* until all of the reacting components are used up; but it comes to a pause when such a condition of equilibrium is established as shall allow some of all four of the reacting components to exist beside each other.

If we start out with equivalent quantities, *i.e.* with those proportions which are convertible into each other, in the sense of the reaction equation, then in both cases we arrive at the same [final] condition of equilibrium. Thus, to select the simplest case, if we mix 1 g.-mol. of alcohol (46 g.) with 1 g.-mol. of acetic acid (60 g.); or if we mix 1 g.-mol. of ethyl acetate (88 g.) with 1 g.-mol. of water (18 g.); then in both cases, as shown by experience, the final composition of the reaction mixture is

$$\tfrac{1}{3} \text{ mol. alcohol} + \tfrac{1}{3} \text{ mol. acetic acid} + \tfrac{2}{3} \text{ mol. ester} + \tfrac{2}{3} \text{ mol. water.}$$

We call a chemical system *homogeneous* when it has the same physical and chemical nature at every point; and when this is not the case, we call it *heterogeneous*. Thus a system consisting of a gas mixture or of a solution, *e.g.* we call *homogeneous*; but if solid substances are also present, or if the liquid is separated into different layers, then the system is *heterogeneous*.

The opinion was formerly entertained that "reversible reactions" were an exceptional class; or, at least, that reactions should be divided into two classes, viz. the *reversible* and the *non-reversible*. But such a sharp distinction as this does not exist; and, moreover, there can be no doubt that by suitable adjustment of the conditions of the experiment, it would be possible to make a reaction take place, now in one direction, now in the opposite—that is, *in principle every reaction is reversible*.

If in following out these considerations we limit ourselves expressly to reversible reactions, then we place a restriction upon ourselves only to this extent, viz. in so far as we make the proviso that the conditions for the reversibility of the reaction in question have been already found. Thus, if the course of a reaction is complete, from a practical standpoint, as, *e.g.*, the union of oxygen and hydrogen to form water, we are not justified, on this basis, in laying down an essential distinction between this and the formation of an ester, where the equivalent proportions of the acid and alcohol react only to $\frac{2}{3}$ of the [theoretically] possible extent. The distinction is only a quantitative one; for hydrogen and oxygen, even in equivalent proportions, are surely unable to unite with each other in absolute completeness; but here also the reaction doubtless comes to a pause before the possible [theoretical] limit of the change is reached, although the quantity of the two gases which remain uncombined in equilibrium with each other at ordinary temperatures, may escape detection, on account of its insignificance.

We will first develop the law of mass-action for reversible reactions which take place in a *homogeneous* system; then there will be no difficulty in developing it later for heterogeneous systems.

According to the usage of van't Hoff, we will introduce into the reaction equation, in place of the sign of equality, two parallel arrows pointing in opposite directions, thus ($\leftrightarrows$), to denote that we are dealing with a reversible reaction.

The law of mass-action will instruct us, not only concerning the *variation* of the condition of equilibrium in a chemical system, with the relative proportions of the reacting components, but also concerning the *velocity* with which this equilibrium is attained. It is the fundamental law of *chemical statics* as well as of *chemical kinetics.*

On the other hand, the law of mass-action has nothing to say regarding the influence of temperature; the law which deals with this will be discussed in Book IV. (Transformation of Energy).

The Kinetic Development of the Law of Mass-Action.

We will assume that the following molecular species, viz. A_1, A_2 A_1', A_2' . . . have been brought together in a *homogeneous system,* where these react on each other, according to the scheme

$$A_1 + A_2 + \ . \ . \ . \rightleftarrows A_1' + A_2' + \ . \ .$$

If we compare this with the universal reaction equation on p. 426, we will notice that here we have a simpler case, because here the n_1, n_2 . n_1', n_2' . . . are all equal to 1, *i.e.* there is *only one molecule* of each substance concerned in the reaction.

The reacting substances may be either gaseous, or may form a liquid mixture, or finally may be dissolved in any selected solvent. In each case we may state the following considerations regarding the course of the reaction.

In order that a decomposition may occur in the direction from left to right, in the sense of the reaction equation, it is of course necessary that the molecules A_1, A_2 . . . shall *collectively* collide at one point; for otherwise a reaction would be impossible, because side reactions must be first excluded. Such a collision as this of course need not necessarily lead to a rearrangement of the atoms in those molecules which finally result from the preceding reaction. Rather must this collision be such as will be favourable to that prerequisite loose union of the atoms, in those particular molecules which must precede the rearrangement. As a result of a large number of collisions of this sort, there will be only a definite percentage, the external conditions being equal, which will cause a rearrangement from left to right in the sense of the equation. But this rearrangement will be greater the more frequent the collisions; and *therefore a direct ratio must exist between these two magnitudes.* Now we observe that these collisions must obviously be proportional to the concentrations of the bodies A_1, A_2 . . ., *i.e.* collectively they must be proportional *to the product of these concentrations.* Therefore the velocity v of the change from left to right, in the sense of the reaction equation, is

$$v = kc_1c_2 \ldots,$$

where c_1, c_2 . . . denote respectively the spatial concentration, *i.e.* the number of g.-mol of the substances A_1, A_2 . . . contained in a litre, and where k is a constant for the given temperature, and which may be called the *velocity coefficient.*

We employ the same consideration for the molecules A_1', A_2' . . . ; here also the change from right to left, in the sense of the reaction equation, will increase with the number of collisions of all these molecules at one point; and the latter, again, will be proportional to the product of their spatial concentrations. If we denote the corresponding proportional factor [velocity coefficient] by k′, then the velocity v′, with which the change occurs from right to left, in the sense of the reaction equation, will be

$$v' = k'c_1'c_2' \ldots;$$

where again c_1', c_2' . . . denote the number of g.-mol of each of the substances A_1', A_2' respectively, contained in 1 litre.

These spatial concentrations are commonly called *the active masses of the reacting components.* The reaction velocity in the sense of the reaction equation, from left to right, or the reverse, is thus proportional to the product of the active masses of the components of the left side, or of the right side, respectively.

We cannot observe either v or v′ alone. The tracing out of the course of a reaction by measurement can merely give us the *difference* between these values. For the *total* reaction velocity, which is actually observed, is made up of the difference between the two *partial* reaction velocities noted above; because the change actually observed, at every moment of time, is equal to the change in one direction, minus the change in the opposite direction, during this moment of time.

Therefore, when the condition of equilibrium has been reached, we are not to conclude that no further change takes place; but we should rather assume merely that the change, in the sense of the reaction equation, from left to right, is compensated by the change, in the sense of the reaction equation, from right to left; and, therefore, that the *total change* to be observed is equal to zero, *i.e.* the system stands in equilibrium. Therefore, here we have the relation

$$v - v' = 0,$$

and, therefore,

$$kc_1c_2 \ldots = k'c_1'c_2' \ldots,$$

which is the fundamental law of chemical statics.

This view must be entertained, viz. not that there is no absolute indifference between the reacting substances in the condition of equilibrium, but rather that the reacting ingredients, with their properties unchanged, persist in their mutual action, and that here the mutual [apparent] change alone has disappeared; this view, moreover, is of fundamental importance for our conception of changes of matter. It is ordinarily expressed by the statement that in this and in all analogous cases the equilibrium is not a static but a *dynamic* one.

This view was derived as an immediate conclusion from the kinetic-molecular mode of explanation, and it has been employed with especial success in the development of the kinetic gas theory.

Thus the equilibrium between water and water vapour, according to Clausius, is not to be conceived of as though there occurred neither the evaporation of liquid water nor the condensation of gaseous water vapour; but rather, when saturated water vapour is in contact with water, both of these processes continue uninterruptedly in equilibrium, *i.e.* through a definite area of the surface of the liquid water, in any interval of time, there are the same number of molecules going in the one direction as in the other (see also p. 216).

This view, as applied to chemical changes, was first advanced by Williamson (1851), and later it was more fully developed by Guldberg and Waage, by Pfaundler, and others.

Regarding the progress of the reaction, *i.e.* the velocity with which the chemical change strives to reach the condition of equilibrium at any moment, this is expressed by the following equation, viz.

$$V = v - v' = kc_1c_2 \ldots - k'c_1'c_2' \ldots$$

This *is the fundamental law of chemical kinetics.*

Of course, the formula for the equilibrium represents only a special case of those so developed; and it is obtained by making the total velocity equal to zero; in the same way, in analytical mechanics, from a specialisation of the general equations of motion, one obtains at once the conditions of equilibrium.

There is no difficulty in generalising the preceding equations, for the reaction which shall occur, according to the scheme,

$$n_1A_1 + n_2A_2 + \ \ldots \ = n_1'A_1' + n_2'A_2' + \ \ldots,$$

where n_1, n_2 . . . n_1', n_2' denote the number of molecules with which each particular substance takes part in the reaction; these numbers (n_1, n_2 . . . n_1', n_2') are necessarily *whole* numbers, and, as a rule, they are not large numbers, that is, *e.g.*, rarely larger than three.

Here also we must put v and v′ proportional to the number of collisions of all the kinds of molecules necessary to the reaction; we must remember however that n_1 molecules A_1, n_2 molecules A_2, and so on, must collide *simultaneously* in order that the reaction may take place from left to right, and similarly n1′ molecules A_1', n_2' molecules A_2', and so on, in order that the reverse reaction may be possible.

If we consider the path of a single molecule arbitrarily chosen for a fixed time, the number of its collisions with other similar molecules will be proportional to c, the concentration of the kind of molecule in question; the number of collisions between two similar molecules of a given kind during the same time is therefore c-times as many, that is, the number of collisions between two similar molecules is proportional to the square of the concentration c, and in general it follows that the number of collisions between n similar molecules of one kind must be considered proportional to c^n.

The number of collisions of n_1 molecules A_1, n_2 molecules A_2, is therefore proportional to $c_1^{n_1}c_2^{n_2}$ . . ., and for the velocity of reaction which is proportional to that number we have

$$v = kc_1^{n_1}c_2^{n_2} \ . \ . \ .,$$

and similarly the velocity of the reverse reaction is

$$v' = k'c_1'^{n_1'}c_2'^{n_2'} \ . \ . \ . ;$$

the total velocity of reaction being again the difference between the partial velocities

$$V = v - v' = kc_1^{n_1}c_2^{n_2} \ . \ . \ . \ - k'c_1'^{n_1'}c_2'^{n_2'} \ . \ . \ .,$$

a formula which represents the most general application of the law of mass-action of *homogeneous systems*. If V is equal to 0, the formula

$$\frac{c_1'^{n_1'} \cdot c_2'^{n_2'} \ . \ . \ .}{c_1^{n_1} \cdot c_2^{n_2} \ . \ . \ .} = \frac{k}{k'} = \text{const.}$$

for the state of equilibrium is arrived at.

An error which is frequently found in the literature of the subject is this, viz. that the active mass of a substance which reacts with n molecules is written c instead of c^n. This oversight, perhaps, does not need to be corrected, inasmuch as the correct notation in this respect has been already used in the writings of Guldberg and Waage. But because, in spite of this, there is still much uncertainty at present, it appears suitable to refer to this in order that there may be no room for doubt; since, moreover, we have proof, both from innumerable experimental facts, and also from the clear kinetic and thermodynamic demonstration, which will be given in the last book.

It should also be emphasised that the kinetic proof,[1] given above, of the law of chemical mass-action, can lay no claim to be in the rank of the laws which are proven with universal completeness; but it can only be considered as a means of making this law appear probable. The thermodynamics of the last book will furnish a stricter theoretical proof; this will find its demonstration in an abundance of facts, part of which are unintelligible without it, and part of which

[1] A more stringent proof will be found in Boltzmann, *Wied. Ann.* **22.** 68 (1884).

were furnished by means of the universal laws [of thermodynamics]. The more recent development of theoretical chemistry leads, more and more convincingly, to recognising that we must regard the formulation of [the law of] chemical mass-action as an expression of a very significant law of nature, which serves as the foundation of a more extended development of the doctrine of affinity.

The History of the Law of Mass-Action.—The first theory of the action of chemical forces is that which was developed by the Swedish chemist, Bergman, in the year 1775. The essential principle of this may be stated in the following law, viz. :—

The magnitude of chemical affinity may be expressed by a definite number; if the affinity of the substance A *is greater for the substance* B *than for the substance* C, *then the latter* (C) *will be completely expelled by* B *from its compound with* A, *in the sense of the equation*

$$AC + B = AB + C.$$

This theory fails entirely to take account of the influence of the *relative masses* of the reacting substances ; and, therefore, as soon as this point was noticed, this statement of the law of course had to be repudiated.

An attempt to consider this factor was made by Berthollet (1801), who introduced into the science the conception of chemical equilibrium. The views of this French chemist may be summed up in the following law :—

Different substances have different affinities for each other ; these exhibit their value only when they are in immediate contact with each other. The condition of equilibrium depends not only upon the chemical affinity, but also essentially upon the relative masses of the reacting substances.

The genuine kernel of Berthollet's idea is to-day the guiding principle of the doctrine of affinity. This holds good, especially for the conception regarding many reactions which, in the sense of Bergman's notion, proceed to a completion, *i.e.* until the reacting substances are all used up ; but only for this reason, viz., that one or more of the products of the reaction either crystallise out or evaporate off from the reaction mixture, and hence the inverse reaction becomes impossible.

Two Norwegian investigators, Guldberg and Waage, clinging to Berthollet's idea, succeeded in formulating the influence of the reacting masses in a simple law, viz. the law of chemical mass-action given above. The results of their theoretical and experimental studies were given in a book[1] which appeared at Christiania in 1867, and which was entitled *Études sur les affinités chimiques.* A new epoch of theoretical chemistry dates from the appearance of this book.

[1] Given in abstract, *Jour. prakt. Chem.* [2], **19.** 69 (1879). Translated into German with notes by Abegg, Ostwald's *Klassiker*, No. 104.

Already, before this, formulæ to describe the progress of certain chemical reactions, and which must be regarded as applications of the law of mass-action, had been described by Wilhelmy (1850) and by Harcourt and Esson (1856). The service of Guldberg and Waage is still undiminished, since they grasped the law in its full significance and applied it logically in all directions.

The treatise of the two Scandinavian investigators remained quite unknown; and so it happened that Jellet (1873), and van't Hoff (1877), and others independently developed the same law.

The *thermodynamic* **basis** of the law of mass-action is due to Gibbs, **and van't Hoff; it** will be dealt with more closely

CHAPTER II

CHEMICAL STATICS—HOMOGENEOUS SYSTEMS

Equilibrium between Gases.—Corresponding to the different states of aggregation, the particular system considered as being in equilibrium, and which must be both physically and chemically homogeneous in all its parts, may be either gaseous, liquid, or solid. In accordance with the old fundamental law, viz. *corpora non agunt nisi fluida* [substances do not react on each other unless they are fluid], the last species of homogeneous systems [*i.e.* the solids], in particular would be excluded from the foregoing consideration. But some facts, though they are but few, lead us not exactly to a repudiation of this law, but rather to limit its universal availability. Thus, for the sake of completeness, a short notice, at all events, must be given to the equilibrium, which is exhibited by homogeneous mixtures existing in the solid state.

In the consideration of gaseous systems, the conception of "active mass" has a very simple and obvious meaning. By the "active mass" of a substance (a molecular species), we mean the number of g.-mols. which are present in 1 litre. But now the partial pressure under which a gas stands in a gas mixture corresponds simply to this value; because, according to Avogadro's law, the pressure of a gas depends only upon the number of molecules which it contains in unit volume. Thus in the equations on p. 431, instead of the concentrations, we may introduce the respective partial pressures of the different molecular species, as they participate in the reaction.

Thus, if a reaction occurs in a gaseous system, in the sense of the equation,

$$n_1A_1 + n_2A_2 + \ . \ . \ . \ \rightleftarrows n_1'A_1' + n_2'A_2' + \ . \ . \ .,$$

and if the partial pressures of the molecular species A_1, A_2, etc., are respectively p_1, p_2, etc., and for the molecular species A_1', A_2', etc., p_1', p_2', etc., then the following relation holds good for the condition of equilibrium: viz.

$$\frac{p_1'^{n_1'} p_2'^{n_2'} \ . \ . \ .}{p_1^{n_1} p_2^{n_2} \ . \ . \ .} = \frac{k}{k'} = K.$$

In this formula we express only the ratio of the two velocity coefficients; we will designate this value as the *reaction constant*, and will denote it by K.

The Formation of Hydrogen Iodide.—The preceding formula will be first applied to the formation of hydrogen iodide from hydrogen and iodine vapour; it occurs in the sense of the equation

$$H_2 + I_2 = 2HI.$$

This equation was first investigated by Hautefeuille,[1] but afterwards, and very thoroughly, by Lemoine.[2] The latter allowed a weighed quantity of iodine to act upon a measured volume of hydrogen in a sealed bulb. After the condition of equilibrium had been reached, the contents of the bulb were introduced into a eudiometer, and the residual hydrogen was measured, while the hydrogen iodide which had been produced was absorbed by the water. At ordinary temperatures, the reaction progressed so very slowly that the two substances seemed absolutely indifferent towards each other; and they could, accordingly, be separated from each other (as by absorption, *e.g.* in Lemoine's work) without any marked displacement of the condition of equilibrium taking place during the separation.

But as the temperature was raised—*and this is quite a universal phenomenon*—the velocity of the reaction increased enormously. At 265° (oil-bath) the time required to bring about the state of equilibrium was counted in *months*; at 350° (boiling mercury) in *days*; at 440° (boiling sulphur) in *hours*.

Moreover, with increasing pressure the velocity of the reaction increased, a fact, by the way, which is in harmony with the considerations advanced on p. 428.

Special researches proved that the final condition of equilibrium reached was the same, whether one started with a mixture of hydrogen and iodine vapour, or with the corresponding quantities of hydrogen iodide with an excess of one of the reaction products: thus the final condition could be reached either by a change from left to right, in the sense of the reaction equation, or in the inverse direction.

If we denote the partial pressure of the hydrogen iodide by p, and that of the hydrogen and of the iodine respectively by p_1 and by p_2, then in the condition of equilibrium we will have

$$\frac{p_1 p_2}{p^2} = K.$$

The total pressure of the gas mixture, according to Dalton's law, of course will be

$$P = p + p_1 + p_2.$$

[1] *C. R.*, **64.** 608 (1867). [2] *Ann. chim. phys.* [5]. **12.** 145 (1877).

We will first investigate the way in which the condition of equilibrium changes with *the external pressure.* If we compress the reaction mixture to the nth part, then the particular partial pressures will rise to n-fold their [original] value. But now we have

$$\frac{np_1np_2}{n^2p^2} = \frac{p_1p_2}{p^2} = K,$$

i.e. the new pressure values meet the requirements of the equilibrium formula, and therefore there is no change in the relative quantities resulting from changes in the pressure.

Thus the condition of equilibrium is independent of the external pressure. This result is general *if there is no change in the number of molecules caused by the reaction.*

Lemoine found for the values of the total pressure P, as given below, the following corresponding values of the decomposition coefficient x (*i.e.* the quantity of the free hydrogen divided by the quantity of the total hydrogen). In all of these researches the glass bulb was filled at first with hydrogen iodide.

P	x	
4·5 Atm.	0·24	
2·3 ,,	0·255	
1·0 ,,	0·27	t=440°
0·5 ,,	0·25	
0·2 ,,	0·29	

The above numbers hardly bring out the influence of pressure on the degree of decomposition of hydroiodic acid, as the numbers show irregular variations. Similarly, the reduction of the other researches of Lemoine (see p. 350 of the first edition of this book) leaves no doubt that the results are affected by some source of error.

In fact, M. Bodenstein,[1] who has recently repeated the investigations of Lemoine, has found that the glass walls absorb considerable quantities of hydroiodic acid, and that consequently less of this gas takes part in the equilibrium than was calculated in Lemoine's experimental arrangement from the amount of free hydrogen; Bodenstein, however, in arriving at the equilibrium, determined besides the volume of free hydrogen, the quantities of free iodine and free hydroiodic acid separately by titration; he was thus able to prove exactly that the decomposition of hydroiodic acid is independent of the pressure, and to arrive at a precise application of the law of mass action of this case.

[1] *Zeitschr. physik. Chem.*, **22.** 1 (1897).

If we express, as above, the degree of decomposition of the pure hydroiodic acid by x, we have

$$p = P(1 - x), \quad p_1 = p_2 = \frac{P}{2}x,$$

and therefore

$$\frac{x^2}{4(1 - x)^2} = \frac{p_1 p_2}{p^2} = K.$$

If in a given volume a mols of iodine and b mols of hydrogen react, and 2γ is the number of mols of hydroiodic acid formed, there remains $a - \gamma$ mols of free iodine, and $b - \gamma$ mols of free hydrogen, so that

$$p = P\frac{2\gamma}{a + b}, \quad p_1 = P\frac{a - \gamma}{a + b}, \quad p_2 = P\frac{b - \gamma}{a + b},$$

and consequently,

$$\frac{(a - \gamma)(b - \gamma)}{4\gamma^2} = K,$$

and this solved for γ gives [1]

$$\gamma = \frac{a + b}{2(1 - 4K)} - \sqrt{\frac{(a + b)^2}{4(1 - 4K)^2} - \frac{ab}{1 - 4K}}.$$

The following table shows the good agreement between calculation and observation; a and b here stand for the number of cubic centimetres of gaseous iodine and hydrogen, reduced to 0° and 760 mm. pressure that were contained in the glass bulb of about 13 cubic centimetres volume; γ is the amount of hydroiodic acid formed expressed in the same manner. The quantities directly determined as mentioned above were $a - \gamma$, $b - \gamma$, and γ.

Heated in Sulphur Vapour: $x = 0{\cdot}2198$, $K = 0{\cdot}01984$

a	b	2γ		Difference.
		Obs.	Cal.	
2·94	8·10	5·64	5·66	+0·02
5·30	7·94	9·49	9·52	+0·03
9·27	8·07	13·47	13·34	−0·13
14·44	8·12	14·93	14·82	−0·09
27·53	8·02	15·54	15·40	−0·14
33·10	7·89	15·40	15·12	−0·28

[1] In solving quadratic equations there can be no doubt as to whether to give the positive or negative sign to the root; only one solution gives a physically possible result. In the above case, for example, the positive sign would yield for γ values higher than a and b, which of course is nonsense.

HEATED IN MERCURY VAPOUR: $x = 0{\cdot}1946$, $K = 0{\cdot}01494$

a	b	2γ		Difference.
		Obs.	Cal.	
2·59	6·63	4·98	5·02	+0·04
5·71	6·22	9·55	9·60	+0·05
10·40	6·41	11·88	11·68	−0·20
26·22	6·41	12·54	12·34	−0·20
23·81	6·21	12·17	11·98	−0·19
22·29	6·51	12·71	12·68	−0·03

The above tables certainly give the most exact confirmation of the law of mass-action for homogeneous gaseous systems that has so far been arrived at.

The Dissociation Phenomena of Gases.—A class of reactions which deserves especial consideration, because of the simplicity and the frequency of their occurrence, is found in the so-called *phenomena of dissociation.* These are distinguished by this fact, viz. that, in the general equation of reaction, the substances standing on one side are reduced to a single molecule, and the change proceeds according to the scheme

$$A = n_1'A_1' + n_2'A_2' + \quad . \ . \ . \ ;$$

here A is the [original] molecular species advancing to the state of dissociation, and A_1', A_2' . . . are the [corresponding] dissociation products. If p denotes the partial pressure of the former [*i.e.* the original molecular species], and p_1', p_2' . . . those of the latter [*i.e.* the resulting molecular species] respectively, then, according to what has preceded, the relation for the condition of equilibrium is

$$\frac{p_1'^{n_1'} p_2'^{n_2'} \ . \ . \ .}{p} = K,$$

where K is designated as the *dissociation constant.*

The number of molecules in the system grows with increasing dissociation. Inasmuch as we have an easy and exact method of learning this [*i.e.* the relative number of molecules], by means of the vapour density determination, there is, therefore, no difficulty here in investigating the condition of equilibrium.

For a simple case, we will consider a gas which dissociates into two new molecules, which latter may be identical with each other, as in the dissociation of nitrogen dioxide, thus

$$N_2O_4 \rightleftarrows NO_2 + NO_2 \ ;$$

or where they may be different, as in the case of the dissociation of phosphorus pentachloride into chlorine and phosphorus trichloride, thus

$$PCl_5 \rightleftarrows PCl_3 + Cl_2.$$

Let δ be the density of the undecomposed gas, as calculated from its [theoretical] molecular weight; at complete dissocation the number of molecules will be doubled, and the vapour density therefore will amount to $\delta/2$. Now, according to p. 346,

$$1 + \alpha = \frac{\delta}{\Delta}, \quad \text{and } \alpha = \frac{\delta - \Delta}{\Delta}.$$

The total pressure P, at which Δ is measured, is made up of the pressures of the undecomposed molecules, and of the products of dissociation; if we denote the former by p, and the latter by p′, then, according to Dalton's law, we will have

$$P = p + p'.$$

Now, since the ratio of the number of the undecomposed molecules to the number of the dissociated ones, is as $1 - \alpha$ to 2α, therefore we obtain

$$p = P\frac{1-\alpha}{1+\alpha} = P\left[2\frac{\Delta}{\delta} - 1\right]$$

$$p' = P\frac{2\alpha}{1+\alpha} = 2P\left[1 - \frac{\Delta}{\delta}\right].$$

Now the law of mass-action gives us the relation

$$p'^2 = Kp,$$

where K denotes the constant of dissociation; then by the substitution of the expressions obtained for the partial pressures, we obtain as the equation of the *isotherm of dissociation*

$$\frac{4[\delta - \Delta]^2 P}{[2\Delta - \delta]\delta} = K \quad . \quad . \quad . \quad . \quad . \quad (1)$$

from which we obtain

$$\Delta = \delta + \frac{K'}{P} - \frac{K'}{P}\sqrt{1 + \frac{\delta P}{K'}} \quad . \quad . \quad . \quad (2)$$

by making

$$K' = \frac{K\delta}{4};$$

that is, the vapour density of a gas existing in a state of dissociation (at constant temperature), changes with the pressure; at very small pressures, it converges towards the lower limit of the vapour density;

at very high pressures, it converges towards the upper limit of the vapour density.

The Dissociation of Nitrogen Dioxide.—The vapour density of nitrogen dioxide has been measured by E. and L. Natanson,[1] and has also been calculated according to the preceding formula. Although there are small deviations between the vapour densities required by the theory and those found experimentally, yet on the whole the results are to be regarded as a satisfactory verification of the theory. This is shown in the following table :—

$t = 49{\cdot}7°$.

P	K′	Δ Obs.	Δ Calc.	γ
0 mm.	...	...	1·590	1·000
26·80 ,,	106	1·663	1·670	0·930
93·75 ,,	112	1·788	1·783	0·789
182·69 ,,	124	1·894	1·906	0·690
261·37 ,,	130	1·963	1·984	0·630
497·75 ,,	121	2·144	2·148	0·493

At a temperature of 49·7° and a pressure 497·75 mm., of 1000 molecules of N_2O_4, 493, i.e. about one-half, are dissociated. At the same pressure, but with increasing temperature, the dissociated fraction becomes greater, a phenomenon observed almost invariably in gaseous dissociation. The dissociation constant K increases with the temperature.

The quantity K′ in the second column of the above table is derived from the former equation (1)

$$K' = \frac{K\delta}{4} = \frac{(\delta - \Delta)^2 P}{2\Delta - \delta},$$

and should be constant according to theory; it shows large but irregular variations due to small errors of observations, especially when Δ is not very different from δ or $\frac{\delta}{2}$. If we take the mean

$$K' = 119$$

and recalculate Δ from equation (2) of the previous section, we arrive at the last column, which is in agreement with the observed values of the vapour density. These errors of experiment show that the conclusions of the theory are thoroughly satisfied.

[1] *Wied. Ann.*, **24.** 454 (1885); **27.** 606 (1886).

As already referred to on p. 347, nitrogen dioxide assumes more colour with increasing dissociation, because the molecules of NO_2 are reddish-brown, while those of N_2O_4 are colourless.

At about 500° the gas begins to decolourise, in consequence of a decomposition into oxygen and nitric oxide, thus

$$2NO_2 = 2NO + O_2.$$

The following vapour densities were measured by Richardson,[1] by the methods of Dumas and of others :—

Temp.	Pressure.	Δ	γ
130°	718·5	1·600	...
184°	754·6	1·551	0·050
279°	737·2	1·493	0·130
494°	742·5	1·240	0·565
620°	760·0	1·060	1·000

At 620°, under ordinary conditions, the decomposition is complete. The fraction α of the decomposed molecules may here be calculated from the equation

$$\alpha = 2\frac{1 \cdot 590 - \Delta}{\Delta}.$$

The Dissociation of the Compound of Hydrochloric Acid and Methyl Ether.—The decomposition of methyl-ether-hydrochloride (a so-called "molecular compound") has been investigated by Friedel.[2] By mixing hydrochloric acid gas with the vapour of methyl ether, a contraction takes place, which, as Friedel has shown, is to be ascribed to the formation of the compound mentioned above, and according to the reaction

$$HCl + (CH_3)_2O = (CH_3)_2O \,.\, HCl.$$

For every 100 volumes of the two gases mixed at the pressure P, there occurred the corresponding contraction of volume x.

P	x	$\frac{P}{x}$	
552	4·6	120	
628	4·9	128	t = 20°
756	6·1	124	

The partial pressure of the undissociated molecules is calculated

[1] *Jour. Chem. Soc.*, **51.** 397 (1887). [2] *Bull. Soc. Chim.*, **24.** 241 (1875)

to be $P\frac{x}{200-x}$, and that of the dissociated ones as $P\frac{100-x}{200-x}$; and therefore we have the relation

$$K = \frac{(100-x)^2 P}{(200-x)x}.$$

Or, since x may be almost neglected on account of its smallness, as compared with 100 and 200, it follows that P and x must be almost proportional to each other, and this is shown by the preceding table.

In the case of a gas which is almost completely dissociated, the undissociated part is proportional to the external pressure.

The Influence of Indifferent Gases.—Experience has shown in a large number of cases that *the condition of dissociation of a gas is not changed when another indifferent kind of gas* (i.e. *a gas which does not react chemically*) *is introduced at constant volume.* This is seen to be in perfect harmony with the equation of the isotherm of dissociation, if one considers that, according to Dalton's law, the presence of a foreign kind of gas does not influence the partial pressures of the reacting ingredients.

This law is, moreover, one of extraordinary importance, and one which throws much light on the general view. Later it will be thoroughly proved from thermodynamic principles.

Of course it is entirely a different case when the mixing with the indifferent gas is attended with an *increase of volume*, the latter acting then as a diluting agent. Then the dissociation increases with the increase of volume, independently of the nature of the kind of gas which is added. If one studies a gas which exists in a state of dissociation by the method of the displacement of air (p. 250), the dissociation is found to increase the more the evaporating substance is diluted with the indifferent gas; and, therefore, results are obtained which depend upon the mutual diffusion, *i.e.* the values of the vapour densities are very irregular. It is erroneous and misleading to speak of the "*dissociating influence*" of other gases.

The Influence of an Excess of the Dissociation Products.—The theory allows us to anticipate the influence of the addition of one of the dissociation products. Thus, for example, let a gas, which gives rise to two different kinds of molecules on decomposition, be brought into equilibrium with its dissociation products; and between the partial pressure p of the undissociated molecules, and that of the dissociation products, which may be p′ for both, let there exist the relation

$$p'^2 = Kp.$$

Now, let one of the dissociation products be added in excess till its pressure is p_0, and let the increase which p experiences be π. Then the partial pressure of the other decomposition product will sink to $(p' - \pi)$, and the pressure of the first decomposition product will amount to $p_0 + p' - \pi$. The calculation of π is obtained by the law of mass-action, from the equation

$$(p' - \pi)(p_0 + p' - \pi) = K(p + \pi).$$

By comparing this equation with the preceding formula, it will be found that π must always have a positive value. That is, *admixture of one of the decomposition products, at constant volume, tends to diminish the degree of dissociation.* This is a very important phenomenon. We shall make repeated application of this rule.

Moreover, research verifies this conclusion. When Friedel added in excess one of the decomposition products of methyl-ether-hydrochloride, the quantity of the latter substance increased. The dissociation of phosphorus pentachloride diminishes when phosphorus trichloride is added in excess.[1] The determination of the vapour density of ammonia by the method of the displacement of air, gives higher values when ammonia or hydrochloric is used as the packing gas [2] [instead of air or some indifferent gas]. We will consider later several ways in which this law is indirectly verified.

If the admixture of the dissociation product occurs with increase of volume, then it also acts as a *diluting agent*, and hence it occasions an *increase* of dissociation. Therefore, according to the special conditions, on the whole, there may result an increase or a decrease of the dissociation. Thus, *e.g.* if any selected volume of gaseous phosphorus pentachloride is mixed with another selected volume of chlorine, or of phosphorus trichloride, so that the resulting mixture occupies a volume which is equal to the sum of the two volumes, then, as can be easily shown, *the condition of dissociation remains unchanged, if the pressure of the added chlorine is exactly as great as its partial pressure in the pentachloride.*

The Frequency of Dissociation Phenomena.—Dissociation phenomena are much more common than was formerly supposed to be the case; and no doubt can be entertained that, under suitable conditions of temperature and pressure, not only all chemical compounds, but also all poly-atomic molecules would be reduced to a state of decomposition more or less considerable. Thus, the di-atomic molecule of iodine, at high temperature and low pressure, decomposes into its two atoms, and a similar thing would doubtless happen in the case of other di-atomic gases, like oxygen, nitrogen, etc., although no such decomposition could be established at 1700°, and at atmospheric pressure.[3]

[1] Würtz, *C. R.*, **76.** 60 (1873).
[2] Neuberg, *Ber. deutsch. chem. Ges.*, **24.** 2543 (1891).
[3] C. Langer and v. Meyer, *Pyrochem. Unters.*, Brunswick (Braunschweig), 1885.

Sulphur, the vapour density of which has been investigated recently by H. Biltz,[1] and Biltz and Preuner,[2] at temperatures varying from 468-606°, behaves exceptionally; the vapour density falls with decrease in the temperature and only gives values agreeing with the formula S_2 at high temperatures. This may apparently be explained by the assumption of molecules S_8 formed by vaporisation of sulphur which partially decompose according to the equation

$$S_8 \rightleftarrows 2S_4,$$

and that the S_4 molecules simultaneously decompose according to the equation

$$S_4 \rightleftarrows 2S_2,$$

therefore the sulphur vapour consists of three kinds of molecules S_8, S_4, S_2, the last of which gains at the expense of the former on decrease of pressure or increase of temperature. The equation of the dissociation isotherm of sulphur vapour is consequently given by the formula

$$K_1p_1 = p_2^2, \quad \text{and } K_2p_2 = p_3^2,$$

where p_1, p_2, p_3 show the partial pressures of the three kinds of molecules, K_1 and K_2 the dissociation constant of the two reactions. Sulphur is accordingly an example of "*dissociation in two stages.*" The present author has found by means of the apparatus described on p. 254 that at 1900-2000° the S_2 molecules are broken up into atoms to the extent of about 45%.

The Decomposition of Carbon Dioxide.—As a further illustration of a reaction progressing in the gaseous condition, we will consider the decomposition of carbon dioxide, which is of great practical interest, and which occurs in the sense of the equation

$$2CO_2 \rightleftarrows O_2 + 2CO.$$

In a limited sense this is no [true] dissociation, because, in order to make the decomposition possible, here two molecules of the compound which is to be decomposed must act together. If we denote the partial pressures of the three preceding molecular species respectively by p_1, p_2, p_3, then, as a direct result of the law of mass-action, we have the relation

$$Kp_1^2 = p_2p_3^2.$$

If one has determined the degree of decomposition at a definite temperature for any one pressure by means of a vapour density method, then it is possible to calculate the degree of decomposition for all pressures by means of this formula. Thus it is shown by Deville that at 3000°, and at atmospheric pressure, about 40 per cent of

[1] *Zeitschr. physik. Chem.*, **2.** 920 (1888). [2] *Ibid.*, **39.** 323 (1902).

the carbon dioxide is decomposed; hence by the aid of these data, by the preceding equation, it is calculated that at this temperature and at 0·001 of an atmosphere, 94 per cent, and at 100 atmospheres 10 per cent [of the CO_2] would decompose.

As will be seen in the last book, in the application of the mechanical theory of heat to the condition of chemical equilibrium, the variation of the equilibrium constant K is so intimately connected with the heat developed in the reaction, that one magnitude can be calculated from the other. Now the heat developed by the combustion of carbon monoxide can be measured at the temperature of a room; and from the difference of the specific heats of the reacting gases, one can calculate this quantity for all other temperatures. Thus there is no difficulty in ascertaining the degree of the decomposition of carbon dioxide under all possible conditions of temperature and pressure.

The number of molecules of carbon dioxide out of 100 molecules which would so decompose is given in the following table:—

Temp.	Pressure=0·001	0·01	0·1	1	10	100 Atm.
1000	0·7	0·3	0·13	0·06	0·03	0·015
1500	7	3·5	1·7	0·8	0·4	0·2
2000	40	12·5	8	4	3	2·5
2500	81	60	40	19	9	4·0
3000	94	80	60	40	21	10
3500	96	85	70	53	32	15
4000	97	90	80	63	45	25

This table was calculated by Le Chatelier.[1] Some remarkable deductions, which are interesting from a practical standpoint, can be drawn from these figures, viz:—

1. **The Smelting Furnace.**—The temperature only reaches about 2000°, and the partial pressure of carbon dioxide only amounts to about 0·2 of an atmosphere; now the carbon dioxide is decomposed only to about 5 per cent, whereby the capacity of the furnace for work is somewhat diminished, though not very much.

2. **Illuminating Flames.**—The incandescent zone in which the combustion of the separated carbon takes place appears to have a temperature of about 2000° in ordinary flames, and a somewhat higher temperature in the regenerative burners. On the other hand, on account of the large quantity of hydrogen in the ordinary illuminating material, the partial pressure [of the carbon dioxide] sinks below 0·1 of an atmosphere. The extent of the decomposition may be more than 10 per cent, and the temperature of the flame may fall the

[1] *Zeitschr. physik. Chem.*, **2.** 782 (1888).

same amount. But the illuminating power, which rises more quickly than the temperature in proportion to it, is also much more strongly diminished than the temperature.

3. **Explosives.**—The temperature of combustion rarely rises above 2500°, and never reaches 3000°. Inasmuch as the pressure of the carbon dioxide is here estimated in thousands of atmospheres, dissociation has no influence.

Decomposition of Water Vapour.—The decomposition of water in the sense of the equation

$$2H_2O = O_2 + 2H_2$$

is of equal importance with that of carbon dioxide, and can be followed by exactly the same formula.

The dissociation of water vapour, as Helmholtz showed, can be calculated from the electromotive force necessary for the reversible decomposition of water, but more precisely by the following method given by the author.[1]

Boudouard,[2] and especially Hahn,[3] have studied the reaction

$$CO + H_2O = CO_2 + H_2 ;$$

it appeared, for example, that at 1000° the constant of the law of mass-action [4]

$$K = \frac{[CO][H_2O]}{[CO_2][H_2]} \quad . \quad . \quad . \quad . \quad (1)$$

is 1·6. If we represent the dissociation constant of water vapour and carbon dioxide at the same temperature by K_1 and K_2, their degree of dissociation with x and y, we have

$$K_1 = \frac{[H_2]^2[O_2]}{[H_2O]^2} = \frac{x^3}{2v(1x)^2} \quad . \quad . \quad . \quad (2)$$

where v is the volume in which a mol of water is contained, and consequently

$$[O_2] = \frac{x}{2v}, \quad [H_2] = \frac{x}{v}, \quad [H_2O] = \frac{1-x}{v}.$$

Similarly, for carbon dioxide we have

$$K_2 = \frac{[CO]^2[O_2]}{[CO_2]^2} = \frac{y^3}{2v(1-y)^2} \quad . \quad . \quad . \quad (3)$$

[1] See here G. Preuner, *Zeitschr. physik. Chem.*, **42.** 50 (1903).
[2] Thèses, Paris, 1901.
[3] *Zeitschr. physik. Chem.*, **44.** 513 (1903).
[4] Here and in the following the concentration of each kind of molecule will be expressed by its chemical formula enclosed in square brackets.

In the equilibrium expressed by equation (1), a certain amount of free oxygen is always present, and this must be in equilibrium with water vapour in agreement with equation (2), and with carbon dioxide in agreement with equation (3). Hence, if we divide equation (2) by (3) as applied to the same system (1), the concentration of oxygen disappears and we obtain

$$K^2 = \frac{K_2}{K_2} = \frac{y^3(1-x)^2}{x^3(1-y)^2}.$$

Using the value of K given above, this equation shows that at 1000° the dissociation of water vapour is 0·73 times that of carbon dioxide.

Until more exact measurements are obtainable, the dissociation of water vapour at higher temperatures may be *approximately* calculated by multiplying the numbers in the table, p. 445, by 0·73. Hence the practical conclusions drawn with regard to carbon dioxide hold also for water vapour.

Equilibrium in Homogeneous Liquid Systems—Esterification.—The dependence of the equilibrium upon the relative quantities of the reacting substances is the same with liquid systems as with gaseous ones; only here we do not have to take account of the partial pressure, but rather of the *concentration* of the reacting substances; and by this is meant the number of g.-mol contained in 1 litre.

The number of special cases which have been investigated here is very large. This is to be explained by the following reasons: firstly, the study of the chemical equilibrium of a liquid system is much easier for the experimenter in many ways than the study of a gaseous system; and secondly, the decompositions which occur in liquid systems are very important, both in the usage of the laboratory, and also in the economy of nature.

We begin with esterification, a reaction to which reference has recently been made (p. 426); the thorough study of this by Berthelot and Péan de St. Gilles[1] has had great influence in defining the conception of chemical equilibrium.

If one mixes any organic acid, as acetic acid, with an alcohol, as ethyl alcohol, there is formed water, and also the corresponding ester; the latter results from the combination of the positive component of the alcohol with the negative component of the acid; thus—

$$\underset{\text{Acetic acid}}{CH_3COOH} + \underset{\text{Alcohol}}{C_2H_5OH} = \underset{\text{Ethyl acetate}}{CH_3COO{\cdot}C_2H_5} + \underset{\text{Water.}}{H_2O}$$

This reaction, which is comparable with the neutralisation of an

[1] *Ann. chim. phys.*, **65.** and **66.** (1862), and **68.** (1863).

acid by a base at ordinary temperatures, progresses very slowly; several days elapse before the state of equilibrium is approximately established, and the reaction comes to a standstill. If this liquid system is heated in a sealed glass tube at 100°, the state of equilibrium is reached after a few hours. The reaction never advances to a completion, *i.e.* until all of the [original] reacting substances are consumed, but only to such a point that all of the four reacting substances stand in equilibrium with each other.

It can be easily ascertained how far the reaction has proceeded at any moment, by titrating the acetic acid which is present, or which has been produced; on account of the slowness of the reaction no marked displacement of the momentary condition need be feared during the operation (of titration).

If equivalent quantities are allowed to react on each other, whether we start with 1 g.-mol. of acetic acid (60 g.) and 1 g.-mol. of alcohol (46 g.), or with 1 g.-mol. of methyl acetate (88 g.) and 1 g.-mol. of water (18 g.); in both cases, after the lapse of a sufficiently long time, there results a homogeneous system, which has the following composition, viz. :—

$$\tfrac{1}{3} \text{ mol. acetic acid} + \tfrac{1}{3} \text{ mol. alcohol} + \tfrac{2}{3} \text{ mol. water} + \tfrac{2}{3} \text{ mol. ester.}$$

This ratio of the relative quantities will remain constant, even if one should wait seventeen years. Moreover, this ratio changes very slightly with the temperature; the reason for this will be given later, Book IV. Chap. III.

Let v be the volume of the preceding reaction mixture, and let it be made up of 1 g.-mol. of acetic acid, with m g.-mol of alcohol, and n g.-mol. of water (or of the ester; it is indifferen which, for the form of the equation). Then in the condition of equilibrium, if x denotes the number of g.-mol. of alcohol decomposed (and of acetic acid also, of course), there exists the relation—

$$k\frac{1-x)(m-x)}{v^2} = k'\frac{(n+x)x}{v^2};$$

here k denotes the velocity with which the alcohol and the acid unite, and k′ the velocity with which the ester and the water unite with each other. The denominator common to both sides of the equation can be neglected.

As in the preceding special case, let $m=1$; $n=0$; and $x=\frac{2}{3}$; then we will have

$$\frac{k}{9}=\frac{4k'}{9},$$

i.e. the constant of equilibrium becomes

$$K=\frac{k}{k'}=4.$$

By introducing this value into the general equation, and then solving it for x, we obtain as the quantity of ester formed—

$$x = \frac{1}{6}\left(4(m+1) + n - \sqrt{16(m^2 - m + 1) + 8n(m+1) + n^2}\right).$$

And this simplified for $n = 0$, becomes

$$x = \frac{2}{3}\left(m + 1 - \sqrt{m^2 - m + 1}\right).$$

The equation is very satisfactorily verified by experiment, as was first shown by Guldberg and Waage, and later and very thoroughly by van't Hoff.[1] Thus Berthelot and Péan de St. Gilles, allowing 1 mol of acid to act upon m mol of alcohol, found that the following quantities of ester were formed, the latter being denoted by x :—

m	x Obs.	x Calc.
0·05	0·05	0·049
0·18	0·171	0·171
0·33	0·293	0·311
0·50	0·414	0·423
[1·00	0·667	0·667]
2·00	0·858	0·845
8·00	0·966	0·945

Results which were fully as satisfactory were obtained by calculation for any casual quantities of water or ester. It is a very important and significant result of the law of mass-action, that such apparently complicated ratios as were found by Berthelot and Péan de St. Gilles in their experimental researches, could be expressed in such simple formulæ.

One learns, moreover, that by the action of a large quantity of acetic acid upon a little alcohol, or by the action of a large quantity of alcohol upon a little acid, the esterification is almost complete; and conversely, by the action of a large quantity of water upon a little ester, the latter is almost entirely decomposed.

Just as in the formation of esters from alcohol and an organic acid, the law of mass-action is applicable in the reaction of hydrochloric acid on alcohol with formation of ethyl chloride and water;[2] and consideration of the equilibrium produced by action of sulphuric acid on water has recently led to remarkable results.[3]

The Influence of the Nature of the Reacting Components.—After it had been ascertained, as already shown, that the

[1] *Ber. deutsch. chem. Ges.*, **10.** 669 (1877).
[2] Cain, *Zeitschr. physik. Chem.*, **12.** 751 (1893).
[3] Zaitschek, *ibid.*, **24.** 1 (1897).

action of acids on alcohols could be expressed in mass and number, this question at once arose—viz. *how does the reaction-capacity depend upon the nature of the particular acid, and the particular alcohol used?* We are indebted to Menschutkin[1] for attacking this problem in a very extended research. He determined the limit of esterification, on bringing together equi-molecular quantities of the most various acids and alcohols; and at the same time he directed the research so as to determine, at least approximately, the velocity with which the limiting state was reached. Of the numerous details of his research, we will merely state that in general—

In homologous series, the relative quantities of ester produced increased with the molecular weight.

In using the same acids, secondary alcohols afforded less ester than primary alcohols, and tertiary alcohols less ester than secondary alcohols.

There was no simple connection observable between the limit of esterification and the velocity with which this was reached.

The Dissociation of Esters.—As Menschutkin observed in his researches, in using tertiary alcohols, the ester produced would decompose [when dissociated] into an acid and a hydrocarbon. This reaction was investigated later by Konowalow,[2] who allowed acids to react upon amylene [pentene]; in this way it appeared that the reaction concerned a dissociation phenomenon, which proceeded according to the scheme

$$CH_3COO(C_5H_{11}) = CH_3COOH + C_5H_{10}.$$

The reaction advances with marked velocity only in the presence of a sufficient quantity of free acid; the pure ester is stable, even at quite high temperatures; and the addition of acid at once occasions dissociation which then advances to a definite limit. The same limit is reached when one starts with free amylene and free acid.

Here we meet *catalytic reactions* for the first time, *i.e.* cases where the presence of a certain substance will strongly accelerate the progress of a reaction, which can take place without it. As the law of mass-action was applicable to this, therefore the writer, together with Hohmann,[3] attempted to prove in a more extended way just how the simple relations which were anticipated were obtained.

If 1 mol of acid is added to 1 mol of amylene, then, according to the law of mass-action, we should obtain

$$\frac{(a - x)(1 - x)}{xV} = K,$$

where x denotes the ester produced, V the volume of the reaction

[1] *Ann. chim. phys.* [5], **20.** 229 (1880); **23.** 14 (1881); **30.** 81 (1883).

[2] *Zeitschr. physik. Chem.*, **1.** 63 (1887); **2.** 6 and 380 (1888).

[3] *Ibid.*, **11.** 352 (1893).

mixture, and K the coefficient of equilibrium. The figures in the following table were obtained in this way, viz. the amylene and trichlor-acetic acid were held in a sealed glass tube at 100° for a sufficiently long time (some hours, up to a day); the quantity of ester produced was found by the decrease in the acidity titration:—

a	V	x Found	K	x Calc.
2·15	361	0·762	0·00120	0·762
4·12	595	0·814	0·00127	0·821
4·48	638	0·820	0·00126	0·826
6·63	894	0·838	0·00125	0·844
6·80	915	0·839	0·00126	0·845
7·13	954	0·855	0·00112	0·846
7·67	1018	0·855	0·00113	0·848
9·12	1190	0·857	0·00111	0·852
9·51	1237	0·863	0·00111	0·853
14·15	1787	0·873	0·00107	0·861

By making $K = 0{\cdot}001205$, and then solving the preceding equation for x, by the help of this mean value, we obtain

$$x = \frac{1}{2}\left(1 + a + KV - \sqrt{(1 + a + KV)^2 - 4a}\right);$$

and by this means we obtain the figures given in the last column, which coincide remarkably well with the results of observation.

As is seen, the quantity of ester produced varies only inconsiderably, whether 4 or 14 mol of amylene are added to 1 mol of acid. One might infer at first glance that the esterification could be carried as far as desired by a sufficient excess of amylene, just as an acid can be completely esterified by an excess of alcohol (p. 449).

But this is not the case; the esterification of amylene and an acid is different from that of an alcohol and an acid; for in the former case only one molecular species is found, viz. the ester; but in the latter case there are two, viz. the ester and the water.

In the formula of equilibrium for the first reaction, the volume of the reaction mixture *does not disappear* as it does in the second case, which thus shows a different behaviour.

Theoretically one could add an infinite quantity of amylene to a limited quantity of acid, without occasioning the esterification of more than 88 per cent of the acid; but as a matter of fact, the acid is soon used up on account of the difficulty in purifying the amylene.

Equilibrium in Solutions.—When one of the pure substances in a homogeneous liquid mixture is present in great excess, we have a *solution*. The great importance which attaches to those reactions which occur in water solutions, both in nature—as in plant and animal organisms—and also in the *rôle* played in the analytical operations of the chemist,—demands a particularly thorough description of the state of equilibrium which is established between substances which are capable of mutual reaction when existing in solution.

We will first consider those cases where the solvent is indifferent, *i.e.* where none of the molecular species which take part in the reaction are from that ingredient which is present in great excess, viz. the solvent. It will be shown later that the ultimate participation of the solvent in the reaction is not essential for the form of the conditions of equilibrium.

At first, great difficulty was found in getting an insight into the condition of equilibrium which prevails in a solution. The methods which were applicable in studying the action of iodine on hydrogen, or of acetic acid on alcohol, and which depended upon the direct analytical determination of one of the molecular species taking part in the reaction, are usually out of the question here, because the point of equilibrium is quickly displaced by the removal of one of the reacting molecular species.

Thus, for example, if one wished to ascertain the state of equilibrium of such a reaction as the following, taking place in water, between the following reacting substances, viz. :—

$$K_2CO_3 + 2CH_3COOH = 2CH_3COOK + CO_2 + H_2O;$$

and if he wished to ascertain it in this way, viz. by removing the free CO_2 by a stream of air, and thus determining it analytically, he would obtain false results; because during this operation [*i.e.* of removing the CO_2 by a stream of air] the condition of equilibrium would be displaced, in the sense of the equation, from left to right.

Thus students were limited almost exclusively to physical methods of determination which, like the vapour-density determination in a gaseous system, can be performed without changing the composition of the reaction mixture; and thus it is only recently that the investigators came into the possession of a method which is almost as universally applicable for solutions as the vapour density determination is for gas mixtures. It is necessary that one should employ, according to the given case, certain physical properties of the solution, such as the power of optical rotation, the heat developed in preparing the solution, the refraction of light, the absorption of light, the colour, etc., in order to infer the composition of the solution; and this should be preceded by familiarising one's self with the dependence of the property in question upon the composition of the system as well as possible.

The Distribution of Hydrochloric Acid between Alkaloids.—As an example of an investigation of the sort just referred to, and one which was conducted with great skill and also with very significant theoretical developments, let us cite the results of a paper communicated by J. H. Jellet[1] on the distribution of hydrochloric acid between alkaloids in alcoholic solution. It appears that Jellet made use of this investigation in the correct formulation and application of the law of mass-action; and although this was subsequent to the work of Guldberg and Waage, yet it was done independently.

The question which Jellet proposed to solve was this, viz.—Such alkaloids, as quinine and codeïne in alcoholic solution, combine with 1 mol of hydrochloric acid; but what will be the distribution when two alkaloids are present in the solution, but when the acid present is not sufficient to saturate the total quantity of alkaloids?

The physical property which was employed in ascertaining the molecular condition of the reacting substances when in equilibrium, was their optical rotatory power when in solution. This, for quinine dissolved in alcohol, was 2·97, for codeïne, 2·63. By the addition of equi-molecular quantities of hydrochloric acid these values increased 1·344 fold and 1·909 fold respectively.

Moreover, since these substances do not mutually influence each other's rotatory power, when in the same solution, except when they react chemically on each other at the same time, therefore it is possible to draw some conclusion regarding the mutual action of the dissolved substances from the rotatory power of the solution.

Thus in a definite volume v of the solution, let n mol. of HCl be added to a_1 mol. of quinine (Ch), and a_2 mol. of codeïne (Cod). Then when there are produced x mol. of quinine "hydrochloride" (ChHCl), and accordingly n − x mol. of codeïne "hydrochloride," the observed rotation α of the plane of polarisation of light will be—

$$\alpha = (a_1 - x)D_{Ch} + (a_2 - [n - x])D_{Cod} + xD_{ChHCl} + (n - x)D_{CodHCl}.$$

The values of D here denote respectively the molecular rotatory power for the compound annexed as the sub-index. By means of this equation, the values of x may be expressed in only those terms which can be directly measured.

The observations gave the following results:—

a_1	a_2	n	x	$\frac{(a_2 - [n - x])x}{(a_1 - x)(n - x)}$
100	104	70·7	42·7	1·91
100	104	91·9	55	2·08
100	104	112·4	66	2·10
100	104	130·3	73	2·02
				Mean 2·03

[1] *Trans. Irish Acad.*, **25.** 371 (1875).

The reaction equation,

$$Ch + CodHCl \rightleftarrows ChHCl + Cod,$$

corresponds to the formula of equilibrium—

$$K = \frac{(a_2 - [n - x])x}{(a_1 - x)(n - x)},$$

since the active masses of the four molecular species participating in the reaction are respectively—

$$\underset{Ch}{\frac{a_1 - x}{v}}, \quad \underset{CodHCl}{\frac{n - x}{v}}, \quad \underset{ChHCl}{\frac{x}{v}}, \quad \text{and} \quad \underset{Cod}{\frac{a_2 - (n - x)}{v}}.$$

The volume v of the solution can be discarded, *i.e.* the condition of equilibrium does not change when the volume of the solution is largely increased by the addition of the solvent, alcohol. And, in fact, in all the mixtures observed above, the rotation angle was proportional to the concentration, and this could not have happened unless the condition of equilibrium remained unchanged with the dilution.

The coefficient of equilibrium K was calculated as in the last column, and showed itself to be a constant as far as was possible with the errors of observation, which were not inconsiderable.

In the same way, Jellet studied the distribution of hydrochloric acid between codeine and brucine; the rotatory power of the latter is 1·66, and it increases 1·291 fold by neutralisation with HCl. He also studied the distribution of HCl between brucine [Bru] and quinine. The following constants of equilibrium were found for these various reactions:—

	K
Ch + CodHCl = ChHCl + Cod	2·03
Cod + BruHCl = CodHCl + Bru	1·58
Bru + ChHCl = BruHCl + Ch	0·32

From the theory we can develop a characteristic relation which must exist between the three constants of equilibrium, and which leads at the same time to a deeper insight into the mechanism of the preceding reaction and also of related reactions. The hydrochloric acid is almost, but not quite entirely, combined with the alkaloids, a very slight quantity remaining free. This amounts to the statement that the alkaloidal hydrochlorides [as substituted ammonium chlorides] are dissociated, at least in traces, into free alkaloid and hydrochloric acid.

If K_1 denotes the dissociation constant of quinine hydrochloride, and K_2 that of codeine hydrochloride, and if ϵ denotes the (very small) quantity of hydrochloric acid remaining free, then according to the law of dissociation, as already stated for gases, and which, according to the law of mass-action, holds good of course also for substances existing in solution, we will have—

$$xVK_1 = \epsilon(a_1 - x),$$

and

$$(n - x)VK_2 = \epsilon(a_2 - [n - x]).$$

Then by division we obtain—

$$\frac{(a_2 - [n - x])x}{(a_1 - x)(n - x)} = K = \frac{K_2}{K_1}.$$

That is, *the constant of equilibrium is equal to the ratio of the dissociation constants of the two compounds between which the hydrochloric acid is divided* (see also p. 446).

Other things being equal, the quinine hydrochloride is dissociated 2·03 times as much as the codeïne hydrochloride. Now this relation also holds good for other reactions. If we denote the dissociation constant of brucine hydrochloride by K_3, then the product of the three constants of equilibrium should be—

$$\frac{K_2}{K_1} \cdot \frac{K_3}{K_2} \cdot \frac{K_1}{K_3} = 1.$$

As a matter of fact, we find that—

$$2{\cdot}03 \times 1{\cdot}58 \times 0{\cdot}32 = 1{\cdot}026.$$

Jellet previously obtained this result in a somewhat different way.

Obviously, in all cases in which the salts of an optically active base have different rotatory power to the base, it may be studied in a similar way, so that the chemical equilibrium may be found even when the second base added is optically inactive. This is a good method of comparing any bases, as was shown by Skraup.[1]

Dissociation in Solutions.—In order to investigate the condition of equilibrium, when it concerns a reaction which involves a change in the number of molecules, as *e.g.* a dissociation, very commonly one may make use of a method which amounts to the same as the determination of the vapour density of a gaseous system. The depresssion of the freezing-point, of the vapour pressure, and of the solubility, experienced by a solvent on the addition of soluble substances, are each directly proportional to the number of molecules of the dissolved substance. This has been shown by the new theory of solutions (see p. 486); and therefore very generally, by the determination of these magnitudes, one may learn the state of equilibrium of those reactions, the progress of which is identified with a

[1] *Monatshefte für Chemie*, **15.** 775 (1895).

change in the number of molecules. This method has, in fact, already given extended information concerning the dissociation phenomena in solutions; but unfortunately it is not sufficiently accurate in all cases, because measurements of this sort are uncertain at small concentrations, and at higher concentration they are disturbed by the irregularities already referred to.

The writer[1] succeeded in showing, in an indirect way, viz. by the study of the distribution of acids between water, and benzene or chloroform, that the equation of the isotherm of dissociation holds good, in the broadest sense, for the decomposition of double molecules into single ones. Thus if c denotes the concentration of the double molecule of the acid, and c_1 that of the normal [single] molecule, then it was shown that—

$$Kc = c_1^2, \quad \text{or } KV(1 - \alpha) = 4\alpha^2,$$

where α denotes the degree of dissociation, and V the volume of the solution containing 2 g.-mol. of the acid.

The preceding formula, of which much use will subsequently be made, can be considerably simplified when the [dissolved] substance in question is either *very highly* dissociated, or when it is *only very slightly* dissociated. In the former case α is almost constant, and nearly equal to 1, and we will have

$$V(1 - \alpha) = \text{constant};$$

i.e. *the concentration of the undissociated molecules* $\left(=\frac{1-\alpha}{V}\right)$, *at extreme dissociation, is proportional to the square of the total concentration* $\left(=\frac{1}{V^2}\right)$.

In the second case, α is small as compared with 1, and therefore $\frac{\alpha^2}{V}$ = constant; i.e. *the concentration of the dissociated molecules* $\left(\frac{\alpha}{V}\right)$, *at very slight dissociation, is proportional to the square root of the total concentration* $\left(=\frac{1}{\sqrt{V}}\right)$.

Such simple laws as these assist greatly in the general treatment of the subject.

Esterification in Benzene.—The reaction described on p. 450, viz. the esterification of amylene and acids, was regularly traced out as the reacting substances were *dissolved in benzene*. When the concentration is not too low, under these conditions the acid is *bi-molecular*, *i.e.* the reaction here progresses according to the scheme

$$S_2 + 2A = 2E,$$

[1] *Zeitschr. physik. Chem.*, **8**, 110 (1891).

where S_2 denotes the double molecule of the acid, and A and E respectively denote the amylene and the ester.

Therefore as before, from the law of mass-action, we obtain

$$\frac{(1-x)(a-x)}{xV} = K,$$

from which the expression

$$\frac{(1-x)(a-x)^2}{x^2V},$$

or else the expression

$$\frac{a-x}{x}\sqrt{\frac{1-x}{V}} = K'$$

must be constant.

The following table contains some results of an investigation conducted with trichlor-acetic acid at 100°. Here V denotes the volume of that reaction mixture in litres, which contains the double equivalent of acid.

a	V	x	K	K′
0·481	3·00	0·181	0·453	0·87
0·963	4·00	0·298	0·392	0·94
0·481	7·77	0·135	0·282	0·85
0·963	13·54	0·197	0·230	0·94

As a matter of fact, in the sense of the theory, and within the limits of the errors of experiment, K′ alone is constant. It should be noticed that if one applies the law of mass-action to the reaction

$$S + A = E,$$

then the active mass of the acid obviously will be

$$\alpha\frac{1-x}{V},$$

when we denote the degree of the dissociation of the acid by α. Therefore the following expression,

$$\alpha\frac{(1-x)(a-x)}{xV},$$

must be constant. But now, according to the equation of dissociation (p. 454), the following expression must be constant, viz.

$$\frac{\alpha^2}{1-\alpha}\cdot\frac{1-x}{V};$$

or because α is very small compared with 1 at slight concentration, then α must be inversely proportional to the expression $\sqrt{\frac{1-x}{V}}$, *i.e.* the square root of the concentration of the acid; and thus we find in this way that the expression, designated above as K′, must be constant.

Thus the requirements of the law of mass-action are found to be unequivocal throughout, as appears from the course of the reaction.

Evidence of Chemical Action from the Osmotic Pressure of Solutions.—Just as the question whether a substance is dissociated in solution or not can be decided by measurement of the osmotic pressure (or the proportional quantities, such as lowering of the freezing-point), so we can investigate whether, in dissolving two or more different substances, a chemical reaction occurs which causes a change in the number of dissolved molecules. If the freezing-point is lowered by each of the substances in the mixture as if the others were not present, no such reaction can have occurred, in the contrary case a reaction is indicated.

As an example of this method, we may quote the measurements of H. C. Jones[1] on solutions of water and alcohol in acetic acid; they produce the same lowering of temperature as if each was present separately, hence there is no formation of alcohol hydrates under these circumstances. But water and sulphuric acid influence each other's depression of the freezing-point, the effect in the mixture being smaller than when the two substances are present alone. The results thus obtained indicate that when a considerable excess of water is present the hydrate $H_2SO_4 \cdot 2H_2O$ is formed, in other cases $H_2SO_4 \cdot H_2O$, but in dilute solutions the latter hydrate is partly dissociated into water and sulphuric acid by the dissociating power of the solvent, acetic acid.

The Influence of the Solvent.—Many investigations have led to the very remarkable result that *the nature of the solvent has a very great influence on the molecular condition of the dissolved substance* (p. 267). Thus the organic acids in *benzene solution*, and at not too low concentration, have *doubled* molecules; in *ether*, *ethyl acetate*, *acetic acid*, and similar solvents, they have their *normal* molecular weight. In *water solution* they suffer *electrolytic dissociation*, more or less strongly. Similar results are found in other cases.

Now, as far as this question can be decided from the material thus far observed, and which is by no means satisfactory in amount, on the whole it appears that the different solutions can be arranged in a series as regards their "power of dissociation" (*dissoziierenden Kraft*, Beckmann).

[1] *Zeitschr. physik. Chem.*, **13**, 419 (1894)

At the head stands *water*, which has the property of splitting many dissolved substances into their respective ions, and of decomposing others at least to their normal molecular size.

The former property [*i.e.* ionisation] is possessed by such solvents as *the alcohols, the phenols, the esters*, and *the ethers*, and also by *acetone*, only in the very slightest degree. On the other hand, substances dissolved in these solvents just enumerated, at not too high concentration, usually have their respective *normal molecular weights.*

Many substances, such as the organic acids, the oximes, the phenols, etc., when dissolved in the following substances, show a tendency, which is more or less strongly expressed, towards the formation of *large molecular complexes*, and especially of *double molecules.* The solvents just referred to which occasion this polymerisation are *anethol, azo-benzene, para-toluidine*, etc., which represent the transition from the preceding group [*i.e.* the slightly ionising group of solvents] to the following group [*i.e.* the group which produces polymerisation], and which includes the *hydrocarbons* (such as *benzene, naphthalene, di-phenyl-methane, di-phenyl*, etc.), *carbon di-sulphide, chloroform, ethylene bromide*, and the like.

Accordingly, we find that acids with amylene, when dissolved in benzene, form more ester, other things being equal, than when dissolved in ether, in which latter solvent the substances do not react to any marked extent. Beckmann called attention to the fact that, excepting acetone, these solvents, which are characterised by great dissociating power,[1] are built on the water type.

We may clearly assume from the observations made hitherto that the high *dielectric constant* of the solvent is favourable not only to dissociation into ions, but also for the dissociation of complex molecules into simpler. This corresponds to the fact that substances in solution are more strongly dissociated than when they exist in a vacuum or in an indifferent gas under the same conditions of temperature and concentration. Thus acetic acid occurs with normal molecular weight in water under circumstances such that, as a gas, it would consist almost entirely of double molecules. Many solutions, moreover, have a greater dissociating force than a vacuum, and hence there follows this practical result, viz. that *the measurements conducted according to Raoult's method usually give more certain information regarding the normal molecular weight than the determinations of the gas density afford*, provided that by the molecular weight we mean the smallest form which the substance in question can assume without a complete dissolution of the molecular union.

The Participation of the Solvent in the Reaction.—Thus far we have always considered the solvent as excluded from taking any part in the reaction, and we will now seek to find out how such a reaction [*i.e.* where the solvent takes part] is to be conceived of.

[1] Beckmann, *Zeitschr. physik. Chem.*, **6.** 437 (1890).

The facility possessed by many substances, especially those containing hydroxyl, of forming double molecules in solutions of small dissociating power, such as benzene or naphthalene, has been thoroughly investigated by Auwers and his students;[1] the regular influence of constitution is shown especially in the phenols and the acetanilides. In both groups of bodies the tendency to the formation of double molecules is reduced or increased by ortho substitution; the aldehyde group CHO acts especially strongly in this direction, then the group CO_2R (carboxalkyl) and CN, more feebly NO_2 and the halogens, and least of all the alkyls. Substitution in the meta or para position is of much less influence; in the phenols such substitution occasionally causes the anomaly of a rise in the freezing-point.

The simplest case is that where only one molecular species, A, occurs in solution, and which reacts on the molecules of the solvent, water, *e.g.* when hydration of the dissolved substance would occur. Then there would take place the following reaction, viz.

$$A(H_2O)_n = A + nH_2O,$$

which also has the form of an equation of dissociation. Let c denote the concentration of the hydrated molecules of the substance A; and c_1 the concentration of the non-hydrated molecules; and c_2 the concentration of the water. Then we obtain the equation

$$Kc = c_1 c_2^n.$$

Now it must be noticed that c_2 is very large in comparison with c and c_1; and therefore, especially when the solution is sufficiently diluted, the active mass of the solvent will only change inconsiderably if the point of equilibrium is displaced in one direction or the other, i.e. *the active mass of the solvent is almost constant.* But then there is a proportionality between c and c_1, and therefore in a dilute solution *the hydrated fraction is independent of the concentration.* This deduction from the law of the mass-action has not been properly noticed hitherto.[2]

But now, since we have no means at present of answering the question, whether or not a molecule of water combines with a molecular species existing in solution, and also since the Raoult-van't Hoff methods of the molecular weight determination are unable to furnish any disclosure concerning this, therefore hitherto the experimental tracing out of the state of equilibrium of such reactions as the preceding or of analogous ones, has eluded us. Yet the experiment described on p. 450, viz. that in an excess of amylene the esterified portion of the acid is almost independent of the quantity of amylene, this speaks in favour of the preceding law.

[1] *Zeitschr. physik. Chem.*, **12.** 689; **15.** 33; **18.** 595; **21.** 337; **32.** 39; **42.** 513.

[2] Thus the view is very commonly found expressed in the literature that hydrates decompose with increasing dilution.

Moreover, a more stringent calculation, on the basis of thermodynamic principles (see Book II. Chap. III.), shows that the active mass of a solvent at constant temperature is proportional to its *vapour pressure.* Now, in keeping with the preceding considerations, this must be regarded as constant in the case of solutions fairly dilute, because it is only very slightly different from that of the pure solvent.

Equilibrium in Solid Systems.—From the van't Hoff method of treatment of "solid solutions" (p. 169), and especially from the apparent capacity of solid substances to diffuse into each other,—which though very slight, yet is undoubtedly true,—from all this there follows immediately the suggestion that mutual chemical reaction may take place even in homogeneous solid systems, and that ultimately a condition of equilibrium will be established. But it is easily conceivable that chemical processes in the solid state of aggregation would usually take place too slowly for one to trace out their progress; and thus, of course, their experimental study would be impossible.

Now, in fact, illustrations of the gradual progress of molecular changes of solids are not wanting. Here would be included the so-called elastic and thermal "effects" (*Nachwirkungen*), which are doubtless to be explained in a more or less extended change of the molecular structure. Also the brittleness of some metals occasioned after a time; as, *e.g.*, that of tin by the cold, or by lively agitation. Also the gradual change of the crystal form, and the change from the so-called amorphous to the crystallised condition, which is illustrated in devitrification.

Moreover, Spring[1] succeeded in one case in demonstrating the establishment of the condition of equilibrium, viz. in the reaction between solid barium carbonate and solid sodium sulphate; here the metathesis into barium sulphate and sodium carbonate, respectively, ceased when about 80 per cent had been so transformed; and conversely, the mutual reaction between barium sulphate and sodium carbonate ceased when about 20 per cent of the equi-molecular mixture had been changed. These reactions are induced by energetic shaking of the finely-powdered substances, and are also accelerated by the application of strong pressures (up to 6000 atmospheres). Their progress was ascertained by extraction with water, and then weighing the insoluble ingredients.

[1] *Bull. soc. chim.*, **46.** 299 (1886).

CHAPTER III

CHEMICAL STATICS—HETEROGENEOUS SYSTEMS

The Kind of Heterogeneity.—The heterogeneity of a chemical system existing in equilibrium obviously cannot consist in any variation of the composition of the liquid, or of the gas mixture, from point to point; because, if this were the case, there would occur a migration of matter, as a result of diffusion, *i.e.* the system would not as yet have reached its state of equilibrium. The heterogeneity must essentially consist only *in the intimate association of different complexes, each of which is homogeneous in itself*, such as solid salts and saturated solutions, mixtures of liquids and of gases, solids and their gaseous dissociation products, etc.; the degree of heterogeneity will be conditioned and estimated by the number of these complexes. The particular system may, of course, be either gaseous, liquid, or solid. There is no limit to the number of solid substances, and of liquids which do not dissolve each other, which participate in material decompositions by a displacement of the point of equilibrium resulting from reactions. Indeed, we know beforehand that, on account of the complete miscibility of gases, there can be present [in any case], *only one gaseous complex.*

The different complexes which are both physically and chemically homogeneous in themselves, which may be either a physical mixture or a chemical compound, and which make up the heterogeneous system, —these complexes are called *the phases of the system,* after the usage of W. Gibbs. Thus, *e.g.* in considering the state of equilibrium between calcium carbonate and its decomposition products, carbon di-oxide and calcium oxide, we must distinguish in this system three phases, two of which, viz. $CaCO_3$ and CaO, are solid, and one, CO_2, is gaseous.

General Law regarding the Influence of the Relative Proportions.—An unusually large number of experiments have led to the following law, which is universal, viz.:—

The condition of equilibrium of a heterogeneous system is independent of the relative quantity by weight in which each phase is present in the system.

Thus, *e.g.* after the state of equilibrium has been established in the aforesaid system, viz.—

$$CaCO_3 \rightleftarrows CaO + CO_2,$$

if there should occur an increase or a diminution in the quantity by weight, in which each of the enumerated substances is present, and if care is taken to preserve the external conditions, temperature and pressure, unchanged, then the state of equilibrium remains unchanged, *i.e.* in the sense of the reaction equation, there is no associated change in either direction, and the composition of each particular phase remains undisturbed.

Among other deductions from the preceding law are these well-known facts, viz. that the vapour tension of a liquid is independent of its quantity; and that the concentration of a saturated solution is the same, whether much or little of the solid salt be present, and the like.

The law may be demonstrated, on the basis of the molecular theory, in the following way. The establishment of chemical equilibrium does not mean that all chemical change has absolutely ceased; but rather that, in the sense of the reaction equation, the change, at any moment and at any point in the one direction, is exactly as great as the change in the opposite direction (p. 431).

Thus if we consider any selected part of a surface which separates two different phases of a system from each other, there will be a continual exchange of molecules in this portion of the surface, and molecules will be continually going out and in at the same time. In order that equilibrium may become established and remain so, this condition must be fulfilled, viz. that the same number of molecules of each species must pass through the portion of the surface in the one direction as in the other. Thus we are considering the necessary extension of the same considerations which earlier (p. 216) led us to conceive of the equilibrium between a liquid and its saturated vapour, as a *dynamic* condition; and there also the state of equilibrium was associated with the condition, that at each portion of the surface separating the liquid and the saturated vapour, at every moment, as many gaseous molecules must condense as shall evaporate from the liquid.

The forces, under the influence of which there occurs the continuous exchange of molecules between two different phases, like all molecular forces, have only very small spheres of action; these forces, at distances which are large enough to be measured, sink to zero. This exchange only takes place as a result of the forces which are active in the immediate vicinity of the limiting surface between molecules which exist in two different phases; and it is indifferent whether the extension of the two phases on both sides of the separating surface is large or not. For the same reason it [*i.e.* the exchange of molecules] is not influenced either by the form[1] or by the extension of the separating surface. This can only mean that *the condition of equilibrium is independent of the relative mass of each of the phases.*

Complete and Heterogeneous Equilibrium.—Among the simplest cases of equilibrium are the so-called "physical," *i.e.* the equilibrium between two different states of aggregation of the same substance; as, *e.g.*, the equilibrium between ice and liquid water; the equilibrium between liquid water and water vapour; and the equilibrium between ice and water vapour, *i.e.* where the reaction—in regard to which an equilibrium has been established—consists in the melting of a solid body, in the evaporation of a liquid, or in the evaporation of a solid (sublimation).

The relations here are very simple. For a definite external pressure there corresponds a definite temperature at which the two systems can exist beside each other; thus ice and water are co-existent at atmospheric pressure at 0°; and liquid water and water vapour at atmospheric pressure and at 100°. If we change the external pressure, at a temperature which is kept constant, or if we change the temperature, at an external pressure which is kept constant—then the reaction advances to a completion in one sense or in the other.

One may become completely familiarised (*orientiert*) with these conditions of equilibrium if he ascertains the dependence of the pressure at which the two respective states of aggregation are capable of existing beside each other—upon the temperature, *i.e.* the dependence of the melting-point upon the external pressure, and the vapour-pressure curve of the liquid or solid substance.

Now, in complete analogy with these foregoing reactions, which are, on the whole, *physical*, there are a number of reactions which are of a purely *chemical* nature. These latter have this peculiarity in common, viz.—

In reactions which occur isothermally, each one of the phases may change its mass but not its composition.

In all reactions of this sort the same thing holds true which was true in the case of the physical reactions, *i.e.* with a certain external pressure, the phases of the system are, collectively, coexistent only at a certain definite temperature; and under all other conditions, the reaction advances in one sense or the other to a completion, *i.e.* till one of the phases is exhausted.

After the example of Roozeboom,[2] whose many-sided experimental

[1] Excepting very great curvatures, where possibly the capillary forces may come noticeably into play (Book IV. Chap. III.).

[2] *Rec. trav. chim. des Pays-Bas*; since 1884; *Z. S. phys. Chem.*; since 1888.

investigations have very largely contributed to clearing up these questions, we will call the equilibrium "*completely heterogeneous.*"

A complete [heterogeneous] equilibrium, *e.g.*, is that between calcium carbonate and its decomposition products. For a definite temperature there corresponds one, and only one, pressure at which the three substances reacting on each other can exist beside each other, in the sense of the reaction equation following, viz.

$$CaCO_3 \rightleftarrows CaO + CO_2.$$

Let us think of the $CaCO_3$ as being at the bottom of a cylinder which is closed with a movable air-tight piston. Then, if we increase the volume by raising the piston, the reaction progresses, in the sense of the preceding equation, from left to right; and if we diminish the volume by depressing the piston, the reaction progresses, in the sense of the preceding equation, from right to left. Equilibrium can exist only at a definite pressure exerted from without upon the piston, viz. the so-called "*dissociation pressure.*" If we make the external pressure only a trifle smaller, always at a temperature which is kept constant, then the reaction advances from left to right until the calcium carbonate is exhausted, *i.e.* until it is *complete.* If we make the external pressure only a trifle greater [than the *dissociation pressure*], there results, conversely, the *complete* reunion of the carbon dioxide with the calcium oxide, in the sense of the equation, from right to left. When the reaction occurs, with the equilibrium pressure and temperature remaining constant, none of the phases changes its composition: this is the necessary pre-existing condition for the occurrence of the reaction under constant equilibrium pressure.

The peculiarity of such reactions is, that they may occur at constant temperature without change in the composition, or in any other property of each of the phases, and with no change except in their total mass; it follows from this, as a very complete generalisation, that for a given composition, there can be but one corresponding external pressure, *and to every definite temperature there corresponds one, and only one, equilibrium pressure.*

In this manner the way is clearly pointed out which must be followed for the experimental investigation of special cases of complete equilibrium. We may become completely familiar with such a case if we ascertain for each temperature the corresponding pressure at which the different phases are coexistent, and if we also ascertain the composition of each of the phases. Sometimes, as was the case with the system,

$$CaCO_3 + CaO + CO_2,$$

the composition may remain unchanged, even with a change in the temperature; but, as a rule, this is not the case.

Let us consider, as an example somewhat similar to the last case, the equilibrium between a solid salt, its saturated water solution, and the water vapour: here, as a sign of complete equilibrium for a definite temperature, the three phases are coexistent only at one pressure, namely, the vapour pressure of the saturated solution; and, moreover, the reaction which occurs isothermally (the evaporation of the water and the precipitation of the salt) does not result in a change of composition of any of the phases. But the composition of the liquid phase (the solution) varies with the temperature as a result of the varying solubility of the salt.

Now, as it is *the influence of the temperature* which especially concerns us in complete equilibrium, the further description of that belongs to the sections on *thermo-chemistry.*

Phases of Variable Composition.—The relations are very different when one or more of the phases changes its composition while the reaction proceeds isothermally. Here, in general, a change of the external pressure establishes a new condition of equilibrium, so that one or more of the phases of the system changes its composition.

An illustration will serve to make the case intelligible.

If we allow pure water to evaporate at the exact pressure of its saturated vapour, then, while the reaction (evaporation) is going on, neither the liquid nor the gaseous phase changes its composition. If we make the external pressure a little smaller, then all the water evaporates; if we increase it, then all the vapour condenses.

Now this is at once changed as soon as a salt is dissolved in the water; then, as is well known, the vapour pressure is lowered in proportion to the concentration [of the solution].

Let a definite volume of water vapour be in equilibrium with the solution. If we now diminish the external pressure a little, not all the water evaporates but only a definite fraction. For, as a result of the evaporation, the concentration of the solution increases, whereby the maximal pressure is still further reduced until it has sunk to the new and lower value of the external pressure, and thus a new state of equilibrium has become established. But now, if the concentration proceeds so far that the solid salt is precipitated, then the concentration remains constant, and equilibrium pressure also remains constant, and the latter does not increase any more with further evaporation.

Thus in those systems in which there occur phases having a variable composition, the equilibrium depends upon *the relative masses* of the reacting components in these phases, and thus the problem arises of formulating this influence.

In what follows, as we advance from the simplest conditions of equilibrium to the experimental examination of more complicated examples which occur, this simple result will be found, viz. *that without the introduction of any new assumptions, the law of mass-action may be also applied to heterogeneous systems.*

We will first consider the case where there occurs *only one phase having a variable composition* in the system, and will then proceed to the more complicated ones where their number is greater than one. This one phase may be either gaseous or liquid. *But solid substances, in contrast to gaseous and liquid phases, do not change their composition by a displacement of the point of equilibrium; and to this extent they occupy an exceptional position, because for this reason they form only phases of constant composition.*

Equilibrium between a Gaseous Phase and a Solid Substance—Sublimation.—According to Dalton's law, the sublimation pressure, *i.e.* the maximal partial pressure under which the solid vaporises into a gas, with which it forms no new compound, is as great as though the sublimation took place in a vacuum. The composition of the gaseous phase is completely determined by the vapour pressure of the corresponding solid substance, and by the quantity of the foreign gases which are present.

The Dissociation of a Solid Substance which produces only one Gas.—This case conducts itself in the same way as the last. Here also, for each temperature, there is a corresponding maximal pressure (*the dissociation pressure*), of the gas resulting from the dissociation, and which is not changed by the presence of indifferent gases. The pressure of dissociation is also independent of the relative quantities of the solid bodies which take part in the reaction.

The classical example of this case is the dissociation of calcium carbonate, the regularity of which was observed by Debray (1867); the reaction equation is

$$\underset{\text{Solid}}{CaCO_3} \rightleftarrows \underset{\text{Solid}}{CaO} + \underset{\text{Gaseous.}}{CO_2}$$

The dissociation pressures of this system have been recently measured very exactly by Le Chatelier;[1] in measuring the temperatures he used a thermopile made of platinum and an alloy of platinum and rhodium.

THE DISSOCIATION TENSIONS OF CALCIUM CARBONATE

t	p	t	p
547°	27 mm. Hg.	745°	289 mm. Hg.
610°	46 ,, ,,	810°	678 ,, ,,
625°	56 ,, ,,	812°	753 ,, ,,
740°	255 ,, ,,	865°	1333 ,, ,,

[1] *C. R.*, **102**. 1243 (1886)

The observation that both the dissociation pressure and the maximal pressure of the saturated vapour increase rapidly with the temperature appears to be quite general; moreover, there is a very close analogy between these phenomena.

The dissociation pressure is independent of the ratio between the relative quantities of solid carbonate and calcium oxide, as also follows if we apply the general law stated on p. 462 to the particular system considered. This is usually expressed as follows: viz. *the active masses of solid bodies as they participate in the chemical equilibrium are constant* (Guldberg and Waage, Horstmann).

The explanation of this behaviour, from the standpoint of the kinetic molecular theory, has thus far presented some difficulty. Thus one might incline to the view that the more CO_2 molecules which were seized and held fast by the mixture of solid carbonate and oxide, the more would the relative quantity of oxide decrease; and the fewer the molecules of CO_2 which were emitted [by the carbonate], the more would the quantity of the carbonate diminish. But in this way the influence of the relative quantities on the dissociation pressure would manifest itself, and this is contrary both to the general law just mentioned and also to experiment.

We have already given on p. 463 a general discussion according to the molecular theory showing that the quantity of the various phases must be without influence on equilibrium; it is worth while however to go through some considerations which show simply and easily the constancy of the dissociation pressure.

Both the calcium oxide and also the carbonate must each certainly possess a certain vapour pressure, or, more correctly, a pressure of sublimation; by this is meant the respective partial pressures of the molecular species CaO and $CaCO_3$, standing in a gas space which is in contact with the oxide and the carbonate. Now, these sublimation pressures are independent of the presence of foreign gases; they remain unchanged when the oxide and the carbonate are present at the same time. We cannot ascertain the magnitudes of these pressures, because they evade direct measurement by reason of their smallness.

In the vapour space, which is in contact with the two solid substances, there occur the three molecular species, viz. $CaCO_3$, CaO, and CO_2; and respecting the reaction,

$$CaCO_3 \rightleftarrows CaO + CO_2,$$

a condition of equilibrium will be established, and to this we may directly apply the law of chemical mass-action.

Let us denote by π_1 and π_2, the sublimation pressures of the oxide and of the carbonate respectively; and by p, the vapour pressure of the carbon-dioxide; this latter, on account of the smallness of the values of π_1 and π_2, will not be very different from the vapour pressure (*i.e* the dissociation pressure) of the whole system. Thus, according to the law of mass-action, it follows that

$$K\pi_2 = p\pi_1,$$

where K denotes the dissociation constant of the *gaseous* $CaCO_3$ molecules. Now, since π_1 and π_2 are independent of the quantity of the solid substances [CaO and $CaCO_3$ respectively], then p, *i.e.* the dissociation pressure at any given temperature, must also be constant. But it will vary greatly with the temperature, because the same also holds good both for K and for π_1 and π_2.

Considered on the basis of the molecular theory, this same proof is the reason for the view that the reaction occurs exclusively in the gaseous phase, and that the solid substances participate in the reaction to the extent of their previous sublimation. This view leads, as an immediate consequence, to the constancy of the dissociation pressure, though, on the other hand, it [*i.e.* the constancy of the dissociation pressure] is not a necessary condition for its derivation.[1]

To the dissociation phenomena of solid bodies there also belong *the dehydration of salts containing water of crystallisation*, resulting from raising the temperature. This has been thoroughly studied by Mitscherlich, Debray, G. Wiedemann, Pareau, Müller-Erzbach, and others. Here also the pressure of decomposition is constant for any given temperature, but increases strongly with increasing temperature. But here *dissociation by stages* is usually found, *i.e.* the salts do not lose *all* their water at a constant pressure, but several stages can be recognised, at which points the different molecules of water suddenly evaporate as the pressure changes.

Thus, for example, copper sulphate ($CuSO_4 + 5H_2O$) gives off the first two molecules of water with constant pressure, the following two also with constant but much lower pressure, and finally, the last molecule of water with the least pressure, so that the dissociation occurs in three stages

$$\begin{aligned} \text{I.}\ & CuSO_4 . 5H_2O = CuSO_4 . 3H_2O + 2H_2O \\ \text{II.}\ & CuSO_4 . 3H_2O = CuSO_4 . H_2O + 2H_2O \\ \text{III.}\ & CuSO_4 . H_2O \ = CuSO_4 + H_2O \end{aligned}$$

each of which possesses its own maximal pressure. A simple and certain method of determining the various stages of hydration is due to Müller-Erzbach.[2] The researches of Andreae[3] show that the *change in question is discontinuous*. The *ammonia compounds of the metal chlorides* behave like salts with water of crystallisation (Isambert [1868], Horstmann [1876]).

[1] The considerations advanced by Horstmann, *Zeitschr. physik. Chem.*, **6.** 1 (1890), and which derived the same result from the same standpoint of "solid solutions," *do not at all* contradict the preceding statements.

[2] See especially *Zeitschr. physik. Chem.*, **19.** 135 (1896).

[3] *Zeitschr. physik. Chem.*, **7.** 241 (1891).

R. Hollmann[1] has shown that crystalline mixtures containing water possess a vapour pressure which is independent of the content of water within certain limits. He showed also that small quantities of an isomorphous mixture in all cases lowered the vapour pressure of the crystal. The discovery that double salts also behave like simple crystals is of great interest; if the vapour pressure of a series of mixtures of two salt hydrates which are isomorphous in all proportions be investigated, it is found that the vapour pressure curve shows a break where the composition corresponds with that of a pure double salt. This is the only method at present known for deciding whether a double salt is found in a complete isomorphous mixture series (p. 120).

Here, of course, each dissociation has its own particular maximal pressure.

The Production of one Gas from several Solid Substances.—It can be shown here, by the same considerations which were advanced above, that for every temperature there is a corresponding *development pressure* (*Entwicklungs-druck*) which is independent of the relative quantities of the solid substances. In fact, Isambert[2] demonstrated that the development of ammonia from lead oxide and ammonium chloride, in the sense of the equation

$$PbO + NH_4Cl = NH_3 + Pb(OH)Cl,$$

took place at the following maximal pressures for the corresponding temperatures:—

t	Pressure.	t	Pressure.
17·5°	296 mm. Mercury	36·3°	599 mm. Mercury
27·0°	420 ,, ,,	48·9°	926 ,, ,,

At about 42° the maximal pressure is equal to the atmospheric pressure, and thus this point is the "boiling-point" of the system.

Another interesting example of the preceding case has been found by Rothmund[3] when calcium carbide is formed according to the equation

$$CaO + 3C = CaC_2 + CO,$$

a single gas is generated from this solid substance, and can therefore be in equilibrium with the system only at one given pressure for a given temperature. In point of fact it is found that above 1620° carbon monoxide is violently evolved with formation of carbide, and that conversely below that temperature carbon monoxide completely decomposes calcium carbide into lime and carbon.

The Dissociation of one Solid Substance which produces several Gases.—When the volatilisation of a solid substance is

[1] *Zeitschr. physik. Chem.*, **37.** 193 (1901). [2] *C. R.*, **102.** 1313 (1886).
[3] *Gött. Nachr.*, 1901, Heft 3.

attended, at the same time, with a more or less complete dissociation, then for a definite temperature there corresponds a definite dissociation pressure; this latter amounts to the same, in the presence of an indifferent gas, as in a vacuum. But where one of the resulting gaseous products of decomposition is present in excess, it constitutes a case requiring special treatment.

Thus ammonium hydrosulphide [NH_4SH] for each temperature has a definitely corresponding vapour pressure; but, as must be inferred from its vapour density, the gas mixture resting above it must be almost completely dissociated into ammonia and hydrogen sulphide, *i.e.* in the sublimation the following reaction occurs, viz.,

$$NH_4SH \rightleftarrows NH_3 + H_2S.$$

At 25·1° the dissociation pressure of the gas amounted to 501 mm., without excess of the decomposition products, *i.e.* the partial pressures of the two gases, NH_3 and H_2S, were equal to each other, each being 250·5° mm.

According as one or the other of the two gases was present in excess, the partial pressures [p_1 and p_2] for the state of equilibrium had the following values:[1]—

NH_3 p_1	H_2S p_2	p_1p_2
208	294	60700
138	458	63200
417	146	60800
453	143	64800
	Mean	62400

As is obvious, the partial pressure of that gas which is not present in excess is diminished by the presence of the other, *i.e. the dissociation pressure has fallen.* This result may be theoretically developed in the following way.

In the vapour of the ammonium hydro-sulphide, aside from the decomposed molecules, there are also some undecomposed molecules, though the number of the latter may be but small: let their partial pressure [*i.e.* of the undecomposed molecules] amount to π. Then by applying the equation of the isotherm of dissociation to the gaseous phase, we obtain

$$K\pi = p_1p_2.$$

But now, according to Dalton's law, at constant temperature, the vapour pressure π of the undecomposed molecules of ammonium hydro-

[1] Isambert, *C. R.*, **92.** 919; **93.** 731 (1881); **94.** 958 (1882).

sulphide must be constant; and, according to the preceding formula, the same thing holds true for the product p_1p_2. Now, denoting the dissociation pressure by P, for the case where $p_1 = p_2$ (*i.e.*, where there is no excess of either of the decomposition products), we obtain

$$K\pi = p_1p_2 = \frac{P^2}{4}.$$

Now, in fact, this equation is found to be established by the preceding table. The value of p_1p_2 varies irregularly, and the mean value 62,400 approximately coincides with the [theoretical] value,

$$\frac{P^2}{4} = \frac{501^2}{4} = 62{,}700.$$

The relations are very similar in the case of ammonium carbamate, a substance studied by Horstmann,[1] to whom we are indebted for the first application of the law of mass-action to the dissociation of solid compounds, which was at the same time one of the most remarkable demonstrations of this. The substance, in sublimation, decomposes almost entirely in the sense of the equation

$$NH_4\text{-}O\text{-}CONH_2 \rightleftarrows 2NH_3 + CO_2.$$

If we denote the partial pressure of the ammonia by p_1, and that of the carbon dioxide by p_2, then the application of the equation of the isotherm of dissociation gives

$$K\pi = p_1^2p_2;$$

here K denotes the dissociation constant of the gaseous ammonium carbamate molecules, and π their partial pressure. Now, since the latter is constant, in the presence of the solid salt, for any given temperature, therefore the right side of the preceding equation must also be constant; and since π, in contrast to the total pressure, has an infinitesimal value, on account of the almost complete dissociation of the ammonium carbamate, therefore we obtain

$$p_1^2p_2 = \frac{4P^3}{27},$$

where P denotes the dissociation pressure at the respective temperature, without excess of either of the dissociation products: and, therefore, two-thirds P denotes the partial pressure of the ammonia; and one-third P, that of the carbon dioxide. The addition of NH_3, then, will depress the dissociation pressure a good deal more than the addition of CO_2. Horstmann, and Isambert[2] later, found the preceding formula to be well established.

[1] *Lieb. Ann.*, **187.** 48 (1877). [2] *C. R.*, **93.** 731 (1881); **97.** 1212 (1883).

Reaction between any Arbitrary Number of Gases and Solid Substances.—This general case can be dealt with, according to what has preceded, without the introduction of any new assumptions.

Thus, let ν_1 molecules of the solid substance a_1, and ν_2 molecules of the solid substance a_2, etc., come together with n_1 molecules of the gas A_1, and n_2 molecules of the gas A_2, etc.; and let them form ν_1' molecules of the solid substance a_1', and ν_2' molecules of the solid substance a_2', etc., and also n_1' molecules of the gas A_1', and n_2' molecules of the gas A_2', etc. Thus the reaction will occur, according to the general scheme,

$$\nu_1 a_1 + \nu_2 a_2 + \ldots + n_1 A_1 + n_2 A_2 + \ldots \rightleftarrows \nu_1' a_1' + \nu_2' a_2' + \ldots + n_1' A_1' + n_2' A_2' + \ldots$$

As above, let the partial pressures of the reacting gases be respectively $p_1, p_2 \ldots, p_1', p_2' \ldots$ Now some of the molecules of the solid substances $a_1, a_2 \ldots, a_1', a_2' \ldots$, will certainly occur in the gaseous system, though, under the circumstances, in very small quantities, since each of these solid substances will have a certain vapour pressure; let this latter amount respectively to $\pi_1, \pi_2 \ldots, \pi_1', \pi_2' \ldots$, etc. Thus the application of the law of mass-action to the gaseous system will give the following condition:—

$$k\pi_1^{\nu_1}\pi_2^{\nu_2} \ldots p_1^{n_1}p_2^{n_2} \ldots = k'\pi_1'^{\nu_1'}\pi_2'^{\nu_2'} \ldots p_1'^{n_1'}p_2'^{n_2'} \ldots$$

Here k and k′ denote the velocity coefficients of the two mutually opposed reactions.

Now if we observe that, according to Dalton's law, the vapour pressures of solid bodies are independent of the presence of other gases, and that they remain constant at constant temperatures, then we can bring the preceding equation to the form

$$Kp_1^{n_1}p_2^{n_2} \ldots = p_1'^{n_1'}p_2'^{n_2'} \ldots,$$

where K denotes a constant, which again we will call the *constant of equilibrium.* We thus obtain the same result, which we previously obtained in a simple way in establishing the conditions of equilibrium between gases which react on each other in the presence of solid substances, and we need to neglect only those molecular species which are present at the same time *in the solid form.*

As has been already noted, Guldberg and Waage expressed this result in the statement, that *the active mass of a solid substance is constant.*

As an example of this, let us consider a reaction which was studied by Deville,[1] viz.

$$3Fe + 4H_2O = Fe_3O_4 + 4H_2.$$

By the reaction of steam at the pressure p upon iron at quite high temperatures, the reaction reached its completion when the partial pressure p′ of the hydrogen produced reached a definite value. The

[1] *Lieb. Ann.*, **157.** 76 (1870).

application of the general equation preceding led to the relation that p^4 and p'^4 must be proportional, and therefore p and p′ must be so also.

In fact, Deville found that, when the pressure of the steam was 4·6 and 10·1 mm., the corresponding pressures of the hydrogen developed amounted respectively to 25·8 and 57·9 mm. The ratios in these two cases are 0·178 and 0·174; the coincidence is more than sufficient, in consideration of the extraordinary difficulties met in prosecuting this research. The figures refer to a temperature of 440°. The pressure of the water vapour was held constant during the research, simply in this way: a tube containing finely divided iron was in communication with a retort containing water, which was kept at 0° the first time, and the second time at 11·5°; thus the pressure of the water vapour in the apparatus would be equal to the saturation pressure of water at these temperatures, namely, 4·6 and 10·1 mm. The partial pressure of the hydrogen developed was measured by the manometer. Other series of experiments therefore show variations not yet explained.

Explanation of the preceding General Case by Means of Sublimation and Dissociation.—When a number of bodies react with a single gas phase we may consider the reaction in general, as has already been shown on p. 470 for the special case of the dissociation of calcium carbide, as occurring in the form of a sublimation of the solid bodies which then react as gases. We may therefore consider the reaction in the gaseous phase as consisting first in a dissociation of the reacting molecules.

The chemical process is thus reduced to sublimation and dissociation, and, however complicated the case, it may be calculated when we know the sublimation pressure and dissociation constants of all the reacting molecules. It would be an important problem for future chemical investigation to determine these quantities for as many gases as possible and to find general rules for their calculations.

This may be elucidated by one or two applications. Ammonium chloride at ordinary temperatures has an exceedingly small sublimation pressure, probably to be reckoned in thousandths of a millimetre of mercury; ammonia and hydrochloric acid in the gaseous phase can remain in equilibrium with solid sal-ammoniac, and the product of the partial pressure of these gases is constant at a constant temperature. If, then, by any means we reduce the partial pressure of hydrochloric acid, that of ammonia must increase, if the partial pressure of hydrochloric acid is made excessively small that of ammonia will become very considerable. If solid lime be added to sal-ammoniac, the hydrochloric acid is almost completely absorbed so that the partial pressure of ammonia must increase, since the product of the partial pressure of hydrochloric acid and ammonia must remain constant so long as solid sal-ammoniac is present. Actually even at atmospheric temperature ammonia is evolved from a mixture of sal-ammoniac and lime with considerable partial pressure, and it is certainly very noteworthy that small sublimation pressures of bodies that dissociate on volatilisation can be raised to considerable amounts by *sufficiently thorough absorption of one of the components.* Thus if sal-ammoniac is mixed with phosphorus anhydride, the pressure of ammonia is made exceedingly small and hydrochloric acid is violently evolved. The formation of carbon dioxide from calcium carbide by acids may be regarded in the same manner. The partial pressure of the calcium oxide in the gaseous phase is extremely reduced, that of the carbon dioxide is consequently considerably raised, since so long as solid calcium carbonate is present the product of the partial pressure of CO_2 and CaO must be constant. Mercury salts are little volatile at low temperatures, but for the same reason if copper filings are brought in contact with them the negative radical of the mercury salt is combined, and the partial pressure of the mercury vapour reaches such an amount that, on moderate warming, enough mercury is given off to serve as the basis of an analytical test for mercury.

These conclusions indicate the way of measuring sublimation pressures which in themselves would be extremely small.

The Equilibrium between a Liquid Phase and Solid Substances—The Solubility of Solid Substances.—The description of the particular cases in which solid substances are in equilibrium with a liquid (solution) should be prefaced by the general remark, that here a behaviour is found which is completely analogous to the equilibrium between a gaseous phase and solid substances. Thus the solution of a solid substance in any solvent is a process which has the greatest similarity to sublimation. In both cases the process is ended as soon as the expansive force of the evaporating or dissolving substance is held in equilibrium by the gas pressure of the vaporised molecules, or by the osmotic pressure of the dissolved molecules, respectively. And, therefore, for the reason already mentioned (p. 146), we call the osmotic pressure of a saturated solution, "the solution pressure" of the respective substance, in order to extend the clear analogy of the vapour pressure to the sublimation pressure.

The solubility varies with the nature of the particular substance, and will always be affected more or less by any change in its chemical composition, or physical properties (as the crystal form, *e.g.*); thus, *e.g.* in general, the different solid hydrates of a salt will have different solubilities.

In determining solubilities, care must always be given to the nature of the solid substances which exist in equilibrium with the solution.

Moreover, the solubility depends also upon the nature of the solvent, and upon the temperature; usually, but not always, it rises with the temperature.

The analogy between the solution, sublimation, and dissociation, respectively, of solid substances is very clearly expressed *in the influence of foreign substances which are present.* The solubility of a solid substance is no more changed by the (moderate) addition of another substance—provided that the foreign substance does not react chemically on the first—than is the sublimation pressure changed by the presence of foreign indifferent gases.

In what follows, as in what has preceded, we will now fix our principal attention upon dilute solutions; their laws have been, thus far, almost exclusively the subject of investigation, and we will consider especially the behaviour of substances which are only *slightly soluble.* When in high concentration,—associated with which there is a deep-seated change in the whole nature of the liquid, and therefore in its solubility also,—the laws which will be developed below suffer considerable modifications; yet these modifications rarely extend so far that the same laws may not serve as a guide to the truth.

The particular "specific" influences which enter here and tend to obliterate the simple regularities of dilute solutions have great interest and deserve a thorough study. It is only recently that the first deliberate step has been made in this direction.[1]

Solubility varies very greatly with the nature of the solvent; this may be described as an effect of "*the nature of the medium,*" but nothing more about it is at present known. If another substance be added to a solvent in a small quantity the nature of the medium is not altered—that is, a dilute solution of a given substance has the same solvent action as the pure solvent. When, however, a considerable quantity of the added substance is present the nature of the medium is altered, and according to present observations this may take place somewhat rapidly—that is, with a higher power (for example, the second) of the concentration of the added component. Thus if alcohol is added to an aqueous solution of cane-sugar, the sugar if precipitated because it is much less soluble in the newly formed medium.

According to recent investigations [2] the solvent action of water is usually much reduced by addition of electrolytes; this is in all probability related to the fact (p. 382) that the density of water is considerably influenced by the presence of other ions.

A good example of the influence of the nature of the medium is to be found in the researches of Villard on the solvent action of compressed gases.[3] In a vacuum or in a *dilute* indifferent gas solids and liquids dissolve in accordance with their vapour pressure; if, however, the indifferent gas is strongly compressed, say at 100 atmospheres, specific solvent action appears; thus bromine evaporates in an atmosphere of

[1] G. Bodländer, *Zeitschr. physik. Chem.*, **7.** 308 and 358 (1891).

[2] See especially Roth, *Zeitschr. physik. Chem.*, **24.** 114 (1897); Rothmund, *ibid.*, **33.** 401 (1900); W. Biltz, *ibid.*, **43.** 41 (1903).

[3] *Journ. de phys.* [3], **5.** 453 (1896); ref. *Zeitschr. physik. Chem.*, **23.** 373 (1897).

compressed oxygen in much greater quantity than in a vacuum; compressed hydrogen has much smaller solvent power.

Solubility of Hydrates.—It may be expected on theoretical grounds that the action of other substances should affect the solubility in water of salts containing water of crystallisation even when the action is too small to alter the nature of the medium. We will not go into the thermodynamic consideration of this influence here; it is easily seen however that the water contained in the dissolving solid is taken up by the solvent the more easily the lower the vapour pressure of the dissolving water, and hence is more strongly dissolved; in other words, the solubility of a hydrate in water is increased by a foreign substance the more the hydrate reduces the vapour pressure of water. For further details see the complete theoretical and experimental investigation of H. Goldschmidt.[1]

Equilibrium between Solids and Solutions.—This case is obviously to be treated in the same way as the equilibrium between solids and gaseous phases. The concentration of each kind of molecule present in the solid state remains constant when the equilibrium is displaced, hence the law of mass-action can be applied to homogeneous solutions in the manner given in the preceding chapter.

The simplest case of this kind is *dissociation on solution*, such as occurs with racemic acid and similar compounds which, on solution, break up into dextro and lævo modifications, and also with many double salts and salts with water of crystallisation which break up into their components (the two single salts or salt and water), and most frequently of all with the solution of salt which dissociates into ions.

The equations for this case are as follows.[2] If u is the concentration of the undissociated component we have

$$u = \text{const.};$$

if u_1 and u_2 are the concentrations of the two components into which the compound breaks up in solution the equation of the dissociation isotherm is

$$Ku = u_1 u_2.$$

So long as the dissociated substance is present in the solid state u is constant, and we have therefore

$$u_1 u_2 = \text{const.}$$

We thus have the same equation as that given on p. 471 for the dissociation of ammonium sulphydrate.

[1] *Zeitschr. physik. Chem.*, **17.** 145 (1895).
[2] Nernst, *Zeitschr. physik. Chem.*, **4.** 372 (1889).

R. Behrend[1] has carried out a study of the foregoïng equation. If anthracene and picric acid are mixed in alcoholic solution, anthracene picrate is formed according to the equation

$$C_{14}H_{10} + C_6H_2(NO_2)_3OH = (C_{14}H_{10}) . (C_6H_2[NO_2]_3OH) ;$$

this becomes evident to the eye from the marked red coloration of the solution, but the compound is only formed to a small extent, *i.e.* it is largely dissociated in solution.

Now the solubility of anthracene and picric acid were determined and found to be 0·176 and 7·452 grms. in 100 parts of solution at 25°. Then a series of solutions of varying composition were analysed, some containing anthracene and some anthracene picrate in solid form; the results of these determinations are given in the following table:—

	1	2	3	4	5	6	7	8	9	
Anthracene a	0·190	0·206	0·215	0·228	0·236	0·202	0·180	0·162	0·151	0·14
Picric acids p	1·017	2·071	2·673	3·233	3·469	3·994	5·087	5·843	6·727	7·51
nthracene diss. u_1	0·176	0·176	0·176	0·176	0·183	0·149	0·127	0·109	0·098	0·09
icric acids diss. u_2	0·999	2·032	2·623	3·166	3·401	3·926	5·019	5·775	6·659	7·44
Picrate u	0·032	0·069	0·089	0·119	0·121	0·121	0·121	0·121	0·121	0·12
$\frac{u_1 \cdot u_2}{u}$	5·5	5·2	5·2	4·7	5·1	4·8	5·3	5·2	5·4	5·9

a and p are the amounts of the two components present in the solution partly free, partly combined. In solutions 1-4 anthracene was in excess, in solutions 6-9 solid anthracene picrate was present; 5 was saturated with both bodies, 10 with the picrate and picric acid.

Solutions 1-4 contain more anthracene than corresponds to its solubility (0·176); the excess must be in the form of picrate in solution; this quantity is given in the fifth line of the table. If the amount of picric acid present in the form of picrate be subtracted from p we obtain the values of u_2 given in the fourth horizontal line.

Solution 5 is saturated both with anthracene and with picrate; the amount of picrate here present, which is equivalent to 0·236 − 0·176 = 0·060 grms. of anthracene, and therefore amounts

[1] *Zeitschr. physik. Chem.*, **15.** 183 (1894).

to $0{\cdot}060\frac{229+178}{178}=0{\cdot}137$ (since 229 and 178 are the molecular weights of picric acid and anthracene), gives the amount of undissociated picrate which must be present in all the solutions containing solid picrate in excess. This amount may however also be calculated from solution 10, which is saturated with picric acid and picrate, and contains $7{\cdot}511-7{\cdot}452=0{\cdot}059$ of picric acid more than corresponds to the solubility of picric acid. This quantity can also only be present as picrate, and we thus find for the amount of undissociated picrate $0{\cdot}059\frac{229+178}{229}=0{\cdot}105$. The mean of these two determinations (0·137 and 0·105) is 0·121; this is taken as the value of u in solutions 5-10 with solid picrate present. u_1 in these solutions is naturally calculated by subtracting u equivalents of anthracene from the total amount of anthracene present as picrate, and u_2 is determined in a similar manner.

The theory requires that for all the solutions the expression

$$K=\frac{u_1u_2}{u};$$

should hold, and, in point of fact, this expression was found constant so far as the errors of observations would allow. In solutions 5-10 on account of the constancy of u

$$u_1u_2=\text{const.},$$

and for the solutions 1-5 on account of the constancy of u_1

$$\frac{u_2}{u}=\text{const.}$$

Several Phases of Variable Composition—The Vapour Pressure of Solutions.—We may now advance a step further and consider the case where *several phases of variable composition occur in a system: these phases may occur in the gaseous or liquid state.*

The state of equilibrium which is established between a dilute solution of a substance which is not noticeably volatile, and the vapour emitted, is completely determined by the formula for the vapour pressure: according to this, *the relative depression of the vapour pressure, which is experienced by the solvent on adding a strange substance, is equal to the ratio of the number of dissolved molecules to the number of molecules of the solvent* (p. 150). For concentrated solutions, this law is at least suitable as a first approximation.

In the sense of the kinetic method of the consideration of chemical equilibrium, we must conceive of this law also in a dynamic way; thus, *e.g.* in the coexistence of a water solution and of water vapour, at every moment, as many molecules of water vapour will be emitted, from every portion of the surface of a solution, as are precipitated by condensation.

We will now consider that case which is in extreme contrast to the last—namely, where only the *dissolved substance* evaporates from the solution, or at least where its volatility is more pronounced than that of the solvent; and where the vapour pressure, exerted by the vapour emitted by the solvent and by the solution, is due almost exclusively to the molecules of the dissolved substance. When the dissolved substance, as such, stands at the same osmotic pressure that its vapour has at the same concentration, or in other words, when its molecular condition does not change on evaporation, then, according to p. 154, a proportionality must exist between the concentrations of the dissolved substance in the two phases of the system considered.

Under these circumstances, Henry's law of absorption holds good. Let us denote the osmotic pressure of the dissolved substance in the solution by π; and by p, the gas-pressure which the evaporated substance has in the space in contact with the solution. Then at constant temperature, we have

$$\pi = Lp,$$

or

$$c = LC,$$

where c and C are the concentrations in the two phases.

The proportional factor L we will call the *solubility coefficient* of the respective substance. It has this simple relation to the *absorption coefficient*, in that it may be obtained from the latter by multiplication by $(1 + 0{\cdot}00366t)$, where t denotes the temperature of the experiment. By the term "absorption coefficient," according to the example of Bunsen,[1] is meant the volume (referred to 0° and 760 mm. pressure) of the gas which is absorbed by unit volume of the solution.

The proportionality between the osmotic pressure and the gas pressure may be easily understood by means of the molecular theory. When the solution is in equilibrium with the vapour in the space above it, then at every moment there are as many molecules of the dissolved substance emitted by the solution as are precipitated upon it from the gas space. But now since the quantity both of the emitted and of the absorbed molecules is proportioned to their number, therefore these quantities must also be proportional respectively to the concentration in the solution and in the gas space; and there must also be a proportionality between the two latter values.

It thus becomes obvious how the *identity of the molecular weights* of the dissolved substance in the solution and in the gas space is a necessary condition for the relevancy of Henry's law; because otherwise the number of the emitted and of the precipitated molecules would not be

[1] *Gasometrische Methoden*, Braunschweig (Brunswick), 1877.

proportional to the concentration. If different molecular species, but such as do not react upon each other, are present, they, of course, do not have any influence on each other, *i.e.* each of the different gases of a mixture is dissolved as though the others were not present (absorption law of Dalton).

The Law of Distribution.—Moreover, the case where several molecular species (which may react chemically upon each other) evaporate at the same time, from any selected solvent,—this may likewise be referred to the preceding case by means of a kinetic treatment. For the method of treatment, which was advanced for the case of one molecular species, may be applied unchanged to each of several molecular species participating in the chemical equilibrium, which has become established between the solution and the vapour space in contact with it.

We thus arrive at the result, that *at a given temperature for every molecular species there exists a constant ratio of distribution between a solvent and its vapour space; and this is independent of the presence of other molecular species, it being a matter of indifference whether the particular molecular species is chemically reactive with the other or not.*[1]

The simultaneous Evaporation of the Solvent and of the Dissolved Substances.—This case, which is of frequent occurrence in nature, may be solved by a combination of the laws just given. It may be stated as follows:—

1. The partial pressure of the solvent in the vapour which stands in equilibrium with the solution is equal to the vapour pressure p of the pure solvent at the respective temperature, diminished by the depression occasioned by the dissolved substance; this depression, according to van't Hoff's vapour-pressure formula, is estimated to be

$$p\frac{n}{N + n},$$

where n denotes the total number of dissolved molecules, in proportion to N molecules of the solvent.

2. The vapour pressures $p_1, p_2 \ldots$ of the dissolved substances can be calculated from the formulæ

$$p_1 = \frac{\pi_1}{L_1}, \quad p_2 = \frac{\pi_2}{L_2} \; \ldots,$$

if $\pi_1, \pi_2 \ldots$ denote the partial osmotic pressures of the particular molecular species existing in the solution, and if $L_1, L_2 \ldots$ denote their respective solubility coefficients.

The partial pressure of each of the particular molecular species in the gas space is, moreover, proportional to the concentration in the

[1] Nernst, *Zeitschr. physik. Chem.*, **8**, 110 (1891); see also Aulich, *ibid.*, **8**, 105.

solution. When no change of the molecular state (*i.e.* neither dissociation nor any other reaction between the dissolved and the evaporating substances) is associated with a change in the concentration, either in the solution or in the gas space, then *the total pressure* of the dissolved substances in the saturated vapour is proportional to the concentration in the solution. But this ceases to be true when there occurs a displacement of the equilibrium existing between the dissolved substances (the displacement being identified with a change in the number of molecules), as a result of a change in the concentration.

In order to prove these requirements of his theory, the writer investigated the equilibrium existing between solutions of acetic acid in benzene and their saturated vapour. Acetic acid, both in solution and in a state of vapour, is a mixture of the molecules (CH_3CO_2H) and $(CH_3CO_2H)_2$; and since the degree of dissociation changes with the concentration, it was anticipated that Henry's law would not hold good for the vapour of acetic acid.

The measurements of the partial pressure of the vapour of acetic acid were accomplished by determining the changes of the boiling-point occasioned by the addition of acetic acid to benzene, and were exactly measured by Beckmann's apparatus.

The partial pressures sought were obtained by the differences between the observed changes in the boiling-point and those calculated from the vapour pressure formula. In the following table, m denotes the number of g. of acetic acid dissolved in 100 g. of benzene; and x its degree of dissociation, as is inferred from the boiling-point determinations of dilute solutions of related non-volatile acids, and calculated for the different concentrations, according to the equation of the isotherm of dissociation, viz.

$$\frac{mx^2}{1-x} = \text{constant.}$$

Here p is derived from the changes in the boiling-point, the partial pressure of the acetic acid being expressed in mm. of mercury.

m	x	p Obs.	p Calc.	Δ	α
0·150	0·20	2·4	2·6	2·24	0·87
0·663	0·10	6·6	6·5	2·44	0·70
1·64	0·065	11·8	11·6	2·61	0·60
1·87	0·061	12·9	12·6	2·63	0·58
2·60	0·055	16·1	15·7	2·71	0·54
4·13	0·042	21·8	21·4	2·81	0·48
5·00	0·038	23·6	23·9	2·83	0·47
6·83	0·033	31·4	31·1	2·96	0·40
7·53	0·031	33·5	33·4	2·99	0·38
8·42	0·029	36·4	36·4	3·02	0·36

Here we are not concerned with the proportionality between m and p which is required by the ordinary conception of Henry's law; but this holds true, nevertheless, for the two particular molecular species, viz. the double and the single molecules, of which the acetic acid vapour is composed. This is shown by the following calculation.

The number of *normal molecules* in the solution is proportional to the value of mx, or also of $\sqrt{m(1-x)}$. The degree of dissociation α for the gaseous state is calculated from the vapour density Δ of acetic acid to be

$$\alpha = \frac{4{\cdot}146 - \Delta}{\Delta},$$

where 4·146 represents the [superior] limiting value of the vapour density of acetic acid, α having the value of zero ($\alpha = 0$). The vapour pressures Δ, given in the fifth column, and corresponding to the temperature of observation (the boiling-point of benzene being 80°), are calculated from the equation of the isotherm of dissociation (p. 440), viz.—

$$\frac{\Delta - 2{\cdot}073}{(4{\cdot}146 - \Delta)^2 p} = K,$$

where, after the example of Gibbs,[1] at 80°, K may be placed at 0·0201.

Now, since the number of normal molecules in unit volume of the vapour is proportional to the product of the mass of the acetic acid vapour in unit volume and its degree of dissociation; therefore, corresponding to the expression

$$\Delta p\alpha = \Delta p \frac{4{\cdot}146 - \Delta}{\Delta} = p(4{\cdot}146 - \Delta),$$

a proportionality must exist between the values,

$$\sqrt{m(1-x)}, \quad \text{and} \quad p(4{\cdot}146 - \Delta).$$

Now, as a matter of fact, we find that the values of p given in the fourth column, and calculated from the formula

$$p = 14{\cdot}4 \frac{\sqrt{m(1-x)}}{4{\cdot}146 - \Delta} \text{mm. Hg.},$$

coincide with the results of observation in a remarkable way. The proportional factor 14·4 corresponds to the solubility coefficients of the CH_3CO_2H molecules.

Similar results are obtained in the measurement of the pressures of ethereal solutions of water, which substance, like acetic acid, tends to form double molecules, in this case $(H_2O)_2$.

[1] *Sill. Journ.* (*Am. J. Sci.*), **18**, 371 (1879).

The Mutual Solubility of Liquids.—Many liquids have the property of mutually dissolving each other to a limited extent, as water and ether, *e.g.* For each temperature there is a corresponding solubility of the one in the other, and conversely of the other in the one. The connection between these two mutual solubilities has not been explained as yet. Only a few facts are known experimentally regarding this.

Regarding the *influence of temperature*, sometimes both of the mutual solubilities increase with increasing temperature. It may happen, as in the case of water and phenol,[1] that the composition of the two solutions may approach nearer and nearer together on raising the temperature, until *complete miscibility* occurs, a phenomenon which reminds one of the *critical temperature.*

Sometimes both solubilities may diminish with increasing temperature.

And, finally, the one may increase and the other may diminish, as is the case with water and ether. Thus if one warms water saturated with ether [the water being the solvent], or cools ether saturated with water [ether being the solvent], in both cases it is noticed that a cloudiness is formed in the originally clear liquids. From this it follows that the solubility of ether in water diminishes with rising temperature; but, inversely, that that of water in ether increases with the temperature.

Two liquids, which are mutually soluble to a limited degree, are in equilibrium with each other when they have dissolved each other to saturation. From this it follows that the *saturated vapours emitted by each of the two layers* [of the two solutions] *have the same pressure and the same composition.* And, moreover, because the two layers of the mixture are in equilibrium with each other, their saturated vapours must be also; for otherwise the condition of equilibrium would be destroyed, resulting from a distillation process which would change the composition of the two layers. This requirement of the theory has been experimentally established by Konowalow (p. 112).

Moreover, the partial pressure of each one of the two ingredients must be smaller than that corresponding respectively to each of the pure solvents, because each of the solvents has experienced a depression of its own particular pressure, by reason of having dissolved some of the other solvent. The magnitudes of these depressions depend of course upon the respective molecular weights, and upon the mutual solubilities. They are very small when the solubility is at a minimun, as is the case, *e.g.*, with water and carbon disulphide. In such cases the resulting vapour pressure is simply [nearly] equal to the sum of the two vapour pressures which each of the two liquids respectively would have independently.

When one of two liquids [A and B] (the mutual solubility of

[1] Alexejew, *Wied. Ann.*, **28.** 305 (1886).

which for each other is limited) dissolves still another substance [C], then the mutual solubility of the former [A or B] for the other [B or A] is diminished, in accordance with the laws of solubility already stated (see p. 141).

The Distribution of a Substance between Two Solvents.— The laws which have already been stated for the evaporation of a substance existing in solution, *i.e.* for the distribution of a substance between a gaseous and a liquid phase, may be applied without further remark to the distribution of a substance between two liquid phases; this is not surprising, in view of the analogies between evaporation and solution, which have been repeatedly emphasised.

Now, by the *distribution coefficients* of a substance between two solvents, we mean the ratio of the spatial concentrations which the one substance has in these two solvents, when the condition of equilibrium shall have been established; and thus, by a simple extension of the law developed on p. 475, we arrive at the following results, viz. :—

1. If the dissolved substance has the same molecular weight in each of the two solvents, then *the coefficient of distribution* is constant for any given temperature (compare also p. 155).

2. In the presence of several dissolved substances, each molecular species distributes itself as though the others were not present.

3. If the dissolved substance does not exist as a unit, but is influenced by dissociation, then law 1 holds good for each of the molecular species resulting from the dissociation. This also follows immediately from the application of law 2.

Thus succinic acid is divided between ether and water, with the following constant *coefficient of distribution* :—

c_1	c_2	$\frac{c_1}{c_2}$
0·024	0·0046	5·2
0·070	0·013	5·2
0·121	0·022	5·4

Here c_1 and c_2 denote the number of g. of acid dissolved in 10 c.c. of water and of ether respectively. This was to be expected, because succinic acid has its normal molecular weight both in ether and also in water, disregarding its very slight electrolytic dissociation in the latter. Similar results were found by Berthelot and Jungfleisch[1] for similar cases.

But, as was to be expected, very different results were found by the

[1] *Ann. chim. phys.* [4], **26**: 396 (1872); Berthelot, *ibid.*, 408.

writer[1] in the study of the distribution of substances which have *different molecular weights* in the two solvents. Thus, c_1 and c_2 denoting the varying concentrations (*i.e.* the number of g. in 10 c.c. of the solvent) of benzoic acid in water and benzene respectively, the following results were found :—

c_1	c_2	$\frac{c_1}{c_2}$	$\frac{c_1}{\sqrt{c_2}}$
0·0150	0·242	0·062	0·0305
0·0195	0·412	0·048	0·0304
0·0289	0·970	0·030	0·0293

Here we are not concerned with the constancy of the quotient $\frac{c_1}{c_2}$, because the acid has its *normal mol. wt.* in water (disregarding again a very slight electrolytic dissociation), while in benzene the *double molecules* preponderate. Now, the number of normal molecules in the latter solvent [benzene], according to the laws of dissociation, is proportional to the square root of the concentration; and thus the law of distribution requires a proportionality between c_1 and $\sqrt{c_2}$; and as a matter of fact we find the values of the expression $\left(\frac{c_1}{\sqrt{c_2}}\right)$, as given in the last column, very constant.

This constancy disappears at extreme dilution, because, in this case the benzoic acid in the benzene is decomposed increasingly into single molecules. The writer calculated the degree of dissociation from the change in $\frac{c_1}{\sqrt{c_2}}$, and thus proved that the equation of the isotherm of dissociation also holds good for the decomposition, in solution, of double molecules [into single molecules].

Hendrixson[2] has more recently made measurements of this kind and thoroughly established the theory. The following table refers to the partition of benzoic acid between benzene and water at 10°; c_1 is the number of grms. of benzoic acid contained in the aqueous phase for 200 grms. of water, c_2 the corresponding quantity for the benzene phase; α is the degree of electrolytic dissociation in water (see the following chapter), $c_1(1-\alpha)$ therefore the quantity of benzoic acid in its normal state in the aqueous phase. Taking as the ratio of

[1] Nernst, *Zeitschr. physik. Chem.*, **8**. 110 (1891).
[2] *Zeitschr. anorg. Chem.*, **13**. 73 (1897).

partition for normal molecules $k=0{\cdot}700$, it follows that the number of normal molecules in the benzene phase is

$$m = \frac{c_1(1-\alpha)}{0{\cdot}700};$$

the number of double molecules in the benzene phase is therefore naturally $c_2 - m$. To test the theory the last column gives the dissociation constant

$$K = \frac{m^2}{c_2 - m},$$

which is found to be constant within the limit of errors of observation; this proves both that the dissociation isotherm holds for the dissociation of double molecules into simple, and that the simple molecules (as also the double molecules) possess a ratio of partition independent of concentration or degree of dissociation.

c_1	c_2	α	$c_1(1-\alpha)$	m	$c_2 - m$	$\frac{m^2}{c_2-m}$
0·0429	0·1449	0·169	0·0357	0·0510	0·0939	0·0277
0·0562	0·2380	0·149	0·0474	0·0677	0·1703	0·0269
0·0823	0·4726	0·125	0·0720	0·1029	0·3697	0·0286
0·1124	0·8843	0·104	0·1007	0·1439	0·7404	0·0279
0·1780	2·1777	0·0866	0·1626	0·2323	1·9454	0·0277
0·2430	4·0544	0·0747	0·2249	0·3213	3·7331	0·0276
0·2817	5·4851	0·0695	0·2621	0·3743	5·1108	0·0274
					Mean K=	0·0277

In the same way it was found that

$$k = 0{\cdot}477 \qquad K = 0{\cdot}122.$$

Application of the Law of Partition to Determination of Chemical Equilibrium.—Just as on p. 475 the solubility was applied to determine the ratio in which the different species of molecules take part in equilibrium, so we may make use of the partition of a substance between two solvents for the same object. This has been done practically in the table in the foregoing paragraph; the following examples which illustrate the application of this method to more complicated cases show that the law of partition has a certain advantage as compared with the principle of constant solubility in application to such cases, for the latter refers only to a single concentration, namely that of saturation, whereas the former is not so limited.

1. Bromine is shared between water and carbon disulphide in a constant ratio, the molecular weight corresponding to the formula Br_2 in both solvents. If potassium bromide is added to the water, it is found that a considerably larger amount of bromine passes into the aqueous layer. This quantity must be used in forming a new species of molecule; Roloff,[1] who, at my instance, first made use of the law of partition to study chemical equilibrium, was able to show that bromine forms, by addition of potassium bromide, the electrolytically strongly dissociated salt KBr_3.

2. The ratio of partition of chlorine between water and carbon tetrachloride varies considerably with the concentration. This, as was shown by A. A. Jakowkin,[2] depends upon the fact that chlorine acts on water according to the equation

$$Cl_2 + H_2O = HCl + HClO.$$

If the amount of unchanged chlorine is used for the calculation, the constant ratio of partition is arrived at; in calculating this the electrolytic dissociation of the hydrochloric acid formed must be taken into account.

The Freezing of Solutions and Crystallisation from Solutions.—The separation of solid substances from solutions, in many respects, may be compared with the process of the evaporation of a mixture, *i.e.* the separation of one of its components in the gaseous form. When this process results in the precipitation of that ingredient of the solution which is present in excess, *i.e. the solvent*, it is called *a freezing of the solution;* and when it results in the separation of *the substance dissolved*, then it is called *a crystallising out of the substance dissolved.* The processes of freezing and of crystallising out are both to be considered from the same point of view; and when we are not dealing with dilute solutions where one ingredient is present in large excess, but with a mixture, where both ingredients are present in about the same proportions, then we would be in actual doubt whether the separation should be regarded as a freezing or a crystallising out.

In fact, it would appear that neither ingredient is separated in an absolutely pure form, but that there probably crystallises out an *isomorphous mixture containing both ingredients*, just as every solution emits a vapour which is a mixture containing *both ingredients.*

Now experiment shows that usually one ingredient preponderates so much in the product which is crystallised out, that we may practically regard it as a separation in a pure form. Then, according to this conception, the two cases which are distinguished above are to be

[1] *Zeitschr. physik. Chem.*, **13.** 341 (1894). The corresponding investigation on the formation of potassium triodide has been carried out by A. A. Jakowkin (see *ibid.*, **20.** 19, 1896).

[2] *Ber. deutsch. chem. Ges.*, **30.** 518 (1897); *Zeitschr. physik. Chem.*, **29.** 613 (1899).

regarded as only two limiting cases of the many which occur in nature; and in fact examples are known, though they are not very numerous, where the solvent and the dissolved substance crystallise out from the solution in an isomorphous mixture (pp. 168 and 272).

The Freezing-Point of Dilute Solutions.—There is no difficulty in solving the case where *the pure solvent freezes out* from a dilute solution. For the condition of equilibrium between the solidified solvent and the solution is completely made known by means of the laws which have been already described (p. 151). These laws may be formulated as follows:—

1. The addition of a [soluble] foreign substance lowers the freezing-point [of the solvent] in all cases.

2. The depression of the freezing-point of the solvent, occasioned by the addition of a foreign substance, is, in most cases, proportional to its concentration (Blagden); and it is so in all those cases where the dissolved substance exists in the solution in its unit molecules, *i.e.* when there is neither dissociation nor polymerisation (van't Hoff).

3. The depression t of the freezing-point, occasioned by the addition of a foreign substance whose mol. wt. is M, is

$$t = E\frac{m}{M};$$

where m denotes the number of grams of the foreign substance in 100 g. of the solvent; E denotes the *molecular depression of the freezing-point, i.e.* the depression occasioned by the addition of 1 g.-mol. of the foreign substance to 100 g. of the solvent; it varies with the particular solvent used (Raoult).

The *molecular depression of the freezing-point* may be theoretically calculated from the *melting-point* T, on the absolute scale, and from the *heat of fusion* r, expressed in g.-cal per g. of substance, from the formula (van't Hoff),

$$E = \frac{0{\cdot}02T^2}{r}.$$

These laws make it possible to ascertain by calculation, from the known concentration of the solution, that temperature-point at which (under atmospheric pressure) the solution and the solidified solvent may exist together in stable equilibrium.

The influence of *pressure* on the freezing-point of a solution has not as yet been experimentally studied; but it may be anticipated that the lowering of the freezing-point experienced by the solvent on the addition of a foreign substance is practically *independent* of the external pressure. The laws formulated above strictly hold good only for *dilute solutions*, and in concentrated solutions (*e.g.* 10 to 20 per cent) they merely serve as tentative standards.

The Crystallising out of Dissolved Substances.—Reference has already been made (p. 488) to the condition of equilibrium which prevails when the dissolved substance is separated in a pure form. Here we will merely show the connection of the preceding section with the new point of view to which we have attained. Thus, *e.g.* we may conceive of such a condition of equilibrium between a solid salt and its saturated water solution, that the freezing-point of the solid salt shall be depressed by the presence of the water to the temperature of the saturated solution. Thus, while on the one hand we may justly parallelise the process of solution with that of evaporation, there is also, on the other hand, an undeniable analogy with the melting of a solidified solvent in presence of its solution.

The so-called "Cryohydrates."—If one cools sufficiently a water solution which is in contact with a solid salt, one finally arrives at the freezing-point of the saturated solution, where, with the separation of ice, there is also associated a separation of the salt existing in the solution. At this temperature-point there is precipitated a mechanical (not a really isomorphic) mass (but a eutectic mixture, p. 126), containing ice and the solid substance, in those proportions which correspond to the particular concentration of the saturated solution. This temperature-point may be found as the intersection-point of the solubility curve of the salt in question, with the curve which represents the dependence of the freezing-point of the solution upon its concentration. Such a solution does not change its composition by fractional freezing, and it must, therefore, have a constant freezing-point, *i.e.* one which is *independent* of the quantity frozen out.[1]

The Equilibrium between Liquid and Solid Solutions.—The most general case is where an isomorphous mixture of the solvent and of the dissolved substance crystallises out of the solution; that is represented to us by a system in which a solid and a liquid solution are in equilibrium with each other. According to the views which were advanced by van't Hoff in his *Theory of Solid Solutions* (p. 168), the study of equilibria of this sort would appear to claim especial attention, because in this way one may perhaps most easily gain an insight into molecular dimensions, and thus also into the molecular structure of particular crystallised substances.

There may be distinguished the following cases respecting the degree of miscibility of solid substances (p. 120), viz.:—

1. The two salts[2] may form isomorphous mixtures in all proportions: then in crystallising out a common solution of both salts, mixed crystals of every composition could be formed.

[1] Amongst the more recent work see especially the investigation of Roloff, *Zeitschr. physik. Chem.*, **17.** 325 (1895).

[2] Of course, what is said above respecting salts may be applied without further remark to other substances.

2. The series of mixtures of the two solid salts has a gap. Then the two inner members must be in equilibrium with each other, in a way similar to the equilibrium between two mutually saturated liquids (p. 484); and, therefore, for exactly the same reason that the two mutually saturated liquids are in equilibrium with the vapour having the same composition, these two inner members must be in equilibrium with the same saturated solution.

In fact, Roozeboom[1] succeeded in verifying this experimentally by mixed solutions of potassium and thallium chlorates. When one causes solutions of thallium chlorate to crystallise out, by adding increasing quantities of potassium chlorate, at first there appear mixed crystals containing the thallium chlorate in excess. The greater the quantity of the potassium salt added, the larger is the proportion of this substance contained in the mixed crystals. But when the proportion of the potassium salt in the mixed crystals has reached 36·3 per cent, then there appear also mixed crystals which contain 98·0 per cent of the potassium salt, *i.e.* the same solution is at the same time in equilibrium with these two inner members of the series of mixtures. If the proportion of the potassium salt in the solution is carried further, then there separate only mixed crystals which contain more than 98 per cent of the potassium salt.

The phenomena observed on evaporating the solution of these two salts are completely analogous to those observed in the evaporation of a mixture, *e.g.* of ether and water.

3. If a double salt is found in the gap of the series of mixtures, then, from solutions of the one salt, by successive and increasing additions of the other, there separate—*firstly*, the mixed crystals on one side [of the gap] of the series of mixtures; then, *secondly*, when a definite concentration is reached, at the same time there appear both the inner end member of the first mixed series and also the double salt [of the gap]; then, *thirdly*, in the interval of certain concentration limits, the double salt alone appears; then, *fourthly*, at a definite point, the double salt, and at the same time the inner end member of the second series of mixtures; and, finally and *fifthly* [with increasing addition of the second salt], the successive mixed crystals of the second series alone. An example of this kind is found in Roozeboom's[2] investigation of the crystallisation of solutions of ferric and ammonium chlorides.

There is thus often a complicated series of stable solutions, and it is necessary in referring to a saturated solution to state the solid substances present.

For the case, where a *dilute solid solution* takes part in the equilibrium, all the experiments thus far made may be transferred to the laws which we have learned in the foregoing sections regarding the

[1] *Zeitschr. physik. Chem.*, **8.** 504 and 530 (1891).
[2] *Ibid.*, **10.** 145 (1892).

equilibrium between phases of variable composition, which are composed of gases or of dilute liquid solutions. Especially the laws regarding the distribution of a substance between two solvents, hold good also for this case.[1]

An interesting application of these laws appears to be possible in the theory of colour processes [dyeing]. According to O. N. Witt,[2] the absorption of the colour by the fibre consists neither in a coarse mechanical arrangement, nor in a chemical compound of the fibre and the colour; but rather, and in opposition to both of these views, against which weighty suspicions have begun to develop, *that the colouring consists in a solution of the dye-stuff in the fibre*, i.e. *in the formation of a solid solution.*

Of the many reasons which Witt has adduced for the plausibility of this view, it should be mentioned that the coloured fibre does not show the colour of the solid [pure] dye-stuff, but of the dissolved dye-stuff: thus, *e.g.*, fibres coloured with fuchsine are not coloured metal-glance-green but red. Rhodamine does not fluoresce in the solid state, but in solution; but silk, coloured with rhodamine, shows a clear fluorescence, which argues for the view that the colouring matter exists in the dissolved state.

In the sense of these views, the colouring process is completely comparable to the shaking any substance out of a water solution by any other solvent, as by ether, carbon disulphide, etc. And thus the laws regarding the distribution of a substance between two solvents are at once applicable to the colour process.

Further, the fact that the same dye-stuff may introduce itself into different fibres, producing different colours, is completely analogous to such cases, as where iodine, *e.g.*, is dissolved in different solvents, producing different colours.

The nature of the so-called "adjective colours" was thus explained by Witt, viz. that the associated mordant is first dissolved by the fibre, and then, in its turn, dissolves the dye-stuff, as a result of a chemical action, as it diffused into the fibre; and thus there results an increase of the solubility of the dye-stuff in the fibre.

Probably when the yarn takes up colouring matter absorption phenomena occur and perhaps even chemical processes, so that it is not a simple case of the partition of a substance between two solvents; this is evidenced by the abnormal values obtained in investigations of the molecular weight of the substances absorbed by the yarn by means of the partition.[3]

[1] See further the literature mentioned on p. 272 and the observations of Muthmann and Kuntze, *Zeitschr. f. Kristallographie*, **23**. 368 (1896).

[2] *Färber-Zeitung*, **1.** (1890-91). Ref. in *Zeitschr. physik. Chem.*, **7.** 93 (1891), also in the *Jahrbuch der Chem.*, **1.** p. 18 (1891), and very thoroughly in the *Chem. Zentralbl.*, **2.** 1039 (1891).

[3] See the literature mentioned on p. 129, and also Zacharias, *Zeitschr. physik. Chem.*, **39.** 468 (1902), and Kaufler, *ibid.*, **43.** 686 (1903).

The carrying down of dissolved salts by precipitation of oxides, sulphides, and the like, so important in analytical chemistry, has been little studied; this phenomenon occurs exclusively with colloidal (amorphous) precipitates. The observations of van Bemmelen,[1] Linder and Picton[2] and Whitney and Ober[3] show that there must be some chemical combination of the salt with the colloidal precipitate.

The Most General Case.—Finally, the following very general case will be treated briefly.

Let there be a reaction between a number of vaporised substances, which are at the same time dissolved in any selected solvent, and let the reaction proceed according to the scheme

$$n_1A_1 + n_2A_2 + \ldots = n_1'A_1' + n_2'A_2' + \ldots$$

That is, let n_1 mol. of a substance A_1, and n_2 mol. of a substance $A_2 \ldots$, etc., unite to form n_1' mol. of a substance A_1', and n_2' mol. of a substance $A_2' \ldots$, etc. Equilibrium will be established when the partial pressures of the particular molecular species are respectively p_1, $p_2 \ldots$, p_1', $p_2' \ldots$, and when their concentrations in the solution amount respectively to c_1, $c_2 \ldots$, c_1', $c_2' \ldots$

Then the application of the Guldberg-Waage law of chemical mass-action gives the two equations—

$$\frac{p_1^{n_1}p_2^{n_2}\ldots}{p_1'^{n_1'}p_2'^{n_2'}\ldots} = K \quad . \quad . \quad . \quad . \quad (1)$$

$$\frac{c_1^{n_1}c_2^{n_2}\ldots}{c_1'^{n_1'}c_2'^{n_2'}\ldots} = K' \quad . \quad . \quad . \quad . \quad (2)$$

Here K and K′, the reaction coefficients, depend only upon the temperature.

The law of distribution gives us several equations, viz.

$$c_1 = p_1k_1, \quad c_2 = p_2k_2 \ldots, \quad c_1' = p_1'k_1', \quad c_2' = p_2'k_2' \quad . \quad . \quad (3)$$

Here k_1, $k_2 \ldots$, k_1', $k_2' \ldots$ denote respectively the solubility coefficients of the particular molecular species; and these again depend only upon the temperature.

From reactions (1) to (3) we conclude that—

$$K = K'\frac{k_1'^{n_1'}k_2'^{n_2'}\ldots}{k_1^{n_1}k_2^{n_2}\ldots} \quad . \quad . \quad . \quad . \quad (4)$$

In most cases the solubility coefficients of a molecular species for any solvent may be directly determined, and this information allows

[1] *Zeitschr. anorg. Chem.*, **23**. 321 (1900). [2] *Chem. Soc. Journ.*, **67**. 63 (1895).
[3] *Zeitschr. physik. Chem.*, **39**. 630 (1902).

one to say beforehand how a number of substances will react on each other in any solvent, provided that their reaction capacity in the gaseous state is known, and conversely.

Of course a similar statement may be made regarding the relation of the distribution coefficients. When *solid substances* take part in the equilibrium, their active mass is constant,[1] and the same is true of reacting molecules, which at the same time play the *rôle* of solvents (p. 459). Therefore, we may state the following general theorem :—

If we know the coefficients of equilibrium of a reaction which progresses to a finish at a definite temperature, and in any selected phase ; and if we know the distribution coefficients of all the molecular species compared with another phase ; then the condition of equilibrium is also known in the second phase at the same temperature.

This theorem should have great practical significance, because it enables us to anticipate, from the distribution coefficients, the reaction capacity in the most various solvents or in the gaseous state, after we have studied it in one particular phase ; and thus the problem con- concerning the "dissociating force" (p. 458), or even concerning the influence of the medium, will be referred to the simpler one of the study of the *distribution coefficients;* and, whenever possible, the attempt will be made to clear up the relation of these [*i.e.* the distribution co-efficients] to the nature of the substance in question, and to the particular phase to which they refer. But this will be the task of the future.

Applications.—The vapour above a mixture of acetic acid, alcohol, water, and ester in equilibrium must also be in equilibrium ; hence the relation

$$\frac{\text{Ester} \times \text{water}}{\text{Alcohol} \times \text{acid}} = \text{const.}$$

must hold for the vapour also ; but the constant in the gaseous phase will in general have a different value to that for the liquid. The experimental verification of this law would be not without interest.

Kuriloff[2] made a very thorough investigation of the above general theorem, the results of which we will calculate in a somewhat different manner to the author. The investigator mentioned determined the equilibrium of solid β-naphthol-picrate in contact first with water and then with benzene, and also the ratio of partition of the reacting molecules, so that all the data necessary for testing the theory are available.

The equilibrium in water was determined by solubility measurements in the same way as in the example of anthracene picrate given on p. 478. The solution contains 6·09 free β-naphthol and 8·80 free

[1] Provided that these form no mixed crystals, double salts, or the like.

[2] *Zeitschr. physik. Chem.*, **25**. 419 (1898).

picric acid, besides 1·20 of picrate when the latter substance was present in the solid state. The numbers are in thousandths of a mol. per litre. The picric acid under these circumstances is electrolytically dissociated to the extent of 94·6 %; hence the product of the free naphthol and the free undissociated picric acid is

$$\mu_1\mu_2 = 6{\cdot}09 \,.\, 8{\cdot}80(1 - 0{\cdot}946) = 2{\cdot}89.$$

Further, the coefficients of partition of the two latter species of molecule between benzene and water are 67 for naphthol and 39 for the undissociated molecules of picric acid. Hence in benzene the value must be

$$\mu_1'\mu_2' = \mu_1\mu_2 \,.\, 67 \,.\, 39 = 7550,$$

and when both species of molecules are present in equivalent quantities their concentrations must be

$$c_0 = \sqrt{\mu_1' \,.\, \mu_2'} = 86{\cdot}9.$$

Now the solubility of the picrate in benzene is 104·5; hence the saturated solution of this substance must be dissociated to the extent

$$\frac{c_0}{104{\cdot}5} = 0{\cdot}83,$$

whilst measurement of the equilibrium by means of solubility, with excess of one or the other component, give 0·64 to 0·85 for the degree of dissociation of the saturated solution.

Thus from the dissociation of the aqueous solution saturated with picrate, and from the ratio of partition of the components, we can calculate how much dissociated substance is contained in a saturated solution of picrate in benzene.

CHAPTER IV

CHEMICAL EQUILIBRIUM IN SALT SOLUTIONS

The Reaction Capacity of Ions.—In the preceding chapter we have become familiar with the general theory, which instructs us regarding the chemical equilibrium in any selected system, and regarding the dependence of this equilibrium upon the relative quantities of the reacting components. But there is one case which we have not as yet considered: viz. *the part played in the reactions by the free ions, i.e.* the study of *water solutions of electrolytes*, or in short, of *salt solutions*. The writer has devoted a special chapter to the consideration of salt solutions, *partly for the purpose of a general view, and partly to show that when one applies the law of mass-action—the universal value of which can be shown by numerous facts—to the study of salt solutions, then the hypothesis of electrolytic dissociation becomes an imperative necessity, at least, according to the present state of our knowledge.*

From the standpoint of the hypothesis of electrolytic dissociation (p. 350), the whole question is at once solved by the simple conclusion, that *the free ions must participate in reactions in proportion to their concentration (their active mass), just like every other molecular species.* Without the introduction of any new hypothesis, we are now in a position to handle the chemical equilibrium between substances, which conduct electrolytically, in just the same simple manner which was employed in considering the reactions between those molecular species which were exclusively electrically neutral.

And thus nothing that is especially new in principle will be given in the following paragraphs, but many new and surprising applications of the law of Guldberg and Waage will be given: these applications will far surpass the former ones in simplicity and also in practical significance, because the salt solutions have always exercised a controlling interest in investigation.

The great honour of giving this point of view its proper value belongs to Arrhenius.

Electrolytic Dissociation.—When a molecular species A, which is electrically neutral, decomposes into ions, thus

$$A = n_1A_1 + n_2A_2 + \ . \ . \ . \ ,$$

the law of mass-action requires that

$$Kc = c_1^{n_1}c_2^{n_2} \ . \ . \ . \ ,$$

where c denotes the concentration of the undissociated part, and c_1, c_2 . . . etc. the respective concentrations of the products (ions) resulting from dissociation, and K, as usual, denotes the *constant of dissociation.* Of course, the ions are always produced in quantities which are electrically equivalent. For a binary electrolyte, we will have

$$Kc = c_1^2 \ ;$$

then, since

$$c = \frac{1 - \alpha}{V}, \text{ and } c_1 = \frac{\alpha}{V},$$

when α denotes the degree of dissociation, and V the volume containing 1 g.-mol of the electrolyte ; we thus obtain

$$KV(1 - \alpha) = \alpha^2,$$

from which, by calculation, we obtain

$$\alpha = \frac{KV}{2}\left(\sqrt{1 + \frac{4}{KV}} - 1\right).$$

Two methods (p. 356) are already known to us for the determination of α ; viz. the measurement of *the osmotic pressure* (the freezing-point, etc.), and the measurement of *the electrical conductivity* ; the latter is much more exact, and gives, as the degree of dissociation, the formula

$$\alpha = \frac{\lambda}{\lambda\infty}.$$

This formula, which was first experimentally proven by Oswald,[1] by application to electrolytic dissociation, was established with the best results for a large number of organic acids. The following table, given by van't Hoff and Reicher,[2] may serve as an illustration :—

[1] *Zeits. phys. Chem.*, **2.** 36 and 270 (1888) ; see also Planck, *Wied. Ann.*, **34.** 139 (1888).

[2] *Zeits. phys. Chem.*, **2.** 779 (1888).

[TABLE

2 K

THE MOLECULAR CONDUCTIVITIES OF ACETIC ACID AT 14·1°.

v	λ	100 α Obs.	100 α Calc.	
0·994	1·27	0·402	0·42	
2·02	1·94	0·614	0·60	
15·9	5·26	1·66	1·67	
18·9	5·63	1·78	1·78	log K = 5·25 − 10
1500	46·6	14·7	15·0	λ∞ = 316
3010	64·8	20·5	20·2	
7480	95·1	30·1	30·5	
15000	129	40·8	40·1	
[∞	316	100	100]	

The coincidence between the degree of dissociation—as determined by the conductivities, and as calculated according to the theoretical formula, where K = 0·0000178—is very remarkable.

Inasmuch as the same form of the isotherm of dissociation holds good for the ordinary binary, and also for the electrolytic, dissociation, therefore the laws developed (p. 456) for electrolytic dissociation amount to this, viz.—

In a binary electrolyte, the concentration of the molecules which are electrically neutral is proportional to the square of the total concentration, when the dissociation is very extensive; and when the dissociation is but slight, the concentration of the ions (and therefore of the conductivity also), is proportional to the square root of the total concentration.

The formula does not exactly fit the observed values, in the case of highly dissociated acids and salts. This is perhaps to be ascribed partly to the fact, that the determination of

$$1 - \alpha = \frac{\lambda\infty - \lambda}{\lambda\infty},$$

is exposed to *great uncertainty*, on account of the smallness of the difference between λ∞ and λ; here, apparently, for reasons unknown at present, the electrical conductivity is not a perfectly exact measure of the degree of dissociation. An explanation of this point would be exceedingly interesting, but it has not been forthcoming as yet.

According to M. Rudolphi,[1] in strongly dissociated electrolytes the expression $\frac{\alpha^2}{(1-\alpha)\sqrt{V}}$ is much more constant than $\frac{\alpha^2}{(1-\alpha)V}$, and for analogous compounds possesses somewhat similar values. Van't Hoff[2] a little later pointed out that the expression $\frac{\alpha^{\frac{3}{2}}}{(1-\alpha)\sqrt{V}}$ or the square

[1] *Zeits. phys. Chem.*, **17.** 385 (1895). [2] *Ibid.* **18.** 300 (1895).

of it $\frac{a^3}{(1-a)^2V}$ is equally or more constant than that of Rudolphi. Considering that

$$c = \frac{1-a}{V}, \quad c_1 = \frac{a}{V},$$

we have

$$\frac{c_1^3}{c^2} = \text{const.},$$

that is, *the third power of the ionic concentration is proportional to the square of the undissociated molecules.* As *F. Kohlrausch*[1] showed *van't Hoff's* expression may also be put in the form that the ratio of the concentration of the undissociated molecules to that of the ions is proportional to the mean distance between the undissociated molecules.

Electrolytic Dissociation and Chemical Nature.—The question now arises,—How does the *degree* of the electrolytic dissociation depend upon the *nature* of the respective electrolyte?—a question which is all the more important because the reaction-capacity depends, in a very pronounced way, upon the degree of dissociation. In what follows, there will be grouped together some of the most important rules thus far recognised in this region of work; a knowledge of these rules will serve, in a remarkable way, to elucidate the general view of chemical equilibrium in salt solutions.

1. *The salts of the alkalies, of ammonium, of thallium, and of silver, with monobasic acids, in dilute solutions and at equivalent concentrations, are dissociated to the same degree; and, moreover, are highly dissociated, as is shown by the figures for potassium chloride, given on p.* 359.

2. On the other hand, the greatest differences are found among *the monobasic acids* and *the monacid bases* [when each is alone in solution]. Thus, some substances, like acetic acid, ammonia, etc., in tenth-normal solutions, are dissociated only to a small percentage; while other substances, such as hydrochloric acid, potassium hydroxide, etc., are as highly dissociated as the salts enumerated above.

3. Certain electrolytes, such as zinc sulphate, cupric sulphate, etc., *which on dissociation are cleaved into only two ions, but each with doubled electrical charges,* are comparatively much less dissociated; zinc sulphate and copper sulphate, in a concentration of 1 g.-mol. per litre, are dissociated only about 25 per cent (see also p. 360).

4. The behaviour of those *electrolytes which cleave into more than two ions,* is much more complicated. According to what is known at present, in general, a *dissociation in stages* seems to take place. Thus sulphuric acid does not decompose all at once into the SO_4 group [sulph-ion] with a double electro-negative charge, and two hydrogen

[1] *Zeits. phys. Chem.*, **18.** 662 (1895).

ions, each with a single electro-positive charge; but the decomposition rather progresses according to the two following equations—

$$\text{I. } H_2SO_4 = \overset{-}{HSO_4} + \overset{+}{H}.$$
$$\text{II. } \overset{-}{HSO_4} = \overset{=}{SO_4} + \overset{+}{H}.$$

The decomposition is similar in the case of such substances as $BaCl_2$, K_2CO_3, etc.

In general this law holds good here, viz., *salts which have an analogous composition, in equivalent solutions are dissociated electrolytically to the same degree.* But this latter law is by no means a rule which holds good without exception; thus the chlorides of calcium, strontium, barium, magnesium, and copper are dissociated nearly to the same extent; but the chlorides of cadmium and of mercury, though of analogous constitution, are cleaved into their respective ions much less strongly.

For dissociation such as $\overset{+}{PbCl} \mid \overset{-}{Cl}$, $\overset{+}{K} \mid \overset{-}{KSO_4}$ etc., it may be assumed with considerable probability that it follows the rule of binary neutral salts, such as potassium bromide. *Von Ende*[1] has recently demonstrated this experimentally in the case of lead chloride.

5. Many polybasic acids, within a wide concentration interval, act like monobasic acids, *i.e.* the equation of the isotherm of dissociation, which was developed on p. 492, for binary electrolytes, is also applicable to them. It is only at extreme dilution that these polybasic acids begin to cleave off the second, the third, etc., hydrogen ions.

The fact that extensive dilution is always required in order to cleave off the last hydrogen ions, would indicate that it is increasingly difficult for the acid residue to assume the last *quanta* of negative electricity.

6. Electrolytic dissociation, as compared with ordinary dissociation, changes with the temperature but slightly; for with rising temperature, it sometimes slowly diminishes, sometimes increases,—in contrast to ordinary dissociation, which always rapidly increases with the temperature.

These rules, as shown by Ostwald,[2] may be applied to the determination of the basicity of acids. Since the state of dissociation varies in a characteristic way, with the basicity of the respective acid of a sodium salt, *e.g.*, therefore the simple study of the conductivity, in its dependence upon the concentration, may be directed to the explanation of this point. Of course the same question may be decided by the measurement of the depression of the freezing-point.

The Strength of the Affinity of Organic Acids.—It is to the

[1] Dissertation, Gottingen, 1900. [2] *Zeits. phys. Chem.*, **1.** 74 (1887); **2.** 901 (1888).

extended researches of Ostwald[1] that we are chiefly indebted for our knowledge of the relations prevailing here; he, in common with his students, has given especial attention to the problem, *how does the capacity of organic acids to dissociate electrolytically, depend upon the structure of the particular radical?* Unfortunately the space is not available to consider in detail the many interesting points which have been developed by research in this region; and it will be merely mentioned that *the dissociation constants*, or as they will also be called, for reasons to be given below, *the affinity strengths* (and which, as shown on p. 496, can be ascertained with the greatest exactness), *these vary in a most outspoken way with the constitution of the acid radical.* This relation to the constitution has not as yet been explained, so as to allow the value of the dissociation constant to be numerically derived from the constitution; for our knowledge thus far is limited to the recognition of the regularity of the influence exerted by the substitution of certain radicals.

According to this the radicals may be divided into *negative* and *positive*, according as they favour the assumption of a negative ionic charge (acids) and therefore hinder the positive charge, or *vice versa*.

The following substances act *negatively:* aromatic radicles (*e.g.* C_6H_5), hydroxyl, sulphur, halogens, carboxyl, nitril and cyanogen; *positively* fatty radicals (*e.g.* CH_3) addition of hydrogen, especially the amido group.

This appears from the affinity constants of the following series of acids:—

	100 K
Acetic acid CH_3COOH	0·00180
α-Volnic acid $CH_2(C_6H_5)COOH$	0·00556
Glycollic acid $CH_2(OH)COOH$	0·0152
Thiacetic acid CH_3COSH	0·0469
Chloracetic acid $CH_2ClCOOH$	0·155
Tri-chlor-acetic acid CCl_3COOH	121
Malic acid $CH_2(COOH)COOH$	0·158
Cyanacetic acid $CH_2(CN)COOH$	0·370
Propionic acid $CH_2(CH_3)COOH$	0·00134
Glycocol $CH_2(NH_2)COOH$	very small

and of *bases*:—

Ammonia NH_4OH	0·0023
Methylamine $NH_3(CH_3)OH$	0·050
Benzylamine $NH_3(CH_2C_6H_5)OH$	0·0024
Aniline $NH_3(C_6H_5)OH$	0·000000011

The *spacial position* of the substituent in the molecule is of considerable influence on its action; the nearer the substituent is to the point at which the ionic charge is taken up the more effective it is; thus we have 100 K in

o-nitrobenzoic acid 0·616 > m-nitrobenzoic acid 0·0345

[1] *Jour. pr. Chem.*, **31.** 433 (1885); *Zeits. phys. Chem.*, **3.** 170 (1889); Bethmann, *ibid.*, **5.** 385; Bader, *ibid.*, **6.** 289 (1890); Walden, *ibid.*, **8.** (1891).

$$\frac{\text{Trichloracetic acid}}{\text{Acetic acid}} = \frac{CCl_3 - COOH}{CH_3 - COOH} \quad \frac{121}{0{\cdot}00180} > \frac{\text{Trichlorlactic acid}}{\text{Lactic acid}}$$

$$= \frac{CCl_3 . CHOH . COOH}{CH_3 . CHOH . COOH} \quad \frac{0{\cdot}465}{0{\cdot}0138}$$

Benzylamine $C_6H_5CH_2NH_3OH$ 0·0024 > Toluidine $CH_3C_6H_4NH_3OH$ circa 10^{-12}

Oxalic acid COOH–COOH 10 > Malonic acid $CO_2H.CH_2.CO_2H$ 0·158

> Tartaric acid $CO_2H.CH_2.CH_2.CO_2H$ 0·00665

We have already met on p. 339 with similar relations in considering the influence of substituting radicals on the absorption of light in the chromophore. It is obvious that the affinity constants are of considerable importance for stereochemistry; they have indeed been repeatedly used with success in that connection.

The behaviour of the *dicarbon acids* is especially interesting: the distance between the two carboxyls is the determining factor between the isomeric maleic and fumaric acids, as in the previous case of oxalic, malonic and tartaric acids; these constants are for

$$\text{Maleic acid} \begin{array}{c} H.C.COOH \\ H.C.COOH \end{array} 1{\cdot}17 > \text{Fumaric acid}, \begin{array}{c} H.C.COOH \\ COOH.C.H \end{array} 0{\cdot}09;$$

on the other hand the dissociation of the hydrogen ion of the second carboxyl group occurs much sooner in fumaric acid than in maleic, which dissociates as a monobasic acid only up to more than 80%. The explanation of this given by *Ostwald*[1] lies in the *electrostatic repulsion* between a negative charge and another of the same sign so that the charge due to dissociation of the first hydrogen ion hinders that of the second more effectively the nearer the two carboxyls stand, since on dissociation they acquire similar charges.

The principle which is at the foundation of the investigations briefly described above, and to which reference has already been made (p. 290), consists in *the study of the mutual influence of different elements or radicals in the molecules, respecting the reaction-capacity.* The measurement of the electrolytic dissociation constant of an acid, as will be shown clearly below, consists merely in the determination of the reaction capacity possessed by the "acid hydrogen" in water solution.

Instead of studying the reaction capacity of "acid hydrogen," one can as well study, as Ostwald has succeeded in doing, the reaction capacity of another element or radical in the molecule, of course with due regard for similarity of results: *but here the problem is to find out a method to determine the reaction-capacity with satisfactory sharpness, and under conditions which are suitably comparable.* Comparable conditions

[1] *Zeits. phys. Chem.*, **9.** 553 (1892).

may be obtained most easily by determining the reaction-capacity in *the same solvent.* The quantitative determination of the reaction-capacity of the radical in question, may be accomplished by the measurement of a suitable chemical equilibruim. Thus it would be very interesting to study in an extended way the question of the capacity of the nitrogen basis to fix hydrochloric acid or any other acid; this could be done very easily by means of a method patterned after that of Jellet (p. 453), since one could allow an optically active base to compete for the acid at the same time with the base to be investigated.

It should be remarked, as a matter of history, that Menschutkin (p. 450) was the first to attack a problem of affinity in an extended way; perhaps the results of his extended investigations would have been greater, if he had studied the state of equilibrium of the esterification under conditions which were more nearly comparable, *e.g.* in a suitable solvent.

Finally, there are here appended some of the most important affinity constants of the acids investigated by Ostwald.

Acid.	100 K.
Malic acid $C_2H_3(OH)(CO_2H)_2$	0·0395
Formic acid $HCOOH$	0·0214
Benzoic acid $C_6H_5 . COOH$	0·0060
Succinic acid $C_2H_4(CO_2H)_2$	0·00665
Butyric acid $C_3H_7 . COOH$	0·00149
Acetic acid CH_3COOH	0·00180
Lactic acid $CH_3CH(OH)COOH$	0·0138
Salicylic acid $C_6H_4(OH) . COOH$	0·102
Tri-chlor-acetic acid $CCl_3 . COOH$	121
Cinnamic acid $CH(C_6H_5) . CH . CO_2H$	0·00355

Walker[1] has measured certain very weak acids, making use of exceedingly pure water for the conductivity (indirect methods for determining the dissociation of extremely weak acids will be given below); acetic acid is taken for comparison:

	$K\ 10^{10}$
Acetic acid $C_2H_3O_2 - H$	180000
Carbon dioxide $HCO_3 - H$	3040
Sulphuretted hydrogen $HS - H$	570
Boric acid $H_2BO_3 - H$	17
Hydrocyanic acid $CN - H$	13
Phenol $C_6H_5O - H$	1·3

The Mixture of Two Electrolytes containing the same Ion.—As the simplest case of equilibrium between several electrolytes, we will first investigate the reaction which occurs *on bringing together any two electrolytes with a common ion, e.g.* two acids having the hydrogen ion in common. The progress of the reaction can be followed without recourse to calculation.

Let a second acid be added to an acid solution keeping the volume

[1] *Zeits. phys. Chem.*, **32.** 137 (1900).

constant, *i.e.* by adding the second acid in the pure form to a dilute water solution of the first acid, so that the concentration of the hydrogen ions will be increased; then the immediate result of this is that the undissociated part of the first acid is no longer able to maintain the equilibrium between the active masses of the hydrogen ion and of the negative ingredient, the masses of which are now increased; *i.e. the dissociation of the acid retrogrades.* We observe a phenomenon similar to this, on adding free chlorine to phosphorus penta-chloride, whereby the dissociation is caused to retreat by the addition of one of the dissociation products (p. 443).

In order to observe the relations in a quantitative way we need employ only the law of mass-action. Let c_1 denote the concentration of the electrically neutral molecules, and c_2 that of the two ions, so that the total concentration C, will be

$$c_1 + c_2 = C,$$

and we will have

$$c_1 K = c_2^2.$$

Now let a second clectrolyte be added, which has one ion the same as one ion of the first electrolyte; and let the concentration of the second electrolyte be c_0. Then when the equilibrium shall have been established, it will be satisfied by the condition,

$$c_1'K = c_2'(c_2' + c_0),$$

where again, of course, $c_1' + c_2'$ must be equal to C. Obviously c_1' will be greater than c_1; and conversely c_2 will be smaller than c_2'; *i.e.* the dissociation of the electrolytes retrogrades, on the addition of a second electrolyte containing an identical ion, and in a ratio capable of exact calculation.

This phenomenon may be very well shown qualitatively by a solution of paranitrophenol; the negative ion of this acid is coloured an intense yellow, whilst the electrically neutral molecule is colourless. If therefore any acid is added to an aqueous solution of this substance the yellow coloration directly vanishes, because the slight dissociation of this weak acid is reduced to almost nothing by even a small addition of hydrogen ions. (See the section on "The Theory of Indicators.")

Arrhenius[1] succeeded in proving this law quantitatively in the following way. Thus, *e g.* sodium acetate was added to a solution of acetic acid, and then he determined the velocity of the inversion of cane sugar contained in the solution; this velocity of sugar inversion (as will be thoroughly demonstrated in the chapter on Chemical Kinetics), is a measure of the number of the free hydrogen ions existing in the solution. Thus, this velocity amounted to 0·74 in a solution

[1] *Zeits. phys. Chem.*, **5.** 1 (1890); also **2.** 284 (1888); and *Wied. Ann.* **30.** 51 (1887).

containing $\frac{1}{4}$ of a mol. of acetic acid to the litre, and when the equivalent quantity of sodium acetate was added, the value sank to 0·0105, the calculated value of the latter being 0·0100.

If any selected volumes of any two electrolytes having ions in common, as two acids, *e.g.*, are mixed, then, in general, the state of dissociation of each of these will change as a result of their mixture; and corresponding to the electrical conductivity of the mixture, another state of dissociation will be established; this state of dissociation is *not* identical with that corresponding to the *mean* of the conductivities of the unmixed components.

But if one selects the concentrations of the two acids so that each shall contain the same number of free hydrogen ions in a litre,—such solutions are called "*iso-hydric,*"—*then there is no change of their state of dissociation resulting from their mixture.* If c and c_1 are the concentrations of the electrically neutral molecules and of the ions, k the dissociation constant, we have the relation

$$kc = c_1^2 \quad . \quad . \quad . \quad . \quad . \quad . \quad . \quad . \quad . \quad . \quad (1)$$

and similarly for the second solution

$$KC = C_1^2, \quad . \quad . \quad . \quad . \quad . \quad . \quad . \quad . \quad . \quad . \quad (2)$$

thus if the volume v of the first solution be mixed with the volume V of the second, the concentration of the electrically neutral molecules and of the not common ions

$$c, c_1 C, C_1$$

will become

$$\frac{cv}{V+v}, \quad \frac{c_1v}{V+v}, \quad \frac{CV}{V+v}, \quad \frac{C_1V}{V+v},$$

whilst the concentration of the ion common to the two solutions reaches the value $\frac{C_1V + c_1v}{V+v}$. Applying the law of mass action to determine the equilibrium in the common solution we have

$$kc = c_1 . \frac{C_1V + c_1v}{V+v}, \quad . \quad . \quad . \quad . \quad . \quad . \quad . \quad . \quad (3)$$

$$KC = C_1 . \frac{C_1V + c_1v}{V+v}; \quad . \quad . \quad . \quad . \quad . \quad . \quad . \quad (4)$$

but the equations (1) to (4) are satisfied if

$$C_1 = c_1,$$

that is, in mixing isohydric solutions no displacement of the dissociation occurs, *hence the conductivity of the mixture must be the mean of the conductivities of the two solutions*, as is shown by experience.

If one mixes solutions of any two electrolytes having an ion in common, and which also have *the same dissociation constant*,—as, *e.g.*, two chlorides of univalent bases,—then these solutions contain the common ion in the same concentrations when the solutions are equivalent. Hence it follows that, in a mixture of such electrolytes, each one is dissociated equally strongly, and just as strongly as though it alone formed a solution in proportions corresponding to the total concentration.

The case of a mixture of two acids in solution can easily be treated by the above method especially the question of how the conductivity of a mixture of two acids changes with increasing dilution. See for particulars A. Wakemann.[1] It is especially interesting to note that the dissociation constant is calculated from the conductivity according to p. 499 and is by no means constant for a mixture of two acids, but varies considerably with the dilution, so that, as has been shown, it can serve as a criterion of the purity of the acid under consideration.

The Equilibrium between any Selected Electrolytes.—The state of equilibrium is much more complicated in the case of a solution containing two binary electrolytes *having no common ion.* Then care must be given to discriminate between the different molecular species in the solution, viz., the four free ions and the four electrically neutral molecules produced by the combination of the former. By a displacement of the point of equilibrium, four reactions may take place beside each other; namely, the dissociation of the (four) electrically neutral molecules into the respective ions, to each of which reactions there always corresponds an equilibrium condition of the form

$$Kc = c_1c_2,$$

where K denotes the corresponding dissociation constant, c the concentration of the molecular species that is electrically neutral, and c_1 and c_2 respectively the concentrations of the two ions.

One may easily convince himself, by a knowledge of the dissociation constants and of the total concentration, the latter being ascertained by chemical analysis, that the equilibrium is definite and unequivocal, and that its ascertainment is only attended by difficulties in the way of pure calculation, which, however, are by no means inconsiderable. But, particularly on the basis of the fact that binary salts, composed of univalent radicals, are in the same state of dissociation, and that a very extended one, the essential generalisations may be collated in the following propositions, as they are formulated by Arrhenius.[2]

1. The degree of dissociation of a weak acid, in the presence of

[1] *Zeits. phys. Chem.*, **15.** 159 (1894). [2] *Zeits. phys. Chem.*, **5.** 1 (1890).

salts of this acid, is inversely proportional to the quantity of salt present.

2. When a weak acid and several strongly dissociated electrolytes exist together in the same solution, their respective degrees of dissociation may be calculated as though the dissociated part of the particular electrolyte were a dissociated part of a salt (as a sodium salt, *e.g.*), of this acid.

There is hardly any special proof needed to show that the case, where any arbitrary number of electrolytes are in a solution, can be solved by the application of the preceding equation. Before advancing any farther, we will now discuss a case which has not been considered as yet, namely, *the participation of water in the equilibrium; i.e.* where the hydrogen ion and the hydroxyl ion react on each other.

Neutralisation.—It may be calculated from the fact that water conducts worse the more carefully it is purified, and accordingly that the traces of conductivity which the most carefully prepared water shows are partly due to traces of salt in solution, that water itself is very slightly dissociated into ions.

It follows directly from this that these two kinds of ions (H and OH), are capable of existing beside each other in water only in the barest traces. Thus if we bring together in a water solution two electrolytes, such that on decomposition one affords a hydrogen ion and the other a hydroxyl ion,—or, in other words, if we mix an acid with a base,—then in all cases the same reaction occurs, viz.,

$$\overset{+}{H} + \overset{-}{OH} = H_2O,$$

and practically in absolute completeness, *i.e.* until one of the reacting components is exhausted.

Now this reaction, the necessity of which, as already shown, was derived theoretically, is in fact well known and of the greatest importance; it is called *the process of neutralisation.*

If the acid and the base are completely dissociated, then the aforesaid reaction will be the only one which takes place; and from this there follows directly a very remarkable conclusion, to which attention was first called by Arrhenius, viz. *the same reactions must correspond to the same evolution of heat ("heat of reaction")*. Therefore by mixing any selected strong base with any selected strong acid in a sufficiently large quantity of water, there always results the same development of heat; this is proved by experiment (see the chapter "Thermo-Chemistry, I." Book IV.).

If, on the other hand, the acid or the base is not completely dissociated, then side reactions other than the aforesaid reaction, take place, namely, the decomposition into the ions; and since this is usually associated with a certain though very slight heat change, therefore under these circumstances one observes a slight variation in the heat of neutralisation.

The law of mass action applied to the dissociation of water causes, since the active mass of the solvent must be constant (p. 460), the relation that in dilute aqueous solutions the product of the concentrations of the hydrogen ions $[\overset{+}{H}]$ and the hydroxyl ions $[\overset{-}{OH}]$ must be constant; if the concentration of each of these ions in pure water is c_0, we have

$$[\overset{+}{H}]\,[\overset{-}{OH}] = c_0^2.$$

Electrolytic dissociation of pure Water.—Although it would seem to be very difficult to determine this exceedingly small dissociation of pure water, the problem has been solved of late years by different investigators in very varied methods, and with good agreement.

1. The electromotive force of the acid alkali cell allows of calculating the concentration of the hydrogen ions in an alkaline solution by means of the osmotic theory of currents (Book IV. Chap. vii.); it was found in this way that at 19° the normal solution of a strongly (80 %) dissociated base it amounts to $0{\cdot}8 \,.\, 10^{-14}$. It follows for such a solution

$$[\overset{+}{H}] = 0{\cdot}8 \,.\, 10^{-14},\ [\overset{-}{OH}] = 0{\cdot}8\text{; therefore } c_0 = 0{\cdot}8 \,.\, 10^{-7} \text{ at } 19^\circ.$$

At a higher temperature it was found in the same way that

$$c_0 = 1{\cdot}19 \,.\, 10^{-7} \text{ at } 25 - 26^\circ.$$

Ostwald[1] and Arrhenius[2] attempted simultaneously to determine c_0 in this manner; the present writer showed shortly afterwards that the calculation was somewhat different to that given by these writers, and that the value obtained is that given above.[3]

2. A second process lies in measuring the *hydrolytic decomposition of salts*, the theory of which is given below on p. 519. Arrhenius[4] found in this way

$$c_0 = 1{\cdot}1 \times 10^{-7} \text{ at } 25^\circ.$$

3. Both hydrogen and hydroxyl ions cause acceleration of the process of saponification of esters dissolved in water; Wiis,[5] at van't Hoff's suggestion, determined the velocity of saponification of methyl

[1] *Zeits. phys. Chem.*, **11.** 521 (1893).
[2] *Ibid.*, **11.** 805 (1893).
[3] *Ibid.*, **14.** 155 (1894).
[4] *Ibid.*, **11.** 805 (1893).
[5] *Ibid.*, **12.** 514 (1893).

acetate in pure water, and calculated according to a theory given by van't Hoff (see the following Chapter) that

$$c_0 = 1{\cdot}2 \,.\, 10^{-7} \text{ at } 25^\circ.$$

4. Finally, Kohlrausch and Heydweiller[1] succeeded in purifying water to such an extent that it showed its own conductivity, that is, was practically free from conductivity due to impurities.

It was known from previous work of Kohlrausch that water conducts the less the more carefully it is purified; it appears, however, that a limiting value can be reached below which the conductivity cannot be reduced, that is that water possesses measurable *conductivity of its own account.* The method of purification is distillation in vacuum. A U-shaped tube, one leg of which ended in a large reservoir and the other in a small resistance cell, was provided with water already very thoroughly purified, and boiled for a long time under the mercury pump. On slightly warming the large vessel the mass of the water was distilled over into the resistance cell and its conductivity measured.

At 18° the conductivity of the purest water was found to be $0{\cdot}0384 \,.\, 10^{-6}$ ohms per cm. (that of ordinarily good water is about 2×10^{-6}); the temperature coefficient at 18° is 5·8 %, much larger therefore than that of salt solutions (2 to 2·5 %) and that of ordinary distilled water (2 %).

The degree of electrolytic dissociation of the water may be calculated from the conductivity found by Kohlrausch and Heydweiller; according to p. 363 the conductivity is

$$\kappa = \eta_0(u + v) = 0{\cdot}0384 \times 10^{-6},$$

where u, the mobility of the hydrogen ion, is 318, and v, that of the hydroxyl ion is 174; hence we may calculate the ionic concentration c_0 of the pure water in g-ions per litre ($= 1000\eta_0$).

$$c_0 = \frac{1000\kappa}{u + v} = 0{\cdot}78 \times 10^{-7} \text{ at } 18^\circ;$$

at 25° this becomes $1{\cdot}05 \times 10^{-7}$; *i.e.*, is in *satisfactory agreement with the values mentioned above, arrived at in entirely different ways.*

Even the surprisingly large temperature coefficient of pure water was predicted by Arrhenius and exactly calculated by him (see Book IV. Chap. iii.).

Water is clearly capable of a second electrolytic dissociation namely

$$\overset{-}{OH} = \overset{=}{O} + \overset{+}{H},$$

that is, water can be treated as a dibasic acid. Since therefore the

[1] *Wied. Ann.* **53.** 209 (1894).

separation of the second hydrogen ion from a dibasic acid is always much harder than that of the first, we may expect that the second stage in the electrolytic dissociation of water is extremely slight, or that the oxygen ions with a double negative charge in water exist in infinitesimal quantities. Nothing more is known at present as to this second dissociation.

The Most General Case of Homogeneous Equilibrium.—Deliberate consideration of the preceding paragraph now finally allows us to remove the last limitation, viz. that the ions of water are not included among the reacting molecular species; and in accordance with this we will develop the general equilibrium of a solution containing any casual electrolytes. This is possible by means of the following propositions.

1. The *total quantity* of each radical which is present in the solution, partly as a free ion and partly combined with other ions,—this is either known from the condition of the experiment, or else may be known by means of analysis.

2. For every combination of ions, we have an equation, according to which the undissociated part per unit of volume is proportional to the product of the active masses of the ions contained in the combination. The proportional factor is the *dissociation constant*, which, according to p. 496, is known for most molecular species, and can be determined by the study of any particular case, by means of the conductivity, by the freezing-point, etc.

3. Hydrogen ions and hydroxyl ions are capable of existing together in only the slightest quantities; their product is (nearly) constant, and its magnitude is extremely small ($0 \cdot 64$ or $1 \cdot 14 \times 10^{-14}$ at $18°$ or $25°$).

By means of the formulæ afforded by the immediate application of these propositions, the state of equilibrium is definitely determined.

Thus one is enabled to state in every case what part of each radical exists in the solution as a free ion, and what part is combined with other ions, provided he knows the dissociation constants of the combinations of all of the ions.

This is a result of the greatest importance. It indicates a partial solution of the problem which is to be regarded as the final goal of the doctrine of affinity, namely, *the expression of the mutual reactions between substances, by means of certain characteristic numerical coefficients.*

To this class of coefficients, belong the dissociation coefficients of electrolytes, a knowledge of which enables us to anticipate the kind of action which takes place between them in dilute solution.

We will later obtain the result, that by means of the solubilities of solid salts, one can also definitely determine the state of equilibrium which is established in dilute solution in the presence of solid (difficultly soluble) salts.

In the following paragraphs, some applications will make the meaning of the preceding results more intelligible.

The Distribution of one Base between two Acids.—We can now answer, in its most general form, a question which was formerly much discussed, and fruitlessly too, in default of the aid of the dissociation theory. This question refers to the distribution of a base between two acids, when the total quantity of the latter is greater than that required for neutralisation; and the distribution of an acid between two bases, when the total quantity of the bases is greater than that required for neutralisation.

The state of equilibrium is definitely determined by the absolute quantities of each of the four radicals of the solution (either two acid radicals, the basic radical, and the hydrogen ion, or else respectively the two basic radicals, the acid radical, and the hydroxyl ion), together with the dissociation constants of the four electrically neutral species which may be produced by the combination of the four radicals aforesaid; and the calculation of this equilibrium offers no difficulties except in the way of pure calculation, which, though usually not inconsiderable, are nevertheless not insuperable.

As an example of such a calculation, we will trace out the following simple case.

Thus let two weak (slightly dissociated) monobasic acids, SH and S′H, compete for a base, *e.g.* NaOH; *and let there be in the volume* V 1 g.-mol. *of each of the three electrolytes.*

Let the quantity of the undissociated part of *the first acid* SH, be x; and therefore the undissociated part of S′H will be $1 - x$. Then the quantity $1 - x$ of the first acid SH, will be concerned with the base; and, moreover, in two ways, for a part of the negative radical S as a free ion, will be electrically neutralised by the equivalent quantity of the positive radical of the base; and a part of the negative radical S will unite to form the electrically neutral molecule SNa. Then let the first of these fractions amount to $\alpha_1 (1 - x)$, and the second to $(1 - \alpha_1)(1 - x)$; here α_1 denotes the degree of dissociation of the salt SNa.

And regarding *the second acid* S′H, the quantity x of this is concerned with the base; and of this quantity x, the quantity $\alpha_2 x$ is in the form of negative ions S′; and the quantity $(1 - \alpha_2)x$ is employed in the formation of electrically neutral molecules having the composition S′Na; here α_2 denotes the degree of dissociation of the salt S′Na.

Moreover, a fraction of the two acids will be electrically dissociated, and let the quantity of *the free H-ions be denoted by* γ. But, according to the proviso, both acids are weak, and by the presence of the sodium salt the dissociation is caused to retreat considerably. Therefore γ will represent an infinitesimal quantity, as compared with x and $1 - x$.

We have now to apply the equation of the isotherm of dissociation to the four following dissociations, viz.,

$$\text{I. } NaS = \overset{+}{Na} + \overset{-}{S}. \qquad \text{III. } HS = \overset{+}{H} + \overset{-}{S}.$$

$$\text{II. } NaS' = \overset{+}{Na} + \overset{-}{S'}. \qquad \text{IV. } HS' = \overset{+}{H} + \overset{-}{S'}.$$

In the case of the first two,—according to the law that binary salts composed of univalent ions are decomposed to the same extent,—we make the dissociation constant as equal to K; and thus we obtain respectively,

$$\text{I. } K(1-x)(1-\alpha_1) = \frac{([1-x]\alpha_1 + x\alpha_2)(1-x)\alpha_1}{V},$$

$$\text{II. } Kx(1-\alpha_2) = \frac{([1-x]\alpha_1 + x\alpha_2)x\alpha_2}{V}.$$

Here $(1-x)(1-\alpha_1)$ and $x(1-\alpha_2)$ are respectively the quantities of the undissociated molecules NaS and NaS′; $(1-x)\alpha_1$ and $x\alpha_2$ are respectively the quantities of the $\overset{+}{S}$ and of the $\overset{-}{S'}$ ions; and finally $(1-x)\alpha_1 + x\alpha_2$ is the quantity of the $\overset{+}{Na}$ ions.

By division of these equations, we obtain

$$\frac{1-\alpha_1}{1-\alpha_2} = \frac{\alpha_1}{\alpha_2}, \text{ or } \alpha_1 = \alpha_2;$$

i.e. *both salts are dissociated to the same extent.*

When K_1 and K_2 denote respectively the dissociation constants of the acids, by the application of the law of mass-action to reactions III. and IV., we obtain respectively

$$\text{III. } K_1 x = \frac{\gamma\alpha_1(1-x)}{V}.$$

$$\text{IV. } K_2(1-x) = \frac{\gamma\alpha_2 x}{V},$$

from which by division, and recollecting the identity of α_1 and α_2, we obtain

$$\frac{K_1}{K_2} = \frac{(1-x)^2}{x^2}.$$

Here $\frac{1-x}{x}$ denotes the *ratio of distribution* of the two acids, and we see at once that *it is independent of the nature of the (mon-acid) base.* If $(1-x) > x$, it denotes that the base claims a larger part of the acid SH than it does of the acid S′H; and we can express this by saying that the first acid has a greater "*affinity*" for the base, or that the first acid is "*stronger*"; but we must take care not to include any more in these expressions than is implied by the preceding considerations. The greater "affinity" or "strength" of the first acid consists in this, and in this solely, viz. *that at the same concentration, the former acid is electrolytically dissociated to a greater extent than the second acid is.*

Now, by the use of the proposition, that salts having an analogous constitution are dissociated to the same extent, the state of equilibrium is definitely determined, in that a greater fraction of the more strongly dissociated acid is concerned with the base than of the acid which is dissociated to a less extent; and *the ratio of distribution* is quantitatively equal to the square root of the ratio of the two dissociation constants.

We may state the result thus. Let us denote the degree of dissociation of the two acids respectively by a_1 and a_2, when each is dissolved alone in volume V; then we have

$$K_1 V = \frac{a_1^2}{1 - a_1}, \text{ and } K_2 V = \frac{a_2^2}{1 - a_2};$$

or, since a_1 and a_2 can be neglected as compared with 1 on account of the slight dissociation of the acids, it follows that

$$\frac{K_1}{K_2} = \frac{(1 - x)^2}{x^2} = \frac{a_1^2}{a_2^2},$$

and, therefore,

$$\frac{1 - x}{x} = \frac{a_1}{a_2}, \text{ and } 1 - x = \frac{a_1}{a_1 + a_2}.$$

That is, the ratio of distribution is accordingly equal to the ratio of the respective degrees of dissociation, at the corresponding dilution.

One method for the experimental determination of the relative distribution of any one base between two acids, was given by Thomsen as early as 1854, long before a clear conception of the process of neutralisation was obtained, and before it was developed as a result of the views regarding the theory of dissociation.

If one equivalent of the base is mixed with one equivalent of each of the acids separately, a certain amount of heat is developed, which in the two cases may amount to a and b, respectively. Now, if we mix one equivalent of the base with one equivalent of each of the acids together, a different quantity of heat will be developed, which may be denoted by c.

If the first acid alone unites with the base, and the second acid has nothing to do with the reaction, then $c = a$; and, conversely, if the second acid alone unites with the base, then $c = b$. Now, in fact, both acids participate in the neutralisation, and therefore c must lie between a and b, provided that there are no disturbing side reactions, such as the formation of acid salts, and the like.

The quantity of the first acid concerned in the neutralisation, must amount to

$$\frac{c-b}{a-b};$$

the nearer c is to a, the greater the value of this fraction will be; and the nearer c is to b, the smaller the value of the fraction. Then, according to the method of notation given above, we shall have

$$1-x=\frac{c-b}{a-b};\ x=\frac{a-c}{a-b};\ \frac{1-x}{x}=\frac{c-b}{a-c}.$$

This method of proof is free from objection, either from the old or the new standpoint. The change in the views merely concerns *the manner* in which the acid and the base are concerned in neutralising each other. This process consists not only in the formation of a salt from the acid and the base, but also at the same time there occurs, and usually in a preponderating degree, according to the circumstances, the formation of the free ions which constitute the salt.

Instead of using the "heat of reaction" to determine the ratio of distribution, one may employ, with just as good or better results, the changes in the volumes or in the refractive powers of the solutions, on neutralisation, as was shown by Ostwald in 1878; and by means of a corresponding method of treatment, one obtains formulæ which are exactly the same in this case as above. In particular, the method of the determination of the changes in volume unites accuracy and simplicity of treatment.

In the following table are given the results of a number of the determinations[1] conducted according to the latter method.

	1 - x Obs.	1 - x Calc.
Nitric acid ; Di-chlor-acetic acid	0·76	0·69
Hydrochloric acid ; Di-chlor-acetic acid	0·74	0·69
Tri-chlor-acetic acid ; Di-chlor-acetic acid	0·71	0·69
Di-chlor-acetic acid ; Lactic acid	0·91	0·95
Tri-chlor-acetic acid ; Mono-chlor-acetic acid	0·92	0·91
,, ,, ; Formic acid	0·97	0·97
Formic acid ; Lactic acid	0·54	0·56
,, ,, ; Acetic acid	0·76	0·75
,, ,, ; Butyric acid	0·80	0·79
,, ,, ; Iso-butyric acid	0·79	0·79
,, ,, ; Propionic acid	0·81	0·80
,, ,, ; Glycolic acid	0·44 (?)	0·53
Acetic acid ; Butyric acid	0·53	0·54
,, ,, ; Iso-butyric acid	0·53	0·54

[1] Ostwald, *Journ. pr. Chem.* [2], **18.** 328 (1878).

The significance of the observed value of $1 - x$ may be illustrated most easily in the following way. If four *equivalent solutions* of SNa, S′Na, SH, and S′H, respectively, are mixed, and in the following proportions, viz.: $1 - x$ volume of SNa, x volume of S′Na, x volume of SH, and $1 - x$ volume of S′H,—then there occurs neither contraction nor dilation. It is indifferent whether one uses, instead of sodium, any other mon-acid base (as is also required by theory). We conclude from this *that acids and salts exist together in those relative proportions which correspond to the equilibrium of their mixture.* For if this were not the case a reaction would take place, which would consist in a change of the degree of dissociation of the four electrolytes, and this would be shown by a change of volume. Therefore the value of $1 - x$ must, as determined experimentally, coincide with the formula as developed above.

As a matter of fact this does occur, as was shown by Arrhenius.[1] In the second column of the preceding table are given the values of $1 - x$, calculated by the formula developed above, viz.

$$1 - x = \frac{a_1}{a_1 + a_2},$$

from the ratio of the degrees of dissociation of the two acids at the dilution employed, which amounted to three litres, inasmuch as the solutions which were studied, resulted from the mixture of the three normal solutions, viz. of the base and of the two acids respectively.

With the exception of the first three values,—where the competition [for the base] was between very strong acids, and, therefore, where the proviso of the theoretical formula is not fulfilled,—and, with the exception of one particular value which probably involved some error,—in general a very good coincidence is established between the values of $1 - x$, as calculated from the changes of volume on neutralisation and those calculated from the respective conductivities of the pure acids. As the value of $1 - x$ is always greater than 0·5, therefore the acid named first is the stronger in all three cases.

In the competition between any two weak acids, and at such dilutions that the salts of the acids may be regarded as completely dissociated, the reaction progresses according to the scheme,

$$SH + \overset{-}{S'} + \overset{+}{Na} = S'H + \overset{-}{S} + \overset{+}{Na},$$

or simply

$$SH + \overset{-}{S'} = S'H + \overset{-}{S}.$$

Now the law of mass-action requires that

$$\frac{\text{Acid I.} \times \text{acid-ion II.}}{\text{Acid II.} \times \text{acid-ion I.}} = \text{a constant.}$$

[1] *Zeits. phys. Chem.*, **5**. 1 (1890)

This equation was found to be established by Lellmann and Schliemann,[1] though, to be sure, they were not very clear as to its meaning; for, on the basis of some misunderstanding, they maintain that their research does not harmonise with the dissociation hypothesis, while the reverse is the case.[2] The method employed by them was in principle that of Jellet (p. 453); only instead of using the rotation of light, they used the absorption of light to analyse the state of equilibrium.

When poly-basic acids—as sulphuric acid, *e.g.*—are employed, then the theoretical treatment of the state of equilibrium is rendered much more difficult by the formation of acid salts and the like: this has not been investigated as yet.[3]

The Strength of Acids and Bases.—It is a very common experience, and one made long ago, that the different acids and bases exhibit very different "intensities" or "strengths" in those solutions where their acid or basic nature, as such, comes into play. But in spite of many strenuous endeavours to accomplish it, it is only recently that it has been possible to find a method for expressing their [relative] strengths numerically, *i.e.* for expressing the numerical coefficient of each particular acid and base, which should make possible the calculation of the degree of their distribution in the reactions which are characteristic of the respective acids and bases.

This problem was first deliberately attacked in a broad way by J. Thomsen (1868); but it was Ostwald (1878-1887) who first succeeded in proving, beyond all doubt, that the property of acids and bases of exerting their action according to the standards of definite coefficients, finds its expression not only in the forming of salts, but also in a large number of other and very different reactions.

Ostwald compared the relative order, in which the acids compete for the same base, according to their strength, as obtained by—

(*a*) Thomsen's thermo-chemical methods, with

(*b*) His own (Ostwald's) volume-chemical methods, and with the relative order in which the acids arrange themselves,—

(*c*) Regarding their capacity to bring calcium oxalate into solution; and,—

(*d*) Regarding the relative velocities with which they convert acetamide into ammonium acetate; and,—

(*e*) Regarding the relative velocities with which they cleave methyl acetate, catalytically, into methyl alcohol and acetic acid; and,—

(*f*) Regarding the relative velocities with which they invert cane sugar; and,—

[1] *Lieb. Ann.*, **270.** 208 (1892).

[2] See Arrhenius, *Zeits. phys. Chem.* **10.** 670 (1892)

[3] See A. A. Noyes, "On the Separation of Hydrogen Ions from Acid Salts, *Zeitschr. phys. Chem.* **11.** 495 (1893).

(*g*) Regarding their relative accelerations of the mutual action of hydriodic acid and bromic acid.

Ostwald showed that in all these various cases investigated, one always obtains *the same order for the relative strengths of the different acids*, as measured by the investigations just mentioned; particular cases will be considered later. It should be noticed that all the decompositions enumerated above, were conducted in *dilute water solution*, and therefore the preceding scale refers only to reaction capacities under these conditions. The order of succession of the acids is apparently independent of the temperature.

Although Ostwald's investigations give undoubted proof regarding *the relative order* of the strength of the particular acids, yet great difficulty was experienced in ascertaining *the quantitative ratios;* and the numerical coefficients, as calculated from the particular reactions, usually exhibit great deviations, although sometimes there are surprising coincidences. In particular the coefficients vary very greatly with the concentration, and in those cases where the concentration of the acid varies in the course of the reaction, the calculation is, of course, entirely untrustworthy.

A similar behaviour was found in the study of bases, though the observations were not so extended.

These seemingly complicated relations may now be explained in one word, by the application of the law of chemical mass-action, as its meaning was recognised by van't Hoff (1885), and as it was thereupon applied by Arrhenius (1887) to electrolytic dissociation, in explaining the exceptional behaviour of substances in water solution. The formulæ to be used here in the calculation of the equilibrium ratios, are of course obtained by a specialisation of the general equations developed on pp. 497 and on.

The peculiarities directly presented by the conduct of acids and bases, which must be presented in the sense of the views developed by Arrhenius,—peculiarities, moreover, which are illustrated both in the old-time distinction between *neutral* solutions, on the one hand, and acid and basic [or rather alkaline] solutions on the other, and also in the recognition of a polar contrast between the two latter [viz. acid and alkaline solutions],—all of these peculiarities, in the light of *the theory of electrolytic dissociation*, are now to be conceived of in the following way.

The reactions which are characteristic of acids existing in solution, and which are common to all acids, and which can only be effected by acids,—these reactions consist in this, namely, that the dissociation of these bodies, *as a class*, results in the production of the same molecular species, the positively charged hydrogen ion $\overset{+}{H}$.

Therefore those chemical actions which are characteristic of acids are to be ascribed to the action of the hydrogen ions.

In the same way, *e.g.*, the chemical actions which are common to the chlorides are to be explained by the action of the free chlorine ions.

And in a similar way, *the reactions which are characteristic of bases existing in solution*, justify the view that the dissociation of this class of substances results in the production of negatively charged hydroxyl ions $\left(\overset{-}{HO}\right)$.

Therefore the specific action of bases is due to the hydroxyl ions.

A solution reacts acid when it contains free hydrogen ions; and alkaline (basic) when it contains free hydroxyl ions. If we bring together an acid solution and an alkaline solution, then, because the positive hydrogen ion and the negative hydroxyl ion are incapable of existing beside each other, the ions unite with each other to form electrically neutral molecules, in the sense of the equation,

$$\overset{+}{H} + \overset{-}{OH} = H_2O,$$

and there results *neutralisation* (p. 507). Thus we find a simple explanation of the polar contrast between acid and basic solutions; it consists simply in this, namely, that *the ions which are respectively characteristic of the acid and of the base, together form the two ingredients of the solvent [water], in which we study the reaction capacities.*

The conception of the "strength" of an acid or of a base now explains itself. If we compare equivalent solutions of different acids, each one will exert the actions characteristic of acids the more energetically, the more free hydrogen ions it contains. This follows as an immediate deduction from the law of chemical mass-action.

The degree of electrolytic dissociation determines the relative strength of the acid; and similar considerations lead to the proposition that—

The strength of bases depends upon the degree of their electrolytic dissociation.

Now the degree of the electrolytic dissociation varies with the concentration, regularly in the way indicated on p. 498. At very extreme dilution, equivalent solutions of the most various acids contain the same number of hydrogen ions, or, in other words, they are equally strong; and a similar thing holds true of bases.

The dissociation decreases with increasing concentration, but the variation changes at a different rate for various substances. Thus the relative strengths of bases and acids must vary with their concentration, as was empirically established by Ostwald.

Now the dissociation constant is the measure of the variation of the degree of dissociation with the concentration; *i.e. we must regard these magnitudes as the measure of the strengths of acids and bases.* Thus by the consideration of this special case, we again obtain the same result as that previously developed in a general way (p. 474), namely, *the dissociation coefficients are the measures of the reaction capacities of all such substances,* e.g. *both acids and bases.*

The order of succession of the acids, as arranged by Ostwald on the basis of the most various reactions, must coincide with the order of succession as developed from their dissociation constants; and also, since the depression of the freezing-point increases with the degree of the electrical dissociation, it must coincide with the order of succession as arranged with reference to the depression of the freezing-point in equivalent solutions. This result is established by experiment.

The degree of dissociation α, of an acid, in a definite concentration at which its molecular conductivity is Λ, according to p. 359, is calculated to be

$$\alpha = \frac{\Lambda}{\Lambda\infty}.$$

The conductivity at very great dilution $\Delta\infty$, according to the law of Koelrausch (p. 363) is found to be

$$\Lambda\infty = u + v.$$

Now since u, the ionic mobility of hydrogen, is more than ten times as great as v, the ionic mobility of the negative radical of the acid, therefore $\Lambda\infty$ has approximately the same value for the most different acids (usually within less than 10 per cent); *and therefore the conductivity of an acid in equivalent concentration corresponds, at least approximately, to the degree of its electrolytic dissociation,* i.e. *to its strength.*

On the whole, the order of succession as arranged in accordance with the conductivities is identical with the order of succession as shown by their acids in their specific reactions. The recognition by Arrhenius and Ostwald (in 1885) of this remarkable parallelism was an important event, for it amounted to the discovery of electrolytic-dissociation.

We will now learn, by a detailed description of these reactions, how the special case of the distribution of acids and bases in the most various reactions, may be calculated quantitatively from their dissociation coefficients, which Ostwald calls the "affinity constants."

Hydrolytic Dissociation.—A very important case in which water as a solvent participates in the reaction, is the so-called *hydrolytic dissociation;* or in short "*hydrolysis,*" *i.e.* the decomposition of a salt into base and acid, with the assumption of the ingredients of water.

The theory of this, according to what has been explained, is very simple. Let any selected quantities of an acid SH, and a base BOH, be dissolved in a large quantity of water. Then, in general, the five following reactions will occur in one sense or the other [*i.e.*, advance or retrograde], by a change in the relative proportions.

$$\text{I. } SB = \overset{-}{S} + \overset{+}{B}.$$

$$\text{II. } SH = \overset{-}{S} + \overset{+}{H}.$$

$$\text{III. } BOH = \overset{+}{B} + \overset{-}{OH}.$$

$$\text{IV. } H_2O = \overset{+}{H} + \overset{-}{OH}.$$

$$\text{V. } SB + H_2O = SH + BOH.$$

Reactions I. to IV. inclusive are cases of electrolytic dissociation; reaction V. is the equation of hydrolytic dissociation. Let K_1 to K_5 inclusive be the respective reaction coefficients, and let the respective concentrations of the reacting molecular species, which are partly electrically neutral molecules, and partly ions, be as follows:—

SB	SH	BOH	$\overset{+}{B}$	$\overset{+}{H}$	$\overset{-}{S}$	$\overset{-}{OH}$
C_1	C_2	C_3	c_1	c_2	c_1'	c_2'.

Now, according to the conditions of the experiment, the total quantity of the radical S, is

$$C_1 + C_2 + c_1' = m;$$

and that of the radical B, is

$$C_1 + C_3 + c_1 = n;$$

and, therefore,

$$c_1 + c_2 = c_1' + c_2';$$

i.e. the solution contains the same number of positive and of negative ions. The active mass of the solvent, *i.e.* of the molecule H_2O, is very nearly constant (p. 460): it is of course indifferent whether water in the liquid state has this molecular weight or another [polymeric].

Now the application of the law of mass-action to reactions I. to V. inclusive, gives respectively the following equations, viz.

$$\text{I. } K_1C_1 = c_1c_1'.$$

$$\text{II. } K_2C_2 = c_1'c_2.$$

$$\text{III. } K_3C_3 = c_1c_2'.$$

$$\text{IV. } K_4 = c_0^2 = c_2c_2'.$$

$$\text{V. } K_5C_1 = C_2C_3.$$

K_1 may, with a satisfactory approach to accuracy, be regarded as the same (p. 499) for all binary electrolytes composed of univalent ions.

K_2 and K_3 may be determined from the conductivity in all cases where the base is not too weak; for those cases where the base is very weak, these values may be most safely determined by a comparison with a stronger base, respecting the competition for the acid by the bases. K_4 *is the electrolytic dissociation constant of water*, the significance of which was first recognised by Arrhenius;[1] its exact determination is of the greatest importance.

By multiplying equations II. and III., and dividing by I. we obtain

$$C_2C_3 = C_1\frac{K_1}{K_2K_3}c_2c_2',$$

and this, by substitution from IV., becomes

$$C_2C_3 = C_1\frac{K_1K_4}{K_2K_3},$$

By comparing this with equation V., it becomes

$$K_5 = \frac{K_1K_4}{K_2K_3},$$

The reaction constant for hydrolysis can therefore be calculated from the dissociation constant of the reacting molecules, that is the degree of hydrolysis can be calculated when we know the strength of the acid and base.

In practically applying the above equation we must consider however, that the value of K is *not* constant (p. 499) for strongly dissociated electrolytes, that is for neutral salts and very strong acids and bases. In this case the above formula may be most practically applied by treating these substances as completely dissociated, and afterwards introducing small corrections on account of the inaccuracy of this assumption.

The strongest hydrolysis is of course met with when both the acid and the base are very weak. It may happen in this case that the salt is entirely decomposed, namely when the acid or base is insoluble and hence the greater part of the substance is precipitated and becomes inactive. Thus white silver borate decomposes on heating, with separation of silver oxide, in the same way ferric acetate decomposes in dilute solution on boiling almost completely into ferric hydroxide and free acid. A case in which the salt decomposes completely in the cold at any attainable dilution is to be found in the behaviour of the salts of the trivalent metals towards carbonates; the hydroxyl is here precipitated at once because the hydrolysis of the carbonate of these metals is almost complete.

[1] *Zeits. phys. Chem.*, **5.** 16 (1890).

Walker[1] investigated quantitatively the hydrolysis of the chlorides of some weak bases (such as aniline) by measuring the amount of free hydrogen ions by the velocity of inversion of methyl acetate (see the following chapter). In this case hydrolysis occurs according to the equation

$$\overset{+}{B} + \overset{-}{Cl} + H_2O = BOH + \overset{-}{Cl} + \overset{+}{H}$$

or more simply

$$\overset{+}{B} + H_2O = BOH + \overset{+}{H};$$

as we may here, on account of the strong dissociation of the hydrochloric acid, regard the concentration of the hydrogen ions as very approximately the same as that of the free acid, and, on account of the strong dissociation of the salt, regard the concentration of the $\overset{+}{B}$-ions as being that of the undissociated salt, and finally, on account of the very feeble dissociation of the base, regard the concentration of BOH as equally that of the free base, we get the equation

$$\frac{[BOH]\,[\overset{+}{H}]}{[\overset{+}{B}]} = \frac{\text{Base} \times \text{Acid}}{\text{Salt}} \quad \text{const.}$$

which Walker found to be confirmed by experiment.

In the case stated by Walker the hydrolysis was considerable despite the strength of the acid because the base was extraordinarily weak; conversely Shields[2] measured the hydrolysis of a number of salts for which the base was strong and the acid very weak. The rate of saponification of methyl acetate was determined, as the amount of free hydroxl ions is directly proportional to it (see the following chapter), and, on acccount of the strength of the base used, this is practically identical with the total concentration of the free base. The degree of hydrolytic dissociation at 25° was found for the following salts in $\frac{1}{10}$ normal solution :—

Potassium cyanide	1·12 per cent.
Sodium carbonate	3·17 ,,
Potassium phenolate	3 05 ,,
Borax	0·5 ,,
Sodium acetate	0·008 ,,

Since in the above cases the base is very strong and the acid very weak the reaction occurs practically according to the equation

$$\overset{-}{CN} + H_2O = \overset{-}{OH} + HCN$$

and by using pure salts, that is by avoiding an excess of acid or base

[1] *Zeits. phys. Chem.*, **4.** 319 (1889). [2] *Ibid.*, **12.** 167 (1893).

the amounts of free acid and base must be equal, so that the law of mass action gives

$$\frac{\text{Acid} \times \text{Base}}{\text{Salt}} = \text{constant, or the base is proportional to } \sqrt{\text{salt}},$$

that is, the degree of hydrolysis is proportional to the square root of the concentration of the undissociated salt, and the latter quantity is not very different from the total concentration of the salt when the hydrolysis is small.

We may calculate the dissociation of water from the fact that 0·1 normal sodium acetate is hydrolysed to the extent of 0·008%. Since acetic acid in presence of its salt is very slightly dissociated, the free base (NaOH) almost completely, we have as the concentrations of the acetic acid and the hydroxyl ions

$$[CH_3COOH] = [\overline{OH}] = 0{\cdot}000008 \frac{\text{mol}}{\text{litre}}.$$

The amount of free hydrogen ions $[\overset{+}{H}]$ is found from the equation

$$K[CH_3COOH] = [\overline{CH_3COO}]\,[\overset{+}{H}],$$

where K, the dissociation constant of acetic acid, reaches 0·0000178 (p. 498), and the concentration of the negative ion of acetic acid is almost precisely that of the salt (0·1). We thus find

$$[\overset{+}{H}] = \frac{0{\cdot}0000178\,.\,0{\cdot}000008}{0{\cdot}1} = 1{\cdot}42 \times 10^{-9}$$

and according to p. 508

$$c_0 = \sqrt{[\overset{+}{H}]\,[\overline{OH}]} = \sqrt{1{\cdot}42\,.\,0{\cdot}8\,.\,10^{-14}} = 1{\cdot}1\,.\,10^{-7}$$

that is, pure water at 25° is 0·11 millionths normal with respect to the hydrogen or hydroxyl ions. Walker (*loc. cit.* p. 503), by means of the values of K given on p. 503, calculated in the same way the hydrolysis of potassium cyanide, potassium phenolate, and borax to be 0·96, 3·0 and 0·3% in striking agreement with the experiments of Shields.

The Theory of Indicators.[1]—Many of the so-called "*colour-reactions*" may be explained by a change in the electrolytic dissociation, whether it is a result of dilution or of the addition of some foreign substance.

We have already seen (p. 384), that each particular ion has its own definite absorption of light, and that in general this changes when

[1] Ostwald, *Lehrb. d. allg. Chim.*, 2nd edit., 1891, Bd. I. p. 799. [See this part translated by P. M. Muir, "Solutions," pp. 268-272. London and New York, 1891.—TR.]; *Grundlagen d. analyt. Chemie.*, chap. vi., 2nd edit., Leipzig, 1897.

the ion unites with another. Thus cupric chloride has a green colour, which is occasioned by the undissociated molecules; and it is only at extended dilution that the blue colour of the cupric ions appears, which is exhibited by all cupric salts dissolved in much water. If one adds hydrochloric acid to a dilute solution of cupric chloride, the dissociation retreats, and the solution again becomes a clear green.

The quantitative study of the gradual change in coloration is an elegant method of investigating chemical equilibrium, which was first introduced by Gladstone (1855). It has been used since by Salet (p. 347), Magnanini,[1] Lellmann (p. 516), and others.

Upon phenomena like this, depends the common use of *indicators* in volumetric analysis, *i.e.* substances which in acid solutions have colours different from those in alkaline [or neutral] solutions. Every weak acid or base [or salt] is suitable for such a purpose when its radical has, as an ion, a colour different from that which it has in an electrically neutral molecule.

The acid or base must be weak so that a very slight excess of hydrogen or hydroxyl ions may cause a great change in coloration. Thus paranitrophenol is an acid indicator; the undissociated molecule of this acid is colourless, its negative ion is coloured an intense yellow. If acid is present in the solution the already slight dissociation of the indicator is almost entirely destroyed and the solution becomes colourless. If, on the contrary, a base is added, the strongly dissociated solution of paranitrophenol is formed and the solution becomes yellow. Another acid indicator of similar chemical character is phenolphthaleine, which is colourless in the undissociated state, that is, in the presence of a trace of hydrogen ions. But as soon as the solution becomes alkaline the strongly dissociated salt of phenolphthaleine is formed, and the intense red of its negative ion appears. Methyl orange is an example of a basic indicator; in acid solutions it is coloured an intense red, in alkaline it is yellowish.[2]

The above considerations show when an indicator is useful, that is, when it gives a sharp change of colour, and when not. If it is a strong or even a moderately strong acid its dissociation would only be destroyed by a considerable excess of hydrogen ions. On the other hand, it must not be too weak an acid, otherwise, with excess of base, the salt formed by the base and the indicator would be hydrolytically decomposed to a considerable extent, and consequently the change of colour would be weakened. The latter circumstance becomes of more importance when the base added is weak. Phenolphthaleine for example is so weak an acid that its ammonia salt is strongly dissociated hydrolytically, hence when ammonia is titrated with phenolphthaleine

[1] *Zeits. phys. Chem.*, **8.** 1 (1891).
[2] See also on this point F. W. Kuster, *Zeits. anorg. Chem.* **13.** 136 (1897), who shows convincingly that the acid function of methyl orange is unimportant as regards change of colour (p. 377).

as indicator the red colour of the ion of phenolphthaleine disappears on addition of acid before the ammonia present has been thoroughly neutralised. Paranitrophenol is a considerably stronger acid, its ammonia salt is less decomposed hydrolytically but the change of colour remains sharp. Hence weak bases may be titrated with paranitrophenol but not with phenolphthaleine as indicator. In titrating strong bases phenolphthaleine gives more exact results than paranitrophenol because the yellow colour of the negative ion of the more strongly dissociated paranitrophenol appears before the neutralisation is completely effected by the alkali added, whilst the much more weakly dissociated phenolphthaleine only shows the red colour of its negative ion when an extremely minute quantity of the strong base is present in excess. The same considerations apply naturally to basic indicators like methyl orange, so that we are led to the following rule as to the use of indicators. *A weak base and a weak acid are not to be used together on account of hydrolysis;* very weak acid indicators are therefore unavailable for titrating weak bases, very weak bases for titrating weak acids. As titrating liquid strong acids and strong bases should always be used (hydrochloric acid, barium hydroxide). If the indicator used is a moderately weak acid or base the change of colour can still be used in titrating weak bases although with a certain loss of sensitiveness.

In multivalent acids it may happen that, on account of the different strengths of its valencies, only one stage of the dissociation may come into evidence on account of the nature of the indicator used. Thus carbonic acid cannot be titrated at all with methyl orange, with phenolphthaleine it appears as a univalent acid ; phosphoric acid behaves as a univalent acid on titration with methyl orange, as a divalent acid on titration with phenolphthaleine. For further details see the study on indicators by Julius Wagner.[1]

The so-called "Hydrate Theory."—In the preceding chapters there has been developed the theory of the state of chemical equilibrium in solution. The way to this was opened by those methods of investigation which were based on the fundamental principles of the doctrine of affinity; and respecting the fruitfulness of these methods, they have been put in evidence, by their abundant results, to the enrichment of our experimental knowledge.

Now in addition to this theory [of chemical equilibrium], another conception has recently shown itself; this, by those who represent it, has been called the "Hydrate Theory of Solutions," although it has provisionally failed utterly to show that it deserves the name of a "theory of solutions."

Mendelejeff[2] supposed that he had discovered that the differential coefficients of the densities of water solutions of sulphuric acid and of

[1] *Zeits. anorg. Chem.*, **27.** 138 (1901). [2] *Zeits. phys. Chem.*, **1.** 273 (1887).

alcohol, vary rectilinearly with the per cent contained: but the dependence upon the concentration is represented not by *one* but by *several* straight lines: breaks [between the curves] were supposed to correspond to definite hydrates.

In fact, Mendelejeff stated the breaks that he supposed he had found, at concentrations corresponding to the following molecular ratios, viz.: $SO_3 + H_2O$; $SO_3 + 3H_2O$; $SO_3 + 7H_2O$; $SO_3 + 151H_2O$. He found similar breaks in the alcohol solutions.

But now it was at once shown[1] by a careful discussion of the material of observation, that irregularities of this sort do not exist. But more impeachable still than the experimental basis, which itself seems to be very untrustworthy, are the considerations which led to the search for "breaks" [in the curve], and to their theoretical interpretation. For even if it were granted that the aforesaid hydrates did exist in the solution, their decomposition would not be manifested by irregularities appearing suddenly; but here, as in general, there would occur a gradual and perfectly continuous change of the equilibrium with the concentration.

Moreover, we have already seen (p. 459) that, according to the requirements of the law of chemical mass-action, any contingent hydration of the dissolved substance would be almost independent of the concentration, at least at extended dilution.

The Mutual Influence of the Solubility of Salts.—We have thus far considered the equilibrium in salt solutions, as though they were *homogeneous* systems. We will now consider the case where *solid salts* participate in the equilibrium. The proposition that a solid substance, which dissociates on solution, has a definite solubility at a definite temperature, like every other solid substance (p. 475), holds good of course for the case where the dissociation is an electrolytic one; and thus the propositions previously developed may be applied, without further remark, to the case in hand. This statement thus enables us to consider, completely, the case where the solid salt participates in the equilibrium, as will be made plain from the following example.

We will first consider the simple case of *a binary electrolyte*, and we will investigate the change in its solubility as it is occasioned by the presence of another binary electrolyte which has a common ion. The process can at once be surveyed in a qualitative way.

The saturated solution of the first electrolyte, of course, is never completely dissociated, for some electrically neutral molecules will be present in the solution. The proposition (p. 475) that the concentration remains unchanged by the presence of other substances in the solution, may be immediately applied to the concentration of these neutral molecules. We will now add to the saturated water solution

[1] Pickering, *Zeits. phys. Chem.*, **6.** 10 (1890).

of the electrolyte, another electrolyte with a common ion; then in exactly the same way that it was found on p. 503, the dissociation of the first electrolyte will retreat, and the quantity of electrically neutral molecules will increase. But this increased quantity prevents the solution pressure of the solid salt from remaining in equilibrium, and therefore a definite part of the dissolved salt will be precipitated from the solution, until the equilibrium is re-established. *Thus the solubility of one salt is depressed in the presence of another having a common ion.*

This proposition may be proved in a *qualitative* way without difficulty. Thus, if one adds to a saturated solution of potassium chlorate a solution of another potassium salt, as *e.g.* potassium chloride, or a solution of another chlorate as *e.g.* sodium chlorate,—and this may be done most easily by adding a few drops of a concentrated solution of the other substance,—he will observe after a moment a copious separation of solid potassium chlorate. A saturated solution of lead chloride gives immediately a white precipitate on addition of a few drops of chloride and so on.

Of course the mutual influence of solubility can also be developed theoretically in a *quantitative* way. Thus let m_0 be the solubility of the solid electrolyte in pure water; and let α_0 be the degree of dissociation corresponding to this concentration (the latter being expressed, as always, in g-equivalents per litre). Then $m_0(1-\alpha_0)$ will denote the undissociated, and $m_0\alpha_0$ the dissociated quantity of the electrolyte. Let its solubility be m, and the corresponding degree of dissociation be α, in the presence of another electrolyte, the free ions of which have the concentration x.

Then the theorem of the constant solubility of the undissociated part, gives

$$m_0(1-\alpha_0) = m(1-\alpha),$$

and the application of the isotherm of dissociation, in the two cases, gives

$$Km_0(1-\alpha_0) = (m_0\alpha_0)^2,$$

$$Km(1-\alpha) = m\alpha(m\alpha + x);$$

and, therefore, there must exist the relation,

$$(m_0\alpha_0)^2 = m\alpha(m\alpha + x),$$

from which we obtain

$$m = -\frac{x}{2\alpha} + \sqrt{m_0^2\left(\frac{\alpha_0}{\alpha}\right)^2 + \frac{x^2}{4\alpha^2}};$$

this equation enables one to calculate the solubility after the addition, from the solubility of the salt in pure water, and from the quantity of salt added.

This preceding law of solubility, which the writer developed in 1889, and also experimentally verified (p. 477), was later subjected to a very careful test by A. A. Noyes;[1] its requirements were established in a striking way. Thus Noyes and others have studied the influence which the solubility of silver bromate experienced in the presence of silver nitrate and potassium bromate.

I. Addition of $AgNo_3$ or of $KBrO_3$ respectively.	II. Sol. of $AgBrO_3$ on the addition of $AgNO_3$.	III. Sol. of $AgBrO_3$ on the addition of $KBrO_3$.	IV. Calc. sol. of $AgBrO_3$.
0·	0·00810	0·00810	0·00810
0·00850	0·00510	0·00519	0·00504
0·0346	0·00216	0·00227	0·00206

The numbers given under II. are the solubilities on addition of silver nitrite, under III. those on addition of potassium bromate; it appears from the table that equivalent mixtures of silver nitrite and potassium bromate displace the solubility of silver bromate equally and by practically the amount that is calculated from the law of mass action. It also appears that a relatively small addition of a foreign substance is sufficient to reduce the solubility [of the $AgBrO_3$] one-fourth. The fact that the calculated values in this case, as well as in the numerous other cases observed by Noyes, are smaller than the observed values, may perhaps be explained by the possibility that the neutral salts are not so highly dissociated electrolytically, as would be indicated by a calculation from the conductivity.

Moreover, Noyes also found that equivalent quantities of chlorides of *univalent* bases, depress the solubility of *thallous* chloride to the same degree; this is a further proof of the proposition (p. 499) that these substances [*i.e.* salts of univalent bases] are dissociated to the same extent when in equivalent solutions.

Moreover, the addition of the chlorides of the *divalent* metals, Mg, Ca, Ba, Mn, Zn, and Cu, have the same effect in depressing the solubility; from which it is to be inferred that these substances, in equivalent solutions, experience the same degree of dissociation. In this way, Noyes found the following values for the degree of dissociation of the chlorides just enumerated.

[1] *Zeits. phys. Chem.*, **6.** 241 (1890); **9.** 603 (1892); **26.** 152 (1898).

[TABLE

Concentration.	Degree of Dissociation.
0·0344	82·2%
0·0567	77·6%
0·1045	69·4%
0·2030	61·5%

These values are calculated as though the salt decomposed according to the scheme

$$CaCl_2 = \overset{++}{Ca} + \overset{-}{Cl} + \overset{-}{Cl};$$

but there is, doubtless, at the same time a decomposition according to the scheme

$$CaCl_2 = \overset{+}{CaCl} + \overset{-}{Cl},$$

although this latter may not amount to very much.

In contrast to the salts just mentioned, $CdCl_2$ is dissociated to a much smaller degree. Thus, in investigating the influence of solubility, we possess, in the suitable choice of substances which are soluble with difficulty, a method, which is applicable to every kind of ion, for the determination of the number of the particular kind of ions contained in a solution.

As an example of the mutual influence of the solubilities of *tenary electrolytes*, Noyes studied the diminution of the solubility of *lead chloride* on the addition of the chlorides of Mg, Ca, Zn, and Mn. As would be expected, these latter substances act with the same strength [of depression of solubility]. The degree of this influence is given by the proposition, that the product of the lead ions and the square of the chlorine ions must be constant.

As before, let m_0 denote the solubility of $PbCl_2$ in pure water, and m its solubility after the addition of x chlorine ions; and let α_0 denote the original dissociation, and α that after the addition. Then the law of mass-action requires that

$$m_0(1 - \alpha_0) = m(1 - \alpha),$$

and

$$(m_0\alpha_0)^3 = m\alpha(m\alpha + x)^2.$$

Then, since $PbCl_2$ is dissociated just as strongly as the added chloride, x is equal to the quantity of chloride added multiplied by α. Thus Noyes found the following results:—

Quantity added.	α	m	
		Found.	Calc.
0·05	0·697	0·0502	0·0522
0·10	0·661	0·0351	0·0351
0·20	0·605	0·0218	0·0176

$$m_0 = 0{\cdot}0777\ ;\ \alpha_0 = 0{\cdot}733.$$

The values of α are taken from the preceding table. The coincidence between calculation and observation is fair.

To precipitate an insoluble salt as far as possible it is therefore convenient to use an excess of the precipitating material in order to displace the solubility of the precipated salt. It is, however, sufficient, especially in the case of very insoluble substances, to use a quite small excess. For example, the concentration of a saturated solution of silver chloride at atmospheric temperature is about $\frac{1}{100000}$ normal; if chlorine ions are added, to precipitate silver, only to the extent of $\frac{1}{1000}$ normal in excess, the concentration of the silver ions, as appears from the preceding formulæ, is reduced to a normality of $\frac{1}{10000000}$.[1] Lead sulphate is appreciably soluble, hence the prescription of analytical chemistry to wash this substance with water acidulated with sulphuric acid rather than with pure water, since the acid reduces its solubility to an amount small as regards analysis.

Moreover, Noyes succeeded in proving, both by theory and by experiment, that *conversely the solubility of a salt must increase* on the addition of a second electrolyte containing no ion in common.

Thus, to follow out a preceding example, if one adds some KNO_3 to $AgBrO_3$, then a number of molecules of $AgNO_3$ and also of $KBrO_3$ will be formed. This will result in a diminution in the number of the molecules of $AgBrO_3$, which must be replaced from the solid salt. In this and in analogous cases the increase of the solubility is of course very slight; but the increase would be very great if one should add, *e.g.*, HNO_3 to a saturated solution of $CH_3 . CO . OAg$; because here, on account of the small dissociation constant of acetic acid, there will be produced a very considerable quantity of undissociated acetic acid molecules as a result of the addition of the HNO_3; and as the product of the silver ions and the acet-ions must regain its former value, a considerable quantity of solid silver acetate must therefore pass into solution.

A similar explanation may be given of the well-known fact that calcium oxalate is soluble in strong acid, for the product of the concentration of the calcium ions and the oxalic acid ions (the so-called

[1] See the interesting study by C. Hoitsema, *Zeits. phys. Chem.*, **20.** 272 (1896).

solubility product) is much diminished when a strong acid is present, since the hydrogen ions of the latter combine with the oxalic ions and raise the concentration of the latter; hence for equilibrium more calcium must go into solution and the amount may become quite large. As another example may be mentioned the solubility of zinc sulphide in strong acids whose hydrogen ions combine with the doubly negative sulphur ions of the zinc sulphide; weak acids do not dissolve it because the concentration of the hydrogen ions is too small. If, therefore, sodium acetate be added to an acid solution of a zinc salt the formation of slightly dissociated acetic acid reduces the concentration of the hydrogen ions to a small amount and the zinc is satisfactorily precipitated by sulphuretted hydrogen.

Anomalies due to Formation of Complex Ions.—There are cases in which the solubility of a salt is raised by addition of another salt containing the same ion; thus potassium nitrate and lead nitrate raise each other's solubility, mercury chloride is more easily dissolved by water containing hydrochloric acid than by pure water. Closer study of these cases however shows that the exception to the general law is only apparent; as has been shown by Le Blanc and Noyes[1] in this and similar cases, new complex molecules are formed, so that the solubility product is not raised by the addition of salt containing the same ion as would otherwise be the case, but reduced.

These anomalous phenomena of solubility are consequently very important for the study of complex salts in solution, a subject which has been very little investigated considering its enormous importance for inorganic chemistry. As an example of the insight to be obtained by means of relatively simple researches we may mention the work of A. A. Noyes and W. R. Whitney.[2] Potassium and sodium hydroxides not only do not reduce the solubility of aluminium hydroxide but take up considerable quantities of this insoluble substance. Yet it was found that the freezing point of the solutions is *not* altered by addition of aluminium hydroxide. From this it may be concluded that on solution the reaction

$$\bar{O}H + Al(OH)_3 = Al\bar{O}(OH)_2 + H_2O$$

takes place, that is, the addition of aluminium hydroxide does not alter the number of molecules in solution. The potassium aluminate has therefore in solution the formula $KAlO(OH)_2$ or $KAlO_2$ (the methods in use at the present day do not allow of distinguishing between these two formulæ any more than between NH_3 or NH_4OH for ammonia in aqueous solution. See the section on "Normal and Anomalous Reactions.")

[1] *Zeits. phys. Chem.*, **6**. 385 (1890). [2] *Ibid.*, **15**. 694 (1894).

The Application of the Law of Mass Action to Strongly Dissociated Electrolytes.—It has already been remarked that the equation of the dissociation isotherm does not hold for strongly dissociated electrolytes such as the neutral salts. The law of mass action gives for binary electrolytes the relation

$$K \cdot c = c_1 \cdot c_2 .$$

A. A. Noyes and Abbot,[1] by systematic determinations of solubility have arrived at the important conclusion that over a certain range of concentration at least for constant c (the concentration of the undissociated molecules), for example, in presence of the solid salt, the product of the two ionic concentrations $c_1 . c_2$ is constant, *i.e.* independent of the relative quantities of the two ions. On the other hand, $c_1 . c_2$ is not proportional to c for strongly dissociated electrolytes, but increases more slowly with increasing concentration than it should, that is, the active mass of the ions is not exactly proportional to the concentration, and the deviations from direct proportionality for all univalent ions appear to be about the same (see also p. 499). It would be of very great importance to make a quantitative investigation of the increase of the active mass with the concentration in order to determine precisely the chemical equilibrium of salt solutions; the usual assumption, adopted for simplicity in the foregoing deductions, that the active mass of the ions, as that of the other molecular species, is directly proportional to their concentration, is only approximately true, although in most cases it is sufficient for practical purposes.

In making exact measurements it is found that salt solutions show small but definite departure from the law of mass action, whilst non-electrolytes in solution obey that law even to a moderately high concentration. It appears not only that, except for extremely dilute solutions, the osmotic pressure of the ions is a more complicated function of the concentration, but that also disturbing action between ions and electrically neutral molecules must exist. In consequence of this, measurements of osmotic pressure (freezing point, etc.), give inaccurate values of the dissociation of strong electrolytes; for an attempt made by the author in connection with calculations of Jahn's to apply the additional terms to the simpler formulæ for osmotic pressure, law of mass action, etc., and obtain a closer agreement with experiment, see *Zeits. phys. Chem.*, **36**. 487 (1901).

It may be taken as certain that the freezing point gives a more correct determination of electrolytic dissociation than conductivity; nothing is known for certain as to the irregularities occurring in electrolytic conductivity which make the equation

$$\alpha = \frac{\Lambda}{\Lambda\infty}$$

inaccurate.

[1] *Zeits. phys. Chem.*, **16**. 125 (1895).

In any case the deviations from the law of mass action are relatively small, and may be neglected for most practical purposes, as may be seen from the numerous examples given in this chapter.

The Reaction between any number of Solid Salts and their Solution.—This general case may also be solved, simply by the theorem, *that the active mass of the solid substances is a constant.*

An example which belongs here, has been already studied by Guldberg and Waage (1867), namely, the state of equilibrium existing between the difficultly soluble $BaSO_4$ and $BaCO_3$, and K_2SO_4 and K_2CO_3 in solution. Here, in the sense of the similar conception considered previously, we have to deal with the reaction

$$\underset{\text{Solid.}}{BaSO_4} + \underset{\text{Dissol.}}{K_2CO_3} \rightleftarrows \underset{\text{Solid.}}{BaCO_3} + \underset{\text{Dissol.}}{K_2SO_4},$$

and therefore the relation obtains that the ratio of the carbonate existing in the solution to the sulphate in solution must be a constant. Guldberg and Waage allowed a equivalents of K_2CO_3, and b equivalents of K_2SO_4, to react on each other in a water solution in the presence of an excess of solid $BaSO_4$; they determined the quantity x of the $BaCO_3$ formed after a sufficiently long interval, and, as a matter of fact, they found the preceding ratio, namely $\frac{a-x}{b+x}$, to be constant, as shown by the following table:—

	a	x	$\frac{a-x}{b+x}$
0	3·5	0·719	3·9
0	2	0·176	4·7
0·25	2	0·200	4·0
0·50	2	0·000	4·0

Now the application of the preceding conception usually leads to results which coincide but poorly with experiment, because it must also take the electrolytic dissociation into account. Moreover, a reaction occurs in the sense of the equation,

$$\underset{\text{Solid.}}{BaSO_4} + \underset{\text{Dissol.}}{\overset{=}{CO_3}} \rightleftarrows \underset{\text{Solid.}}{BaCO_3} + \underset{\text{Dissol.}}{\overset{=}{SO_4}},$$

and the application of the law of mass-action leads to the result that in the condition of equilibrium, the ratio of the SO_4 ions to the CO_3 ions in the solution must be constant. Moreover by a displacement of the point of equilibrium in the solution, there also occur, in the sense of the preceding reaction, the following side reactions, viz.—

$$K_2SO_4 \rightleftarrows 2\overset{+}{K} + \overset{=}{SO_4}, \text{ and } K_2CO_3 \rightleftarrows 2\overset{+}{K} + \overset{=}{CO_3},$$

by means of which the calculation of the equation of the isotherm of dissociation may be accomplished.

But now it should be observed that as these two electrolytes have an analogous constitution, and as they are dissociated to the same extent in a common solution (p. 506), therefore it follows that *the total quantity* of the sulphate existing in the solution must stand in a constant ratio to the carbonate.

Thus the older conception here leads to results which are only partially satisfactory; in other cases—as, *e.g.*, in the explanation of the measurements given in the preceding paragraph—they fail utterly; and the contradictions can only be explained by the aid of the theory of electrolytic dissociation.

But the newer views lead us a step farther. It follows from the laws of solubility (p. 526) that, in the presence of solid $BaCO_3$ and $BaSO_4$ the products of the Ba ions and the CO_3 ions, and of the Ba ions and the SO_4 ions, must be constant. Now the ratio of these products, in the measurements given above, averages about 4·0.

Moreover, since $BaCO_3$ and $BaSO_4$ are doubtless dissociated to the same extent, therefore the value 4·0 is at the same time the ratio of the undissociated quantities of the carbonate and sulphate in their saturated solution.

Finally, since each of the particular solutions of these two substances, considered by itself, is very highly dissociated on account of its difficult solubility, therefore, according to the rules given on p. 498, the value 4·0 is, at the same time, the ratio of the squares of the total concentrations of the two saturated solutions.[1]

If one pours a solution of KBr over solid AgCl, then, as can be proved in a perfectly analogous way, the bromine existing in the solution will be largely replaced by chlorine; because as AgBr *is much less soluble than* AgCl, an equivalent quantity of AgCl will be changed into AgBr. This is also established by experiment. If one knows the solubilities of AgCl and AgBr, then, for a given concentration of KBr, we may state the point of equilibrium which the system strives to reach.

The analogous equilibrium between silver oxide and silver chloride in solutions of hydroxides and chlorides has been completely investigated by A. A. Noyes and Kohr.[2]

When a dissolved salt is so strongly *hydrolysed* that the limit of

[1] In testing this deduction from the theory, it is to be noted that a saturated solution of barium carbonate in pure water is noticeably hydrolysed.

[2] *Zeits. phys. Chem.*, **42.** 336 (1902).

solubility of one of the components (the base or the acid) is exceeded, the solution becomes turbid; thus iron salts give off ferric hydroxide, silicates give silicic acid. To clear such solutions the hydrolysis must be reduced by an excess of acid or, in the second instance, of base.

If sodium acetate is added to a clear, that is a strongly acid solution of ferric chloride, the concentration of the hydrogen ions is much reduced, the hydrolysis is consequently increased, and colloidal ferric hydroxide is precipitated. In the same way silicic acid may be precipitated from alkaline soluble glass by addition of ammonium chloride which reduces the number of free hydroxyl ions with formation of ammonia.

In conclusion, we will collate the following remarks regarding the theoretical treatment of the equilibrium between a salt solution and any number of solid salts.

For every molecular species which can be formed from the ions, there exists a particular *dissociation constant*, which is given by the ratio between the concentration of the particular molecular species and the product of the respective concentrations of the ions of which it is composed.

Further, every such molecular species has a *definite solubility; i.e.* there is a definite value of the concentration beyond which it cannot pass (excluding super-saturation); and this solubility remains unchanged, other things being equal, as long as solid salt remains in contact with the solution.

If one knows the value of the dissociation constants, and also the solubilities of all of the molecular species, then *the equilibrium* in the solution is completely determined, and if the total quantity of each radical is known, one can state *how much of each radical is present as a free ion, and how much of each is combined with other ions, partly in the form of electrically neutral molecules, and partly in the form of solid salt external to the solution.*

The solubilities, as defined above, have a great practical interest and value.

Moreover, the dissociation coefficients determine the number of electrically neutral molecules in the solution, while the solubilities determine the number of those which crystallise out; and although we may succeed in formulating a number of general empirical rules for the values of the dissociated constants, yet this attempt fails entirely in the case of the solubilities. Thus all binary salts composed of univalent ions, exhibit the same degree of dissociation (p. 499), but they do not exhibit the same degree of solubility; moreover, the latter property varies with the different polymorphic [allotropic] modifications of the same salt.

Sometimes, as has been already mentioned (pp. 120 and 491) the salts do not crystallise out from the solution in the pure form, but as isomorphous mixtures. Here the rule holds good that the solubility of each molecular species in a mixture is always smaller than for the particular species when alone. Again, the behaviour of *dilute solid solutions*, when they crystallise out, is very simple. In such cases the principles already developed regarding the equilibrium between phases of variable composition, may be directly applied without further remark. (See the next section.)

Normal and Abnormal Reactions.[1]—The preceding developments now show us, at the same time, the reason for a fact which has long been known, viz., that the [main] reactions of inorganic chemistry, *i.e.* of salt solutions, are characterised by great similarity in classification. Thus, as is well known, we have so-called *typical reactions*, for most radicals [collectively]: thus all acids colour litmus red; all bases colour litmus blue; all chlorides are precipitated by silver salts.[2] These facts are necessary results of the dissociation hypothesis of electrolytes. All acids contain the same hydrogen ion; all bases the same hydroxyl ion, all chlorides the same chlorine ion, etc.; and the reactions which are typical for the classes of substances collectively are the specific reactions of the ion common to them. Thus the behaviour of *electrolytes* regarding their reaction capacity, as also regards their other properties, is *clearly additive.*

Of course it is not necessary for all electrolytes containing a particular radical to exhibit the reactions which are typical for this radical; they *must* do so only when the radical is contained as *a free ion.* Thus sodium acetate [$CH_3 . CO_2Na$] does not exhibit the reactions of hydrogen ions, because the hydrogen is not contained as free ions, but exists in the solution, combined with the negative complex of the salt.

Potassium platinic chloride [K_2PtCl_6] and sodium mono-chloracetate [$CH_2Cl . CO_2 . Na$] do not show the reactions which are typical for chlorine, because the chlorine does not exist there as the free ion, but rather associated with the respective complexes $PtCl_6$ and $CH_2Cl . CO_2$. In this way the contrast between the so-called *normal* and the *abnormal reactions* of certain radicals is explained. The abnormal reactions are those of the new-formed ion complex.

It has already been remarked that potassium bromide can take up bromide forming the salt KBr_3; the solubility of iodine in a solution of potassium iodide is explained in the same way by formation of potassium triiodide, that is, the complex ion $\bar{J}_3$ is formed according to the equation

$$\bar{J} + J_3 = \bar{J}_3;$$

ammonia forms a new complex with silver ions.[3]

[1] Ostwald, *Zeits. phys. Chem.*, **3**, 596 (1889).

[2] *Ibid.*, see also the work quoted on p. 523 by the same author, written in a more popular form, and more suitable for beginners.

[3] Bodländer and Fittig, *Zeits. phys. Chem.*, **39**. 609 (1902).

$$\overset{+}{Ag} + 2NH_3 = (NH_3)_2\overset{+}{Ag};$$

potassium cyanide can take up silver ions to a large extent; a complex ion is formed according to the equation

$$\overset{+}{Ag} + 2\overset{-}{Cy} = Ag\overset{-}{Cy_2}.$$

Naturally these complex ions are more or less dissociated into other components and all possible steps are found. The ion J_3 also is strongly dissociated, the solution of potassium tri-iodide therefore acts like a solution of free iodine. The ion $(NH_3)_2Ag$ is not so strongly dissociated; for a solution of silver nitrate which contains ammonia gives no precipitate of silver chloride on addition of a chloride, that is, the concentration product of the silver ions × chlor-ions remains below the solubility product (10^{-10}). But a precipitate is obtainable on addition of an iodide, that is, the above ion dissociates sufficiently to make the concentration silver ions × iodine ions exceed the solubility product for silver iodide about 10^{-16}. In the complex ion $Ag(Cy)_2$ the silver is retained to an extraordinary extent since it is not precipitated even by addition of iodide; sulphuretted hydrogen however precipitates it on account of the extreme indissolubility of silver sulphide.

In this way we obtain at the same time a strict classification of the *double salts*. The *characteristic double salts* are the isolated points in the series of mixtures, afforded by their components (p. 491). On solution they decompose almost completely into the single salts, and their ions therefore are simply those of their components. The substances sometimes wrongly called "double salts," such as K_2PtCl_6, K_4FeCn_6 and the like, are to be regarded as entirely different; those in solution conduct themselves as *simple electrolytes* (*einheitlicher Electrolyte*), since they afford *only one electrically neutral molecule, and exhibit only one series of ions.* The substances last referred to are simple salts of hydro-chloro-platinic acid [H_2PtCl_6], and of hydro-ferro-cyanic acid [H_2FeCn_6], etc., and therefore they contain no platinum ions, and no iron ions, etc., respectively.

According to present experience the ions, since they have the character of saturated compounds (p. 391) possess in a high degree the capacity for forming molecular compounds.

The systematic study of complex ions therefore tends to throw new light on the nature of compounds which do not fall under the scheme of valency; some rules on this subject have already been given on p. 375.

It must by no means therefore be assumed that all chemical reactions are due to ions; on the contrary every molecular species, ion, or electrically neutral molecule, has its peculiar and typical reaction.

Angeli and Boeris[1] have given a striking example of this. It is well known that an aqueous solution of ammonium nitrite decomposes on heating into water and nitrogen and the more readily the more concentrated it is; in very dilute solutions where only the ions $\overset{+}{N}H_4$ and $\overset{-}{NO_2}$ are present the reaction does not occur; it must therefore be concluded that it is the undissociated molecule NH_4NO_2 that is capable of the reaction

$$NH_4NO_2 = N_2 + 2H_2O;$$

the chemists mentioned have, in fact, shown that on addition of a salt with a similar ion, for example ammonium chloride or sodium nitrite, the production of nitrogen is increased in consequence of the reduced electrolytic dissociation of the ammonium nitrite, whereas the salts which have no ion in common with ammonium nitrite are inactive. In the same way it seems as if the oxidising action of nitric acid was exclusively or mainly due to the molecular species HNO_3 and depends little or not at all on the ions $\overset{+}{H} + \overset{-}{NO_3}$.

Formation and Solution of Precipitates.—Reactions causing precipitation are of exceptional importance in analytical chemistry; the theory of the formation and solution of precipitates is indeed contained in the foregoing sections, but certain points of it may be put together here and illustrated by examples.

A precipitate occurs (unless it is prevented by super-saturation) when an electrically neutral species of molecule exceeds the amount of its solubility product (p. 531); it goes into solution again when the product of the ionic concentrations is reduced below the solubility level.

The latter can only occur when other molecular species, either neutral or ionised, are added, causing an increase in one or more of the ions in question. The following cases arise:—

1. The precipitate to be dissolved is an *acid*; then one of the ions, namely the hydrogen ion, can be removed very completely by addition of a base whose hydroxyl ions conbine with it to form water. If the substance to be dissolved is a base it may be similarly dissolved in acid. (For example benzoic acid dissolves easily in caustic soda, lime in hydrochloric acid, etc.)

If in these cases the acid or the base or both are very weak the solvent action is reduced by hydrolysis; in this way, as was shown by Löwenherz[2] the dissociation of extremely weak acids and bases can be measured.

It may be foreseen that in certain cases the water itself, in consequence of its ionisation, will alter the solubility, namely when the substance to be dissolved is hydrolytically dissociated. Thus, if the

[1] *Acad. Linc.* [1892] [5], **1.** II. 70.
[2] *Zeits. phys. Chem.*, **25.** 385 (1899).

insoluble barium carbonate is brought into contact with water, the hydrogen ions of the water combine with the CO_3 ions to make the very slightly dissociated compound $\overline{HCO_3}$, and the concentration of the hydroxyl ions is accordingly raised in the same way as if hydrogen ions had been added.

2. Addition of hydrogen or hydroxyl ions may, however, occur in the two preceding cases by means of the salts of other weak acids (*e.g.* acetic acid) or a very weak base (*e.g.* ammonia). Examples: the equation

$$[Ca]\,[OH]^2 = \text{const.}$$

holds for the solubility of calcium hydroxide, but in presence of ammonia ions a large addition of hydroxyl ions occurs, since the very slightly dissociated ammonium hydroxide is formed.[1] In the same way the relatively large solubility of magnesium hydroxide in ammonium salts is explained, as well as the fact that magnesium salts are either not all or imcompletely precipitated by ammonia.[2]

3. If the precipitate to be dissolved is the salt of a weak acid its anions can be largely combined with hydrogen ions; example: silver acetate dissolves in acids. In the same way salts of weak bases are dissolved by strong bases.

4. Very often the precipitate dissolves in consequence of the formation of complex ions. Example: silver chloride dissolves in potassium cyanide (p. 537), etc.

It is obvious that the same reagents that dissolve a precipitate prevent its precipitation when they are added before that takes place.

Distribution of an Electrolyte between Water and a Second Phase.—This is a case of the general law of distribution (p. 481) but it must be noted that the ions cannot be separated in appreciable quantity by partition, any more than in diffusion (p. 366).

The simplest method of treatment is to regard the concentration of the electrically neutral molecules as proportional in the two phases. Since also the concentration of the free ions in the gaseous space or in a solvent that will not mix with water is negligible, it follows that the ions must pass over practically completely into the water.

This may be illustrated by some examples (see also the table on p. 487). The partial pressure of hydrochloric acid over its water solution is simply proportional to the number of its undissociated molecules contained in the solution.

If an electrolyte divides itself between water and ether, there must be a proportionality between the electrically neutral molecules in the water and the concentration in the ether. Now the number

[1] Investigated by Noyes and Chapin, *Zeits. phys. Chem.*, **28.** 518 (1899).

[2] See also the thorough investigation by J. M. Lovén, *Zeits. anorg. Chem.*, **11.** 404 (1896).

of electrically neutral molecules in water decreases with increasing dilution much more quickly than proportionally to the concentration; and therefore, at slight concentration, both the vapour pressures of electrolytes, and also their solubilities in another solvent, which is in contact with the water solution, will be very small.

Thus, *e.g.*, it is possible to distil pure water off from a solution of hydrochloric acid.

If one tries to "shake out" with benzene a sufficiently dilute water solution of an organic acid, then only the slightest traces of the acid will go over to the benzene, even though it is much more soluble in benzene that in water.

Hydrocyanic acid which is very slightly dissociated is much more volatile; even dilute solutions show its characteristic smell, so do solutions of the salts of this acid, since according to p. 520, they are markedly hydrolysed, *i.e.* contain free hydrocyanic acid.

The following table given by Kuriloff (p. 494) shows the partition of picric acid between benzene and water; c_1 and c_2 are the concentrations (normalities) in the two solvents, α the degree of electrolytic dissociation.

c_1	c_2	$\frac{c_1}{c_2}$	α	$\frac{c_1}{c_2(1-\alpha)}$
0·09401	0·02609	3·6	0·9027	38
0·0779	0·02080	3·7	0·9104	41
0·06339	0·01963	3·2	0·9138	37
0·06184	0·01882	3·3	0·9164	39
0·0359	0·01320	2·7	0·9353	42
0·01977	0·00973	2·0	0·9463	38
			Mean .	. 39

Thus, although the undissociated molecules of picric acid are thiry-nine times as soluble in benzene as in water, they pass over almost completely into water at great dilutions.

The case of partition of ions between a liquid and a *metallic* phase occurs, *e.g.*, on shaking mercury with dilute silver nitrate solution; mercury ions pass to a certain extent into solution with precipitation of silver. The equilibrium can be treated by means of the law of mass action (see the interesting study by Ogg, *Dissertation*, Göttingen, 1898; *Zeits. phys. Chem.*, **27**. 285 (1898)).[1]

[1] A further series of investigations, in which the methods of this chapter have been applied to the investigation of special problems in inorganic chemistry, will be found in the recent volumes of the *Zeits. f. phys. Chem.*, and the *Zeits. f. anorg. Chem.*

CHAPTER V

CHEMICAL KINETICS

General Observations.—As has been previously emphasised, the hypothesis of Guldberg and Waage is the fundamental law of chemical kinetics. According to this law, the total progress of a reaction occurring *in a homogeneous system*, is determined by the difference between the two velocities with which the decomposition advances from left to right, and conversely from right to left, in the sense of the reaction equation.

Therefore at every instant the velocity of a reaction, *i.e.*, the quantity changed in a moment of time, in the sense of the reaction from left to right, divided by the moment of time,—this velocity is given by the velocity constant of the change in the sense of the equation from left to right, multiplied by the active masses of the molecular species standing on the left side of the equation, diminished by the velocity constant of the change from right to left in the sense of the equation, multiplied by the active masses of the molecular species standing on the right side of the equation.

Thus, *e.g.*, if a homogeneous reaction takes place according to the simple scheme—

$$A_1 + A_2 = A_1' + A_2',$$

and if c_1, c_2, and c_1', c_2' are respectively the concentrations of the four reacting molecular species A_1, A_2, and A_1', A_2'; and also if dc_1 denotes the diminution which c_1 will experience in the moment of time dt, where of course the similar diminution of c_2 is of the same amount,—then the reaction velocity at every moment will be

$$-\frac{dc_1}{dt} = kc_1c_2 - k'c_1'c_2',$$

where k and k′ respectively are the velocity coefficients of the two opposed reactions. If a substance participates in the reaction with n molecules instead of with one, then of course c^n will appear in the equation instead of c.

Now the velocity coefficients are constant at constant temperature; but without exception they increase very strongly with rising temperature. Therefore the application of the fundamental equation preceding is permissible only with the proviso, *that the reaction can progress isothermally, i.e., only when there is no change of the temperature of the system, as occasioned by heat developed or absorbed, in the course of the reaction.*

Now for the time when $t = 0$, let the respective concentrations of the four substances be a_1, a_2, a_1', and a_2'; and in the time t let the quantity of the substance a_1 decomposed be x g.-mol., and accordingly the same amount of a_2. Thus the preceding equation will be written—

$$\frac{dx}{dt} = k(a_1 - x)(a_2 - x) - k'(a_1' + x)(a_2' + x);$$

then by knowing k and k′, and remembering the preliminary condition that, when $t = 0$, then also $x = 0$, by integration one succeeds in obtaining a complete description of the course of the reaction; and the result is similar when the course of the reaction is given for an equation involving any arbitrary number of reacting molecular species.

The ascertaining of the concentrations in the state of equilibrium

$$\left(\frac{dx}{dt} = 0\right),$$

as was thoroughly demonstrated in the second chapter of this book, gives *the ratio of the two velocity constants.*

Now a very important simplification is offered in the case where the reaction advances *almost completely* in one direction in the sense of the equation, *e.g.*, from left to right; this will be applied to the course of most of the reactions thus far investigated. This means that one of the two partial reaction-velocities is very great as compared with the other; or that k is very large as compared with k′. In this case, the right side of the differential equation reduces to the positive term, and the reaction-velocity at every moment is obtained as proportional simply to the product of the active masses of the molecular species standing on the left side of the reaction-equation.

In all cases, the integration of the differential equation of the chemical change gives the result that, strictly speaking, the equilibrium would be reached only after an infinitely long time, *i.e.*—

$$\frac{dx}{dt} = 0,$$

only when $t = \infty$.

According to this, a chemical system, like a pendulum which is highly damped, tends towards an "*aperiodic*" condition of equilibrium. An "over-shooting of the mark" is incompatible with all our views of chemical processes. This would mean that, under certain circumstances the sense of the reaction would depend upon *the previous history* of the system; *i.e.* that in two solutions which are *absolutely identical*, the same reaction might progress in opposite directions, the reaction in one solution approaching one state of equilibrium, while the reaction in the other solution would shoot over towards the other state. As a matter of fact, such has certainly never been observed.

The Inversion of Sugar.—Cane sugar in water solution in the presence of acids is practically absolutely decomposed into dextrose and levulose. The process advances so slowly that it can be followed by easy measurement, as the progress of the reaction can be traced very easily and very sharply by polari-strobometric analysis. The part not inverted turns the plane of polarised light to the right, while the mixture of the products of inversion is laevo-rotatory.

Let α_0 denote the (positive) rotatory angle when the time $t = 0$ which corresponds to an original quantity of sugar a; let α_0' denote the (negative) rotatory angle after complete inversion; and let α be that actually observed after the time t. Then, since all substances rotate the plane in proportion to their concentration, we have

$$x = a\frac{\alpha_0 - \alpha}{\alpha_0 + \alpha_0'}.$$

When the time $t = 0$, then $\alpha = \alpha_0$, *i.e.* $x = 0$. When the time $t = \infty$, after complete inversion, then $\alpha = -\alpha_0'$, *i.e.* $x = a$.

The inversion of sugar in its progress has been investigated by a great number of observers, including Wilhelmy (1850), Löwenthal and Lenssen (1862), Fleury (1876), Ostwald (1884), Urech (1884), Spohr (1885, 1886, and 1888), Arrhenius (1889), Trevor (1892), and others. Therefore, as it plays a very important *rôle* in the history of the doctrine of affinity, it demands a thorough description.

Corresponding to its progress according to the equation

$$C_{12}H_{22}O_{11} + H_2O = 2C_6H_{12}O_6,$$

the law of mass-action shows that *the velocity of the inversion at every moment is proportional to the product of the concentrations of the water and of the cane sugar*; or since the water is usually present in large excess, and suffers only a very slight change in concentration during the course of the reaction, *therefore the velocity must be proportional simply to the concentration of the sugar itself*. That is

$$\frac{dx}{dt} = k(a - x);$$

this is the condition at the start; when $t = 0$, then also $x = 0$. Let k denote *the coefficient of inversion*: then by the integration of the last, we obtain

$$-\ln(a - x) = kt + \text{const.};$$

and as the original condition

$$-\ln a = \text{const.};$$

from this we obtain

$$k = \frac{1}{t}\ln\frac{a}{a - x} = \frac{1}{t}\ln\frac{\alpha_0 + \alpha_0'}{\alpha + \alpha_0'}.$$

This equation was discovered, and proved experimentally, by Wilhelmy, before the law of mass-action was advanced. In fact it is an assumption which follows almost directly from the preceding equation, to suppose that *a constant fraction of the sugar is inverted at every moment of time.* The simple meaning of the inversion coefficient is, that its reciprocal value multiplied by ln2 gives the time required for the inversion of half of the total quantity, as is seen at once if we write

$$x = \frac{a}{2}.$$

How well the results of the preceding equation coincide with those of experiment, is shown by the following table for the inversion of a 20 per cent solution of cane sugar in the presence of a 0·5 normal solution of lactic acid, and at a temperature of 25°:—

t (in minutes)		$\log_{10}\frac{a}{a-x}$
0	34·50°	—
1435	31·10	0·2348
4315	25·00	0·2359
7070	20·16	0·2343
11360	13·98	0·2310
14170	10·61	0·2301
16935	7·57	0·2316
19815	5·08	0·2291
29925	−1·65	0·2330
∞	−10·77	—
		Mean 0·2328

As we are concerned only with the proof of the *constancy* of the expression given in the third column, we can of course introduce the Brigg's logarithms instead of the natural logarithms.

The Catalytic Action of Hydrogen Ions.—The inversion of sugar proceeds with marked velocity only in the presence of an acid the quantity of which remains unchanged during the reaction; such reactions as this have been already called "*catalytic.*"

The ultimate explanation of the preceding case is still unknown to us; but nevertheless some very remarkable results have been obtained regarding the regularity of behaviour which obtains here. Arrhenius[1] succeeded in grouping in one simple orderly arrangement all the abundant material observed concerning this reaction. As the method which is to be pursued here is a typical one, and as it has already led to important results in similar cases, and still promises more, therefore we will give the results observed regarding the dependence of the inversion coefficients upon the nature of the acids and salts which may be present, and we will also give a brief theoretical summary of the same.

The following phenomena were discovered, in a purely empirical way, by the study of the question of the variation of the velocity of inversion, with the *concentration*, with the *nature* of the acid, and in the presence of neutral salts.

The more concentrated the acid, the more quickly is the sugar inverted, although this does not take place in an exact proportion.

In the case of *the stronger acids*, the inverting action occurs somewhat *more quickly* than in proportion to the amount contained, and the *converse* of this is true in the case of *the weaker acids.*

The velocity of the inversion varies very greatly with the *nature* of the acid. Thus the strong *mineral* [*inorganic*] *acids* effect the inversion *most quickly*, and with about the same degree of rapidity; while the *fatty acids*, *e.g.*, exert an inverting action which is much weaker.

In the following table are given some of the results obtained by Ostwald at 25°, and with an acid concentration of 0·5 normal. The figures refer to 1·000 for hydrochloric acid, and are well suited to give a good impression of the variability of the inversion coefficients:—

Hydrochloric acid . . .	1·000	Tri-chlor-acetic acid . .	0·754
Nitric " . . .	1·000	Di-chlor-acetic " . .	0·271
Chloric " . . .	1·035	Mono-chlor-acetic " . .	0·0484
Sulphuric " . . .	0·536	Formic " . .	0·0153
Phenyl-sulphonic acid . .	1·044	Acetic " . .	0·0040

The influence of neutral salts is very remarkable. In the presence of an equivalent quantity of the potassium salt of the respective acid, the inversion velocity is increased about 10 per cent in the case of the stronger acids; but in the case of acids weaker than tri-chlor-acetic acid, the action is diminished, and that the more as the acid is weaker.

In the case of acetic acid this depressing action is perfectly enormous: thus as a result of the presence of an equivalent quantity

[1] *Zeits. phys. Chem.*, **4.** 226 (1889).

of a neutral salt, the inversion velocity sank to $\frac{1}{40}$ of its original value.[1]

The addition of non-electrolytes, in quantities which are not too great, exerts no marked effect.

In order to obtain a general view of the relations aforementioned, which at first sight do not seem to be simple, we will notice at once that it is *all the acids, but only the acids*, which exhibit the characteristic capacity of inverting sugar—*i.e. here we are concerned with a specific action of the free hydrogen ions;* for it is in water solutions of acids, and in these alone, that free hydrogen ions are contained. Now if these hydrogen ions are the actual catalytic agents, then, according to the law of mass-action, it would be expected that the catalytic action of the acids would be respectively proportional to the number of their hydrogen ions, *i.e.* an acid should invert the more strongly accordingly as it is more highly dissociated electrolytically. This anticipation is found to be completely verified in the results of the preceding table, *where the acids are arranged in the same relative order of succession as that required by their respective degrees of electrolytic dissociation.*

But we find only an approximate numerical proportionality between the quantity of the hydrogen ions and the inversion velocity; the reason for this is that the inversion velocity increases more quickly than in proportion to the acid concentration; while the reverse is the case with the hydrogen ions, according to the laws of dissociation. Thus a 0·5 normal solution of HCl inverts 6·07 times more quickly than a 0·1 normal, although the former contains only about 4·64 times as many hydrogen ions as the latter.

Moreover, there is a second action which is very influential here: this was formulated by Arrhenius as follows:—

The catalytic activity of hydrogen ions is greatly stimulated by the presence of other ions.

In this way, which, to be sure, is rather puzzling from a theoretical standpoint, we explain the fact that the inversion velocity of the stronger acids increases more quickly than in proportion to their concentration, because the quantity of the free negative ions, which increases with increasing concentration, stimulates the activity of the hydrogen ions: and thus, in addition, the observed increase of the inverting action of a stronger acid is explained by the presence of its neutral salt.

Nevertheless, although this action of the dissociated part of the neutral salt is very interesting, yet it has rather the character of a side reaction of secondary nature. The action of the hydrogen ions is much more distinct, and thus their inverting action on a solution constitutes a very delicate reaction for the presence of hydrogen ions.[2]

[1] Spohr, *J. pr. Chem.* [2], **32.** 32 (1885).

[2] As shown recently by Trevor (*Zeits. phys. Chem.*, **10.** 321, 1892), this reaction can be employed at quite high temperatures (100°), where the inversion velocity is much greater, in order to detect very small quantities of hydrogen ions.

Now we find in weak acids a simple means for repressing the dissociation to any desired extent, and thus at the same time for diminishing the quantity of free hydrogen ions to any desired degree. According to the laws of dissociation, which were thoroughly discussed on p. 503, the dissociation retreats on the addition of one of the dissociation products, in a way which can be easily calculated. In fact (see above) the researches have always shown a truly enormous diminution of the inversion velocity of the weaker acids in the presence of their neutral sodium salts; and Arrhenius, by a consideration of their side reactions, which are not altogether simple, succeeded in showing that the *quantitative relations* do in fact occur, as required by theory.

At all events, the dissociation does retrograde in the case of the stronger acids, as HCl, *e.g.*, on the addition of another chloride; and although the amount of the diminution of the dissociation may be very slight, yet it does result. The contrary fact, that there is an [apparent] increase of the dissociation, which is by no means inconsiderable, is explained by the fact that the retrograde action is more than counterbalanced by the influence of the neutral salt, as mentioned above.

A very exact investigation by W. Palmaer[1] has yielded the important result that in very dilute solutions, in which there is no longer the action of neutral salts referred to above, the rate of inversion is directly proportional to the concentration of the hydrogen ions. If concentrated solutions of cane sugar are used, the velocity coefficient k increases considerably with the concentration of the sugar, though it should remain constant according to the theory. As E. Cohen has shown, this phenomenon indicates that the volume of the reaction is reduced, hence the number of collisions between the molecules of cane sugar and the hydrogen ions is increased, causing an increase in the velocity of reaction.[2]

Again the decrease in the *rotation of the various kinds of sugar with time* proceeds according to the formula for unimolecular reactions; addition of salts usually accelerates this, hydrogen ions act more strongly, but hydroxyl ions act markedly too.[3]

The Catalysis of Esters.—A phenomenon, which in many respects is very closely related to sugar inversion, is the catalysis of the esters, *i.e.* the accelerating influence of the presence of acids in cleaving an ester in dilute water solution, into the corresponding alcohol and the corresponding acid.

According to the discussion on p. 447 *et seq.*, as a result of the

[1] *Zeits. phys. Chem.*, **22.** 492 (1894). [2] *Ibid.*, **23.** 442 (1897).

[3] See the works of P. Th. Muller (1894), Levy (1895), Trey (1895), also the monograph mentioned on p. 331, by Landolt (p. 238 *et seq.*), and especially Osaka, *Zeits. phys. Chem.*, **35.** 661 (1900), concerning the rotation of milk sugar.

mass-action, the cleaving [saponification] will be complete in the presence of a large excess of water; and therefore we obtain, as the coefficient of the velocity with which the ester and water unite to form alcohol and acid, the same equation as in the case of sugar inversion, viz.

$$k = \frac{1}{t} \log \frac{a}{a - x},$$

provided that in neither case the concentration of the water suffers a marked change; and also provided that by a we understand the quantity of the substance present when the time $t = 0$; and by x, the quantity of substance changed in the time t.

A simple titration shows the progress of the reaction. The velocity of the decomposition at ordinary temperatures is extremely small; but it is very greatly accelerated by the presence of an acid, although this acid may not participate noticeably in the reaction. As was similarly true of sugar inversion, we can arrange the very abundant and also apparently very complicated material here observed, for which we are indebted to Ostwald,[1] in a clear way, under the following simple principles.

1. The velocity with which the ester will be split is at every moment proportional to its concentration, *i.e.* the velocity coefficient remains constant, in the sense of the Guldberg-Waage theory.

2. The catalytic action of an acid increases with the degree of its dissociation, and the velocity coefficient is approximately proportional to the number of hydrogen ions.

3. In a secondary way the catalytic activity of the hydrogen ions is considerably increased, as a result of the presence of neutral salts.

The measurement of the velocity of the saponification of ether also furnishes a method for determining the number of hydrogen ions in a solution. This method was applied by Walker (p. 523) in a very ingenious way to investigate the "*hydrolytic dissociation*" of salts: their respective magnitudes could be ascertained by measuring the velocity with which the methyl acetate added to the solution was catalysed; and this formed a measure of the quantity of the free acid separated from the salt, and thus at the same time the strength of the base of the respective salt (a chloride) was estimated at least approximately.

According to the extensive investigations of B. Lowenherz,[2] the rate of saponification of various esters by the hydrogen ions is nearly independent of the nature of the alcohol contained in the ester, but depends largely on the nature of the acid contained. The behaviour of salicine towards acids is similar to that of cane sugar, it decomposes into dextrose and saligenin with separation of a molecule of water.[3]

[1] *J. pr. Chem.* [2], **28.** 449 (1883); see also Trey, *ibid.* [2], **34.** 353 (1886).

[2] *Zeits. phys. Chem.*, **15.** 389 (1894).

[3] A. A. Noyes and W. J. Hall, *Zeits. phys. Chem.*, **18.** 240 (1895).

Formation of Sulphuretted Hydrogen from its Elements.—There is only a very small number of reactions occurring in the gaseous state which are free from secondary disturbances, such as chemical action or absorption by the walls of the vessel; and their experimental treatment mostly meets with very great difficulties. Bodenstein[1] has recently done a great service in systematically studying gas reactions from the point of view of chemical kinetics, and after discovering a series of sources of error has worked out exact methods for it.

He showed amongst other things that with a sufficient excess of sulphur the formation of sulphuretted hydrogen occurs according to the formula

$$\frac{dx}{dt} = k(a - x),$$

where $a - x$ is the amount of sulphuretted hydrogen present at the time t. The sulphur which occurs as a liquid evaporates fast enough for its active mass (like that of water in the preceding example) to remain practically constant. The following numbers obtained at 310° indicate the satisfactory constancy of $k(a = 1000)$.

t	x	k
720	0·1680	0·000117
1440	0·3049	0·000116
2160	0·4145	0·000114
2880	0·5258	0·000123
4320	0·6610	0·000118
5760	0·7572	0·000118
7200	0·8289	0·000122
8640	0·8494	0·000121
10080	0·9012	0·000123
	Mean .	. 0·000118

It is noteworthy that *catalytic* influences very often occur in gas reactions; thus the formation of seleniuretted hydrogen which is also unimolecular is mainly influenced by the catalytic action of solid selenium.

Unimolecular Reactions.—The same formula for the velocity coefficient, and the same course of the reaction which we met in the case of the inversion of sugar, we find in all cases *where the system suffers an essential change of concentration, as a result of the change of only one molecular species.*

[1] Gas reactions in chemical kinetics, *Zeits. phys. Chem.*, **29.** 147, 295, 315, 429. 665; **30.** 113 (1899).

Thus, according to researches conducted by Harcourt and Esson as early as 1865, potassium permanganate, by oxidising a large excess of oxalic acid which is added, disappears according to the logarithmic formula.

According to the researches of van't Hoff,[1] the same is true of the splitting of di-brom-succinic acid into brom-maleic acid and hydrobromic acid; and also for the decomposition of mono-chlor-acetic acid into glycolic acid and hydrochloric acid, etc.

Following the usage of van't Hoff, we will call reactions of this sort *unimolecular*: their course always corresponds to the differential equation,

$$\frac{dx}{dt} = k(a - x).$$

In an analogous way, we will call those reactions, in the course of which there is a change in the concentration of n molecular species, *n-molecular reactions.*

Bi-molecular Reactions—The Saponification of Esters.—The classical example of this case, where the concentrations of two molecular species are considerably changed in the course of the reaction, is the *saponification of esters.* By bringing together a base with an ester in water solution, there are formed gradually the corresponding alcohol, and the salt from the positive ingredient of the base and the negative ingredient of the ester: the reaction takes place according to the scheme—

C_2H_5-O-O-C_2H_3	+	NaOH	=	CH_3COONa	+	C_2H_5-OH
Ethyl acetate.		Sodium hydroxide.		Sodium acetate.		Ethyl alcohol.

Let a and b represent the original concentrations of the base and of the ester respectively; and let x be the quantity of the ester changed after the time t; this can be easily and sharply determined by the titration of the quantity of the base yet remaining. Thus the reaction velocity for every moment is given by the equation,

$$\frac{dx}{dt} = k(a - x)(b - x),$$

or rearranged,

$$\frac{dx}{a - b}\left(\frac{1}{b - x} - \frac{1}{a - x}\right) = kdt.$$

The integral of this equation is

$$-\frac{1}{a - b}\Big[\ln (b - x) - \ln (a - x)\Big] = kt + \text{const.};$$

[1] *Études de dynamique chimique* (Amsterdam, 1884), pp. 13 and 113.

and, since $x = 0$ when $t = 0$, we have

$$-\frac{1}{a-b}\left(\ln b - \ln a\right) = \text{const.};$$

from which we finally obtain, by subtraction,

$$k = \frac{1}{(a-b)t} \ln \frac{(a-x)b}{(b-x)a}.$$

Saponification was first studied from the standpoint of the law of mass-action, by Warder;[1] and later and more thoroughly by van't Hoff,[2] Reicher,[3] Ostwald,[4] Arrhenius,[5] Spohr,[6] and others.

It appears that the preceding formula harmonises excellently with the values obtained for the stronger bases. Thus the following results were obtained for the action of *sodium hydroxide*, which was present in slight excess, on *ethyl acetate* at 10°; the basic [alkaline] titration of the reaction mixture, and denoted by c, corresponded as follows to the respective times t, counted in minutes.

t (time).	T (titrat).	k
0	61·95	...
4·89	50·59	2·36
10·37	42·40	2·38
28·18	29·35	2·33
∞	14·92	...

The figures under c denote the number of c.c. of a $\frac{1}{23{\cdot}26}$ normal acid solution required to neutralise 100 c.c. of the reaction mixture: in order to reduce these figures to our customary standard of concentration, viz. g.-mol. per litre, we must multiply them by $\frac{1}{23{\cdot}26}$.

Now the quantities a, b, and x in the preceding formula of course correspond thus; viz.: a to the original titration 61·95; b to the original titration minus the final one, *i.e.* $61{\cdot}95 - 14{\cdot}92 = 47{\cdot}03$; and x to $61{\cdot}95 - c$. The formula then becomes

$$k = \frac{2{\cdot}302 \,.\, 23{\cdot}26}{14{\cdot}92 \,.\, t} \,{}^{10}\log \frac{c \,.\, 47{\cdot}03}{61{\cdot}95(c - 14{\cdot}92)}.$$

The factor 2·302 reduces the natural logarithms to the Brigg's logarithms. The values for k, given in the third column of the pre-

[1] *B. B.*, **14.** 1361 (1881).
[2] *Études*, p. 107.
[3] *Lieb. Ann.*, **128.** 257 (1885).
[4] *J. pr. Chem.*, **35.** 112 (1887).
[5] *Zeits. phys. Chem.*, **1.** 110 (1887).
[6] *Ibid.*, **2.** 194 (1888).

ceding table, vary about a mean value, within the limits of error of experiment. The significance of k, considering the fact that the time has been estimated in minutes and the concentration in g.-mol. per litre, is as follows :—

It k denotes the number of g.-mol. of the ester which would be saponified in a minute, if 1 g.-mol. of ester and 1 g.-mol. of sodium hydroxide should react on each other in 1 litre, and provided that one had an apparatus which would constantly remove the resulting reaction products from the system, and as constantly renew the corresponding quantities of base and of undecomposed ester.

If one causes equivalent quantities of ester and of base to react on each other, then the reaction velocity at each moment will be simply

$$\frac{dx}{dt} = k(a - x)^2,$$

which, when integrated, becomes

$$k = \frac{x}{t(a - x)a}.$$

The question regarding *the variation* of the reaction *velocity with the nature of the ester and of the base* has been systematically investigated by Reicher ; his results were as follows :—

1. The saponification of methyl acetate at 9·4°, by the various bases.

	k		k
Sodium hydroxide . .	2·307	Strontium hydroxide . .	2·204
Potassium ,, . .	2·298	Barium ,, . .	2·144
Calcium ,, . .	2·285	Ammonium ,, . .	0·011

2. The saponification of the acetic acid esters of the various alcohols, by means of sodium hydroxide at 9·4°.

	k		k
Methyl alcohol	3·493	Iso-butyl alcohol . .	1·618
Ethyl ,, ...	2·307	Iso-amyl ,, . .	1·645
Propyl ,, ...	1·920		

3. The saponification of the ethyl esters of different acids, by means of sodium hydroxide at 14·4°.

	k		k
Acetic acid . . .	3·204	Iso-butyric acid . . .	1·731
Propionic acid . . .	2·186	Iso-valeric ,, . .	0·614
Butyric acid . . .	1·702	Benzoic ,, . .	0·830

Thus it appears that the strong bases possess a reaction velocity which is about the same; and that as regards the esters, on the whole, their reaction velocity is the smaller, the greater the number of atoms contained.

These numbers show that the nature of alcohol contained in the ester is of less influence than that of the acid; this result has come out even more strongly in recent researches.[1]

The influence of the nature of the base was afterwards very thoroughly investigated by Ostwald, in all the gradations between sodium and potassium hydroxide (which operate most quickly), and ammonia and allyl-amine (which operate very slowly), and he observed a remarkable phenomenon. In the case of the weak bases the preceding formula fails utterly; thus in *the saponification of ethyl acetate by ammonia*, for the times (t), he found the following corresponding velocity coefficients, which are not directly comparable with the preceding ones.

t	k
0	...
60	1·64
240	1·04
1470	0·484

These values of k vary too much to be constant. Ostwald found as the reason for this that *the neutral salt produced* (*ammonium acetate*) *exerted a very strong effect in hindering the progress of the reaction*, which explains the extreme retardation of the saponification.

Thus in *the saponification of ethyl acetate*, when an equivalent quantity of ammonium acetate was added at the outset in addition to the ammonia, the other conditions of the experiment being unchanged, the following values were found:—

t	k
0	...
994	0·138
6874	0·120
15404	0·119

[1] Hemptinne, *Zeits. phys. Chem.*, **13.** 561 (1894): Lowenherz. *ibid.*, **15.** 395 (1894).

As a result of the addition of the neutral salt of ammonium, the reaction velocity becomes considerably smaller, but at the same time the velocity coefficient becomes much more constant; the latter fact is explained by the circumstance that the concentration of the ammonium acetate changes, relatively, much less during the course of the reaction.

This remarkable influence, which is exerted by the presence of neutral salts, was at once investigated by Arrhenius, who, on the basis of very abundant material of observation, succeeded in establishing the following propositions, viz.:—

1. The saponification velocity of the stronger bases in fairly extended dilution is only slightly changed (less than 1 per cent) by the presence of neutral salts.

2. The saponification velocity of ammonia is exceedingly depressed by the presence of ammonium salts; and, moreover, equivalent quantities of the most different salts exert nearly the same effect.

Thus the velocity coefficient k of the action of a $\frac{1}{40}$ normal solution of ammonia on the equivalent quantity of ethyl acetate, in its dependence upon the quantity S of any selected ammonium salt of a mono-basic acid,—this velocity coefficient may be expressed by means of the following formula, which is purely empirical, and which refers to a temperature of 24·7°, viz.:—

$$k = \frac{0{\cdot}1561}{1 + 1241S - 11413S^2}.$$

The Theory of Saponification.—The relations described above, and which seemed at first to be very puzzling, are now shown to be *a necessary result of the law of mass-action, on the assumption of the theory of electrolytic dissociation.*

If we consider the process of saponification in the light of this theory, it appears to consist in the action of the hydroxyl ions upon the ester molecule, in the sense of the equation

$$C_2H_5\text{-}O_2C_2H_3 + \overset{+}{Na} + \overset{-}{OH} = CH_3\overset{-}{C}OO + \overset{+}{Na} + C_2H_5OH,$$

or more simply,

$$C_2H_5\text{-}O_2C_2H_3 + \overset{-}{OH} = CH_3\overset{-}{C}OO + C_2H_5OH.$$

The positive ingredient of the base thus plays a perfectly indifferent *rôle.* Therefore bases subjected to the same degree of dissociation must exert effects of the same degree of strength upon the ester; in fact such is the case with sodium and potassium hydroxides. And the less the base is dissociated, the weaker is the action; as, in fact, is shown by ammonia, or ammonium hydroxide rather, which is cleaved into its ions only in the slightest degree, and which accordingly saponifies with a corresponding slowness. Moreover, this is established in a very striking way by the investigations of Ostwald.

As has been already shown to be necessarily true in analogous cases in "Chemical Statics," so here also *the active mass in the mechanism of the reaction corresponds, not to the total quantity of the base introduced, but only to the quantity which is dissociated.*

Thus let us denote the degree of the dissociation by α, then the formula previously used, viz.

$$\frac{dx}{dt} = k'\alpha(a - x)(b - x),$$

must be corrected. Now the degree of dissociation of the base is given by its dissociation constant, by its concentration, and by the quantity of neutral salt produced from it. For the stronger bases (which are almost as highly dissociated as the corresponding neutral salts produced from them in the reactions), α remains constant during the progress of the reaction. For in a mixture of any two equally dissociated electrolytes having ions in common, *at the same total concentration*, the dissociation is independent of their relative masses (p. 505), and the latter always remains constant during the reaction. Thus if we make

$$k'\alpha = k,$$

then the preceding equation again assumes the original form, which harmonises with experiment.

But the behaviour is quite otherwise in the case of a base, the degree of dissociation of which is very different from that of the resulting salt, *e.g., when the dissociation of the base is much weaker;* this is the case with ammonia and ammonium acetate. Thus it will result that the degree of dissociation of the base will be caused to retreat very much during the progress of the reaction, and also the saponification velocity must diminish much more quickly than it should, if it corresponded to the diminution of the concentration during the reaction; because during the reactions there are produced a relatively large number of ammonium ions. This in fact was found to be true, as above. In this way also the restraining action of the original addition of ammonium acetate is explained.

Now by means of the saponification constants of potassium hydroxide, one may calculate in a quantitative way the saponification constant of ammonium hydroxide in the presence of any quantity of ammonium salts, in the following way.[1]

The saponification constant of potassium hydroxide at 24·7°, and at a concentration of $\frac{1}{40}$ normal, amounts to 6·41 on the scale previously employed; moreover, as shown by both theory and by experiment, it is almost independent of the concentration. The saponification constant of ammonium hydroxide at the same concentration, and with or without the presence of ammonium salts, according to theory,

[1] Arrhenius, *Zeits. phys. Chem.*, **2.** 284 (1888).

must be as much smaller as its dissociation under the same circumstances is less than that of potassium hydroxide, which latter substance, according to its conductivity, is dissociated into its ions to the extent of 97·2 per cent. But now the degree of dissociation of a $\frac{1}{40}$ normal solution of ammonium hydroxide, as calculated on the basis of its conductivity (p. 359), amounts to 2·69 per cent; and its degree of dissociation in the presence of a quantity S of a binary ammonium salt,—which we may regard as being completely dissociated at large dilution, as is the case here,—without noticeable error may be calculated by means of the following equations;—these equations are obtained by a double application of the isotherm of dissociation, firstly to the pure ammonium hydroxide, and secondly to that containing an ammonium salt, thus—

$$\left(\frac{0{\cdot}0269}{40}\right)^2 = \mathrm{K}\frac{1 - 0{\cdot}0269}{40}$$

$$\frac{a}{40}\left(\frac{a}{40} + \mathrm{S}\right) = \mathrm{K}\frac{1 - a}{40};$$

here a denotes the degree of dissociation sought, and K denotes the dissociation constant of ammonia. We thus find the saponification velocity k, in the presence of the quantity S of a neutral salt, to be

$$\mathrm{k} = \frac{a}{0{\cdot}972}6{\cdot}41,$$

and similarly for pure ammonia,

$$\mathrm{k} = \frac{0{\cdot}0269}{0{\cdot}972}6{\cdot}41 = 0{\cdot}177.$$

In the following table there are given the values of k, on the one hand the results being calculated in the way just indicated, and on the other hand the results being calculated according to the formula (p. 555), which was *empirically deduced* by Arrhenius; the latter values of k may therefore be regarded as the immediate expression of a direct observation.

[*N.B.*—In this table the second calc. k is marked " k obs."]

S	a	k Calc.	k Obs.
0	2·69 %	0·177	0·156
0·00125	1·21	0·080	0·062
0·0050	0·71	0·047	0·039
0·0175	0·118	0·0078	0·0081
0·0250	0·082	0·0054	0·0062
0·0500	0·042	0·0028	0·0033

When one considers that the calculation employs as its basis the very much greater values of the saponification coefficients of potassium hydroxide, the coincidence between the last two columns must be regarded as actually surprising ; and this may be expressed as—

The saponification velocity, under conditions which are similar in other respects, is actually proportional with very close approximation to the quantity of free hydroxyl ions present. And also we are able to calculate the saponification velocity of any selected base, from its degree of dissociation.

Measurements of the velocity of saponification of dissolved esters is therefore a means for determining the quantity of hydrogen ions present in a solution. Shields (p. 520) made use of this means in determining the hydrolysis of the salts of strong bases. E. Koelichen showed that the condensation of acetone to di-acetone alcohol in aqueous solution is accelerated by hydroxyl ions and can therefore be used to follow this reaction quantitatively.[1]

Very remarkable results have been found by Wijs in studying the rate of *saponification methyl acetate by pure water.* This process clearly occurs in the following way. If the ester is dissolved in pure water the saponifying action of its hydroxyl ions yields acetic acid and methyl alcohol according to the equation :—

$$CH_3CO_2CH_3 + \overline{OH} = CH_3\overline{COO} + CH_3OH \, ;$$

hence the number of hydroxyl ions is reduced, and that of the hydrogen ions increased. But the hydrogen ions also possess the power of saponifying (p. 545), although to a much smaller degree than the hydroxyl ions ; the comparison between the rate of saponification by acids and alcohols show that the former occurs about 1400 times as slowly as the latter. These considerations show that at first the rate of saponification in pure water of the dissolved methyl acetate must fall very rapidly on account of the reduction in the number of hydroxyl ions, but later, when much acetic acid has been formed, and the water becomes decidedly acid, it will increase again because the catalytic action of the hydrogen has reached a considerable amount. Hence there must be a minimum of the velocity of saponification whose position can be calculated as follows.

Let us suppose the experiment so carried out that the concentration of the ester is kept constant, then the velocity of reaction is

$$\frac{dx}{dt} = k_1[OH] + k_2[H] \qquad . \quad . \quad . \quad (1)$$

where k_1 and k_2 are the velocity coefficients for the saponification by hydroxyl and hydrogen ions respectively and for the constant concentration of the ester.

[1] *Zeitschr. physik. Chem.*, **33.** 129 (1900).

The equation (p. 508),

$$[H][OH] = c_0^2$$

differentiated gives the equation

$$[H]\frac{d[OH]}{dt} + [OH]\frac{d[H]}{dt} = 0 \quad . \quad . \quad . \quad (2)$$

To find the position of the minimum we must differentiate equation (1) with respect to t and put it equal to zero,

$$\frac{d^2x}{dt^2} = k_1\frac{d[OH]}{dt} + k_2\frac{d[H]}{dt} = 0 \quad . \quad . \quad . \quad (3)$$

Equation (3) however is satisfied, as may be seen by comparison with (2), when

$$[H] : [OH] = k_1 : k_2,$$

which determines the position of the minimum ; since hydroxyl ions saponify 1400 times faster than hydrogen ions the time in question must occur when the concentration of the hydrogen ions has become 1400 times as great as that of the hydroxyl ions ; it may also be easily proved that the minimum velocity is 18 times smaller than the initial velocity.

To test these relations experimentally the rate of decomposition of the ester was determined by measuring its electric conductivity ; *it appeared that, in agreement with the theory, the velocity of reaction first fell off, reached a minimum, and then rose again.* To calculate the electrolytic dissociation of water the minimum velocity is the most favourable, so that the values given on p. 508 were determined by calculating the concentration of hydrogen and hydroxyl ions from the observed velocity. The condition that the concentration of the ester should remain constant is easily attained, since the amount of ester decomposed in this early stage of the reaction is only a small fraction of the whole.

Further Bi-Molecular Reactions.—A process which is quite analogous to the saponification of an ester, has been found by van't Hoff[1] *in the action of sodium hydroxide on sodium mono-chlor-acetate ;* this practically advances to a completion, with the formation of sodium glycolate and sodium chloride, according to the equation

$$C_2H_2ClO_2Na + NaOH = C_2H_3O_3Na + NaCl,$$

or in the sense of the more recent conception,

$$\overline{C_2H_2ClO_2} + \overline{OH} = \overline{C_2H_3O_3} + \overline{Cl}.$$

[1] *Études*, p. 19.

In this case, both of the conceptions lead to the same result, viz. that the metathesis must advance according to a formula which will hold good for bi-molecular reactions. As a matter of fact, the measurements gave a satisfactory constancy for the reaction coefficients, which were 0·0128 at 100°, and at 70° amounted to only 0·000822 more. Moreover, the more recent conception would allow us to anticipate that in using different bases, they would be independent of the nature of the positive radical, but that the degree of dissociation would be all important; and that in the case of weak bases the constancy of the velocity coefficients would disappear. The experimental proof of these requirements of the theory is wanting thus far.

The action of *bases upon lactones*, which was investigated by P. Henry,[1] and which results in the formation of salts of the corresponding hydroxy-acids, must likewise progress with a velocity which is proportional to the strength of the base, *i.e.* to the quantity of free hydroxyl ions; this has been thoroughly established by observation.

We may also mention the action of methyl iodide on silver nitrate, which occurs according to the equation

$$C_2H_5I + AgNO_3 = C_2H_5NO_3 + AgI$$

and was studied by V. Chiminello[2] in alcoholic solution.

Tri- and Multi-Molecular Reactions.—When the three molecular species, which vanish from the system in a tri-molecular reaction, are present in equivalent proportions, then the reaction velocity is

$$\frac{dx}{dt} = k(a - x)^3;$$

or, considering that $x = 0$ when $t = 0$,

$$k = \frac{1}{t} \frac{x(2a - x)}{2a^2(a - x)^2}.$$

Noyes[3] has in recent years found the first case of such a reaction in the action of ferric chloride on stannous chloride

$$2FeCl_3 + SnCl_2 = 2FeCl_2 + SnCl_4.$$

To prevent secondary disturbances it was found useful to add a small amount of the products of reaction.

[1] *Zeits. phys. Chem.*, **10.** 96 (1892). [2] *Gazz. chim.*, **25.** [2], 410 (1896). [3] *Zeits. phys. Chem.*, **16** 546 (1895).

	x	a − x	k
2·5	0·00351	0·02149	113
3	0·00388	0·02112	107
6	0·00663	0·01837	114
11	0·00946	0·01554	116
15	0·01106	0·01394	118
18	0·01187	0·01313	117
30	0·01440	0·01060	122
60	0·01716	0·00784	122

In the same way A. A. Noyes and R. S. Wason[1] showed that the reduction of potassium chlorate by ferrous chloride in acid solution is a tri-molecular reaction; the same is true for the reduction of silver salts by silver formate.[2]

Again W. Judson and J. W. Walker[3] succeeded in finding a good example of a quadri-molecular reaction in the action of bromic acid on hydrobromic acid; apparently this occurs (see the section below on "Complication of the Course of Reaction") according to the formula

$$2\overset{+}{H} + \overset{-}{Br} + \overline{BrO_3} = HBrO + HBrO_2.$$

Finally, Donnan and Rossignol[4] have shown that the reaction

$$2KI + 2K_3Fe(CN)_6 = 2K_4Fe(CN)_6 + I_2$$

in neutral solutions is *quinqui-molecular*; the actual process is probably that given by the formula

$$2\overline{\overline{\overline{Fe(CN)_6}}} + 3\overline{I} = 2\overline{\overline{\overline{\overline{Fe(CN)_6}}}} + I_3.$$

That tri-molecular reactions are rare, and those of higher orders still rarer, is evident from the kinetic considerations developed on p. 428; the probability of the simultaneous collision of several molecules is exceedingly small, the velocity of poly-molecular reactions can therefore only be appreciable under quite exceptional conditions.[5] Hence most apparent poly-molecular reactions really take place by means of simpler (uni-, bi- and very rarely tri-molecular) ones, which occur successively. This is confirmed by experience.

The study by O. Knoblauch[6] on the velocity of saponification of esters of polybasic acids is an excellent example of this. If we indicate

[1] *Zeitschr. physik. Chem.*, **22.** 210 (1897).
[2] A. A. Noyes and G. Cottle, *ibid.*, **27.** 579 (1898).
[3] *Journ. Chem. Soc.*, 1898, p. 410. [4] *Trans. Chem. Soc.*, **83.** 703 (1903).
[5] See also Van't Hoff, *Chemische Dynamik*, p. 197, Braunschweig, 1898.
[6] *Zeitschr. physik. Chem.*, **26.** 96 (1898).

by R″ the radical of a dibasic acid its ethyl ester is saponified by caustic soda according to the equation

$$R''(C_2H_5)_2 + 2NaOH = R''Na_2 + 2C_2H_5OH\,;$$

on investigating the course of reaction it appears that this is never a tri-molecular reaction, but that the process occurs in the two stages

$$1.\ R''(C_2H_5)_2 + NaOH = R''(C_2H_5)Na + C_2H_5OH$$
$$2.\ R''(C_2H_5)Na + NaOH = R''Na_2 + C_2H_5OH$$

consisting therefore of two consecutive bi-molecular reactions. In following the process quantitatively it is therefore necessary to apply the law of mass action to the two latter equations, that is, a special velocity constant must be introduced for each of the two practically complete reactions.

It was to be expected that such a reaction process would be found in *the gradual union of gaseous hydrogen and oxygen to form water*; for this occurs with the simultaneous meeting of three molecules, and as van't Hoff[1] discovered, it advances at 440° with a velocity suitable for measurement.

A pear-shaped glass tube, running off into a long narrow tube, was filled with oxygen and hydrogen, and then kept all day long (*tagelang*) in the vapour of boiling sulphur at 440°. The narrow tube into which the "pear" extended at the same time served as a closed manometer, being separated from the gas by a mercury index, and being filled with air. Thus it was possible to show the gradual diminution of pressure in the interior of the "pear," as it was occasioned by the formation of water.

Time in Hours.	Amount of "Knall-gas."
0	1·000
6	0·974
20	0·931
34	0·902
47	0·881
61	0·863

Now, the course of the reaction is not that which is required by the theory, because the calculation of the reaction coefficients furnishes no constant numbers. This is due, as has recently been shown by Bodenstein,[2] to the strong but irregular influence of the glass walls of the containing vessel. When Bodenstein used porcelain vessels he obtained most regular results, and showed the reaction to be *tri-molecular.* The velocity was found to be nearly proportional to the area of the porcelain surface, so that the formation of water did not take place in the interior of the gas, but only by the catalytic action of its surface. It is only at higher temperatures that formation of water occurs in the interior, that is without catalytic acceleration.

[1] *Études*, p. 52; see also V. Meyer, *Lieb. Ann.*, **269.** 49 (1892).
[2] *Zeitschr. physik. Chem.*, **29.** 665 (1899).

The Course and the Mechanism of a Reaction.—We have seen, in the preceding paragraphs, that the course of the reaction always varies in a characteristic way with the number of the molecular species, which experience a considerable change in the course of a reaction in a system considered as homogeneous. This clearly shows the succession of the formulæ, which serve to calculate the velocity coefficients of equivalent quantities of the reacting ingredients. These, in order, are given, namely,

for uni-molecular reactions, by the expression $\frac{1}{t}\ln\frac{a}{a-x}$;

for bi-molecular reactions, by the expression $\frac{1}{t}\frac{x}{(a-x)a}$;

for tri-molecular reactions, by the expression $\frac{1}{t}\frac{x(2a-x)}{2a^2(a-x)^2}$, etc.

These expressions are so different from each other that, if the course of the reaction exhibits a constant velocity coefficient by applying one of the preceding formulæ, this is never the case in the use of one of the other expressions. If we make $x = \frac{a}{2}$, *i.e.* if we calculate *the time required to change one half of the quantity of the substance capable of change*, we find in the *first* case that the time is independent of the original concentration of "a" employed; that in the *second* case it is inversely proportional to this; that in the *third* case it is inversely proportional to the square of this; and that in general, in an n-molecular reaction, this time is inversely proportional to the $(n-1)$ power of this original concentration.

Thus one may decide the question, How many molecular species participate in a reaction?—simply in this way, viz. by starting out with equivalent quantities of the reacting substances, and then determining in two experiments (which differ in concentration) the time required to consume *one half* of the substances capable of reaction.

The honour of the task of first showing the possibility of obtaining a glimpse into the mechanism of a reaction in its course,—this belongs to van't Hoff, who has made some applications of it in his classical treatise, *Études de dynamique chimique* (1884), which has been frequently cited in what has preceded.

Thus arsine and phosphine decompose completely into their respective elements at higher temperatures, and formerly one was inclined to regard the course of the reaction as progressing in this wise, viz.—

$$4AsH_3 = As_4 + 6H_2,$$

and

$$4PH_3 = P_4 + 6H_2;$$

according to this, a collision of two of the four molecules [of AsH_3 or of

PH_3 respectively] would be requisite for the decomposition. Therefore, the calculation of the velocity coefficients, according to a formula good for a *quadri-molecular* reaction, would give constant values.

The measurement of the results of the gradual change of gases in both cases was accomplished by determining the associated change of pressure, which (from the beginning to the end of the reaction) is as 2 : 3.

In the following table are given the values of pressure P, expressed in mm. of Hg, for the corresponding times (in hours) *for the decomposition of arsine*, at a temperature of 310° (boiling di-phenyl-amine) :—

t	P	$\frac{1}{t}\log\frac{P_0}{3P_0-2P}$	$\frac{1}{t}\left[\left(\frac{P_0{}^3}{3P_0-2P}\right)^3-1\right]$
0	784·84	...	...
3	878·5	0·03948	0·422
4	904·05	0·03931	0·491
5	928·02	0·03943	0·581
6	949·38	0·03931	0·683
7	969·08	0·03933	0·814
8	987·19	0·03935	0·975

Now, if the reaction were actually *quadri-molecular*, then the values of the expression, as given in the last column, should be *constant, which is not the case.* But if it is assumed that the reaction takes place in a *uni-molecular* way, the calculation gives the results shown in the third column, where the variations to which they are subject are as small as would be expected. This very remarkable result would seem to indicate that each molecule of arsine decomposes by itself; and then that every pair of hydrogen atoms unites to form a hydrogen molecule; and also that an unknown number of arsenic atoms unite to form the molecules of solid arsenic which separate during the reaction.

A similar result is obtained from the decomposition of phosphine, which was investigated at a temperature of 440° (boiling sulphur), but otherwise under similar circumstances.

As van't Hoff showed,[1] the slow course of such uni-molecular reactions as the decomposition of arsine or phosphine proves that the molecules of a gas are not all in the same condition, otherwise they would either decompose simultaneously or not at all. The occurrence of all stages of velocity of reaction speak strongly in favour of the kinetic views, and especially of Maxwell's conception (p. 209), that the temperature of a single gas molecule oscillates about a mean.

Very often it happens that a reaction will begin simply and smoothly, and it is not until it has advanced to quite a distance, and

[1] *Chem. Dynamik*, p. 187, Braunschweig, 1898.

after the resulting quantity of the reaction products has attained to considerable magnitude, that it is disturbed by side reactions. In these cases one can draw some conclusion regarding the number of molecules of the reacting substances from the dependence of the original velocity upon the original concentration of the reacting substances.

At equivalent concentration c of the reacting ingredients, when n molecules react upon each other, the original velocity is

$$v = kc^n ;$$

now, by observing the original velocities v_1 and v_2 of two concentrations c_1 and c_2, which are very different from each other, we have,

$$n = \ln\frac{v_1}{v_2} \cdot \frac{c_2}{c_1} .$$

Although it is very difficult to ascertain the original velocity directly, and although the results obtained in this way, at all events, are only approximate, yet they are usually sufficiently accurate to discriminate between the results ; for n represents *a whole number* in all cases. Thus van't Hoff[1] found, in studying the action of bromine upon fumaric acid in dilute water solution,—a reaction which leads smoothly to the formation of di-brom-succinic acid only in its first stage,—the result,

$$n = 1{\cdot}87, \text{ instead of } 2 ;$$

this comes sufficiently near to what was expected.

Further applications of this important method are to be found in van't Hoff,[2] Nernst, and Hohmann,[3] and A. A. Noyes.[4] In practice it is more accurate, as the author showed in the work with Hohmann mentioned, to determine the expressions

$$\frac{1}{t}\ln\frac{a}{a-x}, \quad \frac{1}{t}\frac{x}{a-x}, \quad \frac{1}{t}\frac{x(2a-x)}{(a-x)^2}, \text{ etc.}$$

in the initial stage, and see whether they are independent of the volume of the reacting mixture or inversely proportional to the first power or inversely proportional to the second power.

After all this repeated and emphatic proof, it hardly needs any other special demonstration to show that one cannot draw any conclusion by any of the methods indicated respecting the question,—Whether a molecular species which is present in great excess (the solvent, *e.g.*) participates in the reaction or not. The case here resembles that previously encountered (pp. 268 and 459), where it was shown that the hydration of dissolved substances exerts no influence on the depression of their freezing-point.

[1] *Études*, p. 89.

[2] *Chem. Dynamik*, p. 193.

[3] *Zeitschr. physik. Chem.*, **11.** 375 (1893).

[4] *Ibid.*, **18.** 118 (1895).

The Reaction-Velocity and the Constitution.—*The principle of inter-molecular reactions* (p. 287) depends upon the study of the readiness with which the cleavage products of the compounds in question are formed; this has been repeatedly applied to special questions in stereo-chemistry (p. 286).

We have now become familiar with the methods by which the conception of "readiness" (*Leichtigkeit*) (which is doubtless somewhat vague) can be replaced by the knowledge of well-defined magnitudes capable of numerical expression, viz. the *reaction-velocities.* The practical application of this information will, to a large extent, remain a task for the future, but it has already been applied with great methodical skill in one case.

It was Evans[1] who saw a connection between the stereo-chemical constitution of chlor-hydrine and the velocity with which it forms hydrochloric acid in the sense of the reaction,

$$R\begin{matrix} \diagup OH \\ \diagdown Cl \end{matrix} + KOH = KCl + H_2O + R:O;$$

the progress of the reaction as it takes place in dilute solution is ascertained by the titration of the resulting potassium chloride.

In discussing his results, Evans starts out with the view that the distance between the chlorine and the hydroxyl in the molecule [of the chlor-hydrine] conditions the velocity of the formation of the oxide, *i.e.* the velocity will be greater the nearer the two radicals are situated with reference to each other. Space does not allow a thorough discussion of the results obtained regarding the constitution of the seven chlor-hydrines which were investigated.

Further illustrations of chemical kinetics which will obviously throw light on the mechanism of chemical processes are to be found in the important works of A. Hantzsch,[2] and especially H. Goldschmidt's investigation of the formation of the amidoazo compounds,[3] of the azo dyes,[4] and of the anilides.[5] Of investigations touching the relation between constitution and velocity of reaction, we may mention Conrad and Brückner.[6] Study on the determination of affinity coefficients, Wildermann[7] on the velocity of the action of alcoholic potash on the halogen derivatives of the hydrocarbons of the fatty series, H. Goldschmidt[8] on the formation of esters.

[1] *Zeits. phys. Chem.*, **7.** 337 (1891).

[2] *Ibid.*, **13.** 509 (1894).

[3] H. Goldschmidt and Reinders, *Ber. deutsch. chem. Ges.*, **29.** 1369 and 1899 (1896).

[4] Goldschmidt and Merz, *ibid.*, **30.** 670 (1897); Goldschmidt and Buss, *ibid.*, 2075; Goldschmidt and Burckle, *ibid.*, **32.** 355 (1899).

[5] Goldschmidt and Wachs, *Zeitschr. physik. Chem.*, **24.** 353 (1897).

[6] *Zeitschr. physik. Chem.*, **3.** 450; **4.** 273, 450; **5.** 289; **7.** 274, 283 (1889-1891).

[7] *Ibid.*, **8.** 661 (1891).

[8] *Ber. deutsch. Chem. Ges.*, **28.** 3218; **29.** 2208 (1896); see also Donnan, *ibid.*, 2422; and Kellas, *Zeitschr. phys. Chem.*, **24.** 221 (1897).

Catalysis.[1]—We have repeatedly noticed in the foregoing, especially in dealing with the decomposition of esters and the inversion of cane sugar, the striking circumstance that many reactions take place with increase of velocity in presence of certain substances, especially acids. Berzelius gave to this phenomenon the name of catalysis; it means therefore *an increase in velocity of reaction caused by the presence of substances which do not take part in it, although the reaction is capable of taking place without their presence.* Acids and bases seem to exercise a catalytic action on all reactions in which water is affected or split up, and their activity is proportional to the concentration of the hydrogen or hydroxyl ions. One of the longest known and most important catalyses is the accelerated oxidation of sulphur dioxide by oxygen in presence of oxides of nitrogen. Another well-known example of catalysis is the extraordinary increase in velocity of combustion of hydrogen and of sulphur dioxide in presence of finely divided platinum. Finally, we may mention the interesting researches of Dixon and Baker[2] who were led to the conclusion that a number of gas reactions, such as the combustion of carbon monoxide, the dissociation of sal-ammoniac vapour, the action of sulphuretted hydrogen on salts of the heavy metals, cease entirely in the absence of water vapour.

But there is also a negative catalysis, that is a retardation of reaction by addition of another substance. The investigation of Bigelow[3] gives a remarkable example of this. He showed that the oxidation of sodium sulphite by oxygen is very much retarded by small additions of organic substances such as mannite, benzyl alcohol, benzaldehyde, and even that it is practically stopped. The retardation is so enormously great that it is strongly noticeable with solutions, for example, of mannite of the concentration $\frac{1}{160000}$ normal.

The catalysor cannot of course affect the affinity of a process. To do so would be in contradiction to the second law of thermo-dynamics, according to which the affinity of an isothermal process, as measured by the maximal work, depends only on the initial and final states. The activity of the catalysor therefore does not touch the driving force of a reaction but only the opposing resistance, as was early[4] discovered.

Since the catalysor takes no part in the reaction the reaction constant is not altered by its presence. This was seen (p. 434) to be equal to the ratio between the velocity constants of the two reactions in opposite signs. *A catalysor must therefore always affect the velocity of the reverse reaction.* If, for example, any added substance increases the rate of formation of a body, it must equally increase its velocity of

[1] Ostwald has given a sketch of catalytic phenomena in a lecture to the Hamburger Naturforscherversammlung, *Zeitschr. f. Elektrochemie*, **7.** 995 (1891).
[2] *Trans. Roy. Soc.*, **175.** 617 (1884); *Journ. Chem. Soc.*, **49.** 94 and 384 (1886).
[3] *Zeitschr. physik. Chem.*, **26.** 493 (1898).
[4] See, for example, B. Helmholtz, *Erhaltung der Kraft*, p. 25.

decomposition. We find an example of this in the known fact that the presence of acids causes both the formation and the saponification of esters to take place with increased velocity. The observation of Baker that in absence of water vapour gaseous ammonium chloride does not dissociate, and on the other hand dry ammonia does not combine with hydrochloric acid may be explained in the same way.

There is no complete theory of catalytic phenomena at the present moment.

Attempts have often been made to explain it by means of intermediate products formed by the catalysor and the reacting substance. These intermediate products, whose existence has been proved in many cases, are subsequently decomposed into the catalysor and the product of reaction. Thus in the process of manufacturing of sulphuric acid the action of the oxide of nitrogen has been explained by assuming that in the mixture of sulphur dioxide, nitric peroxide and air, nitrosulphuric acid, $SO_2 {-OH \atop -NO_2}$, is first formed, and this, by contact with water, is decomposed into sulphuric and nitrous acids. This method of explanation may be employed when the velocities of such intermediate reactions are greater than the velocity of the total reaction; this theory of catalysis has only so far been proved in a few cases, in many others it seems plausible. In no case does it seem possible to explain all catalytic phenomena from this point of view.

In cases of catalysis in heterogeneous systems, for example the acceleration of gas reactions by platinum, it is probable that the phenomenon is connected with that of solution of the gases in metal.

According to the observations of Bredig,[1] colloidal solutions of metals act in the same way as the metals themselves. In the decomposition of hydrogen peroxide a platinum solution containing only $\frac{1}{70000000}$ of a mol. per litre produces a measureable action. The catalytic activity of a colloidal platinum solution recalls in many respects that of organic ferments, and for this reason Bredig called it an inorganic ferment. The analogy comes out most strikingly in the change of activity by time, by temperature, and by the power of certain substances, which are poisons towards organisms, to poison, that is to destroy, the activity of inorganic ferments also. Such substances are sulphuretted hydrogen, hydrocyanic acid, etc.

Auto-Catalysis.—Acids produce a catalytic action on the transformation of oxy-acids into lactones; the velocity of this has been measured by Hjelt,[2] and especially with regard to catalysis by Henry (p. 559) and Collan.[3] Thus, for example, γ-oxy-valerianic acid in aqueous solution is changed into valero-lactone with separation of

[1] *Anorganische Fermente*, Leipzig, 1900; *Zeitschr. f. physik. Chem.*, **31.** 258 (1899).

[2] *Ber. deutsch. chem. Ges.*, **24.** 1236 (1891).

[3] *Zeitschr. physik. Chem.*, **10.** 130 (1892).

water, and, as usual in such cases, the reaction takes place more quickly in presence of a strange acid; it is only of course the free hydrogen ions that are catalytically active. Now the acid itself is partly electrolytically dissociated, that is in a solution of oxy-valerianic acid without the addition of any strange acid free hydrogen ions occur; it is therefore natural to suppose that these must produce catalysis, so that the acids *catalyse themselves.* This supposition may be easily tested experimentally. If to the acid be added one of its neutral salts, the dissociation is very much reduced in accordance with well-known rules, that is, the number of free hydrogen ions is diminished. Thus the addition of the sodium salt must greatly hinder the conversion of the acid into lactone. Actually it is found that the strength of the acid in the presence of its sodium salt remains unaltered for days. Another deduction from the theory, which is confirmed by experiment, is that the transformation of acid into lactone does not follow the equation applicable to unimolecular reactions, but that the velocity of reaction at any moment is proportional to the product of the concentrations of the undissociated acid and the hydrogen ions. These phenomena, like so many others in chemical kinetics in which actions of electrolytes are concerned, would be inexplicable without the hypothesis of electrolytic dissociation, although in the light of this hypothesis they become almost obvious.

The Progress of Incomplete Reactions.—Finally, we will devote a few words to the general case where the reaction stops at a point where only a small part of the possible decomposition is effected. This occurs in esterification (p. 447); thus, *e.g.* if one mixes 1. g.-mol. of alcohol with 1 g.-mol. of acetic acid, the mutual action comes to a pause after two-thirds of the ester has been formed, which is *in maximo* possible.

The reaction velocity for the time t, when the quantity of ester formed is x, is here given by the equation,

$$\frac{dx}{dt} = k(1 - x)^2 - k'x^2,$$

where k and k′ denote respectively the velocity constants of the two opposed reactions. If we introduce into the preceding equation the value of the ratio of these two constants, viz.

$$\frac{k}{k'} = 4,$$

as ascertained from the equilibrium of the system, then by integration[1] we obtain

[1] For the way of carrying out such calculations see, for example, B. Nernst and Schönflies, *Einführung in die math. Behandl. der Naturwissenschaften*, 2nd edit. p. 14. Munchen, 1898.

$$\frac{4}{3}\left(k - k'\right) = \frac{1}{t} \log \frac{2 - x}{2 - 3x}.$$

The velocity of esterification, under the preceding conditions, was measured at the temperature of a dwelling-room by Berthelot and Péan de St. Gilles; their results were as follows :—

Time t	x Obs.	x Calc.
0 days	0·000	0·000
10 ,,	0·087	0·054
19 ,,	0·121	0·098
41 ,,	0·200	0·190
64 ,,	0·250	0·267
103 ,,	0·345	0·365
137 ,,	0·421	0·429
167 ,,	0·474	0·472
190 ,,	0·496	0·499
∞ ,,	0·677	0·667

The calculated values [1] of x, as given in the third column of the preceding table, are obtained from the theoretical formula, by assuming that

$$\frac{4}{3}\left(k - k'\right) = 0{\cdot}00575.$$

Moreover, the coincidence between experiment and calculation is good throughout, even at the beginning of the series, where some disturbing side reactions seem to have occurred. We will refer to the calculation again at the close of this chapter.

If a small quantity of acid be added to a concentrated solution of water and alcohol, the concentration of the water and the alcohol may be regarded as constant, and we have for the velocity of reaction

$$\frac{dx}{dt} = k_1(a - x) - k_2x,$$

where a is the amount of acid added and x is the amount of ester formed in the time t. This equation of course holds also when a small quantity of ester is added to the alcohol-water mixture, and a is the quantity added to the solution, x is the quantity of ester decomposed in the time t; only that in this case the reaction proceeds in the opposite sense. If in the preceding equation we put the reaction constant

$$K = \frac{k_1}{k_2}.$$

[1] Guldberg and Waage, *J. pr. Chem.* [2], **19.** 69 (1879); Ostwald's *Klassiker*, No. 104.

and integrate, we obtain

$$k_1 + k_2 = \frac{1}{t} \ln \frac{Ka}{Ka + (1 + K)x}.$$

This equation has been tested in a number of experiments by W. Kistiakowsky,[1] he has in particular given an important proof that the value $k_1 + k_2$ obtained is the same whether it is derived from the formation or the decomposition of ester.

The condition of equilibrium is

$$K = \frac{x_0}{a - x_0},$$

and if it be inserted in the above equation we obtain

$$k_1 + k_2 = \frac{1}{t} \ln \frac{a - x_0}{a - x_0 - x}.$$

The following table refers to the formation and decomposition of methyl formate at 25°; the alcohol-water mixture contains 43·9 % of alcohol. To increase the velocity of reaction a small quantity of hydrochloric acid was added, this acted in a partially catalytic manner.

FORMATION OF ESTERS

t	$a - x_0 - x$	$k_1 + k_2$
0	13·74	...
60	10·75	176
100	9·14	177
210	5·92	174

DECOMPOSITION OF ESTERS

t	$a - x_0 - x$	$k_1 + k_2$
0	19·83	...
40	16·91	173
60	15·54	176
90	13·82	174

The means obtained for the two series (176 and 175) are in striking agreement.

[1] *Wied. Beibl.*, 1891, p. 295; *Chemische Umwandlung in homogenen Gebilden.* Petersburg, 1895. See also O. Knoblauch, *Zeitschr. physik. Chem.*, **22.** 268 (1897).

Another example of the incomplete course of a reaction was found by P. Henry (p. 559) in studying the change of *hydroxy-butyric acid into the corresponding lactone.* Here an abundance of free hydrogen ions (in the form of HCl) were added to accelerate the change; therefore the weaker hydroxy-butyric acid could be regarded as entirely undissociated; and, therefore, if "a" denotes the quantity of acid originally present, and x the quantity changed into lactone, then the equation representing the course of the reaction is

$$\frac{dx}{dt} = k_1(a - x) - k_2x.$$

If we introduce the constant of equilibrium, viz.

$$K = \frac{k_1}{k_2},$$

and then integrate, we obtain

$$k_1 + k_2 = \frac{1}{t} \ln \frac{Ka}{Ka - (1 + K)x}.$$

In conducting the research, the original concentration of the acid amounted to $\frac{1}{5 \cdot 66}$ g.-equivalent per litre: at the beginning a standard volume, removed by the pipette, consumed 18·23 c.c. of a barium hydroxide solution, excluding the hydrochloric acid present. After a long time the titration remained constant at 13·28 c.c. Therefore we have

$$K = \frac{13 \cdot 28}{18 \cdot 23 - 13 \cdot 28} = 2 \cdot 68.$$

The series of determinations conducted at 25° were as follows:—

t	x	k_1+k_2
21	2·39	0·0411
50	4·98	0·0408
80	7·14	0·0444
120	8·88	0·0400
220	11·56	0·0404
320	12·57	0·0398
47 hours.	13·28	...

The values in column t denote minutes; the values in column x are the quantities of lactone produced, and expressed in c.c. of the barium hydroxide solution; on the same scale "a" amounts to 18·23. The constancy of the value of $k_1 + k_2$, as calculated according to the preceding equation, is highly satisfactory.

The same equation of reaction holds, as Küster[1] has shown, for the mutual conversion of the two hexachlor-keto-R-pentenes; a mixture of the two isomers tends towards an equilibrium which depends considerably on temperature. It may be noted that the case investigated by Küster was that of a liquid mixture and the two mutually convertible species of molecules without any solvent.

According to Walker and Kay,[2] the formation of urea from ammonium cyanate takes place in accordance with the equation

$$\overset{+}{N}H_4 + C\overset{-}{N}O = CO(NH_2)_2;$$

if a is the quantity of ammonium cyanate present, α the degree of dissociation at the time t, we have

$$\frac{dx}{dt} = k\alpha^2(a - x)^2 - k'x.$$

Multirotation of Milk Sugar.—C. S. Hudson[3] has proved that this is also a case of incomplete reaction, and has completely explained in this way the process of multirotation mentioned on p. 547. Both the hydrate and the lactone of milk sugar slowly change their optical rotation in a freshly prepared solution, for in both cases the reaction

$$C_{12}H_{24}O_{12} \rightleftarrows C_{12}H_{22}O_{11} + H_2O$$

tends to produce equilibrium; since the rotation of both substances can be determined in a freshly prepared solution, the composition of the solution at any phase of the reaction, and finally the equilibrium can be determined polarimetrically. If a mols. of the hydrate are dissolved in a large quantity of water, the equation

$$\frac{dx}{dt} = k(a - x) - k'x,$$

holds, where x is the amount of lactone produced. If r_0 is the rotation of a mols. of the hydrate we obtain by integration, just as on p. 570

$$k + k' = \frac{1}{t} \ln \frac{r_0 - r\infty}{r - r\infty},$$

where r is the variable rotation and $r\infty$ the rotation in equilibrium. Hudson found this equation comfirmed in all cases.

Tautomerism.—Let us consider a mixture of two isomers which are capable of mutual conversion as mentioned above and investigated by Küster, and let us assume that the equilibrium between the two

[1] *Zeitschr. physik. Chem.*, **10.** 161 (1895).
[2] *Journ. chem. Soc.*, 1897, p. 489; *Zeitschr. physik. Chem.*, **24.** 372 (1897).
[3] *Zeitschr. physik. Chem.*, **44.** 487 (1903).

isomers established itself very quickly; if we attempt to separate the components of such a mixture by any chemical method of separation, the other components will be converted into the first in consequence of the displacement of equilibrium, that is, the whole mixture will react as if it consisted of the first component exclusively. If, on the other hand, we apply a chemical reagent which acts only on the second component, the mixture will conversely behave as if it consisted of the second component only. Such a mixture can therefore react according to two constitutional formulæ, that is, we have the phenomenon of *tautomerism* described on p. 291. Thus it has recently been suggested repeatedly [1] that hydrocyanic acid is a mixture of the molecules NCH and CNH which, however, are mutually converted at ordinary temperatures so quickly that separation of the two is impossible or at least very difficult, just as was the case with the mixture studied by Küster at higher temperatures, because one of the isomers passes into the other too quickly. According to this view, reduction of temperature would be the means for isolating the two tautomeric forms.

The same view of tautomerism explains why, as Knorr [2] remarked, this phenomenon is *only found in fluids*, whilst for solids we must always assume a definite structure.

The most thorough test for the accuracy of this view is in separating the isomers and following their mutual conversion. Thus Claisen [3] was able to isolate the enol and keto forms of tribenzoly-methane and similar substances, and Wislicenus [4] those of formyl-phenyl acetic ester. These were distinguished not only by their melting-point but by their chemical behaviour, for example, the acid properties of the enol form and the neutral properties of the keto form. The intensive coloration of the enol form by ferric chloride is especially characteristic, and was used by Wislicenus to follow the course of the inversion and to demonstrate the establishment of equilibrium for both sides. It is to be remarked that the velocity of reaction varies greatly with the nature of the solvent, as in the case of other reactions it is greatest in methyl alcohol, then follow ethyl alcohol and ether, and finally chloroform and benzene.

Hantzsch [5] has made interesting observations on the transformation of tautomeric forms of nitrophenyl-methane and similar bodies, and has shown that the conversion can be followed by measuring the electric conductivity. These substances react sometimes as true nitro bodies, sometimes as isonitro bodies—

$C_6H_5 - CH_2 - NO_2$	$C_6H_5 - CH - NO - OH$
Phenylnitromethane. Neutral. Stable in the free state; in alkaline solution changes into the other form.	Isonitrophenylmethane. Acid. Unstable in the free state; stable as salt.

[1] Similar ideas are to be found in Laar, *Ber. deutsch. chem. Ges.*, **18**. 648 (1885).
[2] *Lieb. Ann.*, **306**. 345 (1899).
[3] *Ibid.*, **291**. 25 (1896).
[4] *Ibid.*, **291**. 147 (1896).
[5] *Ber. deutsch. chem. Ges.*, **32**. 575 (1899).

The passage of the second acid substance into the first neutral one is shown by a decrease in conductivity, which finally disappears altogether. The converse reaction takes place when the first form is dissolved in alkali. Here also a slow decrease in conductivity takes place which is due to the formation of the alkali salt of the isonitrophenyl-methane from the free alkali and the neutral phenyl-nitromethane, producing the ion of this acid in place of the much more rapid hydroxyl ion. The neutralisation, which in the case of real acids is almost instantaneous, takes place here in a measurable time, that required for the conversion of the neutral substance into the acid. Hantzsch describes such compounds which are not acids definitely as pseudo-acids. These are also distinguished by an exceptionally large temperature coefficient of conductivity, and which increases with rising temperature; also by dissociation constants which change abnormally fast when the temperature increases, and by noticeable colour change on change of temperature.

Hollmann[1] has given another excellent example of the equilibrium between the different modifications of a substance in his study on the modifications of acetaldehyde.

According to these investigations it may be finally assumed that tautomerism is nothing less than a kind of isomerism in which the velocity of mutual conversion is very great.[2]

Complications in the Course of the Reaction.—It sometimes happens that the course of a reaction is somewhat different from what it should be, according to the differential equation corresponding to the scheme of the reaction. In most cases it is found that the resulting products of decomposition give rise to side reactions which disturb the simplicity of the relations; thus we saw on p. 553 that the ammonium salt produced in the saponification of ethyl acetate by ammonia acted in a secondary way on the state of dissociation of the base, thus giving rise to certain irregularities; these irregularities at first seemed to be inexplicable; but they were explained after the nature of the disturbance was recognised.

In other cases the discrepancy between calculation and experiment is explained on learning that the reaction equation, which was assumed as the basis of the differential equation, does not correspond to the actual conditions; we have met several examples of this. The mechanism of the reaction must be known, in order to apply the law of mass-action; for in order to write out the expression for the reaction velocity, we need to know not only the *kind* of molecular species, but also the *number* of each kind which participates in the reaction.

It has been already shown in many cases that in those reactions

[1] *Zeitschr. physik. Chem.*, **13.** 190 (1903).

[2] A complete exposition of the views on tautomerism and the experimental material is given by Rabe, *Lieb. Ann.*, **313.** 129 (1900).

in which electrolytes participate, only those formulæ are suited for describing the course of the reaction which are derived in accordance with the [newer] views regarding electrolytic dissociation; while the conception of the mechanism of the reaction which was formerly accepted as practical must be repudiated *in toto.*

This happens in the case of a reaction, which advances to an almost perfect completion, and which fortunately has been recently investigated by Ostwald,[1] viz.

$$HBrO_3 + 6HI = HBr + 3H_2O + 3I_2.$$

The progress of this reaction in water solution can be measured very easily. Let us start with equivalent quantities of the substances on the left-hand side of the equation; thus at the beginning of the reaction let there be a mols. of $HBrO_3$, and 6 a mols. of HI in a litre of the reaction mixture; then in the sense of the old conception, the equation representing the course of the reaction will be

$$\frac{dx}{dt} = k(a - x)^7;$$

here x denotes the quantity of $HBrO_3$ remaining decomposed after the time t.

But it now appears that the equation

$$\frac{1}{(a - x)^6} - \frac{1}{a^6} = kt,$$

which is obtained by integrating the preceding one, does not apply to the course of the reaction. This is clearly intelligible from the standpoint of the theory of electrolytic dissociation. For if we assume that the substances are completely dissociated (which is approximately true in the case of very dilute solutions), then we obtain the following reaction scheme, viz.

$$7\overset{+}{H} + \overline{BrO_3} + 6\overset{-}{I} = \overset{+}{H} + \overset{-}{Br} + 3H_2O + 3I_2.$$

One hydrogen ion, which is common to both sides of the equation, can be neglected. On the whole, the reaction may be dissected into the two following particular reactions, viz.

$$6\overset{+}{H} + \overline{BrO_3} = \overset{-}{Br} + 3H_2O,$$

and

$$6\overset{-}{I} = 3I_2.$$

But these two reactions do not take place together independently of each other; for they must occur at the same time, because the free

[1] *Zeits. phys. Chem.*, **2**. 127 (1888).

[positive] electric charges of the H-ions, at every instant, must be neutralised by those of the Lions.

According to A. A. Noyes,[1] the above-mentioned case is one of a bi-molecular reaction, and is to be explained by the hypothesis that the following reactions occur :—

$$HI + HBrO_3 = HBrO_2 + HIO,$$
$$HI + HBrO_2 = HBrO + HIO,$$
$$HI + HBrO = HBr + HIO$$
$$3HI + 3HIO = 3H_2O + 3I_2 ;$$

only the first bi-molecular reaction requires appreciable time and therefore regulates the velocity of reaction, whilst the other processes take place instantaneously, and consequently no appreciable quantity of intermediate products exist. Walker (p. 560) has shown that a similar assumption probably holds for the action of bromic acid on hydrobromic acid.

Indeed, even without this knowledge, we may predict that the velocity of the decomposition will be accelerated by the addition of acids, *i.e.* by the addition of hydrogen ions, because the hydrogen ions constitute one of the reacting molecular species ; and also that the acceleration occasioned by the addition will be the greater the stronger the acid is, as is clearly to be seen from the results of Ostwald's researches.

Apparently the case is very similar in the behaviour of iodic acid and sulphurous acid in their mutual reaction upon each other, which was investigated by Landolt ;[2] these substances react upon each other in the sense of the equation

$$3SO_2 + HIO_3 = 3SO_3 + HI ;$$

also the hydriodic produced reacts upon the iodic acid in the sense of the equation

$$5HI + HIO_3 = 3H_2O + 3I_2.$$

If one adds thereto some starch paste, then as soon as a sufficient quantity of iodine shall have been set free, there instantly appears a blue colour. The *time* which intervenes between the preparation of the solution and the appearance of the blue colour, and which is usually counted in seconds,—this was measured by Landolt for varying proportions of the reacting substances ; and although this time interval does not represent a simple and direct measure of the velocity of the reaction, but rather has a significance which is somewhat complicated,

[1] *Zeitschr. physik. Chem.*, **18.** 118 (1895).
[2] *B. B.*, **16.** 249 (1885) ; **19.** 1317 (1886).

nevertheless it may at least be stated that the reaction progresses more quickly the shorter this time interval.

Moreover, he succeeded in representing the time interval which elapsed between the preparation of the reaction mixture and the separation of the free iodine (as shown by the blue colour of the added starch), as being in excellent coincidence with all his numerous observations, by means of the empirical formula—

$$t = \frac{524 \cdot 35}{C_S^{0 \cdot 904} C_I^{1 \cdot 642}} \text{ sec.};$$

here C_S and C_I denote respectively the concentration of the sulphurous and of the iodic acid in g.-mol. per cub. cm. The formula holds good for 20°.

The Influence of the Medium.—Strictly speaking, we would expect to find a constant velocity coefficient in the course of the reaction, only in the case of those systems where *the nature of the medium* in which the reaction is completed experiences no essential change as a result of the decomposition of the reacting substances.

This condition is fulfilled, to be sure, only in a faulty way, in the case of esterification previously considered, in the progress of which the nature of the medium is changed considerably; but the condition is apparently very completely satisfied in the case of reactions occurring in gaseous systems, and doubtless also in reactions occurring in dilute solutions. As a matter of fact, we found in the latter a most excellent demonstration of the law of mass-action in its application to chemical kinetics.

The question, how the reaction velocity changes with the nature of the medium in which the reaction occurs has hitherto been attacked only in a desultory way. No researches have been conducted to determine whether the *velocity* of decomposition in gaseous systems is influenced by the presence of a [chemically] indifferent gas or not; whether, *e.g.*, the decomposition of hydrogen arsenide [arsine AsH_3] occurs in the same time in the presence of nitrogen that it would were it alone.

But from the fact that the condition of equilibrium of a gaseous system (*e.g.* the dissociation of a gas) is not displaced by intermixture with foreign indifferent gases, it undoubtedly follows that the two opposed reaction velocities (which in equilibrium exactly compensate each other) would be influenced in the same way, if at all; and until the experimental decision of the question, the most probable provisional assumption is *that both of the two reaction velocities remain unchanged*, i.e. *that [chemically] indifferent gases are without influence on the reaction velocity.* Actually E. Cohen[1] showed that the rate of

[1] *Zeitschr. physik. Chem.*, **25.** 483 (1898).

decomposition of arsine is not altered by the presence of hydrogen or nitrogen.

The problem, and one very interesting for many reasons, respecting the change in the reaction velocity of a chemical process which occurs in a solution, *with the nature of the solvent*,—this was first attacked in a thorough way by Menschutkin.[1] For this purpose he selected the reaction leading to the formation of *tetra-ethyl-ammonium iodide, from tri-ethyl-amine and ethyl iodide, thus*—

$$N(C_2H_5)_3 + C_2H_5I = N(C_2H_5)_4I.$$

The procedure consisted in diluting one volume of an [equivalent] mixture of the two substances with 15 volumes of a solvent; this was maintained for a definite time-interval at 100° in a sealed glass tube; thereafter the quantity of the ammonium base produced was determined by titration, so that the progress of the reaction was ascertained. The reaction progressed, in each one of the twenty-three solvents studied, completely and normally, *i.e.* in accordance with the formula which holds good for bi-molecular reactions; but the value of the velocity coefficient k varied in a very striking way with the nature of the solvent, as is shown by the values given in the following table:—

Solvent.	k	Solvent.	k
Hexane	0·000180	Methyl-alcohol	0·0516
Heptane	0·000235	Ethyl-alcohol	0·0366
Xylene	0·00287	Allyl-alcohol	0·0433
Benzene	0·00584	Benzil-alcohol	0·133
Ethyl-acetate	0·0223	Acetone	0·0608
Ethyl-ether	0·000757	...	...

The presence of a hydroxyl group, and also of an "unsaturated union" in the molecule, according to this, seems to favour the reaction velocity; and, as a rule, in homologous series, the velocity decreases with increasing molecular weight.

But one circumstance would appear to be very remarkable, viz. that, according to this, those solvents which are endowed with a great "*dissociating force*" towards the dissolved substances (see pp. 268 and 458), these, on the whole, are also those which impart the greatest reaction capacity to the dissolved substances. But, as was emphasised by Menschutkin, *the enormous differences between the velocity constants cannot be ascribed to the purely physical action of the solvent*, such as might consist in a difference between the [relative] number of molecular collisions.

Indifferent materials added to the solvent may alter the "nature

[1] *Zeits. phys. Chem.*, **6.** 41 (1890).

of the medium," and consequently exercise an appreciable influence on the velocity of reaction; the action of neutral salts referred to on p. 545 is perhaps to be referred in part to such an influence. According to a thorough investigation by Buchböcks,[1] in the action of foreign substances on the decomposition of carbonyl sulphide dissolved in water

$$COS + H_2O = CO_2 + H_2S$$

some connection with the viscosity of the solution is probable.

According to present experience the velocity of reaction in the gaseous state, at least at ordinary temperatures, is with few exceptions vanishingly small; the gaseous state is therefore a medium which, in accordance with its small dissociating power, imparts to the substance present in it very small capacity for reaction. The fact that oxy-hydrogen gas reacts slowly at the walls of the vessel (p. 540) may be explained by saying that the gas absorbed or dissolved by the walls of the vessel constitutes a medium of greater facility of reaction; it is well known that the facility of reaction is greater still for oxy-hydrogen gas at platinum surfaces. In this way many *catalytic processes* can be, if not explained, at least referred to the apparently less difficult conception of the influence of the medium discussed in this section.

The Kinetics of Heterogeneous Systems.—The velocity of reaction in heterogeneous systems possesses a theoretical interest not inferior to that possessed by the reaction velocity of homogeneous systems; because the former are very largely dependent upon the extent and the nature of the separating surface between the reacting phases, and also upon other circumstances of a secondary nature, such as the diffusion capacity. For these reasons there are difficulties in the way of obtaining, in such processes as the solution of solid substances in a solvent or of metals in acids, figures which shall be constant, and which shall also have a simple meaning, so that they can represent a measure of the velocity of the respective reactions.

Thus, especially in the case of solid substances, the nature of the surface is subject to great irregularities, and it is very difficult to obtain the appropriate material in a condition which is sufficiently uniform.

Thus the velocity of the solubility of pure zinc in acids is enormously increased by admixture with the slightest foreign material. This is simply explained by the electromotive action of these foreign substances upon the zinc, resulting in the formation of innumerable short circuits, and consequently resulting in the opportunity for the electrolytic transference of the zinc into the solution. For this reason

[1] *Zeits. phys. Chem.*, **23**, 123 (1897).

the numerous researches concerning this question have not as yet produced any simple result.[1]

According to the law of mass-action, the following inferences are very direct regarding the solution of metals in acids and analogous processes.

1. The velocity of the decomposition at every instant will be proportional to the extent of the contact surface O between the metal and the acid.

2. It will be proportional to the concentration of the acid.

Thus, let us denote by a the acid titration possessed by the solution at the beginning of the process for the time $t = 0$; and by $a - x$ the acidity after the time t, when x equivalents of the metal shall have gone into solution. Then, according to the assumption made above for the reaction velocity, *i.e.* for the quantity dx of the metal passing into solution in the time dt, we shall have—

$$\frac{dx}{dt} = kO(a - x) \quad . \quad . \quad . \quad . \quad (1)$$

and this, when integrated on the assumption that the surface remains constant during the solution, becomes

$$\ln \frac{a}{a - x} = kOt ;$$

Boguski[2] succeeded in verifying the preceding formula fairly well, by a research concerning the solubility velocity of Carrara marble in acids. A weighed marble plate was immersed in the acid, and after a measured time was removed, washed, dried, and again weighed. The quantity x dissolved was thus ascertained, corresponding to the time t, by the diminution in weight.

As was to be expected, equivalent solutions of HCl, of HBr, and of HNO_3 acted on the marble with the same velocity.

The *influence of temperature* in particular was later investigated by Spring.[3] He showed that an elevation of the temperature from 15° to 35°, and from 35° to 55°, in each case corresponded to a doubling of the solution velocity; and also that the temperature excites a strong accelerating influence on reactions which occur in *heterogeneous* systems, similarly to what has been regularly proved for *homogeneous* systems.

Iceland spar exhibits a dual solution velocity according as the attack of the acid is directed towards the one or the other of its two

[1] De la Rive (1830); Boguski, *Ber. deutsch. chem. Ges.*, **9.** 1646 (1876); Kajander, *ibid.*, **14.** 2050 and 2676 (1881); Spring and van Aubel, *Zeits. phys. Chem.*, **1.** 465 (1887); and others.

[2] *Ber. deutsch. chem. Ges.*, **9.** 1646 (1876); see also Boguski and Kajander, *ibid.*, **10.** 34 (1877).

[3] *Zeits. phys. Chem.*, **1.** 209 (1887).

principal crystallographic directions.[1] The quotient of the reaction velocities in the direction of the basal section and of the longitudinal section respectively, amounted—

At 15° to 1·13
,, 35° ,, 1·15
,, 55° ,, 1·14

Recently A. A. Noyes and W. R. Whitney[2] have established the important fact that the rate of solution is at each moment proportional to the difference between the concentration at the moment and that required for saturation. They showed in the same work, that at the boundary between the crystals and the solution the solid is at each moment saturated; the velocity of solution according to this is conditioned only by the rate of diffusion of the saturated solution of the boundary layer into the interior liquid. By stirring the thickness of the liquid layer adhering to the crystal is diminished, and thus the distance through which the diffusion has to take place reduced.

It is very probable that this rule can be generalised, that in the boundary layer the equilibrium is established with a velocity which, if not infinite, is at any rate so great that the equalisation by diffusion is slow compared with it. According to this view, the velocity of reaction in heterogeneous systems should be calculable from the diffusion coefficient, provided the distance through which the diffusion has to take place is known.

This hypothesis yields, as was shown by the author and E. Brunner,[3] in a complete experimental study, the means of treating theoretically the velocity of reaction in heterogeneous systems. With sufficiently vigorous stirring, which is kept constant throughout the experiment, it may be assumed that the solution has a homogeneous composition, and that the layer of constant thickness adheres to the surface of the solid and causes diffusion. If we consider, for example, the solution of magnesia in acid according to these principles, at the magnesia itself there would be a slight alkalinity, and the concentration of the free acid would therefore be very small; under these conditions the quantity

$$\mathrm{OD}\frac{a - x}{\delta}dt$$

would diffuse to the magnesia, where D is the diffusion constant of the acid, and the equivalent amount of magnesia would go into solution. The reaction constant k therefore, according to equation (1), becomes

$$k = \frac{D}{\delta} \quad . \quad . \quad . \quad . \quad (2)$$

[1] Spring, *Zeits. phys. Chem.*, **2.** 13 (1888). [2] *Zeits. phys. Chem.*, **23.** 689 (1897).
[3] *Dissertation*, Göttingen, 1903.

If the thickness of the adhering layer δ for a given kind and velocity of stirring can once be determined by experiment, the velocity constant k can be calculated in absolute measure. Thus Brunner, after determining δ as 0·03 mm. from a measurement of the rate of solution of benzoic acid, calculated the rate of solution of magnesia in different acids.

Thus it appears that it is not the strength of the acid which regulates the rate at which it attacks magnesia, but merely its diffusion constant, so that acetic acid which diffuses rapidly acts more quickly than benzoic acid, although the latter is much stronger.

Since, according to this, in chemical reactions which occur merely at the boundary between two phases, the phenomenon is essentially one of diffusion, it is useless to try and determine the order of reaction from the rate at which they proceed, as has often been attempted in recent years; this method of argument is only applicable, according to kinetic considerations, to the probability of collisions in homogeneous systems (p. 428), and loses its meaning when applied to heterogeneous systems.

The Linear Velocity of Crystallisation, *i.e.* the velocity with which the formation of crystals situated at a point proceeds, *e.g.* in an undercooled liquid contained in a glass tube, has been studied by Gernez (1882), Moore (1893), and in a very complete manner both theoretically and experimentally by G. Tammann.[1] The general behaviour is that it increases at first with the degree of undercooling, reaches a maximum and, with great undercooling, diminishes; the decrease can go so far that the velocity of crystallisation sinks practically to nothing, so that the undercooled liquid loses its capacity of crystallisation and remains of a glassy character. The maximum rate of crystallisation is, for example, for phosphorus 60,000, azobenzene 570, benzo-phenone 55, salol 4, betol 1 mm. per minute, it varies therefore in an extreme degree from one substance to another.

According to Tammann, the meaning of this remarkable fact is as follows. In the boundary between the solid and fused substance the temperature is that of the freezing-point, so that, when the undercooling is not too great, the velocity of crystallisation measured corresponds to this temperature, although this varies within certain limits with the degree of undercooling. When the undercooling is slight (less than 15°) the velocity is less, partly on account of the impurities always present, and which are very noticeable with slight undercooling, partly because the number of nuclei for crystallisation is too small; this makes the velocity very small if the undercooling is very slight, and causes it at first to increase proportionally to the undercooling before it reaches the maximum. When the under-

[1] Friedlander and G. Tammann, *Zeits. phys. Chem.*, **24.** 152 (1897); Tammann, *ibid.*, **25.** 441; **26.** 307 (1898); **29.** 51 (1899).

cooling is very great, the heat of fusion does not suffice to bring the boundary of the solid and fused substance to the freezing-point; this reduced temperature is the cause of the decrease in the velocity of crystallisation on great undercooling.

It was of importance to establish the fact that slight impurities influence the rate of crystallisation very largely;[1] this phenomenon may be used to decide whether a liquid mixture which solidifies to a homogeneous mass, such as $CaCl_2 + 6H_2O$ or $SO_3 + H_2O$, is a compound or not. In the first case, a slight addition of one of the components would alter the velocity of crystallisation very little, in the latter case very much. See the interesting study by F. A. Lidbury, *ibid.*, **39.** 453 (1902).

The Kinetic Nature of Physical and Chemical Equilibrium.—In concluding this description of the progress of chemical processes, we will turn back again to describe *the condition of equilibrium.*

It has been shown repeatedly that, in the sense of the kinetic molecular theory, *no condition of equilibrium between substances capable of mutual reaction can be regarded as a static, but rather it must be regarded as a dynamic equilibrium;* and this is true whether the action is *physical* or emphatically *chemical*, and also whether the equilibrium is established in a *homogeneous* or in a *heterogeneous* system. According to this view, it is not assumed that the material transformation has entirely ceased in the state of equilibrium; but only that the reaction progresses with the same velocity in the one direction as in the other, and that therefore for this reason in summation no change can be detected in the system.

In this same way we were able to account for the equilibrium—

(*a*) Between water and water vapour (p. 219).

(*b*) Between alcohol and acetic acid on the one side, and ester and water on the other (p. 426).

(*c*) Between the different states of the same substance existing in solution, and in the gaseous state (p. 481).

(*d*) Between the different parts of one substance distributed between two solvents (p. 485), etc.

In all of these cases the state of equilibrium was defined by the statement, that at every instant the amount of the decomposition in the sense of the reaction equation in one direction is the same as that in the other direction.

The question now arises, *How great is this decomposition in each particular case?*—a question which, so far as known to the writer, has previously been neither asked nor answered. It is evident, in any event, that in the sense of the theoretical molecular method of treatment, this question is fully qualified to demand a hearing, although from the nature of the case it may elude a direct experimental decision. It certainly would be very interesting to know how much ester and water

[1] Bogojawlensky, *Zeits. phys. Chem.*, **27.** 585 (1898).

are formed in unit time in the state of equilibrium established between these substances on the one hand, and how much alcohol and acetic acid on the other. Of course the same quantity of ester and water must be formed that is decomposed in the same time into alcohol and water.

As a matter of fact, the answer to this question is possible in all cases where we can measure the reaction velocity of a reaction which does not advance to a completion, *e.g.* it is possible in all of the cases enumerated above.

Thus, again, let us denote by k and k′ (p. 568) the coefficients respectively corresponding to the partial reaction velocities in the two opposite directions of the reaction equation. Then the measurement of the actual velocity gives the difference

$$k - k',$$

and the measurement of the condition of equilibrium gives the quotient

$$\frac{k}{k'},$$

from which both k and k′ can be calculated, and also the opposite [counterbalanced] decomposition in the state of equilibrium.

From the velocity of esterification, which was experimentally measured for equivalent quantities of alcohol and acid, we found that the difference between the coefficients amounted to

$$k - k' = \frac{3}{4}\frac{1}{t}\ln\frac{2 - x}{2 - 3x};$$

and, since the undecomposed quantity x was estimated in equivalents, and the time t in days, this becomes

$$\frac{4}{3}(k - k') = 0{\cdot}00575.$$

Also, according to p. 448,

$$\frac{k}{k'} = 4,$$

from which, by calculation,

$$k = 0{\cdot}00575.$$

Now since in equilibrium, $\frac{1}{3}$ equivalent of alcohol and of acid are present, therefore it follows that the velocity (previously called "partial") of the change in the state of equilibrium is

$$v = 0{\cdot}00575\frac{1}{3} \times \frac{1}{3} = 0{\cdot}00064.$$

Therefore in the system, as it exists in equilibrium, and consisting of

$\frac{1}{3}$ g.-mol. alcohol + $\frac{1}{3}$ g.-mol. acetic acid + $\frac{2}{3}$ g.-mol. ester + $\frac{2}{3}$ g.-mol. water

in the course of a day, 0·00064 g.-mol. of alcohol and acetic acid are transformed ; and of course the same quantity [of ester and water] in the same time is retransformed. From the smallness of this number it is obvious that we are not to conceive of the mutual exchange as always being a very "stormy" one. Of course with increasing temperature (the preceding values refer to the temperature of a dwelling-room), the velocity of exchange will increase in the same proportion as the value $k - k'$, *i.e.* very rapidly. The greater this decomposition in equilibrium is, the more quickly will the system strive to adjust itself to the new condition of equilibrium, when the old one is displaced from some change in the concentration or the like. We may therefore suitably designate its reciprocal value as the *retardation* (*Dämpfung*) *of the system.*

Moreover, particular attention should be called to this, viz. that we must regard the law of mass-action as *an empirical law which is certainly proved,* and therefore one which is independent of every theoretical molecular speculation. If the latter should at some future time come to be regarded as unsatisfactory, and especially if the kinetic conception of physical and chemical equilibrium should have to be relinquished as unreliable,[1] nevertheless the laws developed in this book regarding the decomposition of matter would remain completely undisturbed ; and in addition, it would be the duty of every new theory to give an account in its own way regarding the empirical facts of chemical mass-action.

How far the law of mass-action can be proved in a thermodynamic way, entirely independently of every molecular hypothesis, will be discussed in the last book.

[1] This is all the less likely since chemical kinetics has brought forward a *most striking confirmation* of the kinetic hypothesis.

BOOK IV

THE TRANSFORMATIONS OF ENERGY (THE DOCTRINE OF AFFINITY, II)

CHAPTER I

THERMO-CHEMISTRY I

THE APPLICATIONS OF THE FIRST LAW OF HEAT

General Remarks.—In the preceding books we have considered the transformations of matter, in their dependence upon the relative quantities of the substances which mutually react upon each other. Inasmuch as we invariably considered the displacement of the equilibrium and the progress of the reaction as taking place *isothermally*, and as variations in temperature were thus excluded, disregarding the introduction of electrical energy and the action of light, we conceived the chemical changes as being purely *material*, without reference to their associated *changes of energy*.

Now both the state of equilibrium and the reaction velocity are dependent upon a number of other factors besides the relative quantities [of the reacting substances]; the action of these factors collectively may be regarded as being associated with the introduction of energy into, or the abstraction of energy from, the system considered. These factors are especially *temperature*, *pressure*, *electrification*, and *illumination*.

And, conversely, *a chemical change* is on its part invariably accompanied by *changes of energy*, which are exhibited by a change in one or more of the factors just enumerated.

On the one hand, the action of the pressure and of the temperature upon chemical reaction is by far the most important and far-reaching; and, on the other hand, the development of heat, and the performance of external work by chemical processes, are equally important.

The description of these later relations constitute the subject of *thermo-chemistry*, to which the first five chapters of this book are devoted. In the last two chapters there will be presented the fundamentals of *electro-chemistry* and of *photo-chemistry* respectively.

Heat of Reaction.—As was stated in the introduction, p. 7, in all processes which occur in nature, we can discriminate the following changes of energy, viz. :—

1. The giving off, or the taking on, of heat.
2. The performance of external work.
3. The variation of the internal energy.

Let us consider a chemical system, and that a simple one; and after it has experienced some change in its substance, let it return again to its original temperature, which it had before it had begun to change. Then, according to *the law of the conservation of energy* [JW = A, and U = A − Q], the heat Q developed in the change, increased by the external work A, performed by the system, is equal to the diminution [U] of the system's internal energy.

Now the heat developed by a reaction can be easily measured, by immersing the flask containing the reaction mixture in the water of a calorimeter, and then conducting the reaction in a suitable way. The amount of the heating of the water in the calorimeter, together with "the water value" of the calorimeter itself, corresponds to the heat developed by the reaction. The amount of work which is associated with the reaction consists almost always in overcoming the atmospheric pressure. This work performed, therefore, can be estimated in *litre atmospheres*, from the change in volume associated with the reaction as given in litres; and this by multiplication by 24·25 can be reduced to g.-cal. (p. 12).

The sum of the heat produced in the reaction, and of the external work performed, both of these quantities being expressed in g.-cal. (p. 12), *we will call the heat of reaction in question: of course this can be either positive or negative* (*a*), *according as heat is either produced or absorbed by the reaction; and* (*b*), *according as external work is performed either against the external pressure, or by the external pressure upon the system. This heat of reaction represents the change of the total energy* [U] *of the system.*

Of course, other things being equal, the heat developed and the work performed are proportional to the quantity of the substance which is changed. Wherever nothing is stated to the contrary, *the heat of reaction will always refer to the change of 1 g. equivalent.*

Thus, *e.g.*, it is observed that in the solution of 1 g.-atom of Zn (= 65·4 g.) in dilute H_2SO_4, at 20°, there are developed 34,200 g.-cal. At the same time 1 g.-mol. of H (= 2 g.) is set free, whereby a certain amount of work is performed against the pressure of the atmosphere.

Now since 1 g.-mol. of any gas at 0° occupies 22·42 litres, therefore at the absolute temperature T it occupies

$$22{\cdot}42\frac{T}{273}\text{ litres (p. 40),}$$

and therefore the external work performed by the system amounts to

$$22{\cdot}42\frac{T}{273} = 0{\cdot}0821\text{ T litre atm.};$$

and since 1 litre atm. is equal to 24·25 g.-cal., it amounts to 1·99 T, or in round numbers to 2 T g.-cal.; therefore in the solution of Zn, the work of

$$2(273 + 20^*) = 586\text{ g.-cal.}$$

will be performed *against the pressure of the atmosphere.*

Thus the heat of reaction, or the difference between the values of the internal energy possessed by the system before and after the solution of the Zn, amounts to

$$34{,}200 + 586 = 34{,}786\text{ g.-cal.}$$

Thus it is obvious in this example (where a gas is developed, and where the change of the volume of the system, as a result of the reaction, is very considerable) that the external work performed only plays *the rôle of a correction value:* and moreover in all cases, where the reacting and resulting substances are collectively solid or liquid, the change of volume is of a much smaller order of value, so that in these latter cases all the changes of volume can be disregarded, being infinitesimal in comparison with the unavoidable errors of observation.

In the case of the combustion of hydrogen and oxygen to form liquid water, for every g. of H there results 68,400 cal. Now in this process there disappears 1·5 g.-mol. of the gases; and thus the atmospheric pressure, at the same time, performs the work of

$$586 \times 1{\cdot}5 = 880\text{ cal.,}$$

so that the change in the total energy amounts to

$$68{,}400 - 880 = 67{,}520.$$

The Thermo-chemical Method of Notation.—When a reaction occurs, according to the general scheme,

$$n_1A_1 + n_2A_2 + \ . \ . \ . = n_1'A_1' + n_2'A_2' + \ . \ . \ .,$$

then there will be associated with it a certain heat of reaction, and

* 20 = 20°, the average temperature of a room.—TR.

this will amount to U when n_1 g.-mol. of the substance A_1 unites with n_2 g.-mol. of the substance A_2, etc. Then, according to the law of the *conservation of energy* [First Law], this will amount to $-U$ when n_1' g.-mol. of the substance A_1' unites with n_2' g.-mol. of the substance A_2'. The value of U corresponds to *the energy-difference* between the two systems,

$$n_1A_1 + n_2A_2 + \ldots, \quad \text{and } n_1'A_1' + n_2'(A_2') + \ldots$$

The content of energy of a chemical system is equal to the sum of the contents of energy of the particular components. If we denote the content of energy of 1 g.-mol. of a substance A by the symbol

$$(A),$$

then the content of energy of n-mol. of the substance A will be represented by

$$n(A).$$

The contents of energy of the two systems considered above will be denoted by the symbols

$$n_1(A_1) + n_2(A_2) + \ldots, \quad \text{and } n_1'(A_1') + n_2'(A_2') + \ldots;$$

and

$$U = n_1(A_1) + n_2(A_2) + \ldots - n_1'(A_1') - n_2'(A_2') - \ldots$$

will denote the amount of the heat of reaction taking place between them, because U corresponds to the difference between the contents of energy of the two systems.

When U is *positive*, *i.e.* when the progress of the reaction, in the sense of the equation from left to right, is associated with *a development of heat*, then the reaction is called "*exothermic.*" The opposite reaction, therefore, is associated with *the absorption of heat*, and is called "*endothermic.*"

Thus, *e.g.*, the symbolic equation

$$(S) + (O_2) - (SO_2) = 71,080$$

denotes the union of 32 g. of S with 32 g. of O, and corresponds to the development of 71,080 g.-cal., and the formation of sulphur dioxide from the respective elements is an exothermic reaction.

As a rule the heat of reaction refers to the measurement of substances which react *in dilute water solution.* The energy content of a substance A, *which is dissolved in a large quantity of water*, is denoted by the symbol

$$(A \text{ aq.}),$$

(aq. = *aqua* or water), and therefore the quantity of heat which is developed by the solution of 1 g.-mol. of a substance A, in a large quantity of water, the so-called "*molecular heat of solution,*" corresponds to the expression

$$U = (A) - (A\ aq.).$$

Thus, *e.g.*, the meaning of the thermo-chemical equation,

$$(HCl\ aq.) + (NaOH\ aq.) - (NaCl\ aq.) = 13{,}700,$$

is, that by neutralising 1 equivalent of HCl by 1 equivalent of NaOH *in dilute solution*, there are developed 13,700 g.-cal., the so-called "*heat of neutralisation.*"

It is usually the custom to abbreviate the preceding method of notation in those cases where the resulting condition of the system, after the end of the reaction, can be seen directly from a description of the original condition. In such a case, the difference between the energies contained by the system in its original and in its final condition is denoted by enclosing the formulæ of the reacting substances in a common parenthesis, the formulæ being separated by commas. Thus, *e.g.*, instead of

$$(S) + (O_2) - (SO_2) = 71{,}080,$$

we write the shorter form

$$(S,\ O_2) = 71{,}080\ ;$$

and instead of

$$(HCl\ aq.) + (NaOH\ aq.) - (NaCl\ aq.) = 13{,}700,$$

the shorter form

$$(HCl\ aq.,\ NaOH\ aq.) = 13{,}700\ ;$$

and similarly in other cases. Then, of course, *e.g.*, the formula

$$-(HCl\ aq.,\ NaOH\ aq.)$$

will denote the quantity of heat which will be absorbed by the decomposition of a water solution of NaCl to a water solution of NaOH and of HCl, viz. 13,700 cal.

Similarly the notation

$$(A) + (B) - (AB) = U,$$

is, of course, identical with

$$(A) + (B) = (AB) + U,$$

because the thermo-chemical equations denote simply the summations of energy magnitudes, and accordingly we may apply the ordinary algebraic transformations to them. Thus, *e.g.*, if, from the preceding equation, we substract the following one, viz.

$$(A) + (C) = (AC) + U,$$

we obtain as an immediate result from the two preceding formulæ, the equation

$$(B) + (AC) = (AB) + (C) + U - U';$$

from this latter equation we deduce the result that the substitution of B in the place of C, in the compound AC, corresponds to a heat of reaction of $U - U'$.

We do not know the value of (A) itself, *i.e.* the absolute content of energy of 1 g.-mol. of a substance, although sometimes the kinetic theory of gases leads to a (hypothetical) conception of its magnitude. Thus, according to the accepted views in the light of this theory, the energy content of monatomic mercury vapour consists solely in the translatory energy of its atoms; and this, at the temperature T (p. 204), in "absolute units" amounts to

$$\frac{M}{2}u^2 = 183{,}900^2\frac{T}{273}.$$

But the knowledge of the total energy content of a substance is, for practical purposes, entirely without significance, because we deal only with *the energy differences* of various systems, a knowledge of which is furnished by thermo-chemical measurements.

Finally, it should be noted that, for the sake of brevity, *the atoms* instead of the molecules are denoted in the thermo-chemical formulæ. Thus, *e.g.*, the equation

$$(H_2, O) = 67{,}520$$

does *not* denote the heat of reaction consequent upon the union of *atomic oxygen*, something which is entirely unknown to us, but only the union of 1 g.-atom (16 g.) of ordinary oxygen with [the requisite quantity of] hydrogen. Strictly speaking, it would be more correct to write

$$(2H_2, O_2) = 2 \times 67{,}520,$$

but error is certainly guarded against in such and all similar cases.

The Law of Constant Heat Summation.—If we allow a system to experience various chemical changes, so that it finally returns to its original condition, then the sum of the heat of reaction associated with these processes is equal to zero; for otherwise it would mean a loss or a gain of the total energy, which would contradict the first law of thermodynamics. If we bring the system to the same final condition by means of two different ways, then the heat of reaction must be the same for both methods, *i.e.*—

The energy differences between two identical conditions of the system must be the same, independently of the way by which the system is transferred from one condition to the other.

It is particularly worthy of notice that this theorem, which of course holds good, not only for all chemical processes, but also as a general proposition, was clearly stated by Hess[1] as "*the law of constant heat summation*," and moreover also experimentally proven by him as early as 1840; and this was before "*the law of the conservation of energy*" had emerged from the realm of hazy anticipation, over the threshold and into the consciousness of the scientific world.

The following example will serve to explain the meaning of "*the law of constant heat summation*."

We will consider a system, consisting of 1 g.-mol. of ammonia (NH_3), 1 g.-mol. of hydrochloric acid (HCl), and a large quantity of water in the two following conditions, viz. :—

1. Where the three substances exist separate from each other.
2. Where the three substances form a homogeneous solution of ammonium chloride in a larger quantity of water.

We can cause the system to pass from the first to the second condition in two different ways: thus, *on the one hand*, the two gases [HCl and NH_3] can unite to form solid ammonium chloride, and then this can be dissolved in water; or, *on the other hand*, the two gases can be separately absorbed by water, and then the resulting solutions can be caused to neutralise each other. The respective heats of reaction result as follows :—

The First Way.	The Second Way.
(NH_3, HCl) = +42100 cal.	(NH_3, aq.) = + 8400 cal.
(NH_4Cl, aq.) = − 3900 ,,	(HCl, aq.) = +17300 ,,
	(NH_3, aq., HCl aq.) = +12300 ,,
(NH_3, HCl, aq.) = +38200 cal.	(NH_3, HCl, aq.) = +38000 cal.

Now, in fact, the energy difference between the initial and the final states of the system amounted to the same in both cases, *i.e.* within the limits of error.

The theorem of "the constancy of the heat summation" is very important, and has a many-sided application in practical thermochemistry. Only very few reactions are suited for a direct investigation in the calorimeter, because it is absolutely necessary for exactness of measurement—

(*a*) That the reaction shall be such as can be easily effected.

(*b*) That it shall take place quickly, in order to avoid a large loss of heat from radiation.

(*c*) That the reaction shall be free from side reactions, which cannot usually be taken into consideration.

But even in those cases, where the conditions of [readiness], quickness, completeness, and simplicity of reaction are not well fulfilled, it is usually possible to accomplish the purpose by certain circuitous ways; thus, by the assistance of certain suitable inter-

[1] Ostwald's *Klassiker*, No. 9.

mediate substances, the system can be transferred from the one condition to the other, the energy-difference between which is to be measured. Thus it is not possible to determine directly the energy-difference between charcoal and diamond, because the transformation of one modification into the other has not been accomplished. But if we can change charcoal and diamond into the same compound by means of an intermediate substance, then the difference between these two quantities of heat gives the heat value for the conversion of one modification into the other.

Such an intermediate substance, and one which is very commonly used, is *oxygen*. Thus, *e.g.*, the different modifications of carbon when burnt in "the calorimetric bomb" (see below), gave the following results :—

		Diff.
Amorphous carbon . . .	97650	
		2840
Graphite	94810	
		500
Diamond	94310	

This table means that 2840 cal. would be developed by the conversion of 12 g. of charcoal into graphite; and 500 cal. by the conversion of 12 g. of graphite into diamond; [and therefore 3340 cal. by the conversion of 12 g. of charcoal into diamond].

In a similar way, the "heats of formation" of organic compounds can be ascertained from the respective heat values of their combustion in oxygen. By this process the compound, whose "heat of formation" is unknown, is changed into compounds (CO_2, H_2O, etc.), the heats of formation of which are known.

Thus the energy-difference between (H_2, I_2) and 2(HI) is not obtainable by a direct measurement, because hydrogen and iodine unite with each other only very slowly. But suppose—

(*a*) That we dissolve the hydriodic acid in water.

(*b*) That we neutralise it with potassium hydroxide.

(*c*) That we set the iodine free by means of chlorine.

(*d*) That we decompose the potassium chloride which is formed into potassium hydroxide and hydrochloric acid.

(*e*) That we decompose the hydrochloric acid which is formed again into chlorine and hydrogen.

Then by means of the intermediate substances H_2O, KOH, and Cl_2, we can proceed from gaseous hydriodic acid to free hydrogen and free iodine; and we can do it by means of reactions which progress *quickly and smoothly, in one sense or in the other*, and by reactions each one of which has a heat of reaction capable of good measurement.

In fact, the heat of formation of gaseous hydriodic acid [from the respective elements] has been determined in this way.

The Influence of Temperature upon the Heat of Reaction.—If we allow the same reaction to occur, once at the temperature t_1 and again at the temperature t_2, then the heat of reaction will be different in the two cases; let them amount to U_1 and U_2 respectively.

Now we can imagine the following cyclic process to be carried out. Let the reaction occur at the temperature t_1, whereby the quantity of heat U_1 will be developed; then we raise the temperature of the system to t_2, whereby there is required the introduction of $(t_2 - t_1)c'$ g.-cal. of heat, where c' denotes the heat capacity of the substances *resulting from the reaction.*

Now let the reaction occur *in the opposite sense* at t_2: this process is associated with the quantity of heat U_2; then let the system be cooled to t_1, whereby the quantity of heat $(t_2 - t_1)c$ will be given off, where c denotes the heat capacity of *the reacting substances.* The system has now returned to its original condition.

Now the law of *the conservation of energy* requires the relation, that the heat absorbed by the system shall be the same as that given out; that is, that

$$U_2 + (t_2 - t_1)c' = U_1 + (t_2 - t_1)c,$$

and hence

$$\frac{U_2 - U_1}{t_2 - t_1} = c - c',$$

that is, the excess of the heat capacity of the reacting substances, over the heat capacity of the resulting substances, gives the increase of the heat of reaction per degree of temperature elevation.

Now the specific heats of the substances taking part in the reaction equation are obtainable by direct measurement; therefore the temperature coefficient of the heat of reaction can be determined much more exactly in this indirect way than is possible by a direct measurement of the heat of reaction at two different temperatures. The preceding equation, moreover, follows directly from the application of the universal theorem developed on p. 9.

On p. 176 we derived the proposition that the specific heats of solid compounds constitute an *additive* property; or, in other words, in the union of solid substances to form solid compounds the heat capacities remain unchanged. Then, according to what has preceded, this law can be extended thus:—

The heat of combination of solid substances is independent of the temperature.

The application of the calculation to special cases shows that the temperature coefficient of such substances as iodine and silver, in any case, must be smaller than 0·0001.

Of course in every case, in thermo-chemical statements, reference must be made to the temperature at which the measurements are conducted.

Thermo-Chemical Methods.—In general, the thermo-chemical methods are those of calorimetry, the fundamentals of which will be found described in every text-book of physics. The *water-calorimeter* is used altogether most frequently; although sometimes, in recent work, use has been made of Bunsen's *ice-calorimeter*, especially when the measurements concern very small quantities of heat. The values obtained by means of this last apparatus of course hold good for 0°, and therefore they are not directly comparable with those obtained by means of the water-calorimeter, which usually refer to a temperature somewhere near 18°. In the comparison of different observations, notice should be given to the remarks concerning *the unit of energy* on p. 11.

Here, as in all cases, we assume the calorie referred to water at 18° *as unit.*

Now since, in the case of those chemical reactions which occur very quickly, it takes a certain amount of time before the reaction-heat becomes uniformly distributed through the calorimeter, therefore a correction must be made for the heat absorbed, or that given off by radiation during the reaction. As is well known, this correction is obtained in this way: the course of the thermometer is observed from some time before the beginning of the particular experiment till some time after its close.[1] This correction involves the most dangerous sources of error in thermo-chemical measurements, and therefore the conditions for the investigation must be so arranged that the amount of this correction shall be as small as possible. This end will be obtained by paying attention to the following conditions:—

1. The reaction, the heat of which is to be determined, must take place as quickly as possible.

2. The heat capacity of the calorimeter must be as great as possible, and thus the change of the temperature of the calorimeter, resulting from the heat developed by the reaction, will be made as small as possible.

The second condition requires the use of a sensitive thermometer in order to determine the changes of temperature very accurately, *e.g.* within $\frac{1}{1000}$th part of its amount. The thermometer described on p. 258, and recommended by Beckmann, is very well suited for this purpose; this thermometer shows differences of $\frac{1}{1000}$th of a degree of elevation of temperature, and therefore it allows an exact measurement within this limit.

The accompanying cut (Fig. 31) shows a model of a calorimeter which can be constructed out of simple material, as recommended by Ostwald. The writer can recommend it from experience.[2]

A narrow beaker glass, which is cut down somewhat at the upper

[1] For more particulars see the handbooks of physics, or, *e.g.*, Ostwald, *Allg. Chem.*, 2nd edit., p. 572 (1891).

[2] *Zeit. phys. Chem.*, **2.** 23 (1888).

end, rests on cork supports fastened with shellac in a second larger beaker. This is provided with a wood cover, which rests on the outer beaker. Through this cover there pass a Beckmann thermometer, a thin test-tube in which the reaction is to be effected, and a stirring apparatus made of brass or better of platinum, and provided with a handle of some non-conducting material (as hard gum, wood [cork], or the like).

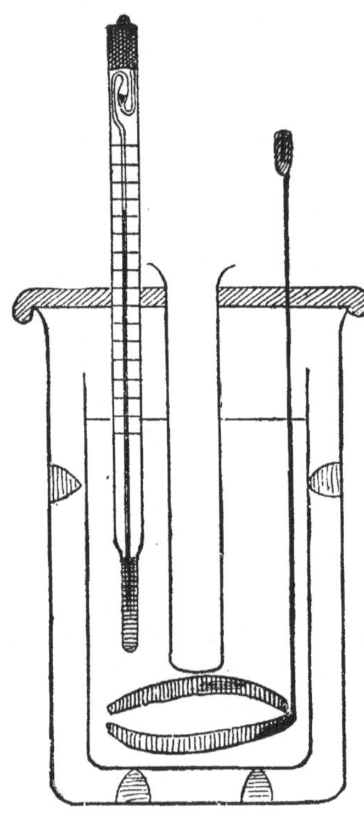

Fig 31.

If it is desired to measure heats of dilution or solution, the substance in question is placed in the test-tube; then after the temperature equilibrium is established, the bottom of the test-tube is punctured with a glass rod.

An advantage of this apparatus is its transparency. Again, if the capacity of the inner beaker is about a litre, then the "water value" of the glass, the stirrer, the thermometer, and all, amounts to only a few per cent of the total heat capacity.

With slight changes, this so-called "universal" calorimeter may be employed also for the measurement of specific heats, of heats of fusion, etc. If one operates with salt solutions, it is important to notice the rule regarding the heat capacity given on p. 106.

Steinwehr has described an apparatus[1] for studying the thermochemistry of dilute solutions capable of measuring very small quantities of heat.

Regarding the determination of the heat of combustion, we will say something more later. A method for the theoretical calculation of the reaction-heat from the change of chemical equilibrium will be considered in the second and third chapters of this book.

As a matter of history, it should be noticed that besides Hess, Andrews, Graham, Marignac, Favre, Silbermann, and others, J. Thomsen in Copenhagen since 1853, and Berthelot in Paris since 1865, especially have subjected the most various reactions to a thorough and systematic chemical examination.[2]

[1] *Zeits. phys. Chem.*, **36.** 185 (1901).

[2] Thomsen has collected his measurements in the work entitled *Thermo-chemische Untersuchungen*, Leipzig, 1882-86; and Berthelot has collected his in his *Essai de mécanique chimique*, Paris (1879). See also Naumann, *Thermochemie*, Braunschweig

Gases and Solutions.—The internal energy of gases, as already shown (p. 42), does not change with their volume; and therefore the heats of reaction are independent of the density of the reacting gases. Also the intermingling of gases is without influence [on the heat of reaction]. Of course these laws hold good only when no external work is performed, such as forcing back the atmospheric pressure. When this happens, then the reaction-heat is diminished by an amount which is equal to the work performed against the atmospheric pressure (p. 587). In the case of strongly compressed gases which are cooled by simple expansion (without the performance of external work), the preceding laws no longer hold strictly true.

A complete analogy is found in the conduct of dilute solutions. Since the energy of dissolved substances is independent of the concentration (p. 148), therefore the latter has no influence on the *reaction-heat* of a dissolved substance. The energy of "(A aq.)" is independent of the dilution of the solution, and herein consists the correctness of the introduction of this symbol (p. 589).

Thus, *e.g.*, if one determines the *heat of solution* of a substance in a large quantity of water, this heat of solution itself is independent of the quantity of water used.

If one determines the *heat of combination* of a metal with an acid, this heat is independent of the concentration of the acid when the latter is dissolved in a large quantity of water, etc.

But this rule ceases to hold good for concentrated solutions when these exhibit marked heat phenomena on dilution. Thus, *e.g.*, the *heat of solution* of zinc in concentrated sulphuric acid is very different from that in the dilute acid.

The *heat of dilution* has already been discussed (p. 114). We will only recall here that the following equation, viz.—

$$(H_2SO_4 + 3H_2O) + 2(H_2O) - (H_2SO_4 + 5H_2O) = 1970,$$

which states that by the addition of two more mol. of H_2O to a solution having the composition $(H_2SO_4 + 3H_2O)$, there will be produced 1970 g.-cal. Moreover, the heat of dilution may be *positive*, as in the case of sulphuric acid; or *negative*, as in the case of saltpetre.

The Changes of the State of Aggregation.—When a substance changes its state of aggregation, whether the change consists in evaporation [condensation], solidification [fusion], sublimation, or finally

(Brunswick), 1882; and H. Jahn, *Thermochemie*, Wien (Vienna), 1892. Thermochemical data collected from different observers will be found in the *Chemiker-kalender*. The figures given in what follows are largely taken from the critical and very complete collection made by Ostwald, *Allg. Chem.*, 2 Aufl. (2nd edit.), II. Leipzig, 1893. [Mention should also be made of the work of Stohmann of Leipzig, *Jour. Prak. Chem.* Probably the best work in English on thermo-chemistry is Muir's *Elements of Thermal Chem.* (1885).—TR.] Berthelot gives a new collection of data in his *Thermochimie*, Paris, 1897.

in a change from one [allotropic] modification to another, the changes are always associated with a change in the content of energy; and therefore the reaction-heat also is changed to the same amount.

This is especially noticed in those reactions where there is a precipitate thrown down and the state of the latter must be considered. Thus, *e.g.*, if *mercuric iodide* is precipitated from a solution of mercuric chloride by a solution of potassium iodide, at first the *yellow* modification comes down; but this quickly changes to the *red* variety, with the necessary accompaniment of a considerable heat of reaction.

The natural supposition that those allotropic changes, which occur *exothermally*, would be associated with an *increase in density*—this does not appear to be established, although the investigation has been thorough.[1]

Thus, *e.g.*, the energy-difference between water at 100°, and water-vapour at the same temperature, is expressed by the equation

$$\underset{\text{Liquid}}{(H_2O)} - \underset{\text{Gas}}{(H_2O)} = 536{\cdot}4 \times 18 - 2 \times 373 = 8910;$$

because the heat vaporisation of 1 g. of water was found to be 536·4, and the external work performed was shown to be $2T = 746$.

The energy-difference between ice and liquid water is

$$\underset{\text{Solid}}{(H_2O)} - \underset{\text{Liquid}}{(H_2O)} = 79 \times 18 = 1422,$$

because 79 cal. are set free in the freezing of 1 g. of water. On account of the slight change of volume, the external work performed here has only a minimal value.

The energy-difference between 1 g.-atom of the orthorhombic and the monoclinic modification of sulphur amounts to

$$(S_r) - (S_m) = 32 \times 2{\cdot}52 = 80{\cdot}6.$$

Thus in all cases where doubt prevails, the thermo-chemical results can afford information respecting the state of aggregation, or the modification respectively, of the reacting substances, or of those produced.

The Heat of Solution.—By the "*heat of solution*" is meant, as already stated, the quantity of heat produced by the solution of 1 g. of a substance in a large quantity of the solvent. If the heat of dilution of a substance is known, then of course it is possible to calculate the quantity of heat [of solution] when dissolved in any quantity of the solvent, however great. The quantity of heat observed when the substance is dissolved to exact saturation is always more or less different from the heat of solution as defined above; and sometimes it has a sign opposite from that of the heat of solution.

[1] Petersen, *Zeits. phys. Chem.*, **8.** 601 (1891).

In the following tables are given[1] the heat of solution of some substances in a large quantity of water at 18-20° :—

HEAT OF SOLUTION OF GASEOUS SUBSTANCES

Substance.	Formula.	Heat of Solution.
Chlorine	Cl_2	+ 4870
Carbon dioxide	CO_2	+ 5880
Ammonia	NH_3	+ 8430
Hydrofluoric acid . . .	HF	+11800
Hydrochloric ,, . . .	HCl	+17310
Hydrobromic ,, . . .	HBr	+19940
Hydriodic ,, . . .	HI	+19210

HEAT OF SOLUTION OF LIQUID SUBSTANCES

Substance.	Formula.	Heat of Solution.
Methyl alcohol	CH_4O	+ 2000
Ethyl ,, ...	C_2H_6O	+ 2540
Propyl ,, ...	C_3H_8O	+ 3050
Ethyl ether	$C_4H_{10}O$	+ 5940
Acetic Acid	$C_2H_4O_2$	+ 420
Sulphuric acid	SO_4H_2	+17850

HEAT OF SOLUTION OF SOLID SUBSTANCES

Substance.	Formula.	Heat of Solution.
Potassium hydroxide . .	KOH	+12500
,, ,, (cryst. hydrate) . .	$KOH + 2H_2O$	− 30
Lithium chloride . . .	LiCl	+ 8440
Sodium chloride . . .	NaCl	− 1180
Potassium chloride . . .	KCl	− 4440
Sodium bromide . . .	NaBr	− 190
,, ,, (hydrate) . .	$NaBr + 2H_2O$	− 4710
Potassium sulphate . . .	K_2SO_4	− 6380
Mercuric chloride . . .	$HgCl_2$	− 3300
Sodium acetate	$NaC_2H_3O_2$	+ 4200
,, benzoate	$NaC_7H_5O_2$	+ 800
Benzoic acid	$C_6H_5CO_2H$	− 6700
Silver chloride	AgCl	−15800
,, bromide	AgBr	−20200
,, iodide	AgI	−26600
Cane sugar	$C_{12}H_{22}O_{11}$	− 800

[1] From Horstmann, *Theoret. Chem.*, p. 502, Braunschweig (Brunswick), 1885.

The figures for gases include the external work: therefore in order to obtain the heat of true solution, the quantity + 580 must be subtracted from the figures given above.

All of the gases thus far examined dissolve with *evolution of heat.* A similar thing is true of liquids, at least as a rule; while solid substances sometimes dissolve with a development of heat, but more often with absorption of heat.

The explanation of this is simple. If one starts out with the assumption that a gaseous substance always has a positive solution-heat, then in the liquid state it will dissolve either with absorption or with evolution of heat, according as its heat of vaporisation exceeds the heat of solution, or not. And similarly, the signs for the solution-heat of substances in the solid state are conditioned by the difference between their heat of sublimation and heat of solution in the gaseous state.

The general rules for the solution-heat of the different states of aggregation as just given therefore mean that, as a rule, the heat of solution in the gaseous state is greater than the heat of vaporisation; but it is usually smaller than the heat of vaporisation plus the heat of fusion, *i.e.* it is smaller than the heat of sublimation.

In general, in comparing substances which are chemically analogous, and soluble with difficulty, the heat of precipitation (= *the negative value of the heat of solution*) *is the greater the more insoluble the substance is.*[1]

We will consider later a method for the determination of the heat of solution of salts which are soluble with difficulty. The difference between the solution-heats of one substance in two solvents would, of course, allow the calculation of the heat phenomena in the distribution of one substance between two solvents (p. 485).

The Heat of Formation.—*By the "heat of formation" of a chemical compound is meant the quantity of heat which is given off in the formation of the compound from its respective ingredients.*

This is *the thermo-chemical characteristic* of the compound in question. If one knows the "heats of formation" of all the substances which participate in any selected chemical reaction, then he knows also the heat of reaction.

Thus we can first imagine the particular substances standing on the left side of the reaction equation to be decomposed into their respective elements; and then imagine the elements to unite to form the substances standing on the right side of the reaction equation. Then in the first of these [hypothetical] stages there would be *absorbed* a quantity of heat which would be equal to the sum of the "heats of formation" of all the substances standing on the left side of the equation; and in the second [hypothetical] stage there would be *developed* a quantity of heat equal to the "heats of formation" of all the substances standing on the right side of the equation. It is a

[1] Thomsen, *Jour. prakt. Chem.* [2], **13.** 241 (1876).

matter of indifference whether the chemical process actually takes place in this way or not; because the change of total energy of any system is independent of the way in which the change is accomplished. We thus develop the proposition that—

The heat of reaction is equal to the sum of the heats of formation of the substances formed, minus the sum of the heats of formation of the substances used up.

Here will be given some "*heats of formation*" which will find application in subsequent calculations. The figures, as always, refer to constant volumes, *i.e.* they are corrected for the external work performed, whatever it may be. The remarks in the last column refer to the condition of the reacting substances.

Reaction.	Heat of Reaction.	Remarks.
$2H+O=H_2O$	+ 67520	Liquid water.
$C+2O=CO_2$	+ 94300	Diamond.
$C+O=CO$	+ 26600	,,
$S+2O=SO_2$	+ 71080	Orthorhombic sulphur.
$H+F=HF$	+ 38600	Gaseous fluorine.
$H+Cl=HCl$	+ 22000	,, chlorine.
$H+Br=HBr$	+ 8400	Liquid bromine.
$H+I=HI$	− 6100	Solid iodine.
$N+3H=NH_3$	+ 12000	...
$N+O=NO$	− 21600	...
$N+2O=NO_2$	− 7700	Dissociated nitric dioxide.
$2N+4O=N_2O_4$	− 2600	Bi-molecular nitric dioxide.
$K+F=KF$	+109500	...
$K+Cl=KCl$	+105600	
$K+Br=KBr$	+ 95300	
$K+I=KI$	+ 80100	...

The *rise of temperature* due to a reaction (*e.g.*, flame temperatures) can be calculated from the heat of reaction and the thermal capacity C of the resulting products; for

$$\int_{t_0}^{t} C dt = q$$

where t_0 is the initial, t the final temperature. The variation of C with temperature must of course be known. See Nernst and Schönflies, *Mathem. Behandl. d. Naturwissensch.*, II. edit. p. 136, Munich, 1898.

The Heats of Combustion of Organic Compounds.—The work of the investigators previously mentioned enables us to infer the heat of reaction of almost all the more important reactions of inorganic chemistry, directly from the table, or else to calculate them indirectly; yet, on the other hand, we do not find ourselves so fortunate regarding many reactions of the carbon compounds. The reason for this is that the number of reactions which take place quickly, and without the formation of side products, and which at the same time are suited for thermo-chemical investigation, is very much smaller in organic than in inorganic chemistry. There is only *one* reaction which occurs quickly and smoothly in all cases, viz. *the combustion of the substance in an excess of oxygen*, whereby all the carbon is oxidised to carbon dioxide, and all the hydrogen to water. Therefore, on the whole, organic thermo-chemistry employs the same reaction which has been long used in elementary analytical analysis, *i.e. combustion.*

Regarding the experimental side, the method followed almost exclusively at present consists in inclosing the substance in a vessel of steel, lined with platinum; then this is filled with oxygen under a pressure of about 20 atm.; and then the substance is ignited by a bit of iron wire made incandescent by an electric current (Berthelot's calorimetric bomb): this whole apparatus is immersed in *a water calorimeter*, which absorbs the heat developed.

In this way one obtains the *heat of combustion at constant volume*, or simply the heat of combustion which corresponds to the change of total energy. The heat of combustion at constant pressure still includes, so to speak, the accidental amount of external work, and is corrected by subtracting the value of "2T" as many times as the number of molecules produced exceeds those which disappear.

The *heat of formation* may be calculated from the *heat of combustion* thus. From the sum of the heats of formation there is substracted the heat of the resulting liquid water (67,500 per mol.) and of the resulting carbon dioxide (94,300 per mol.), and that of the sulphurous acid formed if any (71,100 per mol.); and thus by the use of the figures given in the last table we obtain the heat of formation, from the diamond, from gaseous oxygen and hydrogen, and finally, from orthorhombic sulphur, at the temperature of the experiment.

Now, since the heats of combustion have been determined for the most important organic compounds, the heats of formation are also known; and therefore, according to p. 600, the heat of reaction between these compounds can be calculated. Therefore it is never necessary to calculate the heat of formation, but one can obviously obtain the *heat of reaction* of every reaction, *from the sum of the heats of combustion of the substances which have disappeared, minus the sum of the heats of combustion of the substances formed.*

But the results of such calculations are usually very inexact, because the reaction-heat is the resulting difference between magnitudes which are only slightly different; and of course the errors of observation in these cases amount to a very considerable per cent.

Thus, *e.g.*, the heat of combustion of *fumaric acid* was determined to be 320,300, while that of the geometrical isomer, *maleïc acid*, is

326,900. Therefore the transformation of maleïc acid to the more stable fumaric acid will develop 6600 calories. But this latter number is extremely uncertain, for if the errors of observation in the determination of the heat of combustion amounted to only 0·5 per cent, even then the uncertainty in the difference would amount to over 50 per cent.

Reference has been already made (p. 317) to *the relation between the constitution and the heat of combustion.* We enumerate here the heats of combustion of some important substances; these are expressed in *large calories* (= 1000 g.-cal.), in order to avoid the repetition of ciphers.

Ethyl alcohol	340
Mannite	727
Cellulose	680
Cane sugar	1355
Acetic acid	210
Benzoic acid	772
Ethyl acetate	554
Urea	152

The formation of ethyl acetate from the acid and alcohol (p. 448), according to these figures, would correspond to a heat of reaction of

$$340 + 210 - 554 = -4,$$

which is quite a small amount.

The Thermo-Chemistry of Electrolytes.—The hypothesis of electrolytic dissociation has thrown new light on the meaning of the heat of reactions which take place in salt solutions; and some rules which were previously discovered in an empirical way find a simple explanation as necessary deductions from the theory.

If two solutions of electrolytes, which are completely dissociated, are mixed, there is of course no development of heat, *provided* that the ions of the two electrolytes do not unite to form an electrically neutral molecule nor a new ion complex. This case is illustrated, *e.g.*, in the mixture of most salt solutions.

Experiment establishes the conclusion. The so-called *Law of the thermo-neutrality of salt solutions*[1] is nothing else than an expression of the experience that *no heat phenomena result from the mixture of salt solutions (provided that no precipitate* [and no volatile compound] *is produced).*

Let AB and A′B′ be two salts which obey the "law of thermo-neutrality." Then no noticeable heat of reaction will result from their mixture, thus

$$(\text{AB aq.}) + (\text{A}'\text{B}' \text{ aq.}) = (\text{AB, A}'\text{B}' \text{ aq.}).$$

A similar thing also holds good for the mixture

$$(\text{AB}' \text{ aq.}) + \text{A}'\text{B aq.}) = \text{AB}', \text{A}'\text{B aq.}).$$

[1] Usually called "Hess's Law."—TR.

But now the two resulting solutions, which are expressed by the symbols standing on the right sides of the two equations, are identical; therefore

$$(AB\ aq.) - (AB'\ aq.) = (A'B\ aq.) - (A'B'\ aq.);$$

or, in other words—

The difference between the heats of formation of any two salt solutions having a common ion is a constant which is characteristic for both of the two other radicals, and is independent of the nature of the common ion.

Thus, *e.g.*,

$$\begin{array}{lll} H + Cl + aq. = (HCl\ aq.) & + & 39320 \\ H + I\ + aq. = (HI\ aq.) & + & 13170 \\ \hline Cl + (HI\ aq.) = I + (HCl\ aq.) & + & 26150; \end{array}$$

and

$$\begin{array}{lll} K + Cl + aq. = (KCl\ aq.) & + & 101170 \\ K + I\ + aq. = (KI\ aq.) & + & 75020 \\ \hline Cl + (KI\ aq.) = I + (KCl\ aq.) & + & 26150. \end{array}$$

Thus in fact it does result that

$$(HI\ aq.) - (HCl\ aq.) = (KI\ aq.) - (KCl\ aq.);$$

and thus if one replaces the iodine in [*any*] iodide in a dilute water solution by chlorine, he always observes the same quantity of heat to be developed, viz. 26,150 cal. In this case, of course, one is always concerned with the same reaction, viz. the replacing of the iodine ion by the chlorine ion.

But, on the other hand, if we substitute the iodine in dissolved *potassium iodate* (KIO_3) by chlorine, we observe an entirely different heat of reaction, viz.—the *absorption* of 31,700 cal. instead of the development of 26,150 cal.; for here iodine and chlorine are not ions, and therefore an entirely different reaction occurs.

The case is quite different *where the mixture results in the union of ions to form electrically neutral molecules, which may remain or may be removed.*

We have already met the most important example of this case in *the process of neutralisation* on p. 507. If one mixes a strong acid with a strong base, the *hydrogen ions* and the *hydroxyl ions* unite almost completely to form *molecules of water. The negative ion of the acid and the positive ion of the base remain free* if the salt produced by their union is highly dissociated (as, *e.g.*, is always the case with univalent bases and acids).

We thus develop the important theorem, that—

The neutralisation of a strong acid by a strong base must always exhibit the same heat of reaction.

The following table shows how remarkably the theorem is proved by experiment :—

TABLE OF THE HEATS OF NEUTRALISATION OF ACIDS AND BASES

Acid and Base.	Heat of Neutralisation.
Hydrochloric acid and sodium hydrox. . .	13700
Hydrobromic ,, ,, ,, ,, . .	13700
Nitric ,, ,, ,, ,, . .	13700
Iodic ,, ,, ,, ,, . .	13800
Hydrochloric ,, ,, lithium ,, . .	13700
,, ,, ,, potassium ,, . .	13700
,, ,, ,, barium ,, . .	13800
,, ,, ,, calcium ,, . .	13900

In the case of the di-acid bases [as barium hydrox.], the figures do not of course refer to a molecule, but only to one equivalent of the base. Neglecting the slight correction required by the incomplete dissociation, these figures simply mean that they give the heat developed by the reaction

$$\overset{+}{H} + \overset{-}{OH} = H_2O;$$

or in the language of thermo-chemistry, we obtain

$$(\overset{+}{H}\text{ aq.}, \overset{-}{OH}\text{ aq.}) = 13{,}700.$$

When the acid or the base is only partially dissociated electrolytically, then the heat of neutralisation changes by the amount of energy which comes into play in splitting the molecules into the ions. In fact, the following example shows that the *heat of neutralisation* assumes values which are noticeably different under such circumstances as those just mentioned. Thus *the heats of neutralisation* of the following weak acids, by *sodium or potassium hydroxide*, are as follows :—

Acid.	Heat of Neutralisation.	Heat of Dissociation
Acetic acid	13400	+ 300
Di-chlor-acetic acid . .	14830	− 1130
Phosphoric ,, . .	14830	− 1130
Hydrofluoric ,, . .	16270	− 2570

If these acids were completely dissociated, we should obtain a heat development of 13,700 cal. in the neutralisation. The deviations from this normal value are to be ascribed to the electrolytic dissociation. The *differences* are given in the last column; *these denote the respective quantities of heat which would be required for the electrolytic dissociation of the acids in question.* These figures only claim to be *approximately* correct. Because, on the one hand, the differences between magnitudes which are not very far apart are attended by considerable errors of observation; and, on the other hand, the calculation is based on the faulty assumption that the acid is not at all dissociated, and that the base and the neutral salt are completely dissociated.[1] Nevertheless, these figures, which we will obtain in a perfectly independent way in the following chapter, are well suited to give us a conception of the amount of *the changes of energy which are associated with the separation into ions.*

Of course it happens in all these changes that sometimes heat will be absorbed, and again it will be set free; and conversely, that in the union of ions to form electrically neutral molecules, sometimes heat will be developed, and again it will be absorbed; and that the final absolute amount is not very great in any case.

On neutralising ammonia, a base which is only slightly dissociated, with a strong acid, there results a heat development of 12,300 cal.; therefore the heat used in cleaving the ammonium hydroxide into its ions is estimated to be

$$13{,}700 - 12{,}300 = +1400 \text{ cal.}$$

The calculation of the most general case, viz. *where neither the base, nor the acid, nor the neutral salt formed is dissociated,* is as follows:—

Let the degree of dissociation of the acid SH be a_1, and let that of the base BOH be a_2; now from the mixture of the acid and alkaline solution, each of which contains one equivalent, there results an equivalent of the salt BS; let the degree of dissociation of this salt be a; there also results a molecule of water, and in the sense of the equation

$$SH + BOH = BS + H_2O.$$

Now the heat of neutralisation is made up as follows:—

1. From the heat of dissociation a of water.

2. From the heats of dissociation W_1 of the acid, and W_2 of the base, which participate to the respective amounts of

$$W_1(1 - a_1), \quad \text{and } W_2(1 - a_2).$$

3. From the heat of dissociation W of the salt, which participates to the extent of

$$W(1 - a).$$

Now by observing that the acid and the base are electrolytically

[1] The heat of dilution of some weak acids in dilute solution, from which the heat of dissociation may be calculated, has been measured by E. Petersen (*Zeitschr. phys. Chem.*, **11.** 174, 1893; see also v. Steinwehr, *l.c.*, p. 586.

dissociated in the solution, and conversely, that the water and the undissociated salt are formed by the union of ions, we obtain as the expression for "q" the heat of neutralisation,

$$q = x + W(1 - \alpha) - W_1(1 - \alpha_1) - W_2(1 - \alpha_2).$$

Now let us mix *the slightly dissociated salt of a weak acid with a stronger acid*; then we have a process which is comparable with the neutralisation of a strong acid by a strong base; because here, in consequence of the mixture, two ions unite smoothly to form an electrically neutral molecule.

Thus if we mix, *e.g.*, a sodium salt of the weak hydrofluoric acid with hydrochloric acid, then the reaction progresses almost completely in the sense of the reaction

$$\overset{+}{H} + \overset{-}{F} = HF,$$

and the heat of reaction observed is that of this reaction. Thomsen found that 2360 cal. were associated with this mixture, a value which is only slightly different from the value given above for the heat of dissociation of hydrofluoric acid.

If *the formation of an insoluble precipitate* results from the union of two electrolytes which are completely dissociated, then the negative value of the observed heat of reaction corresponds, of course, to the *heat of solution of the precipitated substance.* Thus, *e.g.*, if we mix a [soluble] silver salt and a chloride, solid silver chloride is precipitated, and, according to Thomsen, a heat development of 15,800 cal. is observed, which corresponds to the simple reaction

$$\underset{\text{Dissolved}}{\overset{+}{Ag} + \overset{-}{Cl}} = \underset{\text{Solid}}{AgCl}$$

Finally, we would observe that it is possible to calculate the heat of reaction on mixing any selected solution, provided that *we know the heats of dissociation, and the heats of solution, of all the molecular species which come into consideration.*

According to the principles developed in Chap. IV. of the preceding book, we can predict the progress of the reaction from the dissociation constants, and from the solubility coefficients, of all the molecular species which are to be considered. But if we know to what extent electrolytic dissociation, and the formation of precipitates, have taken place, and if we know the heat value of each of these processes, then of course we can state the heat of reaction of the total process.

In thermo-chemical relations the heats of dissociation and of solution go hand in hand with the coefficients of dissociation and of solubility. The following chapter will give us the interesting result that these magnitudes are closely related in pairs.

CHAPTER II

THERMO-CHEMISTRY II

TEMPERATURE AND COMPLETE CHEMICAL EQUILIBRIUM

The Application of the Second Law of the Mechanical Theory of Heat: Historical.—The application of the second law of thermodynamics to chemical processes was a step having fundamental significance, because for the first time it afforded an insight into *the relations between chemical energy, heat, and capacity for external* work; and thereby for the first time it was possible to answer the questions, viz. how far is the energy which is set free in chemical processes convertible [into work] without limitation? and is it [the energy] more of the nature of heat, or more of the nature of the kinetic energy of moving masses? and although it is not possible to give a satisfactory answer to these questions in every special case, yet it is possible to outline clearly the way leading to a systematic mode of attack.

The honour of having first considered chemical processes, and dissociation in particular, from the standpoint of thermodynamics, undoubtedly belongs to A. Horstmann,[1] and as a result of his calculations, the fruitfulness of the mechanical theory of heat in this region also was demonstrated.

Then shortly afterwards the problem was treated very thoroughly and in certain special aspects by J. W. Gibbs;[2] but unfortunately the calculations of this author have a character which is entirely too generalised to be capable of a simple and direct application to special cases of investigation.

Thus it came about that independently of Gibbs's work, and a little later, there were discovered a large number of theorems, which may be deduced directly by specialisation of his formulæ; some of these are, *e.g.*—

[1] *Ber. deutsch. chem. Ges.*, **2.** 137 (1869); **4.** 635 (1871); *Lieb. Ann.*, **170.** 192 (1873).

[2] *Trans. Conn. Acad.*, **3.** 108 and 343 (1874-1878); German translation by W. Ostwald, Leipzig, 1892.

(A) The relation between the development of heat and the temperature coefficient in the dissociation of a gas.

(B) The relation between the development of heat and the temperature coefficient of a galvanic element.

The first of these problems in particular has been repeatedly subjected to treatment.[1]

Regarding some of the more recent treatises on the application of chemical processes to thermodynamics, reference should be made to the monographs of Le Chatelier,[2] and especially to that of van't Hoff,[3] the study of which cannot be too highly commended, as leading to a more profound knowledge of the important and difficult problems presented here.

Mathematical deduction has gained greatly in clearness and elegance from the more recent methods of treatment of Planck[4] and Riecke;[5] the latter treatise is based entirely on the application of *the thermodynamic potential*, and therefore has unquestioned advantage for all those who are conversant with the potential theory of physics.

Of course the distinctions between the various methods of treatment are of a *purely conventional nature;* one method can proceed no farther than another. In his own method of presentation, the writer would endeavour to bring all in as close contact as possible with *the results of experiment*, hoping thereby to make the subject in many points more intelligible than it has been made hitherto.

In this and the following chapter we will describe the most important applications made thus far *of the second law of thermodynamics to chemical processes.* We will then consider next *complete chemical equilibrium*, which, according to the developments made on p. 464, is conditioned by the *temperature alone;* and then we will consider *incomplete chemical equilibrium*, which is conditioned not only by the *temperature*, but also by *the relative quantities of the reacting substances.*

Gibbs's "Phase Rule."—The "complete chemical equilibrium" (p. 464) was characterised in this way, viz. for every temperature there existed only one definite pressure at which the different phases of the system were in equilibrium with each other. If we change this pressure at constant temperature, then the reaction advances to a completion in one sense or the other, *i.e.* until one or more of the phases are exhausted. If the external pressure is only changed a very little, then each phase maintains its composition unchanged during the reaction.

[1] van der Waals, *Beibl.*, **4.** 749 (1880); Boltzmann, *Wied. Ann.*, **22.** 65 (1884); and others.

[2] *Recherches sur les équilibres chimiques.* Paris, 1888.

[3] *Études de dynamique chimique.* Amsterdam, 1884.

[4] *Wied. Ann.*, **30.** 562; **31.** 189; **32.** 462 (1887); we may here again refer to the work of Planck quoted on p. 22.

[5] *Zeits. phys. Chem.*, **6.** 268, 411 (1890); **7.** 97 (1891).

We have already become acquainted with numerous examples of chemical equilibrium. In addition to the simplest case of equilibrium between different states of aggregation, there belong in this category, the dissociation of ammonium chloride, excluding an excess of the dissociation products, the dissociation of calcium carbonate, etc.

A very remarkable law regarding *complete heterogeneous equilibrium* was discovered by Gibbs in a theoretical way; this was afterwards thoroughly proved experimentally, and hence it is suited to be used as a safe guide in the investigation of special cases. It may be formulated as follows:—

It is necessary to assemble AT LEAST n *different molecular species, in order to construct a complete heterogeneous equilibrium consisting of* n + 1 *different phases.*

Thus, in order to establish the complete equilibrium,

$$\underset{\text{Liquid}}{H_2O} \rightleftarrows \underset{\text{Vapour}}{H_2O},$$

only *one* [n] molecular species (H_2O) is required, because this consists of two [n + 1] phases.

In order to establish the complete equilibrium between calcium carbonate and its decomposition products (three phases), we need at least the two molecular species, viz. CO_2 and CaO.

By bringing together salt and water, we can establish the three phases of the complete equilibrium between the solid salt, its solution, and the vapour of the solution.

But, on the other hand, if two molecular species, *e.g.* react upon each other in two phases only, the equilibrium will be incomplete, *i.e.* the progress of the reaction is associated with a change in the equilibrium pressure. Thus if we let any mixture of water and alcohol evaporate, the maximal pressure of the mixture changes with the progress of the reaction, in spite of keeping the temperature constant. This reaction would become complete if we should allow the water to freeze, *e.g.*, and a third phase should thus be added to the system.

Of course we can imagine the heterogeneous system in question to be constructed of *more* than "n" molecular species, as, *e.g.*, in the equilibrium between calcium carbonate and its decomposition products, thus, $CaCO_3$, CaO, Ca, CO_2, etc. But we must limit the *minimal value of* "n," and for the preceding system this amounts to 2, whether we imagine the construction of the system to be from CaO and CO_2, or from $CaCO_3$ and CaO, etc. And thus the limitation "at least" is seen to have an essential meaning.[1]

[1] If solid sal-ammoniac be volatilised without excess of its products of dissociation, the system can be built up of one kind of molecule (NH_4Cl) only, so that we may put n = 1, on which point a singular mistake has arisen. See my remarks in *Zeits. phys. Chem.*, **43.** 113 (1903).

The proof of "the phase-rule," as it was accomplished by Gibbs[1] himself, can be shown in a simple way as follows :—[2]

Let the heterogeneous system considered consist of y phases, for the construction of which we need at least n different molecular species. We will select one phase in which all the molecular species are present, there always being one such phase at least. Thus, *e.g.*, each molecular species must occur in every *liquid* phase, because each molecular species has a definite solubility, though it may be perhaps immeasurably small; and hence each species may be present in quantities which are too small to be weighed. A similar thing holds good for the *gaseous* phase of the system, because every molecular species has a finite vapour pressure.

Let the concentration (*i.e.* the number of g.-mol. per litre) of n molecular species in the selected phase be respectively, $c_1, c_2, \ldots c_n$. The composition of the phase will change in a way which is perfectly definite and unequivocal, if we change the external conditions of the system, viz. the temperature T, and the pressure p; and of course if we also change the concentration of each of the particular molecular species. Therefore an equation must exist which so relates the quantities $c_1, c_2, \ldots c_n$, p, and T, with each other, that a variation in the value of one of their magnitudes necessitates a variation in all the others. We thus obtain the equation

$$F_1(c_1, c_2, \ldots c_n, p, T) = 0,$$

where F_1 is the symbol for any selected function of the variable considered.

The fact that such characteristic equations are only very incompletely known at the present time does not impair the stringency of the proof; it is enough to know that there is such an equation in each case. For two molecular species we obtained on p. 242 a characteristic equation of the form

$$f(c_1, c_2, p, T).$$

In the present case the equations must be much more complex, not only because there are more kinds of molecule, but because the relative quantity of each substance may vary with displacement of equilibrium.

Every other phase of the system has its own characteristic equation; but now the composition of one phase unequivocally conditions that of all other phases which may be in equilibrium with it; therefore all phases which are in equilibrium with one phase must be in equilibrium with each other, and this is possible only with perfectly definite ratios of concentration. Thus, *e.g.*, it is evident that from the condition of a

[1] *Trans. Conn. Acad.*, **3.** 108 and 343 (1874-78).

[2] Taken mainly from that of Riecke (*Zeits. phys. Chem.*, **6.** 272), for which see further details.

liquid phase, the composition of the gaseous phase in contact with it is also given.

Therefore it follows that the compositions of all the other phases are definite and unequivocal functions of the same variables upon which the selected phase depends; and also that for every phase there must exist an equation of condition of the form

$$F(c_1, c_2, \ldots c_n, p, T) = 0.$$

We obtain as many equations of condition of this sort as there are phases in our system, *i.e.* "y" in number.

Now the number of the variables, $c_1, c_2, \ldots c_n, p, T$, amounts to $n + 2$; therefore in order that they may be unequivocally determined by the equations of condition, which are "y" in number, it is necessary that there shall be as many equations as there are variables, *i.e.* it follows that

$$y = n + 2.$$

Now this amounts to stating that when n *molecular species react in* $n + 2$ *phases, an equation of condition is possible between all of them, only when the conditions of temperature and pressure are definite and unequivocal; and also when the ratio of concentration of each of the particular phases is perfectly definite.*

Moreover, associated with the coexistence of the $n + 2$ *phases, there is a unique point, the so-called* "**point of transition**" (*Uebergangspunkt*). We will soon learn something more regarding the peculiarities of this point.

Therefore, in order to have a *complete equilibrium, i.e.* when, during a finite interval, for every value of the temperature T, there corresponds a perfectly definite equilibrium pressure p, and, of course, a perfectly definite composition of each of the particular phases,—for this we must have *one equation less* than the number of variables, *i.e.* it must follow that

$$y = n + 1.$$

But this is nothing, except the "phase-rule"; for this result stated that—

The number of phases in a complete equilibrium must be one more than the number of reacting molecular species.

Finally, when

$$y < n + 1,$$

then, with given external conditions of temperature and pressure, $c_1, c_2, \ldots c_n$, remain more or less indefinite, and therefore the composition of all the phases also; here we are dealing with *an incomplete equilibrium.*

The Point of Transition.—The conditions of complete equilibrium require that, within a finite temperature interval, for every temperature point there shall correspond a definite pressure value, at which all of the phases of the system may be coexistent. This temperature interval is always limited, as a result of the fact that one of the phases suddenly ceases to be capable of existence, and hence disappears from the system. Thus, *e.g.*, to take the simplest case imaginable, let us study the equilbrium in the equation

$$\underset{\text{Liquid}}{H_2O} \rightleftarrows \underset{\text{Gaseous}}{H_2O}$$

at different temperatures. Then, on the one hand, according to what was said above, we can follow the measurements of the vapour pressure of liquid water only up to the critical temperature; and, on the other hand, the investigation can be conducted uninterruptedly at lower temperatures, till we arrive at the freezing-point of water (in so far as we deal with the vapour pressure itself); beyond this the liquid water ceases to be capable of existence, except in the labile condition of an under-cooled liquid. But usually another phase appears to take the place of the one which has disappeared. Thus, in the latter case, where we arrive at a new complete equilibrium, thus:

$$\underset{\text{Solid}}{H_2O} \rightleftarrows \underset{\text{Gas}}{H_2O},$$

we deal with the sublimation of ice.

That temperature point, where one phase of a complete heterogeneous equilibrium vanishes, and where another takes its place, we call "*the point of transition.*" Thus at the point of transition, and at the corresponding pressure, there are coexistent beside the others, not only the phases which have begun to disappear, but also those which are beginning to appear.

Now, since there were n + 1 phases composed of n molecular species before the "point of transition" was reached, *therefore, at the point of transition itself, there are* n + 2 *phases, each one of which is, of course, made up of the same molecular species,—a singular state.*

In order to obtain a deeper insight into those relations, the general meaning of which is obvious, we will consider some special cases in what follows.

The Equilibrium between the different Phases of Water.—We can easily determine the "point of transition" at which the *liquid* phase vanishes from the system,—water$\rightleftarrows$water vapour,—only to be replaced by the *solid* phase, ice.

At atmospheric pressure water freezes at 0°; and somewhat below this, under its own vapour pressure, which, to be sure, is very small as compared with the atmospheric pressure, viz. at + 0·0077°, because the freezing-point is raised by this amount by diminishing the pressure one atmosphere (p. 71). The *pressure* corresponding to the "point of transition," according to Regnault's vapour-pressure tables, is 4·57 mm.

Therefore, under these conditions of temperature and pressure [viz. 0·0077° *and* 4·57 mm.], *and under these alone, are the three phases—ice, liquid water, and water vapour, coexistent.*

For the purpose of understanding clearly, not only this case, which is the simplest conceivable, but also for the treatment of the behaviour of complicated equilibria, we may use with advantage a graphic method which represents, in a universal way, the nature of the condition of equilibrium in its dependence upon the external conditions of temperature and pressure.[1]

In a co-ordinate system (Fig. 32), let the abscissæ represent the temperature T (properly on the absolute scale); and let the ordinates represent the pressure p; on this we will plot the curves along which two different phases of water are coexistent. We will call these "the limiting curves" (*Grenz-kurven*) of the system considered.

In general, if we are considering a system composed of a number of phases, and for the construction of which there are requisite n molecular species,—then the meaning of "the limiting curves" is that under the definite and unequivocal conditions of temperature and pressure, n + 1 different phases are capable of existing beside each other.

These "limiting curves" are well known in the case of water; and it is directly evident that there must be *three*,—and each one of these phase-pairs must consist respectively of a combination of two of the coexistent conditions of aggregation of water.

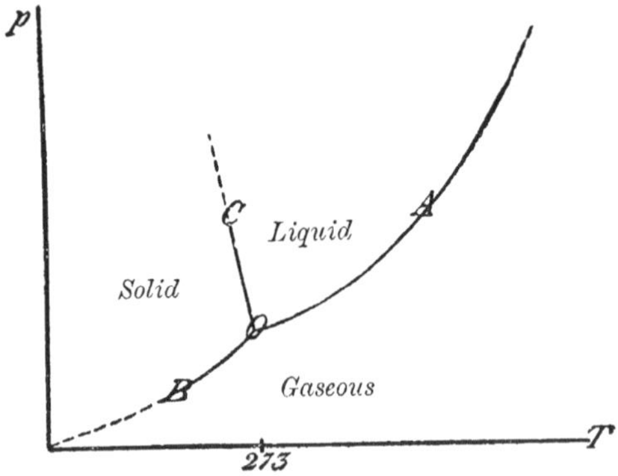

Fig. 32.

Thus we obtain the plotting of the curves shown in Fig. 32. Thus *liquid water* and *water vapour* are coexistent along the curve OA. This is also the "*curve of vapour-pressure*," and it is especially well known in the first part; its upper end is found at T = 273° + 364·3°, which is the "critical temperature" of water.

At the conditions corresponding to the point O (and which we previously determined as T = 273° + 0·0077°, and p = 4·57 mm.), water freezes. Hence the continuation of the curve AO, along the limiting curve OB, represents the conditions under which *ice* [solid H_2O] and water

[1] Roozeboom, *Zeits. phys. Chem.*, **2**. 474 (1888).

vapour are coexistent; this represents the "*vapour-pressure curve*" *of ice* (p. 74). On account of the smallness of the vapour pressure of solidified water, it has been traced only a short distance downwards from O. But from the kinetic treatment of the properties of matter, we may predict with great certainty,—that it intersects the zero-point of our system of co-ordinates,—also that the vapour pressure of ice will become zero at the absolute zero-point of temperature,—and also that at this point [viz. abs. 0°] water vapour ceases to be capable of existence *as such*.

Finally, the "limiting curve" OC represents the conditions for the coexistence of *water* and *ice*. Since the freezing of water is accompanied by an increase of volume, the melting-point of ice sinks with increasing pressure, and by an amount which is 0·0077° per atm., a relatively small amount. And, moreover, since the depression of the melting-point is proportional to the external pressure, at least at pressures which are not too high, OC represents a direction which is only slightly inclined to the p axis. Here also we can follow the course of the curve OC only a slight distance from O; but we may fairly surmise that, on further extension, the curve OC would bend in a convex way towards the p axis, and ultimately would touch it asymptotically at infinity.

The point O, in which all the limiting curves anastomose, and therefore in which all of the three phases are coexistent, is a very remarkable point in the p, T, plane; it has been already called "*the point of transition.*" According to the number of limiting curves which unite at this point, we may call it "three-fold," "four-fold," etc.; or "triple," "quadruple," etc. In the case of water, O is a "triple point."

The limiting curves are, of course, curves of complete equilibrium. If we are considering one point of one of the curves, and if we change the pressure and temperature in a way which is different from what corresponds to the course of the curve, then a complete reaction occurs, and, according to the circumstances, one of the two phases vanishes.

Moreover, the limiting curves divide the plane (p, T) into three fields, each of which corresponds to the existence of one of the three phases. This does not state that the phases cannot exist beyond the limits of their respective fields.

Thus, *e.g.*, it is known that liquid water may exist at a temperature and pressure which are below the point O. But in that case the existence of water is a labile one, *i.e.* it is in an "under-cooled" state.

A similar thing is true of water vapour, which we know can exist at temperatures and pressures where it should have become liquefied, according to the course of the limiting curve.

Moreover, these labile conditions play a much more important *rôle* in nature than we are usually inclined to believe. It has been proved repeatedly that some very stable substances, especially such as usually exist in the solid state of aggregation, may be brought into a state which is comparable to that of an "under-cooled" liquid; and thus, according to the laws of chemical equilibrium, they are for the time being deprived of their right to existence [in their proper phase].

We should not attach to the term "labile" any such notion as that the system necessarily needs only a slight disturbance in order to change it to the more "stable" form.

In this way we can explain why it is that a well-known *fourth* modification of water, *i.e.* "electrolytic gas," has no place in the (p, T) plane; at least there is none in the region thus far considered. "Electrolytic gas" at the ordinary conditions of temperature and pressure is in just as much a labile condition as is under-cooled water, because it, electrolytic gas, can be changed into the stable form of water by several kinds of disturbance; as, *e.g.*, by a suitable elevation of the temperature.

Later, and in an entirely different way, we will also obtain the result that the system

$$2H_2 + O_2$$

is by no means in equilibrium, under ordinary conditions of temperature and pressure, although the mixture may apparently be preserved unaltered for any period, however long.

Moreover, the following circumstance is very remarkable.

Let one in imagination come down on the curve AO at constant volume, removing heat from the system, consisting of water and water vapour, which are enclosed in a suitable vessel. As one arrives at O the water freezes; and now it obviously depends upon the *relative quantities of the phases present*, as to which of the phases, the liquid or the gaseous, will disappear on further cooling, *i.e.* whether one will progress along the curve OB, or along OC, with the continued removal of heat. If the expansion of [or volume occupied by] the gaseous phase is sufficiently great, then all the water will freeze, and one will advance along the curve OB. If, on the other hand, the volume of the liquid water is sufficiently great in comparison with the gaseous phase, then all the water vapour condenses, as a result of the increase of volume resulting from the freezing, and from the associated increase of pressure; thus the melting-point will be strongly depressed, so that one will advance along the curve OC.

The Equilibrium between Water and Sulphur Dioxide.—The system just described could be constructed of *a single molecular species*, viz. H_2O. We will select as a further and more complicated example of complete heterogeneous equilibrium a system composed of H_2O and SO_2, *i.e. two molecular species;* this has been very thoroughly investigated by Roozeboom.[1]

[1] *Zeits. phys. Chem.*, **2.** 450 (1888).

And again, we will plot in a co-ordinate system, the curves along which complete equilibrium can exist.

Now, since we allow two molecular species to meet here, according to Gibbs's "phase-rule," on the limiting curves, *three phases* must be capable of existing beside each other; and it is only at the singular point that *the coexistence of four phases* is possible.

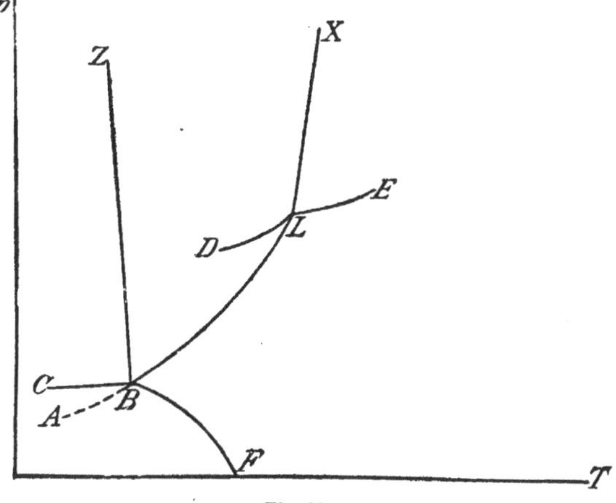

Fig. 33.

By bringing together SO_2 and H_2O, the four following homogeneous systems can be traced in the neighbourhood of the point L (Fig. 33).

1. The *solid hydrate* $SO_2 \cdot (H_2O)$: this can easily be caused to separate out, by cooling a water solution which is saturated with SO_2.

2. A *solution* of SO_2 in water, which we will designate by the symbol fl.$(H_2O + xSO_2)$. Here x denotes the number of g.-mol. which are present in the solution, for 1 g.-mol. of H_2O (18 g.); on account of the excess of water in the solution, x is always less than one. [fl. = fluid or liquid; and similarly in the other symbols.—TR.]

3. A *solution* of water in liquid sulphurous acid, corresponding to the symbol fl.$(SO_2 + yH_2O)$; here, similarly to x, y is always less than one.

4. A *gas mixture*, composed of SO_2 and H_2O, which we will designate by the symbol "gas $(SO_2 + zH_2O)$."

The three phases 2, 3, and 4 are coexistent along the curve LE; the first two are composed of the two liquids, water and sulphurous acid. These do not mix with each other in all proportions, but only partially dissolve each other like ether and water, *e.g.*, thus "x" corresponds to the solubility of SO_2 in water, and "y" to the solubility of water in SO_2. Both of these solubilities vary with the temperature, and of course they must be determined for a number of temperature points, in order to have exact knowledge regarding the aforesaid equilibrium: this has not been done as yet.

The *gaseous* phase consists of the vapour emitted by the two liquids; and the pressure p, which corresponds to any particular point of the curve LE, is the vapour pressure of the two liquids at the corresponding temperature.

Roozeboom did not determine the value of "z"; but it should

be noticed that, by means of the rules already given (p. 484), the composition of the gaseous phase, and therefore the value of z also, may be calculated, at least approximately, from the mutual solubilities, and from the known vapour pressures of the pure solvents at the pressure p_1.

If we cool the system just described, at constant volume, when the quantities of the particular phases are in suitable proportions, we advance along the curve LD; here the water solution of SO_2 disappears, and there appears the solid hydrate No. 1, ($SO_2 . 7H_2O$); while, the two other phases, No. 3, fl.($SO_2 + yH_2O$), and No. 4, gas ($SO_2 + zH_2O$), remain.

The values of y and of z were not measured.

The pressure p of the gaseous phase (the pressure of dissociation of the solid hydrate in the presence of a saturated water solution of sulphur dioxide) is of course independent of the relative quantities of the solid and liquid phases. This pressure p has the following values for the corresponding temperatures T :—

T	p	T	p
273+0·1°	113·1 cm. Hg.	273+11·0°	170·1 cm. Hg.
273+3·05	127·0 ,,	273+11·9	176·2 ,,
273+6·05	141·9 ,,	273+12·1	177·3 ,,
273+9·05	158·2 ,,	...	...

If we advance along the curve DL, by raising the pressure and the temperature, and by a suitable choice of the relative proportions of the particular phases, we can make the gaseous phase disappear at the point L, and thus cause a water solution of sulphur dioxide to appear. We thus arrive at the system composed of the solid hydrate No. 1, ($SO_2 . 7H_2O$), and of the solutions No. 2 and No. 3, viz. fl.($H_2O + xSO_2$), and fl.($SO_2 + yH_2O$); here the hydrate is found in equilibrium with its fusion products, viz. the mutually saturated solutions of water and sulphur dioxide.

At a pressure of 177·3 cm., the melting-point of the hydrate is at T = 273° + 12·1°. This increases with the pressure; at 20 atm. it is at 12·9°, and at 225 atm. it is at 17·1°. Now, since this apparently increases proportionally with the pressure, and indeed only a few degrees in proportion to these enormous changes of pressure, the limiting curve LX advances in a direction only slightly differing from a straight line.

The values of x and y, *i.e.* the mutual solubilities of water and sulphur dioxide in the presence of the solid hydrate, are not known as yet. They should not differ very much from those corresponding to the curve LE, because the mutual solubilities of two liquids vary only slightly with the pressure.

Advancing down along the system of curves LE or LX, we can, by cooling, proceed down along the curve LD, already described; or by suitable proportions of the particular phases, we can proceed down along the curve LB. Here the phases 1, 2, and 4, viz. ($SO_2 . 7H_2O$), fl.($H_2O + xSO_2$), and gas ($SO_2 + zH_2O$), are in equilibrium; here we have to consider a system composed of the solid hydrate, of its saturated solution in water, and of the vapours emitted by these. The concentration of the saturated solution has the following values at the corresponding temperatures:—

T	$x\frac{6400}{18}$	T	$x\frac{6400}{18}$
273+0°	10·4	273+ 7°	17·4
273+2	11·8	273+ 8	19·1
273+4	13·5	273+10	23·6
273+6	16·1	273+12·1	31·0

The numbers denote the parts of SO_2 to 100 parts of H_2O. In order to obtain the value of x (*i.e.* the number of molecules of SO_2 to one mol. of H_2O), they must be multiplied by

$$\frac{18}{6400} = \frac{1}{355 \cdot 5},$$

where 64 = the mol. wt. of SO_2, and 18 = the mol. wt. of water.

The vapour pressure p of the saturated solution was as follows at the corresponding temperatures:—

T	p	T	p
273 − 6°	13·7 cm. Hg.	273+ 4·45°	51·9 cm. Hg.
273 − 4	17·65 ,,	273+ 6·00	66·6 ,,
273 − 3	20·1 ,,	273+ 8·40	92·6 ,,
273 − 2·6	21·15 ,,	273+10·00	117·7 ,,
273 − 2	23·0 ,,	273+11·30	150·3 ,,
273 − 1	26·2 ,,	273+11·75	166·6 ,,
273 − 0	29·7 ,,	273+12·10	177·3 ,,
273 + 2·8	43·2 ,,	...	...

Thus the four limiting curves anastomose in the point L; and although only *three phases* are coexistent at every other point of the respective curves, yet at this point L, and only at this point, there is a coincidence with each other of the four following phases, viz.—

1. The *solid* hydrate, $SO_2 . 7H_2O$.

2. A *water solution* of SO_2, having the composition ($H_2O + 0{\cdot}087SO_2$).
3. A *solution* of H_2O in *liquid* SO_2 (*i.e.* $SO_2 + yH_2O$).
4. A *gas mixture* of SO_2 and H_2O (*i.e.* $SO_2 + zH_2O$).

Inasmuch as each one of these systems can be made from the two molecular species, SO_2 and H_2O, therefore, during a *finite* temperature interval, according to the "phase rule," a complete equilibrium can be established for only *three* of these systems. All four may coexist only at the transition point L, which is in this case a quadruple or fourfold point; it is situated at

$$T = 273^\circ + 12{\cdot}1^\circ, \quad \text{and } p = 177{\cdot}3 \text{ cm.}$$

Roozeboom did not determine the values of y and z for the point L; but we may form a conception of their magnitude in the following way:—

The pressure 177·3 in the gas mixture $SO_2 + zH_2O$ is composed of the partial pressures of the particular gases.

Now, since a water solution of SO_2 is present at the transition point, therefore the partial pressure of the water vapour is equal to the partial pressure of pure water at the temperature in question (12·1°), viz. 1·05 cm., minus the depression experienced by 1 mol. of H_2O on dissolving 0·087 mol. of SO_2; and this latter magnitude, according to the van't Hoff law of vapour pressure, may be estimated in round numbers as equal to

$$1{\cdot}05 \times 0{\cdot}087 \text{ cm.}$$

Thus the partial pressure of the water vapour amounts approximately to 0·9 cm.

Therefore that of the sulphur dioxide is

$$177{\cdot}3 - 0{\cdot}9 = 176{\cdot}4 \text{ cm.}$$

Therefore the value of z is estimated to be

$$z = \frac{0{\cdot}9}{176{\cdot}4} = 0{\cdot}0051.$$

The concentration y of water in sulphur dioxide is found by means of the law for the molecular depression of the vapour pressure. According to Regnault, at 12·1°, pure liquid sulphur dioxide has a vapour pressure of 185 cm. But now, since sulphur dioxide saturated with water is present at the point L, and since the partial pressure of sulphur dioxide so saturated amounts to only 176·4, therefore, in order to occasion the depression of 8·6 cm. [$= 185 - 176{\cdot}4 = 8{\cdot}6$],

$$\frac{8{\cdot}6}{176{\cdot}4} = 0{\cdot}05 \text{ mol. of } H_2O$$

must be dissolved in every molecule of SO_2; and thus the value [0·05 mol. H_2O] corresponds to the value of y.

As stated above, the equilibrium pressures in the system, $SO_2 . 7H_2O$, liq. fl.($H_2O + xSO_2$), and gaseous ($SO_2 + zH_2O$), and which are given by means of the curve LB, these pressures have been measured as far down as $-6°$ (Celsius). But the system is in labile equilibrium beyond the point B (which corresponds to the temperature of $-2·6°$ and to the pressure of 21·1 cm.); because the liquid phase vanishes at once on the appearance of ice, and completely solidifies to ice and solid hydrate. Thus, instead of advancing along BA, the labile extension of the curve LB (and which is figured as a dotted line in Fig. 33), one advances along the curve BC. This curve refers to the three phases forming the system in question; these are—

1. The solid hydrate.
2. Ice.
3. The gas mixture ($SO_2 + zH_2O$).

The vapour pressures of a mixture of ice and the solid hydrate are as follows:—

T	p	T	p
273 – 2·6°	21·15 cm. Hg.	273 – 6°	17·7 cm. Hg.
273 – 3	20·65 ,,	273 – 8	16·0 ,,
273 – 4	19·35 ,,	273 – 9	15·0 ,,

The vapour pressures are considerably greater than would be the case if the system were under-cooled, *i.e.* if the solidification of the water solution of SO_2 had not occurred, as can be readily seen from the curve tracing (Fig. 33). The gas mixture emitted by the ice and the solid hydrate of course consists of H_2O and SO_2. *Now since ice is present, the partial pressures of the water vapour are simply equal to the vapour pressures of ice at the corresponding temperatures;* and these can be taken directly from Regnault's tables, and thus the values of z can be obtained.[1] They are very small in comparison with the total pressures.

Along the curve BF [Fig. 33], ice, the water solution of sulphurous acid, and the gaseous mixture ($SO_2 + zH_2O$), are capable of coexistence. The corresponding values of the pressure and temperature are so correlated that the pressure first conditions the concentration of the sulphurous acid, and then this concentration in turn conditions the lowering of the freezing-point of water. For small pressures, *i.e.* at slight concentration of the sulphurous acid, the curve tends to approach the freezing-point F of pure water at $T = 273°$.

[1] Roozeboom did not draw this conclusion, but its correctness appears to the author to be beyond doubt.

Finally, along the curve BZ, there are capable of coexistence, ice, the solid hydrate, and the melted mixture of the two (*i.e.* a water solution of sulphurous acid). Inasmuch as the melting is attended by a diminution of volume, the curve must be recurrent (*rückläufig*), *i.e.* the temperature of equilibrium must fall with an elevation of the pressure. But this has not been investigated in detail.

Thus the point B represents a second quadruple point; here there co-exist the four following phases, viz. :—

1. Ice.
2. The *solid* hydrate, $SO_2 . 7H_2O$.
3. A *water solution* of SO_2, having the composition $(H_2O + 0{\cdot}024SO_2)$.
4. A gas mixture of SO_2 and H_2O (*i.e.* $SO_2 + zH_2O$).

The co-ordinates of this point are

$$T = 273 - 2{\cdot}6^\circ, \quad \text{and } p = 21{\cdot}1 \text{ cm.}$$

Inasmuch as the vapour pressure of ice at $-2{\cdot}6^\circ$ amounts to 0·38 cm., from this statement *the value of* z amounts to

$$\frac{0{\cdot}38}{20{\cdot}7} = 0{\cdot}0184.$$

The areas in Figs. 32 and 33, enclosed by the curves, constitute *regions of incomplete equilibrium.*

The Hydrates of Ferric Chloride.—As an example of a more extended investigation of those systems, all of whose phases can be formed of *two* molecular species, such as H_2O and Fe_2Cl_6, we will describe Roozeboom's[1] research on the hydrates of ferric chloride. This led to a broadening of our view in several respects.

At the transition points of this system, the four following phases are capable of existing beside each other, viz. :—

1. The solid hydrate, $Fe_2Cl_6 . mH_2O$.
2. The solid hydrate, $Fe_2Cl_6 . nH_2O$.
3. The saturated solution.
4. Water vapour.

For the case where the hydrate *ice* occurs, we can make

$$n = \infty;$$

and for the case where *water-free* ferric chloride separates as a solid phase, we can make

$$n = 0.$$

The four "limiting curves" which anastomose at the transition point are composed of (*a*) and (*b*), the vapour-pressure curves of the

[1] *Zeits. phys. Chem.*, **10.** 477 (1892).

two saturated solutions of the hydrate; (*c*), of the vapour-pressure curve of a *mixture* of the two solid hydrates; (*d*), and of the vapour-pressure curve of a solution which is at the same time saturated with both the hydrates.

By means of these numerous curves, Roozeboom investigated *the solubility curves* of particular hydrates at ordinary pressures; and thus one obtains a fairly complete view of the relations which prevail here.

The following tables contain the results of his measurements:—

The Composition of the Saturated Solutions

n′ = number mol. Fe_2Cl_6 to 100 mol. H_2O,
n″ = " " H_2O " 1 " Fe_2Cl_6.

I. Ice (AB)

	n′	n″
0°	1	∞
−10	1·00	100
−20·5	1·64	61
27·5	1·90	52·6
40	2·37	42·2
55	2·75	36·4

II. $Fe_2Cl_6 \cdot 12H_2O$(BCD)

t	n′	n″
−55°	2·75	36·4
−41	2·81	35·6
−27	2·98	33·6
0	4·13	24·2
10	4·54	22·0
20	5·10	19·6
30	5·93	16·9
35	6·78	14·8
36·5	7·93	12·6
37	8·33	12·0
36	9·29	10·8
33	10·45	9·57
30	11·20	8·92
27·4	12·15	8·23
20	12·83	7·80
10	13·20	7·57
8	13·70	7·30

III. $Fe_2Cl_6 . 7H_2O$(DEF)

t	n′	n″
20°	11·35	8·81
27·4	12·15	
32	13·55	7·38
32·5	14·29	7·00
30	15·12	6·61
25	15·54	6·47

IV. $Fe_2Cl_6 . 5H_2O$(FGH)

t	n′	
12°	12·87	7·77
20	13·95	
27	14·85	6·73
30	15·12	6·61
35	15·64	6·40
50	17·50	5·71
55	19·15	5·22
56	20·00	5·00
55	20·32	4·92

V. $Fe_2Cl_6 . 4H_2O$(HJK)

t	n′	n″
50°	19·96	5·01
55	20·32	4·92
60	20·70	4·88
69	21·53	4·64
72·5	23·35	4·28
73·5	25·00	4·00
72·5	26·15	3·82
70	27·90	3·58
66	29·20	3·43

VI. Fe_2Cl_6 – Anhydride (KL)

t	n′	n″
66°	29·20	3·43
70	29·42	3·40
75	28·92	3·46
80	29·20	3·43
100	29·75	3·33

In order to explain this according to Roozeboom's method, we will employ another graphic delineation, Fig. 34, which differs slightly from

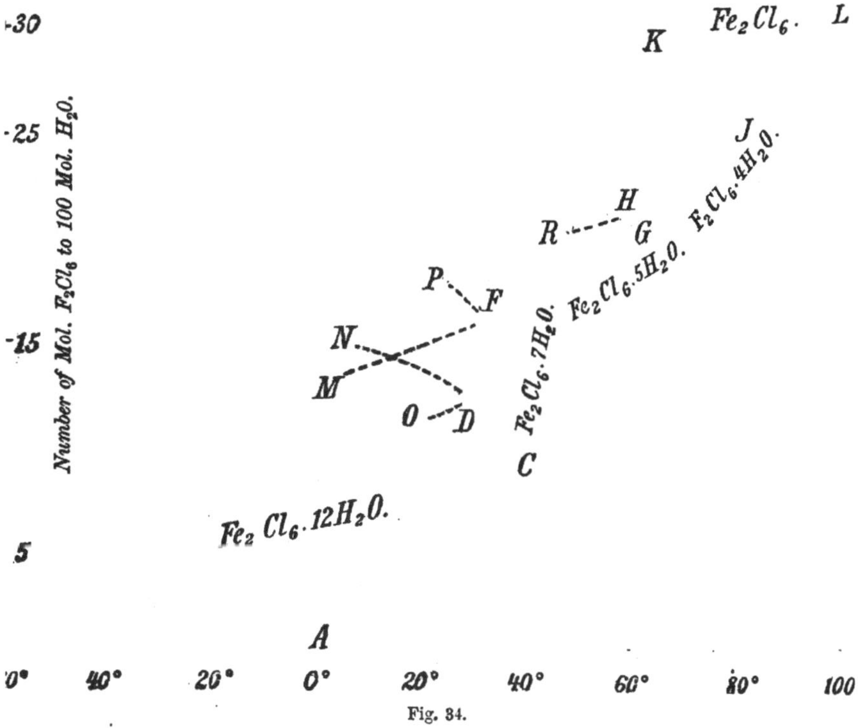

Fig. 34.

those in the preceding cases. Inasmuch as the *concentration* of the solutions here commands our chief interest, therefore we will express the *abscissæ* in terms of *temperature* (on the ordinary scale); and the *ordinates* in terms of the composition, expressed in molecules of Fe_2Cl_6 to 100 mol. of H_2O (*i.e.* the n′ of the preceding tables).

Fig. 34 gives a good general view regarding the behaviour of the equilibria.

2 S

If we start out from the equilibrium, water + ice, and add thereto ferric chloride, we obtain the curve AB (Table I. of the preceding), *i.e.* the curve of the depression of the freezing-point by adding the salt. At about $-55°$ we reach the point B, which corresponds to the saturation point of the hydrate of $12H_2O$; here there separates the so-called "cryo-hydrate," *i.e. a mechanical mixture of ice and solid salt* (p. 490).

A further addition of ferric chloride causes the ice to disappear, and we advance along the curve BC, which is *the solubility curve* of the hydrate with $12H_2O$.

At 37° the concentration of the saturated solution becomes equal to that of the solid hydrate, and at this temperature either the solution $Fe_2Cl_6 + 12H_2O$ congeals smoothly to the solid hydrate, or else this solidified hydrate is changed smoothly to a homogeneous liquid. Thus 37° is the *melting-point* of this hydrate.

Now, if we add anhydrous ferric chloride to this melted hydrate, we will advance along the curve CDN. Of the two branches which diverge from C, the one, CB, corresponds to the curve of the depression of the freezing-point occasioned by adding H_2O; the other, CDN, corresponds to the curve of the depression of the freezing-point of the hydrate, occasioned by the addition of Fe_2Cl_6. Thus below the melting-point [37°] of the pure hydrate [$Fe_2Cl_6 + 12H_2O$] it is possible to prepare *two* saturated solutions, one of which contains more H_2O, and the other less than the hydrate in question, as it stands in equilibrium with the solution. We will return to this remarkable phenomenon a little later.

In a perfectly similar way, one finds for the hydrate with $7H_2O$ the curve DEF; for the hydrate with $5H_2O$, the curve FGH; and for the hydrate with $4H_2O$, the curve HJK. At K begins the almost straight solubility curve of *anhydrous* Fe_2Cl_6. The melting-points of the respective hydrates are, at C (37°); at E (32·5°); at G (56°); and at J (73·5°).

The respective curve fractions DN, FM, DO, FP, and HR, correspond to certain labile conditions. Ice and the hydrate containing most water are in equilibrium at the point B. At the points D, F, and H the respective contiguous hydrates are in equilibrium. And finally, at K, the hydrate containing the least quantity of water is in equilibrium with the anhydrous Fe_2Cl_6.

The *composition* of the respective solutions at each of these points lies between the composition of the two respective solid bodies [hydrates], because at these respective points [D, F, H, and K] the second branch of a hydrate containing more water intersects the first branch of one containing less water. These respective points [of the intersection of the respective curves] are, at $-55°$ [B]; at 27·4° [D]; at 30° [F]; at 55° [H]; and at 66° [K]; and these points also represent the temperatures at which there solidify mixtures of the two respective [contiguous] hydrates.

In order to obtain a clear picture of the relations which prevail here, let one imagine the concentration and temperature of a solution of ferric chloride to be given by a point which lies in the right-hand side of the region bounded by the multiple curves ABCDEFGHJKL. On cooling, the solution at first traverses a horizontal line, the composition remaining constant; and at a definite temperature, its path will cut one of the portions of the curve, say FGH, *e.g.* If we suppose supersaturation to be excluded at the moment of the intersection, there will occur the separation of a solid substance having the composition, *e.g.* $Fe_2Cl_6 . 5H_2O$ [to correspond to the curve FGH]. By more extended cooling the curve will retreat towards lower temperatures until it reaches its final point, where another solid substance appears, and complete solidification will result.

If the solution should have the same composition as its hydrate, it would solidify completely at its melting-point. If the solution should have a composition corresponding to the intersection of any two of the contiguous curves [as at the points B, D, F, H, or K], then it would solidify completely at the temperature of the point in question.[1]

Most interesting phenomena would be observed if one should evaporate down a dilute solution of ferric chloride, and the results would be most striking between 30° and 32°. Here, by the removal of water vapour, the dilute solution would first dry down to $Fe_2Cl_6 . 12H_2O$. It would then liquefy, and then dry down to $Fe_2Cl_6 . 7H_2O$. It would then liquefy again, and a third time dry down, this time to $Fe_2Cl_6 . 5H_2O$.

The whole series of these phenomena corresponds to the respective stable condition. This very remarkable behaviour would be inexplicable were it not a necessary result conditioned by the relations which are illustrated in the curves shown in Fig. 34.

Regarding the curve branches, *e.g.* BCD and DEF [of one curve], it follows that, within the limits of a certain temperature interval, there are two saturated solutions having a different composition, which are in equilibrium with the respective solid hydrates. One of these always contains more water, the other less than the solid hydrate.

The second kind of saturated solution was fortunately discovered by Roozeboom in an investigation[2] on *the hydrates of calcium chloride.* It should be particularly emphasised that both of these solutions [*i.e.* on both branches of any one curve] are stable throughout, and nowhere supersaturated.

Supersaturation only occurs in a solution when its melting-point lies in the region to the left of the curve ABCDEFGHJKL. Supersaturation may be produced by adding to a solution a piece of its respective hydrate in the solid state, when, according to circumstances, the proportion of ferric chloride will be either diminished or increased,

[1] These two cases are good examples of the "eutectic and dystectic" mixtures described by the writer on p. 127.

[2] *Zeits. phys. Chem.*, **4**. 31 (1889).

accordingly as the supersaturated solution belongs to the first or the second category [*i.e.* to the first or second branch of any particular curve].

The writer has described this particular investigation at this length, because, in his judgment, it is entitled to claim very high methodical value. Before the work of Roozeboom, only one of the hydrates of ferric chloride, viz. the highest one, was certainly recognised, the others being entirely unknown. But the investigation of the equilibria led at once and of necessity to the discovery of the others. While Roozeboom was studying the "solubility curve" of the hydrate with $5H_2O$, he found certain irregularities which he suspected to be due to the existence of another hydrate, and thus he was led to the discovery of the hydrate $Fe_2Cl_6 . 7H_2O$. Now the stable part of its solubility curve extends only over the portion DEF, *i.e.* from 27·4° to 32·5°, and from 30° to 32·5°.

Thus this hydrate would hardly have been discovered without a systematic investigation of this sort.

Roozeboom and his students[1] have studied, with similar thoroughness, a number of other conditions of equilibrium of the systems composed of two molecular species.

Detection of Chemical Compounds by Means of Melting Curves.—A curve of melting-points may always be used to detect chemical compounds of two substances in the same way as for water and ferric chloride in the example given above; each compound is indicated by a peak, since an excess of either component lowers the melting-point of the compound. This assumes that the compound crystallises pure, and not as an isomorphous mixture, otherwise an excess of one component does not necessarily lower the melting-point.[2] The question whether the same compound, *e.g.* a hydrate of ferric chloride, exists in the liquid state, can of course not be settled by this method.

This method is especially useful in studying racemic mixtures; on account of the exact similarity of the dextro and laevo forms, the melting curve must be symmetrical about the racemic compound. According to the extensive investigations of Adriani,[3] three cases are to be distinguished:—

1. A continuous (convex or concave) curve, indicating mixed crystals.

2. Two curves, cutting at the composition of the racemic mixture, give a eutectic point (no compound).

3. Three curves, the central one with a maximum, indicating a racemic compound; eutectic points lie symmetrically on each side (mixture of the compound with the dextro- or laevo-body).

[1] *Zeits. phys. Chem.*, **4.** 31 (1889).
[2] See especially Roozeboom, *Zeits. phys. Chem.*, **30.** 385 (1899).
[3] *Zeits. phys. Chem.*, **33.** 453 (1900).

Systems of Three Components.—The investigation of such systems is, of course, much more complex, as the multiplicity of forms involved in complete equilibrium has already reached an extraordinarily high degree. For the same reason their experimental investigation is very incomplete; the only systems thoroughly studied so far are $CuCl_2$, KCl, H_2O[1] and PbI_2, KI, H_2O,[2] as well as those due to mixing K_2SO_4, $MgSO_4$, H_2O[3] and $FeCl_3$, HCl, H_2O.[4] The complex relations discovered in these systems cannot be described here, but may be found in van't Hoff's "chemical dynamics" already quoted. In the same book the still more complex behaviour of the system of four components, $MgSO_4$, K_2SO_4, KCl, H_2O, investigated by Löwenherz[5] is described.

The phase rule, it must be noted, is only a scheme into which complete heterogeneous equilibrium fits, and which, therefore, any experimentalist in this region must be familiar with, just as the analytical chemist must bear in mind the conservation of mass during his operations. But when B. Roozeboom, very naturally overrating the rule that guided him in his fine experimental work, says that there is no better point of view in heterogeneous equilibrium[6] than Gibbs's phase rule, the statement is as misleading as if the analytical chemist made it with respect to the law of conservation of mass. It must not be forgotten that for the more frequent, and by far the more important heterogeneous equilibria, viz. the incomplete, the phase rule has nothing to say. But apart from this, research must not be too limited in its object, and chemistry would almost be reduced to a triviality if it followed Roozeboom's suggestions. Molecular theory, thermodynamics of incomplete equilibrium, and especially the law of mass action, are far more important guides, and of deeper meaning, than the scheme of the phase rule, useful and even indispensable as the latter is to chemical research.[7] Roozeboom expresses these views in reviewing a book, *The Phase Rule*, by Bancroft (Ithaca and Leipzig, 1897), which contains a summary of the work on the subject, of which he says, "The chemist will be astonished at the important development this branch of his science has reached, and what a precious guide the phase rule is;" whilst Ostwald[8] much more correctly remarks, "No reader can get through the 248 pages of the book without a certain feeling of tiredness." We may refer also to the good and brief exposition, *Die Phasenregel und ihre Anwendungen*, by W. Meyerhoffer (Leipzig and

[1] Meyerhoffer, *Zeits. phys. Chem.*, **5.** 97 (1890); **9.** 641 (1892).
[2] Schreinemakers, *ibid.*, **9.** 57 (1892).
[3] Van der Heide, *ibid.*, **12.** 416 (1893).
[4] Roozeboom and Schreinemakers, *ibid.*, **15.** 588 (1894).
[5] *Ibid.*, **13.** 459 (1894); and **23.** 95 (1897).
[6] *Journ. phys. Chem.*, **1.** 559 (1897).
[7] Thus Van't Hoff remarked, "It is a pity that, valuable as the contents of the phase rule is, there has been a certain exaggeration of its scope (*Ber. deutsch. chem. Ges.*, **35.** 4252 (1892).
[8] *Zeits. phys. Chem.*, **23.** 706 (1897).

Wien, 1893), to the paper of van't Hoff quoted on p. 481 and the monograph, *Zinn, Gips, und Stahl* (München, Oldenbourg, 1901). It is with especial pleasure that one notes that Roozeboom has begun to put his researches on the phase rule into book form (the first volume published by Vieweg, Brunswick, 1901).

The Thermodynamics of Complete Equilibrium.—If we allow a complete equilibrium to pass beyond "the limiting curve" (p. 614), one phase will disappear and a new one will appear; both these phases are coexistent in the limiting curve itself.

Let us denote by Q the quantity of heat *absorbed* in the transition, by V_0 the associated increase of volume of the system, and by A the maximal external work which the system sustains as a result of the designated change. Thus the second law of thermodynamics gives the equation (p. 22),

$$Q = T\frac{dA}{dT}.$$

Both Q and A refer to a *definite* change of volume V_0 of the system, *e.g.* an increase of volume of 1 c.c. If we denote by p the pressure at that point of the limiting curve where the transition occurs, then it follows that

$$A = V_0 p, \quad \text{and } dA = V_0 dp;$$

and we find that

$$Q = T\frac{dp}{dT}V_0 \quad . \quad . \quad . \quad . \quad . \quad (I.)$$

This equation embraces all that the second law can teach us concerning a chemical system which stands in complete equilibrium. It is obvious that the thermodynamic formulæ derived for *evaporation* (p. 59), for *sublimation* (p. 73), and for the *process of melting* (p. 70), are special cases of equation (I.); because, as is shown on p. 464 and following, the conditions of equilibrium between different states of aggregation are to be enumerated among the cases of complete equilibrium.

Now, since we have seen that the important applications of equation (I.), cited in the places just mentioned in Book I., and their manipulation in complicated cases, offer no particular difficulties,—therefore we may be excused from describing here any further special applications, and especially so, because we shall have frequent dealings with the preceding equation.

Equation (I.) gives simple results when it is applied to *the passage across the limiting curves in the immediate neighbourhood of their points of intersection*, i.e. *of their transition points.*

Let us imagine the system to be conducted around the transition point in a very small curve; then all of the n limiting curves will be crossed, and we will obtain n equations of the form of (I.), viz.—

$$Q = T\frac{dp}{dT}V_0, \quad \text{or } TV_0 = \frac{dT}{dp}Q.$$

If we add these n equations, it follows that

$$\Sigma Q = \Sigma T\frac{dp}{dT}V_0, \quad \text{and } \Sigma TV_0 = \Sigma\frac{dT}{dp}Q\,;$$

now ΣQ is the sum of the heat developed in circling around the transition point, and accordingly, as is true in every reversible and isothermal heat process (p. 22), is equal to zero; and a similar thing is true of ΣV_0, because the system returns to its original volume; and of course ΣTV_0 is also equal to zero, because T changes only to an infinitesimal amount during the passage around the circle. And thus we find that

$$\Sigma\frac{dp}{dT}V_0 = \Sigma\frac{dT}{dp}Q = 0.$$

These phenomena, which are observed in passing from one phase to another, are the relations which must exist between the values of the trigonometrical tangents of the angle $\frac{dp}{dT}$ (with which the limiting curves anastomose in their common intersection point), and their latent heats on the one hand; and the changes of volume on the other.

Reference should be made to the investigation of Riecke[1] for several more broadly generalised relations, which are required by thermodynamics for the particular factors of the complete equilibrium.

Tammann (monograph quoted on p. 99) has discovered the following theorem on the position of curves that terminates in a triple point: "The prolongation of each curve of equilibrium must lie between the other two curves." The proof of this theorem rests on the consideration that on a temperature entropy diagram a triangle corresponds to the triple point; now the corresponding statement necessarily holds for the perpendiculars to the three sides, and this leads to the above theorem.

Condensed Systems.—Although the cases enumerated thus far are subject to the universal laws of complete equilibrium, yet the complex cases (which van't Hoff[2] calls "*condensed systems*") offer certain peculiarities which justify us in giving them a special description. *Such are the heterogeneous systems, all the reacting components of which are in the solid or liquid state, but not in the gaseous state.*

[1] *Zeits. phys. Chem.*, **6.** 268, 411 (1890). [2] *Études de dyn. chim.*, p. 139.

The simplest type of this sort of reactions is *the melting of a solid body.* The condition of equilibrium with which we are here concerned, and which consists in the coexistence of a solid body and its melted product,—this is a "complete equilibrium"; because for any definite temperature there is only one pressure at which both phases of the system are stable beside each other. This pressure changes with the temperature, and in such a way that it can be calculated from Thomson's formula, with the aid of the change of volume in the melting, and the heat absorption associated therewith (p. 70). The temperature at which the two phases can coexist at atmospheric pressure is called the "melting-point" of the solid body.

In contrast to those reactions, where a gaseous phase either appears or disappears in the "condensed system," the change of volume occasioned by the change of phase is *relatively very small;* and therefore, according to equation (I.) (p. 630), the influence of the pressure on the equilibrium is so much the smaller. Thus while the boiling-point varies greatly with the external pressure, the melting-point changes to only a relatively small extent.

And therefore we must consider this as the characteristic peculiarity of the reactions of "condensed systems," viz. in that they group themselves in a purely quantitative contrast to those reactions whose substances are volatilised.

Thus for most purposes it is immaterial whether we study the "condensed systems" at atmospheric pressure, or at some other pressure not very far removed from it; and it is still less important to notice the slight variations of the atmospheric pressure.

That temperature at which all phases of a condensed system can exist beside each other is called the "temperature of transformation" (*Umwandlungs-temperatur*). Below this the reaction will advance to a completion (*i.e.* until one phase at least is completely exhausted) *in one direction or in the opposite.*

The "transformation temperatures" of the systems described on pp. 616 and 622 can be found by searching out the temperature points corresponding to the atmospheric pressure on the limiting curves which separate the region of liquid and solid phases. Moreover, these transformation temperatures, as will be more thoroughly shown later, usually lie in the immediate vicinity of the intersections of the limiting curves, *i.e.* near the transition points.

If we allow the transformation in a condensed system to complete itself in one sense, a little below the transformation temperature, and if we then warm it a little above the transformation temperature, so that the reaction will play itself out in the reverse sense—then after cooling down to the ordinary temperature, the system is again in its original condition. These two contrasted transformations complete themselves spontaneously, and are able to perform a certain amount of external work, although this is usually very slight.

Therefore, in this cyclic process, heat must fall from a higher to a lower temperature. That is, *the transformation below the temperature of transformation must occur with a development of heat; and above this temperature with absorption of heat; or the system which is stable at higher temperatures is formed from the system which is stable at lower temperatures with an absorption of heat.*

As is well known, it always takes an *elevation* of temperature to melt a solid substance; this fact necessarily requires that there shall be an introduction of heat in order to melt a substance.

Allotropic Transformation.—An important example of condensed systems is found in *the equilibrium between two modifications of the same substance.*

The transformation of *orthorhombic into monoclinic sulphur* has been well studied. At atmospheric pressure these two phases are in equilibrium at 95·6°. If the temperature is maintained constant, then above this temperature the orthorhombic sulphur goes over to the monoclinic variety, while below this temperature, conversely, the monoclinic variety changes to the orthorhombic; in both cases, the reaction occurs without change of composition of one of the phases, and therefore it occurs completely, as is the case in all such reactions.

The transformation temperature, which is completely analogous to the melting-point, varies with the external pressure; and also, as is directly obvious from the considerations of the preceding section, it occurs according to a method given by the same formula of Thomson, and which makes disclosures regarding the dependence of the melting-point upon the external pressure.

Thus if dT denotes the elevation of the transformation temperature corresponding to the increase of pressure dp; and if σ and τ denote respectively the specific volumes of the monoclinic and orthorhombic sulphur, at the transformation temperature T, on the absolute scale; and if r denotes the amount of heat in g.-cal. absorbed by the transformation of 1 g. of sulphur, then, according to p. 70, we have

$$\frac{dT}{dp} = 24{\cdot}25\frac{T(\sigma - \tau)}{1000\ r},$$

and since

$$T = 273 + 95{\cdot}6^\circ, \quad \sigma - \tau = 0{\cdot}0126 \text{ c.cm.}, \quad \text{and } r = 2{\cdot}52 \text{ cal.},$$

therefore

$$\frac{dT}{dp} = 0{\cdot}045^\circ.$$

That is, the temperature of transformation must rise 0·045°, on account of the elevation of the external pressure [per atm.]. The experimental test of this theoretical prediction, as conducted by

Reicher,[1] confirms quantitatively that the elevation of the transformation temperature amounts to about 0·05° per atm.

Similarly one always obtains the same kind of results in a large number of analogous cases, *i.e.* when we find two modifications of a dimorphous substance, there is always a temperature, corresponding to the melting-point, above which only the one variety can correspond to the state of equilibrium (at atmospheric pressure), and below which only the other variety can so correspond; and by passing beyond this temperature, there occurs a *total* transformation of one modification into the other.[2] This regularity of behaviour has been called by van't Hoff "the incompatibility of condensed systems."

Although the influence of pressure on the temperature of transformation of allotropic modifications may be very slight, yet it really manifests itself in all cases, and may become considerable when the action of enormous pressures is concerned.[3] This point is of great importance in mineralogy. In rocks which are cooling under enormous pressure, certain modifications may be formed, the preparation of which has not been accomplished as yet in the laboratory, because it has not been possible to establish the conditions suitable for their formation.

In general, the study of the relations of the equilibrium between the allotropic forms of the same substance is usually rendered impossible by reason of the *slowness* of the transformation. Thus we do not know certainly which of the modifications of carbon is the more stable, or where the temperature of transformation lies, etc. Moreover, the conclusion that diamond must be more stable than graphite at higher temperatures, because, according to p. 593, it can be formed from graphite with absorption of heat; this conclusion is not inevitable; for the heat of transformation changes with the temperature, on account of the difference between the specific heats of the two modifications; and therefore the heat of transformation may vary with the temperature of transformation, on which it alone depends; and even the signs prefixed may be different from those referred to ordinary temperatures.[4]

As to the velocity of the change there are the greatest differences. In some cases, *e.g.* Tetrabrommethane, it is as rapid as the solidification of a liquid; in others so slow that the two modifications can be kept year-long without the transformation taking place. Then it is hardly possible to tell which is the stable form, or where the transition temperature lies, *e.g.* between graphite and diamond, between quartz and tridymite. In fusion, so far, equilibrium has only been over-

1 *Zeit. f. Krist.*, **8.** 593 (1884).

2 For other similar examples, see O. Lehmann, *Molecular Physik*, **1.** Leipzig, 1888.

3 See especially the researches and calculations in Tammann's work quoted on p. 99.

4 On the formation of diamond see the very interesting work of Moissan, *The Electric Furnace.*

stepped in one sense by under-cooling the liquid; it has not been found possible to overheat the solid without melting it; but in solid transitions both things are possible, so that a body can often be heated well above its transition point without being turned into the stable form. The transition is usually hastened by contact with already transformed material, just as an undercooled liquid can be made to freeze by contact with a crystal of the solid. E. Cohen[1] has studied a good instance of this in tin. The white form in which tin is usually known is really only stable above 20°, but can be very much undercooled without conversion into the grey form. In very cold winters spontaneous conversion sometimes takes place, and the mass of tin falls into powder. If the white tin is "infected" with a trace of grey tin, the conversion into the grey form, "the tin disease," goes on at ordinary temperature.

The Melting of Salts containing Water of Crystallisation.—Among the other phenomena which belong here, and in contrast to those preceding which, although of a physical nature, have been included in the chemical section, is the so-called *melting of salts containing water of crystallisation.* The fact that their fusion is usually accompanied by the separation of a salt containing less water of crystallisation, shows that here we are not dealing with a case of simple fusion, *i.e.* with the uniform change of a solid into a liquid.

Thus from the fusion of Glauber's salt, there not only results a liquid product (which is a saturated water solution of Glauber's salt, $SO_4Na_2 . 10H_2O$), but also the solid anhydrous salt, SO_4Na_2.

Now it is at once evident that here we are dealing with a "complete equilibrium"; for in the liquefaction of Glauber's salt we must distinguish the three phases, viz. $SO_4Na_2 . 10H_2O$ (solid); SO_4Na_2 (solid); and $H_2O + xSO_4Na_2$ (saturated solution). And it is also evident that for the construction of the three phases, there are required at least the two molecular species, H_2O and Na_2SO_4.

Therefore, according to the phase-rule, for a given pressure there can correspond one, and only one, temperature at which all three phases can coexist beside each other. Below this temperature the reaction,

$$SO_4Na_2 . 10H_2O \rightleftarrows SO_4Na_2 + 10H_2O,$$

will advance to a completion in the sense of the equation from right to left; above this temperature, completely from left to right.

This temperature (33°), which corresponds to the atmospheric pressure, is called the "temperature of transformation" (*Umwandlungstemperatur*). On account of the very slight changes in volume associated with the reaction, this temperature varies to only an inconsiderable degree with the external pressure.

[1] *Zeits. phys. Chem.*, **30.** 601 (1899); **33.** 57 (1900); **35.** 588 (1900).

The Formation of Double Salts.—Now there are known[1] to us a number of transformation temperatures of condensed systems which consist of *four phases;* therefore, according to the phase-rule, they must be constructed of at least *three molecular species.*

These reactions include the *formation of double salts,* such as blödite (*astrakanite, simonite*), $(SO_4)_2Na_2Mg . 4H_2O$, from the sulphates of sodium and magnesium. This reaction occurs according to the equation

$$SO_4Na_2 . 10H_2O + SO_4Mg . 7H_2O \rightleftarrows (SO_4)_2Na_2Mg . 4H_2O + 13H_2O.$$

The four phases coexistent in equilibrium are formed from the three solid salts and their saturated solution. The temperature of equilibrium is 21·5°; above this, only the phases occurring on the right side of the reaction equation are coexistent; below this temperature only those on the left. This will appear from the following considerations.

If a mixture of finely pulverised blödite and water, in the proportions given above, is prepared *below* 21·5°, the mixture, which first was a thin paste, hardens in a short interval to a completely dry and solid mixture of the first two sulphates. *This does not happen above* 21·5°.

Conversely, if one mixes the two sulphates of sodium and magnesium in molecular proportions, and reduces them to a fine powder, then above 21·5° blödite is formed after a longer or shorter interval, and the water which is set free occasions a partial liquefaction of the blödite; but below 21·5° the mixture of the finely powdered sulphates of sodium and magnesium remains unchanged.

A still further complication should be noticed here, viz. the liquefaction of the Glauber's salt, which, as mentioned above, in the absence of other salts, occurs at 33°; but here, as a result of the presence of the magnesium sulphate in molecular proportions, the melting-point experiences a depression of 7°, just as the melting-point of water is depressed by the presence of dissolved salts.

The relations are completely analogous in *the formation of sodium-ammonium-racemate,* a double salt, which can be formed above 27° by rubbing together a dry mixture [in molecular proportions] of right and left sodium-ammonium-tartrates; but the double salt [*i.e.* the racemate] cannot be so formed below this temperature.

Here, at the temperature of transformation, the following *four phases* are coexistent, viz. the three solid salts and their saturated solution.

At every other temperature the reaction

$$2C_4O_6H_4NaNH_4 . 4H_2O \rightleftarrows (C_4O_6H_4NaNH_4)_2 . 2H_2O + 6H_2O$$

advances to a completion either in one sense or the other. Thus, if one attempts to cleave the optically inactive mixture of the right and

[1] Van't Hoff and van Deventer, *Zeits. phys. Chem.*, **1.** 170 (1887).

left compound, by the method described on p. 334, he will find it absolutely necessary to work on the appropriate side of the temperature of transformation; in this case, *e.g.*, below it.

The formation of *calcium-copper-acetate* ($CaCu(Ac)_4 . 8H_2O$) from the two salts ($Ca(Ac)_2 . H_2O$ and $Cu(Ac)_2 . H_2O$), and the corresponding quantity of water, $6H_2O$,—this exhibits some peculiarities,[1] inasmuch as the formation takes place easily at lower temperatures, but not above 76°; also, by raising them above this temperature, the double salts are decomposed, instead of forming the single salts described above; and also this decomposition of the double salts occasioned at higher temperatures is accompanied by a very considerable contraction, and by a very apparent change in colour, as the double salt is blue, the copper acetate is green, and the calcium acetate is colourless.

The formation of *cupric-potassium-chloride* ($CuCl_2KCl$), and also that of *cupric-di-potassium-chloride* ($CuCl_22KCl . 2H_2O$), has been thoroughly investigated by Meyerhoffer.[2] There are two corresponding reactions, viz.—

$$CuCl_22KCl2H_2O \rightleftarrows CuCl_2KCl + KCl + 2H_2O;$$
$$CuCl_22KCl2H_2O + CuCl_22H_2O \rightleftarrows 2CuCl_2KCl + 4H_2O;$$

and to these there correspond respectively the transformation temperatures of 92° and 55°; below these temperatures, only the systems on the left side are coexistent; above them, only those on the right side. In both cases, at the respective transformation temperatures, there are *four* different phases which are coexistent, and in both cases the same three molecular species (viz. H_2O, KCl, and $CuCl_2$) are required to construct the four respective phases.

The systems standing on the left differ by the presence of a molecule of cupric chloride; and the fact that this molecule occasions a depression of about 37° in the transformation of the cupric-potassium-chloride, reminds us again of the depression of the freezing-point of a solvent by the addition of a foreign substance.[3]

The Double Decomposition of Solid Salts.—In conclusion, some condensed systems are known where *five* phases are coexistent in the equilibrium, and hence where *four* molecular species are required to construct the phases. This is the case in *the double decomposition of solid salts*; as, *e.g.* of magnesium sulphate and sodium chloride to form the double salt blödite (*a sodium-magnesium-sulphate*) and magnesium chloride. This occurs in the sense of the following equation, viz.,

$$2NaCl + 2SO_4Mg . 7H_2O \rightleftarrows (SO_4)_2MgNa_2 . 4H_2O + MgCl_2 . 6H_2O + 4H_2O.$$

[1] Reicher, *Zeits. phys. Chem.*, **1.** 221 (1887).
[2] *Ibid.*, **3.** 336 (1889), and **5.** 97 (1890).
[3] For further examples, see Van't Hoff, *Bildung und Spaltung von Doppelsalzen* (Leipsic, 1897).

The transformation temperature is at 31° ; below this temperature, if one mixes finely pulverised blödite with magnesium chloride and water, in the aforesaid proportions, the thin paste at once solidifies to a solid mass, which is completely dry, and which consists of sodium chloride and magnesium sulphate ; but above 31° the mixture remains unchanged. And conversely, an equi-molecular mixture of NaCl and $MgSO_4 . (H_2O)_7$ changes into blödite and $MgCl_2$ only above 31°, accompanied by a partial liquefaction, which is occasioned by the water which is eliminated.

At the temperature of transformation there are *five* phases coexistent together, which are composed of the four solid salts and their saturated solution.

The relations are precisely analogous in the reaction

$$SO_4Na_2 . 10H_2O + 2KCl \rightleftarrows SO_4K_2 + 2NaCl + 10H_2O,$$

the transformation temperature of which[1] is 3·7°.

The Vapour Pressure and the Solubility at the Temperature of Transformation.—In all of the condensed systems enumerated above, we have considered the cases where—

a. The n molecular species react in n + 1 phases, and where

b. The phases were in complete equilibrium.

But now we can imagine another *new* phase to be added to the system in all of these cases, viz. *the gaseous phase ;* because we can think of the system as standing under its own vapour pressure instead of that of the atmosphere. Of course, as a result of this pressure, the transformation temperature will be displaced ; but this displacement, on account of its slight dependence upon the external pressure, is relatively small, and rarely amounts to a tenth of a degree, for this independence of the external pressure is a prominent characteristic of condensed systems.

Under these circumstances we will have to consider the system as containing n + 2 phases, for the construction of which we need only n molecular species. That point at which the n + 2 phases can coexist is the "transition point" (*Uebergangs-punkt*), in the sense defined on p. 612, and, moreover, it is an (n + 2)-fold point.

The space above the n-systems will be filled by the vapours emitted by each of the solid or liquid phases ; and although the vapour pressure of each one of the substances (as of the water-free salts, *e.g.*) may be extremely small, nevertheless it is not absolutely zero ; and this total vapour pressure therefore has a definite, though small, amount.

Now, let us consider, on the one hand, the phases standing on the left side of the reaction equation, and on the other hand, those standing on the right side ; these are mutually convertible into each other.

[1] Van't Hoff and Reicher, *Zeits. phys. Chem.*, **3.** 482 (1889).

Now, from the fact that they are capable of co-existing beside each other in stable forms, at the respective conditions of temperature and pressure, therefore it necessarily follows that, at the temperature of equilibrium, the *separate vapours emitted by the two particular groups of phases must have the same composition and density.*

Therefore *the vapour-pressure curves* of two systems which are mutually convertible into each other, as *e.g.*, ice and water, orthorhombic and monoclinic sulphur, Glauber's salt and the saturated water solution of $Na_2SO_4 . H_2O$, etc. etc.,—*these curves must cut each other* at the transition point, and also approximately at the adjacent temperature of transformation: this inference has been most conclusively proven in all the cases thus far investigated.

Now it is usually possible to investigate the vapour pressures of the two systems separately from each other, *i.e.* the one above and the other below the transformation temperature, where each respectively is in a labile condition. Then, of course, the vapour pressure of that system which is the stable one [relatively] will be the smaller; and the vapour pressure of that which is the labile one will be the greater. And, as a matter of fact, the measurements made thus far have given this result, viz. that before the transformation temperature is reached, the stable system does show the smaller vapour pressure, but that above this point the same system, which has now become the labile one, exhibits the greater vapour pressure.

An analogous inference is drawn for the solubility of the two groups of phases which are convertible into each other in the sense of the respective reaction equations.

Thus let us consider the two systems in any solvent (which is chemically indifferent) at the temperature of transformation; then at this point the two solutions must have the same composition and concentration. For otherwise, if the two solutions could communicate with each other by means of diffusion, an equalisation of the difference in composition would take place; and during this adjustment, wherever the two groups of phases are in contact with each other by means of the intermediate solution present, the constancy of the concentration would be [adjusted and] maintained by continuous solution or crystallisation, and thus a process of this sort would necessarily occur, until one or more of the phases would disappear, *which is impossible;* hence the two systems of phases must be in equilibrium.

This conclusion has been experimentally verified in a great number of cases by van't Hoff and his students.

Thus, *e.g.*, a saturated water solution of blödite at the transformation temperature, 21·5°, has the same concentration as a saturated solution of sodium and magnesium sulphates.

As before, the mixture of the two sulphates, as the more stable system, had the lesser solubility, and the solution of blödite, as the labile system, the greater solubility.

Thus the solution of the less stable system partakes of the character of a supersaturated solution, as follows, in fact, from the crystallisation occasioned by contact with the ingredients of the other systems.

Above the transition point the two systems exchange their rôle, and the solubility curves intersect each other at the transition point.

Amorphous substances (p. 98), since they are in an unstable state, must have greater vapour pressure and solubility than crystalline.

The Determination of the Temperature of Transformation. —In most cases it is easy to obtain an upper and a lower limit for the temperature of transformation, because it is easy to find two temperature points, at one of which the reaction advances in one direction, while at the other point the reaction advances in the other. But on account of the slowness of the reactions, it is only in the rarest cases that a very exact determination can be made in this way.

Yet this can be almost always accomplished by means of the following methods, devised by van't Hoff, and which, in part, remind us of the melting-point determination (p. 328):—

1. The fact that the change of the condensed systems of the one group into the other group is almost always associated with more or less change in volume can be applied in the following way.

The ingredients of the one group, intimately mixed, are placed in a dilatometer, which, it should be noticed, is filled with an indifferent liquid, as oil. The bulb of the dilatometer is then placed in a water bath, the temperature of which is very gradually elevated through the requisite interval; the oil level in the projecting capillary tube then varies very gradually. But in the immediate neighbourhood of the point of transformation temperature there is exhibited a sudden and very significant displacement of the oil level, occasioned by the complete transformation, and which is sharply contrasted with the otherwise gradual change.

It is advisable, as a rule, in order to expedite the occurrence of the reaction, and to avert any delay of the system in the labile condition, to mix with the substances a little of the ultimate products of decomposition.

2. The fact that the change is almost always accompanied by either a development or an absorption of heat; this can be used to determine the transformation temperature in a way similar to that in which the development of heat in freezing allows an exact determination of the freezing-point.

Those systems which are changed into others with a development of heat become under-cooled; then when the reaction occurs, the temperature mounts to the transformation temperature, which may be read off by means of a thermometer which dips into the mixture.

3. One may study the vapour-pressure curves, or the solubility

curves, of the two systems, and may thus search out the intersection point which corresponds to the temperature desired.

In order to measure the very slight differences of vapour pressures which are met here, it is advantageous to use the *differential tensimeter*.[1]

Thus, in ascertaining the temperature of transformation of the system,

$$SO_4Na_2 . 10H_2O \rightleftarrows SO_4Na_2 + 10H_2O,$$

the results showed an identity of vapour tensions at 32·5° – 32·6°; while the interpolation of the following results of Loewel, in determining the solubilities of the two solid salts, showed an identity of solubility at 32·65°.

Temperature.	Solutions of	
	SO_4Na_2	$SO_4Na_2 . 10H_2O$
31·84°	40	50·37
32·73°	50·76	49·71

Thus the two determinations of the temperature coincide with each other remarkably, and also sufficiently well with the so-called melting point, 33°, which is directly determined.

[1] Bremer, *Zeits. phys. Chem.*, **1**. 424; Prowein, *ibid.*, **1**. 10 (1887).

CHAPTER III

THERMO-CHEMISTRY III

TEMPERATURE AND INCOMPLETE EQUILIBRIUM

The Thermodynamics of Incomplete Equilibrium.—Although the influence of temperature on *a complete equilibrium* is always such that, at constant pressure, the slightest change of temperature is sufficient to cause one of the phases to vanish completely, thus occasioning an entire rearrangment of the system, yet a similar change of the temperature has an entirely different effect in the case of an *incomplete equilibrium.* Here a slight change in the temperature never occasions more than a very slight change in the equilibrium; because the relative proportions of the reacting ingredients are always so changed by a displacement of the equilibrium in one sense or the other, that there is a compensation for the slight change in the reaction coefficient occasioned by a slight change in the temperature.

The results of a variation in the external pressure, at constant temperature, are exactly similar to the preceding; in the former case [that of complete equilibrium], one observes the complete elimination of one of the phases; in the latter case [that of incomplete equilibrium], one observes only a very slight displacement of the equilibrium [by a change of the external pressure at constant temperature].

If we denote by

$$\frac{\partial p}{\partial T}dT,$$

the change experienced by the pressure of a reaction mixture, from the temperature elevation dT, at constant volume; and by

$$\frac{\partial Q}{\partial V}dV,$$

the heat absorption which takes place if we increase the volume of the reaction mixture by DV, at constant temperature, then, according to p. 24, the equation

$$\frac{\partial Q}{\partial V} = T\frac{\partial p}{\partial T} \quad . \quad . \quad . \quad . \quad \text{(II.)}$$

holds good here.

This equation is a special case of equation (I.) on p. 630, and which was obtained when Q and V were proportional to each other; *it contains all that the second law of thermodynamics can state regarding the incomplete equilibrium.*

Equation (II.) is applicable both to *gaseous* systems and also to solutions, provided that in the former cases the equilibrium pressure p is measured by means of an ordinary manometer, and in the latter case by means of an osmotic apparatus. It also continues to be applicable even when the gases or the dissolved substances react in a fairly high state of concentration,[1] or when various solid substances participate in the equilibrium.

By applying the preceding formula to the equilibrium between a solution and its vapour, one may easily convince himself that he will be led to the equation of Kirchhoff, which was developed on p. 118, by a method identical in principle with the preceding.

But equation II. is sometimes inconvenient on account of its very universality, so that the desire is aroused for a more convenient form. Such a form can be obtained if again, as in the consideration of incomplete equilibrium in the previous book, we limit ourselves to this condition, viz.—*that the phases of systems of variable composition, shall be either gaseous and at not too high pressure, or else solutions at not too high concentration.*

The Reaction Isotherm, and the Reaction Isochore.—It was van't Hoff who succeeded in reducing to a very simple form, the equations which are obtained by the application of the second law of thermodynamics to the special chemical processes described above.

Let us consider a chemical system consisting of one phase which has a variable composition (as a gas mixture or a dilute solution), and of any arbitrary number of phases of constant composition; then when we are considering reactions which are in progress, the following scheme holds good, viz.

$$\nu_1 a_1 + \nu_2 a_2 + \ldots + n_1 A_1 + n_2 A_2 + \ldots \rightleftarrows \nu_1' a_1' + \nu_2' a_2' + \ldots + n_1' A_1' + n_2' A_2' + \ldots$$

Here $a_1, a_2, \ldots a_1', a_2' \ldots$ denote the respective solid substances reacting, with the respective number of molecules, $\nu_1, \nu_2, \ldots \nu_1', \nu_2' \ldots$; and we

[1] Compare, *e.g.*, van Deventer and van der Stadt, *Zeits. phys. Chem.*, **9.** 43 (1891).

also suppose that the reaction progresses according to the equation for the law of mass-action, viz.

$$K = \frac{c_1'^{n_1'} c_2'^{n_2'} \dots}{c_1^{n_1} c_2^{n_2} \dots} \quad . \quad . \quad . \quad \text{(III.)}$$

where c_1, c_2, . . . c_1', c_2' . . . denote the concentrations of the molecular species, A_1, A_2, . . . A_1', A_2' . . .

The equilibrium coefficient K *is constant at a given temperature*, i.e. *it is independent of the relative proportions of the reacting substances ; its change with the temperature is determined by the following equation deduced by van't Hoff ;* viz.

$$\frac{d \ln K}{dT} = -\frac{q}{RT^2};$$

here, as always, ln denotes the natural logarithm ; q the heat of reaction measured at the abs. temp. T ; and R denotes the gas constant.

This equation has a very great importance and a many-sided application, as will appear in the following sections ; and he who would penetrate deeper into the relations between heat and chemical energy, must be familiar with its meaning and manipulation.

While, on the one hand, the equation of Guldberg and Waage, viz.

$$K = \frac{c_1'^{n_1'} c_2'^{n_2'} \dots}{c_1^{n_1} c_2^{n_2} \dots},$$

is able to instruct us regarding the influence of changes in concentration *at constant temperature ;* on the other hand, the van't Hoff equation, viz.

$$\frac{d \ln K}{dT} = -\frac{q}{RT^2},$$

instructs us regarding the influence of temperature upon the equilibrium of a system *at constant volume.* Therefore I suggest that the former be called *the equation of the reaction isotherm,* and the latter *the equation of the reaction isochore.*

The latter is a differential equation ; its integral is

$$\ln K = \frac{q}{RT} + B,$$

where B represents the integration constant.

If the value of K, at the temperatures T_1 and T_2 respectively, amounts to K_1 and K_2, then we obtain

$$\ln K_1 = \frac{q}{RT_1} + B,$$

and

$$\ln K_2 = \frac{q}{RT_2} + B;$$

and by subtraction we obtain

$$\ln K_2 - \ln K_1 = \frac{q}{R}\left(\frac{1}{T_2} - \frac{1}{T_1}\right).$$

In effecting the integration, it was provided that q should not change with the temperature, which is, in fact, only approximately fulfilled. But, if the changes are but slight, and if the temperatures T_1 and T_2 are not too far removed from each other, then the preceding equation will be useful in all cases, and will give a value of q corresponding to the temperature

$$\frac{T_1 + T_2}{2}.$$

The Derivation of the Reaction Isotherm.—The equation of the reaction isotherm, which we have learned to regard as the universal expression of the law of chemical mass-action, can be made to appear correct, at least in all probability, by means of kinetic considerations, as we have already seen on p. 430. Moreover, as we have already thoroughly shown in the third book, a widely extended experience has shown us that the law of mass-action is to be regarded as a reality well grounded on facts.

Yet, in spite of all this, it will not be superfluous for the real importance of this fundamental chemical law, to ask a further question, viz. how it stands as regards the requirements of thermodynamics?

Let us consider the following process. There is given a system, in equilibrium, and which is gaseous, or else consists of a dilute solution; and moreover, let it be in contact with various solid substances.

Let the reaction, the equilibrium of which has been established, progress according to the scheme designated above (p. 643); viz.

$$\nu_1 a_1 + \nu_2 a_2 + \ldots + n_1 A_1 + n_2 A_2 \ldots \rightleftarrows \nu_1' a_1' + \nu_2' a_2' + \ldots + n_1' A_1' + n_2' A_2' + \ldots$$

and again we denote the concentration of the molecular species A by c (of course adding the corresponding indices). Also, let all the molecular species be in the free state; and let those denoted by a be in the solid state, as they also participate in the equilibrium; and let those denoted by A be present in unit degree of concentration, either as gases if the phase of variable composition of the system considered is gaseous, or else dissolved in the respective solvent if the variable phase forms a solution.

The following process we will consider thermodynamically in this section and the following one. *We will suppose the ν or n molecules of the substances occurring on the left side of the equation, to be introduced into the mixture; and at the same time the ν′ or n′ molecules of the substances on the right side of the equation, to be removed from the reaction mixture.*

On account of its simplicity, we can suppose this process to be so conducted that the reaction mixture shall preserve its composition unchanged, that at every moment the same equivalent quantity of A molecules shall be driven in, as of A′ molecules that are driven out, so that the change in the reaction mixture shall continually complete itself from left to right, without the occurrence of any marked change in its relative concentration.

In this way it is clearly possible to conduct the reaction isothermally and reversibly: it remains to calculate the work so performed.

It will be simpler first to take a particular case, viz. the formation of water vapour according to the equation

$$2H_2 + O_2 = 2H_2O.$$

Let c_1, c_2, c_1' be the concentrations of the reacting molecular species in equilibrium, and suppose the substances originally in the free state with concentrations C_1, C_2, C_1'.

We will first deal with the problem of calculating the work involved in transferring a mol. of gas from a space at concentration C to one at concentration c; and for simplicity we assume that both spaces are so large that the addition or withdrawal of a mol. makes no appreciable difference to the concentration. If the pressure and volume [of one mol.] in the two spaces are PV and pv respectively, the mol. must first be withdrawn from space I, by which the amount of work PV is gained, then expanded to volume v, yielding work $= RT \ln \frac{v}{V}$ (see p. 52) and finally introduced under the constant pressure p into space II, which requires work pv; hence the work to be calculated is

$$PV + RT \ln \frac{v}{V} - pv,$$

or since

$$PV = pv,$$

simply

$$RT \ln \frac{v}{V} = RT \ln \frac{C}{c},$$

since

$$V : v = c : C.$$

In conveying n mols. the work is, of course,

$$nRT \ln \frac{C}{c}.$$

We can now calculate the maximum work done when hydrogen and oxygen of concentration C_1, C_2 are converted isothermally and reversibly into water vapour of concentration C_1'. If two mols. of hydrogen are taken from a space of concentration C_1 into a space where the concentration is c_1, and there is equilibrium with the other molecular species, we gain work $2RT \ln \frac{C_1}{c_1}$; for a mol. of oxygen $RT \ln \frac{C_2}{c_2}$, and to remove the water vapour formed, requires the work $2RT \ln \frac{C_1'}{c_1'}$. Hence the work to be gained is

$$A = 2RT \ln \frac{C_1}{c_1} + RT \ln \frac{C_2}{c_2} - 2RT \ln \frac{C_1'}{c_1'}$$

or

$$A = RT \ln \frac{C_1^2 C_2}{C_1'^2} + RT \ln \frac{c_1'^2}{c_1^2 c_2}.$$

The maximal work must, however, be independent of the nature of the reaction mixture, which only plays the part of an intermediary which suffers no appreciable change during the process. This is only possible if, at constant temperature, the expression $RT \ln \frac{c_1'^2}{c_1^2 c_2}$, and therefore $\frac{c_1'^2}{c_1^2 c_2}$ remains constant; this is, however, nothing else than the law of mass-action.

There is now no difficulty in dealing with the most general case. We have only to calculate the work done in introducing into the mixture the substances on the left of the equation, and removing those on the right, the process being, of course, isothermal and reversible.

The work done in introducing or removing the molecular species a and a′ which are solid, is, of course, nothing; hence the two parts of the total work are

$$n_1 RT \ln \frac{C_1}{c_1} + n_2 RT \ln \frac{C_2}{c_2} \ldots$$

and

$$-\left(n_1' RT \ln \frac{C_1'}{c_1'} + n_2' RT \ln \frac{C_2'}{c_2'} + \ldots\right),$$

and the total is

$$A = RT \ln \frac{C_1^{n_1} C_2^{n_2} \ldots}{C_1'^{n_1'} C_2'^{n_2'} \ldots} + RT \ln \frac{c_1'^{n_1'} c_2'^{n_2'} \ldots}{c_1^{n_1} c_2^{n_2} \ldots}.$$

Since A does not depend on the nature of the reaction mixture

$$K = \frac{c_1'^{n_1'} c_2'^{n_2'} \ldots}{c_1^{n_1} c_2^{n_2} \ldots}$$

is constant; *i.e. we have the law of mass-action in its most general form.*

If the concentrations C_1, C_2, ... C_1', C_2' are all unity we have simply

$$A = RT \ln K;$$

so that the maximal work can be very simply derived from the constant K.

Assuming that the process described is realisable in all cases, then the proof given above may claim strict validity, *i.e.* the law of chemical mass-action is a necessary conclusion from thermodynamics. We can hardly doubt its possibility of realisation in those cases where there are several molecular species on both the left and the right sides of the equation.

But when this is not the case, as when we are given a case of dissociation to consider, then the suspicion arises that the respective molecular species cannot be removed from the reaction mixture in a pure state, but that they are dissociated—that is, that if we had a mixture of PCl_5, of PCl_3, and of free Cl_2, and if we should remove the first molecular species PCl_5, it would at once dissociate.

Now, regarding this suspicion, we find help in the remark that the molecular species in question may be removed from the reaction mixture so quickly and may be brought to such a state that dissociation cannot take place; thus, *e.g.*, it may be brought to a state of strong condensation, or into a solvent which does not occasion dissociation; and thus the process described above may be modified in a way which is perfectly immaterial for the final result. If we assume the possibility of working *with sufficient rapidity*, then the process described is realisable *in all cases, and we find that the law of mass-action is a strict postulate of thermodynamics.*

Now, the maximal work F is not only independent of the relative concentration, but is also independent of the whole nature of the reaction mixture, *e.g.* it is independent of the nature of the solvent in which the equilibrium of the reaction considered is established.

Moreover, we find from simple considerations, that *if we know the equilibrium constant K for any particular phase, and also if we know the distribution coefficients of the reacting molecular species as compared with any other selected phase, then we are also able to state the condition of equilibrium of this phase.* This result has been already obtained in an entirely different way (p. 494).

If the *solvent* is to be counted in as one of the reacting molecular species, then it may be easily shown (p. 116), that for the transference of n molecules, there is required the work

$$n \ln \frac{c}{c_0},$$

where c and c_0 denote respectively the concentrations at the temperature in question, of the saturated vapour (of the solution containing the reacting substances), and of the pure solvent. Now, c_0 is constant at a given temperature; whence it follows that the solvent participates with the active mass c; *i.e. we may regard the active mass of a solvent as proportional to the concentration of the vapour emitted by it*, as was proved by the proposition given on p. 459. This result could not be deduced from kinetic considerations, and was first discovered by the writer in the aforesaid way. Thermodynamics is here able to conduct us farther than hitherto, and it should be emphasised that, *in the consideration of concentrated reaction mixtures*, it alone at present is able to undertake the management of the theoretical treatment. If we were in the possession of rules for the vapour pressure of mixtures of any arbitrary degree of concentration, then we would be able to manipulate the reactions of these systems with the same completeness as in the case of the reactions of dilute solutions.

Finally, in order to consider briefly the case of *electrolytic dissociation* from the standpoint of thermodynamics, it necessarily follows, from Chap. IV. of the preceding book, that the thorough application of the law of mass-action leads to this purely experimental result, viz., for the compression of an electrolyte which is dissociated into n ions, n times as much work is required as in the case of the undissociated substance.

The Derivation of the Reaction Isochore.—*The equation of the reaction isochore* is at once given by the application of the fundamental equation on p. 23, viz.

$$A - U = T\frac{dA}{dT}.$$

We found the maximal work A to be

$$A = RT \ln \frac{C_1^{n_1} C_2^{n_2} \ldots}{C_1'^{n_1'} C_2'^{n_2'} \ldots} + RT \ln K,$$

whence

$$\frac{dA}{dT} = R \ln \frac{C_1^{n_1} C_2^{n_2} \ldots}{C_1'^{n_1'} C_2'^{n_2'} \ldots} + R \ln K + RT\frac{d \ln K}{dT},$$

and for the decrease in total energy U we have the heat of reaction

$$U = q;$$

by q we mean the quantity of heat which is *evolved when the process occurs without the performance of external work.*

Therefore it follows on introducing the value of A, U and $\frac{dA}{dT}$ that

$$q = -RT^2\frac{d \ln K}{dT}.$$

This is *the equation of the reaction isochore.*

Now it should be expressly emphasised that, as also doubtless follows from the derivation of the equation of the reaction isochore, that here we are dealing *with the concentration and not with the partial pressure* of the particular molecular species. It was immaterial in the application of the equation of the reaction isotherm whether we worked with one or with the other of these magnitudes (p. 438); but this is not the case with the reaction isochore, since at constant volume the concentration of a substance remains constant but not its pressure.

In what follows we will make several applications of the integrated form of the preceding equation, viz.

$$\ln K_2 - \ln K_1 = \frac{q}{R}\left(\frac{1}{T_2} - \frac{1}{T_1}\right).$$

If we express the heat of reaction q, as is usual in calorific measure, then R amounts to 1·991. If we work with *the ordinary logarithms* instead of the natural, then finally we will have

$$q = -\frac{4{\cdot}584(\log K_2 - \log K_1)T_1T_2}{T_2 - T_1} \text{ g.-cal.}$$

Vaporisation.—We have found for the equilibrium between a simple liquid and its saturated vapour, the relation

$$K = c = \frac{p}{RT};$$

i.e. for every temperature there corresponds a definite concentration of the saturated vapour. Now if we denote, by p_1 and p_2 respectively, the values of the vapour pressure corresponding to the temperatures T_1 and T_2 (which are only slightly different from each other), then we obtain from the preceding equations the expression

$$\ln \frac{p_2}{T_2} - \ln \frac{p_1}{T_1} = -\frac{q}{R}\left(\frac{1}{T_1} - \frac{1}{T_2}\right).$$

From the figures which Regnault obtained for water, viz.,

$$T_1 = 273°, \qquad p_1 = 4{\cdot}54 \text{ mm.},$$
$$T_2 = 273° + 11{\cdot}54°, \qquad p_2 = 10{\cdot}02 \text{ mm.},$$

it follows that $q = -10{,}050$. Now the molecular heat of evaporation λ, at 5·77°, was found to be 10,854; and if we subtract from this latter value the external work, $2T = 558$, then it follows that

$$q = -10{,}296,$$

a value which is in satisfactory accord with the preceding.

We have previously (p. 62) found that

$$\lambda = RT^2 \frac{d \ln p}{dT}.$$

Now if we compare this equation with that obtained by the application of the reaction isochore, viz.

$$-q = RT^2 \frac{d \ln \frac{p}{T}}{dT},$$

we obtain, in harmony with the preceding, the equation

$$\lambda + q = RT^2 \frac{d \ln T}{dT} = RT.$$

The Dissociation of Solid Substances.—The *heat of sublimation of a solid substance* can be calculated from its vapour pressure at two different temperatures, in exactly the same way that the heat of vaporisation is calculated for a liquid. Therefore here we will consider only the case where *the sublimation is attended with dissociation.*

Let us suppose that the solid substance is broken up into n_1 mol. of the substance A_1, and n_2 mol. of the substance A_2, and that the partial pressures of the particular molecular species amount respectively to p_1, p_2, etc. Then, according to p. 473,

$$K = c_1^{n_1} c_2^{n_2} \ldots = \frac{p_1^{n_1} p_2^{n_2} \ldots}{T^{n_1+n_2+} \ldots}.$$

If the decomposition products in the vapour space in contact with the solid bodies are present in the same proportion in which they are produced by the reaction, then we shall have

$$p_1 = P \frac{n_1}{n_1 + n_2 + \ldots}, \quad p_2 = P \frac{n_2}{n_1 + n_2 + \ldots},$$

where P denotes the total pressure (dissociation pressure) of the gases. If this, for the two temperatures, T_1 and T_2 respectively, amounts to P_1 and P_2, then it is calculated that

$$\ln K_1 - \ln K_2 = (n_1 + n_2 + \ldots)\left(\ln \frac{P_2}{T_2} - \ln \frac{P_1}{T_1} \right) = \frac{q}{R}\left(\frac{1}{T_1} - \frac{1}{T_2} \right).$$

In the *dissociation of ammonium sulphhydrate*, $n_1 = 1$, and $n_2 = 1$; thus

$$NH_5S = NH_3 + H_2S.$$

And therefore we will find that

$$\ln \frac{P_1}{T_1} - \ln \frac{P_2}{T_2} = \frac{q}{2R}\left(\frac{1}{T_1} - \frac{1}{T_2} \right).$$

And from the figures

$$T_1 = 273 + 9{\cdot}5^\circ, \quad P_1 = 175 \text{ mm.},$$
$$T_2 = 273 + 25{\cdot}1^\circ, \quad P_2 = 501 \text{ mm.},$$

it follows that $q = -21{,}450$ g.-cal.

Now the thermo-chemical measurements give 22,800 as the molecular heat of sublimation of NH_4SH; and if we subtract from this value the amount of the external work, *i.e.*

$$4T = 1160,$$

we obtain as the value of q observed, $= -21{,}640$ g.-cal.[1]

It is of some historical interest to revert to the fact that, by a calculation which is completely analogous to the preceding, Horstmann[2] in 1869 ascertained theoretically the heat of sublimation of *ammonium chloride*, and thus for the first time applied the second law of the mechanical theory of heat [thermodynamics] to chemical processes; and Horstmann and his successors have pushed on the investigations in this field, with clear and brilliant results.

Of course a similar calculation can be made for the *dissociation of compounds containing water-of-crystallisation;* and thus the combining heat of water-of-crystallisation can be theoretically calculated from the change of the dissociation pressure with the temperature. Horstmann has already called attention to this, and Frowein[3] has performed the task with good results.

Moreover, *the curve of the dissociation pressure of calcium carbonate*[4] has made possible the calculation of the heat of formation of this substance from CO_2 and CaO.

The Dissolving of Solid Substances.—The analogy between the processes of dissolving in any solvent, and of evaporation, is shown in this, viz.—that the same formula which allows us to find the heat of vaporisation from the change of the density of the saturated vapour with the temperature, allows the calculation of *the heat of solution from the change of solubility with the temperature.* Thus, since every substance has a definite solubility in a definite solvent at a definite temperature, we have simply

$$K = c;$$

here c denotes the concentration of the saturated solution at the temperature T. Let us suppose that c has the respective values c_1 and c_2, at the temperatures T_1 and T_2; then for the heat q absorbed

[1] van't Hoff, *Études*, p. 139.
[2] *B.B.*, **2.** 137 (1869); more thoroughly **14.** 1242 (1881).
[3] *Zeits. phys. Chem.*, **1.** 5 (1887).
[4] Le Chatelier, *l.c.* 98.

in the solution, *i.e.* the negative value of the heat of solution of 1 g-mol. of the dissolved substance, we obtain

$$\ln c_1 - \ln c_2 = \frac{q}{2}\left(\frac{1}{T^1} - \frac{1}{T^2}\right).$$

From the solubility of succinic acid in water with the following values, viz.

$$c_1 = 2{\cdot}88 ; \quad T_1 = 273^\circ ;$$

$$c_2 = 4{\cdot}22 ; \quad T_2 = 273^\circ + 8{\cdot}5^\circ,$$

van't Hoff[1] calculated the value of q to be -6900, while the direct measurement by Berthelot gave -6700.

The values of the solubility denote the weight per cent. By the term g.-mol. is meant of course that quantity of the dissolved substance which exerts, at the same volume and temperature, the same [osmotic] pressure as the pressure of 1 g-mol. of an ideal gas. And therefore the applicability of the preceding formula gives us information regarding the molecular condition [of the substance in question] in the solvent used. In the special case just considered (succinic acid), it was dissociated electrolytically to only a slight extent, and therefore had the normal mol. wt.

Moreover, conversely, from the comparison of the observed heat of solution with that calculated, it is, of course, possible to draw an inference as to the size of the molecule in the respective solvent.

The correction for the external work performed, a correction which must be considered in the case of the heat λ of vaporisation (see above), can be neglected in the case of the heat of solution, because no considerable external work is associated with the solution of a solid substance.

See also p. 145, where Q denotes the heat of solution under the osmotic pressure of the saturated solution. This differs from q by the amount of the external work, viz. 2T.

The Dissociation of Solid Substances by Dissolving.— This case is, of course, treated in a way exactly similar to that for the dissociation of solid substances in vaporisation (p. 651).

As an example we will consider the dissolving of silver chloride; thus,

$$AgCl = \overset{+}{Ag} + \overset{-}{Cl}.$$

If the solubility of this substance at T_1 and T_2 is c_1 and c_2 respectively, then it follows that

$$\ln c_1 - \ln c_2 = \frac{q}{2R}\left(\frac{1}{T_1} - \frac{1}{T_2}\right).$$

[1] *Lois de l'équilibre*, etc., p. 37 (1885).

F. Kohlrausch and F. Rose[1] have measured the conductivity of saturated silver chloride solutions at a number of different temperatures, and give as the solubility c_t at $t°$ the formula

$$c_t = c_{18}[1 + 0{\cdot}049(t - 18) + 0{\cdot}00089(t - 18)^2]$$

for temperatures above 18°.

Applying the unintegrated equation

$$q = -2RT^2\frac{d\ln c}{dT}$$

which is suitable we get

$$\frac{d\ln c}{dT} = \frac{0{\cdot}049 + 2 \cdot 0{\cdot}00089(t - 18)}{1 + 0{\cdot}049(t - 18) + 0{\cdot}00089(t - 18)^2}.$$

From this may be calculated at 22° that $q = -16,000$, whilst on p. 607 we found $-15,800$. The agreement is remarkable, c_{18} is $1{\cdot}05 \times 10^{-5}$ g. equiv. per litre.

The Dissolving of Gases.—For gases we have

$$K = \frac{c'}{c},$$

where c denotes the concentration in the gaseous state, and c′ that in solution (Law of Henry). Let us denote the Bunsen coefficient by α; then, according to p. 480,

$$K = \alpha\frac{T}{273};$$

and therefore

$$q = RT^2\frac{d\ln K}{dT} = \frac{RT^2}{\alpha} \cdot \frac{d\alpha}{dT} + RT.$$

The heat of solution Q, amounts to

$$Q = RT - q,$$

and therefore, according to Kirchhoff (1858)

$$Q = -\frac{RT^2}{\alpha} \cdot \frac{d\alpha}{dT}.$$

Thus Naccari and Pagliani[2] found for carbonic acid, that

$$\alpha = 1{\cdot}5062 - 0{\cdot}03651\,t + 0{\cdot}000292\,t^2;$$

from which, for $t = 20$ (*i.e.* $T = 273° + 20°$), Q is calculated to be 4820; while Thomsen by a direct measurement found 5880. It is

[1] *Wied. Ann.*, **50.** 136 (1893); see also F. Kohlrausch, *Zeits. phys. Chem.*, **44.** 197 (1903).

[2] *N. Cim.* [3], **7.** 71 (1880).

probable that the measurements of the absorption coefficients are not accurate enough for this calculation; thus, for instance, the older measurements of Bunsen are entirely unsuited for this purpose.

The Dissociation of Gaseous Substances.—Let us suppose that a molecular species A, whether it exists as a gas or in dilute solutions, decomposes according to the universal reaction equation of dissociation, viz.

$$A = n_1A_1 + n_2A_2 + \ldots;$$

then the condition of equilibrium is denoted by the equation

$$Kc = c_1^{n_1}c_2^{n_2} \ldots;$$

where c, c_1, c_2 . . . are respectively the concentrations of A, A_1, A_2 . . .

If the dissociation products are present in equivalent proportions, and if x denotes the dissociation coefficient, then, if 1 g.-mol. occupies the volume v, we have

$$c = \frac{1 - x}{v}, \quad c_1 = \frac{n_1x}{v}, \quad c_2 = \frac{n_2x}{v}, \ldots$$

and therefore

$$K = \frac{n_1^{n_1}n_2^{n_2} \ldots x^{n_1+n_2+\ldots}}{(1 - x)v^{n_1+n_2+\ldots-1}}.$$

Let us suppose that the g.-mol. considered, at the temperatures T_1 and T_2, occupies the volumes v_1 and v_2 respectively, and that it is dissociated to the fractional extent of x_1 and x_2 respectively. Then the equation of the reaction isochore, in the calculation of the heat of dissociation, gives the relation

$$\ln\frac{x_2^{n_1+n_2+\ldots}}{(1 - x_2)v_2^{n_1+n_2+\ldots-1}} - \ln\frac{x_1^{n_1+n_2+\ldots}}{(1 - x_1)v_1^{n_1+n_2+\ldots-1}} = -\frac{q}{2}\left(\frac{1}{T_1} - \frac{1}{T_2}\right).$$

Let us apply this equation to the dissociation of nitrogen dioxide, viz.

$$N_2O_4 = 2NO_2;$$

then

$$n_1 = 2, \quad n_2 \ldots = 0,$$

and we obtain

$$\ln\frac{x_2^2}{(1 - x_2)v_2} - \ln\frac{x_1^2}{(1 - x_1)v_1} = -\frac{q}{2}\left(\frac{1}{T_1} - \frac{1}{T_2}\right).$$

Let the vapour density, at the temperatures T_1 and T_2, and at atmospheric pressure, be Δ_1 and Δ_2 respectively; then, according to p. 440, we shall have

$$x_1 = \frac{3{\cdot}179 - \Delta_1}{\Delta_1}, \quad \text{and } x_2 = \frac{3{\cdot}179 - \Delta_2}{x_2};$$

here 3·179 denotes the [theoretical] vapour density without dissociation, as calculated from its molecular weight, $N_2O_4 = 96$.

Now the volumes occupied by 1 g.-mol. of N_2O_4, in the two cases, are respectively

$$v_1 = 0{\cdot}0821T_1\frac{3{\cdot}179}{\Delta_1}, \quad \text{and } v_2 = 0{\cdot}0821T_2\frac{3{\cdot}179}{\Delta_2};$$

because the g.-mol. of an ideal gas at T, and at atmospheric pressure, occupies the volume, 0·0819T litre; and the volume of 1 g.-mol. of N_2O_4, as a result of the partial decomposition, is increased in the ratio $\frac{3{\cdot}179}{\Delta}$. Moreover, we also observe that

$$1 + x = \frac{3{\cdot}179}{\Delta},$$

and we finally obtain

$$\ln\frac{x_2^2}{T_2(1-x_2^2)} - \ln\frac{x_1^2}{T_1(1-x_1^2)} = -\frac{q}{2}\left(\frac{1}{T_1} - \frac{1}{T_2}\right).$$

From the values

$$T_1 = 273 + 26{\cdot}7^\circ,\ \Delta_1 = 2{\cdot}65,\ x_1 = 0{\cdot}1996,$$
$$T_2 = 273 + 111{\cdot}3^\circ,\ \Delta_2 = 1{\cdot}65,\ x_2 = 0{\cdot}9267,$$

it follows that

$$q = -12{,}900.$$

The dissociation of 96 g. of N_2O_4 then consumes the very considerable quantity of heat of 12,900 g.-cal., provided that it occurs without the performance of external work; as, *e.g.*, in this way, viz., when a vessel filled with "nitrogen dioxide" [N_2O_4] is connected with another one which is exhausted. During the equalisation of the pressure, as the enclosed gas mass assumes an enlarged volume, a certain fraction of it, which can be calculated from the equation of the isotherm of dissociation, is cleaved into the single molecules [NO_2].

No experiments have as yet been conducted leading to the direct measurement of q; but by means of the mean specific heat of nitrogen dioxide (as measured by Berthelot and Ogier (p. 347), at atmospheric pressure between 27° and 150°), van't Hoff[1] calculated the heat of dissociation in this way; viz., from the quantity of heat required for the elevation of temperature, he subtracted the energy required for the simple warming of the gas, and also the energy consumed in the performance of the external work, and thus he obtained the energy employed in the dissociation. It coincides well with the calculated value[2]

$$q = -12{,}500.$$

[1] *Études*, p. 133.

[2] See also Boltzmann, *Wied. Ann.*, **22.** 68 (1884).

All the vapour-density determinations of nitrogen dioxide are in harmony with the preceding formula; moreover, the variation of the specific heat of this gas, as observed by Berthelot and Ogier, can be completely calculated in a theoretical way, as was recently and thoroughly shown by Swart.[1]

The heats of dissociation of several other gases have been calculated as follows, from the variation of the degree of dissociation with the temperature:—

Vapour of iodine . . .	28,500
,, ,, acetic acid . . .	20,000
Methyl-ether-hydro-chloride[2] .	8,600

Thus it appears that in general the quantities of heat associated with the dissociation of gases are quite considerable; their variation with the temperature is obtained from the differences of the specific heats of the dissociation products (referred to constant volume), as compared with the specific heats of the undissociated molecules. On account of the magnitude of the heat of dissociation, this variation can be neglected, especially if it does not involve too great a temperature interval, for it numbers only a few calories and rarely exceeds 10.

Therefore we may regard q as being independent of the temperature, in the equation for the reaction isochore, and thus can use the integrated form

$$\ln K = \frac{q}{RT} + \text{a const.}$$

Now if we substitute the value of K, viz.,

$$K = \frac{4x^2}{(1-x)v} = \frac{4x^2P}{(1-x^2)RT},$$

(where P = the total pressure of the gas in the preceding equation), we obtain

$$\ln\frac{(1-x^2)T}{x^2P} = -\frac{q}{RT} + \text{a const.};$$

or after introducing the theoretical vapour density δ, and that observed Δ, we obtain

$$\ln\frac{(2\Delta-\delta)T}{(\delta-\Delta)^2 P} = -\frac{q}{RT} + \text{a const.}$$

This is a general equation of condition for gases existing in the state of binary dissociation.

In order to determine q and the integration constant, use is made of the measurements of the vapour density at different temperatures; and then by means of the two preceding formulæ, one can calculate

[1] *Zeits. phys. Chem.*, **7.** 120 (1891). See also Schreber, *ibid.*, **24.** 651 (1897).

[2] See p. 441.

the degree of dissociation, and also the vapour density, for any arbitrary condition of temperature and pressure.

It should be observed, in making calculations of this kind, that the constants in the two equations have different values, as is apparent from the derivation. It is more convenient to manipulate the first form, especially when one has calculated a table for $\ln \frac{1 - x^2}{x^2}$ for every case.

It was shown by Gibbs[1] (1879) that all the observations on the vapour densities of formic acid, acetic acid, and phosphorus pentachloride, can be satisfactorily represented by means of the preceding equations.

The Dissociation of Dissolved Substances.—The formulæ developed in the preceding section can be applied to this case without further remark.

Thus let v_1 and v_2 denote respectively the volumes of solutions containing 1 g.-mol. [of the solid substance in question], at the temperatures T_1 and T_2; and let x_1 and x_2 be the respective degrees of dissociation under these conditions. Thus, as a special formula for the case of a binary electrolyte, we have

$$\ln \frac{K_2}{K_1} = \ln \frac{x_2^2 (1 - x_1) v_1}{x_1^2 (1 - x_2) v_2} = -\frac{q}{2}\left(\frac{1}{T_1} - \frac{1}{T_2}\right);$$

on account of the very slight expansion of water solutions from heat, one may, without hesitation, make $v_1 = v_2$, if the same solution is being compared at two different temperatures.

Unfortunately there are, as yet, no satisfactory numerical data at hand for the calculation of the *heat of ordinary dissociation in solution* (p. 455). Inasmuch, however, as the measurement of the electrical conductivity affords a very exact determination of the degree of electrolytic dissociation, the data for the calculation of the heat of dissociation are furnished by the very simple and exact *measurement of the temperature coefficients of the conductivity.*[2]

In this way Arrhenius[3] ascertained the heats of dissociation for

Electrolyte.	q
Acetic acid	− 28
Propionic acid	183
Butyric acid	427
Phosphoric acid	2103
Hydrofluoric acid	3200

[1] *Sill. Jour.*, **18.** 277 (1879); see also p. 483.

[2] The heat of dissociation of benzoic acid in benzene in 8710 cal. according to Hendrixson's (p. 486) experiments on the distribution between two solvents.

[3] *Zeits. phys. Chem.*, **4.** 96 (1889), **9.** 339 (1892); see also Petersen, *ibid.*, **11.** 174 (1893).

the respective electrolytes, as given in the preceding table : the figures refer to a temp. of 21·5°.

At higher temperatures the heat of dissociation increases, *i.e.* the specific heat of the electrically neutral molecules is greater than that of the free ions; and thus the heat of dissociation of acetic acid at higher temperatures is negative.

In this way the result was ascertained, that *the dissociation of a substance into its ions is usually attended with a development of heat.*

The value for hydrofluoric acid [3200] coincides in a very interesting way with the value 2570, as found on p. 607, especially when we consider that the latter can only be estimated in a general way.

A stricter calculation shows that the coincidence between the heats of dissociation, as calculated from the conductivities, with the thermo-chemical measurements of Thomsen, is most excellent; this is shown by the following results, which are calculated according to the more exact formula of Arrhenius (p. 607); here x, the dissociation heat of water, is put at 13,210; and the values of W, W_1, and W_2, are ascertained from the conductivities.

Acid.	Obs.	Calc.
HCl	13,700	13,740
HBr	13,760	13,750
HNO_3	13,810	13,680
C_2H_5COOH	13,400	13,450
HF	16,120	16,270

The figures refer to the heat developed on neutralising the acid with sodium hydroxide, when the concentration of acid and base is $\frac{1}{3\cdot6}$ normal.

Attention should be especially directed to the fact that *the value of the heat of neutralisation does not stand in any direct relation to the strength of the acid.* Thus propionic and hydrofluoric acid are both weak acids; yet the neutralisation heat of the former is smaller than, and that of the latter is greater than, the heat of neutralisation of either of the three strong acids, HCl, HBr, or HNO_3. This shows, therefore, that we cannot detect any simple relation between the heat of dissociation of acids and their strength. And therefore it is not allowable to do, as is often wrongly done, viz., to measure the affinity between an acid and a base by the quantity of heat developed on neutralising them.

If an electrolyte develops heat by decomposing into its ions, then its dissociation must diminish with rising temperature. Therefore if a solution of such an electrolyte is warmed, then the number of conducting molecules is diminished, which, of course, causes its *conductivity to diminish.* But, on the other hand, the conductivity increases considerably as a result of the diminished ion friction; and as the latter influence preponderates, therefore usually an increase of the conductivity of water solutions is observed with rising temperature.

It was Arrhenius who, guided by such considerations as the preceding, first discovered those electrolytes, where, conversely, the restraining influence of the retrograding dissociation preponderates. Thus phosphoric and hypo-phosphorous acid at the respective temperatures of 54° and 75°, show *maxima of conductivity*, and above these respective temperatures they have negative temperature coefficients. No one before this had suspected the existence of such electrolytes.

Electrolytic Dissociation of the Solvent.—We found on p. 508 for the dissociation of water

$$K = c_0^2;$$

whence the heat of dissociation is

$$q = -RT^2 \frac{2}{c_0} \frac{dc_0}{dT}.$$

In this equation q is known, so that the temperature coefficient of the dissociation can be calculated. Since the temperature coefficient of the ionic friction is known, that of the conductivity may be arrived at theoretically, and Kohlrausch and Heydweiller[1] calculated it as 5·81 % at 18°; this is unusually large, because the increase in ionic mobility is joined to a rapid rise of dissociation with temperature. These observers found that with increased purity the temperature coefficient increased from 2·4 % as shown by ordinarily "pure water" to a maximum of 5·32 %; it appears, then, that the theoretical temperature coefficient is nearly, but not quite reached, so that the most completely purified water is probably not quite pure. They estimated the remaining trace of impurity from the difference between the observed and theoretical temperature coefficient, and in this way obtained a fairly accurate value for the dissociation of water. They found at

$$\begin{array}{llllll} t = & 0° & 10° & 18° & 34° & 50°, \\ c_0 = & 0·35 & 0·56 & 0·80 & 1·47 & 2·48; \end{array}$$

where c_0 is the amount in mols dissociated in 10,000,000 litres.

The Most General Case.—The calculation of the heat of reaction, which is associated with the displacement of the condition of equilibrium between any arbitrary number of gaseous or dissolved substances, according to what has preceded, does not offer any more difficulties in any case; for the calculation can be accomplished by a knowledge of the equilibrium at two temperatures which do not lie

[1] *Wied. Ann.*, **53**. 209 (1894).

too far apart. Therefore we devote a little space to the discussion of such examples: thus we have already (p. 444) discussed the case of the decomposition of carbon dioxide from the standpoint of the law of mass-action; thus,—

$$2CO_2 = O_2 + 2CO.$$

We found the equation for this case to be

$$Kc_1^2 = c_2c_3^2;$$

where the concentrations of the molecular species, CO_2, O_2, and CO amount respectively to c_1, c_2, and c_3. We will assume no excess of [either of] the decomposition products, and will denote the total pressure by P, and the dissociation coefficient by x, corresponding to the pressure P and the temperature T. Then we will obtain

$$c_1 = \frac{P\frac{2-2x}{2+x}}{RT}, \quad c_2 = \frac{P\frac{x}{2+x}}{RT}, \quad c_3 = \frac{P\frac{2x}{2+x}}{RT};$$

and therefore

$$K = \frac{P}{RT} \cdot \frac{x^3}{(2+x)(1-x)^2}.$$

Now if the degrees of dissociation, at the temperatures T_1 and T_2, and at the pressures P_1 and P_2, amount respectively to x_1 and x_2, then we obtain

$$\ln\frac{P_2x_2^3}{T_2(2+x_2)(1-x_2)^2} - \ln\frac{P_1x_1^3}{T_1(2+x_1)(1-x_1)} = -\frac{q}{R}\left(\frac{1}{T_1} - \frac{1}{T_2}\right).$$

By means of the heat of combustion of carbon monoxide—as measured directly at lower temperatures, and capable of being calculated for higher temperatures from the difference of the specific heats—from these Le Chatelier ascertained the heats of dissociation given on p. 445. They coincide well with those observed. Of course both in the preceding formula and in all other similar cases, q refers to the mean temperature between T_1 and T_2.

With increasing temperature the heat of dissociation $(-q)$ diminishes, and in all probability, at a temperature (5000°), which certainly lies too high for experiment, q becomes 0; above this temperature q would assume a negative value. But a positive value of q would correspond to a *decrease of the dissociation coefficient with the temperature*, and this leads us to the unexpected result *that at very high temperatures the dissociation of carbon dioxide must retrograde.* Of course this result—obtained as it is by an extrapolation of the specific heats of reacting gases at lower temperatures, to very high temperatures—this is not to be adopted beyond all doubt. But we have no reason for regarding a retrograding of the gaseous dissociation with the temperature as impossible; for this is certainly the case with electro lytic dissociation [*i.e.* within the limits attainable by experiment.—Tr.]. Moreover, Troost and Hautefeuille in the dissociation of *silicon chloride*, and Ditte in the dissociation of *hydrogen selenide*, both observed a *dissociation maximum*; and probably many molecular species unknown to us are formed at high temperatures.

The Process of the Blast Furnace.—In conclusion, an example taken from the monograph of Le Chatelier, will show us how the equations of the reaction isotherm and of the reaction isochore find immediate application in practice.

The production of iron by means of the blast furnace occurs according to the equation

$$Fe_2O_3 + 3CO = 2Fe + 3CO_2.$$

Experience shows that an equilibrium is established between the reacting substances, and also that there escapes from the blast furnace a mixture of carbon monoxide and carbon dioxide. Therefore it was hoped that a larger product would be obtained by building furnaces of very large dimensions. But the results of the undertaking were disappointing, even with the great expense involved in the work.

Now, on the one hand, the law of mass-action teaches that the ratio of the two gases [viz. CO and CO_2], at constant temperature, must be constant; and on the other hand, as the heat of reaction of the process amounts to very little, that therefore the temperature exerts a very slight influence relatively on the amount of the product. Moreover, as the furnaces of moderate dimensions nearly succeeded in reaching the limit below which the ratio of carbon monoxide and dioxide cannot be lowered, therefore, as was to be expected, a further enlargement of the furnace cannot succeed in increasing the relative amount of the product.[1]

The Law of the Mutuality (*Vertretbarkeit*) **of Phases.**—*If two phases, respecting a certain definite reaction, at a certain temperature, are in equilibrium with a third phase, then at the same temperature and respecting the same reaction, they are in equilibrium with each other.*

This obvious generalisation is an immediate conclusion from the considerations advanced on p. 612.

Moreover, it also necessarily follows from the proposition given on page 19, viz. that it is not possible to construct an apparatus which shall, at constant temperature, be able to furnish external work continuously.

For if two phases, A and B, were in equilibrium with a third, C,

[1] Of course this does not affect the positive result, that large furnaces may handle a large product more cheaply than a small furnace can handle a correspondingly small product, with the same ratio of gaseous products in both, an important item from a business standpoint.—Tr.

but not with each other, then in the combination arranged in a cyclic way as follows,

C
A B,

at first there would occur a displacement between A and B, but this would disturb the equilibrium with C. And thus, at constant temperature, the respective reactions between A and B, between B and C, and between C and A, would take place continuously, without ever establishing an equilibrium [*which is absurd, hence the validity of the preceding proposition.*—TR.].

We have already made repeated applications of this law, which clearly illustrate its fruitfulness.

Thus, for example, let us consider, as a reaction, the mutual removal of different phases from water; then the preceding law teaches that two liquid or solid phases which are in equilibrium with water-vapour of the same pressure, must also be in equilibrium with each other.

If we consider a liquid with one solid salt, then the latter will absorb more water, the smaller the dissociation pressure of its water-of-crystallisation (p. 469). This observation gives us the theory of the method of drying [as by a desiccator].

If we have once investigated the equilibrium between water vapour and ether containing water,—then conversely the quantity of water which will be extracted from the ether by a salt capable of combining with water-of-crystallisation,—this will give the dissociation pressure of the salt, as was shown by Linebarger.[1]

Moreover, the identity of the law of the relative depression of the solubility (p. 152), with the law of the relative depression of the vapour pressure, is an immediate deduction from the preceding principle.

According to p. 471, the product of the partial pressures of ammonia and hydrochloric acid over sal-ammoniac is constant; according to the above law the product must be the same over a saturated solution of sal-ammoniac. Since, however, free ammonia is present in the solution, in consequence of hydrolysis, but hardly any free hydrochloric acid, so that the partial pressure of the ammonia is the greater, and on evaporating, the distillate is found to be alkaline, the residue acid.

Influence of the Extent of a Phase.—The general law was stated on p. 464, and has been applied since, that equilibrium in heterogeneous systems is independent of the extent of the phases, *e.g.* the vapour pressure of a liquid is independent of the amount of liquid, the solubility of a solid independent of the amount of solid. A simple

[1] *Zeits. phys. Chem.*, **13.** 500 (1894).

thermodynamic theorem shows, however, that this is only true when the extent of the phase does not fall below a certain minimum.

If from a given heterogeneous system we allow unit mass of one phase to disappear, by means of a change in pressure, the system does a certain amount of work A_0. If now we first suppose this unit mass divided into n equal parts, separate in space, and that to perform this separation requires A_n of work, then we must have

$$A = A' - A_n,$$

where A′ is the work done when the unit of mass is removed in n portions.

So long as the subdivision is not pushed too far, A_n remains infinitesimal, so that the law as originally stated is true. If the work of subdivision becomes appreciable, so that A′ differs from A, the active mass is changed; for the modified equilibrium we have simply

$$A = RT \ln K = RT \ln K' - A_n,$$

where K′ is the new reaction constant; the latter may be calculated when A_n is known.

This can easily be found for a liquid from the surface tension; in this case the equation becomes identical with the formula given by Lord Kelvin in 1870 for the vapour pressure over small drops.

The work of subdivision for solids is unknown; but G. A. Hulett[1] showed experimentally that very fine powder is more soluble than large crystals; *e.g.* barium sulphate, ground into powder of about 10^{-5} cm. diameter, is nearly twice as soluble as larger crystals of the same salt. This is a fact of importance for the theory of spontaneous crystallisation (p. 242).

The Influence of the Temperature, and of the Pressure, upon the State of Chemical Equilibrium.—Thus far, on the whole we have employed our fundamental thermo-chemical formula, viz.,

$$\frac{d \ln K}{dT} = -\frac{q}{RT^2},$$

in order to calculate, from the displacement of a chemical equilibrium, the heat of reaction which is associated with that reaction, the equilibrium of which we are investigating. But, of course, we can, conversely, reason from the heat of reaction to the influence of the temperature, and then obtain the following law—

If we heat a chemical system, at constant volume, then there occurs a displacement of the state of equilibrium, and in that direction towards which the reaction advances with absorption of heat.

[1] *Zeits. phys. Chem.*, **37.** 385 (1901).

If q > 0, K decreases with increase of temperature, and the equilibrium is displaced on increase of temperature in the sense of the chemical equation from right to left, *i.e.* in the sense in which the reaction proceeds with absorption of heat.

Of course, the preceding formula primarily proves this law only for those systems for which it is valid, *i.e.* only for those systems in which the phase, having a variable composition, is represented by a gas mixture or a solution.

We have already seen that it [*i.e.* the preceding formula] is also applicable to condensed systems. *Thus it is universally valid.*

Those chemical forces which condition a development of heat, will always be weakened by an increase of temperature; and conversely, those which condition an absorption of heat will be strengthened by such an increase in temperature; and it is this fact which, primarily, gives the preceding proposition its universal validity.

This law, which van't Hoff called the "*principle of mobile equilibrium*" (*principe de l'équilibre mobile*), assists us in a wonderful way in becoming acquainted with the subject. Thus it shows us at once that, *e.g.*, the pressure of a gas, the vapour pressure, the degree of dissociation, etc., must each and all increase with the temperature; because heat absorption is respectively associated with the expansion of gases, with evaporation, and with the decomposition of complex molecules into more simple ones.

Thus the reaction of acetic acid and alcohol in forming water and an ester, is not accompanied by a marked development of heat (p. 603), and therefore the state of equilibrium between these substances is independent of the temperature (p. 448).

Substances formed with absorption of heat, such as ozone, acetylene, hydrogen peroxide, are unstable in considerable concentration at ordinary temperature, but must gain in stability on heating.

Completely parallel with the preceding law we may state another, which gives an explanation of the influence of *pressure* upon a chemical equilibrium, viz.—

If we compress a chemical system at constant temperature, there follows a displacement of the equilibrium in that direction which is associated with a diminution of volume.

This proposition can be easily derived for gaseous systems, from the law of mass-action (see p. 436). Indeed it holds good universally.

Thus the solubility of a salt in water, *e.g.*, will increase with the pressure, provided that the dissolving is associated with a contraction *of the solution plus the salt;* and conversely, the solubility will decrease if the separation of the salt [from the solution] is associated with a diminution of the volume of the system.[1]

Moreover, those chemical forces are strengthened by compression, which condition a diminution of volume; and those chemical forces

[1] F. Braun, *Wied. Ann.*, **30**, 250 (1887); *Zeits. phys. Chem.*, **1**. 259 (1887).

are weakened by compression, which condition an increase in volume.

The influence of pressure on equilibrium in dilute solution is obtained as follows.[1] If $d\nu$ mols of a reactive mixture are converted, the work $d\nu RT \ln K$ is done (p. 648); if the mixture is at pressure p, and suffers a change in volume dv, the work pdv is given out. Hence

$$dA = RT \ln K \,.\, d\nu + pdv;$$

but this equation is of the form

$$dA = \mathfrak{K}_1 d\mathfrak{w}_1 + \mathfrak{K}_2 d\mathfrak{w}_2 \text{ [T const.]},$$

so that, according to p. 26,

$$\frac{\partial \mathfrak{K}_1}{\partial \mathfrak{w}_2} = \frac{\partial \mathfrak{K}_2}{\partial \mathfrak{w}_1},$$

or in the special case considered,

$$\frac{\partial RT \ln K}{\partial v} = \frac{\partial p}{\partial \nu},$$

or

$$RT \frac{\partial \ln K}{\partial p} = \frac{\partial v}{\partial \nu} = V_0,$$

whence V_0 is the change in volume due to conversion of one mol.

Fanjung[2] applied this equation to the influence of pressure on the dissociation of weak acids. He found the dissociation constant under different pressures by means of the conductivity, and found, *e.g.*, for acetic acid,

$$p_1 = 1 \text{ atmo. } \log^{10} K_1 = 0{\cdot}254 - 5,$$

$$p_2 = 260 \text{ atmo. } \log^{10} K_2 = 0{\cdot}305 - 5.$$

Since K and p vary proportionally within wide limits we have

$$\frac{d \ln K}{dp} = 2{\cdot}302 \frac{\log^{10} K_2 - \log^{10} K_1}{p_2 - p_1},$$

and

$$V_0 = 0{\cdot}0821\,(273 + 18)\,2{\cdot}302 \frac{0{\cdot}305 - 0{\cdot}254}{259} = 0{\cdot}0108.$$

Since p is expressed in atmos. and R (= 0·0821) in litre-atmospheres, we get V_0 in litres; *i.e.* when the ions of acetic acid combine to form undissociated molecules there is an increase in volume of 10·8 c.c. per mol; according to p. 381, the molecular volume of acetic acid when ionised is 40·5, and when undissociated 51·1, an increase of 10·6 c.c.

[1] Planck, *Wied. Ann.*, **32.** 494 (1887).

[2] *Zeits. phys. Chem.*, **14.** 673 (1894).

This is as good an agreement as could possibly be expected; Fanjung found equally good results for the other acids studied.

As electrolytic dissociation always, so far as known, involves contraction, the dissociation must, as with acetic acid, always be favoured by increase of pressure.

It is possible, following Le Chatelier,[1] to unite the two preceding laws in the following principle :—

Every change of one of the factors of an equilibrium occasions a rearrangement of the system in such a direction that the factor in question experiences a change in a sense opposite to the original change.

This reminds us of the mechanical principle of ACTION and REACTION.

Effect of non-uniform Pressure.—So far we have always assumed that the external pressure (as measured by a manometer) is the same in all parts of the chemical system. Le Chatelier[2] has considered the interesting case of a fluid or a solid compressed by a piston that is not gas tight, so that one phase of the system may have quite a different pressure to a second phase in contact and equilibrium with it. The thermodynamic treatment of this case also can be carried out exactly.

If $d\nu$ mols. of a solid or liquid of vapour pressure π are distilled isothermally into a phase of pressure π_0, the work

$$RT \ln \frac{\pi}{\pi_0} . d\nu$$

is gained; if at the same time the system which is under pressure p contracts by dv, the external work pdv is done; so that

$$dA = RT \ln \frac{\pi}{\pi_0} d\nu - pdv;$$

and as on p. 666 we find

$$\frac{\partial RT \ln \pi}{\partial v} = -\frac{\partial p}{\partial \nu},$$

or

$$\frac{RT}{\pi} . \frac{\partial \pi}{\partial p} = -\frac{\partial v}{\partial \nu},$$

the percentage change which the vapour pressure suffers as a consequence of the external pressure is therefore

$$-\frac{100}{RT} V_0,$$

where by $\frac{\partial v}{\partial \nu} = V_0$ is meant the increase in volume that the solid or

[1] *Équilibres*, p. 210.

[2] *Zeits. phys. Chem.*, **9.** 335 (1892). See also the more complete treatment by E. Riecke, *Gott. Nachr.* (1894), No. 4.

liquid suffers by evaporation of one mol., and which naturally is always negative. Thus for ice $V_0 = -19{\cdot}65$ c.c. $RT = 22{,}420$ c.c. if p is reckoned in atmospheres. Hence $\frac{d\pi}{\pi}$ per atmo. increase of pressure is 0·00088, *i.e.* the vapour pressure of ice is increased by pressure to the extent of 0·088 % per atmo. This is, moreover, equal to the fall in vapour pressure produced by an equal osmotic pressure in the interior of a liquid. Similar considerations show that the solubility of solid is increased if whilst it is in contact with the solvent it is compressed by a sieve-like arrangement. Also a consideration of the diagram Fig. 15 (p. 142) shows at once that compression of ice (or other solid) must lower its melting point by a calculable amount if it is in contact with the uncompressed liquid; for the melting point is the point at which solid and liquid have the same vapour pressure.

It is well known that ice can be cut by pressing a wire through it. The solidification of moist precipitated powders by pressure is another phenomenon that falls into this category.

Thermodynamic Potential.—If we consider two phases in contact, work can be done if one gains at the cost of the other; if the volume increases by dv, work is done to the extent pdv where p is the pressure on both phases. On the other hand, components of the one phase may pass over into the other, and then for $d\nu$ mols. the work done must be expressed as $(\mu_1 - \mu_2)\,d\nu$, where μ_1 and μ_2 are factors peculiar to the phases and components in question. We find therefore that

$$dA = pdv - \Sigma(\mu_1 - \mu_2)d\nu \text{ (T const.)} \quad . \quad . \quad . \quad . \quad . \quad (1)$$

where the sum is to be taken for all the components. The reason for the negative sign of $d\nu$ is that we understand by ν the amount of the given component in the first phase, and this after the reaction becomes $\nu - d\nu$.

$(\mu_1 - \mu_2)$ is called, after Gibbs, the difference in thermodynamic potential for the given component; μ_1 and μ_2 are therefore the *thermodynamic potentials* of the component in the two phases.

For equilibrium, according to p. 28,

$$\delta A = 0 \quad . \quad . \quad . \quad . \quad . \quad . \quad (2)$$

for all possible isothermal changes consistent with the conditions of the system. If the system is kept at constant volume

$$\delta A = \Sigma(\mu_1 - \mu_2)\delta\nu,$$

and since this relation must hold for all the (infinitesimal) variations $\delta\nu$, it follows that

$$\mu'_1 = \mu'_2,\ \mu''_1 = \mu''_2, \text{ etc.} \quad . \quad . \quad . \quad . \quad . \quad (3)$$

i.e. *if two phases maintained at constant temperature and volume are to be in equilibrium, the thermodynamic potential of each component must be identical in the two phases.*

Example.—If a molecular species in the gaseous state or in dilute solution changes in concentration from c to c_0, the work done is $RT \ln \frac{c}{c_0}$; hence

$$\mu d\nu = (RT \ln c - A)d\nu,$$

where

$$A = RT \ln c_0.$$

Hence the thermodynamic potential of this molecular species is

$$\mu = RT \ln c - A$$

for two coexistent phases in equilibrium

$$RT \ln c' - A' = RT \ln c'' - A'',$$

or

$$\frac{c'}{c''} = e^{\frac{A' - A''}{RT}}.$$

Since, however, the expression on the right is constant for any given molecular species, at a given temperature, it follows that for equilibrium there must be a constant ratio of partition between the two phases; this is the law enounced on p. 481.

In equilibrium the different states of aggregation have, according to the above, equal thermodynamic potentials (but not equal free energy). For further applications of the thermodynamic potential see the references on pp. 608 and 609; it should be remarked, however, that the theory of thermodynamic potential is essentially equivalent to the thermodynamic treatment adopted in this book.

THERMO-CHEMISTRY IV

TEMPERATURE AND THE REACTION VELOCITY

The Acceleration of Chemical Reactions by Means of an Elevation of the Temperature.—In the last two chapters we have sought to formulate the influence of the *temperature upon chemical equilibrium*, but in this chapter we return again to *chemical kinetics.*.

The problem for investigation here is perfectly clear. In the last chapter of the preceding book, we became acquainted with certain equations which enable us to calculate the course of reactions which occur at constant temperature, and therefore they enable us to express the influence of temperature in *numerical values of the velocity-coefficients.*

Here we are to consider the solution of *the problem of penetrating into the nature of these temperature functions.*

The following result has been obtained in a purely empirical way, viz.—

All experimental measurement has shown that the velocity with which a chemical system strives to reach its state of equilibrium, increases enormously with the temperature.

This appears to be a universal phenomenon, and such is its importance for the course of chemical decomposition, and such also is its significance for the so-called "stormy reactions" (decrepitation, explosion, etc.) that it will be described at once.

As an example of this we will give the figures obtained for the *velocities according to which cane-sugar is inverted* at the respective temperature t, the other conditions being unchanged:—

[TABLE

t	Inversion Coefficient.
25°	9·67
40	73·4
45	139
50	268
55	491

It appears that an elevation of only 30° is sufficient to increase the reaction velocity fifty-fold; and the increase is similarly rapid in many other cases which have been investigated.[1]

It is very easy, from the standpoint of the molecular theory, to account for the fact that substances react much more quickly in homogeneous, gaseous, or liquid systems, the higher the temperature rises, because the activity of the heat movement, and therefore also the relative number of the collisions of the reacting substances, increases with the temperature. But when we consider that the velocity of the molecular movement in gases—and in all probability in liquids also—is proportional to the square root of the absolute temperature, and also that at the temperature of a dwelling-room it increases by only about $\frac{1}{6}$ of one per cent per degree, and that the number of collisions must increase in proportion to the velocity of the molecule—when all this is considered, one would suppose that the increase of reaction velocity actually observed (and which usually amounts to 10 to 12 per cent per degree), would be impossible, provided that one had regard to calculation only, in considering the increased heat movement. Regarding the explanation of this, only conjectures have been expressed hitherto.[2]

When we consider various systems at the same temperature we find the greatest conceivable divergences of reaction velocity.

Thus, *e.g.*, while in neutralising an acid by a base, the reacting ingredients unite so quickly with each other as to evade all estimation of the reaction velocity, yet, on the other hand, hydrogen and oxygen unite with each other at ordinary temperatures, so slowly that it has thus far been impossible to measure the velocity. It is only by a depression of the temperature in the former case, and an elevation of temperature in the latter case, that it is sometimes possible to adapt the experiment to conditions favourable for observation.

Non-reversible Decomposition.—The relations prevailing here are of the greatest importance for the problem of the determination of chemical equilibrium. Such a problem can of course be studied only

[1] A rise of 10° usually doubles or trebles the velocity of reaction; see the collection of data in Van't Hoff, *Chem. Dynamik*, p. 225.

[2] See Arrhenius, *Zeits. phys. Chem.*, **4.** 232 (1889).

when the progress of the reaction in question is sufficiently rapid to allow the equilibrium to be established in observable time. When this is not the case, then one must work at higher temperatures; and when one shall have accomplished the measurement at two different temperatures, then, by means of the thermodynamic principle given in the two preceding chapters, he can calculate the equilibrium for those lower temperatures for which a direct observation is impossible.

Sometimes even this method is debarred; as, *e.g.*, when the reaction velocity does not become sufficiently great until such temperatures are reached that the equilibrium has already been so displaced on one side or the other, that a measurement is impossible for this very reason.

It is possible that this is the case in *the direct formation of ammonia from nitrogen and hydrogen.* At lower temperatures neither do hydrogen and nitrogen react upon each other, nor does ammonia decompose into nitrogen and hydrogen, at least in observable time intervals. At high temperatures, such as those produced by the electric spark, ammonia decomposes with practical completeness; and therefore the determination of the real equilibrium between ammonia and its decomposition products cannot be made.

The fact that the reaction velocity in chemical systems is usually extraordinarily small, no matter how far removed from the point of equilibrium, is of the very greatest importance for our knowledge of chemical compounds. Probably nine-tenths, or rather ninety-nine hundredths, of all organic compounds would never have seen the light of day, if they had proceeded to their stable conditions with a greater velocity. The many polymeric hydrocarbons of the formula C_nH_n could not all exist at the same time if they all tended to go at once to the system of greatest stability corresponding to the formula C_nH_n. Thus the fact that, in the sense of the preceding, *organic chemistry is peculiarly the region of unstable compounds*, and that they either go over to a more stable form, very slowly in measurable time, or else not at all—all this finds its explanation in *the inertia* of the carbon union (p. 287).

Those chemical systems which are far removed from the stable form usually change with an increase of temperature, when this gives a sufficient value to the velocity with which they strive for the state of equilibrium. Thus consider the innumerable decompositions, charrings, decrepitations, etc., of organic compounds on heating, or consider the combustibility of many compounds in oxygen. In most of these cases heat only accelerates a reaction—whether a decomposition or a union—which would take place spontaneously, though, to be sure, perhaps only during the lapse of, say, a thousand years.

If the decomposition shall have once taken place, then, of course, it cannot be made reversible by cooling, because now the system is in a more stable condition than before the cooling.

In this way we can explain the existence of the many non-reversible reactions: these are essentially different from the dissociation phenomena proper; and also many of the reactions which progress in only one sense are similarly explained.

This fact brought it about that until very recently the real nature of chemical equilibria escaped the attention of the chemist; and in his estimation of nature he too largely regarded chemical processes as belonging to the phenomena of non-reversible processes. Just as the physicist could not study successfully the laws of the so-called "physical reactions" of the changes of aggregation by means of under-cooled vapours or liquids, so the chemist could not study the laws of chemical processes by starting out with the investigation of the non-reversible decompositions, as, *e.g.*, the dissociation phenomena of explosives.

The Application of Thermodynamics.—Strictly speaking, the doctrines of thermodynamics have nothing to teach regarding the *velocity* of a process; because the velocity always depends, not only upon the driving force, but also upon the magnitude and *the nature of the friction*, and these lie wholly beyond the domain of thermodynamics. Nevertheless, we will venture some theoretical conclusions regarding the much-discussed formula which has already done so much good service, viz.—

$$\frac{d \ln K}{dT} = -\frac{q}{RT^2},$$

which represents the dependence of the reaction velocity upon the temperature.

Let us recall the meaning of K in the sense of the theory of Guldberg and Waage (p. 431), according to which this quantity is the ratio of the velocities of the mutually opposed reactions, the difference between which conditions the total reaction velocity,

$$K = \frac{k}{k'}.$$

Thus, in the first place, we arrive at the result that when $q = 0$, then K is independent of the temperature, and *k and k′ represent the same temperature function.* Thus, *e.g.*, the velocity with which alcohol and acetic acid unite to form water and an ester, increases in exactly the same way as the velocity with which the ester and water unite to form acetic acid and an alcohol.

The equation

$$\frac{d \ln \frac{k}{k'}}{dT} = -\frac{q}{RT^2},$$

can be decomposed into the two more general forms

$$\frac{d \ln k}{dT} = \frac{A}{T^2} + B,$$

and

$$\frac{d \ln k'}{dT} = \frac{A'}{T^2} + B.$$

Here

$$A' - A = \frac{q}{R};$$

and B may be an arbitrary temperature function. But van't Hoff[1] finds that in many cases

$$B = 0,$$

because usually the coefficient k of the reaction velocity from the equation

$$\frac{d \ln k}{dT} = \frac{A}{T^2},$$

by integration, becomes

$$\ln k = -\frac{A}{T} + C.$$

Here C denotes a constant (not a temperature function), which can be calculated with excellent results from observation, by a suitable choice of the values for A and C. The expression resembles the interpolation formula, which was used to represent the vapour pressure curve, and the enormous increase of the reaction velocity reminds us of the increase of the vapour pressure with the temperature.

Since chemical equilibrium is established aperiodically, it is a process of the same kind as the movement of a material point under great friction (p. 14), the displacement of ions in a solvent (p. 360), or the diffusion of dissolved substances (p. 365). In all these cases the velocity of the process is at each instant proportional directly to the deriving forces, and inversely to the frictional resistance; so that we conclude that an equation of the form

$$\text{velocity of reaction} = \frac{\text{chemical force}}{\text{chemical resistance}}$$

similar to Ohm's law must hold. The "chemical force" at any moment may be derived from the change in free energy (pp. 26 and 27); of the chemical resistance we know little, but it is not impossible that it may be measured directly. In that case the problem of calculating chemical velocities of reaction would be similar to that of calculating the rate of diffusion of electrolytes in absolute measure (p. 369). The

[1] *Études*, p. 114.

attempt of Helm in his *Grundzugen der mathematischen Chemie* (Leipsic, 1894) is wrong in principle and leads to results incompatible with experience.

According to all experience the chemical resistance increases rapidly with fall of temperature, and becomes infinitely great at the absolute zero (in accordance with kinetic views). At the absolute zero, therefore, all chemical reaction would cease, since the denominator of the above fraction is infinite.

The velocity of reaction in heterogeneous systems is zero at the transition point, because then the chemical force is nil. With fall of temperature it must first increase on account of the increase in the numerator (chemical force) on departure from the temperature of equilibrium. When the temperature is sufficiently lowered, however, it decreases on account of the enormous increase in the denominator of the above fraction. Examples of such behaviour are to be found in the researches on velocity of crystallisation (p. 582) and those of E. Cohen on velocity of transition.[1]

Explosions and Combustions.—The degree of the influence of temperature upon the velocity of a chemical change, in the first place, gives us this practical rule—viz. that care must be taken in measuring the course of a reaction to maintain the system at a constant temperature, which can be accurately measured. This can be most easily accomplished by putting the system in a thermostat.

But it is not usually possible to avoid a slight excess or insufficiency of the proper temperature, on account of the heat of reaction associated with the chemical change; as when the change takes place with a velocity which is greater than the equalisation of the temperature of its surroundings can follow. The tracing and measuring of such a process offers very considerable difficulties.

Reactions with Development of Heat.—Let us first consider the case where the reaction progresses in the sense which is associated with the *development of heat*. As a result of the progress of the reaction, there ensues an elevation of temperature, which in turn accelerates the velocity. But this accelerated velocity means a quicker decomposition, and therefore in turn occasions an increased development of heat, which again reacts to hasten the decomposition. Thus it is evident how a very extraordinary acceleration of the reaction velocity may take place under favourable circumstances. In this way we can explain the "stormy reactions." It will be found invariably that these are associated with a development of heat.

The reaction velocity in many systems at the ordinary temperature is very slight, and perhaps may have no appreciable value. In such cases the mutual action of the acceleration of the reaction velocity

[1] *Zeitschr. f. Elektrochemie*, **6.** 85 (1899).

upon the development of heat and the converse, as in the phenomenon just described—this does not exhibit its full value, because the slight amount of heat developed is conducted away by the environment before a very marked elevation of temperature follows.

Thus, *e.g.*, this is the case with *electrolytic gas;* doubtless a mutual reaction with the corresponding development of heat takes place between oxygen and hydrogen at all times; but as this occurs at ordinary temperatures, with most extreme slowness, it reaches no appreciable amount; and therefore the elevation of the temperature of the electrolytic gas mixture above its surroundings is too small to be measured. But it is quite otherwise at 530° to 600°; here the reaction velocity reaches a magnitude sufficient to occasion such a lively development of heat as to accelerate the union of the two gases enormously, and this results in a combustion or explosion of the system.

Moreover, it is by no means necessary to bring the *whole* system to that temperature at which the reaction velocity shall reach an amount sufficient for the ignition; but it is sufficient if there occurs a local heating to a sufficient degree, as can be done by means of the electric spark.

Let us consider again a simple case of a homogeneous system as an electrolytic gas mixture; then in every case, at that point where the temperature reaches a sufficient limit, the reaction between the two gases will progress more quickly, and therefore the temperature of the point will rise. But now this may result in either one of two ways: thus—

Firstly, the heat developed at the point may be taken away into the environment by radiation and conduction more quickly than it can be generated anew, and therefore after a short time there will occur a sinking of the temperature, so that the reaction velocity will quickly return to a minimal value: or—

Secondly, the heat developed at the point considered may be sufficiently great to heat the surroundings to a temperature of lively activity; thus with the higher temperature, the rapid reaction between the gases spreads over the whole system; then a combustion takes place, resulting in the almost complete union of all the gases in the system, which are capable of reaction.

That temperature, to which a point of the system must be heated in order to cause combustion, is called the "ignition temperature." The degree of this temperature, as is obvious from the preceding, is conditioned by a large number of factors, such as the heat of the reaction, the thermal conductivity, the capacity for diffusion possessed by the gas, and the dependence of the reaction velocity upon the temperature; and it will also vary, especially with the temperature of the surroundings, and with the pressure of the system.[1]

[1] Actually v. Meyer and Freyer found that an exact determination of the tempera-

Thus, the ignition temperature has quite a secondary nature; as is clear from what precedes, it cannot claim to be the point where the mutual action of the gases *begins*. That would be as contrary to fact, as though one should regard the boiling temperature as the point where vaporisation takes its beginning.

All of these considerations may be applied to explosions in heterogeneous systems, as the ignition of gunpowder and the like.

Reactions with Absorption of Heat.—But the relations are entirely different in those cases where the progress of the reaction is associated with an *absorption of heat*, *i.e.* with a cooling down. Here the temperature sinks during the course of the reaction, and the chemical change will be retarded the more quickly the temperature sinks.

The phenomenon of a retardation occasioned by the chemical change itself may be observed in the process of *evaporation*, which process is associated with a strong absorption of heat; as a result of the cooling occasioned thereby, the vapour pressure falls very rapidly.

The fact that gunpowder is an explosive substance, while solid carbon dioxide is not, and this in spite of the further fact that both are capable of the same reaction, viz. conversion into gaseous products; this is all explained by observing that *in the first case* [gunpowder], when the reaction has once been started, it spreads and gains acceleration from the lively development of heat; but in the second case [solid CO_2], on the other hand, the action is reduced to a standstill almost immediately by the cooling down.

The Reaction Capacity of Gases.—The considerations advanced in the preceding section will be applied to gaseous systems in some further respects. It is a striking fact that many gases which are able to combine with each other with a lively development of heat—as oxygen and hydrogen, carbon monoxide and chlorine—do nevertheless approach so slowly to that state of equilibrium (which involves almost complete combination), that these gases may be regarded as chemically indifferent towards each other at ordinary temperatures. Yet, as has been repeatedly emphasised, there can be no doubt that a reaction *does* take place at ordinary temperatures; but here the reaction velocity is so extremely slow that the amount changed in the lapse of years would be smaller than that produced in a fraction of a second by an elevation of a few hundred degrees. We can regard this as only another example of the enormous influence of temperature upon the reaction velocity.

The tracing of the course of a reaction in a gaseous system by

ture of ignition is not possible (*Ber. deutsch. chem. Ges.*, **25.** 622, 1892). See also Bodenstein's work on combination of oxygen and hydrogen referred to on p. 561.

accurate measurement has unfortunately been accomplished thus far in only a few cases.

The determination of the ignition temperature, and the investigation of the dependence of the explosion capacity of gases upon foreign intermixtures, as explained above, are very interesting and important in many respects, and they bring surprising facts to light; yet the phenomena of the subject are too complicated to be accessible by a simple theoretical treatment.

A similar thing is true of the numerous investigations [1] which have been conducted regarding the distribution of oxygen between hydrogen and carbon monoxide, or of hydrogen between oxygen and chlorine, when the latter gases respectively are in excess, and the gas mixture is caused to ignite. An application of the law of mass-action to such observations as these is *directly invalid;* because, in the first place, we have *no certain guaranty* that an equilibrium *is* established at the moment of ignition; and secondly, if such were the case, one would even then be ignorant of the relations of the equilibrium to the variations of the temperature.

It is a fact that the mutual action of gases can be brought to an almost complete standstill by cooling; this fact can be employed to give an insight into the condition of equilibrium at higher temperatures.

This fact was first applied by Deville (1863) in the construction of his "cold-hot tube" (*kalt-warmen Röhre*); by means of this apparatus he succeeded in showing the decomposition of CO_2, of SO_2, and of HCl. The gases were conducted through an incandescent porcelain tube, through the axis of which passed a silver tube conducting a stream of cold water. As soon as the decomposition products produced at high temperatures diffused from the hot wall of the porcelain tube towards the cold wall of the silver tube, they were suddenly cooled, and thus a reunion to form the original compound was hindered, at least to a partial extent.

It is possible, in a similar way, to detect the dissociating action of strong electric sparks. A part of the gas is raised to a very high temperature, and thus is decomposed to a greater or less extent; then the decomposed products are partially separated from each other by diffusion, and are cooled before they can reunite with each other completely.

Thus it is possible, as found by A. W. Hofmann,[2] to demonstrate the dissociation of CO_2, or of water vapour, in this way: the gases are led through a glass-tube in which are developed the sparks from a strong induction apparatus, intensified by being in connection with the

[1] See, *e.g.*, Bunsen, *Lieb. Ann.*, **85.** 137 (1853); Horstmann, *ibid.*, **190.** 228 (1878); Bötsch, *ibid.*, **210.** 207 (1881); Schlegel, *ibid.*, **226.** 133 (1884); Hautefeuille and Margottet, *Ann. chim. phys.* [6], **20.** 416 (1890), etc.

[2] *Sitzungsber. der Berl. Akad.*, 1889, p. 183; *Ber. deutsch. chem. Ges.*, **23.** 3303 (1890).

poles of a Leyden jar. Of course the decomposition cannot pass beyond a certain limit, for there would then occur the explosive reunion of the gaseous products of decomposition. Thus Hofmann and Buff (1860) observed that by means of a constant spark stream, and at suitably chosen proportions, carbon dioxide, which was enclosed in a eudiometer, could be alternately converted partially into carbon monoxide and oxygen, which again united with a weak explosion. Here the decomposition only goes so far as to be [partially] capable of explosion, and thus by the ignition the decomposition products again unite, only to begin the cycle anew.

Deville's method can, with suitable alteration, give valuable service in investigating qualitatively equilibrium in gaseous systems at high temperatures. If the gaseous mixture to be studied is led into a bulb heated to a steady temperature, equilibrium will be established in accordance with the dimensions of the apparatus; if the gas is allowed to flow out quickly through a narrow capillary, so as to be rapidly cooled, the change in equilibrium during cooling may be negligible; and if, further, the equilibrium changes only slowly at atmospheric temperature, an analysis of the outflowing gas will show what is the state of equilibrium at the temperature of the bulb.

Equilibrium may be hastened by suitable catalysors, such as platin-asbestos; by this artifice Knietsch[1] followed the equilibrium of the system

$$2SO_3 = 2SO_2 + O_2$$

at comparatively low temperature.

Reactivity of Oxygen.—Oxygen, at high temperatures a highly reactive element, is strikingly inert at ordinary temperature, not because its affinity, but because its reactivity, is small. Only a few bodies—the spontaneously oxidisable or "autoxidisable" bodies—are capable of combining more or less energetically with oxygen at ordinary temperature. To this class belong the alkaline metals, especially those of high atomic weight, rubidium, cæsium, the compounds of sulphurous acid and sulphuretted hydrogen, finely divided metals, such metallic oxides as are capable of higher stages of oxidation; but most of all, certain organic bodies, alkyl compounds of phosphorus, arsenic, antimony, zinc, the aldehydes, many ethereal oils like turpentine, etc.[2]

In the spontaneous oxidation of these substances it is observed that they induce oxidation of others not spontaneously oxidisable. We may therefore attribute to them the power of putting oxygen into a more active state. This fact has received practical applications, *e.g.* the bleaching of textiles and paper by turpentine.

[1] *Ber. deutsch. chem. Ges.*, **34.** 4069 (1901).

[2] See Bodlander, *Langsame Verbrennung*, Stuttgart, 1899.

These processes have been specially studied by Schönbein,[1] Brodie,[2] Clausius,[3] Löw,[4] Hoppe-Seyler,[5] Baumann,[6] M. Traube,[7] and in more recent times by van't Hoff,[8] Jorissen,[9] and especially by Engler and Wild.[10] Most early writers thought the activation was due to conversion of the oxygen molecule into ozone and a hypothetical "autozone," or decomposition into free atoms. M. Traube showed that in such processes of oxidation, especially of finely divided metals, in presence of water, hydrogen peroxide is formed, and this causes further oxidation. Van't Hoff and Jorissen investigated the facts quantitatively and found, as Schönbein[11] and Traube[12] had partly before, that the autoxidisable substances activates as much oxygen as it requires for its own oxidation; in other words, that the autoxidisable and the other substance take up equal amounts of oxygen. They tried to explain this by a decomposition of oxygen into opposite electrically charged atoms. According to the important investigations of Engler and Wild, the phenomena attending autoxidation are to be explained by combination not with single atoms but with half-broken oxygen molecules —O—O—, forming peroxides of the type of hydrogen peroxide, *i.e.* of constitution

$$\begin{array}{l} \dot{R}-O \\ \quad\ \ | \\ \dot{R}-O \end{array} \qquad \text{or} \qquad \ddot{R}\!\left\langle \begin{array}{c} O \\ | \\ O \end{array} \right. .$$

These peroxides like hydrogen peroxide can give up an atom of oxygen to other oxidisable substances, being converted themselves into the simple oxides $\dot{R}_2O$ and $\ddot{R}O$. The active oxygen is therefore not in the form of free atoms, but combined to form easily decomposable substances. The stability of the peroxides varies according to the nature of the radicle R. Some are easily isolated (peroxides of alkali metals, sodium, rubidium, etc., of hydrogen—in palladium-hydride—of aldehydes, acetyl peroxide, propionyl peroxide, benzoyl peroxide), others are less stable. Probably too the autoxidation of phosphorus yields a very easily decomposed peroxide, which gives off spontaneously an atom of oxygen, which combines with a molecule of oxygen to form ozone.

The action of light on autoxidation is marked, causing an extra-

1 *Verh. Bas. naturw. Ges., N. F.*, **1.** 467; **2.** 113.
2 *Phil. Trans.*, 1850, p. 759; *Jahresber. f. Chem.*, 1850, p. 248.
3 *Pogg. Ann.*, **103.** 644 (1858); **121.** 256 (1864).
4 *Zeits. f. Chem., N. F.*, **6.** 610.
5 *Zeits. phys. Chem.*, **2.** 24; *Ber. deutsch. chem. Ges.*, **12.** 1551 (1879).
6 *Zeits. phys. Chem.*, **5.** 244.
7 *Ber. deutsch. chem. Ges.*, **15.** 644 (1888).
8 *Zeits. phys. Chem.*, **16.** 411 (1895); *Chem. Ztg.*, 1896, 807.
9 *Zeits. phy. Chem.*, **22.** 34-59 (1897).
10 *Ber. deutsch. chem. Ges.*, **30.** 1669 (1897); see also Engler, *ibid.*, **33.** 1109 (1900).
11 *Journ. prakt. Chem.*, **93.** 25 (1864).
12 *Ber. deutsch. chem. Ges.*, **26.** 1471 (1893).

ordinary acceleration, as may be observed especially in the oxidation of organic bodies such as aldehyde, turpentine, etc. The bleaching of dyes may very probably be due to the action of peroxides, accelerated by illumination. It may be imagined that light dissociates the closed oxygen molecules O=O into reactive complexes —O—O—, an hypothesis however that has not yet been verified.

Van't Hoff and Jorissen's rule as to the quantities of oxygen taken up by the autoxidisable and other substance holds only when the lower oxide formed from the peroxide is itself stable, and does not oxidise further itself. If the lower oxide can give up its oxygen the original autoxidisable body is reformed, and by a kind of catalytic action is capable of oxidising an indefinite quantity of the other substance. Oxidation in the animal body very likely takes place in this way. Hæmoglobin has two stages of oxidation, oxyhæmoglobin and metoxyhæmoglobin, and the investigations of Schützenberger and others show that in oxyhæmoglobin half the oxygen is more easily split off than the other half. The first is to be regarded as peroxide oxygen, the second as oxide oxygen.

According to Haber,[1] autoxidation occurring in aqueous solution may most clearly be conceived as the process in a voltaic cell; at the anode a mol. of oxide of the substance in question; simultaneously, but in this case at a separate place, a mol. of hydrogen peroxide is found at the cathode, which must be covered with oxygen.

The Catalytic Action of Water Vapour.—Another very striking phenomenon is this, viz. that the presence of *the very slightest traces of water vapour* are of the utmost importance for the ignition capacity of certain explosive gas mixtures. Thus Dixon[2] discovered that a *completely dry mixture of carbon monoxide and oxygen* cannot be made to explode by means of the electric spark, or at most only with the greatest difficulty; but that the mixture is made capable of explosion by the introduction of the very slightest traces of water vapour.

Quantitative data are at hand regarding the increase of the velocity of the explosions, etc., which show the perfectly enormous influence of traces of water-vapour. Thus "carbon monoxide electrolytic-gas" [*i.e.* $2CO + O_2$], when dried with P_2O_5 or with H_2SO_4, and then when saturated with water-vapour at the temperature of 10°, 35°, and 60°, showed for the velocity of explosion in these five cases the respective results of 36, 119, 175, 244, and 317 m. per sec.

In the cases where *other foreign gases* besides water were mixed with the carbon monoxide mixture [$2CO + O_2$], the following results were obtained:—

1. Explosion occurred on passing the electric spark with traces of

[1] *Zeit. phys. Chem.*, **35.** 81 (1900).

[2] *Trans. Roy. Soc.*, **175.** 617 (1884); *Jour. Chem. Soc.*, **49.** 94 and 384 (1886).

H_2S, C_2H_4, H_2CO_2, NH_3, C_5H_{12}, or HCl (all of which contain hydrogen), but—

2. No explosion occurred when traces of SO_2, CS_2, CO_2, N_2O, C_2N_2, or CCl_4 (none of which contain hydrogen) were introduced into the carbon monoxide mixture.

It is the view both of Dixon and of L. Meyer[1] that the action of the water-vapour, requisite for the combustion of the carbon monoxide mixture [$2CO + O_2$], is, that the water-vapour is reduced [to hydrogen] by the carbon monoxide, and then the mixture of hydrogen and oxygen burns with a velocity sufficient for explosion, and at much lower temperatures than those at which the combustion of the carbon monoxide occurs; according to this, the water-vapour acts as a carrier of oxygen, in the sense of the two following equations, viz.:—

$$(1) \quad CO + H_2O = CO_2 + H_2,$$

and

$$(2) \quad 2H_2 + O_2 = 2H_2O,$$

which together lead to the result

$$(3) \quad 2CO + O_2 = 2CO_2.$$

Thus, in reaction (3) it only requires a much higher elevation of temperature than in (1) and (2), for the course of the reaction to advance with a velocity sufficient to cause an explosion.

In other cases this explanation fails, as has been shown by Baker,[2] in numerous examples; hydrochloric acid and ammonia form no ammonium chloride, and vapour of ammonium chloride does not dissociate when they are made very thoroughly dry. The peculiar catalytic action of water in these instances is suggestive of the high dissociating power of liquid water.

The Velocity of the Transmission of Explosions.—The velocity with which the explosion of an inflammable gas mixture can be transmitted has for a decade or more been a subject of investigation, and especially by Berthelot, Vieille, Mallard, and Le Chatelier.

Berthelot in particular showed that this velocity is independent of the pressure, of the diameter of the tubes containing the explosive gas mixture, and also of the material composing the tubes, but that *every gas mixture exhibits a characteristic constant*, which Berthelot called its "explosive wave," the determination of which offers great interest.

In the following table are arranged some of the figures of Berthelot, as well as some of the more recent results of Dixon:[3]—

[1] *Ber. deutsch. chem. Ges.*, **19.** 1099 (1884).
[2] *Journ. Chem. Soc.*, 1894, p. 611; *Chem. News*, **69.** 270 (1894).
[3] *Chem. News*, **64.** 70 (1891).

	Berthelot.	Dixon.	Calc.
H_2+O	2810	2821	2830
H_2+N_2O	2284	2305	2250
CH_4+4O	2287	2322	2427
C_2H_4+6O	2210	2364	2517
C_2H_2+5O	2482	2391	2660
C_2N_2+4O	2195	2321	2490

[The figures in this table refer to metres per second.—Tr.]

The coincidence of the values of Dixon with the older observations of Berthelot is very satisfactory. Moreover, they are not too far removed from Berthelot's calculated values, *which identified the mean molecular velocity of the combustion products with the velocity of the transmission of the explosion*, and this is derived from the temperature at the instant of the explosion; and this last is in its turn obtained from the heat of combustion, and the heat capacity of the exploding gas mixture —provided that one disregards the small quantity of heat lost during the explosion and some slight dissociation.

Although it has been completely shown by the investigations thus far that the explosion velocity of gaseous systems is a characteristic magnitude in just the same way that the sound velocity is for a gas, yet this question of velocity of explosion is still an open one for explosives in other states of aggregation. Berthelot[1] has very recently attacked the question—*whether liquid and solid explosives show similar relations.* It is conceivable that the quantitative investigation of some explosives with which we are acquainted might not be entirely attractive even for a bold experimenter, and it is not without astonishment that we learn of some of the experiments conducted by this great French investigator. Regarding the explosion, two of its special characteristics are the development of heat and the change of volume.

The following table gives these values for some explosives; the figures in the "volume" column denote the volume occupied by the decomposition products of 1 g. of the respective explosive at 0°; the associated development of heat is given in the last column.

Explosive.	Volume.	Heat Devel.
Methyl nitrate	870 c.cm.	1431 Cal.
"Nitro-glycerine" . .	713 ,,	1459 ,,
Nitro-mannite	692 ,,	1427 ,,
Gun-cotton	859 ,,	1010 ,,

[1] *Compt. rend.*, **112.** 16; thoroughly in *Ann. chim. phys.* [6]; **23.** 485 (1891).

When the preceding explosives were exploded in their own volume, a rough estimate of the pressure showed that it mounted to about 10,000 kilograms per sq. cm.

The measurement of the velocity with which the explosion is transmitted in *liquid methyl nitrate*, which was contained in long tubes of 1 mm. in diameter, was conducted by means of the chronographic method, which had been much used by Berthelot. The velocity in rubber tubes was 1616 m. per sec.; in glass tubes, 1890-2482 m. per sec.; in tubes made of Britannia-metal, 1230 m. per sec.; and in steel tubes, 2100 m. per sec. In all cases the tubes were burst by the explosion, and usually in long strips; the glass tubes of course were ground to dust. In general, the stronger the tube, the more rapid the explosion. Whether one would obtain constant numbers if the tubes had not burst, could not be decided, because such tubes could not be prepared. But Berthelot suspected that in such a case [*i.e.* with infinitely strong tubes] the velocity of the explosion of liquid substances would be approximately constant, and would mount to about 5000 m. per sec.

The velocity for nitro-glycerine in lead tubes amounted to 1300 m., for dynamite 2500 m.; thus the physical nature of the explosive is a very important factor. Parallel with these results may be given the great velocities of *solid explosives;* nitro-mannite, 7700; picric acid, 6500; gun-cotton (by heavy loading), up to 5400 m. per sec.

THERMO-CHEMISTRY V

HEAT AND CHEMICAL ENERGY

The "Principle of Maximal Work," an Erroneous Interpretation of Heat of Reaction.—In what precedes we have described the most important laws which have been developed regarding *the application of the doctrine of energy to chemical change.*

Thus the *law of the constancy of the heat summation* (the expression of the first law of thermodynamics), and *the equation of the reaction isochore* (the simplest expression of the second law of thermodynamics),—these must be regarded as results of investigation which rest on the surest foundation; for, on the one hand, they have served to build up an extensive system of natural science; and, on the other hand, they have received confirmation from experiments of all sorts, and thus these laws may be regarded as among the safest possessions of natural science.

There is no doubt that every future doctrine of affinity in connection with the law of mass-action (the reaction isotherm) must recognise these two preceding inferences from the doctrine of energy as guiding principles.

In fact—as the history of theoretical chemistry in the last two decades shows—a faulty consideration of these principles has led to persistent error.

This error owes its origin to a mistaken notion which has in many ways invaded the most different regions, and which has been conquered at the cost of much trouble; the notion is this, viz. that the *heat of reaction* (*i.e.* the development of heat plus the external work), *according to which a process occurring in nature completes itself, must be regarded as the measure of the force which urges the system in question into its new condition.*

In the sense of this view we would have to regard the heat of reaction associated with the reaction as the measure of the mutual affinity which exists between reacting substances; and therefore we would have to conclude that "*every chemical change gives rise to the production of those substances which occasion the greatest development of heat.*" This theorem (*Satz*) was advanced in 1867 by that highly-gifted (*genial*[1]) experimenter, Berthelot, although Thomsen had previously, 1854, expressed an opinion similar to this. Berthelot later[2] claimed that this was not only the guiding principle of thermochemistry, but even of all chemical mechanics.

For a long time this theorem found the same unquestioning recognition in foreign lands [*i.e.* in countries other than Germany.—TR.] which it enjoyed till recently in the school of Berthelot. Then it was shown in the most different ways, by Horstmann, Rathke, Helmholtz, Boltzmann, and others, that the theorem was untenable, both from a theoretical and an experimental standpoint.

Since the "after pains" which resulted from the birth of this unhappy child of theoretical chemistry have not been controlled even in Germany until recently, we will subject to a short critique the theorem, called by Berthelot "*the third principle*," or "*the principle of maximum work*"; and this we will do in order to discuss the intimate relations between affinity and *heat of reaction*, to which the modern doctrine of energy has led with unquestionable logic.

The Standpoint of Thermodynamics.—In the first place, in order to test the deductions from Berthelot's theorem by the standpoint of thermodynamics—let us notice that, according to the principle stated on p. 16, all processes which occur without the introduction of external energy—such as all spontaneously occurring chemical reactions—such reactions can advance only in that sense which is associated with the production of external work.

If in every case the sense of the reaction were entirely dependent upon the heat of reaction q, then the measure of the maximal work which could be obtained by a complete change, in an isothermal and reversible way, would be represented by A, *i.e.* it must follow that

$$A = q.$$

Now from the application of the equation on p. 23, viz.

$$A - U = T\frac{dA}{dT},$$

[1] Why can we not recover the use of the word *genial*, as used by Sir T. Browne, Hare, and others, in the sense of *being endowed with genius?* The word *genial* means more than *talented* or *gifted*, and we need such a word.—TR.

[2] *Essai de mécanique chimique*, Paris, 1878.

it follows that

$$\frac{dq}{dT} = 0 ;$$

and we thus arrive at the necessary (but not sufficient) condition for the validity of Berthelot's theorem, viz. that the *heat of reaction should be independent of the temperature.*

Now this law is by no means established by experiment; but, on the contrary, in all those cases where *liquids* and *gases* participate in the reaction, the heat developed varies very noticeably with the temperature (p. 593); this is on account of the difference between the specific heats of the reacting substances and of those produced, a difference which is very considerable. And thus we conclude that the heat of reaction which is associated with a chemical change does not at all correspond to the maximal external work to be obtained by the isothermal progress of a reaction.

The case is, of course, different where *solid* substances unite to form new complexes which exist in the *solid* state; here the heat of reaction is found to be practically independent of the temperature; and in this case, although it is by no means necessary, yet by what precedes, the possibility is not excluded that by a suitable utilisation of the reaction process, the energy changed into heat under ordinary circumstances might be obtained entirely in the form of mechanical energy.

If A, the maximal work, is independent of the temperature, then *necessarily* must

$$A = q.$$

This latter condition is identical with the principle of the greatest work, and is itself self-evident. But there is another proposition which follows as a necessary conclusion from the preceding formula; and this is that there can be an identity of the maximal work A, and the heat of reaction q, only when the temperature coefficient of A, *i.e.* $\frac{dA}{dT}$, is equal to zero; whereby it results that the other factor T, of the right-hand number of the equation, vanishes.

At absolute zero,[1] *i.e.* at $-273°$, the maximal work and the heat of reaction are identical. At this temperature the theorem of Berthelot holds good absolutely, because here only *complete and exothermic reactions* are possible. It is to be expected that the farther we are removed from this point, the greater the probability that we will meet with endothermic reactions.

As a matter of fact, and on the whole, the preponderating chemical reactions at lower temperatures are the combinings (associations) which take

[1] That q should become equal to zero (and A also) at about the same limit is more than improbable.

place with a development of heat; while the reactions preponderating at higher temperatures are the cleavings (dissociations) which take place with the absorption of heat.[1]

The Standpoint of Kinetics.—As viewed from the standpoint of the kinetic molecular theory, the question assumes in brief the following aspect.

Let us suppose, *firstly*, that the chemical forces which are active between the atoms in the molecule and also between molecules of different kinds do not vary with the temperature; this is the case, *e.g.*, with the Newtonian gravitation, and also with the electrostatic or electro-dynamic mutual action. And, moreover, let us suppose, *secondly*, that the storehouse of potential energy (which conditions the very existence of the aforesaid forces) experiences in chemical changes that diminution which we measure thermo-chemically as the development of heat. Then in that case, as a matter of fact, the heat developed would be equal to the maximal work which could be obtained by the chemical decomposition; and, moreover, the fulfilment of these provisos would, as a matter of fact, result in the absolute validity of Berthelot's principle.

But now the second of the two preceding conditions is certainly *not* fulfilled. For in all probability the atoms are quite complicated structures, and certainly the atoms in the molecule possess considerable quantities of kinetic energy. And therefore in no case can the heat developed by a chemical process correspond simply to the work performed by the molecular forces; as is the case of the fall of a stone to the earth, where the diminution of the potential energy of the system (*i.e.* the stone and the earth) appears in the form of heat which is developed by the work performed by the mutual attractive forces.

Let us consider the union of two molecules to form a new and larger molecule; here, in addition to the quantity of heat set free by the work of their mutual action, in general there will occur still other changes of energy; these latter consist of changes in the kinetic energy of the atoms, and nothing can be predicted regarding the amount of the heat of reaction to be observed.

The Results of Experiments.—Now as a matter of fact, a critical comparison of the thermo-chemical data with the course of the reaction—a comparison which must be conducted with sufficient precaution, this shows that the direction of the chemical change does not necessarily coincide with the direction in which the reaction progresses exothermally.[2]

Thus, on the one hand, let us bring equivalent quantities of

[1] See Van't Hoff, *Études*, p. 174.

[2] See especially Rathke, *Abhandl. der naturforschenden Gesellschaft zu Halle*, **15.** (1881); also *Beibl. z. Wied. Ann.*, **5.** 183.

gaseous HCl and of gaseous NH_3 into a given space: then a part of the gases will unite to form solid NH_4Cl, and the production of this salt will extend to the point corresponding to the dissociation pressure at the temperature in question. On the other hand, let us bring the NH_3 and HCl, united in the form of solid NH_4Cl, into a sufficiently large space at the same temperature; then the same substance which was formed in the first case is decomposed in the second case. But in the first case we meet with an exothermic reaction; in the second case with an endothermic one.

And thus in general we may say that every single one of the numerous examples of reversible reactions is enough to disprove the universal validity of Berthelot's principle. For if the course of a reaction in the one sense is exothermic, then the reverse direction must be endothermic; and if only the former were possible, then the reaction would be one of those which advance to a completion, and therefore could not establish a chemical equilibrium. And after what has been shown in Book III. regarding this, we must absolutely agree with Ostwald[1] when he characterises the advancement of Berthelot's principle as a retreat to Bergmann's doctrine of affinity.

Moreover, Berthelot himself has not failed to notice the inadmissibility of his principle in the illustration described above. And therefore he has added this limitation, viz., that a chemical system strives towards the final condition which has the greatest diminution of total energy as contrasted with the original state, only in those cases where the interference of foreign energy does not tend to disturb things. We will spend no more time here in considering the failure of the attempt to ascribe every process which occurs with an absorption of heat, to the interference of some foreign (not chemical) force,—a force which in Berthelot's reasoning usually plays the part of a *deus ex machina*.

Regarding the energetic defence of the *universality* of the principle of Berthelot, we should not neglect to call attention to the fact that, on the whole, the occurrence of those reactions which develop heat, would seem more probable than endothermic reactions; *and also that very frequently the sense of the chemical forces coincides with that in which a chemical process occurs with a development of heat.*

Therefore the claim of this rule to be an absolute law of nature must be rejected; and yet the rule does hold too often for us to ignore it entirely. It would be as absurd to give it complete neglect, as to give it absolute recognition.

Now it is never to be doubted in the investigation of nature, that a rule which holds good in many cases, but which fails in a few cases, contains a genuine kernel of truth,—a kernel which has not as yet been "shelled" from its enclosing hull. And so in this case it is quite possible that, in a clarified form, Berthelot's principle may some time

[1] *Allg. Chem.*, II. 614 (1887).

come to have some value. Inasmuch as we are dealing here with a question of great significance—a significance which is rarely met in chemical investigation—we will try to specialise the somewhat too general form of Berthelot's rule, and apply it in some simple cases.

And, in the first place, we must be on our guard against a confusion which is liable to arise in this kind of reasoning. Since, on the one hand, the development of heat resulting from the associated elevation of temperature always increases the reaction velocity, and since, conversely, heat absorption decreases it, then exothermic reactions must exert a stimulating influence, and endothermic reactions must, conversely, exert a restraining action, unless the reaction is artificially maintained at constant temperature. The phenomenon already described on p. 677 has obviously *an entirely secondary nature* for the question regarding the relations between heat and chemical energy.

Both in chemical work which is analytical, and also in that which is preparative, those reactions which progress smoothly and completely are to be preferred. These complete reactions are usually explained by the fact that some substances escape from the reaction mixture, whether gaseous or liquid, and thus by leaving the battle-ground give their victorious opponents complete possession of the field. And it is this kind of reactions which usually speaks in favour of Berthelot's principle. If the proper chemical equilibrium should be applied to such reactions, and if the reacting ingredients should adjust themselves without allowing the escape of any, then the rule of Berthelot would be repudiated still more frequently.

But now, those substances, *the vapour pressures or the solubilities of which have the lowest values*, are most liable to escape from the reaction mixture. The rule mentioned on p. 327 can now be formulated thus, viz., that on the whole, the vapour pressure will be smaller,[1] the greater the heat of condensation. And we also learned (p. 607) that the heat of precipitation is greater the smaller the solubility. These two rules may be combined in the statement, *that, other things being equal, there is the more chance that a substance can be formed, the greater its heat of condensation.* It appears to the writer that the essential reason for the customary value of Berthelot's rule lies in the preceding statement.

Thus let us convert silver chloride to silver bromide by pouring a solution of a bromide over the former; then, according to Berthelot, it would be explained by the greater heat of precipitation of the latter substance (20,300) as compared with that of the former (15,800). But, on the other hand, we have already seen (p. 653), that the necessity of this reaction is derived, independently of all thermo-chemical data, from the equilibrium in the solution and the slighter solubility of the bromide.

To be sure, the application of Berthelot's theorem is usually much

[1] According to Le Chatelier, *Compt. rend.*, **102.** 1243 (1886), a similar law holds good for the dissociation pressure and the dissociation heat of *solids*.

simpler, because the data available for calculation are much more complete; but it usually gives unreliable results, and it can never predict anything as to the relative quantity of the products of the reaction in question. Thus it cannot explain why, in the preceding example, the silver chloride is *not completely* changed to silver bromide; for this is known to be true, not only from the requirements of theory, but this residue can also be determined quantitatively. There can be no doubt as to which of the methods has the greater *scientific value.*[1]

As the final statement of our argument, let us advance the following. A law of nature lies hidden in this "principle of maximal work," the further development of which is very important. In its form as thus far applied, it sometimes undoubtedly gives very false results; but more often, and especially so in the case of the so-called "non-reversible reactions," the chemical processes advance in accordance with its requirements.

No objections can be raised against the careful use of Berthelot's theorem as *a rule which usually agrees with experiment;* one may put about the same confidence in it that he does when he expects bad weather because the barometer falls. But it was an entire mistake to elevate this theorem to be the guiding principle of thermo-chemistry.

The Measure of Affinity.—Proof has been given in the preceding section that the heat of a reaction will not serve at all as a measure of the force which occasions the reaction, nor as a measure of the affinity which the reacting substances have for each other. The *excessive estimation* which has been ascribed to the heat developed in chemical processes in its significance for chemical mechanics, is responsible for this in particular, viz., that neither the experimental study of thermo-chemistry nor its theoretical valuation has produced results for the doctrine of affinity which are at all worthy of the industry and sagacity which have been given to this field of work.

But it appears from all the futile attempts made thus far, that we have at least one certain result, in that we are no longer in doubt as to the right pathway to follow in order to arrive the sooner at a deeper knowledge of the relations between material changes and the associated changes of energy.

No special proof, in addition to what has been given, is needed to show that the *velocity* of the course of a reaction is no measure of the chemical affinity. For the velocity depends, so to speak, on the accidental resistant friction which opposes the course of the reaction. Then the supposition that iodine has a greater affinity for hydrogen than oxygen has at 400°, because the former substance reacts much

[1] It is noteworthy that in concentrated solutions (p. 166), the thermal phenomena commonly coincide with the change in free energy, and consequently follow Berthelot's rule.

more quickly than the latter,—this supposition is just as inadmissible as though one should measure the capacity of two motors for work by the relative number of revolutions.

Or, consider the interesting experiment of Raoul Pictet,[1] in which metallic sodium had no action on dilute alcohol at $-80°$; we are not to consider from this that the [real] chemical affinity of sodium for water is *less* than at ordinary temperatures. Such a conclusion would not be admissible until it could be shown that hydrogen when conducted into a solution of sodium in dilute alcohol would precipitate metallic sodium (which certainly is not the case). The most obvious meaning of this experiment is that the reaction velocity is correspondingly depressed by the enormous depression of the temperature; and that water and sodium at $-80°$ only show an (apparent) indifference, similar to that shown by oxygen and hydrogen at ordinary temperatures.[2]

Since every chemical process, like every process of nature (p. 16), can advance without the introduction of external energy only in that sense in which it can perform work; and since also for a measure of the chemical affinity, we must presuppose the absolute condition, that every process must complete itself in the sense of the affinity;—then on this basis we may without suspicion regard *the maximal external work of a chemical process* (i.e. "*the change of free energy*"), *as the measure of affinity.*

Therefore the clearly defined problem of thermo-chemistry is to measure the amounts of the changes of free energy associated with chemical processes, with the greatest accuracy possible, and as far as possible in the same limits within which the changes of the total energy[3] have been studied by means of the measurement of the heat of reaction. When this problem shall have been solved, it will be possible to predict whether or not a reaction can complete itself under the conditions. All reactions advance only in the sense of a diminution of free energy; *i.e.* only in the sense of affinity, as defined above.

In order to ascertain the change of free energy which is associated with a chemical reaction, we must cause the reaction to occur isothermally and reversibly; and thereby we can obtain directly the desired information respecting the maximal amount of external work which can be obtained from the chemical change. Let us suppose that, under the conditions described, the change may occur in any one of several ways; even then the change of free energy would always be the same. For otherwise we could complete the change *in one way* and then reverse it *in another*; and thus we could establish a reversible, isothermal, cyclic process by means of which any arbitrary

[1] *Chem. Zentralbl.*, 1893, I. p. 458.

[2] This explanation, given by me immediately after the publication of Pictet's experiments (*Jahrb. d. Chem.*, II. 41, 1893), is confirmed by Dorn and Vollmer (*Wied. Ann.* **60**. 468, 1897).

[3] See Van't Hoff, *Kongl. Svenska Akad. Hand.*, 1886, p. 50.

amount of external work could be performed at the cost of the h the environment. But this is contrary to the second law of th dynamics. And thus we obtain the theorem, that—

The change of the free energy of a chemical process is independent of the way in which the change is completed, and is determined solely by the initial and final states of the system.

This theorem is analogous to the law of the "constancy of the heat summation" (p. 591).

It follows directly from this that one may employ the changes of free energy in calculation exactly as the total energy is used. Thus, *e.g.*, the change of the free energy of a chemical process is equal to the sum of the "free energies of formation" of the newly formed molecules, minus the "free energies of formation" of the molecular species decomposed. By the "*free energy of formation*" (*freie Bildungs-energie*) of a compound, we mean the maximal work capable of being obtained by the union of the respective elements to form the compound in question. This magnitude [*i.e.* "*the free energy of formation*"] plays a *rôle* in the chemistry of the changes of free energy, which is similar to that which belongs to "*the heat of formation*" in thermo-chemistry, and particular importance must be attached to ascertaining its dimensions.

Methods for the Determination of Affinity.—We have already found, *in the determination of the equilibrium between the reacting substances*, a method of very universal applicability for the determination of the affinity of reactions; this refers to the change of free energy A (p. 648); *i.e.*

$$A = RT \ln K.$$

If this refers to a dissociation, then K denotes the dissociation constant, and the negative value of A denotes the "free energy of formation" of the compound. If all the dissociation constants of all of the reacting compounds are known, then we know the affinities of all the reactions; this theorem finds an excellent illustration [1] in the treatment of the reactions between electrolytes, as given in Chap. IV. of Book III.

Another method for the measurement of affinity, which is very simple and exact, will be given in the following chapter: it depends on the *determination of the electromotive force of galvanic combinations.* A noticeable application is given by St. Bugarszky.[2]

Whatever importance is possessed by such affinity determinations as those just described, may be illustrated by the single following example instead of many.

As is well known, the combustion of carbon is the reaction which furnishes the work capacity of the driving power of most of our

[1] Compare van't Hoff, *Zeits. phys. Chem.*, **3.** 608 (1889)
[2] *Z. S. anorg. Chem.*, **14.** 145 (1897).

motors. Now we do not know the affinity of this reaction, *i.e.* the maximal external work which 1 g. atom of carbon (= 12 g.) can furnish in its combustion to carbon dioxide; and therefore we do not know the work capacity (*Leistungs-fähigkeit*) of an ideal machine working at its maximum efficiency, when it is fed with coal.

The following method may be suited for the solution of this problem. We know the equilibrium between carbon dioxide, carbon monoxide, and oxygen (pp. 444 and 661), *i.e.* we know the affinity A_1, of the reaction

$$2CO + O_2 = 2CO_2 + A_1,$$

and, moreover, at all temperatures. And, on the other hand, Rathke (p. 688) states that, according to his observation, in subjecting carbon dioxide to glowing coal, it is *not possible to effect the complete reduction to carbon monoxide.* Thus a single quantitative investigation of the equilibrium between carbon monoxide, carbon dioxide, and carbon, would give us for all temperatures, exactly as in the preceding case, the affinity A_2, of the reaction

$$C + CO_2 = 2CO + A_2.$$

Then the addition of these two energy equations would give us, indirectly,

$$C + O_2 = CO_2 + A_1 + A_2;$$

that is, we would obtain the affinity sought for the combustion of carbon: this evades a direct determination on account of the very limited extent to which carbon dioxide is dissociated into oxygen and carbon.

In a recent work Boudouard [1] has found that at 1000° atmospheric pressure 99·3% CO and 0·7% CO_2 can exist in presence of solid (amorphous) carbon; this allows of determining the affinity of carbon and oxygen in accordance with the theory put forward above. We may content ourselves here with an approximate calculation. According to p. 445 at 1000° carbon dioxide is at atmospheric pressure dissociated to the extent of 0·06%. The quantity of oxygen x which coexists with carbon monoxide at 0·993 and carbon dioxide at 0·007 atmospheres is therefore according to the law of mass action

$$K.(1)^2 = (0{\cdot}0006)^2.(0{\cdot}0003)$$
$$K.(0{\cdot}007)^2 = (0{\cdot}993)^2.x,$$

and therefore $x = 5{\cdot}4 . 10^{-15}$ atmospheres. In order to combine carbon and oxygen at atmospheric pressure we may suppose, according to the cyclic process on p. 643, oxygen introduced into the system described

[1] *Compt. rend.*, **128.** 842; *Bull. soc. chim.*, from 5th August 1899 and 5th March 1900.

by Boudouard and carbon dioxide removed. This gives as the affinity in question

$$A = A_1 + A_2 = RT \ln \frac{1}{x} - RT \ln \frac{1}{0{\cdot}007}$$

or (p. 647)

$$A = 4{\cdot}58 \,.\, 1273 \,.\, \log \frac{0{\cdot}007}{x} = 70{,}625 \text{ cal.}$$

The heat of reaction at atmospheric temperature is, according to p. 593 97,650 ; but as the system $C + O_2$ on the one hand, and that of CO_2 on the other, have nearly the same thermal capacity, the heat of combustion at the absolute zero must differ very little from 97,650, hence we obtain a second value for A, for that quantity is equal to the heat of combustion at the absolute zero. By interpolation we find at atmospheric temperature

$$A = 97{,}650 - \frac{(97{,}650 - 70{,}625)291}{1273} = 91{,}470,$$

that is, the *heat of combustion of coal at atmospheric temperature is almost completely convertible into external work.* Further calculations and considerations of the same kind are to be found in Bodländer.[1]

[1] *Zeits. f. Elektrochemie,* **8.** 833 (1902).

CHAPTER VI

ELECTRO-CHEMISTRY I

GENERAL FACTS

The Conveyance of Electrical Energy.—A chemical system may absorb or give up energy in the form of the kinetic energy of moving matter; viz. in the form of heat (*i.e.* the irregular movement of the molecules), or in the form of external work (*i.e.* the orderly movement of the molecules); the action of these forms of energy upon chemical change makes up the principal contents of the preceding chapter. But a chemical system may also absorb or give up energy in two other forms, viz., on the one hand, in the form of vibrations of a hypothetical medium, which we call the *luminiferous ether;* and, on the other hand, in the form of electrical energy, the conveyance of which is apparently effected by means of the same medium. The consideration of the influence exerted upon the condition of a chemical system, by the absorption or the surrender of electrical energy, constitutes the subject of *electro-chemistry;* the similar influence exerted by the [luminiferous] vibrations of the ether constitutes the subject of "*photo-chemistry.*". This and the following chapter will, respectively, be devoted to these two branches of theoretical chemistry.

For complete expositions of electro-chemistry in recent times we refer to Le Blane, *Lehrbuch der Elektrochemie* (Leipzig, 3rd edit., 1903), and H. Jahn, *Grundriss der Elektrochemie* (Wien, 1895); in the first of these two excellent works electro-chemistry is treated more from the chemical, in the second from the mathematical-physical standpoint. A short but full account of electro-chemistry is that of W. Loeb, *Grundzüge der Elektrochemie* (Leipzig, 1897). Whilst for more complete information we may refer to the second volume of Ostwald's *Lehrbuch der Allgemeinen Chemie*, 2nd edit. We may further note the striking work of R. Lüpke, *Grundzüge der Elektrochemie*, 4th edit. (Berlin, 1903), which contains a series of instructive electrical experiments.

The *conveyance of electrical energy* may be effected in two essentially different ways.

Firstly, let us bring a body having an *electrostatic charge* into the neighbourhood of a system which does not conduct the galvanic current. Then the system will be changed into a certain state of tension which is called *dielectric polarisation*. The influence of such a tension upon the state of chemical equilibrium, should certainly occur, but it is probably very slight.[1]

Thus far at least there are no researches known which give positive results. But some phenomena have long been known which are to be regarded as *rapidly alternating dielectric polarisations;* and in particular, the *transformation of oxygen into ozone*, thus

$$3O_2 = 2O_3.$$

This change is effected in the ozoniser under the influence of the so-called silent discharge.

Secondly, much more exact research has been devoted to the material changes effected by the conveyance of the galvanic current from some source of electricity (as a galvanic element, a thermo-pile, or a dynamo machine), through a material aggregate.

While the former mode of conveyance of electrical energy [electrostatic], presupposed a non-conducting system, here of course the system must be *a conductor of electricity.*

This change, which is effected by the passage of an electric current, is called *electrolysis, in so far as the electricity exhibits its action, as such.*

The changes which are occasioned by simple heat, and which are always associated with the transport of electricity, cannot of course be designated as electrolysis. Such, *e.g.*, is the decomposition of carbon dioxide or of other gases, under the influence of the hot electric spark.

Electrolytic Conduction. — The conveyance of electricity in conducting substances may happen in two different ways, viz. *with or without* the associated transportation of matter. The latter happens in the case of *metallic* conductors [first class]; the former in *electrolytic* conductors [second class]; hence these are called conductors of the "*first*" and "*second*" *classes*, respectively.

The nature of metallic conduction is little known.

On the other hand, we have very thorough conceptions regarding the nature of electrolysis, which, in the history of physical and chemical discipline, has so often played the *rôle* of a linking band. It consists in a chemical decomposition which overcomes the strongest affinities, by means of the electric force, the study of which has long been the favourite occupation of students of physics.

In considering the theory of electrolytic dissociation (p. 353), we

[1] This should hold good for the action of magnetisation upon a chemical system; but thus far no certain action has been observed, either upon the equilibrium or upon the reaction velocity; see, *e.g.*, M. Loeb, *Am. Chem. Journ.*, **13.** 145 (1891). A. Bucherer, *Wied. Ann.* **58.** 568 (1896); Hemptinne, *Z. S. phys. Chem.* **34.** 669 (1900).

saw how the process of the conduction of the current, as a result of the electric forces, consisted in the displacement of the *free* ions in the solution; by free ions we mean those which are not united with each other to form electrically neutral molecules; the positive ions migrate in a direction, from anode to cathode; the negative ions, in the opposite direction.

A solution conducts electricity the better, the more numerous the ions, and the smaller the friction which the ions encounter in their migration.

This conception may now be applied, unchanged, to every substance which conducts electrolytically,—whether gaseous, liquid, or solid,—whether simple or a mixture.

The electrolytic charges of the ions are equally great and equivalent, whether they occur in solution, or in a substance having a simple composition; this would be anticipated, because the fundamental electrolytic laws of Faraday hold good both for water solutions, and also for fused salts.

All that is most important for electrolytic conductivity *in water solution*, has been already treated in discussing the theory of electrolytic dissociation, and in the doctrine of affinity (p. 497).

The conductivities of simple substances, as *e.g.* of fused salts, as is the case with solutions, vary directly with the degree of dissociation, and inversely with the friction of the ions. But it has not been possible as yet to reduce the observed conductivities to these two factors; because here of course there can be no occurrence of the phenomenon, analogous to the changes of concentration near the electrodes, as a result of the Hittorf transportation in the solutions. And thus we cannot state how large a fraction of the total molecules present in a fused salt is decomposed into ions, and the solution of this question seems to be dependent upon circuitous means.[1]

Electrolysis.—When a system composed of conductors of the *first class* is traversed by a galvanic current, *thermal* action (Peltier's effect) is developed on the surface between different conductors; and also heat (Joule's heat) is developed in the whole circuit; but there is no migration of matter associated with the passage of the electricity.

But, on the other hand, when the galvanic current passes through conductors of the *second class*, then, *in addition to the phenomena enumerated above, there occurs a transportation of matter* (migration of the ions); and also on the limiting surfaces between the conductors of the first and second classes, there occur peculiar chemical processes; these latter consist primarily, in the solution of the electrodes or in the separation of the ions from the electrolytes; but they are usually

[1] For water, which is at ordinary temperature an electrolyte, though a very poor one, the problem has actually been solved indirectly; see p. 508.

complicated by secondary reactions between the electrolyte and the separated products.

Let us consider two similar metallic electrodes to be dipped into an electrolyte (a solution, or a fused, or a solid, salt); then if the current is J, in an electrolytic cell the resistance of which is W; and from a source of electricity the electromotive force of which is E, and the [internal] resistance of which is w; then the strength of the current, according to Ohm's law, is

$$J = \frac{E}{W + w}.$$

This assumes that there are no changes, produced by the current in the electrolytic cell, of such a nature as to affect it as regards the electromotive forces. But it usually happens in the process of electrolysis, that either the nature of the surface of one electrode is changed, whether by having another metal precipitated upon it or by dissolving (occluding) the separated gases, or else the composition of the electrolyte bathing the electrode is changed in some way.

In all such cases, one can observe the occurrence of a resistant electromotive force in the cell, the so-called *galvanic polarisation;* if we denote the value of this by ϵ, then the strength of the current at once falls to

$$J = \frac{E - \epsilon}{W + w}.$$

We may obtain information respecting the quantity of ions separated by the current, from the law discovered by Faraday in 1833; according to this, the quantity of the ion separated in unit of time, upon the electrode, is proportional to the intensity [strength] of the current; and the same quantity of electricity will, in the most various electrolytes, electrolyse chemically equivalent quantities of ions.

In those cases where the chemical value of the ions is capable of changing, of course the meaning of *chemical equivalence* changes; thus the same current which separates 200 g. of mercury from a solution of $HgNO_3$, will separate 100 g. of mercury from a solution of $Hg(CN)_2$.

The quantity of electricity that suffices to deposit one electro-chemical gram equivalent is

$$\frac{1}{1{\cdot}035} \cdot 10^5 = 96{,}540 \text{ coulombs (ampere seconds)}$$

and will be expressed by F (*Faraday*).

The Development of Electrical Energy by Chemical Systems.—Several different forms of apparatus are known which are capable of developing the galvanic current. Among these are those for the application of heat, to develop electrical energy (thermo-piles); those for the application of mechanical energy (dynamo machines); and those for the application of chemical energy (galvanic elements). Only the current-producing apparatus of the last category will be considered here.

A chemical system in which the changes of energy, associated with the changes of matter, succeed in producing electromotive activity, is called a *galvanic element.* Inasmuch as the galvanic current is associated with changes of matter only in the case of electrolytes—conductors of the second class, and moreover, since conversely changes of matter tend to produce galvanic currents only in these electrolytes, therefore the galvanic elements must contain electrolytic conductors. Hitherto these have been almost exclusively water solutions (hydro-elements = *hydro-ketten*), and molten salts,[1] which have found application.

Galvanic elements can be constructed either by means of substances which conduct electrolytically, or else by the use of conductors of the first class (*i.e.* electrodes of metal or carbon). In the first category are included the so-called "liquid cells" (*Flüssigkeits-ketten*), which are formed by arranging in series, water solutions of electrolytes (acids, bases, and salts); in this sort of galvanic circuit, the electromotive forces developed on the contact surfaces between the metals and liquids, are completely eliminated; these liquid circuits have been the subject of repeated investigation, by du Bois-Reymond (1867), Worm-Müller (1870), Paalzow (1874), and others.

These circuits have recently excited great interest because, by means of the newer theory of solution, they have afforded an insight into the mechanism of the development of the current (see below); here the active electromotive process consists simply in the mingling of the various solutions.

But in the case of other galvanic elements, in most instances the nature of the process producing the current is simple and clear. The current occasions certain changes in the element producing it, which changes can almost always be predicted from a knowledge of the nature of the electrode, and of the liquid which bathes it.

Thus in the cell of Volta, *e.g.*,

$$Zn \mid H_2SO_4aq. \mid Cu,$$

the metal of the negative electrode, zinc, goes into solution and hydrogen is developed on the positive copper pole.

In the Daniell element, thus,

$$Zn \mid ZnSO_4aq. \mid CuSo_4aq. \mid Cu,$$

the process which develops the current can be expressed by the equation,

$$Zn + CuSO_4 = Cu + ZnSO_4.$$

[1] See, *e.g.*, Fabinyi and Farkas, *Compt. rend.*, **106.** 1597 (1888); also Poincaré *Ann. chim. phys.* [6], **21.** 289 (1890).

The Clark element,

$$Zn \mid ZnSO_4aq. \mid Hg_2SO_4 \mid Hg,$$

consists of mercury covered with [solid] mercurous sulphate, a saturated solution of zinc sulphate, and metallic zinc ; here zinc from the negative pole goes into solution, and is then separated primarily on the positive pole ; but instead of forming an amalgam with the mercury, the solid mercurous sulphate is reduced according to the equation

$$Zn + Hg_2SO_4 = ZnSO_4 + 2Hg.$$

All of these chemical processes occur in accordance with Faraday's law.

If the same quantity of electricity passes through the most various galvanic elements, the decompositions are all in the ratios of (electrical) equivalents.

The experimental determination of electromotive forces is most easily carried out by comparison with a standard cell; as such the Clark cell suffices; it can be easily constructed for one's self, and it has a potential which is reliable to within one in a thousand; its value in international volts according to Jäger and Kahle is[1]

$$E_t = 1{\cdot}4292 - 0{\cdot}0012\ (t - 18).$$

Recently the Weston cell has been preferred; this is constructed in the same way as the Clark cell, but instead of zinc and zinc sulphate cadmium and cadmium sulphate are used. The electromotive force of this cell (see the preceding references) is

$$E_t = 1{\cdot}0187 - 0{\cdot}000035\ (t - 18),$$

that is, for most purposes may be regarded as independent of temperature. For very exact measurements of electromotive force it is desirable to use both cells, as they can be constructed without difficulty. For details of the measurements see Kohlrausch, *Leitf. d. Phys.*, 9th edit., and especially W. Jäger, *Normalelemente*, Halle, 1902.

Special Electrochemical Reactions.—In the electrolysis of hydrochloric acid the ions are directly deposited as gaseous hydrogen and gaseous chlorine, but more often this is not the case, but complications occur from the action of the ions on each other, on the metal of the electrode, or on the solvent, or, finally, on other dissolved substances. The following reactions which are intelligible without further explanations may be taken as examples :—

$$2\overline{O}H + 2\oplus = H_2O + \frac{1}{2}O_2$$

$$2CH_3\overline{C}OO + 2\oplus = C_2H_6 + 2CO_2$$

[1] *Zeitschr. f. Instrumentenkunde*, 1898, Heft 6, p. 161; *Wied. Ann.*, **65.** 926 (1898).

$$\overset{-}{Cl} + \oplus + Ag = AgCl$$

$$\overset{++}{Zn} + 2\ominus + 2H_2O = Zn(OH)_2 + H_2$$

$$\overset{-}{Cl} + \oplus + FeCl_2 = FeCl_3$$

$$2\overset{+}{H} + 2\ominus + H_2O_2 = 2H_2O.$$

All these reactions may be regarded as due to the action of positive and negative *electrons* on the electrolyte; the introduction of negative electrons into the reaction indicates that the reaction takes place at the cathode, where the positive ions (cations) are deposited, and similarly the occurrence of positive electrons indicates that the reaction takes place at the anode. The number of electrons occurring in the reaction gives at the same time the number of F's which are required for the electrochemical reaction in question.

For more precise knowledge of the nature of electrochemical principles it is indispensable to know the most important electrochemical reactions; for this we may refer to the excellent work of W. Borchers, *Elektrometallurgie*, 3rd edit., Leipzig, 1903, Hirzel, F. Haber, *Grundriss der technischen Elektrochemie auf theoretischer Grundlage*, R. Oldenbourg, München, Leipzig, 1904, and B. Neumann, *Theorie und Praxis der analytischen Elektrolyse der Metalle*, W. Knapp, Halle, 1897. For the experimental study we may note as introductions to practical electrochemical work the writings of W. Loeb (Leipzig, 1899), R. Lorenz (Göttingen, 1901), and K. Elbs (Halle, 1902).

CHAPTER VII

ELECTRO-CHEMISTRY II

THERMODYNAMIC THEORY

Electrical Work.—Electrical work is given by the product of the potential (volts) and the quantity of electricity (coulombs); the unit of electric work is the volt-ampere-second, or, more briefly, the watt second, and is performed when a current of one ampere flows under a difference of potential of one volt. In absolute measurement the volt is 10^8, the ampere 10^{-1}, so that the watt second is

$$10^7 \text{ absolute units } (\text{cm}^2 \text{ g sec}^{-2}),$$

or according to p. 12.

$$\frac{10^7}{41{,}770{,}000} = 0{\cdot}2394 \text{ cal.}$$

This amount of heat is evolved, for example, when a current of 1 ampere flows for one second through a resistance of 1 ohm.

To deposit one electrochemical gram equivalent is needed 96,540 coulombs, which number we have already designated F. When a galvanic cell of electromotive force E yields so much current as suffices for the chemical conversion of one gram equivalent the work done is

$$\text{EF Watt seconds} = 96{,}540 \times 0{\cdot}2394 \times \text{E} = 23{,}110 \text{ E cal.}$$

If an electrolytic cell has an electromotive force ϵ the work needed to decompose one gram equivalent of an electrolyte is ϵF.

Application of the First Law of Thermodynamics.—If a galvanic cell of electromotive force E and internal resistance W is closed by a resistance w, then, according to Ohm's law, the current is

$$i = \frac{E}{W + w}$$

and the heat evolved in the external circuit is

$$i^2wt = E^2\frac{w}{(W + w)^2}t.$$

If W is very small compared with w all the electrical energy given out by the cell is spent on the external circuit and we have

$$i^2wt = \frac{E^2}{w}t = \nu FE,$$

where the number of F's yielded by the element is given by

$$it = \nu F.$$

If q is the heat of reaction of the process that yields the current per gram equivalent of chemical conversion, a cell which is short circuited and enclosed in a calorimeter will give the thermal effect νq, whilst ν gram equivalents are converted; for, according to the law of conservation of energy, the thermal effect is the same in whatever way the chemical transformation may take place. But if the electrical circuit is placed outside the calorimeter, the heat developed in the calorimeter will be less by the amount i^2wt, and if w is great compared with W the conversion of ν equivalents will yield in the calorimeter the quantity of heat

$$\nu q - i^2wt = \nu(q - FE)$$

or per gram equivalent in the cell the heat

$$H = FE - q$$

will be *absorbed*. H may conveniently be described as the latent heat of the cell.

If we may make the supposition that H is negligible, it follows that

$$E = \frac{q}{F},$$

that is, the electromotive force of the cell can be directly calculated from the heat of reaction. Experience shows that in many cases, especially in those in which q is a large value, this assumption holds; it is not permissible, however, to make the assumption tacitly.

In the following we will describe the heat of reaction in electrochemical processes by q, bearing in mind that in practical applications the systems of units in which E and q are respectively expressed, must be taken into consideration.

It was long believed and is still occasionally stated that in a galvanic cell the decrease in total energy which is associated with the chemical change, that is the heat of reaction, is the direct measure of the electro-motive force, or, in other words, that the chemical energy is simply converted into electrical; this assumption would involve the equation

$$E = q;$$

if q is measured in calories and E in volts we should have, according to p. 703—

$$E = \frac{q}{23,110} = 0{\cdot}00004327\, q \text{ volt.}$$

This relation was put forward by v. Helmholtz (1847) and W. Thomson (1851), and is commonly known as Thomson's law; *it would follow from the law of conservation of energy if a galvanic cell neither rose nor fell in temperature when yielding current*, that is, if it neither gave heat to its surroundings nor withdrew heat from them, but though this is often approximately the case it is not accurately true. It was long considered that the difference between calculation and observation was due to errors of experiment, this view was completely disproved by the experimental investigations of Thomson,[1] and especially Braun,[2] and by the thermodynamic treatment of v. Helmholtz as given below, which led to new and very careful measurements, and finally controverted Thomson's rule both experimentally and theoretically.

The unqualified recognition which this rule long enjoyed was largely due to the circumstance that in the first case to which it was applied it very approximately holds; in the Daniell cell the electromotive force has almost precisely the value which was calculated from the heat of reaction. The heat of formation of one equivalent of zinc sulphate from metal, oxygen and very dilute sulphuric acid is

$$\frac{1}{2}(\text{Zn, O, SO}_3\text{, aq.}) = 53,045,$$

and for copper sulphate the corresponding amount is

$$\frac{1}{2}(\text{Cu, O, SO}_3\text{, aq.}) = 27,980.$$

The difference between these two values,

$$53,045 - 27,980 = 25,065,$$

gives the change of energy associated with the transport of one F of electricity from the cell and consequently the deposition of 1 gramme equivalent of copper from the solution of its sulphate by zinc; according to Thomson's rule therefore the electromotive force of the Daniell cell should be

$$E = 0{\cdot}00004327 \times 25,065 = 1{\cdot}085 \text{ volt,}$$

whilst the direct measurement gives 1·09 to 1·10. Similarly good agreement is found for combinations of the type of the Daniell cell

[1] *Wied. Ann.* **11.** 246 (1880). [2] *Wied. Ann.* **17.** 593 (1882).

in which silver replaces copper and cadmium replaces zinc in the solutions of their salts.

Thomson's rule, however, fails altogether in liquid cells and concentration cells; here the current yielding process consists simply in the mixing of solutions of different concentrations, and if the solutions are sufficiently dilute the heat of reaction is nothing, whilst the electromotive force can reach a considerable amount. We shall subsequently make calculations with a whole series of galvanic combinations in which there are large differences between the measured and calculated electromotive force.

Just the same considerations are attributable to electrolytic processes, since instead of electromotive force E of the galvanic combination we have the opposing electromotive force of polarisation ϵ (p. 699). Here also only approximate agreement is to be found; for example, to electrolyse dilute hydrochloric acid requires about 1·3 volt, whilst the heat of reaction is, according to the equation

$$\frac{1}{2}(H_2, Cl_2) = 22{,}000$$

$$(HCl, aq.) = 17{,}310$$

$$\text{Total} = 39{,}310$$

and therefore yields

$$\epsilon = \frac{39{,}310}{23{,}110} = 1{\cdot}71;$$

it must be remarked, however, that q is independent of the concentration of the hydrochloric acid, whereas ϵ is not so (see Chap. VIII.). The galvanic polarisation of a common salt solution calculated from the thermodynamic equation

$$\frac{1}{2}(H_2, Cl_2) = 22{,}000$$

$$(HCl, aq.) = 17{,}310$$

$$(HCl\ aq., NaOH\ aq.) = 13{,}700$$

$$\text{Total} = 53{,}010$$

gives

$$\epsilon = \frac{53{,}010}{23{,}110} = 2{\cdot}30,$$

whereas experiment shows it to be 2·0 volt.

Reversible Elements.—When we send the same quantity of electricity through a galvanic cell, now in one direction, and then in the opposite, we must distinguish between two cases. The element may return to the original state or not. An example of the first case [*i.e.* where the cell may be reversed to the original state] is given by the Daniell cell,

$$Zn \mid ZnSO_4 \mid CuSO_4 \mid Cu.$$

Here, if we conceive the quantity of electricity 1 (measured electrochemically) to be sent through the cell in the direction from left to right, then, firstly, an equivalent of zinc goes into solution at one pole; and secondly, an equivalent of copper is precipitated upon the other pole. Then if we conceive the same quantity of electricity to be sent through the cell from right to left, conversely, firstly, an equivalent of copper is dissolved by the solution; and secondly, an equivalent of zinc is precipitated from the solution. Thus the system returns again to its original condition.

An example of the second case is given by the cell of Volta, combined according to the scheme,

$$Zn \mid H_2SO_4 \mid Cu.$$

Here, let the quantity 1 of electricity pass through the cell from left to right; at the zinc pole an equivalent of zinc passes into solution, and an equivalent of hydrogen is set free at the copper pole. Then when the same quantity of electricity passes back from right to left, at the copper pole an equivalent of copper passes into solution; and at the zinc pole an equivalent of hydrogen is set free. Thus, at the end of the experiment, an equivalent of copper and an equivalent of zinc have passed into the solution, and two equivalents of hydrogen have been set free from the system.

In the first case, the heat developed during the cyclic process, and also the external work performed, are both equal to zero. Therefore the element must possess the same electromotive force, when the current flows in one direction as when in the opposite.

But, in the second case, there must be a compensation for the chemical changes produced in the course of the experiment; this consists simply in the fact that the sum of the work (expended in the passage of the quantity of electricity 1, now in one direction and then in the other, through the cell) has a finite amount: *i.e.* that *the current in the two respective directions has a different electromotive force.* This is possible only when an opposite electromotive force is developed by the passage of electricity, that is when the element is polarised.

We therefore call these two classes of cells respectively, *polarisable* and *non-polarisable*, or *non-reversible* and *reversible*. A kind of intermediate link between these two groups is found in such elements as Grove's cell (platinum, nitric acid, sulphuric acid, and zinc); these are non-polarisable in one direction, viz., that direction in which a current is afforded; but they become polarised by leading the current through in the opposite direction.

Strictly speaking, all reversible batteries can be used as *accumulators*, and, conversely, all good accumulators can act reversibly. Only the reversible elements work rationally, *i.e.* with the maximum efficiency; the cells which are non-reversible are comparable to badly built steam engines, which work with leaky pistons and valves. This fact explains why here we deal only with cells which work ideally, *i.e.* with those which are reversible.

In the first place, it is indispensable, for the reversibility, that the *processes taking place at the electrodes shall be reversible.* A metal plate which dips into a solution of one of its salts *is a "reversible electrode"* of this sort; because, by the passage of the current from the electrode into the solution, metal only passes into the solution, and when the current passes in the opposite direction the metal passes back from the solution to the metal plate. Thus the change is reversible. Such electrodes may be called "*reversible electrodes of the first class.*"

Moreover, the electrodes just described can also be suitably called "*electrodes which are reversible as regards the cation,*" because the transfer of electricity is limited exclusively to the *cation.*

And, moreover, similarly, we may call the variety, to be described presently, "*reversible electrodes of the second sort,*" or "*electrodes which are reversible with reference to the anion.*"

Thus, *e.g.*, if we cover a piece of silver with a layer of AgCl, and then dip the electrode so prepared into a solution of a chloride, as KCl, then the required conditions are fulfilled. The passage of electricity from the electrode to the electrolyte can only take place in such a way, that either chlorine ions go into the solution, being set free from the AgCl by the precipitated potassium; or else, when the current is in the opposite direction, the chlorine ions will be precipitated upon the electrode, uniting with the silver to form AgCl. In both cases the electro-negative ions are the ones which accomplish the transport of the electricity from the electrode to the solution. Therefore the electropositive silver cannot participate in the transport of the electricity, at least not to any considerable extent, because silver ions are not capable of existence in a solution of a chloride on account of the extreme insolubility of AgCl.

Thus the electrode described behaves as though it were a modification of chlorine, which conducts like a metal; and, therefore, it completely satisfies the condition of reversibility.

In general, every metal when covered with one of its insoluble salts, in which it forms the basic ingredient, and when placed in a solution of another salt containing the same electro-negative ingredient as the first salt, represents an "*electrode which is reversible as regards the anion.*"

For this purpose, it is preferable to use *mercury* as the metal, because, on the one hand, being a liquid, it presents a surface which behaves in a constant way; and also, on the other hand, mercury forms a large number of insoluble compounds with negative radicals.

Since the latter render the electrode unpolarisable they are called *depolarisators*.

Therefore the reversible galvanic elements may be divided into the three following classes—

1. Those composed of two reversible electrodes of the first class, as, *e.g.*, the Daniell element.

2. Those composed of one reversible electrode of the first class, and of one reversible electrode of the second class, as, *e.g.*, the Clark "normal" element.

3. Those composed of two reversible electrodes of the second sort; no use has thus far been made of any element of this last type.

The Law of the Conversion of Chemical Energy into Electrical Energy.—We will make use of a cyclic process in order to arrive at the relations between *the heat development* of the current-generating process of a reversible galvanic cell, and the external work performed by this element, *i.e. its electromotive force.*

We will allow the galvanic element having the electromotive force E to perform the work E, at the temperature T, whereby an equivalent of the positive metal is precipitated upon the positive pole, and an equivalent of the metal of the negative pole goes into solution. Let q denote the heat of reaction associated with these chemical processes; this can be calculated from the thermo-chemical data. In the Daniell element, *e.g.*, it corresponds to the difference between the heats of formation, thus

$$(Zn, SO_4, aq.) - (Cu, SO_4, aq.).$$

Therefore the heat developed in the element is equal to the heat of reaction minus the external work, *i.e.*

$$q - E = -H.$$

We will now bring the element from the temperature T to T + dT, whereby the electromotive force will change from E to E + dE; and we will make the chemical change in the element reversible in this way, viz., that by sending in the quantity of electricity 1 in the opposite direction, there is required the expenditure of the work E + dE, and there occurs an absorption of heat amounting to

$$q + dq - E - dE.$$

After cooling to T the system comes again to its original condition.

Now, during this reversible cyclic process, the work dE is performed from without; and at the same time the quantity of heat, E – q, is taken from T + dT to T.

But now, according to the law regarding the conversion of heat into external work, it follows that

$$dE = (E - q)\frac{dT}{T},$$

or

$$E - q = T\frac{dE}{dT}$$

According as the electromotive force of the element increases or decreases with the temperature,

$$E > q, \quad \text{or else } E < q.$$

That is, the electromotive force will, respectively, be greater or smaller than the heat of reaction of the chemical processes which produce the galvanic current.

The preceding equation may be obtained directly from the fundamental equation on p. 23, if we make the change of the free energy A equal to E, and the change of the total energy U equal to q; it was first derived by H. v. Helmholtz,[1] and was soon after subjected to an experimental test by Czapski;[2] its correctness was thus shown by these experimental results, together with those of Jahn.[3]

Now experience shows that there are both reversible galvanic elements with positive temperature coefficients, and also those with negative; and therefore in these cases there may be obtained, from the chemical processes producing the current, sometimes a greater and sometimes a smaller amount of external work than is equivalent to their respective heat of reaction.

In the recent and very exact researches of Jahn (*l.c.*) the heat values were measured directly by means of the Bunsen ice calorimeter. These heat values, developed in the total closed circuit of a galvanic element and corresponding to q, show that the difference E – q is by no means always equal to zero, and that at most it can hardly be neglected in comparison with the total change of energy.

In the following table there are grouped the values of the electromotive force E, for a number of combinations, expressed both in volts and also in g.-cal.; there are given also the heat of reaction q, of the chemical changes referred to 1 g. equivalent.

It was found for the Daniell cell at 0°

$$E = 25{,}263, \quad q = 25{,}055, \quad E - q = 208;$$

on the other hand, its temperature coefficient is

$$\frac{dE}{dT} = +0{\cdot}00034,$$

[1] *Sitzungsber. der Berl. Akad.* for Feb. 2 and July 7, 1882.
[2] *Wied. Ann.*, **21.** 209 (1884); see also Gockel, *ibid.*, **24.** 618 (1885).
[3] *Ibid.*, **28.** 21 and 491 (1886). See also Jahn, *Elektrochemie*. Wien, 1895.

from which, according to Helmholtz's formula,

$$E - q = 0{\cdot}00034 . 23{,}110 . 273 = 213 \text{ cal.}$$

The agreement is amply sufficient, considering the uncertainty of the slight difference between E – q.

Combination.	E expressed in		q	E – q	
	Volts.	g.-cal.		Obs.	Cal.
Cu, $Cu(C_2H_3O_2)_2$ aq. Pb, $Pb(C_2H_3O_2)_2 + 100\,H_2O$	0·470	10,842	8,766	+2076	+2392
Ag, AgCl Zn, $ZnCl_2 + 100\,H_2O$	1·015	23,453	26,023	–2570	–2541
Ag, AgCl Zn, $ZnCl_2 + 50\,H_2O$	1·001	23,146	24,456	–1310	–1305
Ag, AgCl Zn, $ZnCl_2 + 25\,H_2O$	0·960	22,166	23,493	–1327	–1255
Ag, AgBr Zn, $ZnBr_2 + 25\,H_2O$	0·828	19,138	19,882	–644	– 663

A combination found by Bugarszky,[1]

$$Hg \mid HgCl - KCl - KOH - Hg_2O \mid Hg,$$

is especially interesting. This cell yields an electromotive force of 7566 cal., and the current given by it is from right to left, so that the current producing process is—

$$HgCl + KOH = \frac{1}{2}Hg_2O + \frac{1}{2}H_2O + KCl.$$

This process, however, occurs with *absorption of heat* (q = – 3280), that is an endothermic reaction is here electromotively active. $T\frac{dE}{dT}$ according to Bugarszky = + 11276, it is therefore very considerable, and according to Helmholtz's theory yields the value $q = E - T\frac{dE}{dT} = -3710$.

Finally, we may note the application of thermodynamics to the *Grove* gas battery (see Smale);[2] here at 20° E = 1·062 volt, q = 34,200 cal. = 1·480, and therefore E – q = – 0·418 volt, whilst $T\frac{dE}{dT} = -293 . 0{\cdot}00142 = -0{\cdot}416$. The remarkable agreement shows *that*

[1] *Zeitschr. Anorg. Chem.*, **14.** 145 (1897).
[2] *Jahrb. d. Elektrochemie*, 1894, p. 36.

the decomposition or formation of water is a reversible process with voltage of 1·06.

Another remarkable application of the Helmholtz equation is due to van 't Hoff, Cohen and Bredig.[1] If there is a point at which $E = 0$, it follows that $-q = T\frac{dE}{dT}$, that is the temperature-coefficient can be calculated from the electromotive force itself, and that $E_{T+\Delta T} = \Delta T\frac{dE}{dT}$. Hence if the variation of q with temperature is known it is possible to determine E at all temperatures by means of the heat of reaction.

Since the electromotive force E, of a reversible galvanic element, is the measure of the external work which the current-developing process is capable of performing *in maximo*, therefore it gives directly the "degree of affinity" (p. 693) of the reaction in question. If, as required by the "principle of Berthelot," the change of the free energy of a chemical reaction were identical with the total change of energy q, then would

$$E = q;$$

i.e. in that case the "rule of Thomson" would be fulfilled.

The formulation both of "Thomson's rule" and of "Berthelot's principle" sprang from views which are similar, and which lay very near the truth, but which are now shown to be inadmissible. The fact that we may find

$$E > q,$$

as well as

$$E < q,$$

again shows that the heat of reaction is no [accurate] measure of the work which the chemical forces are able to perform at their maximal efficiency, *i.e.* in a reversible reaction.

Moreover, on the other hand, there is the fact that "Thomson's rule," *i.e.* that the electromotive force and the heat of reaction are not *very* different from each other; this rule usually is in the best of coincidence with the result to which one is led by a consideration of the spontaneously occurring processes. That is, *in many cases the heat of reaction is at least an approximate measure of the affinity.*

We have already found (p. 687), that it is an indispensable proviso for this, that the heat of reaction shall be independent of the temperature. As a matter of fact, from the equality of E and q, viz.,

$$E - q = T\frac{dE}{dT} = 0,$$

[1] *Zeitschr. physik. Chem.*, **16.** 453 (1895).

it also follows that

$$\frac{dE}{dT} = \frac{dq}{dT} = 0.$$

That is, the heat of reaction of a current-developing process is independent of the temperature, if it coincides with the maximal work.

Galvanic Polarisation.—If we electrolyse a system which is standing in equilibrium, there occurs a deformation of the equilibrium as a result of the electrolytic disturbance. This of course needs the application of a certain amount of external work which is performed by the galvanic current. It follows necessarily from this, *that the current which is transmitted through the system has to overcome an opposing electromotive force.* And from this there follows this theorem; viz., *if we electrolyse a chemical system, which is kept at constant temperature, one observes a reaction in that sense which is opposed, in an electromotive way, to the current conducted through the system.* This principle is included under the more general one of *action and reaction*, stated on p. 667.

As remarked on p. 699, the current in a circuit which contains the electromotive force E and an electrolytic cell is given by the expression

$$i = \frac{E - \epsilon}{W + w};$$

if the polarising force is equal to the back electric force we shall get equilibrium, that is the electrolytic cell forms a galvanic element of electromotive force $E = \epsilon$. Hence the formula

$$\epsilon - q = T\frac{d\epsilon}{dT}$$

will, according to p. 710, apply to it. It is to be remarked however that q, the heat of reaction of the process yielding the current, can rarely be determined with certainty, because polarisation is commonly due to extremely small quantities of substances occluded by the electrodes or by changes of concentration in the immediate neighbourhood of the electrodes, which are difficult to define. For these reasons the above formula has only been applied in a few cases.

Moreover, ϵ is usually considerably less than E on account of certain irreversible processes, such as convection, diffusion, and the like; see especially von Helmholtz (*Sitzungsberichte der Berliner Akademie*, 1883, p. 660; *ges. Abh.* vol. iii.).

The thermal accompaniments of polarisation have been very thoroughly studied by H. Jahn.[1]

Thermodynamic Calculation of Electromotive Forces.—Since the electromotive force of a reversible galvanic cell measures the

[1] *Zeitschr. physik. Chem.*, **18**. 399 (1895); **26**. 385 (1898); **29**. 77 (1899).

maximal work which the current yielding process can give, and since at any given temperature this amount of work is perfectly defined, *we can calculate the electromotive force of any combination if we know the affinity of the current yielding reaction.* This has been done, in the first place, with so called concentration cells in which the current yielding process consists simply in the mixing of solutions of different concentrations. The calculation was first carried out in 1877 by Helmholtz;[1] if two copper electrodes are dipped into two communicating cells of a copper salt, copper is dissolved in the dilute solution and precipitated from the more concentrated; as current yielding process we have the equalisation of concentration of the solutions round the two electrodes. The maximal work to be obtained in this case can be calculated in various ways, most simply by means of isothermal distillation, as shown on p. 115, and it follows that this work must be equal to the electromotive force, as is confirmed by experiment.[2] Later the author[3] considerably simplified the theory of Helmholtz for dilute solutions by application of van 't Hoff's law, and by making use of reversible electrodes of the second kind (see the following chapter) was enabled to apply it to the relatively exact experiments.

A different kind of concentration cell was subsequently studied by G. Meyer,[4] after Türin[5] had given a theoretical discussion of it. In an element which follows the scheme

Concentrated amalgam	Solution of a salt of the metal dissolved in the amalgam	Dilute amalgam

the current which it yields transfers metal from the more concentrated to the dilute amalgam; the electromotive force of the cell is, however, since the process is reversible, again the measure of the maximal work which can be obtained from the current yielding process. On the other hand, the osmotic pressure of the metal dissolved in the mercury measures the same quantity, hence by determining the electromotive force of cells constructed according to the above type the osmotic pressure of the dissolved metal may be obtained. In all the cases experimented on (Zn, Cd, Pb, Sn, Cu, Na) it appeared that (1) the osmotic pressure is proportional to the concentration of the metal dissolved in mercury, and (2) that its value (to within a small percentage) is that calculated from the temperature (mostly 18-20°) and concentration of the amalgam on the assumption *that the molecular weight of the dissolved metal is identical with its atomic weight;* the latter result is in complete agreement with other determinations of the osmotic pressure of metals dissolved in mercury (see p. 402).

[1] *Wied. Ann.*, **3.** 201 (1877); *Ges. Abhandl.*, 1. p. 840.
[2] J. Moser, *Wied. Ann.*, **14.** 61 (1881).
[3] *Zeitschr. physik. Chem.*, **4.** 129 (1889).
[4] *Ibid.*, **7.** 477 (1891).
[5] *Ibid.*, **5.** 340 (1890).

The preceding considerations show that at the melting point of a metal, electrodes in the *solid and liquid states* must possess *the same electromotive activity;* the force of such an element therefore suffers no change when the electrode melts, which is in agreement with the experiments of Miller.[1]

If we were able to determine the equilibrium of the reaction

$$Zn + CuSO_4 = Cu + ZnSO_4$$

in any way, we could determine the electromotive force of a Daniell cell; conversely, from the known electromotive force of this cell we could arrive at the equilibrium which exists between the vapour of the above four reacting substances.

Application of Thermodynamics to the Lead Accumulator.—The thermodynamic treatment given above has been applied to the lead accumulator in a very interesting paper by F. Dolezalek.[2] If we assume that the current yielding process is, as suggested by Planté, expressed by the equation

$$PbO_2 + Pb + 2H_2SO_4 = 2PbSO_4 + 2H_2O,$$

the electromotive force must clearly increase when the concentration of the sulphuric acid is increased and that of the water decreased. If two lead accumulators filled with acid of different concentrations are opposed to one another the current yielding process of the combination is simply that for two Fs two mols. of sulphuric acid are transported from the first accumulator to the second, and at the same time two mols. of water from the second accumulator to the first. If p_1 and p_2 are the vapour pressures of the sulphuric acid, P_1 and P_2 those of the water in the two accumulators, the work done per F, which is accordingly the electromotive force ΔE of the combined cell, is

$$\Delta E - RT\left(\ln \frac{p_1}{p_2} + \ln \frac{P_2}{P_1}\right) \quad . \quad . \quad . \quad . \quad (1)$$

The vapour pressures of water over aqueous sulphuric acid are known (p. 167); those of sulphuric acid are impossible to measure on account of their smallness. According to the considerations on p. 117 we can calculate the vapour pressure of one of the components of a mixture when we know that of the other as a function of the composition. We may therefore express $\ln \frac{p_1}{p_2}$ in equation (1) by means of the vapour pressure P of water; we found (p. 117)

[1] *Zeitschr. physik. Chem.*, **10.** 459 (1892).

[2] *Zeitschr. f. Elektrochemie*, **4.** 349 (1898); *Weid. Ann.*, **65.** 894 (1898); see also especially the monograph by the same author, *Theorie des Bleiakkumulators*, Halle, 1901, Knapp.

$$\ln \frac{p_0}{p} + x \ln \frac{P_0}{P} = \int_0^x \ln \frac{P_0}{P} dx$$

and this may be applied to the two solutions, of which one contains x_1 and the other x_2 mols. of water for one mol. of sulphuric acid, and yields

$$\ln \frac{p_1}{p_2} = x_2 \ln P_2 - x_1 \ln P_1 - \int_{x_1}^{x_2} \ln P \, dx,$$

hence

$$\Delta E = RT\left(x_2 \ln P_2 - x_1 \ln P_1 - \int_{x_1}^{x_2} \ln P \, dx + \ln \frac{P_2}{P_1}\right).$$

In order to obtain ΔE in volts we must set

$$R = 0{\cdot}861 \, . \, 10^{-4}.$$

Dolezalek calculated the electromotive force of two accumulators opposed to one another also from the formula

$$\Delta E = q + T \frac{\partial \Delta E}{\partial T};$$

where q can be calculated from the heat of dilution of sulphuric acid, and $\frac{\partial \Delta E}{\partial T}$ is given by the measurements of Streintz[1] on the variation of the temperature coefficient of the accumulator with the concentration of the acid. Moreover $T \frac{\partial \Delta E}{\partial T}$ is very small (0·02 volt as the maximum), which is in complete agreement with the considerations on p. 167, according to which the heat of dilution of a sulphuric acid mixture is almost the same as the maximal work done in dilution. The following table contains the results of calculation, and shows the striking agreement between the electromotive force calculated in different ways and measured by different observers.

[1] *Weid. Ann.*, **46.** 454 (1892).

[TABLE.

ELECTROMOTIVE FORCE OF ACCUMULATORS OF DIFFERENT DENSITIES OF ACID

No.	Density of acid.	% H_2SO_4	x	Vapour pressure P mm. Hg.	Electromotive force E in volts (0°).			
					calculated		measured	
					from P.	from q.	Dolezalek.	Streintz.
1	1·553	6·45	3	0·431	2·383	2·39	2·355	...
2	1·420	52·15	5	1·297	2·257	2·25	2·253	2·268
3	1·266	35·26	10	2·975	(2·103)	(2·10)	2·103	(2·103)
4	1·154	21·40	20	4·027	2·000	2·06	2·008	1·992
5	1·035	5·16	100	4·540	1·892	1·85	1·887	1·891

During the charge of an accumulator the acid becomes, according to the above equation, more concentrated, during discharge more dilute; accordingly during charge the electromotive force of the accumulator must rise, during discharge it must fall. If the charge and discharge are carried out with considerable current density, relatively large changes of concentration will occur at the plates, and accordingly the accumulator will show a noticeably higher electromotive force during charge and a noticeably smaller during discharge than corresponds to the normal value of 1·95 volt; this is confirmed by experiment. Accordingly we have in the accumulator a polarisation of a peculiar kind, namely a change in the concentration of acid which takes place in the same sense at both the positive and negative plates, although to a greater extent at the positive. The accumulator, which of all galvanic cells is by far the most important, is also, as Dolezalek has shown in the monograph mentioned above, one of the most interesting combinations from the thermodynamic point of view; the thermodynamic treatment has yielded the important result that *the accumulator works reversibly in the sense of the above reaction.*

Electromotive Force and Chemical Equilibrium. — The maximal work that a chemical process can yield is, according to p. 647,

$$A = RT \ln \frac{C_1^{n_1} C_2^{n_2} \dots}{C_1'^{n_1'} C_2'^{n_2'} \dots} + RT \ln K; \quad (1)$$

if E is the electromotive force of a galvanic combination, in which the current yielding process is given by a reaction to which equation (1) refers, we have

$$E = A. \quad (2)$$

To express E in volts it is necessary, according to p. 703, to put

$$R = \frac{1 \cdot 991}{23,110} = 0 \cdot 8615 \times 10^{-4}.$$

This reaction has recently been confirmed experimentally by C. Knüpffer [1] for the reaction

$$TlCl + KSCN \text{ aq.} = TlSCN + KCl \text{ aq.},$$

who, at the suggestion of Bredig, measured both the chemical equilibrium in the above reaction and the electromotive force of the combination

Tl-amalg. | TlCl – KCl – KSCN – TlSCN | Tl-amalg.

again, G. Preuner [2] has succeeded in calculating the electromotive force of a gas battery from the dissociation of water vapour; and Danneel [3] has investigated electrically and chemically the equilibrium between silver and hydroiodic acid.

The Galvanic Element considered as a Chemical System.—A galvanic element represents a heterogeneous chemical system, and has certain peculiarities; but, on the whole, as regards the relations of the equilibrium, both to the relative proportions of the ingredients and also to the temperature, it is subject to the same laws which have been already developed for ordinary heterogeneous chemical systems.

The different complexes, which are both physically and chemically homogeneous, and which, by their interpenetration, form heterogeneous systems,—these are called "the phases of the system"; we have already seen (p. 609) that complete equilibrium can be established only when the number of phases is one greater than the number of the reacting molecular species. We can easily convince ourselves, by an example, that the same rule holds good for a galvanic element.

Thus, *e.g.*, let us consider the Clark element—

$$Hg \mid \underset{\text{Solid}}{Hg_2SO_4} \mid \underset{\text{Liquid}}{H_2O + xHg_2SO_4 + yZnSO_4} \mid \underset{\text{Solid}}{ZnSO_4} \mid Zn;$$

thus we distinguish *five* different phases; these include the two metals, their two solid sulphates, and the saturated solution of these sulphates at the respective concentrations of x and y. In order to construct all of these *five* phases, we must bring together at least *four* different molecular species, viz.:—

$$Hg, Zn, H_2O, SO_4;$$

and we thus conclude that here we have a complete chemical equilibrium.

Now we know, as a matter of fact, that (at constant pressure) for every definite temperature there corresponds a certain definite

[1] *Zeitschr. physik. Chem.*, **26.** 255 (1898).
[2] *Ibid.*, **42.** 50 (1903).
[3] *Ibid.*, **33.** 415 (1900).

electromotive force of the cell; and that the phases are in equilibrium only when we oppose the element with an electromotive force of the same magnitude. But, on the other hand, if the opposed electromotive force is greater or smaller than that of the element, at the temperature in question, then the chemical change advances in the sense of the equation,

$$Zn + Hg_2SO_4 \rightleftarrows 2Hg + ZnSO_4,$$

and completely, *i.e.* until the exhaustion of one of the phases, in one direction or the other; and it also advances in such a way that none of the phases changes its composition during the reaction.

We thus find here all the criteria which were characteristic of *a complete equilibrium.*[1]

But the case is quite otherwise when there are only *four* phases in the cell in question, *i.e.* as soon as one of the two solid salts disappear. In that case, although the element continues to furnish a current, yet the composition of the solution, and the electromotive force of the element also, changes. Thus the occurrence of phases of variable composition is characteristic of *an incomplete equilibrium.*

One peculiarity which characterises galvanic batteries above other chemical systems is this, that a conducting [outside] union is needed between the two metals in order to make the reaction possible. We may have a Clark cell in open circuit for an indefinite length of time without observing a chemical change; but we cannot conclude from this that the system is in equilibrium, any more than we were at liberty to draw a similar conclusion regarding a "knall-gas" mixture (p. 677).

The reaction velocity in an "open" element, if not absolutely *nil*, is at least practically so, and it does not attain a measurable value until a conducting [outside] union is established between the two poles. This reaction velocity can be made to vary as desired, by changing the resistance in the circuit, and corresponds directly to the intensity of the current; for according to Faraday's law, the intensity of the current and the chemical change are directly proportional to each other.

Thus we can speak of having an equilibrium, only when the two poles are connected by a conducting union, and when also we compensate the potential difference in some way by means of an opposing electromotive force. This opposing electromotive force is completely analogous to the opposing pressure which we found necessary to apply in the case of solid NH_4Cl, *e.g.*, in order to guard against its complete sublimation and the attendant separation into its two dissociation products.

[1] For standard cells evidently those with $n+1$ phases are the best; it is also desirable that E and q shall differ as little as possible in order that the temperature may be of small influence.

CHAPTER VIII

ELECTRO-CHEMISTRY III

OSMOTIC THEORY

The Mechanism of Current-Production in Solutions.—The considerations thus far advanced rest essentially upon a thermodynamic basis. It is inherent in the nature of this method of investigation, they are undoubtedly correct when applied with precaution; but they do not give obvious results. In particular, the mechanism of galvanic-current production has remained, for the present, entirely beyond the domain of our consideration. But it now appears that the more recent views of the ion theory enable us to take an important step forward. Therefore the last section of this chapter will be devoted to the presentation of *a special theory regarding the electromotive activity of ions;* this was developed in 1888 and 1889, and is now generally accepted.

We have already developed the view, p. 366, that in the contact between two solutions of an electrolyte of different concentrations, there is developed an electromotive force which so acts between them, that the one ion strives to pass by the other. In this way there was obtained for the first time a mechanical explanation of the potential difference between any two substances; and it also became possible to calculate this value in absolute measure.

If we solve the differential equation on p. 367.

$$dS = -Uqdz\left(\frac{dp}{dx} + \eta\frac{dP}{dx}\right) = -Vqdz\left(\frac{dp}{dx} - \eta\frac{dP}{dx}\right),$$

for $\frac{dP}{dx}$ we find

$$\frac{dP}{dx} = -\frac{U-V}{U+V}\frac{1}{\eta}\frac{dp}{dx}$$

or, bearing in mind equation (2) p. 368,

$$\frac{dP}{dx} = -\frac{u-v}{u+v}\frac{1}{\eta}\frac{dp}{dx}.$$

Inserting the value of the osmotic pressure,

$$p = \eta RT$$

and integrating, we have

$$P_1 - P_2 = \frac{u-v}{u+v} RT \ln \frac{\eta_2}{\eta_1} \quad . \quad . \quad . \quad . \quad (1)$$

where $P_1 - P_2$ is the potential difference between two solutions of the same kind, consisting of two univalent ions and assumed to be completely dissociated; η_1 and η_2 are the concentrations of the two solutions.

The preceding formula may also be deduced thermodynamically in the following way, as was shown by the author.[1] Suppose the quantity of electricity F to pass across the surface of contact of the two solutions in the direction from the concentrated to the dilute solution, $\frac{u}{u+v}$ equivalents of cations will migrate in the same direction, $\frac{v}{u+v}$ anions in the opposite. Since the cations pass from the concentration η_1 to the concentration η_2, they are capable, according to p. 52, of doing the work of expansion $\frac{u}{u+v} RT \ln \frac{\eta_1}{\eta_2}$, whilst conversely the anions need an amount of work $\frac{v}{u+v} RT \ln \frac{\eta_1}{\eta_2}$; the total work performed by the process is, therefore,

$$A = \frac{u-v}{u+v} RT \ln \frac{\eta_1}{\eta_2},$$

and thus the expression is equal to the electromotive force $P_2 - P_1$. If $\eta_1 > \eta_2$ and $u > v$, we have $A > 0$, there is an electromotive force tending to produce a current from the concentrated to the dilute solution. In the case of incompletely dissociated solutions the ionic concentrations must be used instead of η.

We can explain in an entirely similar mechanical way,[2] the potential difference occasioned by *the contact of the solutions of any two different electrolytes.*

Thus let us bring into contact with each other a solution of HCl and of Li Br; then, on the one hand, more hydrogen ions than chlorine ions will diffuse from the first solution into the second, and therefore the second solution [Li Br] will receive a positive charge; and on the other hand, more bromine ions than lithium ions will diffuse from the second solution into the first, because of the greater

[1] *Zeitschr. physik. Chem.*, **4.** 129 (1889).

[2] Nernst, *Zeitschr. physik. Chem.*, **2.** 613; **4.** 129 (1889); *Wied. Ann.*, **45.** 360 (1892.)

mobility of the bromine ions; and thus the positive charge of the second solution will be increased.

Moreover, these electromotive forces can be calculated in absolute units of measurement, from the gas laws and the ion mobilities, for which purpose Planck,[1] by integration, has developed the general equations given by the writer.

In this way the general problem has been solved regarding the calculation of the electromotive forces of any liquid cells, provided that only dilute solutions are used; and this has been done theoretically by means of the gas laws and the ion mobilities; and, moreover, the mechanism, according to which these batteries produce the current, has been searched out, even to all the details.

The electromotive forces between different solutions are nearly always small and usually counted only by hundredths or thousandths of a volt, but of course in exact calculations must be taken into account, the more so as in most cases they can be estimated with sufficient accuracy. In dealing with dilute solutions there is a very simple means to reduce them almost to the vanishing point; this is to use as solvent, not pure water, but a solution of a suitable indifferent electrolyte (see p. 728). Under these circumstances the current flows almost entirely through the indifferent electrolyte and consequently no appreciable osmotic work is done, and there is no appreciable potential difference between the solutions.

The gas constant R may be expressed in the electrochemical system of units as follows. We have

$$p = \eta RT = \eta \frac{p_0}{273} T,$$

where p_0 is the osmotic pressure in a space of unit concentration. Expressing the concentration in mols. per c.cm. we have

$$p_0 = 22420 \text{ atm.} = 22420 \,.\, 1033{\cdot}3 \,.\, 981 \text{ abs. units.}$$

Now the unit quantity of electricity in absolute measure is associated with $1{\cdot}035 \,.\, 10^{-4}$ mols. of a univalent ion; accordingly if we take as the unit of concentration the corresponding number $1{\cdot}035 \,.\, 10^{-4}$,

$$p_0 = 22420 \,.\, 1033{\cdot}3 \,.\, 981 \,.\, 1{\cdot}035 \,.\, 10^{-4},$$

we find, as on p. 716,

$$R = 0{\cdot}861 \,.\, 10^4 \text{ abs. units};$$

but in order to express potential differences in volts the preceding number must be multiplied by 10^{-8}.

Hence

$$R = 0{\cdot}861 \,.\, 10^{-4}.$$

[1] *Wied. Ann.*, **40.** 561 (1890).

We have therefore

$$RT \ln \frac{c_1}{c_2} = 0{\cdot}861 \,.\, 10^{-4}\ T \ln \frac{c_1}{c_2} = 0{\cdot}0001983\ T \log^{10} \frac{c_1}{c_2} \text{ volt.}$$

Or, putting $T = 273 + 18$, we get

$$RT \ln \frac{c_1}{c_2} = 0{\cdot}0577 \log^{10} \frac{c_1}{c_2} \text{ volt.}$$

The Dissolving of Metals.—In order to account for the activity of galvanic elements, we will subject the current-producing process to a closer examination; now, in this process, nothing is obviously of more importance than the respective dissolving and precipitating of the metals of the electrode. And this at once leads us to the following observation.

The metals throughout are characterised in that they receive only positive charges; i.e. *they are able to go into solution only in the form of positive ions.*

Thus, whether, in the electrolysis of a solution of a nitrate, we convert the silver serving as the anode into silver nitrate, or whether we change zinc into zinc sulphate by simply dipping it into sulphuric acid, under all such circumstances the ions of the metals appear in solution in such a way that they either remain free, or else unite with the negative ion of the salt contained in the solution, to form electrically neutral molecules.

Thus the process of the dissolving of metals appears in a very peculiar light. We see that *forces of an electrical nature* play a very prominent part in these chemical processes. And, in fact, if we accept the aid of the conceptions which we have already developed (pp. 147 and 475) regarding the process of dissolving, we can, by means of the views advanced above, explain the electromotive activity of the metals, simply and without constraint.

Let us, therefore, ascribe to each metal a *certain solution pressure regarding the water;* for every substance must have such a pressure, as regards the selected solvent; and, as before, we imply the expansive force, which tends to drive the molecule of the substance into the solution. In the case of electrically neutral molecules, this expansive force will be reduced to an equilibrium by the osmotic pressure of the saturated solution.

But now this circumstance is a very characteristic mark of the metals, viz. *that those molecules which they strive to bring into solution, by dint of their solution pressure, are charged in an electro-positive way.* Therefore their solution pressure is suitably called "*electrolytic.*"

It is now easy to examine the processes which will take place on dipping a metal into pure water or into any solution.

We will first consider the case where the ions of the metal in question, are either not at all present in the solution, or, at most, to only a very slight extent. It will happen that a number of ions, driven by the solution pressure of the metal, will pass into solution. And the immediate result of this will be that the solution becomes positively charged and the metal negatively. But, just as appeared in the diffusion of electrolytes (p. 366), so here these electrical charges occasion a force component, which, on the one hand, opposes the further passage of the metallic ions into the solution; and, on the other hand, seeks to drive the metallic ions already in the solution back on to the metal.

On account of the enormous electrostatic capacities of the ions, this force attains an extremely high value long before a weighable quantity has passed into solution. Now this may result in one of two ways.

1. Either the action of the solution pressure of the metal will be exactly compensated by the electrostatic charges, and a state of equilibrium will be established; in this case no more of the metal will pass into the solution, neither will any more positive ions be driven out of the solution. This happens, *e.g.*, when silver is dipped into a solution of NaCl. We should not conclude that the solution pressure of this metal has no marked value, just because the silver is not dissolved under these circumstances. It is much more conceivable that it is to be estimated in thousands of atmospheres, and that it is compensated in its action by the electrostatic charge of the solution as it comes into contact with the metal.

2. Or the electrostatic charges may reach such an amount, as a result of the magnitude of the solution pressure, that the positive ions which are contained in solution are driven out on to the metal. We observe an illustration of this case, *in the precipitation of one metal by another;* thus, *e.g.*, when iron is dipped into a solution of a copper salt, the iron ions go into the solution and the electrically equivalent quantity of copper ions is driven out of the solution, by electrostatic repulsion of the solution, and by the electrostatic attraction of the metal; and thus the copper ions are precipitated upon the iron.

The case is similar in the *development of hydrogen.*

The foregoing considerations have lately received a striking confirmation by the researches of W. Palmaer.[1] If, as shown by Fig 35, mercury is dropped into a solution which contains only a few mercury ions, some of these ions will be deposited on each drop, because mercury, being a noble metal, possesses an extremely small solution pressure. If the drops fall into a mass of mercury below, their charge disappears, that is, they give the mercury ions back to the solution. We see then that the point where the new mercury surface is formed, that is, at the opening of the mercury funnel, the solution becomes poorer in mercury salt, whilst where the drops unite and there is a diminution of surface

[1] *Zeitschr. physik. Chem.*, **25.** 265 (1898).

between mercury and electrolyte the solution must become richer in salt. It has long been noticed that a potential difference arises between dropping mercury and still mercury; it was pointed out by the author (see my note on contact electricity)[1] that this arrangement may be regarded as a concentration cell, and Palmaer proved experimentally that changes of concentration do occur in the sense expected; this was demonstrated first by measurements of potential difference as compared with small inserted mercury electrodes, and afterwards[2] by purely chemical means.

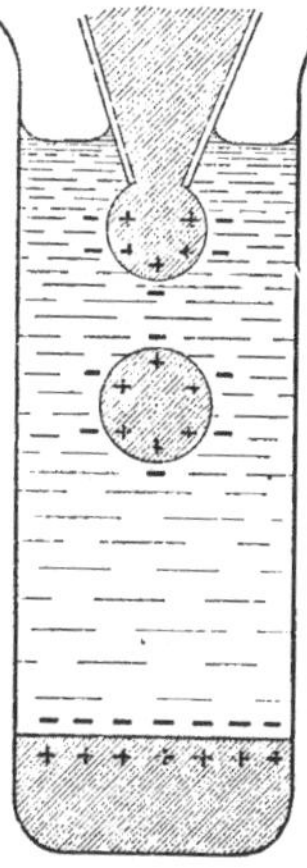

Fig. 35.

The Theory of the Production of the Galvanic Current.—The considerations advanced in the preceding section, led directly to an insight into the production of galvanic currents from systems possessed of electromotive activity, in so far as these systems are dependent on the use of metals. Here we will limit ourselves to reversible galvanic elements, and in order to fix the idea, we will consider one particular example, namely the Daniell element.

Let a zinc rod be dipped in a solution of zinc salt (for example, $ZnSO_4$), as in Fig. 36, and a copper rod in a solution of copper salt ($CuSO_4$); then the zinc, in consequence of its greater solution pressure, will send out a quantity of positive ions into the solution, whilst conversely, if the osmotic pressure p_2 of the copper ions is sufficiently great, copper ions will be deposited on the copper rod and the copper will thus receive a positive charge.

As long as the circuit is *open*, neither zinc ions nor copper ions will go over into the solution, because the electrolytic solution pressures of these metals are held in equilibrium by the opposing charges, which both the metals and the solutions have received from the inappreciable passage of ions; and therefore chemical change is excluded from the element.

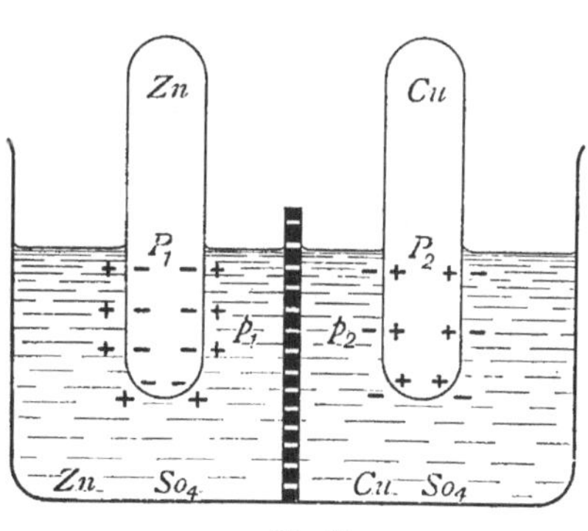

Fig. 36.

But the case is changed as soon as the circuit between the two poles of the element is *closed* by a conducting union; then the reaction can advance, because it is possible to equalise the electric

[1] Supplement to *Wied. Ann.* 1896, Heft, Nr. 8.
[2] *Ibid.*, **28**, 257 (1899).

charges associated with the solution and the precipitation respectively, of the two metals; and thus the reaction takes place in such a way, that that metal having the greater solution pressure drives its ions into the solution; and conversely, that metal having the smaller solution pressure is precipitated from the solution.

In the Daniell element the zinc is dissolved, and the correspondingly equivalent quantity of copper is driven out of the solution, because the force tending to bring the zinç ions into solution is greater than that tending to bring the copper ions into solution.

The passage of the zinc into the solution and of the copper out of the solution, necessarily results in a movement of positive electricity in the external circuit, from copper to zinc; *i.e. the production of a galvanic current* ensues in the direction indicated.

We have already seen that the osmotic pressure of the ions of a metal tends to oppose its solution pressure. Thus the force which brings the zinc ions into solution will be the smaller, the greater the concentration of the zinc sulphate solution; and likewise the force tending to separate the copper ions will be the greater, the stronger the concentration of the copper sulphate. Thus the electromotive force of the Daniell element will increase by diluting the zinc sulphate solution; and it will diminish by diluting the copper sulphate solution; and both of these statements are completely verified by experiment.

The electromotive force of a Daniell cell may be expressed by means of the formula given below—

$$E = \frac{RT}{2}\left(\ln \frac{P_1}{p_1} - \ln \frac{P_2}{p_2}\right);$$

here the small potential difference at the contact of the two electrolytes is neglected.

Concentration Cells.—The above considerations lead at once to a very simple expression for the potential difference between the metal and the solution which contains a larger or a smaller amount of ions in question. If A is the work that can be performed by the solution of one electro-chemical gramme equivalent of the electrode metal in the electrolyte considered, then the osmotic pressure of the univalent ions of the metal amounts to p; thus, clearly,

$$A = E, \quad . \quad . \quad . \quad . \quad (I.)$$

where E is the potential difference required. If p becomes p + dp, A becomes A + dA, and accordingly E becomes E + dE. Let us now dissolve one electro-chemical gramme equivalent; then dA is equal to the work that must be spent if the electro-chemical equivalent is to be brought from pressure p + dp to pressure p. This amounts, therefore, to p . dv (where v is the volume which one electro-chemical equivalent assumes in the solution), hence we have

$$dE = dA = p \, . \, dv = -RT\frac{dp}{p}$$

or, integrating,

$$E = -RT \ln p + \text{const.}$$

that is

$$E = RT \ln \frac{P}{p}.$$

In this equation clearly E = 0, when P = p, hence P indicates the electrolytic solution pressure of the metal in question. Instead of using the pressure p we may use the ionic concentration c, which is proportional to it, and obtain

$$E = RT \ln \frac{C}{c} \quad . \quad . \quad . \quad . \quad (II.)$$

where c is the concentration corresponding to the osmotic pressure p. In what follows we shall describe P simply as the solution pressure.

The equation

$$A = E = \text{const.} - RT \ln c$$

may also be derived immediately from the theory of the thermodynamic potential given on p. 668.

If the ion whose solution is considered is n-valent, we have

$$dA = -\frac{RT}{n} \cdot \frac{dp}{p},$$

and accordingly

$$E = \frac{RT}{n} \cdot \ln \frac{C}{c}.$$

This equation, given by the author in the year 1889, contains both the theory of galvanic current production and that of galvanic polarisation.

We will first consider the case of two electrodes of a univalent metal dipping in two solutions of a salt of that metal in different concentrations; we then have a concentration cell, for example

$$Ag \left| \begin{array}{c} AgNO_3 \\ c_1 \end{array} \right| \begin{array}{c} AgNO_3 \\ c_2 \end{array} \left| Ag. \right.$$

In these galvanic combinations we have three contacts of dissimilar conductors, and accordingly three potential differences, which may be separately calculated and whose sum gives the electromotive force of the combination. At one of the metal-electrolyte contacts we have the force $RT \ln \frac{C}{c_1}$, at the contact of the two solutions, according to p. 721, $\frac{u-v}{u+v} RT \ln \frac{c_1}{c_2}$, at the other contact between metal and electrolyte the force $RT \ln \frac{C}{c_2}$; hence the total electromotive force of the cell will be

$$E = RT \ln \frac{C}{c_1} + \frac{u-v}{u+v} RT \ln \frac{c_1}{c_2} - RT \ln \frac{C}{c_2} = -\frac{2v}{u+v} RT \ln \frac{c_1}{c_2}.$$

Assuming, for example, $c_1 = 0{\cdot}1$, $c_2 = 0{\cdot}01$ mol. per litre, we get, at atmospheric temperature, according to p. 723,

$$E = 2 \,.\, 0{\cdot}522 \,.\, 0{\cdot}0577 = 0{\cdot}0604 \text{ volt}$$

$\left(\frac{v}{u+v} = 0{\cdot}522\right.$, according to p. $\left.364\right)$; experiment gave 0·055 volt, *i.e.* a somewhat smaller value. Bearing in mind, however, that the electrolytic dissociation in the solutions used was incomplete, the theory gives 0·057, which is in complete agreement with the experiment. If two metal electrodes dip into the same solution of an indifferent electrolyte, and a small quantity of the ions of the electrode metal are added to produce concentration c_1 in the first solution and c_2 in the second, where c_1 and c_2 are small in comparison with the concentration of the indifferent electrolyte, the electromotive force of contact between the two electrolytes vanishes; then under these circumstances the current is conducted almost entirely by the indifferent electrolyte, so that practically no osmotic work is spent in the electrolyte, and for concentration cells of this type we have the simple formula

$$E = \frac{RT}{n} \ln \frac{c_2}{c_1}.$$

The electromotive force of such cells can occasionally become very large, viz. when one of the concentrations (c_1 or c_2) becomes extraordinarily small; thus the electromotive force of the cell

$$Ag \mid 0{\cdot}1\ AgNO_3 \mid 1{\cdot}0\ KCl \mid AgCl \mid Ag$$

has been found 0·51 of a volt; this relatively large electromotive force of a cell with two identical electrodes is due to the fact that the osmotic pressure of the silver ions in the solution of silver nitrate is considerable, but in the potassium chloride solution surrounding the silver chloride is excessively small; not only on account of the very small solubility of silver chloride but because it is further reduced by the presence of chlorine ions. This conception can be carried out quantitatively; c_1 is about 0·1 gramme ion per litre, since the solution of silver nitrate was deci-normal; c_2, according to p. 654, at 20° would be $1{\cdot}1 \times 10^{-5}$, if there were no potassium chloride present, and, as the latter forms a normal solution, the solubility sinks, according to p. 530, to $(1{\cdot}1 \times 10^{-5})^2 = 1{\cdot}21 \times 10^{-10}$ Hence the electromotive force of the cell becomes

$$0{\cdot}058\,(9 - \overset{10}{\log}\ 1{\cdot}21) = 0{\cdot}52 \text{ volt},$$

which is in very good agreement with the direct measurements (0·51); bearing in mind that the electrolytic dissociation of the KCl and $AgNO_3$ is not complete, the calculated value should be a trifle reduced.[1]

In the case of negative ions, such for example as a platinum electrode charged with chlorine, the potential difference is of the opposite sign, and in this case

$$E = -RT \ln \frac{C_1}{c_2} = RT \ln \frac{c_1}{C_1} \qquad \text{(III.)}$$

If mercury is covered with the very slightly soluble mercurous chloride the passage of a current through the electrode either forms or reduces calomel according to its direction, in other words, ions either come out from the solution or enter it. Such an electrode behaves therefore electrolytically, as if it were made of a metallically conducting modification of chlorine, and equation III. is applicable to it. Such electrodes, which consist of a metal covered with one of its insoluble salts and a solution which contains the same anion, are known as *reversible electrodes of the second kind* (p. 708), where electrodes to which equation II. is applicable are called *reversible electrodes of the first kind.* Electrodes of the second kind may be reduced to the first kind, as may be seen from the foregoing calculation, with the electrode Ag(AgCl).

In a concentration cell such as

$$\text{Hg} \mid \text{HgCl} \mid \text{chloride } c_1 \mid \text{chloride } c_2 \mid \text{HgCl} \mid \text{Hg},$$

the electromotive force may be determined by means of equations I. and III. exactly in the same way as on p. 727 for concentration cells with electrodes of the first kind. We find in this way

$$E = \frac{2u}{u+v} RT \ln \frac{c_1}{c_2}.$$

For example the author found:—[2]

Chloride	c_1	c_2	E	
			Obs.	Cal.
HCl	0·1	0·01	0·0926	0·0939
KCl	0·1	0·01	0·0532	0·0542
LiCl	0·1	0·01	0·0354	0·0336

[1] See also Goodwin, *Zeitschr. physik. Chem.*, **13.** 577 (1894).
[2] *Zeitschr. physik. Chem.*, **4.** 161 (1889).

In the first of these three combinations the electromotive force at the contact of the two electrolytes adds to the total effect, in the third it reduces it, in the second it is almost nothing.

Normal and Abnormal Potentials.—According to the equation

$$E = \frac{RT}{n} \ln \frac{C}{c}$$

the potential difference between the metal and electrolyte varies at atmospheric temperature by $\frac{0{\cdot}058}{n}$ volt when the concentration of the n-valent ions of the metal is altered tenfold; the change of electromotive force is consequently small when dealing with moderate concentrations, and even if the ionic concentration is changed several fold the electromotive force of the combination suffers, as a rule, only a small percentage modification. It is otherwise of course when the ionic concentration is reduced by several times tenfold; then there are very considerable alterations in electromotive force.

Thus for example silver in solutions of a silver salt of moderate concentration has always approximately the same potential; if by any means silver ions are very completely removed from the solution the potential is considerably altered. This can be done in two different ways; either by adding a reagent (for example a chloride) to precipitate the silver ions, or by adding a reagent such as potassium cyanide, which forms a complex salt with the silver ions and so equally removes them from the solution. By these two ways it is possible to reduce the concentration of the silver ions enormously, accordingly it is found that metals in such solutions give quite different so called "anomalous" potentials in opposition to the "normal" potentials observed in solutions containing larger quantities of their ions.

If a metal is dipped in a solution which originally contains no ions of it, the potential observed is, unlike that in the preceding cases, variable and uncertain; traces of the metal in larger or smaller quantity, according to the circumstances (for example under the action of the atmospheric oxygen), pass into solution, and these accidental quantities of course give accidental potential differences.

As an example of the great change in potential of a metal against an electrolyte that can be produced by precipitation, we have already mentioned the concentration cell with silver electrodes, which has an unusually large electromotive force (p. 537). If to the copper of a Daniell cell a sufficiently concentrated potassium cyanide solution is added, the copper ions are so completely taken up by it that the electromotive force of the combination even changes its sign, *i.e.* in such combinations copper goes into solution and reduces the zinc. A series of further examples were found by Hittorf,[1] who first carefully studied cases of this kind, and by Ostwald,[2] who gave an explanation of these anomalous electromotive forces by means of the author's formula.

Gas Batteries.—Platinum charged with hydrogen behaves electrolytically like the electrode of a conducting variety of hydrogen, charged with oxygen as if it were a metallically conducting form of oxygen. If these electrodes are immersed in a solution of an electrolyte we have a galvanic combination of the electromotive force

$$E = RT \ln \frac{C}{c} + \frac{RT}{2} \ln \frac{C_1}{c_1} = \frac{RT}{2} \ln \frac{C^2 . C_1}{c^2 . c_1};$$

here C and C_1 are the solution pressures of the two electrodes, c and c_1 the concentrations of the hydrogen ions and the doubly charged oxygen ions. Since in dilute aqueous solutions

$$c^2 \cdot c_1 = \text{const.}$$

it follows that E is independent of the nature of the dissolved substances; at atmospheric temperature it amounts to 1·08 volt, a value which is of much importance in electro-chemical calculations.

The potential of a hydrogen electrode is very different in acid and alkaline solutions (about 0·8 volt), because the concentration of the hydrogen ions in the two cases is, according to p. 508, enormously different, and the same is of course true of the oxygen electrode.

In a similar way the theory of all gas batteries which contain dilute solutions of an electrolyte may be studied by means of the osmotic theory. It must be noted however that the solution pressure of an electrode charged with gas depends on the degree of saturation of the gas in the electrode, and naturally increases with the concentration of the dissolved gas.

Oxidation and Reduction Cells.—Chemically an oxidising material is characterised by its power of giving off oxygen, a reducing material by its power of giving off hydrogen. In some cases this power extends to the visible evolution of gas; thus hydrogen peroxide gives off oxygen violently on a platinum surface, chromous solutions give off hydrogen, etc. Clearly the oxidising or reducing power is the greater the higher the pressure the evolution of gas can reach. If thus we bring platinum electrodes into solutions which contain an oxidising or reducing agent they will be charged with oxygen or hydrogen; by combination we get a cell according to the scheme

Pt | oxidising medium | indifferent solution | reducing medium | Pt,

[1] *Zeitschr. physik. Chem.*, **10.** 593 (1892).
[2] *Lehrbuch der allg. Chem.*, 2nd edit., vol. ii.

we have, therefore, an *oxyhydrogen cell*, but with this difference, that the oxygen or hydrogen charge can be, according to the nature of the oxidising or reducing medium, greater or smaller than if the charge were immediately produced by oxygen or hydrogen at atmospheric pressure, as in the ordinary gas cell. In other words, the values C_1 and C in the formula of the previous section depend upon the nature of the reagent used.

Hence it follows that (for constant charge of the platinum) the potential difference of each electrode against the solution depends on the concentration of the hydrogen or hydroxyl ions, and that if several oxidising agents are present at once in the solution the strongest determines the potential difference, that is the one which charges the electrode with oxygen most strongly. The same of course is true for the reducing materials.

If an oxidising and a reducing material are present simultaneously in the solution it often, but not always, happens that chemical action will take place; this must, however, happen as soon as platinum is brought into the solution and the charge of gas reaches sufficient magnitude, because oxygen and hydrogen occluded by platinum react violently.

In order to calculate the dependence of the potential difference on the concentration of the various reagents we must write the reaction for the evolution of oxygen and hydrogen in each case. Thus, ferrous sulphate charges platinum according to the equation

$$\overset{++}{\mathrm{Fe}} + \overset{+}{\mathrm{H}} = \overset{+++}{\mathrm{Fe}} + \frac{1}{2}\mathrm{H}_2 ;$$

hence the hydrogen charge is proportional to the concentration of the ferrous and hydrogen ions directly, and inversely to that of the ferric ions. See on this point the work by Peters,[1] which is full of interesting observations. The numbers of Bancroft[2] must be received with caution, as later measurements have shown.[3] See further the original study, *Oxydations und Reduktionsketten*, by K. Ochs, Dissertation, Basle, 1895. Ostwald has pointed out that measurements of electromotive force determine the oxidising or reducing action of a reagent.[4]

Theory of Electrolysis.—If we consider an electrolytic cell with two unalterable electrodes, on electrolysis the cations must be deposited at the cathode, the anions at the anode; if the electromotive force necessary for the first of these processes be called ϵ_1, for the second ϵ_2, we have for the back electromotive force of polarisation

$$E = \epsilon_1 + \epsilon_2 .$$

1 *Zeitschr. physik. Chem.*, **26.** 193 (1898).
2 *Ibid.*, **10.** 387 (1892).
3 See the same author, *ibid.*, **26.** 215.
4 See *Allg. Chem.*, 2nd edit., p. 883, etc., Leipzig, 1893.

Now the cation is the more easily deposited at the cathode the higher its concentration and the more difficultly the more dilute it is; by considerations exactly similar to this on p. 727 we find that

$$\epsilon_1 = \frac{RT}{n} \ln \frac{C_1}{c_1}$$

and

$$\epsilon_2 = \frac{RT}{n} \ln \frac{C_2}{c_2}$$

If we have now several kinds of cations and anions present *together* in the solution, a case which always happens when we are dealing with aqueous solutions, in which besides the ions of the dissolved substance there is water present, *the electrolysis will begin when the electromotive force E, conveniently known as the decomposition potential, is large enough to discharge one of the cations present and one of the anions.*

The great service involved in bringing out clearly this point, following on the later researches of Helmholtz, Berthelot, and others, is due to Le Blanc,[1] who investigated electrolytic decomposition in a very thorough manner. Le Blanc showed further that the conceptions and formulæ given by the author for galvanic production of current can be applied without change to electrolysis, and so laid the foundation for the osmotic theory of electrolysis. This experimentalist has also made many important applications of his law.

Thus it became possible,[2] by applying different potentials, to arrive at a method for separating electrolytically the various metals; it is clearly not the current density which primarily determines the electrolytic process, but the *potential difference between the electrodes.*

In the following table Wilsmore[3] has determined the potential differences of the most important electrodes from a critical summary of the measurements so far obtained:—

DECOMPOSITION POTENTIALS

For Normal Concentrations

ϵ_1 (Cations)		ϵ_2 (Anions)
Mg + 1·482		J − 0·520
Al + 1·276	H = ± 0	Br − 0·993
Mn + 1·075		O − 1·08*
Zn + 0·770		Cl − 1·353
Cd + 0·420		OH − 1·68*
Fe + 0·344		SO_4 − 1·9
Co + 0·232		HSO_4 − 2·6
Ni + 0·228		
Pb + 0·151		
Cu − 0·329		
Hg − 0·753		
Ag − 0·771		

[1] *Zeitschr. physik. Chem.*, **8.** 299 (1891); **12.** 333 (1892).

[2] This was first shown by M. Kiliani (1883); the further working out of it is due to H. Freudenberg, *Zeitschr. physik. Chem.*, **12.** 97 (1893), on the suggestion of Le Blanc.

[3] *Zeitschr. physik. Chem.*, **35.** 291 (1900).

These numbers, in accordance with the preceding formulæ, refer to normal concentrations of the ions; a reduction of tenfold in the concentration increases the numbers by $\frac{0\cdot058}{n}$ volt (where n = the number of charges or the chemical valency of the ion). The potential of hydrogen is taken as zero; for there is always both anode and cathode, and therefore an arbitrary term to be added to all the above numbers, that is, it is necessary to make an arbitrary standard of some one of them. The value of O and OH (distinguished by an*) refer to a *concentration which is normal with respect to the hydrogen ions.* In order to discharge O or OH from normal concentration of OH the voltage needed is less by 0·8, in order to set free hydrogen from the same solution 0·8 volt more is required than in acid solution, as may be calculated from the concentration of the ions of water.

A number of important conclusions may be drawn from the above numbers. In the first place, they yield the decomposition potentials of all ionic combinations. For example, zinc bromide requires for decomposition 0·99 + 0·77 = 1·76 volt, if its ions are present in normal solution. The decomposition of hydrochloric acid requires 1·353 + 0 = 1·353 volt, and so on. We saw that it is easy to separate silver from copper electrolytically, because the difference between their solution pressures is almost half a volt; but similarly the electrolytic separation of iodine from bromine and bromine from chlorine can be carried out satisfactorily.[1] The electrolytic decomposition of silver iodide in normal solution would, according to the above figures, not only require no force but should, on the contrary, yield 0·26 of a volt (0·52 − 0·78 = − 0·26). Silver iodide itself, however, cannot be obtained in such concentrations on account of its extreme insolubility in water, and indeed we may conclude from the above numbers that at usual temperatures silver iodide to be *stable* must be very insoluble, a conclusion which of course can very easily be generalised.

Naturally also the above numbers give at the same time the electromotive force of cells combined with the electrodes in question; thus the Daniell cell gives 0·770 + 0·329 = 1·099 volt.

An important step in this direction has been taken by G. Bodländer.[2] The decomposition potential of a saturated solution of an electrolyte is

$$E = \epsilon_1 + \epsilon_2 + \left(\frac{1}{n_1} + \frac{1}{n_2}\right) \ln c_0,$$

where n_1 and n_2 are the valencies of the anions and cations and c_0 the concentration of the saturated solution. Now E is the free energy of formation of the solid salt which, according to Chap. V. of this book,

[1] See Specketer, *Zeitschr. f. Elektrochem.*, **4.** 539 (1898).
[2] *Zeitschr. physik. Chem.*, **27.** 55 (1898).

is mostly not very different from the heat of formation. If then we put, according to p. 705,

$$E = \frac{q}{23,110} = \epsilon_1 + \epsilon_2 + \left(\frac{1}{n_1} + \frac{1}{n_2}\right) \ln c_0,$$

the value of c_0 can be calculated from the decomposition potentials in the above table and from thermo-chemical data. Bodländer found, in fact, in many cases an agreement, at least as to order of magnitude with the solubility of salts, a very noticeable result, *for it is the first means that has been found to calculate solubilities even approximately.*

As a further example which gives rise to interesting speculations we may take the electrolysis of sulphuric acid. It is well known that in this case we get hydrogen at the cathode and oxygen at the anode, assuming of course that the electrodes are unalterable. We concluded from the above table that electrolysis will only take place if the potential difference is greater than 1·08 volt; with this potential difference hydrogen ions can be discharged at the cathode and doubly charged oxygen ions at the anode; it is actually found that if a small platinum point be used as cathode and a large platinised plate as anode with 1·1 volt, hydrogen is evolved violently at the point, and the electrolysis can be kept up indefinitely by means of this electromotive force. We have here the reverse of the gas battery on p. 711. But if a large platinum plate charged with hydrogen is used as cathode and opposite it a small platinum point as anode, oxygen bubbles do not form until the potential has reached 1·66 volt, that is when the decomposition potential of hydroxyl ions is exceeded and the electrolysis only takes place freely with still higher electromotive force, by means of which the SO_4 ions are discharged. The doubly charged oxygen atoms are present in such small quantity that it is not possible to keep up any considerable electrolysis by their means, whereas hydrogen ions are abundant and consequently can be discharged in large quantities as soon as the potential rises sufficiently high. The circumstances are more favourable in the case of the hydroxyl ions, whose concentration is much greater than that of the doubly charged oxygen ions, but in order to electrolyse freely the potential must be raised till both at the cathode and anode ions present in large quantities can be discharged.

In electrolysing *lead salts* it was observed that at the cathode lead is precipitated, at the anode lead peroxide. According to the views of Liebenow [1] this is a case of primary electrolysis, since he supposes lead salt to be hydrolytically dissociated according to the equation

$$\overset{++}{Pb} + 2H_2O = \overset{--}{PbO_2} + 4\overset{+}{H}.$$

As a matter of fact compounds like PbO_2Na_2, PbO_2Ca, and so on, are

[1] *Zeitschr. f. Elektrochem.*, **2.** 120 (1895-96).

known. If a solution of lead salt is electrolysed in presence of a copper salt, copper comes down on the cathode instead of lead, because copper has a much smaller decomposition potential; all lead finally appears at the anode in the form of peroxide (Luckow's process for the electrolytic determination of lead). If, on the other hand, oxalic acid is added to lead salt, its negative ions are more easily discharged at the anode than the lead peroxide ions, so that all the lead appears in metalic form on the cathode.

In the *accumulator* on charge, according to Liebenow's view, lead ions are deposited primarily on the cathode, $\overset{--}{PbO_2}$ ions on the anode. Although on account of the small solubility of lead sulphate very little of these ionic species are present in solution, they are continuously replaced by means of the lead sulphate on the electrode. It is only when the lead sulphate is used up that the sulphuric acid is electrolysed, and hydrogen appears at the cathode and oxygen at the anode.

A very noticeable phenomenon was investigated by Caspari,[1] who found that visible electrolytic separation of hydrogen at a platinised plate takes place at a potential of practically zero (0·005 volt), but with other metals requires a special excess-potential. He thus found the values

Au 0·02, Pt (blank) 0·09, Ag 0·15, Cu 0·23, Pd 0·48, Sn 8·53, Pb 0·46, Zn 0·70, Hg 0·78 volts.

In consequence of this excess-potential, hydrogen does not appear on the cathode of an accumulator during charge, but lead is separated, and it is only when the lead sulphate is reduced and the potential consequently raised that gaseous hydrogen is evolved. It can therefore be shown[2] that with small currents lead sulphate is much more easily reduced at a lead electrode than at a platinum electrode, and also that by means of electrodes with high excess-potentials (especially mercury and zinc) it is possible to obtain reduction products that would otherwise be very difficult to prepare.[3] Apparently hydrogen can only give off bubbles when the electrodes have occluded an appreciable amount; in metals which occlude very little hydrogen, considerable quantities of gas must first be generated by the potential before any formation of bubbles can take place. A similar phenomenon may be expected in the case of other gases which can be produced electrolytically; oxygen has been investigated by Coehn and Osaka[4] in this respect. Nickel behaves in a remarkable manner in this respect, since oxygen can be evolved from alkaline solution in gaseous form at about 1·3 volt against hydrogen, whilst on platinum 1·7 is required.

[1] *Zeitschr. physik. Chem.*, **30.** 89 (1899).
[2] Nernst and Dolezalek, *Zeitschr. f. Elektrochemie*, **6.** 549 (1899-1900).
[3] See Tafel, *Ber. deutsch. chem. Ges.*, **33.** 2209 (1900).
[4] *Zeitschr. anoragan. Chem.*, **34.** 68 (1903).

Chemical Application of the Osmotic Theory.—It has long been customary to draw chemical deductions from the galvanic potential series of the metals, which is given in the table (p. 733), and which is now seen to find quantitative expression in the solution pressures or the decomposition pressure so determined; it must not, however, be assumed that copper is always precipitated by zinc. On the contrary, the ionic concentration is a second important factor, and we have indeed already mentioned (p. 730) the experiment with the Daniell cell in which, under suitable conditions, zinc is reduced by copper.

Similar conclusions may be drawn as to the decomposition potential of the anions; thus it is known that bromine throws out iodine from solution of iodide, and the chlorine similarly precipitates bromine quickly and very completely (how complete the precipitation is may be easily calculated from the solution pressure according to principles already given). We have, in fact, the well-known simple reactions—

$$Br_2 + 2\overline{J} = J_2 + 2\overline{Br};$$
$$Cl_2 + 2\overline{Br} = Br_2 + 2\overline{Cl}.$$

We saw, further, that chlorine must be capable of evolving oxygen from acid solution, but not so bromine or iodine. It is known, however, that the evolution of oxygen by chlorine takes place very slowly, unlike the rapid deposition of bromine by chlorine. This need not surprise us in view of what has gone before; for the chlorine, in order to pass to the ionic state, must replace the $\overline{\overline{O}}$ ions that are present in excessively small quantities, since the $\overline{OH}$ ions, which are more abundant, do not give up their negative charge except by means of a potential 0·3 volt higher than that of chlorine. So far, it has not been possible to obtain oxygen from it.

One of the most interesting reactions is the *decomposition of water by metals with formation of hydrogen;* the conditions for this process can easily be deduced from the foregoing considerations (p. 732). The electrical forces in question act not only on the ions of the metals in question, but also on all the positive ions present; for example, on the hydrogen ions which always exist in aqueous solution. The separation of hydrogen ions must occur as soon as *the osmotic pressure of the hydrogen ions and the electrostatic attraction are sufficient to overcome the electrolytic solution pressure of hydrogen at atmospheric pressure*, that is, we must have $\epsilon_1 > \epsilon_2$ or $\sqrt[n_1]{\frac{C_1}{c_1}} > \frac{C_2}{c_2}$, where the index 1 refers to the metal and 2 to the hydrogen, and n_1 is the chemical valency of the metal in question.

We see, therefore, that the favourable conditions for decomposition of water are:—

1. Large osmotic pressure of the hydrogen ions.

2. Large electrostatic force, that is great solution pressure of the metal and small opposition pressure of the ions of this metal.

Potassium decomposes water violently under all circumstances on account of its enormously great electrolytic solution pressure, for we can neither make the osmotic pressure of the hydrogen ions small enough, nor that of the potassium ions large enough to hinder the solution. *Zinc* possesses large enough solution pressure to decompose water in acid solution, but it is incapable of doing so when the concentration of the zinc ions is sufficiently great, or that of the hydrogen ions sufficiently small; for example, when zinc is dipped into a neutral solution of zinc sulphate. In strongly alkaline solutions it is again capable of decomposing water rapidly, although the concentration of the hydrogen ions is exceedingly small, because in this case, through formation of zincates, the concentration of the zinc ions is very greatly reduced. *Mercury*, in spite of its small solution pressure, evolves hydrogen from strong hydrochloric acid because the coneentration of the hydrogen ions is large, and that of the mercury ions very small, owing to the insolubility of mercurous chloride and the presence of considerable quantities of chlor ions. *Copper*, whose ions, as we have already seen, are very strongly dissolved by platinum cyanide, gives off hydrogen violently in such solution, despite its alkalinity.

The preceding considerations apply, of course, only to reversible formation of hydrogen, which, however, we can always produce when the metal in question is surrounded with platinum wire (best platinised); otherwise formation of hydrogen only takes place when

$$\epsilon_1 > \epsilon_2 + \eta,$$

where η is the gas potential given on p. 736. The sponge-like electrode of an accumulator in moderately concentrated sulphuric acid gives off no hydrogen, although here $\epsilon_1 > \epsilon_2$, it does so as soon as it is touched with a platinum wire. In very concentrated acid it boils spontaneously, because the above inequality is satisfied.

These considerations for the formation of hydrogen by metals can be at once applied to the *electrolytic separation* of metals. If $\sqrt[n_1]{\frac{C_1}{c_1}} < \frac{C_2}{c_2}$, the metal will be deposited, if $\sqrt[n_1]{\frac{C_1}{c_1}} > \frac{C_2}{c_2}$, hydrogen is more easily evolved electrolytically. Hence for galvanic separation of metals it is necessary to make first the concentration of the metallic ions as large as possible, second, that of the hydrogen ions as small as possible. Now in aqueous solution the concentration of the hydrogen ions is inversely proportional to that of the hydroxyl ions; hence the product of the concentration of the metallic and hydroxyl ions must be lessened as much as possible. But, according to the laws of solubility, a limit is set to it by the solubility of the hydroxide of the metal; hence the impossibility of separating aluminium or magnesium from aqueous solution is due, not only to the great solution pressures of those metals, but to the insolubility of their *hydroxides*.[1]

At the electrodes the elements or radicals appear deprived of their electric charges, and their affinities, which in the ionic state are saturated by the electric charges, and in neutral dissolved molecules by those of the other constituent, are now unsatisfied. This leads most frequently to the combination of two similar ions: 2Cl gives Cl_2, 2H gives H_2, $2HSO_4$ gives per sulphuric acid $H_2S_2O_8$ (according to F. W. Küster's views), a case which has recently been investigated very thoroughly by Elbs, $2KCO_3$ gives, according to Hansen and Constam,[2] potassium per carbonate $K_2C_2O_6$. Many other so-called secondary reactions, that is, chemical actions between ions deprived of their electrical charges, are known. The application of electrolysis to oxidation, reduction, chlorination, and so on, belongs here; the methods for this have recently been carried out with many important results by Gattermann, Elbs, Loeb, and others.[3]

The pressure under which the ion is given off in the gaseous form, or the concentration in which it dissolves, depend essentially on the potential with which it is electrolysed; in other words, *the active mass can be varied arbitrarily by applying varying electromotive force of polarisation.* We can thus cause chlorine to appear at the anode in a state of more than homœopathic dilution, or under pressures which are to be reckoned by millions of atmospheres, and employ it *chemically*. It cannot be doubted, therefore, that all possible stages of chlorination may be reached; for example, in organic preparations, by changes of potential. It is true that the current density, to which attention was formerly paid, changes with the potential, and is closely related to that quantity, but it depends also upon the form of the electrodes and the specific resistance of the electrolyte, and can therefore not be regarded as a measure of the capacity of a current to chlorinate, oxidise, reduce, etc.

The principles given above are found in important application in the work of Haber;[4] he showed that the reducing action of electrolytic hydrogen on nitro-benzene depends solely on the potential at the cathode. Thus he was able to stop the reduction at the stage of azo-oxy-benzene by keeping the cathode potential constantly under a certain fixed value. Another application is to be found in a paper by Dony-Hénault,[5] who oxidised quantitatively alcohol to aldehyde, that is, according to Faraday's law, by keeping the anode potential during

[1] See also Glaser, *Zeitschr. f. Elektrochemie*, **4.** 355 (1898).

[2] *Zeitschr. f. Elektrochemie*, **3.** 137, 445 (1897).

[3] See also especially the collection of Loeb, *Zeitschr. f. Elektrochemie*, **2.** 293.

[4] *Zeitschr. f. Elektrochemie*, **4.** 506 (1898).

[5] *Ibid.*, **6.** 533 (1900).

electrolysis under a certain critical value; otherwise acetic acid and higher products of oxidation of alcohol appear.

Theory of Galvanic Polarisation.—A quantity of electricity passed through a voltameter gives changes of concentration in all cases; the theory of galvanic polarisation is therefore simply that of concentration cells.[1]

The changes of concentration producing electromotive force can consist either in a change in the concentration of the ions of the electrode metal or in a change in the substance occluded by the electrodes. We have an instance of the former in the electrolysis of sulphuric acid with mercury electrodes. If a quantity of electricity is passed through this voltameter the concentration of the mercuro-ions at the cathode is reduced, at the anode increased, hence we have a potential difference amounting to (p. 730)

$$E = \frac{RT}{n} \ln \frac{c_1}{c_2}.$$

Since $\frac{c_1}{c_2}$ can be made extremely large, considerable back electromotive force may be produced in such cases, and for a given quantity of electricity the change of concentration produced will be greater the smaller the concentration of the ions of the electrode metal in the first place. We have an instance of the second case in the electrolysis of sulphuric acid between platinum plates; since these are charged with oxygen through the air they may be regarded as reversible with respect to oxygen; and since the concentration of the oxygen ions in the electrode is not appreciably altered by small quantities of electricity, the most important concentration change is that in the occluded oxygen. The back electromotive force is therefore

$$E = \frac{RT}{4} \ln \frac{c_1}{c_2},$$

where c_1 and c_2 are the concentrations (active masses) of oxygen in the electrodes; the factor $\frac{1}{4}$ depends upon the fact that the oxygen molecule is electrochemically quadri-valent.

Further details on this subject belong to the region of pure physics.

For further literature on the subject of the osmotic electrochemical theory see the *Jahrbuch der Elektrochemie*, edited by Danneel, published by Knapp, from 1895 onwards.

General Theory of Contact Electricity.[2]—The general prin-

[1] See also Warburg, *Wied. Ann.*, **38.** 321 (1889).

[2] See my report, "Ueber Beruhrungselektrizitat," supplement to *Wied. Ann.*, 1896, Heft 8, where a collection of literature is to be found.

ciple by means of which, as we have shown in this chapter, we have calculated the potential differences between substances may be formulated as follows. We attribute to the ions the same properties as to electrically neutral molecules ; if we now consider any phenomenon which involves a change of place of molecules (*molecular phenomenon*), then the same process applied to free ions will usually have the consequence of separating anions and cations ; this causes a potential difference. We may of course calculate the latter if we know the laws of the molecular phenomenon in question. On account of the enormous electrostatic capacity of the ions the quantities that are actually separated are too small to weigh.

Thus, for example, the theory of non-electrolytes (molecular phenomena) leads to the theory of potential difference between dilute solutions, since the general laws of diffusion are applicable to the diffusion of electrolytes (ionic phenomena). The comparison of the solubility of ordinary substances with the solubility of metals leads to the much used formula for the potential difference between metal and electrolyte.

A further example may be briefly noted. If a solution has different temperature in different parts the dissolved substance will travel along the temperature gradient, a phenomenon observed by Soret.[1] If this molecular phenomenon be applied to the solution of electrolytes, and we suppose that Soret's formula is different for the different ions as it is for the different species of molecules, we arive at the result that there must be potential differences in a solution of varying temperature.

It may easily be shown, as on p. 367, that in order that equal masses of positive and negative ions should move along the temperature gradient the equation

$$U\left(\frac{dp}{dx} + c\frac{dP}{dx} + ck'\frac{dT}{dx}\right) = V\left(\frac{dp}{dx} - c\frac{dP}{dx} + ck''\frac{dT}{dx}\right)$$

must be satisfied, where T is the variable temperature and k' and k'' the forces which, apart from osmotic action, drive the ions along the temperature gradient. The theory which van't Hoff[2] gave for Soret's phenomenon puts $k' = k'' = 0$, which however is not always true. The above equation can of course be utilised for the case of any number of dissolved substances. (See the thorough investigation of electrolytic thermo-elements, by W. Duane.[3])

If we consider further the division of a substance between two solvents (p. 485), and attribute, as we may according to all analogies, to each of the ions a *specific coefficient of distribution between two solvents*,

[1] This was observed before Soret (1881) by Ludwig, as early as 1856.

[2] *Zeitschr. physik. Chem.*, **1.** 487 (1887).

[3] *Wied. Ann.*, **65**, 374 (1898).

it follows that the ions will not be divided between the phases in electrically equivalent quantities, if no other force is added to that arising from the existence of coefficients of distribution. But ions in the interior of a homogeneous phase must be present in electrically equivalent quantities; if this is to be so some other force must exist, and it is easily seen that it is again of an electrostatic character, that is, there must be in general a potential difference between two homogeneous phases. Such phenomena must be found in the *occlusion of gases* by electrodes, and in the *precipitation of one metal on another*, and may help to explain the phenomena of polarisation as well as the behaviour of inconstant cells.

The above theory of contact electricity may be applied without change to non-conductors (for example, to the frictional electricity between glass and silk and the like), since, according to our present knowledge, these substances are merely bad electrolytes; there is, however, the difficulty that we know nothing of the ions in such cases. From the observations so far given, however, Coehn [1] has arrived at an important conclusion, *that substances of high dielectric constant become positive by contact with substances of lower dielectric constant.*

It is only a short step to apply the same treatment to the *electrons* which we assume on p. 407 in the explanation of metallic conduction. If the metals are regarded as solvents containing positive and negative electrons in varying concentration, then a consideration of the changes of position which they undergo in various circumstances would, like the ionic theories already developed, lead to a theory of potential difference in different metals or metals of differing temperature.

[1] *Wied. Ann.*, **64.** 217 (1898).

CHAPTER IX

PHOTO-CHEMISTRY

The Action of Light.—When the ether vibrations pass by a material system, they occasion two results which are essentially different. Thus, on the one hand, they raise the temperature of the system, their energy being partially converted into heat; and, on the other hand, they occasion chemical changes, which of course occur at the expense of a certain amount of the energy of the vibration.

We have already considered the first class of phenomena (p. 336), under the subject of the *absorption of light.*

The description of the second class of phenomena, which may be called *the photo-chemical absorption of light*, will form the subject of this chapter.

The ordinary absorption of light is a very general phenomenon. Every substance, in a way which varies largely with its own particular nature, and with the wave-length of light, can change the energy of the ether vibrations partially into heat; and this can be done completely if the layer permeated has a sufficient thickness.

But the *chemical action of light*, so-called, takes place only in exceptional cases, since it is rather rare that illumination is able to exert an influence on the reaction velocity of a system which is in process of change, or on the state of equilibrium of a system which is in chemical repose.

Of course it is not impossible that the photo-chemical action may be very general; and that it may usually be too slight to be noticed under the ordinary conditions of research.

The chemical action of sunlight, such, *e.g.*, as that shown in the bleaching process, in the production of the green colour of plants, in its destructive effect on certain colours of the artist,—all this has been known since antiquity. But only recent investigation has taught us that numerous compounds are sensitive to light, and has convinced us that here we are dealing with a mutual action between ether vibrations and chemical forces, which is deserving of the greatest interest.

It would take too much space to enter into the detailed enumeration and description of the particular phenomena belonging here, and therefore we will only refer to the very complete bibliography prepared by Eder.[1] But it should be emphasised that *gases* (as, *e.g.*, the explosive mixture of hydrogen and chlorine), and *liquids* (as, *e.g.*, chlorine water, which gives up oxygen, under the influence of light), and *solids* (as, *e.g.*, white phosphorus, which changes to the red modification in the light; or cinnabar, which turns black),—all these gases, liquids, and solids respond to the ether vibrations.

Also the photo-chemical process may consist either in the *production* of a compound, as is the case with "chlorine knall-gas"; or it may consist in the *decomposition* of a compound, as is seen in the decomposition of hydrogen phosphide, with the separation of phosphorus.

Although, on the one hand, this kind of light action exhibits the greatest variety in the photo-chemical processes, according to the nature of the system illuminated, — and in contrast to ordinary absorption which always develops heat,—yet like ordinary absorption is highly dependent upon the wave-length of the light used. Thus we know photo-chemical reactions which are, on the whole, occasioned by ultra-violet rays, or by the visible rays, or by the ultra-red rays of the spectrum, respectively; and in all cases the intensity of the photo-chemical action depends in the highest degree upon the wave-length of light, a fact which should be given proper attention in researches of this sort.

After a discussion of all the material up to date, Eder[2] comes to some general empirical laws, the essentials of which will be given in the following statements.

1. Light of every wave-length, from the infra-red to the ultra-violet, is capable of some sort of photo-chemical action.

2. Only those rays are effective which are absorbed by the system; so that the chemical action of the light is closely associated with the optical absorption; but conversely, optical absorption does not always necessitate chemical action.

3. According to the nature of the substance absorbing the light, every kind of light may act in an oxidising or in a reducing way; indeed it may be said, in general, that *red* light has usually an *oxidising* effect, and *violet* light a *reducing* effect, on the metals. The case where red light may also exert a reducing effect, occurs in the *latent light action*[3] of silver salts. Thus far no oxidising action of violet rays on metallic compounds has been observed with certainty.

Violet and blue light usually act most strongly on the compounds

[1] *Handbuch der Photographie*, Halle, 1884, p. 10; Fehling's *Handwörterbuch*, under *Chemische Wirkungen des Lichtes*. See further M. Roloff, *Zeitschr. physik. Chem.*, **26**. 337 (1898).

[2] *l.c.* p. 28; and *Beibl. zu Wied. Ann.*, **4**. 472 (1880).

[3] For the various applications of the sensitiveness of silver salts, see especially the text-books of photography, by Eder (Halle, 1884 and 1885), and by H. W. Vogel (Berlin, 1878, Supp. 1883.

of the metalloids with each other: as, *e.g.*, on "chlorine knall-gas," on nitrous acid, on sulphurous acid, and on hydriodic acid [1]; yet hydrogen sulphide solution is decomposed more quickly by red light.

The light action is partly oxidising and partly reducing,[2] according to the nature of the substance. In most cases, violet light exerts the strongest oxidising action on organic compounds, especially the colourless ones. Colourless substances are oxidised most strongly by those light rays which they absorb.

4. Not only does the absorption of the light rays, by the illuminated substance itself, play an important part; but also the absorption of light by *foreign* substances mixed with the principal substance, is important; for the sensitiveness of the main substance can be stimulated for those rays which are absorbed by the admixed substance (*optical sensitisation*).

5. A substance sensitive to light, admixed with the main substance, and which unites with one of the products resulting from the photo-chemical reaction (as oxygen, bromine, and iodine), tends to accelerate the reaction velocity to such an extent that a reversal is impossible. This may be also regarded as a consequence from the law of mass-action (chemical sensibilisation).

In many cases, as Roloff observed (*Z. S. phys. Chem.*, **13.** 327, 1894), the action of light consists in the transport of ionic charges: see the paper of the same author quoted on p. 744.

Actinometry.—The action of the light upon a chemical system is the greater, the more intensive the ether vibrations underlying this influence. In the quantitative prosecution of any selected photo-chemical process, we possess a means for measuring the intensity of the chemically active rays.

All those pieces of apparatus which are designed to measure the photo-chemical intensity of light, and all of which, collectively, depend upon the observed changes which are experienced by substances which are sensitive to light, and when under the influence of the ether vibrations,—all such pieces of apparatus are called *actinometers*. Inasmuch as all the empirical laws of photo-chemistry thus far discovered, have been ascertained by the aid of actinometers, therefore the most important of these will be enumerated in the pages immediately following.

But first let us preface a general remark regarding the estimation to be put upon the figures of the actinometer for the intensity of the light action.

[1] Not all of the illustrations are good, as hydrogen is certainly a gaseous *metal*, both physically and chemically.—Tr.

[2] As oxidation and reduction are relative terms, *both* occur where either does; and when we say that a substance is oxidised or reduced, we only mean that we are ignoring the other substances present.—Tr.

The data obtained from all kinds of actinometers are to be considered, respectively, as having a purely individual nature.

Thus it may be shown in two ways that they serve to give only a *relative* measurement of the intensity of light.

For, on the one hand, even when using the same kind of light, the nature and reaction velocity of the chemical process occasioned in each particular case, will vary according to the varying behaviour of the system which is subjected to the action of light.

And, on the other hand, when light is used which consists of rays of very different wave-lengths, the data of the same actinometer will, by no means, be proportional to the intensity of light; because the action of different kinds of light varies greatly according to the wave-length.

And, moreover, *the eye* is an actinometer, having an individual nature; because apparently its sensitiveness to the ether vibrations depends upon certain photo-chemical processes which are occasioned thereby.

The results of the [visual] photometric measurement of light are not parallel with the results measured by means of the actinometer to be described below; and neither of these are parallel with the results of the *thermometric* measurement of the intensity of light, and the latter method is usually regarded as the absolute measure of radiation. It would perhaps be more correct to regard the diminution of free energy —which, to be sure, is unknown,—associated with the change of radiant energy into heat, as the measure of the intensity of light.

It has been proved that the results obtained by several actinometers are at least approximately proportional to each other. Although one is not justified in drawing even an approximate conclusion regarding the photo-chemical activity of two light sources which have been studied only in an optical-physiological [visual] way, yet, on the whole, the "chlorine knall-gas" actinometer and the "silver chloride" actinometer give results which correspond with each other.

The Chlorine "Knall-gas" Actinometer.—This depends upon a discovery of Gay-Lussac and Thénard in 1809, regarding the action of light on the combining of chlorine and hydrogen; in strong light this action advances with a velocity which results in an explosion, but in weak light it progresses gradually and steadily. This actinometer was first constructed by Draper in 1843, but it was brought by Bunsen and Roscoe[1] to an improved form suited for exact measurement. The method consists in measuring the diminution of a volume of "chlorine knall-gas" (standing over water and kept at constant pressure), as a result of the formation of hydrochloric acid which is absorbed by the water.

Inasmuch as the manipulation of this apparatus makes unusually

[1] *Pogg. Ann.*, **96.** 96 and 373; **100.** 43 and 481; **101.** 255; **108.** 193 (1855-1859).

large requirements on the patience and skill of the observer, therefore Bunsen and Roscoe later turned their attention to—

The Silver-Chloride Actinometer.[1]—In this the *time* required to darken photographic paper until a definite "normal" shade is reached, serves as a measure of the light intensity.

The Mercury-Oxalate Actinometer.—A solution of mercuric chloride and ammonium oxalate will remain unchanged an indefinitely long time in the dark; but in the light, CO_2 and mercurous chloride are developed in the sense of the equation,

$$2HgCl_2 + C_2O_4(NH_4)_2 = Hg_2Cl_2 + 2CO_2 + 2NH_4Cl.$$

Either the quantity of CO_2 set free, or else the amount of mercurous chloride precipitated, may serve as a measure of the intensity of light; the latter gives much more exact results.

According to Eder,[2] it is best to mix two litres of water containing 80 g. of ammonium oxalate, with 1 litre of water containing 50 g. of mercuric chloride; some of this is then poured into a beaker glass of about 100 c.c. capacity, which is "light tight" on all sides, but which has an opening in its cover.

Inasmuch as the concentration of the sensitive solution changes during the illumination, therefore the separation of the mercurous chloride takes place with an increasing slowness, and thus does not directly correspond to the light energy introduced; and therefore it is necessary to apply a correction, the amount of which can be taken from the table furnished by Eder.

Moreover, elevation of the temperature is favourable to the action of light, and this must be considered in quantitative research.

This apparatus is chiefly sensitive to the ultra-violet rays.

Instead of mercuric chloride and ammonium carbonate, in a similar way, we may make use of the oxalate of iron or uranium.

The Electro-Chemical Actinometer.—Let two silver electrodes which have been chlorinised or iodised, be dipped into dilute sulphuric acid; then, as observed by Becquerel in 1839, an electromotive force will be established between them, as long as *one* electrode is illuminated; *the current will flow in the solution from the unlighted to the lighted pole.*

The strength of a current, as read by means of a sensitive galvanometer, will serve to determine the intensity of light. The results obtained by this actinometer are approximately parallel with those obtained in photometric ways. The most useful form is that of Rigollet,[3] which consists simply of two copper plates, slightly oxidised in the Bunsen flame, and immersed in 1 per cent solution of an alkaline haloid, only one plate being exposed to the light.

[1] *Pogg. Ann.*, **117.** 529; **124.** 353; **132.** 404. Regarding the numerous modifications which this has experienced in order to adapt it to the needs of practical photography, see Eder, *Handbuch der Photographie*, **1.** 174 and *ff.*

[2] *Wiener Sitzungsber.*, **80.** 1879; *Handbuch*, **1.** 169.

[3] *Journ. de Phys* [3], **6.** 520 (1897).

Photo-Chemical Extinction.—Now the light which is chemically active, performs a certain amount of work; and therefore it is to be expected—other conditions being the same—that the light will be absorbed to a greater degree when it occasions or accelerates a chemical process, than when such is not the case.

In fact, Bunsen and Roscoe[1] found that light which had passed through a layer of chlorine knall-gas, was much more weakened in its chemical activity (as measured by the chlorine knall-gas photometer) than when it had only passed through a layer of pure chlorine of the same thickness, and thus had had no opportunity to form hydrochloric acid.

In both cases the light is weakened by absorption by the chlorine; the absorption occasioned by the hydrogen can be neglected. But in the first case the absorption is due purely to optical activity; and therefore the loss of energy of the light reappears in the heat which is developed. But in the second case, an additional fraction of the light-energy is consumed in performing chemical work, which thus occasions a stronger absorption.

This phenomenon was called by Bunsen and Roscoe, "*photo-chemical extinction*"; it is apparently a very common phenomenon, and it has surpassing interest for our conceptions regarding the mechanism of the chemical action of light.

Photo-Chemical Induction.—Another very remarkable fact, for the discovery of which we are likewise indebted to the classical investigations of Bunsen and Roscoe, is the so-called "*photo-chemical induction*"; by this is meant the phenomenon where light usually acts very slowly at first, and attains its full activity only after a lapse of some time.

Thus, with the illumination of a constant kerosene lamp, the quantities (S) of hydrochloric acid produced in single minutes (and as measured by the displacement of the water filament in the scale tube of the chlorine knall-gas actinometer), are given in the following table; the time t denotes the minutes in order.

t	S	t	S
1	0·0	7	14·6
2	1·6	8	29·2
3	0·5	9	31·1
4	0·0	10	30·4
5	0·5	11	32·4
6	2·1	...	...

As is obvious, the action [which at first is very slow] begins to be constant only after about nine minutes; after this the hydrochloric

[1] *Pogg. Ann.*, **101.** 254 (1857).

acid produced is proportional to the product of the time and the light intensity; and only then does the actinometer become suitable for measurements. But if one afterwards allows the actinometer to remain some time in the dark, then it requires a renewed illumination, although a shorter one, in order to bring it again to the condition where it gives [constant] results which are proportional to the product of the intensity of the light and the duration of the illumination. But if the apparatus stands as long as half an hour in the dark, then the influence of the preceding illumination vanishes completely.[1]

Pringsheim has recently succeeded in showing the probability of the view, that the phenomenon of induction is to be ascribed *to the formation of intermediate compounds.*[2] Thus the fact that chlorine knall-gas is very much more sensitive to light when *moist* than when *dry*; and also that it is extremely difficult, even by the action of intense light, to explode the mixture when it is very carefully dried,—these facts make it appear probable that the union of chlorine and hydrogen, under the influence of light, does not take place according to the equation,

$$H_2 + Cl_2 = 2HCl\ ;$$

but rather that the union of the reacting substances with water to form *intermediate compounds*, plays an important part in the process.

Although it is extremely difficult to obtain any exact information regarding the nature of these intermediate compounds (which, at all events, are only produced in the slightest quantities), yet by means of the mere assumption that such compounds are produced by the action of light, and that they take part in the reaction, it is possible to account for the phenomenon of induction.

Thus let us suppose, for illustration, that the chlorine, under the action of light, *first* acts upon the water vapour, according to the equation

$$\text{I.} \qquad H_2O + Cl_2 = Cl_2O + H_2\ ;$$

and then, in the second phase of the reaction, that the resulting compound of chlorine and oxygen in turn decomposes in the sense of the equation

$$\text{II.} \qquad 2H_2 + Cl_2O = H_2O + 2HCl.$$

In this way we can explain, on the one hand, the catalytic action of the water; and, on the other hand, the phenomenon of induction. Thus the production of hydrochloric acid at first should be small; and not until the hypothetical intermediate compound shall have been found in sufficient quantity, can the second phase of the reaction leading to the production of hydrochloric acid become prominent.

[1] Marked induction is found in this photo-chemical reaction, $Cl_2 + CO = COCl_2$ (Wildermann, *Z. S. phys. Chem.*, **42.** 257, 1903).

[2] *Wied. Ann.*, **32.** 384 (1887).

Moreover, the state of the chlorine knall-gas, after the action of the light has passed by the preliminary stage of induction, is characterised by the stationary concentration of the hypothetical intermediate substance; for at every moment it would be produced, according to equation I., with the same velocity at which it would be decomposed according to equation II.

Dixon[1] has put forward the very suggestive hypothesis that free oxygen atoms are the intermediate body, according to the equations

$$\text{I.} \qquad Cl_2 + H_2O = 2HCl + O;$$

$$\text{II.} \qquad O + H_2 = H_2O.$$

Moreover, there is the well-known fact, that a very slight preliminary exposure of photographic plates makes their sensitiveness much greater; and also, in harmony with this, that "under-exposed" plates can be strengthened by a slight subsequent exposure. It is highly probable that these facts must be regarded as analogous to the photo-chemical induction of chlorine knall-gas. Both of the facts just mentioned indicate that the photographic action of light is relatively slow in the first moments of exposure; and that it is some time before the state of maximum sensitiveness is reached.[2]

The precipitation of calomel in Eder's photometer also suffers an initial retardation; the cause of this photo-chemical induction is, however, of a secondary nature, as Eder showed; enough calomel must be formed for saturation before any is precipitated.

The Latent Light-Action of Silver Salts.[3]—The so-called "*latent light action*" of silver salts (p. 744) has both great theoretical and also great practical interest; it offers, in particular, much that is puzzling, but it has long found practical application in photography.

All of the photographic methods, from the "positives" of Daguerre to the "collodion emulsion" and the "dry-plate" methods (which latter find almost universal application at present)—all of these depend on this principle, viz., the light is not allowed to act upon the plate until a visible picture is formed, but the light action is interrupted long before this; the image is then "developed" or called forth, by suitable treatment of the plate in the dark.

The photographic action of the light does not consist in a marked material change of the exposed plate, but only to such an extent that the different parts of the plate, having met more or less light respectively, react with correspondingly greater or lesser velocity in the subsequent treatment.

[1] *Z. S. phys. Chem.*, **42.** 313 (1903).

[2] See the researches of Abney, Eder's *Jahrbuch*, 1895, pp. 123 and 149; Englisch, *Arch. wiss. Phot.*, **1.** 117 (1899).

[3] See, in connection with this section, the attractively written monograph, "Chem. Vorgange in der Photographie," by R. Luther, Halle (Knapp) 1899.

The action of the developer produces visible changes; a picture appears, and the chief art of the photographer consists in the proper choice of the time of "exposure," and in interrupting the process of "development" at the right moment.

In order to preserve the picture, the remaining substance, which is sensitive to light, and which has not been fixed by the developer, must be removed ("fixing"). For this purpose use is commonly made of some suitable solvent which removes the undecomposed silver salt.

A distinction must be made between *physical* and *chemical development.*

The *first* found application, *e.g.* in the daguerreotype; this physical development consists in the fact that mercury vapour is most quickly precipitated upon those portions of a silver plate which have been exposed to the light, the plate having been previously slightly iodised on the surface.

The image is formed in the same way in the collodion process, as silver is separated from a solution of a silver salt with addition of reducing agents, where illumination occurs. The chemical development is applied in the modern negative process, in which a gelatine film impregnated with silver bromide is illuminated and is then treated with reducing agents (aqueous solution of potassium ferro-oxalate, alkaline salts of amido- or polyphenols, hydroquinone, pyrogallol, p-amidophenol, etc.); the silver haloid in the plate is then reduced to silver most quickly at the illuminated spots. The distinction between the two methods is a purely external one, for in all cases the precipitation of the solid produced by developing (mercury, silver) takes place more rapidly in the places where the product of reaction of light exists, and about proportionally to the amount of this product.[1] Only the material for depositing silver is taken from the photographic layer in the "chemical" process, and is added from outside in the "physical."

Many views have been put forward as to the chemical nature of the latent action of light. There is no doubt that tho changes that silver haloids undergo on illumination is a reduction with formation of free halogen; but the nature of the reduction product is not in all cases known. Direct chemical investigation and isolation of the reduced substance has not been accomplished, and is not promising, on account of the minimal quantity. Further, the nature of the binding material—collodion, gelatine, albumen, etc.—in which the silver haloid is embedded, seems to affect the reduction. It has recently been shown by Luther,[2] for films of silver chloride and bromide without any binding material, that the products of reduction are the subchloride Ag_2Cl and subbromide Ag_2Br.

Under ordinary conditions of illumination of silver halide the photo-chemical reaction is indefinite, inasmuch as the halogen set free

[1] Abegg, *Arch. wiss. Phot.*, **1.** 109 (1899).
[2] *Z. S. phys. Chem.*, **30.** 628 (1899); and *Arch. wiss. Phot.*, **2.** 35, 59 (1900).

is at an arbitrary potential, on account of thickness, resistance to diffusion, moisture, and especially the chemical nature of the film. Luther showed first that sufficient exposure, with any intensity of illumination, produces a definite equilibrium, that can be reached from either side, involving, besides the solid phases of unchanged and reduced silver salt, a definite halogen potential, increasing with the illumination. This shows that the action of light on the silver haloids is reversible. Further, Luther showed that the *latent* developable image produced by short illumination and the *visible* image that takes longer to produce are permanent in solutions of a certain halogen potential, and are destroyed in one of a higher potential, so that they apparently consist of the same substance. Finally, this halogen potential of the reduction product agrees nearly with that which exists over silver subchloride and subbromide, so that these substances must be regarded as the latent and visible products of reaction of light when no binding material is used.

The following view has long been taken of the process of development. On the illuminated spots of the plates small particles of metallic silver are deposited by reduction, and with density increasing with the intensity of the light; but always in such small quantity that no visible change occurs in the substance of the plate. When the plate is put in the developer those invisible silver particles act as nuclei for precipitation of silver, just as a small crystal brings about crystallisation in a supersaturated solution. The denser the silver particles at any spot the denser will be the deposit of silver during development.

This theory requires slight modification, on the view that the latent image does not consist of metallic silver. It may be supposed that the silver sub-halide is easily reduced by the developer, and the silver nuclei thus produced serve for further deposit.

There is much uncertainty in the chemical theory of ordinary photographic preparations, which contain the silver haloid in various binding materials, although steps towards an explanation have been made by Luggin.[1]

Moreover, a further very remarkable discovery was made by H. W. Vogel (1878). According to this, photographic plates may be made much more sensitive by intermixture with slight traces of organic colouring substances, and the plates are usually especially sensitive for the kinds of light absorbed by the particular colouring substances (*optical sensitisation*). And thus, as desired, plates may be prepared sensitive to yellow or red light, etc.

Thus far, no theoretical explanation has been given for this phenomenon. A thorough investigation of this subject by E. Vogel[2] has led to the result that, of the eosin colours, erythrosin and di-iodo-

[1] *Zeits. phys. Chem.*, **14**. 385 (1894); **23**. 577 (1897).
[2] *Wied. Ann.*, **43**. 449 (1891).

fluorescein work best; and that, in general, those substances "*sensitise*" best which are themselves most sensitive to light. The sensitising action increases in a striking way with the diminution of *fluorescence*.

The Laws of Photo-Chemical Action.—Apparently there is no difference between the visible rays, and those which are chemically active, and probably also the rays of great wave-length (such as those produced by electrical agitation of the ether, Maxwell and Hertz), except the differences of wave-length; all of these must be regarded as occasioned by the transmission of disturbances developed in the luminiferous ether. And therefore there can be no doubt that the chemically active rays should be refracted, reflected, and polarised, like all other rays; that their intensity should diminish as the reciprocal of the square of the distance from the point of origin (*i.e.* origin of the light); and that if an absorbing substance should be placed in the path of these rays, their photographic action should be weakened according to the same laws as those for optical rays, etc., etc. And, as a fact, the test of these laws has given the anticipated results.[1]

Moreover, abundant research has led to the result, that the action in the illumination of a photographic system, is conditioned solely by the amount of the impinging light; and it is immaterial whether the *time*, in which the same number of similar vibrations are introduced to the system, is the same or not. This law is usually formulated thus.

When light of the same kind is used, the photo-chemical action depends solely upon the product of the intensity and the duration of exposure.

Thus Bunsen and Roscoe[2] proved in the very sharpest way that the time required for the development of the normal colour on their sensitive paper was proportional to the number of light waves which struck the paper in a second; by changing the cross-section of the aperture through which the sunlight entered, it varied precisely according to the way indicated. A corresponding proof was given by Goldberg[3] for the oxidation of quinine by chromic acid.

The Theory of Photo-Chemical Action.—Thus far, only vague surmises have been expressed regarding that *mechanism*, by dint of which the energy of the vibrations of the luminiferous ether is applied to the performance of chemical work. But, in consideration of the fact that, in the sense of the later views, the light vibrations are produced by electric agitations, therefore it is an immediate inference to suppose that in the chemical action of light we are dealing with phenomena which are not far removed from the formation and decomposition of chemical compounds under the influence of the galvanic

[1] See especially the investigations of Bunsen and Roscoe.
[2] *Pogg. Ann.*, **117.** 536 (1862).
[3] *Z. S. physik. Chem.*, **41.** 1 (1902).

current. But it is possible that the time to advance from this standpoint to hypotheses of a special nature, will not arrive until our views regarding the mechanics of the light process shall have become clearer themselves, and also shall have been worked out with more completeness.

But the *principles* which *the formal mathematical description of the course of the reaction* taking place under the influence of light, must follow,—these *principles* may be ascertained.

Let us consider any homogeneous system, whether liquid or gaseous, in which a reaction occurs according to the general scheme, p. 430, and where the corresponding reaction velocity is given by the expression

$$V = kc_1^{n_1}c_2^{n_2} \ldots - k'c_1'^{n_1'}c_2'^{n_2'} \ldots$$

Thus the action of light simply reduces to this, viz., *that the velocity coefficients* k *and* k′ *depend upon the intensity of light;* and it is an immediate inference to suppose, as some experimental facts indicate,[1] *that, for the same kind of light, the changes of these coefficients follow in proportion to the intensity of the light.*

This, as a matter of fact, is the mathematical description of the photo-chemical process. But it must be observed in applying this calculation, that the light intensity of the system varies from point to point, on account of the [local] optical and photo-chemical absorption of the light rays; and therefore k and k′ become functions of place.

But this requires, as a still further complication, that, as a result of the varying reaction velocities, there must also result differences of concentration in the system, which will be equalised by diffusion—a point which should be particularly observed in the theory of liquid actinometers.

The relations are very simple, if the light intensity of the system is regarded as constant; and also if the reaction advances almost to a completion, so that $k' = 0$. These conditions were fulfilled in Wittwer's[2] researches, who investigated *the velocity of the action of dissolved chlorine upon water, under the influence of light.* In this special case, the preceding general equation reduced simply to

$$V = -\frac{dc}{dt} = kc.$$

Here c denotes the concentration of the chlorine; and in fact it coincided very well with the values observed.[3]

According to the change experienced by the velocity coefficients of the two opposed reactions, the action will take place either *with* or *against* the sense of the chemical forces which occasion the reaction.

1 See Wildermann's measurements (p. 748).

2 *Pogg. Ann.*, **94.** 598 (1855).

3 See also Lemoine, *Compt. rend.*, **112.** 936, 992, 1124 (1891).

In the former case, the light will have a decisive tendency ; *i.e.* it will tend to accelerate the velocity with which the system strives to reach the state of equilibrium. In the second case, the state of equilibrium will be displaced, and therefore *a certain amount of work will be performed against the chemical forces.*

An important step has been taken by Luther (p. 751) in the experimental study of the latter case, which is also of great theoretical importance.[1]

The foregoing considerations do not, of course, exclude the supposition that the mechanism of light action may be something quite different to a spontaneous chemical process. M. Bodenstein[2] has given a good example of this. The decomposition of hydriodic acid in the light is a unimolecular reaction, whereas its spontaneous decomposition is bimolecular. In the first case the reaction is

$$HJ = H + J,$$

in the second

$$2HJ = H_2 + J_2.$$

Another important difference between photo-chemical and ordinary reactions is that the velocity of the former increases but little with rise of temperature, while that of the latter increases enormously.[3] We must therefore not regard the light action as a direct loosening of the atoms in the molecule, like that effected by heating; rather the primary effect must be some action on the luminiferous ether, and suggests that ionisation, as mentioned in Book III. Chap. IX., plays a part in photo-chemical processes.

That chemical equilibrium must be affected by illumination follows from the change in the thermodynamic potential of the components by illumination, as may be most clearly deduced from the electromagnetic theory of light; it was shown on p. 697 that electrification and magnetism alter the thermodynamic potential and the action of light waves is, according to the theory, that of rapidly alternating electric (or magnetic) fields. These actions are, however (on account of the smallness of the pressure of radiation), too minute to affect equilibrium, or accelerate reaction, sensibly. This seems to exclude the possibility of referring photo-chemical action to them, as Luther[4] attempted; other experiences also contradict this view. Clearly in photo-chemical reaction we have to do with phenomena of resonance, possibly similar to those of optical absorption.

The question regarding the amounts of the light which are effective in each particular case, in one sense or the other, these, for the most

[1] Work is, of course, always done against chemical forces when the light-action is reversed in the dark ; see examples by Marckwald (*Z. S. physik. Chem.*, **30.** 140, 1899).
[2] *Z. S. physik. Chem.*, **22.** 23 (1897).
[3] See the work of Goldberg, mentioned on p. 753.
[4] *Z. S. physik. Chem.*, **30.** 628 (1899).

part, still await their solution. The great importance of all this may perhaps best be made clear by the hint that perhaps the only processes by means of which the energy of the sun's rays can be stored up as utilisable work (the product of which constitutes the object of strife on the part of the animal world) are the photo-chemical processes.

And therefore, as emphasised by Boltzmann,[1] the strife is not for the component *substances*, for these component substances of all organisms, as air, water, and earth, are abundant; neither is the strife for *energy as such*, for this occurs in abundance, as the heat content of the matter of our environment; but the strife is for *the free energy available for the performance of work.* This is accumulated in the products of the plant world from the sunlight, just as electrical energy is stored in an accumulator.

The problem of the relations between heat, and electrical and chemical energy, has been treated in the two preceding chapters of this book; and now it may be ventured, as a final remark regarding *the capacity of radiant energy for transformation into chemical energy*, that this doubtless has great significance, and its accomplishment will mean a long step forward. And when this shall have once been accomplished by theoretical chemistry, then side by side with *the doctrine of the material changes of nature* (which has for the most part claimed the interest and research of chemists) will be recognised, in complete equality, *the doctrine of the transformations of energy.*

[1] *Der zweite Hauptsatz der mech. Wärmeth.* (the second law of thermodynamics). *Vortrag.*, Wien, bei Gerold, 1886, p. 21.

INDEX OF SUBJECTS

Abnormal reactions, 536
Absolute temperature, 41
  units, 4
  zero, 204
  zero, reaction at, 687
Absorbed gas, catalytic action of, 579
Absorption of gases, 480
  of light, 336
  law of, 153
Accumulator, lead, 715, 736
Acetic acid, conductivity of, 498
  freezing point, 71
  and sodium acetate, 505
  solutions in, 269
  solution in benzene, 482
Acid-alkali cell, 508
Acids and bases, strength of, 516
  effect of pressure on, 666
Actinometry, 745
Action and reaction, 667
Active mass of bases, 555
  of solvent, 460
Additive properties of salt solutions, 380
  property, 105, 342
Adiabatic expansion, 50
  process, 27
Adsorption, 129
Affinity, 129, 278, 423, 432, 689, 691, 713
  of acids and bases, 501, 519
After effects in solids, 461
Alcohol and acetic acid, 447
Alkaloids, relative strength of, 452
Allotropic transformation, 633
Allotropy, 273
Alloys, molecular weight in, 102
Ammonia compounds, dissociation, 469
  dissociation of, 364, 672
  saponification by, 556
Ammonium carbamate, 472
  chloride, 652
  cyanate, 572
  hydrosulphate dissociation, 471
  nitrite, decomposition of, 538
Amorphous body, 75
  state, 98
Analogous reactions, 287
Anthracene picrate, 478, 494
Aqueous solutions, dissociation, 352
Areometric method, 301
Arsine decomposition, 562
Associated liquids, 271, 324, 459
Atom, 32
Atomic heat, 175
  refraction, 309
  volume, 189, 302
  weight, 33, 173
  weight, calculated from gas density, 255
  weights, regularity of, 182, 189
Atomistic theory of electricity, 390
Autocatalysis, 567
Autoxidation, 679
Avogadro's law, 39, 174, 205, 247, 256
Axis, crystallographic, 81
  of symmetry, 80

Band spectrum, 194, 198
Baroscope gas, 255
Bases, saponifying power of, 552
Basicity of acids, 500
Batho-chromic groups, 338
Benzene, curve of fusion, 72
  heat of vaporisation, 59
  theory, 288
  vapour pressure of, 61, 74
Benzoic acid, in water and benzene, 486
Benzophenone, transition of, 97
Beryllium, atomic weight of, 193
Bi-molecular reactions, 550
Binary dissociation, 498, 657
Blast furnace, process of, 662
Boiling point, 65
  as corresponding temperature, 303
  curves, 325
  elevation of, 262
  regularities of, 321
  relations of, 63
  of solutions, 139
Boyle's law, 38

Cadmium cell, 701

Calcium carbide formation, 470
 carbonate, 652
 dissociation, 463, 465, 467, 610
Calomel electrode, 729
Caloric, 10, 595
Calorimeter, thermo-chemical, 596
Calorimetric bombs, 593
Cane sugar, inversion of, 543
Carbon compounds, 283
 dioxide, dissociation, 444, 661, 679
 dioxide, isotherms of, 218
 free energy of oxidation of, 694
 heat of combustion, 593
Carbonyl sulphide and water, 579
Catalysis, 566
Cathode rays, 392
Centre of symmetry, 80
Change of state, 598
Characteristic equation, 240
 (Van der Waals'), 212
Chemical energy, 689
 forces, 275
 kinetics, law of, 430
 resistance, 674
Chloral hydrate, 351
Chlorhydrines, velocity of reaction, 565
Chlorides, dissociation of divalent, 529
Chromophore, 340
Clark cell, 701, 720
Cleavage, 93
Coexistence of liquid and vapour, 227
Cold-hot tube, 678
Colligative properties, 343
Colloidal solutions of metals, 567
Colloids, 408.
Colour and dissociation, 347
 of salt solutions, 384
Combining weight, 172
Common ions, effect of, 504, 527
Complete equilibrium, thermodynamics of, 630
Complex ions, 531, 536
Compound, chemical, 29
Compressibility of gases, 54
 laws of, 55
 of liquids, 226
Concentration cell, 714, 726
 effect on potential, 734
Condensed system, 631
Conduction, electrolytic, 697
Conductivity of alloys, 405
 electric, 358
 maxima of, 660
 and temperature, 658
 thermal, 348
Configuration, 294
Conservation of energy, 8, 587
Constant boiling point of mixtures, 113
 heat summation, law of, 591
 and multiple proportions, law of, 30
Constitution of molecule, 274
Constitution and reaction velocity, 565
Constitutive properties, 342
Contact potential, 721, 740
Continuity of state, 67
Continuous spectrum, 194
Convergence temperature, 260
Copper sulphate, hydrates of, 469
Corresponding states, 224, 230
Critical coefficient, 326
 data, 231, 326
 density, 272
 phenomena, 66
 point, 221
 point of mixtures, 119, 484
Cryohydrate, 126, 490
Cryohydrates, use of, 259
Crystal form, 340
Crystalline indices, 77
 state, 74
 structure of alloys, 405
 systems, 83
Crystallisation, and molecular theory, 241
 separation by, 145
 velocity of, 99, 582
Crystallography, 75
Crystals, liquid, 97
 physical properties, 88
 properties of mixed, 122
Current, production of, 725
Cyclic process, 21

Daniell cell, 700, 705, 710, 725
Density of gases, 248
 of molecules, 419
 of salt solutions, 380
 of solids, 301
 of vapours, 248
Depolariser, 709
Dialysis, 414
Diamond and graphite, 634
Dielectric constant, 307, 313, 419, 459, 742
 constant and dissociation, 373
Diffusion, 156
 coefficient, 157
 of electrolytes, 365
 of gases, 103
 of gas ions, 397
 in jellies, 413
 of liquids, 244
 in solids, 169
 and viscosity, 210
Dilution, maximal work of, 131
Dispersion, atomic, 310
Displacement, vapour density by, 249, 253
Dissociating force of water, 268
 substance, vapour density of, 251
Dissociation, 277, 345
 constant, 438, 503
 degree of, 357
 by dissolving, 653

Dissociation, heat of, 651
of solids, 467
in solution, 350, 455, 477
in stages, 500
Distillation, fractional, 112
Distribution of a base between acids, 511
between two solvents, 267
of electrolyte between water and a second phase, 539
law of, 481, 485, 494, 741
Double bond, 289, 295, 302
decomposition, 533
decomposition of solid salts, 637
salts, 120, 537
salt, formation of, 636
Draper's law, 194
Dropping electrode, 724
Dualistic and unitary view, 279
Dyeing, theory of, 338, 492
Dystectic mixture, 127

Effusion, 346
Elastic forces, 69
molecules, 201
Elasticity, 92
Electro-affinity, 377
chemical equivalent, 699
chemistry, 696
Electrode, reversible, 708
Electrolysis, laws of, 356
theory of, 732
Electrolytes with a common ion, 503
Electrolytic conduction, 353
dissociation, 496
dissociation, character of, 365
gas, 616
Electromotive force, 705, 709, 713
force and free energy, 693
Electron, 391, 702
Electrostriction, 382, 407
Elements, 6, 33
Ellipsoid of dilation, 89
Enantiomorphic, 80
Endothermic cell, 711
reaction, 589
reactions at high temperature, 687
Energy of a gas, 42
indestructibility of, 6
internal, 7
Entropy, 27
Equations of motion, 12
Equilibrium, chemical, 425
and electromotive force, 717
condition of, 28
and dissociation constants, 455
Erg, 10
Esterification, 426, 447, 450, 494, 570, 585
in benzene, 456
Esters, catalysis of, 547
dissociation of, 450

Esters, reduced co-ordinates of, 235
Etched figures, 88
Ether and water, mutual solubility, 267
Ethyl acetate, heat of formation, 603
iodide, velocity of reaction of, 578
Ethylene, pressure and volume of, 213
Eutectic mixture, 126, 627
Evaporation, 216, 677
separation of dissolved substance by,
separation of solvent by, 136
Excess of dissociation products, 442
potential, 736
Expansion of a gas, 19
Explosions, 446, 675, 683

Faraday, 699
Ferments, inorganic, 567
Ferric chloride and water, 622
Ferrocyanide membrane, 132
Flame, 446
conductivity of, 378
Fluor-benzene and ether, 225
Fluorescence, 340, 387, 753
Free energy, 26, 691
Freezing mixture, 126
point, depression, 257
point of dilute solutions, 489
point of solutions, 142, 152, 170
Friction, 13
Frictional resistance to diffusion, 158
Fusible alloys, 124

Galvanic cells, 700
Gas battery, 711, 731
constant, 40
constant, exact value of, 215
Gaseous laws, deviation from, 210
Gases, 37
density of, 41
dissociation of, 655
equilibrium in, 434
evolution from solids, 470
expansion of, 38
heat of solution, 654
kinetic theory of, 201
and solids, reaction between, 473
specific heat of, 177
thermodynamics of, 46, 52
Gas ions, 394
ions, condensation by, 399
ions, number of molecules in, 399
mixtures, 100
pressure on kinetic theory, 203
Gay-Lussac's law, 38
Gelatination, 412
Geometrical isomerism, 295
Glass, 99
Gold, diffusion in lead, 169
leaf, thinness of, 416
Graphical method, 21

Hæmoglobin, 681
Heat capacity, 44
  of combustion, 317
  of dilution, 597
  of dissociation of acids, 605
  of dissociation of gases, 656
  of dissociation in solution, 658
  of evaporation, 326
  of explosion, 683
  of formation, 600
  of hydration, 652, 413
  of mixture, 118
  of neutralisation, 603
  of reaction, 587, 645, 650, 687, 704
  of reaction and thermal capacity, 594
  of solution, 598, 653
  of sublimation, 73
  of transition, 97
  production in galvanic cell, 710
Heating baths, 251
Hemihedral, 82
Hemimorphic, 82
Heterogeneity, 462
Heterogeneous equilibrium, 464
  system, kinetics, 579
Hexachlor-keto-R-pentene, 572
Holohedral, 82
Homogeneity, 75
Homogeneous equilibrium, 510
  system, 427
Hydrate, 460, 469
  theory, 525
Hydrated salts, melting of, 635
Hydrates of chromium chloride, 376
  solubility of, 477
Hydriodic acid, action of light on, 755
  acid, formation of, 276, 435
  acid, heat of formation, 593
Hydrocarbons, heat of combustion, 318
Hydrochloric acid, vapour pressure of, 165
  energy of formation, 706
Hydrocyanic acid, volatility, 540
Hydrogen, deviation from Boyle's law, 213
  electrode, 731
  first spectrum of, 195
  ions, acid property, 517
  ions, catalytic effect of, 544
  limiting density, 214
  permeability of iron for, 169
  production of, 724
Hydrolysis, 352, 508, 519, 524, 543
Hydroxides, insolubility of, 739
Hydroxyl, anomalous dispersion of, 315
  ions, basic, 518

Ice, effect of pressure, 71
Iceland spar, velocity of reaction, 581
Ideal concentrated solutions, 166
Ignition temperature, 676
Incomplete equilibrium, 466
  equilibrium, thermodynamics of, 642
Incomplete reactions, 568
Indestructibility of matter, 5
Index-ellipsoid, 91
Indicators, theory of, 521
Indifferent gas, influence of, 442
Indium, chlorides of, 281
Inertia of carbon compounds, 285, 672
Intermolecular action, 287
Internal friction (viscosity), 331
  pressure of liquids, 216, 223
  salt, 377
Inversion of cane sugars, velocity of, 671
Iodic acid and sulphur dioxide, 576
Ion, 355, 391, 698
  absolute charge of, 421
Ionic conductivity, 364
  friction, 371
Ionisation, 353
Ionising agents, 395
Ions, diffusion of, 720
  reactivity of, 496
Iridium vessel for vapour density, 254
Iron and water, reaction between, 473
Isochore of reaction, 643, 649
Isochores, linear, 237
Iso-dimorphism, 121, 123, 180
Iso-hydric solutions, 505
Isomeric ions, 387
Isomerism, 96, 273
Isomers, boiling point of, 323
  heat of formation, 602
  number of, 287
Isomorphism, 177
Isomorphous mixtures, 120, 171, 470, 490
  mixtures, thermodynamics of, 128
Isosmotic solutions, 133, 156
Isotherm of a condensing vapour, 218
  of dissociation, 439
  of reaction, 643, 645
Isothermal distillation, 115, 158
  process, 18, 23
Isothermals, projection of reduced, 226
Isotropy, 8, 90

Kilogrammeter, 10
Kinetic energy of molecules, 202
  equilibrium, 583
  theory, 200
Kinetics, chemical, 541

Lactone, formation of, 571
Latent heat, 23, 598, 630
  heat of evaporation, 238
  light action, 744
Law of distribution, Maxwell's, 209
Lead salts, electrolysis of, 736
Limiting curves, 614
Line spectra, 194
Liquid cells, 700, 706
  mixtures, 105
  mixtures, thermodynamics of, 114

Liquids, 56
  kinetic theory of, 215
  molecular weight of, 270
Litre-atmosphere, 12, 587
Luminescence, 194

Magnetic rotation of polarised light, 316
Magnetism, atomic, 317
Manocryometer, 71
Mass action, application of law of, 468, 471
  action, law of, 395, 423, 428, 432, 644, 647
  of electrons, 393
Maximal work, 24, 647, 692
  work, principle of, 685
Mean free path, 202, 208, 244
  square velocity, 209
Mechanical equivalent of heat, 7
Medium, influence on velocity of reaction, 577
Melting curves, detection of compounds by, 628
  point, 69, 328, 632
  point of alloys, 404
  point, effect of pressure on, 70
  point of elements, 190
  point of mixtures, 124
Mercury, conductivity of, 358
  salts, action of copper on, 475
  vapour pressure of, 64
  vapour, specific heats of, 207
Metal-ammonia compounds, 375
Metallic conduction, 401
Metals, decomposition of water by, 738
  diffusion of, 403
  solution of, 723
  and metalloids, 186
Metamerism, 275
Methyl ether hydrochloride, dissociation, 441
  iodide, 559
  orange, 522
Migration of ions, 354, 360
Milk sugar, rotation of, 572
Miscibility of crystals, 180
Mixed acids, dissociation of, 506
Mixture, 100
  of constant boiling point, 484
  kinetic theory of, 242
  refraction of, 109
  thermodynamics of, 243
  vapour pressure of, 272
Mobile equilibrium, principle of, 665
Mobility of ions, 361, 387
  of gas ions, 396
  of ions and periodic system, 387
Modulus of density for ions, 383
  of refractive power for ions, 384
Molecular attraction, 211
  compounds, 282
Molecular compounds, Werner's theory of, 374
  conductivity, 363
  fractional composition, 262
  heat, 42
  heat of solids, 176
  hypothesis, 31
  refraction, 306
  state of dissolved bodies, 155
  surface energy, 270
  volume, 300, 302
  volume of fluids, 222, 234
  weight, 41, 247
  weight of colloids, 409, 411
  weight determination, 258
  weight from osmotic pressure, 256
  weight in mercury solutions, 714
Molecules, absolute size of, 416, 420
  space occupied by, 418
  volume of, 211, 303, 306, 312
Multimolecular reactions, 560
Multiple point, 615, 619, 622
Mutual solubility of liquids, 111, 484

Neutral elements, 187
Neutralisation, 507, 518
  heat of, 590, 592, 603
Neutrality of carbon, 284
Neutral salts, effect on action of acids, 545
Neutron, 392
Nitril and isonitril, 290
Nitrogen dioxide, dissociation, 440, 656
Non-aqueous solutions, 372
Non-reversible processes, 673
Notation, thermo-chemical, 588
Nuclei of crystallisation, 242, 582
Number of reacting molecules, 564

Occluded oxygen, 740
Ogdohedral, 82
Ohm's law, 360, 674, 699
One-sided property, 109
Optical isomerism, 293, 334
Optically active crystals, 90
Optic axis, 91
Organic compounds, heat of combustion 601
Organisms, action on optic isomers, 335
Osmometer, 131
Osmotic pressure, 131
  pressure and chemical action, 458
  pressure at high concentration, 159
  pressure, kinetic explanation, 244
  pressure, laws of, 146
  pressure of colloids, 409
  pressure of mixture, 159
  pressure of strong solutions, 246
  theory, 720
  theory, application of, 737
  work, 162, 245
Oxides, composition of, 185

Oxygen, reactivity of, 679
  refractive power of, 310
Ozone, formed by silent discharge, 697

Palladium, possessability of, 103
Paranitrophenol, 522
Partial pressure, 58, 110
Periodic system, 182, 191
Permanganates, absorption spectra, 385
Perpetuum mobile, 15
Phase, influence of the extent of. 663
  rule, 609
Phases, chemical, 462
  in galvanic cell, 718
  law of equilibrium between, 662
Phenol and water, mutual solubility, 267
Phenolphthalein, 522
Phosphine decomposition, 562
Photo-chemical absorption, 743
  action and temperature, 755
  extinction, 748
  induction, 748
Photography, 750
Physical methods in analysis, use of, 452
Picric acid, in water and benzene, 540
Piezo-electricity, 93
Plane of symmetry, 80
Plasmolysis, 134
Platin-ammonia compounds, 376
Point system, 86
Poisons to colloidal platinum, 567
Polarisation, galvanic, 699, 705, 707, 713, 740
  of light, 90, 92
Polari-strobometer, 331
Polymerisation, 170, 275
  of solvent, 374
Polymorphism, 95
Positive and negative, 280
  and negative elements, 399
Potassium chloride, conductivity, 359
  chloride, osmotic pressure of, 164
Potential, 13
  anomalous, 730
  decomposition, 733
  thermodynamic, 668
Precipitates, carrying down of salts by, 493
  formation and solution of, 538
Precipitation of silver chloride, 530
Preparation of carbon compounds, 286
Pressure, effect on equilibrium, 436, 665
  effect of non-uniform, 667
Prout's hypothesis, 187
Pseudo-symmetric, 87
Pycnometer, 300
Pyro-electricity, 93

Racemates, formation of, 636
Racemic acid, 334
Radicals, atomic, 279
Radio-active substances, 393
Radium, 393
Ratio of specific heats, 49
Rationality, law of, 78
  of compound ratios, 78
  of indices, law of, 77
Reaction, constant, 542, 644, 648
  constant and co-efficient of distribution, 648
  practically irreversible, 542
  reversible, 426
  subsidiary, 575
Rectilinear diameter, law of, 232
Reduced characteristic equation, 223, 228
Reduction, electrolytic, 739
Refractive power of salt solutions, 383
  power, specific, 107
Refractometer, 305
Retrogression of acid dissociation, 504
Reversible cell, 707
  electrode, 729
  process, 17
Rings, atomic, 281
Rotation of polarised light, 331, 338, 386
Rotatory power of alkaloids, 453
Ruby gold glass, size of particles in, 417

Salammoniac, action of lime on, 474
Saponification of esters, 548, 550, 554
Saturated layer round solid, 581
Second law of thermodynamics, 205, 608, 643
Semi-permeable partition, 101
Sensitisation, 745, 752
Separation, electrolytic, 738
Silver bromate solubility, 528
  bromide precipitation, 690
  chloride, heat of precipitation, 607, 654
  chloride solubility, 728
  nitrate and mercury, equilibrium, 540
  nitrate concentration cell, 727
  salts, actions of light on, 750
Smelting furnace, process in, 445
Sodium, action at low temperatures, 692
  monochloracetate, 558
  sulphate, transition of, 635, 641
Solids, 69
  density of, 304
  equilibrium in, 461
  kinetic theory of, 240
  molecular weight, 272
  refractive index of, 306
  vapour pressure of, 72
Solid solutions, 168, 171, 272, 490
Solubility, 535
  calculation of, 735
  in compressed gas, 476
  influence of size of crystals on, 664
  lowering of, 266
  mutual influence of salts, 526
  product, 531
  selective, 133, 141, 145

Solubility and temperature, 652
at transition point, 639
Solution, 30
of metals in acid, 580
pressure, 475, 723, 737
Solutions, electromotive force between, 722
equilibrium in, 451
freezing of, 488
kinetic theory of, 243
Solvent, influence of, 244, 458, 476, 648
molecular constitution of, 268
nature of, 154
Sound, velocity of, 50
Space-grating, 87
Spatial arrangement of atoms, 287
Specific heat and dissociation, 347
heat of gases, 42, 47
heat of liquids, 236
heats, ratio of, 206
volume, 300
Spectra of elements, 193
Spectral series, 197
Sphere of action of molecular forces, 416
Sphero-crystals, 98
Standard cell, 719
Steam, specific volume of, 60
Stereochemistry, 291
of nitrogen, 296
Strong electrolytes, dissociation of, 499
electrolytes, law of mass action applied to, 532
Structural chemistry, 201
Sublimation 172, 467
and dissociation, 474
heat of, 651
Substitution, effect on boiling point, 322
Sulphur dioxide, catalytic oxidation, 566
dioxide, critical point, 68
.dioxide and water, 616
dissociation, 444
transition of, 97, 633
trioxide, dissociation, 679
Sulphuretted hydrogen, formation of, 549
Sulphuric acid solutions, 167
Supersaturation, 627
Surface energy, 88
tension, 57
tension and chemical equilibrium, 664
Symmetry, 75, 79

Tartaric acids, rotation of, 333
Tautomerism, 572
Temperature, effect on heat of reaction, 594
variation of specific heat, 207
Tetartohedral, 82
Thallous chloride solubility, 528
Thermal conductivity, 209
Thermodynamic classification, 35
functions, 22
Thermodynamics, application to reactions, 673
Thermodynamics, laws of, 16, 22
Thermo-electric effects in alloys, 406
Thermo-element, electrolytic, 743
Thermometer, differential, 259
Thermo-neutrality, law of, 603
Thomson's rule, 705
Three components, system of, 629
Tin, transition of, 635
Total energy of chemical system, 587
Transition point, 96, 612, 675
temperature, 632
temperature, determination of, 640
temperature, effect of pressure, 633
Translation, kinetic energy of, 204, 591
Tri-molecular reactions, 559
Triple point, 615
Turbid solutions due to hydrolysis, 534
Twin elements, 189
Twinning, 87

Ultra violet, methods of study of, 337
Undercooled water, 615
Undercooling, 74, 98, 127
Unilateral absorption, 337
Unimolecular reactions, 549
Unstable compounds in organic chemistry, 672
fluid state, 220
Unsymmetrical carbon compounds, 292, 332

Valency, 277, 280
Vaporisation, heat of, 58, 650
Vapour densities, abnormal, 344
density at reduced pressure, 254
density at very high temperature, 251
pressure, 58
pressure of concentrated solutions, 164
pressure curve, 61, 615
pressure of dilute sulphuric acid, 716
pressure, lowering, 261
pressure of mixtures, 110
pressure of solutions, 137, 479, 481, 487
pressure of solvent, 049
pressure at transition point, 638
Velocity of explosion, 681, 682
of reaction, 429, 542
of reaction and equilibrium, 584
of reaction and temperature, 435, 670, 674
Very dilute solutions, 259
Viscosity, 209
Volatile substances, boiling point of solutions of, 266
Volume, specific, 105

Water, conductivity of, 509, 660
electrolysis of, 735, 737
of crystallisation, 121
heat of formation, 588
heat of vaporisation of, 59
infra-red dispersion, 315
ionisation of, 507, 521, 558

Water, phases of, 613
saponification by, 557
specific heat of, 11, 236
vapour, catalytic action of, 567, 681, 749
vapour, decomposition of, 446
vapour, formation of, 646, 676
velocity of formation, 561

Weight, 41
Work done in chemical reaction, 588
electrical, 703

Zeeman phenomenon, theory of, 393
Zinc sulphide, solubility in acids, 531
Zone 76

INDEX OF AUTHORS

Abbot, 532
Abegg, 144, 246, 260, 261, 371, 377, 432, 751
Abney, 750
Abraham, 393
Adie, 133
Adriani, 628
Alexandrow, 410
Alexejew, 484
Amagat, 54, 55, 212, 226, 227
Ames, 196
Anderlini, 310
Andreæ, 469
Andrews, 54, 66, 67, 218, 219, 221, 596
Angeli, 538
Appleyard, 129
Arago, 331
Arons, 314
Arrhenius, 136, 354, 359, 363, 371, 378, 496, 504, 506, 508, 515, 517, 519, 521, 543, 545, 547, 551, 554, 658, 660, 671
Arzruni, 179
Askenasy, 158
Aston, 271, 374
Aubel, van (see *van Aubel*)
Austen, 169
Auwers, 257, 274
Avogadro, 30, 39, 174, 352

Babo, 147
Bader, 501
Baeyer, 296, 321, 329
Baker, 566, 682
Bakker, 238
Balmer, 195, 197
Bancroft, 629, 732
Barnes, 11
Barrell, 114
Barth, 394
Bartoli, 11
Batelli, 71, 152
Batschinski, 327
Baud, 271
Baumann, 680
Baynes, 213
Beckmann, 138, 152, 170, 258, 262, 263 265, 351, 458, 482, 595
Becquerel, 393, 400, 747
Bedson, 175
Behn, U., 371
Behrend, R., 478
Beilstein, 330
Bel, Le (see *Le Bel*)
Bellati, 169
Bemmelen, 129
Bender, 382, 384
Bérard, 43
Bergman, 423, 432
Bernoulli, Daniel, 201
Berthelot, D., 44, 152, 154, 255, 319, 347, 447, 449, 485, 569, 596, 653, 656, 682, 686, 687, 690, 712, 733
Berthollet, 157, 423, 432
Berzelius, 33, 274, 279, 374, 424, 566
Bethmann, 501
Bigelow, 566
Bijlert, van (see *van Bijlert*)
Biltz, H., 187, 251, 444, 476
Biot, 331
Blagden, 150, 489
Blanc, Le (see *Le Blanc*)
Bleier, 254
Bodenstein, M., 436, 549, 561, 677, 755
Bodlander, 81, 123, 171, 255, 377, 402, 404, 476, 536, 679, 734
Boeris, 538
Bogojawlensky, 583
Boguski, 580
Bohl, 416
Boisbaudran (see *Lecoq de B.*)
Bois-Reymond, du (see *du Bois-R.*)
Boltzmann, 102, 104, 201, 203, 205, 210, 239, 243, 244, 419, 431, 609, 656, 686, 756
Borchers, 702
Borelli, 423
Bornstein, 41
Bose, 371
Bötsch, 678

Boudouard, 446
Bourdon, 43
Boyle, 38, 52, 214
Braun, 705
Bravais, A., 81, 86
Bredig, 191, 262, 315, 387, 513, **712**, 718
Bremer, 641
Brigg, 544, 551
Brillouin, 227
Brodie, 680
Brown, F. D., 114
Browne, Sir T., 686
Brückner, 565
Brühl, 268, 309, 311, 313
Bruner, 152
Brunner, E., 581
Buchbocks, 579
Buchkremer, 108
Buchner, 296
Buckingham, 387
Buff, 679
Bugarszky, 711
Bunsen, 71, 194, 305, 347, 480, 595, **654**, 678, 746, 747, 748, 753
Bürckle, 565
Buys-Ballot, 341

Cagniard de la Tour, 66, 67
Cailletet, 54, 232, 270
Cain, 449
Callendar, 11
Cannizzaro, 344
Carnegie, 685
Carnot, 15, 16, 22
Carrara, 310, 374, 382, 580
Caspari, 736
Cauchy, 308, 315
Celsius, 42, 218, 620
Chaperon, 136
Chapin, 539
Chappuis, 129
Chatelier, Le (see *Le Chatelier*)
Chiminello, V., 559
Ciamician, 171
Claisen, 573
Clapeyron, 21
Clark, 718, 719
Clausius, 15, 16, 17, 22, 27, 60, 62, **65**, 140, 145, 201, 209, 217, 228, 238, 309, 316, 327, 379, 419, 420, 430, 680
Clement, 49
Coehn, 736, 742
Cohen, E., 547, 577, 635, 675, 712
Cohn, 254, 314
Collan, 567
Conrad, 565
Conrady, 309
Constam, 739
Cooke, J. P., 417
Coppet, 151
Cornu, 196
Cottle, 560
Couper, 283
Crafts, 65, 248, 253
Curie, 81, 88, 393
Czapski, 710

Daguerre, 750
Dale, 107, 307
Dalton, 30, 31, 32, 63, 153, 173
Dammer, 303
Daniel, 368
Daniell, 18, 730, 734
Danneel, 718, 740
Davy, 279
De Boisbaudran (see *Lecoq de B.*)
Debray, 467, 469
De la Rive, 580
Delaroche, 43
Demerliac, 71
Demuth, 254
Deslandres, 198
Desormes, 49
Deventer, van (see *van Deventer*)
Deville, 251, 345, 347, 444, 473, 678
De Visser, 71
De Vries, 134, 257
Dewar, 67
Dieterici, 11, 164, 167
Dixon, 566, 681, 682, 750
Dobbie, 338
Doebereiner, 183
Dolezalek, 119, 166, 715, 716, 717, 736
Donnan, 560, 565
Dony-Henault, 739
Draper, 746
Drude, P. 90, 314, 382, 417
Duane, W., 741
Du Bois-Reymond, 700
Dufet, 123
Duhem, 119
Duhring, 63
Dulong, 50, 175, 313
Dumas, 248, 251, 254, 441
Dutoit, 374

Earp, 324
Eder, 744, 747, 750
Elbs, 702, 739
Engler, 680
Enke, F., 289
Eötvös, 270
Esson, 433, 550
Evans, 565
Ewan, Th., 164, 385
Exner, 346
Eykman, 152, 254, 257, 310, 351

Fabinyi, 257, 700
Fanjung, 667

Faraday, 274, 316, 353, 356, 391, 416, 698, 719
Farkas, 700
Favre, 596
Fedorow, 86
Fehling, 322, 329, 744
Ferche, 74
Ferrantini, 171
Fick, 157
Fischer, W., 74
Fittig, 536
Fleury, 543
Fock, 123
Forster, F., 402
Fourier, 156
Fournier, 50
Frankenheim, 86
Frankland, 195, 283
Fresnel, 91
Freudenberg, H., 733
Freyer, 676
Friedel, 441
Friedlander, 582
Friedrich, 215, 222, 362
Frowein, 641, 652
Fuchs, G., 258, 263

Gadolin, A., 81
Galitzine, 416
Garelli, 171
Gattermann, 739
Gauss, 4, 78
Gautier, 175
Gay-Lussac, 38, 42, 174, 249, 746
Geissler, 393
Georgewics, 129
Gernez, 332, 582
Gernhardt, 265
Gibbs, W., 26, 433, 462, 483, 608, 610, 617, 658, 668
Gladstone, 107, 307, 383, 412, 521
Glaser, 739
Glauber, 635, 637
Gockell, 710
Goldberg, 753, 755
Goldschmidt, 274, 348, 477, 565
Gomberg, 281
Goodwin, 729
Gouy, 136
Graham, 156, 408, 411, 413, 415, 596
Grebe, G., 338
Griffiths, 12, 59
Groshans, 63
Groth, 340
Grotrian, 60
Grove, 99, 711
Gubser, 376
Guldberg, 143, 303, 395, 430, 432, 449, 453, 468, 473, 493, 496, 533, 548, 569, 644, 673
Gussler, 195
Guthrie, 126
Guye, 215, 222, 270, 271, 272, 326, 336

Haber, 681, 702, 739
Habermann, 254
Hahn, 446
Hall, 548
Hallock, 128
Hallwachs, 301, 381
Hamburger, 134
Hansen, 739
Hantzsch, A., 291, 297, 565, 573
Harcourt, 433, 550
Harden, 173
Hare, 686
Hartley, 338
Hartmann, W., 386
Hausrath, 261
Hautefeuille, 435, 662, 678
Hauy, 77, 79
Hedin, 134
Heilborn, 67, 326
Heintz, 124
Helm, 675
Helmholtz, H. von, 4, 9, 15, 26, 28, 115, 118, 315, 379, 391, 446, 566, 686, 705, 710, 711, 712, 713, 714, 733
Hemmelmayer, v., 418
Hemptinne, 553
Hendrixson, 486
Henrichsen, 317
Henry, L., 291, 322
Henry, P., 153, 559, 567, 571
Heraeus, 254
Hertz, 63, 390, 753
Hess, 592, 596
Hesse, 322
Hessel, J. F. C., 81
Heycock, 402, 404
Heydweiller, 509, 660
Hibbert, 412
Hirzel, 702
Hittorf, 360, 362, 302, 698, 731
Hjelt, 567
Hoff, Van't (see *van't Hoff*)
Hofmann, 249, 251, 254, 678, 679
Hohmann, 450, 564
Hoitsema, C., 272, 530
Holborn, 358
Hollemann, 257, 470, 574
Hoppe-Seyler, 680
Horstmann, 302, 433, 468, 469, 599, 608, 652, 678, 686
Hudson, C. S., 572
Hulett, G. A., 664
Huyghens, 90

Isambert, 469, 471

Jäger, 317, 701
Jahn, 315, 386, 597, 696, 710, 713

Jakowkin, A. A., 488
Jamin, 313
Jellet, 433, 453, 503, 516
Jolly, 41
Jones, H. C., 458
Jordan, Camille, 86
Jorissen, 80
Joule, 9, 11, 42, 211, 214, 237
Judson, W., 560
Jungfleisch, 154, 485

Kahlbaum, 61, 63, 326
Kahle, 701
Kajander, 580
Kannonikoff, 309
Kaufler, 492
Kaufmann, W., 392
Kawalki, 373
Kay, 572
Kayser, 194, 196, 337
Kekule, 283, 289, 319, 344
Kellas, 565
Kelvin, Lord (see *Thomson, Sir W.*)
Kiliani, M., 733
Kirchhoff, 74, 92, 115, 118, 194, 643
Kistiakowsky, W., 570
Kleber, 321
Klemencic, 419
Klobukow, 257
Knapp, 702
Knietsch, 679
Knoblauch, O., 560, 570
Knorr, 573
Knüpffer, C., 718
Koch, E. 338
Koch, R., 414
Koelichen, E., 557
Kohlrausch, 5, 301, 304, 358, 362, 363, 368, 374, 381, 499, 509, 654, 660, 701
Kohr, 534
Konowalow, 112, 450, 484
Kopp, 176, 179, 302, 304, 331, 344
Koppe, 134
Krapiwin, 374
Krause, 254
Kroenig, 201
Krüss, 338
Kuenen, 120
Kundt, 50, 207
Kuntze, 492
Kuriloff, 494, 540
Küster, 125, 171, 242, 267, 272, 377, 522, 572, 739

Laar, 573
Ladenburg, 387
Lagergren, 129
Landauer, 194
Landolt, H., 5, 41, 107, 110, 307, 309, 315, 328, 331, 334, 336, 386, 547, 576
Langer, 44, 253, 443
Langevin, 394, 396, 399
Laplace, 50
Laszczynska, St. v., 374
Lea, C., 191
Lebedew, 419
Le Bel, 291, 296
Le Blanc, 306, 383, 531, 696, 733
Le Chatelier, 43, 46, 445, 467, 609, 652, 661, 667, 682, 690
Lecoq de Boisbaudran, 196
Leduc, 41, 215, 248
Lehfeldt, 119
Lehmann, O., 98, 120, 181, 634
Lellmann, 516, 521
Lemery, 423
Lemoine, 435, 754
Lenssen, 543
Lenz, 330
Levi, 382
Levy, 547
Lidbury, F. A., 583
Liebenow, C., 405, 407, 735, 736
Liebermann, C., 340
Liebig, 274
Liebisch, T., 81, 88, 92
Linder, 493
Linebarger, 272, 315, 410, 413, 663
Ljubavin, 412
Loeb, M., 134, 363, 697, 702, 739
Loewel, 641
Longinine, 327
Lorentz, H. A., 109, 241, 243, 244, 307, 393
Lorenz, L., 109
Lorenz, R., 189, 702
Loschmidt, 421
Lottermoser, 415
Loven, J. M., 539
Löw, 680
Löwenherz, 538, 548, 553, 629
Löwenthal, 543
Lowy, 129
Luckow, 736
Ludeking, 413
Lüdin, 11
Ludwig, 741
Luggin, 752
Lunge, 255
Lupke, R., 696
Lussana, 43, 169
Luther, 751, 752, 755

Mach, 9
Maey, E., 405
Magnanini, 348, 521
Magnus, 158
Malagnini, 348
Malfatti, 254
Mallard, 43, 682
Maneuvrier, 50

Marbach, 80
Marckwald, 289, 322, 334, 755
Margottet, 678
Margules, M., 55, 119
Marignac, 106, 596
Mariotte, 38, 52
Markownikoff, 329
Marshall, 59, 327
Mathias, 232, 238, 270
Maxart, 313
Maxwell, 201, 208, 209, 228, 390, 398, 420, 563, 753
Mayer, J. R., 9, 15, 49
Mazotto, 127
M'Clelland, 395
M'Kenzie, 334
Meier, Fr., 253
Mendelejeff, 67, 183, 185, 186, 197, 525
Mensching, 253
Menschutkin, 450, 503, 577
Merz, 565
Meslin, 227
Meyer, C., 251
Meyer, G., 404, 714
Meyer, K., 240
Meyer, L., 175, 182, 185, 187, 191, 331, 682
Meyer, O. E., 201, 209
Meyer, R., 340
Meyer, St., 317
Meyer, V., 44, 249, 251, 253, 254, 274, 443, 561, 676
Meyerhoffer, W., 629, 637
Miculescu, 11
Miller, 715
Minnegerode, B., 81, 92
Miolati, 124, 375
Mitscherlich, E., 96, 177, 179, 182, 469
Mohr, 300, 345
Moissan, 175, 339, 634
Moore, 582
Morley, 41, 248
Moser, J., 714
Mossotti, 307, 316, 419
Muir, M. M. P., 175, 521, 597
Müller, P. Th., 547
Muller-Erzbach, 469
Muthmann, 492

Naccari, 654, 659
Nasini, 310
Natanson, E. and L., 440
Natterer, 54
Naumann, 80, 280, 596
Nernst, 60, 133, 141, 150, 154, 160, 166, 253, 260, 314, 354, 363, 365, 373, 382, 392, 477, 481, 486, 564, 568, 601, 714, 721, 729, 736
Neuberg, 255, 443
Neumann, B., 702
Neumann, F. E., 76, 176
Neville, 402, 404
Newlands, 187
Newton, 423
Nicol, 331
Nietzki, 339
Nilson, 175, 193, 251, 281
Nobel, A., 414
Norremberg, 94
Noyes, A. A., 516, 527, 529, 531, 532, 534, 539, 548, 559, 560, 564, 576, 581

Ober, 493
Ochs, K., 732
Oeconomides, 338
Oeholm, W., 370
Ogier, 347, 656
Oldenbourg, R., 702
Onnes, Kammerlingh, 239
Ortloff, 191
Osaka, 547, 736
Ostwald, 9, 16, 26, 33, 97, 105, 112, 127, 172, 178, 259, 287, 301, 316, 331, 342, 358, 362, 365, 385, 387, 432, 497, 500, 502, 508, 514, 516, 518, 519, 521, 536, 543, 545, 551, 553, 566, 569, 575, 592, 595, 597, 608, 629, 689, 696, 731, 732
Otto, G., 178, 302
Oudeman, 386
Overbeck, A., 417

Paalzow, 700
Pagliani, 654, 659
Palmaer, W., 547, 724, 725
Pareau, 469
Parrot, 156
Pasteur, 92, 384
Pawlewski, 67, 119
Péan de St. Gilles, 447, 449, 569
Pebal, 346
Pekar, 272
Peltier, 406, 698
Perkin, W. H., 316
Pernet, 11
Perrot, 251
Peters, 732
Petersen, 329, 598, 606
Petit, 175
Pettersson, 152, 175, 193, 251, 281
Pfaundler, 430
Pfeffer, 132, 134, 144, 146
Philip, J., 315
Pickering, 526
Pictet, Raoul, 692
Picton, 493
Piloty, 281
Planck, 22, 422, 497, 609, 666, 722
Planté, 715
Pockels, F., 69, 123
Poincaré, 700
Ponsot, 144

Poynting, J. H., 245
Freuner, G., 444, 446, 718
Pringsheim, 50, 749
Frout, 187
Pulfrich, 108, 304

Rabe, 574
Ramsay, 63, 65, 67, 68, 74, 152, 185, 187, 231, 237, 270, 271, 325, 327
Raoult, 144 148, 150, 153, 160, 169, 257, 261, 262, 266, 269, 351, 402, 459, 460, 489
Rathke, 686, 688
Ratz, F., 315
Raveau, C., 226, 227
Rayleigh, Lord, 41, 102, 201, 208, 248
Recoura, 269
Reformatzky, 413
Regnault, 41, 54, 59, 60, 61, 118, 140, 167, 176, 213, 218, 248, 405, 620, 650
Reicher, 256, 359, 497, 552, 634, 637, 638
Reinders, 565
Reinganum, 159, 240, 407
Reinitzer, 98
Reinold, 417
Remsen, 175, 287
Retgers, 121, 180, 301, 340
Riccard, 329
Richardson, 441
Riecke, 199, 236, 243, 392, 609, 611, 631, 667
Rigollet, 747
Rive, De la (see *De la Rive*)
Roberts, 169
Rodger, 331
Roloff, 160, 164, 259, 488, 490, 744, 745
Rontgen, 50, 353, 396, 404, 416, 422
Roozeboom, 127, 464, 491, 614, 616, 620, 623, 627, 629
Roscoe, 173, 746, 747, 748, 753
Rose, F., 654
Rossignol, 560
Roth, 476
Rothmund, 470, 476
Rowland, 11, 197
Rucker, 417
Rudolphi, M., 498
Rüdorff, 150
Ruhlmann, 308
Runge, 196
Rutherford, 395
Rydberg, 189, 191, 197

Sabanejeff, 410, 412
Salet, 347, 521
Sarreau, 43
Savoisier, 5
Schall, 255
Schaum, K., 97
Scheffer, 157, 370
Schenk, R., 98, 271
Schiff, 141
Schlegel, 678
Schliemann, 516
Schmidt, G. C., 119, 129, 321
Schoenfliess, A., 60, 81, 86, 568, 601
Schönbein, 680
Schoop, 254
Schreinemakers, 629
Schumann, 63
Schutt, 108
Schutze, M., 338, 340
Schwarz, W., 97
Seubert, K., 187
Shields, 270, 524, 557
Siedentopf, 417
Siemens, 23
Silbermann, 596
Silliman, 410
Silow, 314
Skraup, 455
Smale, 711
Sohncke, 86
Sommerfeldt, E., 128
Soret, 741
Soxhlet, 263
Specketer, 734
Spohr, 543, 546, 551
Sprengel, 259, 300
Spring, 128, 169, 404, 461, 580, 581
Stadt, van der (see *van der Stadt*)
Staigmuller, 187
Stark, J., 394
Stefan, 209
Steinwehr, 596, 606
Steno, 76
Steudel, 331
Stillman, 152
Stock, 170
Stohmann, 319, 321, 597
Strauss, 346
Streintz, F., 402, 716
Strenger, P., 110
Summer, 50
Swain, 152
Swart, 657

Tafel, 736
Tammann, G., 72, 98, 118, 134, 241, 402, 410, 582, 631
Thénard, 746
Thiele, J., 289, 321
Thomas, 114, 235
Thomsen, J., 106, 114, 168, 186, 189, 317, 319, 513, 516, 596, 600, 607, 654, 659, 686
Thomson, James, 70
Thomson, J, J., 395, 399, 422
Thomson, Sir W., 42, 69, 71, 74, 173 211, 214, 237, 418, 632, 664, 705 712

Thorpe, 333
Tolloczko, 267
Töpler, M., 189, 191
Townsend, 395, 398
Traube, H., 92
Traube, J., 381
Traube, M., 132, 134, 680
Trevor, 543, 546
Trey, 547
Troost, 251, 662
Trouton, 327
Turin, 714

Urech, 543

Valson, 382, 384
van Aubel, 580
van Bemmelen, 493
van Bijlert, 124, 170
van der Heide, 629
van der Stadt, 146, 643
van der Waals, 63, 115, 200, 201, 210, 212, 217, 219, 221, 227, 236, 239, 245, 255, 269, 271, 303, 325, 417, 419, 421, 609
van Deventer, 146, 636, 643
van Kuenen, 243
Van't Hoff, 115, 124, 131, 138, 146, 148, 151, 153, 156, 160, 168, 169, 256, 262, 266, 268, 284, 286, 291, 294, 334, 350, 352, 364, 402, 428, 433, 460, 461, 481, 489, 498, 509, 517, 550, 558, 560, 561, 564, 629, 631, 634, 636, 638, 652, 656, 664, 671, 680, 688, 692, 712, 741
Vernon, 324
Vieille, 43, 44, 682
Vieweg, 630
Vignon, 124
Villard, 68, 476
Visser, de (see *de Visser*)
Vogel, E., 338, 752
Vogel, H. W., 337, 744, 752
Voigt, W., 93, 239
Voigtlander, 413
Vries, de (see *de Vries*)

Waage, 395, 430, 432, 449, 453, 468, 473, 493, 496, 533, 548, 569, 644, 673
Waals, van der (see *van der Waals*)
Wachs, 565
Wagner, J., 191, 523
Wakemann, A., 500
Walden, 501
Walker, 129, 172, 186, 316, 387, 503, 523, 525, 548, 560, 572, 576
Wallach, 340
Warburg, 207, 740
Warder, 551
Wason, R. S., 560
Waterston, J. J., 201
Watt, 23, 27
Weber, H. F., 175
Werner, A., 283, 297, 374, 375, 376
Wheatstone, 314, 358
Whitney, 493, 531, 581
Wiebe, 10, 60
Wiedemann, E., 43, 46, 413
Wiedemann, G., 387, 469
Wijs, 509, 557
Wild, 680
Wildermann, 565, 749, 754
Wilhelmy, 433, 543, 544
Will, 262
Williams, G. H., 81, 86
Williams, 175
Williamson, 430
Wilsmore, 733
Wilson, 395
Winck, G., 191
Winkelblech, 377
Winkelmann, 194, 196
Winkler, 175
Wirkner, V., 61, 63
Wislicenus, 274, 291, 294, 573
Witt, O. N., 340, 492
Wittwer, 754
Wladimiroff, 135
Wohler, 274
Wollaston, 31
Wood, 254
Worm-Muller, 700
Wroblewski, 213
Wullner, 60
Wurfel, 272
Wurtz, 443

Young, 57
Young, S., 61, 63, 65, 67, 68, 74, 230, 235, 237, 248, 325

Zacharias, 492
Zaitschek, 449
Zanninovitch-Tessarin, 374
Zawidzki, 119
Zeemann, 393
Zeleny, 395, 399
Zunino, 348
Zsigmondy, 417

THE END